London Mathematical Society Lecture Note Series, 149

Solitons, Nonlinear Evolution Equations and Inverse Scattering

M.J. Ablowitz
Program in Applied Mathematics, University of Colorado at Boulder
and
P.A. Clarkson
Department of Mathematics, University of Exeter

CAMBRIDGE UNIVERSITY PRESS
Cambridge
New York Port Chester Melbourne Sydney

Published by the Press Syndicate of the University of Cambridge
The Pitt Building, Trumpington Street, Cambridge CB2 1RP
40 West 20th Street, New York, NY 10011–4211, USA
10 Stamford Road, Oakleigh, Victoria 3166, Australia

First published 1991

Printed in Great Britain at the University Press, Cambridge

Library of Congress cataloguing in publication data available

British Library cataloguing in publication data available

ISBN 0 521 38730 2

LONDON MATHEMATICAL SOCIETY LECTURE NOTE SERIES

Managing Editor: Professor J.W.S. Cassels, Department of Pure Mathematics and Mathematical Statistics, University of Cambridge, 16 Mill Lane, Cambridge CB2 1SB, England

The books in the series listed below are available from booksellers, or, in case of difficulty, from Cambridge University Press.

34 Representation theory of Lie groups, M.F. ATIYAH *et al*
36 Homological group theory, C.T.C. WALL (ed)
39 Affine sets and affine groups, D.G. NORTHCOTT
40 Introduction to H_p spaces, P.J. KOOSIS
46 p-adic analysis: a short course on recent work, N. KOBLITZ
49 Finite geometries and designs, P. CAMERON, J.W.P. HIRSCHFELD & D.R. HUGHES (eds)
50 Commutator calculus and groups of homotopy classes, H.J. BAUES
57 Techniques of geometric topology, R.A. FENN
59 Applicable differential geometry, M. CRAMPIN & F.A.E. PIRANI
62 Economics for mathematicians, J.W.S. CASSELS
66 Several complex variables and complex manifolds II, M.J. FIELD
69 Representation theory, I.M. GELFAND *et al*
74 Symmetric designs: an algebraic approach, E.S. LANDER
76 Spectral theory of linear differential operators and comparison algebras, H.O. CORDES
77 Isolated singular points on complete intersections, E.J.N. LOOIJENGA
79 Probability, statistics and analysis, J.F.C. KINGMAN & G.E.H. REUTER (eds)
80 Introduction to the representation theory of compact and locally compact groups, A. ROBERT
81 Skew fields, P.K. DRAXL
82 Surveys in combinatorics, E.K. LLOYD (ed)
83 Homogeneous structures on Riemannian manifolds, F. TRICERRI & L. VANHECKE
86 Topological topics, I.M. JAMES (ed)
87 Surveys in set theory, A.R.D. MATHIAS (ed)
88 FPF ring theory, C. FAITH & S. PAGE
89 An F-space sampler, N.J. KALTON, N.T. PECK & J.W. ROBERTS
90 Polytopes and symmetry, S.A. ROBERTSON
91 Classgroups of group rings, M.J. TAYLOR
92 Representation of rings over skew fields, A.H. SCHOFIELD
93 Aspects of topology, I.M. JAMES & E.H. KRONHEIMER (eds)
94 Representations of general linear groups, G.D. JAMES
95 Low-dimensional topology 1982, R.A. FENN (ed)
96 Diophantine equations over function fields, R.C. MASON
97 Varieties of constructive mathematics, D.S. BRIDGES & F. RICHMAN
98 Localization in Noetherian rings, A.V. JATEGAONKAR
99 Methods of differential geometry in algebraic topology, M. KAROUBI & C. LERUSTE
100 Stopping time techniques for analysts and probabilists, L. EGGHE
101 Groups and geometry, ROGER C. LYNDON
103 Surveys in combinatorics 1985, I. ANDERSON (ed)
104 Elliptic structures on 3-manifolds, C.B. THOMAS
105 A local spectral theory for closed operators, I. ERDELYI & WANG SHENGWANG
106 Syzygies, E.G. EVANS & P. GRIFFITH
107 Compactification of Siegel moduli schemes, C-L. CHAI
108 Some topics in graph theory, H.P. YAP
109 Diophantine Analysis, J. LOXTON & A. VAN DER POORTEN (eds)
110 An introduction to surreal numbers, H. GONSHOR
111 Analytical and geometric aspects of hyperbolic space, D.B.A.EPSTEIN (ed)

113 Lectures on the asymptotic theory of ideals, D. REES
114 Lectures on Bochner-Riesz means, K.M. DAVIS & Y-C. CHANG
115 An introduction to independence for analysts, H.G. DALES & W.H. WOODIN
116 Representations of algebras, P.J. WEBB (ed)
117 Homotopy theory, E. REES & J.D.S. JONES (eds)
118 Skew linear groups, M. SHIRVANI & B. WEHRFRITZ
119 Triangulated categories in the representation theory of finite-dimensional algebras, D. HAPPEL
121 Proceedings of *Groups - St Andrews 1985*, E. ROBERTSON & C. CAMPBELL (eds)
122 Non-classical continuum mechanics, R.J. KNOPS & A.A. LACEY (eds)
124 Lie groupoids and Lie algebroids in differential geometry, K. MACKENZIE
125 Commutator theory for congruence modular varieties, R. FREESE & R. MCKENZIE
126 Van der Corput's method for exponential sums, S.W. GRAHAM & G. KOLESNIK
127 New directions in dynamical systems, T.J. BEDFORD & J.W. SWIFT (eds)
128 Descriptive set theory and the structure of sets of uniqueness, A.S. KECHRIS & A. LOUVEAU
129 The subgroup structure of the finite classical groups, P.B. KLEIDMAN & M.W.LIEBECK
130 Model theory and modules, M. PREST
131 Algebraic, extremal & metric combinatorics, M-M. DEZA, P. FRANKL & I.G. ROSENBERG (eds)
132 Whitehead groups of finite groups, ROBERT OLIVER
133 Linear algebraic monoids, MOHAN S. PUTCHA
134 Number theory and dynamical systems, M. DODSON & J. VICKERS (eds)
135 Operator algebras and applications, 1, D. EVANS & M. TAKESAKI (eds)
136 Operator algebras and applications, 2, D. EVANS & M. TAKESAKI (eds)
137 Analysis at Urbana, I, E. BERKSON, T. PECK, & J. UHL (eds)
138 Analysis at Urbana, II, E. BERKSON, T. PECK, & J. UHL (eds)
139 Advances in homotopy theory, S. SALAMON, B. STEER & W. SUTHERLAND (eds)
140 Geometric aspects of Banach spaces, E.M. PEINADOR and A. RODES (eds)
141 Surveys in combinatorics 1989, J. SIEMONS (ed)
142 The geometry of jet bundles, D.J. SAUNDERS
143 The ergodic theory of discrete groups, PETER J. NICHOLLS
144 Introduction to uniform spaces, I.M. JAMES
145 Homological questions in local algebra, JAN R. STROOKER
146 Cohen-Macaulay modules over Cohen-Macaulay rings, Y. YOSHINO
147 Continuous and discrete modules, S.H. MOHAMED & B.J. MÜLLER
148 Helices and vector bundles, A.N. RUDAKOV et al
149 Solitons, nonlinear evolution equations and inverse scattering, M.J. ABLOWITZ & P.A. CLARKSON
150 Geometry of low-dimensional manifolds 1, S. DONALDSON & C.B. THOMAS (eds)
151 Geometry of low-dimensional manifolds 2, S. DONALDSON & C.B. THOMAS (eds)
152 Oligomorphic permutation groups, P. CAMERON
153 L-functions in Arithmetic, J. COATES & M.J. TAYLOR
154 Number theory and cryptography, J. LOXTON (ed)
155 Classification theories of polarized varieties, TAKAO FUJITA
156 Twistors in mathematics and physics, T.N. BAILEY & R.J. BASTON (eds)
157 Analytic pro-p groups, J.D. DIXON, M.P.F. SU SAUTOY, A. MANN & D. SEGAL
158 Geometry of Banach spaces, P.F.X. MÜLLER & W. SCHACHERMAYER (eds)
159 Groups St Andrews 1989 Volume 1, C.M. CAMPBELL & E.F. ROBERTSON (eds)
160 Groups St Andrews 1989 Volume 2, C.M. CAMPBELL & E.F. ROBERTSON (eds)
161 Lectures on block theory, BURKHARD KÜLSHAMMER
162 Harmonic analysis and representation theory for groups acting on homogeneous trees, A. FIGA-TALAMANCA & C. NEBBIA
163 Topics in varieties of group representations, S.M. VOVSI
166 Surveys in Combinatorics, 1991, A.D. KEEDWELL (ed)
167 Stochastic analysis, M.T. BARLOW & N.H. BINGHAM (eds)

Contents

Preface

An exciting and extremely active area of research investigation during the past twenty years has been the study of Solitons and the related issue of the construction of solutions to a wide class of nonlinear equations. Indeed there have been a few books written which serve to review aspect of this burgeoning field. A book coauthored by one of us (MJA) exactly ten years ago, discussed many of the relevant viewpoints as well as a variety of applications. Certain important and novel subareas of research such as the the application of the Inverse Scattering Transform (I.S.T.) to solve nonlinear wave equations on the infinite interval, in one spatial and one temporal dimension $(1+1)$, were described in detail. At that time the complete inverse scattering methodology had been carried out primarily for those nonlinear equations related to second order scattering problems. Although it was known that certain nonlinear evolution equations in one and two spatial dimensions were related to suitable (higher order and two dimensional) linear scattering problems, and special techniques were available, nevertheless it was not yet clear that a unified and effective procedure could be applied to all of these nonlinear equations. The main purpose of this book is the description of how the I.S.T. technique can be applied to these situations.

Our presentation begins with a "state of the art" introduction. Here we list as many integrable systems and relevant scattering problems that we are familiar with — though it is still possible that we have missed some interesting ones. In order to help the interested reader and to establish our point of view, in Chapter Two we have reviewed the I.S.T. technique as it applies to the famous Korteweg-de Vries (KdV) equation on the infinite line. In this case the KdV equation is related to the classical second order scalar time-independent Schrödinger scattering problem. Chapter Three discusses the inverse scattering associated with $N \times N$ systems and higher order systems of scattering problems; the 2×2 case is reviewed. The I.S.T. technique is then applied to various systems of partial differential equations. Chapter Three contains a brief discussion of discrete problems as well. Chapter Four details the inverse transform as it applies to a rather novel class of scattering problems which arise in the solution of singular integro-differential evolution equations such as the intermediate long wave and Benjamin-Ono equations. Chapter Five involves the inverse transform for $2+1$-dimensional nonlinear wave equations, and discusses the "DBAR" ($\bar{\partial}$) method of inverse scattering. Key solvable nonlinear wave equations include the Kadomtsev-Petviashvili, Davey-Stewartson and three wave equations in $2+1$-dimensions. Chapter Six analyzes certain multidimensional inverse problems (generalised Schrödinger and $N \times N$ systems) via the $\bar{\partial}$ method of inverse scattering. It is especially worth noting that Faddeev's inverse formulation for the time-independent Schrödinger problem in three(or more dimensions) follows naturally from the $\bar{\partial}$ method in a limiting case. There are few nonlinear wave equations in more than 2 + 1-dimensions which are now known to be solvable by I.S.T.. However the few that are known are discussed with special emphasis placed on the self-dual Yang-Mills (SDYM) system. Indeed the SDYM system plays a rather central role in integrable systems as the reader will understand. Chapter Seven details some related questions involving the properties and solution of certain nonlinear ordinary differential equations. Especially important amongst these equations are the classical equations of Painlevé and coworkers. Indeed the ideas of Painlevé lead naturally to the study of more general equations which possess the so called Painlevé property. Moreover there is a connection between nonlinear equations solvable by I.S.T. and such equations. Indeed recent work has shown that such properties can be exploited to provide useful *a priori* tests to determine when a given partial differential equation might be integrable. These ideas are also discussed in Chapter Seven. Finally we conclude with some remarks and brief discussion of some important open problems. A large and we hope reasonably complete bibliography is included. Given the rate at which this field has developed and the numerous applications it is natural that some references will be missed or forgotten. We apologise for such situations whenever they occur.

In this book we have not considered solutions to nonlinear wave equations with periodic boundary conditions, or the various methods of construction of special solutions (e.g., Hirota's direct method, pole expansions, dressing techniques etc.), or the in depth functional analysis which underlies a considerable portion of the theory. Nor have we considered the rather substantive issue of the underlying algebraic connections that serve to endow these nonlinear wave equations with considerable inherent structure. The main reason for this approach is that doing otherwise would have entailed a much larger project and hence a considerably expanded book. Since there is so much to cover and explain even with the reduced scope of this effort we felt that it would have been simply unwise and likely detrimental to the reader to have included much more than we did.

These notes review many years of work, partially supported by the mathematics divisions of the Air Force Office of Scientific Research, the National Science Foundation and the Office of Naval Research. Their support is gratefully appreciated. We sincerely acknowledge the collaboration of our many colleagues and note especially M.D. Kruskal, for innumerable valuable discussions and insights, A.S. Fokas who collaborated on many of the projects, A. Nachman who collaborated on the multi-dimensional inverse scattering, and J.B. McLeod for introducing one of us to this exciting field of research. We also thank A.P. Bassom for help in proof-reading the manuscript and A.C. Hicks and S. Hood for assistance in producing the figures. Of course we are deeply indebted to our wives Enid and Kim who endured many hours away from their husbands while this book was written.

Chapter One

Introduction.

1.1 Historical Remarks and Applications.

"Solitons" were first observed by J. Scott Russell in 1834 [1838, 1844] whilst riding on horseback beside the narrow Union canal near Edinburgh, Scotland. There are a number of discussions in the literature describing Russell's observations. Nevertheless we feel that his point of view is so insightful and relevant that we present it here as well. He described his observations as follows:

> "I was observing the motion of a boat which was rapidly drawn along a narrow channel by a pair of horses, when the boat suddenly stopped — not so the mass of water in the channel which it had put in motion; it accumulates round the prow of the vessel in a state of violent agitation, then suddenly leaving it behind, rolled forward with great velocity, assuming the form of a large solitary elevation, a rounded, smooth and well-defined heap of water, which continued its course along the channel apparently without change of form or diminution of speed. I followed it on horseback, and overtook it still rolling on at a rate of some eight or nine miles an hour, preserving its original figure some thirty feet long and a foot to a foot and a half in height. Its height gradually diminished, and after a chase of one or two miles I lost it in the windings of the channel. Such, in the month of August 1834, was my first chance interview with that rare and beautiful phenomenon which I have called the Wave of Translation ... ."

Subsequently, Russell did extensive experiments in a laboratory scale wave tank in order to study this phenomenon more carefully. Included amongst Russell's results are the following:

1. he observed solitary waves, which are long, shallow, water waves of permanent form, hence he deduced that they *exist*; this is his most significant result;
2. the speed of propagation, c, of a solitary wave in a channel of uniform depth h is given by $c^2 = g(h + \eta)$, where η is the amplitude of the wave and g the force due to gravity.

Further investigations were undertaken by Airy [1845], Stokes [1847], Boussinesq [1871, 1872] and Rayleigh [1876] in an attempt to understand this phenomenon. Boussinesq and Rayleigh independently obtained approximate descriptions of the solitary wave; Boussinesq derived a one-dimensional nonlinear evolution equation, which now bears his name, in order to obtain his result.

These investigations provoked much lively discussion and controversy as to whether the inviscid equations of water waves would possess such solitary wave solutions. The issue was finally resolved by Korteweg and de Vries [1895]. They derived a nonlinear evolution equation governing long one dimensional, small amplitude, surface gravity waves propagating in a shallow channel of water

$$\frac{\partial \eta}{\partial \tau} = \frac{3}{2}\sqrt{\frac{g}{h}}\frac{\partial}{\partial \xi}\left(\tfrac{1}{2}\eta^2 + \tfrac{2}{3}\alpha\eta + \tfrac{1}{3}\sigma\frac{\partial^2 \eta}{\partial \xi^2}\right), \qquad \sigma = \tfrac{1}{3}h^3 - Th/(\rho g), \tag{1.1.1}$$

where η is the surface elevation of the wave above the equilibrium level h, α an small arbitrary constant related to the uniform motion of the liquid, g the gravitational constant, T the surface tension and ρ the density (the terms "long" and "small" are meant in comparison to the depth of the channel). The controversy was now resolved since equation (1.1.1), now known as the Korteweg-de Vries (KdV) equation, has permanent wave solutions, including *solitary wave solutions* (see §1.3 for details). Equation (1.1.1) may be brought into nondimensional form by making the transformation

$$t = \tfrac{1}{2}\sqrt{g/(h\sigma)}\,\tau, \qquad x = -\sigma^{-1/2}\xi, \qquad u = \tfrac{1}{2}\eta + \tfrac{1}{3}\alpha.$$

Hence, we obtain

$$u_t + 6uu_x + u_{xxx} = 0, \tag{1.1.2}$$

where subscripts denote partial differentiations. Henceforth, we shall consider the KdV equation in this form (1.1.2) (note that any constant coefficient may be placed in front of any of the three terms by a suitable scaling of the independent and dependent variables). (1.1.2) may be thought of as the simplest "nonclassical" partial differential equation since it has the minimum number of independent variables, two; the lowest order not considered classically, that is three; the fewest terms of that order, one; the simplest such term, an unmixed derivative; the fewest number of terms containing the other derivative, which is of first order; the simplest structure for these terms, linear; and the simplest additional term to make the equation nonlinear, quadratic. (It might be thought that a simpler nonlinear term would be u^2, however the KdV equation has an extra symmetry with uu_x (Galilean invariance) and if u is interpreted as a velocity, then the convective derivative, which arises in continuum mechanics and is familiar to physicists and engineers, is $du/dt = u_t + uu_x$.)

As Miura [1976] points out, despite this early derivation of the KdV equation, it was not until 1960 that any new application of the equation was discovered. Gardner and Morikawa [1960] rediscovered the KdV equation in the study of collision-free hydromagnetic waves. Subsequently the KdV equation has arisen in a number of other physical contexts, stratified internal waves, ion-acoustic waves, plasma physics, lattice dynamics, ... (for details and further references see, for example, the articles by Jeffrey and Kakutani [1972]; Scott, Chu and McLaughlin [1973]; Miura [1976] and monographs by Ablowitz and Segur [1981]; Calogero and Degasperis [1982]; Dodd, Eilbeck, Gibbon and Morris [1982]; Lamb [1980]; Novikov, Manakov, Pitaevskii and Zakharov [1984] — for a discussion concentrating primarily on the pre-1965 history of the KdV equation, see Miles [1980b]).

It has been known for a long time that the KdV equation (1.1.2) possesses the solitary wave solution

$$u(x,t) = 2\kappa^2 \operatorname{sech}^2\left\{\kappa(x - 4\kappa^2 t - x_0)\right\}, \tag{1.1.3}$$

where κ and x_0 are constants (in fact this solution was known to Korteweg and de Vries). Note that the velocity of this wave, $4\kappa^2$, is proportional to the amplitude, $2\kappa^2$; therefore taller waves travel faster than shorter ones. Zabusky and Kruskal [1965] discovered that these solitary wave solutions have the remarkable property that the interaction of two solitary wave solutions is elastic, and are called *solitons* (see §1.4 details).

The solitons observed by Russell were small amplitude surface waves. There have been several investigations examining the validity of the KdV equation (1.1.2) as a model of the evolution of small amplitude water waves as they propagate in one direction in shallow water. These studies have compared the solutions of (1.1.2) with experimental results (see, for example, Hammack and Segur [1974, 1978]). In physical terms the KdV equation arises if the water waves are strictly one-dimensional, that is one spatial dimension and time (for a derivation of the KdV equation see, for example, Chapter 4 of Ablowitz and Segur [1981]).

In many physical situations, internal waves can arise at the interface of two layers of fluid due the gravitational effects in a stably stratified fluid. Several theoretical models exist which govern the evolution of long internal waves with small amplitudes in a stably stratified fluid including the KdV equation (1.1.2), the intermediate long wave (ILW) equation

$$u_t + \delta^{-1} u_x + 2uu_x + \mathrm{T}u_{xx} = 0, \tag{1.1.4a}$$

where $\mathrm{T}u$ is the singular integral operator

$$(\mathrm{T}f)(x) = \frac{1}{2\delta} ⨍_{-\infty}^{\infty} \coth\left\{\frac{\pi}{2\delta}(y-x)\right\} f(y)\,\mathrm{d}y, \tag{1.1.4b}$$

with $⨍_{-\infty}^{\infty}$ the Cauchy principal value integral (Joseph [1977]; Kubota, Ko and Dobbs [1978]), and the Benjamin-Ono (BO) equation

$$u_t + 2uu_x + \mathrm{H}u_{xx} = 0, \tag{1.1.5a}$$

where $\mathrm{H}u$ is the Hilbert transform

$$(\mathrm{H}f)(x) = \frac{1}{\pi} ⨍_{-\infty}^{\infty} \frac{f(y)}{y-x}\,\mathrm{d}y \tag{1.1.5b}$$

(Benjamin [1967]; Davies and Acrivos [1967]; Ono [1975]). In the shallow water limit, as $\delta \to 0$, (1.1.4) reduces to the KdV equation

$$u_t + 2uu_x + \tfrac{1}{3}\delta u_{xxx} = 0, \tag{1.1.6}$$

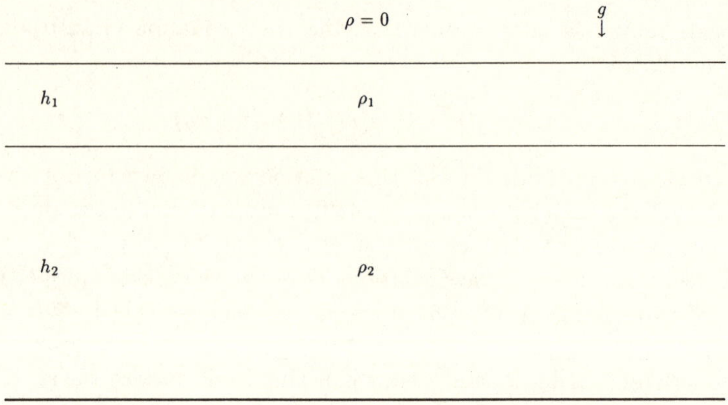

FIGURE 1.1.1 *The two-layer configuration.*

and in the deep-water limit as $\delta \to \infty$, to the BO equation. Therefore the ILW equation (1.1.4) may be thought of as being an equation intermediate between (1.1.5) and (1.1.6).

Consider two incompressible, immiscible fluids, with densities and depths h_1, h_2 ($h := h_1 + h_2$) with the lighter fluid, of height h_1, lying over a heavier fluid of height h_2, in a constant gravitational field (Figure 1.1.1). The lower fluid rests on a horizontal impermeable bed, and the upper fluid is bounded by a free surface.

Suppose the the characteristic wave amplitude is denoted by a and the characteristic wavelength by $\lambda = k^{-1}$. The basic assumptions for the derivation of the KdV equation (1.1.6), the ILW equation (1.1.4) and the BO equation (1.1.5), as models for internal waves are (see Chapter 4 of Ablowitz and Segur [1981] for details).

Korteweg de-Vries equation:

(A1) the waves are long waves in comparison with the total depth, $kh \ll 1$;
(A2) the amplitude of the waves is small, $\varepsilon = a/h \ll 1$;
(A3) the two effects (A1) and (A2) approximately balance, i.e., $kh = O(\varepsilon)$;
(A4) viscous effects may be neglected.

Intermediate-Long-Wave equation:

(B1) there is a thin (upper) layer, $\varepsilon = h_1/h_2 \ll 1$;
(B2) the amplitude of the waves is small, $a \ll h_1$;
(B3) the two effects (B1) and (B2) balance, i.e., $a/h_1 = O(\varepsilon)$;
(B4) the characteristic wavelength is comparable to the total depth of the fluid, $kh = O(1)$;
(B5) the waves are long waves in comparison with the thin layer, $kh_1 \ll 1$ [this is implied by (B1) and (B4)];
(B6) viscous effects may be neglected.

Note that in the ILW equation (1.1.4), the parameter δ is effectively kh. For derivations of the ILW equation (1.1.4) in which the fluid is confined between two rigid walls see Kubota, Ko and Dobbs [1978]; Segur and Hammack [1982] (these two derivations are slightly different).

Benjamin-Ono equation:

(C1) there is a thin (upper) layer, $h_1 \ll h_2$;

(C2) the waves are long waves in comparison with the thin layer, $kh_1 \ll 1$;

(C3) the waves are short in comparison with the total depth of the fluid, $kh \gg 1$;

(C4) the amplitude of the waves is small, $a \ll h_1$;

(C5) viscous effects may be neglected.

Under these assumptions one obtains the KdV, ILW and BO equations, after a suitable scaling. Segur and Hammack [1982] have examined the validity of the KdV and ILW equations as models for internal waves by comparing the theoretical solutions of the KdV and ILW equations with experimental results.

Recently there has been considerable interest in the observation of what scientists think might be solitons in the oceans. The availability of photographs taken from satellites and spacecraft orbiting the earth have greatly assisted in these observations. Peculiar striations, visible on satellite photographs of the surface of the Andaman and Sulu seas in the Far East, have been interpreted as secondary phenomena accompanying the passage of "internal solitons", solitary wavelike distortions of the boundary layer between the warm upper layer of sea water and cold lower depths. These internal solitons are travelling ridges of warm water, extending hundreds of meters down below the thermal boundary, and enormous energies which they carry are presumed to be the cause of unusually strong underwater currents experienced by deep-sea drilling rigs. In order to continue deep-sea drilling for oil in areas where these internal solitons occur, the drilling rigs will have to be built to withstand these forces.

Osborne and Burch [1980] (see also Osborne [1990]) investigated the underwater currents which were experienced by an oil rig in the Andaman sea, which was drilling at a depth of 3600ft (one drilling rig was apparently spun through ninety degrees and moved one hundred feet by the passage of a soliton below). Satellite photographs had shown that there were 100km long striations on the Andaman sea, separated by 6 to 15km and grouped in packets of typically 4 to 8. The average time between the arrival of these packets at the research vessel was 12 hours 26 minutes, suggesting that this was some kind of tidal phenomenon. Osborne and Burch spent four days measuring underwater currents and temperatures. The striations seen on satellite photographs turned out to be kilometer-wide bands of extremely choppy water, stretching from horizon to horizon, followed by about two kilometers of water "as smooth as a millpond". These bands of agitated water are called "tide rips", they arose in packets of 4 to 8, spaced about 5 to 10km apart (they reached the research vessel at approximately hourly intervals) and this pattern was repeated with the regularity of tidal phenomenon.

Osborne and Burch [1980] found that the amplitude of each succeeding soliton was less than the previous one, which, as we shall see in §1.7 below, is precisely what is expected for solitons (recall that the velocity of a solitary wave solution (1.1.3) of the KdV equation increases with amplitude). Thus if a number of solitons are generated together, then we expect them eventually to be arranged in an ordered sequence of decreasing amplitude. From the spacing between successive waves in a packet and the rate of separation calculated from the KdV equation, Osborne and Burch were able to estimate the distance the packet had travelled from its source and thus identify possible source regions. They concluded that the solitons are generated by tidal currents off northern Sumatra or between the islands of the Nicobar chain that extends beyond it.

Underwater measurements showed a rapid circulation of water associated with the solitary waves. The interaction of this internal circulation with ordinary surface waves produces the tide rip and appear to explain the origin of the forces experienced by the oil drilling rigs. Osborne and Burch [1980] conclude that, despite the irregular geometry of the Andaman sea, their observations have good general agreement with the predictions for internal solitons as given by the KdV equation.

Apel and Holbrook [1980] (see also Holbrook, Apel and Tsai [1980]) undertook a detailed study of internal waves in the Sulu sea. Satellite photographs had suggested that the source of these waves was near the southern end of the Sulu sea and their research ship followed one wave packet for more than 250 miles over a period of two days — an extraordinary coherent phenomenon. These internal solitons travel at speeds of about 8 kilometers per hour (5 miles per hour), with amplitude of about 100 meters and wavelength of about 1700 meters.

Further observations and studies of solitons in oceans include the Strait of Messina (Alpers and Salustri [1983]; Santoleri [1983]); the Strait of Gibraltar (Lacombe and Richez [1982]); off the western side of Baja California (Apel and Gonzalez [1983]); the Gulf of California (Fu and Holt [1984]); the Archipelago of La Maddalena (Manzella, Bohm and Salustri [1983]) and the Georgia Strait (Hughes and Gower [1983]).

Rossby waves (Rossby [1939]) are long waves between layers of the atmosphere, created by the rotation of the planet. As one might suspect, there is an analogy between internal waves and Rossby waves under suitable conditions. The KdV equation has been proposed as a model for the evolution of Rossby waves (Long [1964]; Benney [1966]; Redekopp [1977]; Redekopp and Weidman [1978] — see also Miles [1980b] and the references cited therein). There has been been a conjecture (Maxworthy and Redekopp [1976]) that Jupiter's Great Red Spot might be a solitary Rossby wave, though recent work seems to indicate that the red spot is a result of strong differential rotation (cf. Meyers, Sommeria and Swinney [1989]).

We have noted that the KdV equation (1.1.2) can describe the evolution of small amplitude water waves as they propagate in shallow water; the evolution described being weakly nonlinear and weakly dispersive, where these effects are of the same order. These solitary waves are usually regarded as propagating in a uniform rectangular channel. There have been several studies on the problem of quasi one-dimensional solitary waves in channels of varying width, depth or shape of the channel (for a review see Miles [1980b] and the references cited therein).

A further restriction in the application of the KdV equation as a practical model for water waves, is that the KdV equation is strictly only one-dimensional (that is one spatial-dimension plus time), whereas the surface is two-dimensional. A two-dimensional generalization of the KdV equation is the Kadomtsev-Petviashvili (KP) equation

$$(u_t + 6uu_x + u_{xxx})_x + 3\sigma^2 u_{yy} = 0, \tag{1.1.7}$$

where $\sigma^2 = \pm 1$ (Kadomtsev and Petviashvili [1970]). Whereas the KdV equation (1.1.2) describes the evolution of long water waves of small amplitude if they are strictly one-dimensional, the KP equation (1.1.7) describes their evolution if they are weakly two-dimensional (in §1.2 below, we give a derivation of the KP equation as a model for surface waves). The evolution described by the KP equation is weakly nonlinear, weakly dispersive and weakly two-dimensional with all three effects being of the same order; the choice of sign depends on the relevant magnitude of gravity and surface tension (cf. §1.2). The KP equation has also been proposed as a model for surface waves and internal waves in straits or channels of varying depth and width (see, for example, Santini [1981]; David, Levi and Winternitz [1987a, 1989]).

The KdV equation (1.1.2), the ILW equation (1.1.4), the BO equation (1.1.5) and the KP equation (1.1.7) have been extensively studied in recent years, primarily by mathematicians and physicists. We have already seen that they arise in the description of physically interesting phenomena, however much of the interest in these nonlinear evolution equations is due to the fact that they are thought of as being *completely integrable*, or *exactly solvable* equations. This terminology is a consequence of the fact that the initial value problem for each of these nonlinear equations can be solved exactly by a method employing *inverse scattering*, which we shall study in depth in these notes. The inverse scattering method was discovered by Gardner, Greene, Kruskal and Miura [1967] as a method of solving the initial value problem for the KdV equation (on the infinite line), for initial values that decay sufficiently rapidly at infinity. Subsequently numerous other physically interesting equations in one spatial dimension were solved by generalizations of this technique, which is now referred to as the *Inverse Scattering Transform* (I.S.T.), for example the nonlinear Schrödinger, Sine-Gordon, three-wave interaction, Modified KdV and Boussinesq equations. The I.S.T. scheme for the ILW (1.1.4) and the BO equation (1.1.5), which are integro-differential equations, was derived by Kodama, Ablowitz and Satsuma [1982] and

Fokas and Ablowitz [1983b], respectively. For the KP equation (1.1.7), the I.S.T. scheme is dependent upon the choice of sign for σ^2. The scheme for the KP equation with $\sigma^2 = -1$ (which is known as KPI) was derived by Manakov [1981] and Fokas and Ablowitz [1983d]; and the scheme in the case $\sigma^2 = 1$ (which is known as KPII), by Ablowitz, BarYaacov and Fokas [1983]. (As we show in Chapter 5, the I.S.T. schemes for KPI and KPII are fundamentally different.)

The organization of these notes is as follows. In Chapter 1 we give an introduction to several of the most important and significant aspects in the development of the I.S.T. schemes for nonlinear evolution equations. In Chapters 2 and 3 we discuss I.S.T. schemes for one-dimensional equations; discussing the KdV equation (1.1.2) in Chapter 2 and more general I.S.T. schemes in one dimension in Chapter 3 (including the I.S.T. schemes for differential-difference and partial-difference equations). In Chapter 4 we discuss the I.S.T. scheme for integro-differential equations using the ILW equation (1.1.4) and the BO equation (1.1.5) as the prototype examples. In Chapters 5 and 6 we discuss multidimensional I.S.T. schemes; in Chapter 5 discussing inverse scattering in two spatial dimensions respectively, using the KP equation (1.1.7) as prototype example and in Chapter 6 discussing inverse scattering in higher dimensions. In Chapter 7 we discuss several topics concerning integrability associated with the Painlevé equations, which are six second order, nonlinear ordinary differential equations; in particular, the so-called Painlevé tests and inverse scattering schemes for the Painlevé equations using the second Painlevé equation

$$\frac{\mathrm{d}^2 y}{\mathrm{d}x^2} = 2y^3 + xy + \alpha, \tag{1.1.8}$$

where α is a constant, as the prototype example. Finally, in Chapter 8, we make some concluding remarks and mention some difficult open problems.

1.2 Physical Derivation of the KP Equation.

In this section we present a physical derivation of the KP equation (1.1.7) as a model for surface waves.

The classical problem of water waves is to find the irrotational motion of an inviscid, incompressible, homogeneous fluid, subject to a constant gravitational force g. The fluid rests on a horizontal and impermeable bed of infinite extent at $z = -h$ and has a free surface at $z = \eta(x, y, t)$. Since the fluid is irrotational and incompressible, then it has a velocity potential ϕ satisfying

$$\nabla^2 \phi = 0, \qquad -h < z < \eta(x, y, t). \tag{1.2.1a}$$

It is subject to the boundary conditions on the bottom $z = -h$

$$\phi_z = 0 \tag{1.2.1b}$$

(since the bed is impermeable), and on the free surface $z = \eta$

$$\frac{\mathrm{D}\eta}{\mathrm{D}t} \equiv \eta_t + \phi_x \eta_x + \phi_y \eta_y = \phi_z \tag{1.2.1c}$$

(kinematic condition) and

$$\phi_t + g\eta + \tfrac{1}{2}|\nabla\phi|^2 = \frac{T}{\rho}\,\frac{\eta_{xx}(1+\eta_y^2) + \eta_{yy}(1+\eta_x^2) - 2\eta_{xy}\eta_x\eta_y}{(1+\eta_x^2+\eta_y^2)^{3/2}} \tag{1.2.1d}$$

(dynamic condition), where T is the surface tension and ρ is the density of the fluid. Boundary conditions in (x, y) and initial conditions are also required. For isolated waves, then $\nabla\phi$ and η should vanish as $(x^2 + y^2) \to \infty$. (In other problems, periodic boundary conditions in x and y may be appropriate.)

This problem, first posed by Stokes [1847], is nonlocal, highly nonlinear and not surprisingly remains unsolved in its general form. To make further progress, we need to impose additional assumptions on the solutions to equations (1.2.1). The first such assumption is that the wave amplitudes should be small. If we interpret small to mean infinitesimal, then we may linearize equation (1.2.1) about $\nabla\phi = 0$, $\eta = 0$, and seek solutions of the linearized equations proportional to $\exp\{\mathrm{i}(kx + my - \omega t)\}$ (see Lamb [1932,§§228,266,267]). The result is the linearized dispersion relation

$$\omega^2 = (g\kappa + \kappa^3 T/\rho)\tanh(\kappa h), \tag{1.2.2}$$

where $\kappa^2 := k^2 + m^2$. From this one computes the group velocity and shows that the linearized problem is dispersive at most wave numbers, but not as $\kappa \to 0$ (i.e., long waves, or shallow water waves), where it is only weakly dispersive. The KdV and KP equations arise as models of the water wave problem in this weakly dispersive limit $\kappa h \ll 1$.

We orient the horizontal coordinates so that the x-direction is the principal direction of wave propagation. To derive the KP equation (1.1.7) we assume that:

1. wave amplitudes are small, $\varepsilon = |\eta|_{\max}/h \ll 1$;
2. the water is shallow relative to typical horizontal wavelengths, $(\kappa h)^2 \ll 1$;
3. the waves are nearly one-dimensional, $(m/k)^2 \ll 1$;
4. these three effects all have comparable influence (i.e., they balance), $(m/k)^2 = O((\kappa h)^2) = O(\varepsilon)$.

These assumptions imply a certain scaling of the original equation (see below); an evolution equation which follows may then be found using a multiple scales method.

At leading order, the equations are linear (from 1), nondispersive (from 2) and one-dimensional (from 3), the result being the linear wave equation

$$\eta_{tt} - gh\eta_{xx} = 0. \tag{1.2.3}$$

At this order every wave has permanent form, not because it is soliton but because we are solving the one-dimensional, linear wave equation.

When the perturbation expansion is continued to second order, the one-dimensional, linear wave equation has homogeneous (forcing) terms representing weak nonlinearity, weak dispersion and weak two-dimensionality. Each of these effects contributes to a secular term at second order. Define

$$r = \varepsilon^{1/2}\,\frac{x - \sqrt{gh}\,t}{h}, \quad s = \varepsilon^{1/2}\,\frac{x + \sqrt{gh}\,t}{h}, \quad \zeta = \varepsilon\,\frac{y}{h}, \quad \tau = \varepsilon^{3/2}\,\frac{\sqrt{gh}\,t}{6h}, \tag{1.2.4a}$$

and seek solutions of the form

$$\eta(x, y, t; \varepsilon) = \tfrac{2}{3}\varepsilon h\,[u(r, \zeta, \tau) + v(s, \zeta, \tau)] + O(\varepsilon^2 h). \tag{1.2.4b}$$

Since we are interested in problems where the initial disturbances are localized, then it is convenient to assume *a fortiori* that the physical quantities have compact support initially. In this case it is easily shown that u and v in (1.2.4b) have compact support as well. No secular terms arise at second order provided that $U(r, \zeta, \tau)$ and $U(s, \zeta, \tau)$ satisfy

$$[U_\tau + 6UU_r + (1 - \hat{T})U_{rrr}]_r + 3U_{\zeta\zeta} = 0, \tag{1.2.5a}$$
$$[V_\tau - 6VV_s - (1 - \hat{T})V_{sss}]_s - 3V_{\zeta\zeta} = 0, \tag{1.2.5b}$$

where $\hat{T} := T/(3\rho g h^2)$ is the dimensionless surface tension. Thus the left and right running waves each evolve according to their own KP equation, which describe how the two sets of waves each interact with themselves over a long time scale

$$\tau = \tfrac{1}{6}\varepsilon^{3/2}\,\sqrt{g/h}\,t = O(1).$$

To make the model quite explicit, we also write the KP equation for the right-going waves in its dimensional form for $\eta(x, y, t)$:

$$\frac{\partial}{\partial x}\left[\frac{1}{\sqrt{gh}}\eta_t + \eta_x + \frac{3}{2h}\eta\eta_x + \left(\frac{3h^2\rho g - T}{18\rho g}\right)\eta_{xxx}\right] + \tfrac{1}{2}\eta_{yy} \sim 0. \tag{1.2.6}$$

For most cases of interest in water waves $1 - \hat{T} > 0$ (in fact usually $\hat{T} \ll 1$ and so $\hat{T}$ may be neglected), which corresponds to $\sigma^2 = 1$ in equation (1.1.7) (i.e., KPII). Thus from equation (1.2.2), the linearized phase speed is a (local) maximum at $\kappa = 0$. In this case, equation (1.2.6) is equivalent to equation (1.1.7) with $\sigma = +1$, i.e.

$$(u_t + 6uu_x + u_{xxx})_x + 3u_{yy} = 0, \tag{1.2.7}$$

which is usually called KPII.

Obviously every solution of the KdV equation (1.1.2) is also a solution of the KPII equation (1.2.7). More generally, Satsuma [1976] showed that the KPII equation (1.2.7) has N line-soliton solutions, with the N line-solitons travelling in N different directions and interacting obliquely. In particular, a one line-soliton solution is

$$u(x,y,t) = 2\kappa^2 \operatorname{sech}^2 \left\{\kappa[x + \lambda y - (4\kappa^2 + 3\lambda^2)t + \delta_0]\right\}. \tag{1.2.8}$$

Essentially these are one-dimensional solutions travelling at an angle to the y-axis. A two line-soliton (written in Hirota's notation — see §2.6.5 below) is

$$u(x,y,t) = 2\frac{\partial^2}{\partial x^2} \ln F(x,y,t), \tag{1.2.9a}$$

with

$$F(x,y,t) = 1 + \exp(\eta_1) + \exp(\eta_2) + \exp(\eta_1 + \eta_2 + A_{12}), \tag{1.2.9b}$$

where

$$\eta_i = 2\kappa_i[x + \lambda_i y - (4\kappa_i^2 + 3\lambda_i^2)t] + \delta_i, \tag{1.2.9c}$$

$$\exp(A_{12}) = \frac{4(\kappa_1 - \kappa_2)^2 - (\lambda_1 - \lambda_2)^2}{4(\kappa_1 + \kappa_2)^2 - (\lambda_1 - \lambda_2)^2}. \tag{1.2.9d}$$

One and two line-soliton solutions of the KPII equation are illustrated in Figures 1.2.1 and 1.2.2, respectively. Note that in Figure 1.2.2, away from the interaction region, each wave is essentially a KdV-type soliton (as expected, there is a phase shift due to the interaction). We remark that solutions such as (1.2.8) and (1.2.9) do not decay as $(x^2+y^2)^{1/2} \to \infty$ in all directions.

For very thin sheets of water $\hat{T} > 1$ (when surface tension dominates gravity), then equation (1.2.6) is equivalent to equation (1.1.7) with $\sigma^2 = -1$, i.e.

$$(u_t + 6uu_x + u_{xxx})_x - 3u_{yy} = 0, \tag{1.2.10}$$

which is usually called KPI. Here one-dimensional solitons are unstable (Kadomtsev and Petviashvili [1970]). (We remark that the KP equation with $\hat{T} > 1$ also applies if we neglect gravity, remove the horizontal bed (and the viscous boundary layer it creates) on which the thin sheet of water lies and restrict out attention to the symmetric modes so that $\phi_z = 0$ holds.) In this case, there exist "lump" solutions of the KPI equation,

$$u(x,y,t) = 4\,\frac{\left\{-\left[x + \lambda y + 3(\lambda^2 - \mu^2)t\right]^2 + \mu^2(y + 6\lambda t)^2 + 1/\mu^2\right\}}{\left\{\left[x + \lambda y + 3(\lambda^2 - \mu^2)t\right]^2 + \mu^2(y + 6\lambda t)^2 + 1/\mu^2\right\}^2}, \tag{1.2.11}$$

which decay algebraically as $(x^2+y^2)^{1/2} \to \infty$ (cf. Manakov, Zakharov, Bordag, Its and Matveev [1977]; Satsuma and Ablowitz [1979]).

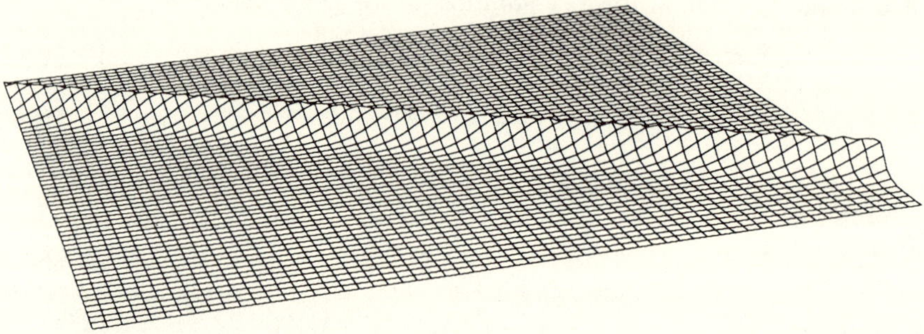

FIGURE 1.2.1 *One line-soliton solution of the KPII equation* (1.2.7).

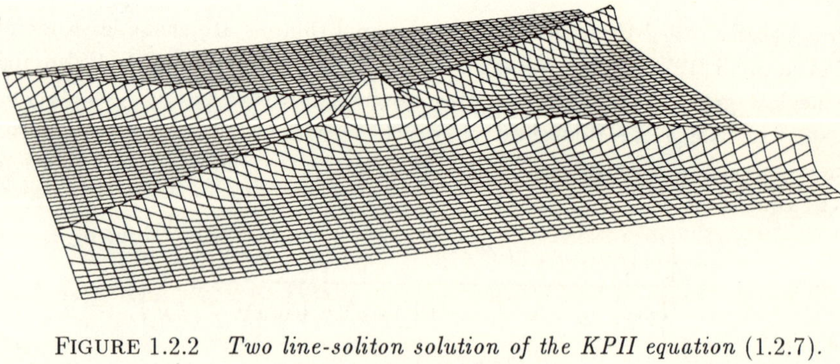

FIGURE 1.2.2 *Two line-soliton solution of the KPII equation* (1.2.7).

1.3 Travelling Wave Solutions of the KdV Equation.

The first interesting property of the KdV equation is the existence of permanent wave solutions, including solitary wave solutions.

DEFINITION 1.3.1

A solitary wave solution of a partial differential equation

$$\Delta(x,t,u)=0,$$

where $t\in\mathbb{R}$, $x\in\mathbb{R}$ are temporal and spatial variables and $u\in\mathbb{R}$ the dependent variable, is a travelling wave solution of the form

$$u(x,t)=w(x-\gamma t)=w(z), \tag{1.3.1}$$

whose transition is from one constant asymptotic state as $z\to-\infty$ to (possibly) another constant asymptotic state as $z\to\infty$. (Note that some definitions of solitary waves require the constant asymptotic states to be equal — often to zero.)

To obtain travelling wave solutions of the KdV equation, we seek a solution in the form (1.3.1) which yields a third order ordinary differential equation for w

$$w'''+6ww'-\gamma w'=0, \tag{1.3.2}$$

where $'\equiv \mathrm{d}/\mathrm{d}z$. Integrating this twice gives

$$\tfrac{1}{2}(w')^2=f(w):=-\tfrac{1}{2}(2w^3-\gamma w^2-Aw-B), \tag{1.3.3}$$

with A, B constants. Since we are interested in obtaining real, bounded solutions for the KdV equation, then we require that $f(w)\geq 0$ and so we study the zeros of $f(w)$. There are two cases to consider: (1) when $f(w)$ has only one real zero and (2) when $f(w)$ has three real zeros (for which there are three subcases).

Case 1. If $f(w)$ has only one real zero, a, then it is one of the two forms shown in Figure 1.3.1. If $w'(0)<0$, then $f(w)>0$ for all $z>0$, and so w decreases monotonically to $-\infty$ as $z\to\infty$. If $w'(0)>0$, then w increases until it reaches a, at say z_1 (that is $w(z_1)=a$), which is a simple maximum of w. Therefore $w'(z)<0$, for $z>z_1$ and thereafter w decreases monotonically to $-\infty$ as $z\to\infty$. Hence there are no bounded solutions in this case.

Case 2. If $f(w)$ has three real zeros, a, b and c, then we may assume that $a\leq b\leq c$, and so we may write

$$f(w)=-(w-a)(w-b)(w-c), \tag{1.3.4}$$

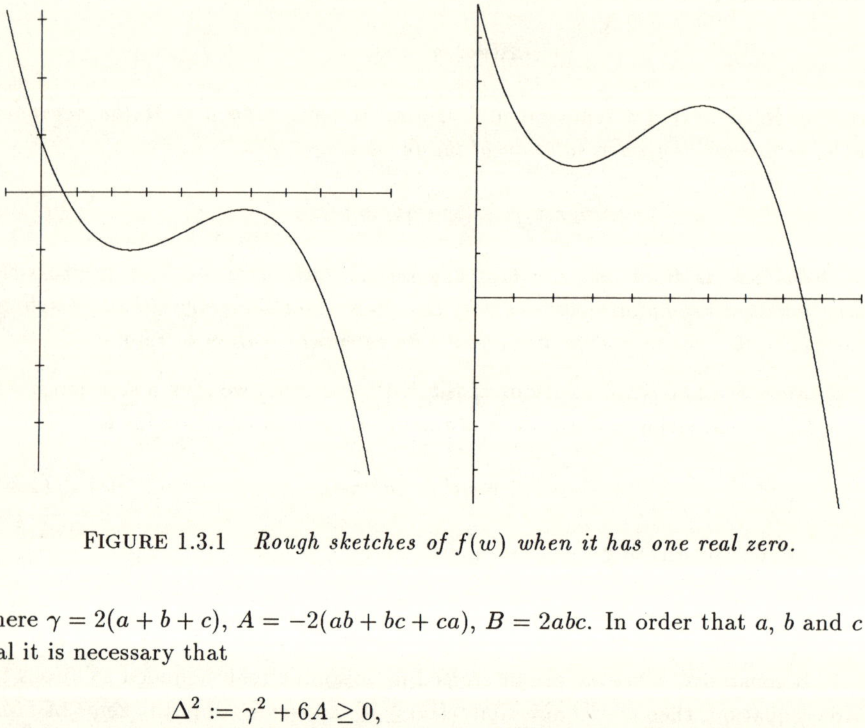

FIGURE 1.3.1 *Rough sketches of $f(w)$ when it has one real zero.*

where $\gamma = 2(a+b+c)$, $A = -2(ab+bc+ca)$, $B = 2abc$. In order that a, b and c are real it is necessary that

$$\Delta^2 := \gamma^2 + 6A \geq 0,$$
$$f\left(\tfrac{1}{3}(\gamma+\Delta)\right) \geq 0, \qquad \text{and} \qquad f\left(\tfrac{1}{3}(\gamma-\Delta)\right) \leq 0.$$

The possible behaviors of the function $f(w)$ against w when it has three real zeros is shown in Figure 1.3.2.

The general behavior of the function $f(w)$ against w is shown in Figure 1.3.2a. In this case all the roots are distinct and the real solution of (1.3.3) represents a nonlinear oscillation between b and c. There are two special cases; one when $b = a$ (Figure 1.3.2b), which corresponds to the solitary wave solution, and the other is when $b = c$ (Figure 1.3.2c), which gives rise to a constant solution related to linear sinusoidal waves. If all three roots coalesce (Figure 1.3.2d), then again we obtain a constant solution as the only bounded solution (see below). We shall now consider the four cases corresponding to Figures 1.3.2a, 1.3.2b, 1.3.2c and 1.3.2d.

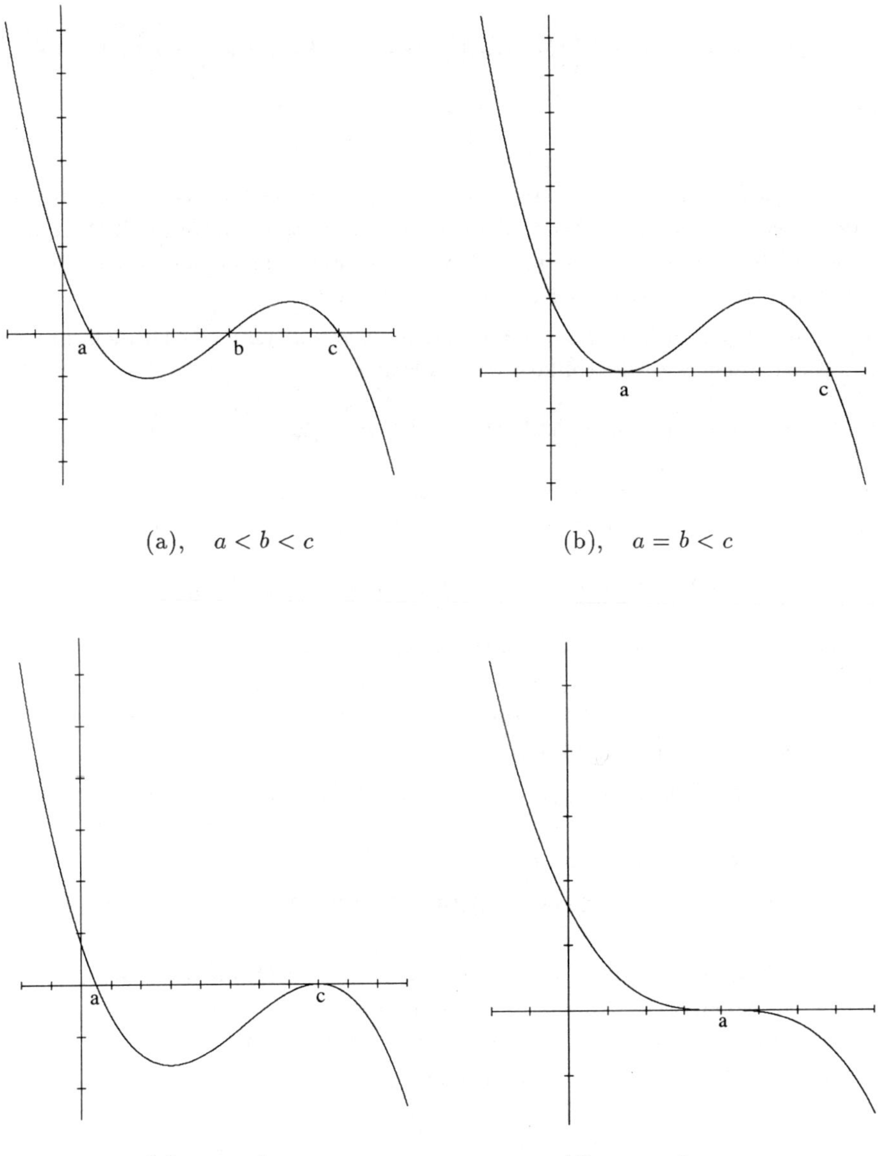

(a), $a < b < c$ (b), $a = b < c$

(c), $a < b = c$ (d), $a = b = c$

FIGURE 1.3.2 *Rough sketches of $f(w)$ when it has three real zeros.*

Case 2a. If a, b, c are distinct, then the solution of the KdV equation is of the form of ***cnoidal waves***, which may be expressed in terms of the Jacobian elliptic function $\mathrm{cn}(z;m)$

$$u(x,t) = b + (c-b)\,\mathrm{cn}^2\left\{\left[\tfrac{1}{2}(c-a)\right]^{1/2}[x-(a+b+c)t+x_0]\,;m\right\}, \tag{1.3.5a}$$

with

$$m = \frac{c-b}{c-a}, \qquad \text{that is} \qquad 0 < m < 1, \tag{1.3.5b}$$

where x_0 is a constant and m denotes the modulus of the Jacobian elliptic function (cf. Whittaker and Watson [1927]). These cnoidal wave solutions of the KdV equation were originally found by Korteweg and de Vries themselves and appear their celebrated paper [1895] (in fact they coined the term cnoidal wave).

Case 2b. $a = b \neq c$. If in (1.3.5) we take the limit $b \to a$ (i.e. $m \to 1$), then, provided that $b \neq c$, it reduces to the solitary wave solution

$$u(x,t) = a + (c-a)\,\mathrm{sech}^2\left\{\left[\tfrac{1}{2}(c-a)\right]^{1/2}[x-(2a+c)t+x_0]\right\}, \tag{1.3.6}$$

since $\mathrm{cn}(z;1) = \mathrm{sech}(z)$ (Abramowitz and Stegun [1972, p.571]). If in (1.3.6) we now set $u_0 = a$ and $\kappa^2 = \frac{1}{2}(c-a)$ then we obtain the solitary wave solution

$$u(x,t) = u_0 + 2\kappa^2\,\mathrm{sech}^2\left\{\kappa[x-(4\kappa^2+3u_0)t+x_0]\right\}. \tag{1.3.7}$$

From this we can easily make the following observations.

1. The speed of the wave relative to the uniform state u_0 is proportional (in this case by a factor of two) to the amplitude (that is, if $u_0 = 0$, then the wave speed is twice the amplitude).
2. The width of the solitary wave $2\pi/\kappa$ is inversely proportional to the square root of the wave amplitude and therefore taller solitary waves are narrower in width and travel faster than shorter ones.
3. The amplitude is independent of the uniform state u_0, the limit of $u(x,t)$ as $|t| \to \infty$.

[We could have derived the solitary wave solution (1.3.7) directly from (1.3.3) by imposing the boundary conditions $w \to u_0$, $w' \to 0$, $w'' \to 0$, as $|z| \to \infty$; note that these compare favorably with the observations of Russell mentioned above.]

Case 2c. If $a \neq b = c$, then the general solution of (1.3.4) is

$$w(z) = b - (b-a)\sec^2\left\{\left[\tfrac{1}{2}(b-a)\right]^{1/2}(z+z_0)\right\},$$

where z_0 is a constant. Therefore the only bounded real solution of the KdV equation is the constant solution $u(x,t) = b$, which may be obtained by taking the limit $c \to b$, that is $m \to 0$, in (1.3.5). Suppose m is very small, and define $\kappa^2 := \frac{1}{2}(c-a)$, then (1.3.5) becomes

$$u(x,t) = b + 2\kappa^2 m\,\mathrm{cn}^2\left\{\kappa[x-(a+b+c)t+x_0];m\right\}.$$

Since $\mathrm{cn}(z;m) = \cos(z) + O(m)$ for small m (Abramowitz and Stegun [1972, p.573]), it follows that m sufficiently small

$$u(x,t) \sim \tfrac{1}{2}(b+c) + \kappa^2 m \cos\left\{2\kappa[x-(a+b+c)t+x_0]\right\},$$

so that the case $c \to b$ may be thought of as the limiting case of a sinusoidal wave with finite period.

Case 2d. If $a = b = c$, then the general solution of (1.3.4) is

$$w(z) = a - \tfrac{1}{2}(z+z_0)^{-2},$$

where z_0 is a constant. Therefore, as for Case 2c, the only bounded real solution of the KdV equation is the constant solution $u(x,t) = a$ [which may have been obtained by taking the limit $c \to a$ in (1.3.6)].

For further details on travelling wave solutions of the KdV equation, see Drazin and Johnson [1989].

1.4 The Discovery of the Soliton.

The physical model which motivated the recent discoveries associated with the KdV equation was the Fermi-Pasta-Ulum (FPU) problem of a one-dimensional anharmonic lattice of equal masses coupled by nonlinear strings which was studied numerically by Fermi, Pasta and Ulum [1955]. They felt that any smooth initial state would eventually relax into equilibrium due to the nonlinear couplings.

The model Fermi, Pasta and Ulum [1955] considered consisted of identical masses connected by nonlinear springs with the force law $F(\Delta) = -K(\Delta + \alpha\Delta^2)$. The equations of motion governing this lattice are

$$m\frac{\partial^2 y_n}{\partial t^2} = K(y_{n+1} - 2y_n + y_{n-1})[1 + \alpha(y_{n+1} - y_{n-1})], \tag{1.4.1}$$

for $n = 1, 2, \ldots, N-1$, with $y_0 = 0 = y_N$ and initial condition $y_n(0) = \sin(n\pi/N)$, $y_{n,t}(0) = 0$ (typically N was taken to be 32). Here y_n measures the displacement of the nth mass from equilibrium.

A great surprise was encountered. Rather than equilibrating, the energy eventually recurred. That is, after flowing back and forth amongst the low order modes, the energy recollected into the lowest mode to within an accuracy of one or two percent, and from there on the process approximately repeated.

In order to understand this phenomenon, Zabusky and Kruskal [1965] considered a continuum model. Calling the length between springs h, $t' = \omega t$ ($\omega = \sqrt{K/m}$), $x' = x/h$ with $x = nh$ and expanding $y_{n\pm1}$ in a Taylor series, equation (1.4.1) reduces to (dropping the primes)

$$y_{tt} = y_{xx} + \varepsilon y_x y_{xx} + \tfrac{1}{12}h^2 y_{xxxx} + O(\varepsilon h^2, h^4), \tag{1.4.2}$$

where $\varepsilon = 2\alpha h$. A further reduction is possible by looking for an asymptotic solution of the form (unidirectional waves)

$$y \sim \phi(\xi,\tau), \qquad \xi = x - t, \qquad \tau = \tfrac{1}{2}\varepsilon t,$$

whereupon (1.4.2) gives

$$\phi_{\xi\tau} + \phi_\xi \phi_{\xi\xi} + \delta^2 \phi_{\xi\xi\xi\xi} + O(h^2, h^4\varepsilon^{-1}) = 0, \tag{1.4.3}$$

where $\delta^2 = \frac{1}{12}h^2\varepsilon^{-1}$ and the $O(h^2, h^4\varepsilon^{-1})$ terms are small. By setting $u = \phi_\xi$ and ignoring small terms, (1.4.3) reduces to the KdV equation

$$u_\tau + uu_\xi + \delta^2 u_{\xi\xi\xi} = 0 \tag{1.4.4}$$

(for further details of this derivation see, for example, Kruskal [1975]).

We saw in §1.3 that the KdV equation possesses a solitary solution

$$u(x,t) = 2\kappa^2 \operatorname{sech}^2\left\{\kappa(x - 4\kappa^2 t - x_0)\right\}, \tag{1.4.5}$$

where κ and x_0 are constants [set $u_0 = 0$ in (1.3.6)]. This special solution of the KdV equation had been well known for a long time, though it was not until Zabusky and Kruskal [1965] did extensive numerical studies for the KdV equation (1.4.4) that the remarkable properties of these solitary waves were discovered. Zabusky and Kruskal considered the initial-value problem for the KdV equation (1.4.4) with the initial condition

$$u(\xi,0) = \cos(\pi\xi), \qquad 0 \le \xi \le 2, \tag{1.4.6}$$

and u, u_ξ, $u_{\xi\xi}$ periodic on $[0,2]$ for all t; they took $\delta = 0.022$ (the periodic problem is more complicated than the one on the infinite line, but is better suited to numerical computations). Zabusky and Kruskal discovered that after a short time the wave steepens and almost produces a shock, but the dispersive term ($\delta^2 u_{xxx}$) then becomes significant and a balance between the nonlinear and dispersion terms ensues. Later the solutions develops a train of eight well-defined waves, each like sech^2 functions, with the faster (taller) waves continually catching up and overtaking the slower (smaller) waves. When two solitary waves of the form (1.4.5) (with different wave speeds), are initially well separated, with the larger one to the left (the waves are travelling from left to right), then the faster, taller wave overlaps the slower, smaller one, and the waves interact nonlinearly. After the interaction, the waves separate, with the larger one on the right, having regained their initial amplitude and velocity and the only effect of the interaction are phase shifts, that is the centers of the waves are at different positions than where they would have been had there been no interaction. Finally, after a long time the initial profile recurs.

At the center of these observations is the discovery that these nonlinear waves can interact elastically and continue afterwards almost as if there had been no interaction at all. Because of the analogy with particles, Zabusky and Kruskal named these special waves as *solitons*.

DEFINITION 1.4.1

A soliton is a solitary wave which asymptotically preserves its shape and velocity upon nonlinear interaction with other solitary waves, or more generally, with another (arbitrary) localized disturbance.

Kruskal and Zabusky's remarkable numerical discovery demanded an analytical explanation and detailed mathematical study of the KdV equation. However the KdV equation is nonlinear and at that time no general method of solution for nonlinear equations existed. (We remark that Lax and Levermore [1983a,b,c] and Venakides [1985a,b] have reinvestigated (1.4.4) with δ small.)

1.5 Fourier Transforms.

Before discussing the inverse scattering method of solution for the KdV equation we shall discuss a general method for solving *linear* partial differential equations.

Consider the partial differential equation

$$u_t = \Delta(x, t, u(x,t)), \tag{1.5.1}$$

where $\Delta(x,t,u)$ is a function of $u(x,t)$ and its spatial derivatives, with $t \in \mathbb{R}$, and $x \in \mathbb{R}$, the temporal and spatial variables, respectively. A typical problem associated with equation (1.5.1) is to solve it subject to the given initial condition

$$u(x,0) = f(x), \tag{1.5.2}$$

provided that the problem is well posed for f in some Banach space $\mathcal{B}$ (usually either a space of functions vanishing sufficiently rapidly as $|x| \to \infty$ or a space of periodic functions). If (1.5.1) is a linear partial differential equation, then it is usually solvable by Fourier transform methods (on the infinite line, provided that $f(x)$ decays sufficiently rapidly as $|x| \to \infty$, for example $f \in L^1(\mathbb{R})$). Given the linear partial differential equation in the form

$$u_t = -\mathrm{i}\omega(-\mathrm{i}\partial_x)u, \tag{1.5.3}$$

where $\partial_x \equiv \partial/\partial x$ (usually ω is a polynomial function with constant coefficients, referred to as the dispersion relation); its solution, given the initial condition (1.5.2), is obtained by use of the Fourier transform pair

$$u(x,t) = \frac{1}{2\pi}\int_{-\infty}^{\infty} \hat{u}(k,t)\,\mathrm{e}^{\mathrm{i}kx}\,\mathrm{d}k, \tag{1.5.4}$$

$$\hat{u}(k,t) = \int_{-\infty}^{\infty} u(x,t)\,\mathrm{e}^{-\mathrm{i}kx}\,\mathrm{d}x. \tag{1.5.5}$$

Assuming the validity of the interchange of derivative and integral, by taking the Fourier transform of (1.5.3) we obtain a linear ordinary differential equation for $\hat{u}(k,t)$

$$\frac{\mathrm{d}\hat{u}}{\mathrm{d}t} = -\mathrm{i}\omega(k)\hat{u}, \tag{1.5.6}$$

which has general solution

$$\hat{u}(k,t) = \hat{u}(k,0)\exp\{-\mathrm{i}\omega(k)t\}, \tag{1.5.7a}$$

where

$$\hat{u}(k,0) = \hat{f}(k) = \int_{-\infty}^{\infty} f(x)\,\mathrm{e}^{-\mathrm{i}kx}\,\mathrm{d}x. \tag{1.5.7b}$$

Therefore from (1.5.4)

$$u(x,t) = \frac{1}{2\pi}\int_{-\infty}^{\infty} \hat{f}(k)\exp\{\mathrm{i}[kx - \omega(k)t]\}\,\mathrm{d}k. \tag{1.5.8}$$

Conceptually the Fourier transform method may be viewed as follows.

1. The initial data is transformed into Fourier space by means of (1.5.7b).
2. The time evolution in Fourier space is particularly simple since it satisfies (1.5.6).
3. $u(x,t)$ is recovered from the inverse transform (1.5.4) and is given by (1.5.8).

Schematically this may be written as:

$$\begin{array}{ccc}
u(x,0) & \xrightarrow{\text{Fourier Transform}} & \hat{u}(k,0) \\
 & & \Big\downarrow \ \omega(k)\text{: dispersion relation} \\
u(x,t) & \xleftarrow{\text{Inverse Fourier Transform}} & \hat{u}(k,t)
\end{array}$$

Although (1.5.8) is a solution in quadrature form, useful information may be obtained from it in the asymptotic limit $t \to \infty$. If the system is conservative ($\omega(k)$ is real for real k) and dispersive ($\omega''(k) \neq 0$) then the initial data decays into wave packets which move with their group velocity $\omega'(k)$ and decay algebraically as $t \to \infty$ (in general, asymptotic formulae can be given to describe this see also below). Hence the asymptotic behavior of these linear problems is relatively simple.

EXAMPLE 1.5.1 *The linearized Korteweg-de Vries equation*

A simple example of the method of Fourier transforms is given by the linearized KdV equation

$$u_t + u_{xxx} = 0, \tag{1.5.9a}$$

with

$$u \to 0, \qquad \text{as} \quad |x| \to \infty, \qquad u(x,0) = f(x) \in L^1(\mathbb{R}). \tag{1.5.9b}$$

The dispersion relation is given by $\omega(k) = -k^3$ and so the Fourier transform solution of (1.5.9) is

$$u(x,t) = \frac{1}{2\pi}\int_{-\infty}^{\infty} \hat{f}(k)\exp\left\{\mathrm{i}(kx + k^3 t)\right\} \mathrm{d}k.$$

Asymptotically the solution has the representation as $|t| \to \infty$

$$u(x,t) \sim (3t)^{-1/3}\,\mathrm{Ai}(z)\left(\frac{\hat{f}(y)+\hat{f}(-y)}{2}\right) + (3t)^{-2/3}\,\mathrm{Ai}'(z)\left(\hat{f}(y) - \frac{\hat{f}(-y)}{2\mathrm{i}k}\right),$$

where $z = x/(3t)^{1/3}$ and $y = (-x/(3t))^{1/2}$, and $\mathrm{Ai}(z)$ is the Airy function given by

$$\mathrm{Ai}(z) = \frac{1}{2\pi}\int_{-\infty}^{\infty} \exp\left\{\mathrm{i}\left(sz + \tfrac{1}{3}s^3\right)\right\} \mathrm{d}s$$

(see Ablowitz and Segur [1981, p.363] for further details).

1.6 An Infinite Number of Conserved Quantities.

An important stage in the development of the general method of solution for the KdV equation was the discovery that KdV had an infinite number of independent conservation laws.

DEFINITION 1.6.1

For the partial differential equation

$$\Delta(x,t,u(x,t)) = 0, \tag{1.6.1}$$

where $t \in \mathbb{R}$, $x \in \mathbb{R}$ are temporal and spatial variables and $u(x,t) \in \mathbb{R}$ the dependent variable, a conservation law is an equation of the form

$$\mathrm{D}_t T_i + \mathrm{D}_x X_i = 0, \tag{1.6.2}$$

which is satisfied for all solutions of the equation (1.6.1), *where $T_i(x,t)$, the conserved density, and $X_i(x,t)$, the associated flux, are, in general, functions of x, t, u and the partial derivatives of u; D_t denotes the total derivative with respect to t; and D_x the total derivative with respect to x.*

If additionally, $u \to 0$ as $|x| \to \infty$ sufficiently rapidly, then

$$\frac{\mathrm{d}}{\mathrm{d}t}\int_{-\infty}^{\infty} T_i(x,t)\,\mathrm{d}x = 0.$$

Therefore

$$\int_{-\infty}^{\infty} T_i(x,t)\,\mathrm{d}x = c_i,$$

where c_i, constant, is the *conserved density*. For the KdV equation, the first three conservation laws are

$$\begin{aligned}
&(u)_t + (3u^2 + u_{xx})_x = 0,\\
&(u^2)_t + (4u^3 + 2uu_{xx} - u_x^2)_x = 0,\\
&\left(u^3 - \tfrac{1}{2}u_x^2\right)_t + \left(\tfrac{9}{2}u^4 + 3u^2u_{xx} - 6uu_x^2 - u_xu_{xxx} + \tfrac{1}{2}u_{xx}^2\right)_x = 0.
\end{aligned}$$

The first two conservation laws correspond to conservation of momentum and energy, respectively. The third (and less obvious one) was discovered by Whitham [1965].

The fourth and fifth conservation laws for the KdV equation were found by Kruskal and Zabusky [1963]. Subsequently four additional ones were discovered and it was conjectured that there was an infinite number. Despite rumors that only nine existed, Miura worked out the tenth. After studying the conservation laws of the KdV equation, and those associated with another partial differential equation, the Modified Korteweg-de Vries (mKdV) equation

$$v_t - 6v^2v_x + v_{xxx} = 0, \tag{1.6.3}$$

Miura [1968] discovered the following transformation, now known as *Miura's transformation*. If v is a solution of the mKdV equation (1.6.3), then

$$u = -(v^2 + v_x), \tag{1.6.4}$$

is a solution of the KdV equation (1.1.2). This is readily seen from the relation

$$u_t + 6uu_x + u_{xxx} = -(2v + \partial_x)(v_t - 6v^2 v_x + v_{xxx}), \tag{1.6.5}$$

where $\partial_x \equiv \partial/\partial_x$. Note that every solution of the mKdV equation (1.6.3) leads, via Miura's transformation (1.6.4), to a solution of the KdV equation, but the converse is not true (that is not every solution of the KdV equation can be obtained from a solution of the mKdV equation — see, for example, Ablowitz, Kruskal and Segur [1979]). Miura's transformation leads to many other important results related to the KdV equation. Initially it formed the basis of a proof that the KdV and mKdV equations have an infinite number of conserved densities (Miura, Gardner and Kruskal [1968]), which we outline below. However, even more significant was the motivation given by Miura's transformation (1.6.4) in the development of the *inverse scattering method* for solving the initial value problem for the KdV equation (1.1.2), which we shall discuss in §1.7.

Early on there was an ingenious proof of the existence of an infinite number of conservation laws (Miura, Gardner and Kruskal [1968]). Define w by the relation

$$u = w - \varepsilon w_x - \varepsilon^2 w^2, \tag{1.6.6}$$

which may be thought of as generalization of Miura's transformation (1.6.4). Then the equivalent relation to (1.6.5) is

$$u_t + 6uu_x + u_{xxx} = \left(1 - \varepsilon\partial_x - 2\varepsilon^2 w\right)\left(w_t + 6(w - \varepsilon^2 w^2)w_x + w_{xxx}\right). \tag{1.6.7}$$

Therefore u, as defined by (1.6.6), is a solution of the KdV equation provided that w is a solution of

$$w_t + 6(w - \varepsilon^2 w^2)w_x + w_{xxx} = 0 \tag{1.6.8}$$

(as with Miura's transformation, the transformation (1.6.6) is only one way). Since the KdV equation does not contain ε, then it's solution u depends only upon x and t, however w, a solution of equation (1.6.8), depends on x, t and ε. Then seek a formal power series solution of (1.6.6), in the form

$$w(x, t; \varepsilon) = \sum_{n=0}^{\infty} w_n(x, t)\varepsilon^n. \tag{1.6.9}$$

Since the equation (1.6.8) is in conservation form, then

$$\int_{-\infty}^{\infty} w(x, t; \varepsilon)\,\mathrm{d}x = \text{constant},$$

and so

$$\int_{-\infty}^{\infty} w_n(x, t)\,\mathrm{d}x = \text{constant},$$

for each $n = 0, 1, 2, \ldots$. Substituting (1.6.9) into (1.6.6) and equating coefficients of powers of ε and solving recursively gives

$$w_0 = u, \tag{1.6.10a}$$

$$w_1 = w_{0,x} = u_x, \tag{1.6.10b}$$

$$w_2 = w_{1,x} + w_0^2 = u_{xx} + u^2, \tag{1.6.10c}$$

$$w_3 = w_{2,x} + 2w_0w_1 = u_{xxx} + 4uu_x, \tag{1.6.10d}$$

$$w_4 = w_{3,x} + 2w_0w_2 + w_1^2 = u_{xxxx} + 6uu_{xx} + 5u_x^2 + 2u^3, \tag{1.6.10e}$$

etc.. Continuing to all powers of ε gives an infinite number of conserved densities. The corresponding conservation laws may be found by substituting (1.6.9-10) into equation (1.6.8) and equating coefficients of powers of ε (odd powers of ε actually give no useful information since they are the derivative of the previous even power, however the even powers give independent conservation laws for the KdV equation).

1.7 The Associated Linear Scattering Problem and Inverse Scattering.

1.7.1 The Inverse Scattering Method. Many physical problems are modelled by nonlinear partial differential equations for which, unfortunately, the Fourier transform method fails to solve the problem. There was no unified method by which classes of nonlinear partial differential equations could be solved, and the solutions were often obtained by rather *ad hoc* methods. Therefore a most significant result was the development by Gardner, Greene, Kruskal and Miura [1967, 1974] of a method for the exact solution of the initial-value problem for the KdV equation (1.1.2) for initial values which decay sufficiently rapidly, through a series of *linear* equations. Prior to this work, the only known exact solutions of the KdV equation were the solitary wave and cnoidal wave solutions described in §1.3. The aim is to solve the KdV equation (1.1.2) for $(x,t) : x \in \mathbb{R}$, $t > 0$, subject to the initial condition

$$u(x,0) = f(x), \tag{1.7.1}$$

where $f(x)$ decays sufficiently rapidly as $|x| \to \infty$.

The basic idea is to relate the KdV equation to the time-independent Schrödinger scattering problem

$$\mathrm{L}v := v_{xx} + u(x,t)v = \lambda v, \tag{1.7.2}$$

which has been extensively studied by mathematicians and physicists. The motivation for this equation came from studying the Miura transformation relating solutions of the KdV and mKdV equations. Recall that if $U(x,t)$ is a solution of the mKdV equation

$$U_t - 6U^2U_x + U_{xxx} = 0,$$

then

$$u = -(U^2 + U_x), \tag{1.7.3}$$

is a solution of the KdV equation

$$u_t + 6uu_x + u_{xxx} = 0. \tag{1.7.4}$$

(1.7.3) may be viewed as a Riccati equation for U in terms of u, and it is well known that it may be linearized by the transformation $U = v_x/v$, which yields

$$v_{xx} + uv = 0. \tag{1.7.5}$$

Since the KdV equation is Galilean invariant, that is invariant under the transformation

$$(x, t, u(x,t)) \rightarrow \left(x - ct, t, u(x,t) + \tfrac{1}{6}c\right),$$

where c is some constant, then it is natural to consider (1.7.2) rather than (1.7.5), in which t plays the role of a parameter and $u(x,t)$ the potential. For (1.7.2), the eigenvalues and the behavior of the eigenfunctions as x determine the *scattering data*, $S(\lambda, t)$, which depends upon the potential $u(x,t)$. The *direct scattering problem* is to map the potential into the scattering data. The *inverse scattering problem* is to reconstruct the potential from the scattering data.

The time dependence of the eigenfunctions of (1.7.2) is given by

$$v_t = (\gamma + u_x)v - (4\lambda + 2u)v_x, \tag{1.7.6}$$

where γ is an arbitrary constant. Assuming that $\lambda_t = 0$, then from (1.7.2) and (1.7.6) we obtain

$$v_{txx} = [(\gamma + u_x)(\lambda - u) + u_{xxx} + 6uu_x]\,v - (4\lambda + 2u)(\lambda - u)v_x, \tag{1.7.7a}$$
$$v_{xxt} = [(\lambda - u)(\gamma + u_x) - u_t]\,v - (\lambda - u)(4\lambda + 2u)v_x. \tag{1.7.7b}$$

Therefore (1.7.2) and (1.7.6) are compatible (i.e., $v_{xxt} = v_{txx}$), if and only if u satisfies the KdV equation (1.7.4). Similarly, if (1.7.4) is satisfied, then necessarily the eigenvalues must be time independent (i.e., $\partial\lambda/\partial t = 0$).

The solution of (1.7.4), corresponding to $u \rightarrow 0$ as $|x| \rightarrow \infty$, proceeds as follows:

1. Direct problem. At time $t = 0$, given $u(x,0)$ we solve the direct scattering problem. The spectrum of the Schrödinger equation (1.7.2) consists of a finite number of discrete eigenvalues, $\lambda = \kappa_n^2$, $n = 1, 2, \ldots, N$, for $\lambda > 0$ and a

continuum, $\lambda = -k^2$, for $\lambda < 0$. The eigenfunctions corresponding to these eigenvalues may be computed and their asymptotic behavior written as: for $0 < \lambda = \kappa_n^2$,

$$v_n(x,t) \sim c_n(t)\exp(-\kappa_n x), \qquad \text{as} \quad x \to \infty, \tag{1.7.8a}$$

with

$$\int_{-\infty}^{\infty} v_n^2 \, \mathrm{d}x = 1, \tag{1.7.8b}$$

and for $0 > \lambda = -k^2$,

$$v(x,t) \sim \mathrm{e}^{-\mathrm{i}kx} + r(k,t)\,\mathrm{e}^{\mathrm{i}kx}, \qquad \text{as} \quad x \to \infty, \tag{1.7.9a}$$

$$v(x,t) \sim a(k,t)\,\mathrm{e}^{-\mathrm{i}kx}, \qquad \text{as} \quad x \to -\infty, \tag{1.7.9b}$$

where $r(k,t)$ is the *reflection coefficient* and $a(k,t)$ the *transmission coefficient*. Therefore we have the scattering data at time $t = 0$;

$$S(\lambda, 0) = \left(\{\kappa_n, c_n(0)\}_{n=1}^{N}, r(k,0), a(k,0)\right).$$

2. Time evolution. From equation (1.7.6) we can determine the time evolution of the scattering data. It may be shown that

$$\kappa_n = \text{constant}, \qquad n = 1, 2, \ldots, N \tag{1.7.10a}$$

$$c_n(t) = c_n(0)\exp(4\kappa_n^3 t), \qquad n = 1, 2, \ldots, N \tag{1.7.10b}$$

$$a(k,t) = a(k,0), \tag{1.7.10c}$$

$$r(k,t) = r(k,0)\exp(8\mathrm{i}k^3 t), \tag{1.7.10d}$$

so we have the scattering data at time t;

$$S(\lambda, t) = \left(\{\kappa_n, c_n(t)\}_{n=1}^{N}, r(k,t), a(k,t)\right).$$

3. Inverse Problem. Given the scattering data at the initial time, it is possible to determine its time evolution. The inverse scattering problem is to reconstruct from knowledge of the scattering $S(\lambda, t)$ data, the potential $u(x,t)$ which is the required solution of the KdV equation. This problem was considered by Gel'fand and Levitan [1955] (see also Faddeev [1963] — for a recent discussion see Deift and Trubowitz [1979]; later, we shall view this techniques in a different way — as a Riemann-Hilbert boundary value problem). The results may be summarized as follows. First, using the scattering data (1.7.10), define the function

$$F(x;t) = \sum_{n=1}^{N} c_n^2(t)\exp(-\kappa_n x) + \frac{1}{2\pi}\int_{-\infty}^{\infty} r(k,t)\,\mathrm{e}^{\mathrm{i}kx}\,\mathrm{d}k. \tag{1.7.11}$$

Then solve the linear integral equation

$$K(x,y;t) + F(x+y;t) + \int_x^\infty K(x,z;t)F(z+y;t)\,\mathrm{d}z = 0, \tag{1.7.12}$$

called the *Gel'fand-Levitan-Marchenko equation.* Finally the potential is reconstructed by the relation

$$u(x,t) = 2\frac{\partial}{\partial x}\left[K(x,x;t)\right]. \tag{1.7.13}$$

This method is therefore conceptually analogous in many ways to the Fourier transform method for solving linear equations, except however that the final step of solving the inverse scattering problem is highly nontrivial. Schematically this may be written as

$$\begin{array}{ccc}
u(x,0) & \xrightarrow{\text{direct scattering}} & S(\lambda,0) = \left(\{\kappa_n, c_n(0)\}_{n=1}^N, r(k,0), a(k,0)\right) \\
 & & \big\downarrow\ \omega(k)\text{: dispersion relation} \\
u(x,t) & \xleftarrow{\text{inverse scattering}} & S(\lambda,t) = \left(\{\kappa_n, c_n(t)\}_{n=1}^N, r(k,t), a(k,t)\right)
\end{array}$$

The scattering data plays the role of the Fourier transform and the inverse scattering problem the inverse Fourier transform.

1.7.2 Reflectionless Potentials. If the reflection coefficient is zero (i.e., $r(k,0) = 0 = r(k,t)$), then we obtain the special soliton solutions. If $r(k,t) = 0$ in (1.7.11), then the kernel and inhomogeneous terms in the Gel'fand-Levitan-Marchenko integral equation are reduced to finite sums over the discrete spectrum. This is solvable by separation of variables (cf., Kay and Moses [1956] and Gardner, Greene, Kruskal and Miura [1974]).

If $r(k,t) = 0$ in (1.7.11) then

$$F(x;t) = \sum_{n=1}^N c_n^2(t)\exp(-\kappa_n x), \tag{1.7.14}$$

with $c_m(t) = c_m(0)\exp(4\kappa_j^3 t) > 0$ and distinct $\kappa_m > 0$, $m = 1, 2, \ldots, N$ and so the Gel'fand-Levitan-Marchenko equation becomes

$$\begin{aligned} K(x,y;t) &+ \sum_{n=1}^N c_n^2(t)\exp\{-\kappa_n(x+y)\} \\ &+ \int_x^\infty K(x,z;t)\sum_{n=1}^N c_n^2(t)\exp\{-\kappa_n(z+y)\}\,\mathrm{d}z = 0. \end{aligned} \tag{1.7.15}$$

The solution of this equation now takes the form

$$K(x,y;t) = \sum_{n=1}^{N} c_n(t) v_n(x) \exp(-\kappa_n y),$$

then (1.7.15) becomes

$$v_m(x) + \sum_{n=1}^{N} \frac{c_m(t)\, c_n(t)}{\kappa_m + \kappa_n} \exp\{-(\kappa_m + \kappa_n)x\} v_n(x) = c_m(t) \exp(-\kappa_m x), \qquad (1.7.16)$$

for $m = 1, 2, \ldots, N$. This is a system of algebraic equations which can be written in the form

$$(\mathbf{I} + \mathbf{C})\boldsymbol{v} = \boldsymbol{f}, \qquad (1.7.17)$$

where $\boldsymbol{v} = (v_1, v_2, \ldots, v_N)$, $\boldsymbol{f} = (f_1, f_2, \ldots, f_N)$ with $f_m := c_m(t) \exp(-\kappa_m x)$, $m = 1, 2, \ldots, N$, $\mathbf{I}$ is the $N \times N$ identity matrix and $\mathbf{C}$ is a symmetric, $N \times N$ matrix with entries

$$C_{mn} = \frac{c_m(t)\, c_n(t)}{\kappa_m + \kappa_n} \exp\{-(\kappa_m + \kappa_n)x\}, \qquad m, n = 1, 2, \ldots, N.$$

A sufficient condition for (1.7.17) to have a unique solution is that $\mathbf{C}$ is positive definite. Consider the quadratic form

$$\begin{aligned} \boldsymbol{\xi}^{\mathrm{T}} \mathbf{C} \boldsymbol{\xi} &= \sum_{m=1}^{N} \sum_{n=1}^{N} \frac{c_m c_n \xi_m \xi_n}{\kappa_m + \kappa_n} \exp\{-(\kappa_m + \kappa_n)x\} \\ &= \int_x^{\infty} \left[\sum_{n=1}^{N} c_n \xi_n \exp(-\kappa_n y) \right]^2 \mathrm{d}y, \end{aligned}$$

where $\boldsymbol{\xi} = (\xi_1, \xi_2, \ldots, \xi_N)$, which clearly positive and equal to zero only if $\boldsymbol{\xi} = \mathbf{0}$, and so $\mathbf{C}$ is positive definite. The unique solution to the KdV equation in this case is

$$u(x,t) = 2 \frac{\partial^2}{\partial x^2} \left\{ \ln \det(\mathbf{I} + \mathbf{C}) \right\}, \qquad (1.7.18)$$

This is the N-soliton solution for the KdV equation, which corresponds to N waves which asymptotically as $t \to \pm\infty$ have the form

$$u_n(x,t) \sim 2\kappa_n^2 \operatorname{sech}^2 \left\{ \kappa_n (x - 4\kappa_n^2 t + x_n) \right\}, \qquad n = 1, 2, \ldots, N,$$

where $x_n = \ln c_n^2(0)$ is a constant. There is a one-to-one relationship between the number of discrete eigenvalues and the number of solitons which emerge asymptotically (see below). As mentioned previously above, these waves interact in such a way as to preserve their identities in the limit asymptotically, however by using the inverse scattering method to solve the KdV equation, we are able to mathematically confirm the numerical observations of Zabusky and Kruskal.

If in (1.7.18) we set $N = 1$, then we obtain the one-soliton solution

$$u(x,t) = 2\kappa_1^2 \operatorname{sech}^2\left\{\kappa_1(x - 4\kappa_1^2 t + x_1)\right\},$$

as derived in equation (1.3.7) above.

For $N = 2$, we write

$$\begin{aligned}\Delta :&= \det(\mathbf{I} + \mathbf{C})\\ &= 1 + \exp(2\eta_1) + \exp(2\eta_2) + \exp(2\eta_1 + 2\eta_2 + A_{12}),\end{aligned} \tag{1.7.19a}$$

with

$$\eta_n = -\kappa_n(x - 4\kappa_n^2 t + \ln c_n^2(0)), \qquad A_{mn} = 2\ln\left(\frac{\kappa_m - \kappa_n}{\kappa_m + \kappa_n}\right) \tag{1.7.19b}$$

(below we explicitly give the two-soliton solution of the KdV equation). From equation (1.7.19) it is easily shown that the only effect of the interaction of two solitary waves is a phase shift (see Figure 1.7.1). Consider the trajectory $\eta_1 =$ constant, and assume that $\kappa_1 > \kappa_2 > 0$. Then

$$\Delta \sim 1 + \exp(2\eta_1), \qquad \text{as} \quad t \to -\infty, \tag{1.7.20a}$$

$$\Delta \sim \exp(2\eta_2) + \exp(2\eta_1 + 2\eta_2 + A_{12}), \qquad \text{as} \quad t \to \infty. \tag{1.7.20b}$$

Therefore from (1.7.18) it follows that for fixed η_1

$$\begin{aligned}u(x,t) &= 2\frac{\partial^2}{\partial x^2}\left\{\ln \Delta\right\},\\ &\sim 2\kappa_1^2 \operatorname{sech}^2(\eta_1 + \delta_1^{\pm}), \qquad \text{as} \quad t \to \pm\infty,\end{aligned} \tag{1.7.21a}$$

with

$$\delta_1^+ = \tfrac{1}{2}A_{12}, \qquad \delta_1^- = 0. \tag{1.7.21b}$$

Similarly, for fixed η_2

$$u(x,t) \sim 2\kappa_2^2 \operatorname{sech}^2(\eta_2 + \delta_2^{\pm}), \qquad \text{as} \quad t \to \pm\infty, \tag{1.7.22a}$$

with

$$\delta_2^+ = 0, \qquad \delta_2^- = \tfrac{1}{2}A_{12}. \tag{1.7.22b}$$

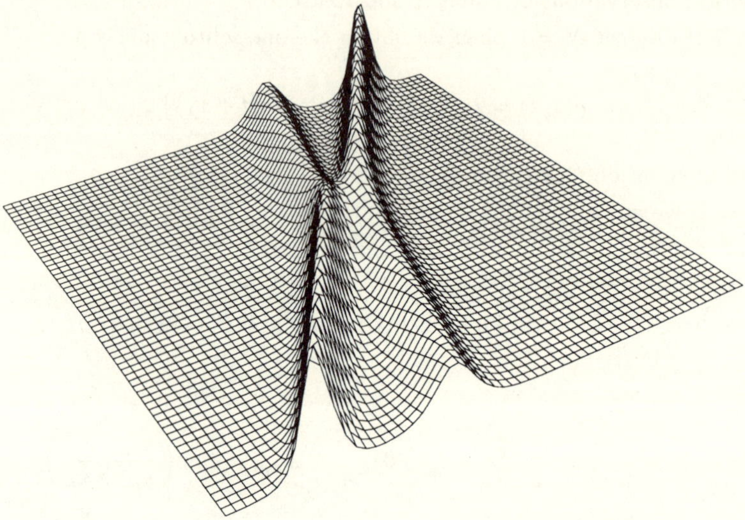

FIGURE 1.7.1 *Two-soliton solution of the KdV equation.*

For large negative time, the larger soliton lies to the left of the smaller soliton, and vice-versa for large positive time (the solitons are travelling from right to left). From (1.7.21,22) it follows that as a consequence of the interaction, the larger and smaller solitons undergo phase shifts given by

$$\delta_1 = \delta_1^+ - \delta_1^- = \tfrac{1}{2}A_{12} = \ln\left(\frac{\kappa_1 - \kappa_2}{\kappa_1 + \kappa_2}\right) < 0, \tag{1.7.23a}$$

$$\delta_2 = \delta_2^+ - \delta_2^- = -\tfrac{1}{2}A_{12} = -\ln\left(\frac{\kappa_1 - \kappa_2}{\kappa_1 + \kappa_2}\right) > 0, \tag{1.7.23b}$$

respectively. Therefore, after the interaction, the larger soliton has been shifted to the right of where it would have been had there been no interaction, and the smaller shifted to the left by the same amount.

A similar calculation can be done in the N-soliton case (see Gardner, Greene, Kruskal and Miura [1974]; Zakharov [1971]; Wadati and Toda [1972]; Tanaka [1972]). Assume that $\kappa_1 > \kappa_2 > \ldots > \kappa_N > 0$, then for fixed η_n

$$u(x,t) \sim 2\kappa_n^2 \operatorname{sech}^2(\eta_n + \delta_n^{\pm}), \qquad \text{as} \quad |t| \to \infty. \tag{1.7.24}$$

Therefore the nth soliton undergoes a phase shift given by

$$\delta_n = \delta_n^+ - \delta_n^- = \sum_{m=n+1}^{N} \ln\left(\frac{\kappa_n - \kappa_m}{\kappa_n + \kappa_m}\right) - \sum_{m=1}^{n-1} \ln\left(\frac{\kappa_m - \kappa_n}{\kappa_m + \kappa_n}\right). \tag{1.7.25}$$

It is clear from this expression that the total phase shift experienced by the nth soliton is equivalent to the sum of phase shifts that arise from pairwise interactions with every other soliton. The general question of the interaction of N-soliton and continuous spectra has been examined by Ablowitz and Kodama [1980]; Alonso [1985a,b,c]; Schuur [1986].

To demonstrate the relationship between the number of discrete eigenvalues and the number of solitons, suppose that the initial condition is given by

$$u(x,0) = N(N+1)\operatorname{sech}^2 x.$$

In this case the scattering problem, with $\lambda = \kappa^2$, is

$$v_{xx} + \left\{N(N+1)\operatorname{sech}^2 x - \kappa^2\right\} v = 0. \tag{1.7.26}$$

If we make the transformation $\mu = \tanh x$, then (1.7.26) becomes

$$(1-\mu^2)\frac{\mathrm{d}^2 v}{\mathrm{d}\mu^2} - 2\mu\frac{\mathrm{d}v}{\mathrm{d}\mu} + \left(N(N+1) - \frac{\kappa^2}{1-\mu^2}\right) v = 0, \tag{1.7.27}$$

which is the ***Associated Legendre equation*** (cf. Abramowitz and Stegun [1972, Chapter 8]). Equation (1.7.27) has N distinct eigenvalues $\kappa_n = n$, $n = 1, 2, \ldots, N$ and bounded associated eigenfunctions $v_1(x), v_2(x), \ldots, v_N(x)$, given by

$$v_n(x) = \gamma_n P_N^n(\tanh x) \sim c_n \exp(-nx), \qquad \text{as} \quad x \to \infty,$$

where $P_N^n(\mu)$ is the Associated Legendre polynomial defined by

$$P_N^n(\mu) = (\mu^2 - 1)^{n/2}\frac{\mathrm{d}^n}{\mathrm{d}\mu^n}\left(\frac{1}{2^N N!}\frac{\mathrm{d}^N}{\mathrm{d}\mu^N}(\mu^2-1)^N\right),$$

and $c_n(0)$ is determined from the normalization condition (1.7.8b), and so from equation (1.7.10b)

$$c_n(t) = c_n(0)\exp(4n^3 t).$$

Therefore, from equation (1.7.11), the functions $F(x;t)$ in the Gel'fand-Levitan-Marchenko equation is given by

$$F(x;t) = \sum_{n=1}^{N} c_n^2(t)\exp(-nx),$$

and so, from we equation (1.7.18) we obtain the N-solution of the KdV equation

$$u(x,t) = 2\frac{\partial^2}{\partial x^2}\left\{\ln\det(\mathbf{I} + \mathbf{C})\right\}, \tag{1.7.28a}$$

where

$$C_{mn} = \frac{c_m(t)\, c_n(t)}{m+n} \exp\{-(m+n)x\}. \tag{1.7.28b}$$

In particular, the two-soliton solution of the KdV equation is [see also (1.7.19)]

$$u(x,t) = 12 \left\{ \frac{3 + 4\cosh(2x - 8t) + \cosh(4x - 64t)}{[3\cosh(x - 28t) + \cosh(3x - 36t)]^2} \right\}. \tag{1.7.29}$$

If we introduce $\xi = x - 16t$ and $\eta = x - 4t$, then the two-soliton solution can be expressed as

$$u(x,t) = 12 \left\{ \frac{3 + 4\cosh(2\xi + 24t) + \cosh(4\xi)}{[3\cosh(\xi - 12t) + \cosh(3\xi + 12t)]^2} \right\}, \tag{1.7.30}$$

and

$$u(x,t) = 12 \left\{ \frac{3 + 4\cosh(2\eta) + \cosh(4\eta - 48t)}{[3\cosh(x - 24t) + \cosh(3\eta - 24t)]^2} \right\}. \tag{1.7.31}$$

Expanding (1.7.30,31) as $t \to \pm\infty$, keeping ξ and η fixed respectively, shows that as $t \to \pm\infty$

$$u(x,t) \sim 2\,\mathrm{sech}^2\left(\eta \pm \tfrac{1}{2}\ln 3\right) + 8\,\mathrm{sech}^2\left(2\xi \mp \tfrac{1}{2}\ln 3\right).$$

(The structure of the two-soliton solution of the KdV equation during the interaction has been discussed analytically by a number of authors, e.g., Lax [1968]; Ablowitz and Kodama [1980]; Moloney and Hodnett [1986]; LeVeque [1987]; Hodnett and Moloney [1989].)

For a discussion of the case when the reflection coefficient is nonzero see for example Ablowitz and Kodama [1980]; Ablowitz and Segur [1981, §1.7]; Alonso [1985a,b,c]; Schuur [1986] and the references therein.

1.8 Lax's Generalization.

Lax [1968] put the inverse scattering method for solving the KdV equation into a more general framework which subsequently paved the way to generalizations of the technique as a method for solving other partial differential equations. Consider two operators L and M, where L is the operator of the spectral problem and M is the operator governing the associated time evolution of the eigenfunctions

$$\mathrm{L}v = \lambda v, \tag{1.8.1a}$$

$$v_t = \mathrm{M}v. \tag{1.8.1b}$$

Now take $\partial/\partial t$ of (1.8.1a), giving

$$\mathrm{L}_t v + \mathrm{L} v_t = \lambda_t v + \lambda v_t,$$

hence using (1.8.1b)

$$\begin{aligned} \mathrm{L}_t v + \mathrm{LM}v &= \lambda_t v + \lambda \mathrm{M}v, \\ &= \lambda_t v + \mathrm{M}\lambda v, \\ &= \lambda_t v + \mathrm{ML}v, \end{aligned}$$

Therefore we obtain

$$[\mathrm{L}_t + (\mathrm{LM} - \mathrm{ML})]\, v = \lambda_t v,$$

and hence in order to solve for nontrivial eigenfunctions $v(x,t)$

$$\mathrm{L}_t + [\mathrm{L}, \mathrm{M}] = 0, \tag{1.8.2}$$

where

$$[\mathrm{L}, \mathrm{M}] := \mathrm{LM} - \mathrm{ML},$$

if and only if $\lambda_t = 0$. Equation (1.8.2) is called *Lax's equation*, and contains a nonlinear evolution equation for suitably chosen L and M. For example if we take

$$\mathrm{L} := \frac{\partial^2}{\partial x^2} + u, \tag{1.8.3a}$$

$$\mathrm{M} := (\gamma + u_x) - (4\lambda + 2u)\frac{\partial}{\partial x}, \tag{1.8.3b}$$

then L and M satisfy (1.8.2) provided that u satisfies the KdV equation

$$u_t + 6uu_x + u_{xxx} = 0. \tag{1.8.4}$$

Therefore, the KdV equation may be thought of as the compatibility condition of the two linear operators given by (1.8.3). As we shall see below, there is a general class of equations which are associated with the Schrödinger operator (1.8.3a).

If a nonlinear partial differential equation arises as the compatibility condition of two such operators L and M, then (1.8.2) is called the *Lax representation* of the partial differential equation and L and M are the *Lax pair*. Lax [1968] indicates how given L, an associated M may be constructed so that (1.8.2) is nontrivial. A major difficulty is that for a given partial differential equations there is to date no completely systematic method of determining whether or not it has a Lax representation and, if so, how to determine the associated operators L and M. Work has usually relied on inspired guesswork, either fixing the form of the operators L and M and seeing what partial differential equations result or looking for operators of a certain form given the partial differential equation (the so called prolongation structure method due to Wahlquist and Estabrook [1975, 1976], which we discuss in §2.6.8 below — see also Kaup [1980c]; Dodd and Fordy [1983]).

1.9 Linear Scattering Problems and Associated Nonlinear Evolution Equations.

Following the development of the method of inverse scattering to solve the initial value problem for the KdV equation by Gardner, Greene, Kruskal and Miura [1967], it was then of considerable interest to determine whether the method would be applicable to other physically important nonlinear evolution equations. The method of inverse scattering is highly nontrivial and was thought by some to be a fluke, a clever transformation analogous to the Cole-Hopf transformation (Cole [1951]; Hopf [1950]) which linearizes Burgers' equation

$$u_t + 2uu_x - u_{xx} = 0.$$

If we make the transformation $u = -\phi_x/\phi$, then $\phi(x,t)$ satisfies the linear heat equation $\phi_t - \phi_{xx} = 0$. (We remark that Forsyth [1906, p101] first pointed out the relationship between Burgers' equation and the linear heat equation.)

However, Zakharov and Shabat [1972] proved that the method indeed was no fluke by extending Lax's ideas in order to relate the nonlinear Schrödinger equation

$$\mathrm{i}u_t + u_{xx} + \kappa u^2 u^* = 0, \tag{1.9.1}$$

where $*$ denotes the complex conjugate and κ is a constant to a certain linear scattering problem. They showed that if

$$\mathrm{L} = \mathrm{i}\begin{pmatrix} 1+k & 0 \\ 0 & 1-k \end{pmatrix}\frac{\partial}{\partial x} + \begin{pmatrix} 0 & u^* \\ u & 0 \end{pmatrix}, \tag{1.9.2a}$$

$$\mathrm{M} = \mathrm{i}k\begin{pmatrix} 1 & 0 \\ 0 & 1 \end{pmatrix}\frac{\partial^2}{\partial x^2} + \begin{pmatrix} \dfrac{-\mathrm{i}uu^*}{1+k} & u_x^* \\ -u_x & \dfrac{\mathrm{i}uu^*}{1-k} \end{pmatrix}, \tag{1.9.2b}$$

with $\kappa = 2/(1-k^2)$, then L and M satisfy Lax's equation

$$\mathrm{L}_t + [\mathrm{L},\mathrm{M}] = 0, \tag{1.9.3}$$

if and only if $u(x,t)$ satisfies the nonlinear Schrödinger equation (1.9.1). Using the operators (1.9.2), Zakharov and Shabat were able to solve (1.9.1), given initial data $u(x,0) = f(x)$ (provided that $f(x)$ decays sufficiently rapidly as $|x| \to \infty$). Shortly thereafter, Wadati [1972] gave the method of solution for the Modified KdV equation

$$u_t - 6u^2 u_x + u_{xxx} = 0, \tag{1.9.4}$$

and Ablowitz, Kaup, Newell and Segur [1973a], motivated by several important observations by Kruskal, solved the Sine-Gordon equation

$$u_{xt} = \sin u. \tag{1.9.5}$$

Ablowitz, Kaup, Newell and Segur [1973b, 1974] developed a procedure, which showed that the initial value problem for a remarkably large class of physically interesting nonlinear evolution equations could be solved by this method. Because of the analogy between the Fourier transform method for solving the initial value problem for linear evolution equations and the inverse scattering method for solving the initial value problem for nonlinear evolution equations, they termed the inverse scattering technique the *Inverse Scattering Transform* (I.S.T.). (Referred to briefly in §1.1, by the I.S.T. scheme for a nonlinear evolution equation, we shall mean the inverse scattering method for solving the initial value problem of the equation.)

Consider two linear equations

$$\boldsymbol{v}_x = \mathbf{X}\boldsymbol{v}, \tag{1.9.6a}$$

$$\boldsymbol{v}_t = \mathbf{T}\boldsymbol{v}, \tag{1.9.6b}$$

where $\boldsymbol{v}$ is an n-dimensional vector and $\mathbf{X}$ and $\mathbf{T}$ are $n \times n$ matrices. If we require that (1.9.6) are compatible, that is requiring that $\boldsymbol{v}_{xt} = \boldsymbol{v}_{tx}$, then $\mathbf{X}$ and $\mathbf{T}$ must satisfy

$$\mathbf{X}_t - \mathbf{T}_x + [\mathbf{X}, \mathbf{T}] = 0. \tag{1.9.7}$$

This equation (1.9.7) and Lax's equation (1.9.3) are similar; (1.9.6) is somewhat more general as it allows eigenvalue dependence other than $\mathrm{L}v = \lambda v$.

As an example, consider the 2×2 scattering problem [a generalization of (1.9.2a)] given by

$$v_{1,x} = -ikv_1 + q(x,t)v_2, \tag{1.9.8a}$$

$$v_{2,x} = ikv_2 + r(x,t)v_1, \tag{1.9.8b}$$

and the most general linear time dependence is given by

$$v_{1,t} = Av_1 + Bv_2, \tag{1.9.9a}$$

$$v_{2,t} = Cv_1 + Dv_2, \tag{1.9.9b}$$

where A, B, C and D are scalar functions of $q(x,t)$, $r(x,t)$ and k, independent of (v_1, v_2). Essentially, we just specify that

$$\mathbf{X} = \begin{pmatrix} -ik & q \\ r & ik \end{pmatrix}, \qquad \mathbf{T} = \begin{pmatrix} A & B \\ C & D \end{pmatrix}, \tag{1.9.10}$$

in (1.9.6). Note that if there were any x-derivatives on the right-hand side of (1.9.9) then they can be eliminated by use of (1.9.8). Furthermore, when $r = -1$, then (1.9.8) reduces to the Schrödinger scattering problem

$$v_{2,xx} + (k^2 + q)v_2 = 0, \tag{1.9.11}$$

which is equivalent to (1.7.2), with $k^2 = -\lambda$. It is interesting to note that the most physically interesting nonlinear evolution equations arise from this procedure when either $r = -1$ or $r = q^*$ (or $r = q$ if q is real).

This procedure provides a simple technique which allows us the find nonlinear evolution equations expressible in the form (1.9.7). The compatibility of equations (1.9.8-9), that is requiring that $v_{j,xt} = v_{j,tx}$, for $j = 1,2$, and assuming that the eigenvalue k is time-independent, that is $\mathrm{d}k/\mathrm{d}t = 0$, imposes a set of conditions which A, B, C and D must satisfy. Therefore

$$\begin{aligned} v_{1,xt} &= -\mathrm{i}kv_{1,t} + q_t v_2 + q v_{2,t}, \\ &= -\mathrm{i}k(Av_1 + Bv_2) + q_t v_2 + q(Cv_1 + Dv_2), \\ v_{1,tx} &= A_x v_1 + Av_{1,x} + B_x v_2 + Bv_{2,x}, \\ &= A_x v_1 + A(-\mathrm{i}kv_1 + qv_2) + B_x v_2 + B(\mathrm{i}kv_2 + rv_1). \end{aligned}$$

Hence by equating the coefficients of v_1 and v_2, we obtain

$$A_x = qC - rB, \tag{1.9.12a}$$
$$B_x + 2\mathrm{i}kB = q_t - (A - D)q, \tag{1.9.12b}$$

respectively. Similarly

$$\begin{aligned} v_{2,xt} &= \mathrm{i}kv_{2,t} + r_t v_1 + r v_{1,t}, \\ &= \mathrm{i}k(Cv_1 + Dv_2) + r_t v_1 + a(A_1 + Bv_2), \\ v_{2,tx} &= C_x v_1 + Cv_{1,x} + D_x v_2 + Dv_{2,x}, \\ &= C_x v_1 + C(-\mathrm{i}kv_1 + qv_2) + D_x v_2 + D(\mathrm{i}kv_2 + rv_1), \end{aligned}$$

and equating the coefficients of v_1 and v_2 we obtain

$$C_x - 2\mathrm{i}kC = r_t + (A - D)r, \tag{1.9.13a}$$
$$(-D)_x = qC - rB. \tag{1.9.13b}$$

Therefore, from (1.9.12a) and (1.9.13b), without loss of generality we may assume $D = -A$, and hence it is seen that A, B and C necessarily satisfy the compatibility conditions

$$A_x = qC - rB, \tag{1.9.14a}$$
$$B_x + 2\mathrm{i}kB = q_t - 2Aq, \tag{1.9.14b}$$
$$C_x - 2\mathrm{i}kC = r_t + 2Ar. \tag{1.9.14c}$$

We now solve (1.9.14) for A, B and C (thus ensuring that (1.9.8) and (1.9.9) are compatible). In general, this can only be done if another condition (on r and q) is satisfied, this condition being the evolution equation. Since k, the eigenvalue, is a free parameter, we may find solvable evolution equations by seeking finite power series expansions for A, B and C:

$$A = \sum_{j=0}^{n} A_j k^j, \qquad B = \sum_{j=0}^{n} B_j k^j, \qquad C = \sum_{j=0}^{n} C_j k^j. \tag{1.9.15}$$

Substituting (1.9.15) into (1.9.14) and equating coefficients of powers of k, we obtain $3n+5$ equations. There are $3n+3$ unknowns, A_j, B_j, C_j, $j = 0, 1, \ldots, n$, and so we also obtain two nonlinear evolution equations for r and q. Now let us consider some examples.

EXAMPLE 1.9.1 $n = 2$

Suppose that A, B and C are quadratic in k, that is

$$A = A_2 k^2 + A_1 k + A_0, \tag{1.9.16a}$$

$$B = B_2 k^2 + B_1 k + B_0, \tag{1.9.16b}$$

$$C = C_2 k^2 + C_1 k + C_0. \tag{1.9.16c}$$

Substitute (1.9.16) into (1.9.14) and equate powers of k. The coefficients of k^3 immediately give $B_2 = C_2 = 0$. At order k^2, we obtain $A_2 = a$, constant, $B_1 = \mathrm{i}aq$, $C_1 = \mathrm{i}ar$. At order k^1, we obtain $A_1 = b$, constant, for simplicity we set $b = 0$ (if $b \neq 0$ then a more general evolution equation is obtained), then $B_0 = -\frac{1}{2}aq_x$ and $C_0 = \frac{1}{2}ar_x$. Finally, at order k^0, we obtain $A_0 = \frac{1}{2}arq + c$, with c a constant (again for simplicity we set $c = 0$). Therefore we obtain the following evolution equations

$$-\tfrac{1}{2}aq_{xx} = q_t - aq^2 r, \tag{1.9.17a}$$

$$\tfrac{1}{2}ar_{xx} = r_t + aqr^2. \tag{1.9.17b}$$

If in (1.9.17) we set $r = \mp q^*$ and $a = 2\mathrm{i}$, then we obtain the nonlinear Schrödinger equation

$$\mathrm{i}q_t = q_{xx} \pm 2q^2 q^*. \tag{1.9.18}$$

In summary, given the scattering problem (1.9.8) and the associated time dependence (1.9.9), then these are compatible provided (1.9.14) holds. In this example, setting $r = -q^*$, then we find that

$$A = 2\mathrm{i}k^2 \pm \mathrm{i}qq^*, \tag{1.9.19a}$$

$$B = 2qk + \mathrm{i}q_x, \tag{1.9.19b}$$

$$C = \mp 2q^* k \pm \mathrm{i}q^*_x, \tag{1.9.19c}$$

satisfy (1.9.14) provided that $q(x,t)$ satisfies the nonlinear Schrödinger equation (1.9.18).

EXAMPLE 1.9.2 $n = 3$

If we substitute the third order polynomials in k

$$A = a_3k^3 + a_2k^2 + \tfrac{1}{2}(a_3qr + a_1)k + \tfrac{1}{2}a_2qr - \tfrac{1}{4}\mathrm{i}a_3(qr_x - rq_x) + a_0, \tag{1.9.20a}$$
$$B = \mathrm{i}a_3qk^2 + \left(\mathrm{i}a_2q - \tfrac{1}{2}a_3q_x\right)k + \left[\mathrm{i}a_1q - \tfrac{1}{2}a_2q_x + \tfrac{1}{4}\mathrm{i}a_3(2q^2r - q_{xx})\right], \tag{1.9.20b}$$
$$C = \mathrm{i}a_3rk^2 + \left(\mathrm{i}a_2r + \tfrac{1}{2}a_3r_x\right)k + \left[\mathrm{i}a_1r + \tfrac{1}{2}a_2r_x + \tfrac{1}{4}\mathrm{i}a_3(2r^2q - r_{xx})\right], \tag{1.9.20c}$$

in (1.9.14), with a_3, a_2, a_1 and a_0 constants, then we find that $q(x,t)$ and $r(x,t)$ satisfy the evolution equations

$$q_t + \tfrac{1}{4}\mathrm{i}a_3(q_{xxx} - 6qrq_x) + \tfrac{1}{2}a_2(q_{xx} - 2q^2r) - \mathrm{i}a_1q_x - 2a_0q = 0, \tag{1.9.21a}$$
$$r_t + \tfrac{1}{4}\mathrm{i}a_3(r_{xxx} - 6qrr_x) - \tfrac{1}{2}a_2(r_{xx} - 2qr^2) - \mathrm{i}a_1r_x + 2a_0r = 0. \tag{1.9.21b}$$

For special choices of the constants a_3, a_2, a_1 and a_0 in (1.9.21) we find physically interesting evolution equations. If $a_0 = a_1 = a_2 = 0$, $a_3 = -4\mathrm{i}$ and $r = -1$, then we obtain the KdV equation

$$q_t + 6qq_x + q_{xxx} = 0.$$

If $a_0 = a_1 = a_2 = 0$, $a_3 = -4\mathrm{i}$ and $r = q$, then we obtain the mKdV equation

$$q_t - 6q^2q_x + q_{xxx} = 0.$$

[Note that if $a_0 = a_1 = a_3 = 0$, $a_2 = -2\mathrm{i}$ and $r = -q^*$, then we obtain the nonlinear Schrödinger equation (1.9.18).]

We can also consider expansions of A, B and C in inverse powers of k.

EXAMPLE 1.9.3 $n = -1$

Suppose that

$$A = \frac{a(x,t)}{k}, \qquad B = \frac{b(x,t)}{k}, \qquad C = \frac{c(x,t)}{k}, \tag{1.9.22}$$

then the compatibility conditions (1.9.14) are satisfied if

$$a_x = \tfrac{1}{2}(qr)_t, \qquad q_{xt} = -4\mathrm{i}aq, \qquad r_{xt} = -4\mathrm{i}ar. \tag{1.9.23}$$

Special cases of these are (i),

$$a = \tfrac{1}{4}\mathrm{i}\cos u, \qquad b = c = \tfrac{1}{4}\mathrm{i}\sin u, \qquad q = -r = -\tfrac{1}{2}u_x, \tag{1.9.24}$$

then u satisfies the Sine-Gordon equation

$$u_{xt} = \sin u,$$

and (ii),

$$a = \tfrac{1}{4}\mathrm{i}\cosh u, \qquad b = c = \tfrac{1}{4}\mathrm{i}\sinh u, \qquad q = r = \tfrac{1}{2}u_x, \tag{1.9.25}$$

where u satisfies the Sinh-Gordon equation

$$u_{xt} = \sinh u.$$

These three examples only show a few of the many nonlinear evolution equations which may be obtained by this procedure. We saw above that when $r = -1$, the scattering problem (1.9.8) reduced to the Schrödinger equation (1.9.11). In this case we take an alternative associated time dependence

$$v_t = Av + Bv_x. \tag{1.9.26}$$

By requiring that this and

$$v_{xx} + (k^2 + q)v = 0,$$

are compatible and assuming that $\mathrm{d}k/\mathrm{d}t = 0$, yields equations for A and B analogous to equations (1.9.14), then by expanding in powers of k^2, we obtain a general class of equations.

There have been numerous applications and generalizations of this method. Early on Ablowitz and Haberman [1975b] generalized the 2×2 case to the $n \times n$ system (see §1.10 below). Other generalizations includes that of Wadati, Konno and Ichikawa [1979a,b] (see also Shimizu and Wadati [1980]; Ishimori [1981, 1982]; Wadati and Sogo [1983]; Konno and Jeffrey [1984]). For example, instead of (1.9.8), consider the scattering problem

$$v_{1,x} = -f(k)v_1 + g(k)q(x,t)v_2, \tag{1.9.27a}$$

$$v_{2,x} = f(k)v_2 + g(k)r(x,t)v_1, \tag{1.9.27b}$$

where $f(k)$ and $g(k)$ are functions of the eigenvalue k, and the time dependence is given (as previously) by

$$v_{1,t} = Av_1 + Bv_2, \tag{1.9.28a}$$

$$v_{2,t} = Cv_1 - Av_2, \tag{1.9.28b}$$

The compatibility of (1.9.27) and (1.9.28) requires that A, B and C satisfy

$$A_x = g(qC - rB), \tag{1.9.29a}$$

$$B_x + 2fB = gq_t - 2Agq, \tag{1.9.29b}$$

$$C_x - 2fC = gr_t + 2Agr. \tag{1.9.29c}$$

As earlier, postulating that A, B, C, f and g have finite power series expansions in k (where the expansions for f and g have constant coefficients), then one obtains a variety of physically interesting evolution equations.

EXAMPLE 1.9.4

Choosing

$$f(k) = \mathrm{i}ak^2 - \sqrt{2b}\,k, \qquad g(k) = ak + \mathrm{i}\sqrt{b/2}, \tag{1.9.30a}$$

$$A = -2\mathrm{i}a^2k^4 + 4a\sqrt{2b}\,k^3 + (4\mathrm{i}b - \mathrm{i}a^2rq)k^2 + \sqrt{2b}\,arqk + \mathrm{i}rq\sqrt{b/2}, \tag{1.9.30b}$$

$$B = 2a^2qk^3 + 3\mathrm{i}\sqrt{2b}\,aqk^2 + (-2bq + \mathrm{i}aq_x + a^2rq^2)k + (-q_x + \mathrm{i}arq^2)\sqrt{b/2}, \tag{1.9.30c}$$

$$C = 2a^2rk^3 + 3\mathrm{i}\sqrt{2b}\,ark^2 + (-2br - \mathrm{i}ar_x + a^2r^2q)k + (r_x + \mathrm{i}ar^2q)\sqrt{b/2}, \tag{1.9.30d}$$

where a and b are constants, then the compatibility conditions (1.9.29) are satisfied if r and q are solutions of the evolution equations

$$r_t + \mathrm{i}r_{xx} - a(r^2q)_x + \mathrm{i}br^2q = 0, \tag{1.9.31a}$$

$$q_t - \mathrm{i}q_{xx} - a(rq^2)_x - \mathrm{i}brq^2 = 0. \tag{1.9.31b}$$

If in (1.9.31) we set $r = q^*$, then q satisfies the generalized nonlinear Schrödinger equation

$$q_t = \mathrm{i}q_{xx} + a(q^2q^*)_x + \mathrm{i}bq^2q^*, \tag{1.9.32}$$

which reduces to the nonlinear Schrödinger equation if $a = 0$, and to the derivative nonlinear Schrödinger equation

$$q_t = \mathrm{i}q_{xx} + a(q^2q^*)_x, \tag{1.9.33}$$

if $b = 0$, which was solved by inverse scattering in this way by Kaup and Newell [1978a]. (In fact, if $a \neq 0$, then the generalized nonlinear Schrödinger equation (1.9.32) can be transformed into the derivative nonlinear Schrödinger equation (1.9.33) — see Clarkson and Cosgrove [1987].)

EXAMPLE 1.9.5

Another interesting class of equations can be obtained by choosing

$$f(k) = g(k) = k, \tag{1.9.34a}$$

$$A = -\frac{4\mathrm{i}k^3}{(1-rq)^{1/2}} + \frac{(rq_x - qr_x)k^2}{(1-rq)^{3/2}}, \tag{1.9.34b}$$

$$B = \frac{4qk^3}{(1-rq)^{1/2}} + \frac{2\mathrm{i}q_xk^2}{(1-rq)^{3/2}} - k\left[\frac{q_x}{(1-rq)^{3/2}}\right]_x, \tag{1.9.34c}$$

$$C = \frac{4rk^3}{(1-rq)^{1/2}} - \frac{2\mathrm{i}r_xk^2}{(1-rq)^{3/2}} - k\left[\frac{r_x}{(1-rq)^{3/2}}\right]_x, \tag{1.9.34d}$$

for which the compatibility conditions (1.9.29) are satisfied provided that q and r satisfy

$$q_t + \left[\frac{q_x}{(1-rq)^{3/2}}\right]_{xx} = 0, \tag{1.9.35a}$$

$$r_t + \left[\frac{r_x}{(1-rq)^{3/2}}\right]_{xx} = 0. \tag{1.9.35b}$$

If we take the special case $r = -q$, then we obtain

$$q_t + \left[\frac{q_x}{(1+q^2)^{3/2}}\right]_{xx} = 0, \tag{1.9.36}$$

which has so-called "loop solitons" (Konno and Jeffrey [1984]). Also, if in (1.9.35), $r = -1$ and $q = u - 1$, then we obtain the Harry-Dym equation (Kruskal [1975])

$$u_t = 2(u^{-1/2})_{xxx}. \tag{1.9.37}$$

The Harry-Dym equation is related to both the KdV equation (1.1.2) and the mKdV equation (1.6.3) through hodograph transformations (i.e., transformations which involve the interchange of dependent and independent variables — see Levi, Ragnisco and Sym [1984]; Kawamoto [1985]; Clarkson, Fokas and Ablowitz [1989]). Additionally, both Ishimori [1982] and Wadati and Sogo [1983] have shown that the inverse scattering schemes of equations (1.9.8,9) and (1.9.27,28) may be related by a sequence of gauge transformations (see also Rogers and Wong [1984]).

EXAMPLE 1.9.6 *Landau-Lifschitz equation*

The Landau-Lifschitz equation

$$\boldsymbol{S}_t = \boldsymbol{S} \wedge \boldsymbol{S}_{xx} + \boldsymbol{S} \wedge \mathbf{J}\boldsymbol{S}, \tag{1.9.38a}$$

where

$$\mathbf{J} = \operatorname{diag}(J_1, J_2, J_3), \qquad J_1 < J_2 < J_3, \tag{1.9.38b}$$

$$\boldsymbol{S} = (S_1, S_2, S_3), \qquad |\boldsymbol{S}|^2 = S_1^2 + S_2^2 + S_3^2 = 1, \tag{1.9.38c}$$

describes nonlinear spin waves in an anisotropic ferromagnet (Sklyanin [1979]; Mikhailov [1982]; Rodin [1984, 1987]). It possesses a linear eigenvalue problem with elliptic function dependence in the scattering parameter

$$\boldsymbol{U}_x(x,t;k) = -\mathrm{i}\left(\sum_{j=1}^{3} S_j(x,t)W_j(k)\boldsymbol{\sigma}_j\right)\boldsymbol{U}(x,t;k), \tag{1.9.39a}$$

where $\boldsymbol{\sigma}_j$, $j = 1,2,3$ are the Pauli spin matrices given by

$$\boldsymbol{\sigma}_1 = \begin{pmatrix} 0 & 1 \\ 1 & 0 \end{pmatrix}, \qquad \boldsymbol{\sigma}_2 = \begin{pmatrix} 0 & -\mathrm{i} \\ \mathrm{i} & 0 \end{pmatrix}, \qquad \boldsymbol{\sigma}_3 = \begin{pmatrix} -1 & 0 \\ 0 & 1 \end{pmatrix}, \tag{1.9.39b}$$

and

$$W_1(k) = \rho \frac{1}{\mathrm{sn}(k;m)}, \qquad W_2(k) = \rho \frac{\mathrm{dn}(k;m)}{\mathrm{sn}(k;m)}, \qquad W_3(k) = \rho \frac{\mathrm{dn}(k;m)}{\mathrm{sn}(k;m)}, \tag{1.9.39c}$$

with

$$k := \left(\frac{J_2 - J_1}{J_3 - J_1}\right)^{1/2}, \qquad 0 < k < 1, \qquad \rho := (J_3 - J_1)^{1/2}, \tag{1.9.39d}$$

and $\mathrm{sn}(k;m)$, $\mathrm{dn}(k;m)$, $\mathrm{cn}(k;m)$ are the Jacobian elliptic functions (cf. Whittaker and Watson, [1927]).

We remark that in the special case $\mathbf{J} = \mathbf{0}$, equation (1.9.38) reduces to the Heisenberg Ferromagnet equation

$$\boldsymbol{S}_t = \boldsymbol{S} \wedge \boldsymbol{S}_{xx}, \tag{1.9.40a}$$

where

$$\boldsymbol{S} = (S_1, S_2, S_3), \qquad |\boldsymbol{S}|^2 = S_1^2 + S_2^2 + S_3^2 = 1, \tag{1.9.40b}$$

which is considered by Lakshmanan [1977] and Takhtajan [1977] and is equivalent to the nonlinear Schrödinger equation.

1.10 Generalizations of the I.S.T. in One Spatial Dimension.

An obvious question is: *what is the most general class of nonlinear evolution equations*? Here we use the 2×2 scattering problem

$$v_{1,x} = -\mathrm{i}kv_1 + q(x,t)v_2, \tag{1.10.1a}$$

$$v_{2,x} = \mathrm{i}kv_2 + r(x,t)v_1. \tag{1.10.1b}$$

Ablowitz, Kaup, Newell and Segur [1974] answered this question by considering the equations (1.9.12) associated with the functions A, B and C which arise in the time dependence (1.9.9) of the eigenfunctions. Under certain restrictions, the general relation was found which directly gives a class of solvable nonlinear equations. This relationship depends upon the dispersion relation of the linearized form of the nonlinear equation and a certain integro-differential operator. Suppose that q and r vanish sufficiently rapidly as $|x| \to \infty$, then the general evolution equation can be shown to be

$$\begin{pmatrix} r \\ -q \end{pmatrix}_t + A_0(\mathrm{L}) \begin{pmatrix} r \\ q \end{pmatrix} = 0, \tag{1.10.2a}$$

where $A_0(k) = \lim_{|x|\to\infty} A(x,t,k)$ ($A_0(k)$ may be the ratio of two entire functions), and L is the integro-differential operator given by

$$\mathrm{L} = \frac{1}{2\mathrm{i}} \begin{pmatrix} \partial_x - 2r(\mathrm{I}_- q) & 2r(\mathrm{I}_- r) \\ -2q(\mathrm{I}_- q) & -\partial_x + 2q(\mathrm{I}_- r) \end{pmatrix}, \tag{1.10.2b}$$

where $\partial_x \equiv \partial/\partial x$ and

$$(\mathrm{I}_- f)(x) := \int_{-\infty}^{x} f(y)\,\mathrm{d}y. \tag{1.10.2c}$$

Note that L operates on (r, q), and I_- operates both on the functions immediately to its right and also on the functions to which L is applied. Equation (1.10.2) may be written in matrix form

$$\boldsymbol{\sigma}_3 \boldsymbol{u}_t + A_0(\mathrm{L})\boldsymbol{u} = 0, \tag{1.10.3a}$$

where

$$\boldsymbol{\sigma}_3 = \begin{pmatrix} 1 & 0 \\ 0 & -1 \end{pmatrix}, \qquad \boldsymbol{u} = \begin{pmatrix} r \\ q \end{pmatrix}. \tag{1.10.3b}$$

It is significant that $A_0(k)$ is closely related to the dispersion relation of the associated *linearized* problem. If $f(x)$ and $g(x)$ vanish sufficiently rapidly as $|x| \to \infty$, then in the limit $x \to -\infty$,

$$f(x)(\mathrm{I}_- g)(x) \equiv f(x) \int_{-\infty}^{x} g(y)\,\mathrm{d}y \to 0,$$

and so for infinitesimal q and r, L is the diagonal differential operator

$$\mathrm{L} = \frac{1}{2\mathrm{i}} \frac{\partial}{\partial x} \begin{pmatrix} 1 & 0 \\ 0 & -1 \end{pmatrix}.$$

Hence in this limit (1.10.2) yields

$$r_t + 2A_0\left(-\tfrac{1}{2}\mathrm{i}\partial_x\right) r = 0, \tag{1.10.4a}$$

$$-q_t + 2A_0\left(\tfrac{1}{2}\mathrm{i}\partial_x\right) q = 0. \tag{1.10.4b}$$

Equations (1.10.4) are linear (decoupled) partial differential equations which may be solved by Fourier transform methods. Moreover, substituting the wave solutions $r(x,t) = \exp\{\mathrm{i}(ikx - \omega_r(k)t)\}$, $q(x,t) = \exp\{\mathrm{i}(ikx - \omega_q(k)t)\}$, into (1.10.4), we obtain the relationship

$$A_0(k) = \tfrac{1}{2}\mathrm{i}\omega_r(2k) = -\tfrac{1}{2}\mathrm{i}\omega_q(-2k). \tag{1.10.5}$$

Therefore the general evolution equation is

$$\begin{pmatrix} r \\ -q \end{pmatrix}_t = -\mathrm{i}\omega(2\mathrm{L}) \begin{pmatrix} r \\ q \end{pmatrix}, \tag{1.10.6}$$

or in matrix form

$$\boldsymbol{\sigma}_3 \boldsymbol{u}_t + \mathrm{i}\omega(2\mathrm{L})\boldsymbol{u} = 0. \tag{1.10.7}$$

Hence, the form of the nonlinear evolution equation is characterized by the dispersion relation of its associated linearized equations and an integro-differential operator (for further details see Ablowitz, Kaup, Newell and Segur [1974]).

EXAMPLE 1.10.1 *The nonlinear Schrödinger equation*

For the nonlinear Schrödinger equation

$$\mathrm{i}q_t = q_{xx} \pm 2q^2 q^*, \tag{1.10.8}$$

the associated linear equation is

$$\mathrm{i}q_t = q_{xx}. \tag{1.10.9}$$

Therefore the dispersion relation is $\omega_q(k) = -k^2$, and so from (1.10.5), $A_0(k)$ is given by

$$A_0(k) = 2\mathrm{i}k^2. \tag{1.10.10}$$

The evolution equation is obtained from either (1.10.2) or (1.10.6), therefore we have

$$\begin{pmatrix} r \\ -q \end{pmatrix}_t = -4\mathrm{i}\mathrm{L}^2 \begin{pmatrix} r \\ q \end{pmatrix} = -2\mathrm{L} \begin{pmatrix} r_x \\ q_x \end{pmatrix} = \mathrm{i} \begin{pmatrix} r_{xx} - 2r^2 q \\ q_{xx} - 2q^2 r \end{pmatrix}. \tag{1.10.11}$$

Therefore, when $r = \mp q^*$, both of these equations are equivalent to the nonlinear Schrödinger equation (1.10.8).

Note that equation (1.10.10) is in agreement with equation (1.9.19) in Example 1.9.1 (that is $\lim_{|x|\to\infty} A = 2\mathrm{i}k^2$, since $\lim_{|x|\to\infty} r(x,t) = 0$). This explains *a posteriori* why the expansion of A, B and C in powers of k are related so closely to the dispersion relation. Indeed, the fact that the nonlinear Schrödinger equation is related to (1.10.10) implies that an expansion starting at k^2 will be the judicious choice. Similarly, the Modified KdV equation and the Sine-Gordon equation can be obtained from the operator equation (1.10.6) using the dispersion relations $\omega(k) = -k^3$ and $\omega(k) = k^{-1}$ respectively. These dispersions relations suggest expansion commencing in powers of k^3 and k^{-1} respectively which indeed we saw to be the case (Examples 1.9.2 and 1.9.3).

The derivation of (1.10.6) required that q and r tend to 0 as $|x| \to \infty$ and therefore we cannot simply set $r = -1$ in order to obtain the equivalent result for the Schrödinger scattering problem

$$v_{xx} + (k^2 + q)v = 0. \tag{1.10.12}$$

However the essential ideas are similar and Ablowitz, Kaup, Newell and Segur [1974] showed that the general evolution equation in this case is

$$q_t + \gamma(\mathrm{L})q_x = 0, \tag{1.10.13a}$$

where

$$\mathrm{L} := -\tfrac{1}{4}\partial_x^2 - q + \tfrac{1}{2}q_x \mathrm{I}_+, \tag{1.10.13b}$$

$$(\mathrm{I}_+ f)(x) := \int_x^\infty f(y)\,\mathrm{d}y. \tag{1.10.13c}$$

$$\gamma(k^2) = \frac{\omega(2k)}{2k}, \tag{1.10.13d}$$

with $\partial_x := \partial/\partial x$ and $\omega(k)$ odd. As above, $\omega(k)$ is the dispersion relation of the associated linear equation with $q = \exp\{\mathrm{i}(kx - \omega(k)t)\}$.

EXAMPLE 1.10.2 *The Korteweg-de Vries equation*

For the KdV equation

$$q_t + 6qq_x + q_{xxx} = 0, \tag{1.10.14}$$

the associated linear equation is

$$q_t + q_{xxx} = 0. \tag{1.10.15}$$

Therefore $\omega(k) = -k^3$ and so $\gamma(k^2) = -4k^2$, thus $\gamma(\mathrm{L}) = -4\mathrm{L}$. Hence (1.10.13) yields

$$q_t - 4\mathrm{L}q_x = 0,$$

thus

$$q_t - 4\left(-\tfrac{1}{4}\partial_x^2 - q + \tfrac{1}{2}q_x\mathrm{I}_+\right)q_x = 0,$$

therefore

$$q_t + q_{xxx} + 4qq_x + 2qq_x = 0,$$

which is just the KdV equation!

Associated with the operator L there is a hierachy of equations (sometimes referred to as the Lenard hierarchy) given by

$$u_t + \mathrm{L}^k u_x = 0, \qquad k = 1, 2, \dots\,. \tag{1.10.16}$$

The first two higher order equations in the KdV hierarchy are given by

$$u_t - \tfrac{1}{4}\left(u_{5x} + 10uu_{3x} + 20u_xu_{xx} + 30u^2u_x\right) = 0, \tag{1.10.17}$$

$$\begin{aligned} u_t + \tfrac{1}{16}\big(&u_{7x} + 14uu_{5x} + 42u_xu_{4x} + 70u_{xx}u_{3x} + 70u^2u_{3x} \\ &+ 280uu_xu_{xx} + 70u_x^3 + 140u^3u_x\big) = 0, \end{aligned} \tag{1.10.18}$$

where $u_{nx} = \partial^n u/\partial x^n$.

In the above we have encountered two different scattering problems, namely the classical time-independent Schrödinger equation

$$v_{xx} + (u + k^2)v = 0, \tag{1.10.19}$$

which was used to solve the KdV equation, and the 2×2 scattering problem

$$\boldsymbol{v}_x = k\begin{pmatrix} -\mathrm{i} & 0 \\ 0 & \mathrm{i} \end{pmatrix}\boldsymbol{v} + \begin{pmatrix} 0 & q \\ r & 0 \end{pmatrix}\boldsymbol{v}, \tag{1.10.20}$$

which was used to solve the nonlinear Schrödinger, mKdV and Sine-Gordon equations (as we have seen, (1.10.19) can be interpreted as a special case of (1.10.20), however it is more convenient, from the point of view of possible generalizations, to regard them as different scattering problems).

A third order generalization of the scattering problem (1.10.19) has been considered by several authors (see Chapter 3).

We see that there are two particularly important classes of one dimensional linear scattering problems, associated with solvable nonlinear evolution equations; scalar equations and systems. Immediate generalizations are:

(I) *Scalar Nth order Sturm-Liouville scattering problem*

Gel'fand and Dikii [1977] proposed the Nth order generalization of the scattering problem (1.10.19)

$$\frac{\mathrm{d}^N v}{\mathrm{d}x^N} + \sum_{j=2}^{N} u_j(x) \frac{\mathrm{d}^{N-j} v}{\mathrm{d}x^{N-j}} = \lambda v, \tag{1.10.21}$$

and investigated many of the *algebraic* properties of equations solvable through this scattering problem. The general inverse problem associated with this scattering problem is treated in detail in the monograph of Beals, Deift and Tomei [1988] — see also Beals [1985]. [We discuss scattering and inverse scattering associated with the operator (1.10.21) further in §3.1.5.]

(II) $N \times N$ *Systems*

Ablowitz and Haberman [1975b] proposed an $N \times N$ matrix generalization of the scattering problem (1.10.20)

$$\frac{\partial \boldsymbol{v}}{\partial x} = \mathrm{i}k\mathbf{J}\boldsymbol{v} + \mathbf{Q}\boldsymbol{v}, \tag{1.10.22a}$$

where $\mathbf{Q}(x) \in \mathrm{M}_N(\mathbb{C})$ [the space of $N \times N$ matrices over $\mathbb{C}$] with $Q^{ii} = 0$, $\mathbf{J} = \mathrm{diag}(J^1, J^2, \ldots, J^N)$, with $J^i \neq J^j$ for $i \neq j$, $i,j = 1,2,\ldots,N$, and $\boldsymbol{v}(x,t)$ is an N-dimensional vector eigenfunction. The time dependency of the vector eigenfunction is chosen such that

$$\frac{\partial \boldsymbol{v}}{\partial t} = \mathbf{T}\boldsymbol{v}, \tag{1.10.22b}$$

where $\mathbf{T}$ is also an $N \times N$ matrix. The compatibility of (1.10.22) yields

$$\mathbf{T}_x = \mathbf{Q}_t + \mathrm{i}k[\mathbf{J}, \mathbf{T}] + [\mathbf{Q}, \mathbf{T}]. \tag{1.10.23}$$

As for the 2×2 case discussed in §1.9, associated nonlinear evolution equations may be found by postulating an expansion for $\mathbf{T}$ in powers or inverse powers of the eigenvalue k

$$\mathbf{T} = \sum_{j=0}^{n} k^j \mathbf{T}_j, \tag{1.10.24}$$

Ablowitz and Haberman [1975b] established that this scattering problem can be used to solve several physically interesting equations such as the three-wave interaction equations in one spatial dimension (with $N = 3$ and $n = 1$) and the Boussinesq equation (with $N = 3$ and $n = 2$). The recursion operators generating all equations solvable through (1.10.22) have been obtained by Newell [1978] and Fokas and Anderson [1982]. The inverse problem associated with the scattering problem (1.10.22) in the general $N \times N$ case has been rigorously studied by Beals and Coifman [1984, 1985a] (see also Caudrey [1982]; Mikhailov [1979, 1981]; Zhou [1989]). We discuss scattering and inverse scattering associated with the operator (1.10.22) further in §3.1.2.

The scattering problems (1.10.19) and (1.10.20), and their generalizations (1.10.21) and (1.10.22), are the primary scattering problems used in connection with the solution of nonlinear partial differential equations in 1 + 1-dimensions (that is one spatial and one temporal dimension). As mentioned earlier, there exist several variations on these scattering problems, including systems of Schrödinger scattering problems and scattering problems with more complicated dependence on the scattering parameter.

In §1.7 we saw that the inverse scattering problem for the KdV equation, that is the reconstruction of the solution of the KdV from the scattering data, was solved using the Gel'fand-Levitan-Marchenko equation, that is by using a linear integral equation. However, as we discuss in Chapter 2, it is more convenient (and relevant to generalizations) to view the inverse scattering problem for the KdV equation as a *Riemann-Hilbert boundary value problem in scattering space* (an introduction to Riemann-Hilbert problems is given in Appendix A). We also show in Chapter 2 that (when it can be obtained at all), the Gel'fand-Levitan-Marchenko equation can be obtained as a special case of the Riemann-Hilbert problem which therefore may be thought of as more fundamental.

In Chapter 2 we shall primarily by concerned with the solution of the initial value problem on the infinite line (that is for $x \in \mathbb{R}$), for initial conditions which decay sufficiently rapidly as $|x| \to \infty$.

The periodic problem for the KdV equation is to solve

$$u_t + 6uu_x + u_{xxx} = 0, \tag{1.10.25}$$

with periodic boundary conditions of period T,

$$u(x,t) = u(x+T,t), \tag{1.10.26a}$$

given suitable initial condition

$$u(x,0) = f(x). \tag{1.10.26b}$$

This problem has also attracted much attention, these studies include Novikov [1974]; Dubrovin and Novikov [1975]; Its and Matveev [1975]; Lax [1975]; McKean and van Moerbeke [1975]; McKean and Trubowitz [1976]; Dubrovin, Matveev and Novikov [1976] (see also the review by Dubrovin [1981]; §2.3 of Ablowitz and Segur [1981]; Novikov, Manakov, Pitaevskii and Zakharov [1984]). In the course of these studies, significant results on abelian integrals, theta functions, etc., were uncovered from as early as the 1920's and many new results were discovered. Fundamental knowledge of and expertise in algebraic geometry were found to be crucial in the understanding of solutions of partial differential equations for the first time.

1.11 Classes of Equations Solvable by Inverse Scattering.

The scattering operators considered in §1.10 were associated with nonlinear evolution equations in $1+1$ dimensions. In this section we give the scattering operators associated with nonlinear evolution equations in other dimensions. Additionally we list different classes of equations which are solvable by inverse scattering, in order of dimensionality.

1.11.1 Ordinary Differential Equations. The Painlevé equations are six classical nonlinear ordinary differential equations (called $\mathrm{P_I} - \mathrm{P_{VI}}$), which were discovered and studied by Painlevé and his colleagues at the turn of this century. They arise naturally in the study of integrable equations and Chapter 7 of these lecture notes is devoted to this problem.

Significant progress in establishing the solutions of $\mathrm{P_{II}} - \mathrm{P_V}$ has been been made. Ablowitz and Segur [1977] showed that many of the classical Painlevé equations were connected with integrable partial differential equations in $1+1$-dimensions, and for $\mathrm{P_{II}}$ found a new one-parameter family of solutions by a Gel'fand-Levitan integral equation (see also Ablowitz, Ramani and Segur [1980b] and Clarkson and McLeod [1991] for similar results of other Painlevé equations). By considering similarity reductions of the MKdV equation (1.6.3), Flaschka and Newell [1980] showed that the second Painlevé equation $\mathrm{P_{II}}$

$$\frac{\mathrm{d}^2 y}{\mathrm{d}x^2} = 2y^3 + xy + \alpha, \tag{1.11.1}$$

arises as the compatibility condition $v_{j,kx} = v_{j,xk}$, $j = 1, 2$, of the equations

$$v_{1,k} = -\mathrm{i}\left(4k^2 + x + 2y^2\right) v_1 + \left(4ky + 2\mathrm{i}y' + \alpha/k\right) v_2, \tag{1.11.2a}$$

$$v_{2,k} = \left(4ky - 2\mathrm{i}y' + \alpha/k\right) v_1 + \mathrm{i}\left(4k^2 + x + 2y^2\right) v_2, \tag{1.11.2b}$$

$$v_{1,x} = -\mathrm{i}kv_1 + y(x)v_2, \tag{1.11.2c}$$

$$v_{2,x} = y(x)v_1 + \mathrm{i}kv_2. \tag{1.11.2d}$$

Solving the initial value problem for the second Painlevé equation (1.11.1) amounts to solving an "inverse problem" for $\boldsymbol{v}(k,x) = (v_1, v_2)$. As with inverse scattering, the analytical properties of $\boldsymbol{v}(k,x)$ (in k with $k \in \mathbb{C}$) allows one to show that given appropriate monodromy data, one can, in principle, reconstruct $\boldsymbol{v}$. Obtaining monodromy data is classical and analogous to the direct problem in I.S.T.. Fokas and Ablowitz [1983a] have expressed the inverse problem in terms of a Riemann-Hilbert boundary value problem on a complicated contour (six semi-rays intersecting at the origin). They mapped this Riemann-Hilbert problem through a series of transformations to three different standard Riemann-Hilbert problems (each defined on two rays, which provide the continuation of each other). Each of these Riemann-Hilbert problems can be solved through a system of two Fredholm integral equations. Fokas, Mugan and Ablowitz [1987] point out the striking similarity of the inverse monodromy method to I.S.T. and refer to it as I.M.T.. In some sense the method is the analogue of the Laplace method for linear ordinary differential equations just as I.S.T. is the analogue of Fourier analysis for linear problems. We discuss the inverse scattering and the I.M.T. for ordinary differential equations, in particular for the second Painlevé equation (1.11.1), in further detail in §7.3.) Rigorous results regarding the inverse problem associated with P_{II} and the fourth Painlevé equation P_{IV} (see below) have been obtained by Fokas and Zhou [1991].

Other Painlevé equations which have been considered by I.M.T. include: the third Painlevé equation P_{III}

$$\frac{\mathrm{d}^2u}{\mathrm{d}x^2} = \frac{1}{u}\left(\frac{\mathrm{d}u}{\mathrm{d}x}\right)^2 - \frac{1}{x}\frac{\mathrm{d}u}{\mathrm{d}x} + \frac{\alpha u^2+\beta}{x} + \gamma u^3 + \frac{\delta}{u},$$

with α, β, γ and δ constants (Flaschka and Newell [1980]; Morris and Dodd [1980]; Kitaev [1989]); the fourth Painlevé equation P_{IV}

$$\frac{\mathrm{d}^2u}{\mathrm{d}x^2} = \frac{1}{2u}\left(\frac{\mathrm{d}u}{\mathrm{d}x}\right)^2 + \frac{3}{2}u^3 + 4xu^2 + 2(x^2-\alpha)u + \frac{\beta}{u},$$

with α and β constants (Fokas, Mugan and Ablowitz [1987]); and the fifth Painlevé equation P_{V}

$$\frac{\mathrm{d}^2u}{\mathrm{d}x^2} = \left(\frac{1}{2u} + \frac{1}{u-1}\right)\left(\frac{\mathrm{d}u}{\mathrm{d}x}\right)^2 - \frac{1}{x}\frac{\mathrm{d}u}{\mathrm{d}x} + \frac{(u-1)^2}{x^2}\left(\alpha + \frac{\beta}{u}\right) + \frac{\gamma u}{x} + \frac{\delta u(u+1)}{u-1},$$

with α, β, γ and δ constants (Jimbo [1979]; Morris and Dodd [1980]; Fokas, Mugan and Ablowitz [1987]).

Another class of interesting problems are many body problems which arise when one considers the motion of pole solutions of partial differential equations. The prototype problem is:

$$\frac{\mathrm{d}^2u_j}{\mathrm{d}x^2} = 8\sum_{\substack{k=1\\k\neq j}}^{n}\frac{1}{(u_j-u_k)^3}, \qquad j=1,2,\ldots,n$$

(Calogero [1971]; Moser [1975]); a number of pole type solutions can be found in Ablowitz and Segur [1981].

1.11.2 Partial Differential Equations in One Spatial Dimension. Many partial differential equations in 1 + 1-dimensions have been studied (some have already been mentioned), we list for completeness, known ones below. The I.S.T. for 1 + 1-dimensional problems is discussed in Chapters 2 and 3.

The Korteweg-de Vries equation

$$u_t + 6uu_x + u_{xxx} = 0$$

(Gardner, Greene, Kruskal and Miura [1967]).

The Modified Korteweg-de Vries equation

$$u_t - 6u^2 u_x + u_{xxx} = 0$$

(Wadati [1972]).

Other third order equations include:

$$u_t + u_{xxx} - \tfrac{1}{8}u_x^3 + u_x\left(ae^u + be^{-u} + c\right) = 0$$

(Calogero and Degasperis [1981]; Fokas [1980]);

$$u_t = u_{xxx} - \tfrac{3}{2}\frac{u_x u_{xx}^2}{1+u_x^2} - \tfrac{3}{2}\wp(u)(u_x^2+1)u_x + \gamma u_x,$$
$$u_t = u_{xxx} - \tfrac{3}{2}u_{xx}^2 u_x^{-1} + \alpha u_x^{-1} - \tfrac{3}{2}\wp(u)u_x^2 + \gamma u_x,$$

where α, β and γ are arbitrary constants, and $\wp(u; g_2, g_3)$ is the Weierstrass elliptic function (Svinolupov and Sokolov [1982]; Svinolupov, Sokolov and Yamilov [1983]); the cylindrical Korteweg-de Vries equation

$$u_t + 6uu_x + u_{xxx} + \frac{1}{2t}u = 0$$

(Calogero and Degasperis [1978]).

Coupled partial differential equations with third order derivatives:

(a),
$$u_t + u_{xxx} + 6uu_x - 6vv_x = 0,$$
$$v_t - 2v_{xxx} - 6uv_x = 0$$

(Hirota and Satsuma [1981]);

(b),
$$u_t + u_{xxx} + 6uu_x + 2vv_x = 0,$$
$$v_t + 2(uv)_x = 0$$

(Ito [1982]);

(c),
$$u_t + (u_{xx} - 3vu_x + 3uv^2 - 3u^2)_x = 0,$$
$$v_t + (v_{xx} + 3vv_x + 3v^3 - 6uv)_x = 0$$

(Kaup [1975]; Kuperschmidt [1984, 1985]; Antonowicz and Fordy [1987, 1989]) a model for dispersive water waves;

(d),
$$u_t + 3vv_x = 0,$$
$$v_t + 2v_{xxx} + 2uv_x + u_x v = 0,$$

(Drinfel'd and Sokolov [1981]; Wilson [1982]).

Klein-Gordon equations:

(a), the Sine-Gordon equation
$$u_{xt} = \sin u$$

(Ablowitz, Kaup, Newell and Segur [1973a])

(b),
$$u_{xt} = \mathrm{e}^{2u} - \mathrm{e}^{-u}$$

(Mikhailov [1979, 1981]; Fordy and Gibbons [1980b])

(c),
$$u_{xt} = \mathrm{e}^{2u} - \mathrm{e}^{-u}\cosh v,$$
$$v_{xt} = 3\mathrm{e}^{-u}\sinh v$$

(Fordy and Gibbons [1980b]);

(d),
$$\theta_{xt} = \mathrm{e}^{\theta}\cosh\phi - \mathrm{e}^{-\theta}\cosh\psi,$$
$$\phi_{xt} = \mathrm{e}^{\theta}\sinh\phi,$$
$$\psi_{xt} = \mathrm{e}^{-\theta}\sinh\psi$$

(Fordy and Gibbons [1980b]);

(e),
$$u_{xx} - u_{tt} \pm \sin u\cos u + (v_x^2 - v_t^2)\cot u\,\mathrm{cosec}^2\,u = 0,$$
$$v_{xx} - v_{tt} = 4(u_x v_x - u_t v_t)\,\mathrm{cosec}\,2u$$

(Pohlmeyer [1976]; Lund and Regge [1976]).

Other interesting systems include:

(a), the Reduced Maxwell-Bloch equations
$$E_t - v = 0,$$
$$v_x - \omega r - Eq = 0,$$
$$q_x + Ev = 0,$$
$$r_x + \omega v = 0,$$

with ω a constant (Gibbon, Caudrey, Bullough and Eilbeck [1973]).
(b), the self-induced-transparency (SIT) equations

$$E_t + E_x = \langle p \rangle = \int_{-\infty}^{\infty} p(x,t,\alpha)g(\alpha)\,\mathrm{d}\alpha,$$
$$p_t + 2\mathrm{i}\alpha p = E\eta,$$
$$\eta_t = -\tfrac{1}{2}(Ep^* + E^*p).$$

(Eilbeck, Gibbon, Caudrey and Bullough [1973]; Lamb [1973]; Ablowitz, Kaup and Newell [1974]) with $g(\alpha)$ being a probability density (if $g(\alpha) = \delta(\alpha - \alpha_0)$, then the SIT equations reduce to the Sine-Gordon equation — see Ablowitz and Segur [1981, p.322]).

"Boussinesq" systems:
(a), the Boussinesq (1871) equation

$$u_{tt} - u_{xx} + (u^2)_{xx} \pm u_{xxxx} = 0$$

(Zakharov [1974]; Ablowitz and Haberman [1975b]);

(b), $$3u_{tt} - 6u_t u_{xx} - 6u_x^2 u_{xx} + u_{xxxx} = 0$$

(Quispel, Nijhoff and Capel [1982]) a modified Boussinesq equation;

(c), $$u_t + (uv)_x + v_{xxx} = 0,$$
$$v_t + u_x + vv_x = 0$$

(Kaup [1975]; Kuperschmidt [1984, 1985]; Antonowicz and Fordy [1987, 1989]).

Fifth order equations:

(a), $$u_t + u_{xxxxx} + 10u_x u_{xx} + 10uu_{xxx} + 20u^2 u_x = 0$$

(Sawada and Kotera [1974]; Caudrey, Dodd and Gibbon [1976])

(b), $$u_t + u_{xxxxx} + 10u_x u_{xx} + 25uu_{xxx} + 20u^2 u_x = 0$$

(Kaup [1980b]).

Nonlinear Schrödinger-like equations:
(a), the nonlinear Schrödinger (NLS) equation

$$\mathrm{i}u_t + u_{xx} \pm 2|u|^2 u = 0$$

(Zakharov and Shabat [1972]);

(b), $$\mathrm{i}u_t + \frac{\mathrm{i}}{2t}u + u_{xx} \pm 2|u|^2 u = 0$$

(Leclert, Karney, Bers and Kaup [1979]) a "cylindrical" nonlinear Schrödinger equation;

(c),
$$\mathrm{i}u_t + u_{xx} \pm 2\mathrm{i}(|u|^2u)_x = 0$$

(Kaup and Newell [1978a]) a derivative nonlinear Schrödinger equation;
(d), the Landau-Lifshitz equation

$$\boldsymbol{S}_t = \boldsymbol{S} \wedge \boldsymbol{S}_{xx} + \boldsymbol{S} \wedge \mathbf{J}\boldsymbol{S},$$

where $\mathbf{J} = \mathrm{diag}(J_1, J_2, J_3)$ with $J_1 < J_2 < J_3$, $\boldsymbol{S} = (S_1, S_2, S_3)$ with $|\boldsymbol{S}|^2 = S_1^2 + S_2^2 + S_3^2 = 1$ (Sklyanin [1979]; Mikhailov [1982]; Rodin [1984, 1987]) which has as a special case $\mathbf{J} = \mathbf{0}$ the Heisenberg Ferromagnet equation

$$\boldsymbol{S}_t = \boldsymbol{S} \wedge \boldsymbol{S}_{xx}$$

(Lakshmanan [1977]; Takhtajan [1977]);
(e), the coupled Schrödinger (NLS) equation

$$\mathrm{i}u_t + u_{xx} + 2(|u|^2 \pm |v|^2)u = 0,$$
$$\mathrm{i}v_t \pm v_{xx} + 2(|v|^2 \pm |u|^2)v = 0$$

(Manakov [1975]; Zakharov and Schulman [1982]);
(f), a vector nonlinear Schrödinger equation

$$\mathrm{i}\boldsymbol{u}_t + \boldsymbol{u}_{xx} \pm 2|\boldsymbol{u}|^2\boldsymbol{u} = 0,$$

where $\boldsymbol{u} = (u_1, u_2, \ldots, u_n)$ (Zakharov and Manakov [1976]; Zakharov and Schulman [1982]; Fordy and Kulish [1983]);
(g), a vector derivative nonlinear Schrödinger equation

$$\mathrm{i}\boldsymbol{u}_t + \boldsymbol{u}_{xx} \pm 2\mathrm{i}(|\boldsymbol{u}|^2\boldsymbol{u})_x = 0,$$

where $\boldsymbol{u} = (u_1, u_2, \ldots, u_n)$ (Fordy [1984]);
(h), a long-short-wave interaction system (Langmuir waves)

$$\mathrm{i}u_t + u_{xx} - uv = 0,$$
$$v_t + v_x + (|u|^2)_x = 0$$

(Yajima and Oikawa [1976]);

(i),
$$\mathrm{i}u_t + u_{xx} - 2\kappa|u|^2u + \mathrm{i}uv_x + uv^2 = 0,$$
$$v_t - 2\kappa(|u|^2)_x = 0$$

(Newell [1978]).

The principal SU(n) chiral field equation

$$\left(\mathbf{U}^{\dagger}\mathbf{U}_x\right)_t + \left(\mathbf{U}^{\dagger}\mathbf{U}_t\right)_x = 0,$$

where $\mathbf{U}(x,t)$ is a unitary matrix, $\mathbf{U}^{\dagger}\mathbf{U} = \mathbf{I}$, of SU($n$) (Zakharov and Mikhailov [1978]).

The Ernst equations

$$F\nabla^2 F = (\nabla F)^2 - (\nabla G)^2,$$
$$F\nabla^2 G = (\nabla F)\bullet(\nabla G),$$

where ∇ is the axisymmetric gradient operator (Ernst [1968]; Hauser and Ernst [1980]; Cosgrove [1980]), which are equivalent to the stationary axisymmetric vacuum Einstein equations. ($F = F(r,z)$, $\nabla^2 F = F_{rr} + F_r/r + F_{zz}$, $|\nabla F|^2 = F_r^2 + F_z^2$.)

The massive Thirring model

$$\mathrm{i}u_x + v + u|v|^2 = 0,$$
$$\mathrm{i}v_t + u + v|u|^2 = 0$$

(Mikhailov [1976]; Kuznetsov and Mikhailov [1977]; Villarroel [1991]).

The O(n) model

$$\boldsymbol{u}_{xt} + (\boldsymbol{u}_x \bullet \boldsymbol{u}_t)\boldsymbol{u} = 0,$$

where $\boldsymbol{u} = (u_1, u_2, \ldots, u_n)$ (Pohlmeyer [1976] — see also Orfanidis [1980]).

The so called Harry-Dym equation

$$u_t = 2(u^{-1/2})_{xxx}$$

(Kruskal [1974]; Wadati, Konno and Ichikawa [1979b]).

Two highly nonlinear equations

$$u_t + \left[u_x(1+u^2)^{-3/2}\right]_{xx} = 0,$$
$$\mathrm{i}u_t + \left[\frac{u}{(1+|u|^2)^{1/2}}\right]_{xx} = 0$$

(Wadati, Konno and Ichikawa [1979b]).

The three-wave interaction equations

$$\frac{\partial u_i}{\partial t} + a_i \frac{\partial u_i}{\partial x} = b_i u_j^* u_k^*, \tag{1.11.3}$$

where a_i, b_i are constants and $i,j,k = 1,2,3$ are cyclically permuted (a_i are the three group velocities associated with the underlying waves; Zakharov and Manakov [1973, 1976]; Ablowitz and Haberman [1975b]; Kaup [1976]).

The N-wave interaction equations

$$\frac{\partial Q^{ij}}{\partial t} = \alpha_{ij}\frac{\partial Q^{ij}}{\partial x} + \sum_{k=1}^{n}(\alpha_{ik} - \alpha_{kj})Q^{ik}Q^{kj}, \tag{1.11.4}$$

where $Q^{ij} = \sigma_{ij}Q^{ji*}$, $Q^{ii} = 0$, with $\sigma_{ik}\sigma_{kj} = -\sigma_{ij}$ for $j \neq k \neq i$ and α_{ij} are real constants, which are $\frac{1}{2}n(n-1)$ equations (Ablowitz and Haberman [1975b]). In the case $n = 3$, equations (1.11.4) reduce to the three-wave interaction equations (1.11.3) under a suitable scaling of the variables. For example, choosing

$$\begin{aligned} &Q^{12} = \mathrm{i}u_3, \qquad && Q^{23} = \mathrm{i}u_2, \qquad && Q^{31} = \mathrm{i}u_1, \\ &\alpha_{12} = -a_3, && \alpha_{23} = -a_2, && \alpha_{31} = -a_1, \\ &\sigma_{12} = \gamma_3, && \sigma_{23} = \gamma_2, && \sigma_{31} = \gamma_1, \end{aligned}$$

with $\gamma_1\gamma_2\gamma_3 = -1$, $\gamma_i = \pm 1$ for $i = 1,2,3$, then we obtain (1.11.3) with

$$b_1 = \mathrm{i}\gamma_2\gamma_3(a_2 - a_3), \qquad b_2 = \mathrm{i}\gamma_3\gamma_1(a_3 - a_1), \qquad b_3 = \mathrm{i}\gamma_1\gamma_2(a_1 - a_2).$$

1.11.3 Differential-Difference Equations. Some well known differential-difference equations are listed below. Aspects of their solution are discussed in §§3.2.1,3.2.2 below.

The scattering operators associated with the Toda Lattice

$$\frac{\partial^2 u_n}{\partial t^2} = \exp\{-(u_n - u_{n-1})\} - \exp\{-(u_{n+1} - u_n)\},$$

(Toda [1967, 1975, 1976]) and the discrete nonlinear Schrödinger equation

$$\mathrm{i}\frac{\partial u_n}{\partial t} = (u_{n+1} + u_{n-1} - 2u_n) \pm |u_n|^2(u_{n+1} + u_{n-1}),$$

are

$$a_n v_{n+1} + a_{n-1}v_{n-1} + b_n v_n = v_n, \tag{1.11.5a}$$

with

$$a_n = \tfrac{1}{2}\exp\left\{-\tfrac{1}{2}(u_n - u_{n-1})\right\}, \qquad b_n = -\tfrac{1}{2}u_{n-1,t}, \tag{1.11.5b}$$

and

$$v_{1,n+1} = kv_{1,n} \mp u_n^* v_{2,n}, \tag{1.11.6a}$$

$$v_{2,n+1} = k^{-1}v_{2,n} + u_n v_{1,n}, \tag{1.11.6b}$$

respectively. The I.S.T. schemes for the Toda Lattice and the discrete nonlinear Schrödinger equation were developed by Flaschka [1974a,b]; Ablowitz and Ladik [1975,

1976a] (which we discuss in further detail in §§3.2.1,3.2.2). Further nonlinear differential difference equations which are solvable by I.S.T. include the nonlinear ladder equation, or discrete KdV equation

$$\frac{\partial u_n}{\partial t} = \exp(u_{n+1}) - \exp(u_{n-1})$$

(Manakov [1975]; Kac and van Moerbeke [1975] — see also Bogoyalevlensky [1988] for other integrable discretizations of the Korteweg-de Vries equation), the discrete Sine-Gordon equation

$$\frac{\partial u_{n+1}}{\partial t} - \frac{\partial u_n}{\partial t} = \sin(u_{n+1} + u_n)$$

(Orfanidis [1978]; Pilloni and Levi [1982]), the discrete mKdV equation

$$\frac{\partial u_n}{\partial t} = (1 \pm h^2 u_n^2)(u_{n+1} - u_{n-1})$$

(Ablowitz and Ladik [1977]) and the nonlinear self-dual network equations

$$\frac{\partial u_n}{\partial t} = (1 \pm u_n^2)(v_n - v_{n-1}),$$
$$\frac{\partial v_n}{\partial t} = (1 \pm v_n^2)(u_{n+1} - u_n)$$

(Ablowitz and Ladik [1975]) — see §2.2 of Ablowitz and Segur [1981] for further details.

Similarly partial difference equations can be solved by the I.S.T. method; two are listed below. (1), A partial difference equation which reduces to the nonlinear Schrödinger equation as $\Delta t \to 0$ and $\Delta x \to 0$

$$\begin{aligned}
\mathrm{i}\,\frac{\Delta^m u_n^m}{\Delta t} = \frac{1}{2(\Delta x)^2}&\left[\left(u_{n+1}^m - 2u_n^m + u_{n-1}^m \prod_{k=-\infty}^{n-1} \Lambda_k^m\right)\right. \\
&\left.+ \left(u_{n+1}^{m+1} \prod_{k=-\infty}^{n} \Lambda_k^m - 2u_n^{m+1} + u_{n-1}^{m+1}\right)\right] \\
\pm\frac{1}{4}&\left[u_n^m\left(u_n^{m*}u_{n+1}^m + u_n^{m+1*}u_{n+1}^{m+1}\right) + u_n^{m+1}\left(u_{n-1}^m u_n^{m*} + u_{n-1}^{m+1}u_n^{m+1*}\right)\right. \\
&\left.+ 2u_n^m u_n^{m*} u_{n+1}^{m+1} \prod_{k=-\infty}^{n} \Lambda_k^m + 2u_n^{m+1} u_n^{m+1*} u_{n-1}^m \prod_{k=-\infty}^{n-1} \Lambda_k^m\right] \\
&- u_n^m \sum_{k=-\infty}^{n} \Delta^m S_k^m - u_n^{m+1} \sum_{k=-\infty}^{n-1} \Delta^m S_k^{m*},
\end{aligned}$$

with

$$\Lambda_k^m = \frac{1 \pm u_k^{m+1} u_k^{m+1*}}{1 \pm u_k^m u_k^{m*}}, \qquad S_n^m = u_k^m u_{k-1}^{m*} + u_{k+1}^m u_k^{m*}, \qquad \Delta^m u_n^m = u_n^{m+1} - u_n^m$$

(Ablowitz and Ladik [1976b, 1977]).
(2), A totally discrete Toda Lattice

$$\Delta_t^2 \ln(1 + u_n(t)) = v_{n+1}(t) + v_{n-1}(t) - 2v_n(t),$$

with

$$v_n(t) = \delta^{-2} \ln(1 + \delta^2 u_n(t)),$$

and where

$$\Delta_t^2 f(t) = \delta^{-2} \left[f(t+\delta) + f(t-\delta) - 2f(t) \right]$$

(Hirota [1977b]).

1.11.4 Singular Integro-Differential Equations. Significantly certain nonlinear evolution equations with singular integral operators can be solve by I.S.T. (see Chapter 4). The best known ones are as follows: (a), the Intermediate Long Wave (ILW) equation

$$u_t + \delta^{-1} u_x + 2uu_x + \mathrm{T}u_{xx} = 0, \tag{1.11.7a}$$

where $\mathrm{T}u$ is the singular integral operator

$$(\mathrm{T}f)(x) = \frac{1}{2\delta} \fint_{-\infty}^{\infty} \coth\left\{\frac{\pi}{2\delta}(y-x)\right\} f(y)\,\mathrm{d}y \tag{1.11.7b}$$

(with $\fint_{-\infty}^{\infty}$ the Cauchy principal value integral). The ILW equation reduces as $\delta \to 0$ to the KdV equation

$$u_t + 2uu_x + \tfrac{1}{3}\delta u_{xxx} = 0;$$

(b), the Benjamin-Ono (BO) equation

$$u_t + 2uu_x + \mathrm{H}u_{xx} = 0, \tag{1.11.8a}$$

where $\mathrm{H}u$ is the Hilbert transform

$$(\mathrm{H}f)(x) = \frac{1}{\pi} \fint_{-\infty}^{\infty} \frac{f(y)}{y-x}\,\mathrm{d}y, \tag{1.11.8b}$$

is the limit as $\delta \to \infty$ of the ILW equation.

The scattering problem associated with the ILW equation (1.11.7) is given by

$$\mathrm{i}v_x^+ + (u - \lambda)v^+ = \mu v^-, \tag{1.11.9}$$

where $\lambda = -\frac{1}{2}k\coth(k\delta)$ and $\mu = \frac{1}{2}k\,\mathrm{cosech}(k\delta)$ (Satsuma, Ablowitz and Kodama [1979]). In (1.11.9), $v^{\pm}(x) := \lim_{\mathrm{Im}(z)\to 0\pm} v^{\pm}(z)$, analytic in the horizontal strips between $\mathrm{Im}(z) = 0$ and $\mathrm{Im}(z) = \pm 2\delta$, and periodically extended vertically. The scattering problem (1.11.9) defines a Riemann-Hilbert boundary value problem in a strip of the complex z-plane (where z is the complex extension of x). Kodama, Satsuma and Ablowitz [1981] have developed the I.S.T. scheme for solving (1.11.7) (see also Kodama, Ablowitz and Satsuma [1982]). In the limit as $\delta \to 0$, the scattering problem (1.11.9) tends to the Schrödinger equation

$$\delta\left(w_{xx} + \tfrac{1}{4}k^2 w\right) + uw = 0,$$

where $w(x) = \lim_{\delta \to 0} v(x)$, and as $\delta \to \infty$

$$iv_x^+ + (u+k)v^+ = kv^-, \tag{1.11.10}$$

where $v^{\pm}(z)$ are analytic in the upper and lower half planes. For the BO equation (1.11.8), the associated scattering problem (1.11.10) was obtained by Bock and Kruskal [1979] and Nakamura [1979] and defines a Riemann-Hilbert boundary value problem in the complex z-plane (where z is the complex extension of x). The Riemann-Hilbert problem for the BO equation is nonlocal, as opposed to local for the KdV and ILW equations, which is also a feature of certain multidimensional equations. (multidimensional equations have been solved either via nonlocal Riemann-Hilbert problems or a more general method called the DBAR technique — see §1.11.5.) The I.S.T. scheme for the BO equation was developed by Fokas and Ablowitz [1983b] and provided a conceptual pivot between inverse scattering in one spatial dimension and inverse scattering in two spatial dimensions. We also note that Anderson and Taflin [1985] constructed an alternative inverse scattering problem to (1.11.10); Coifman and Wickerhauser [1990] have rigorously investigated this inverse problem.

Degasperis and Santini [1983] have generated a class of integro-differential equations associated with the scattering problem

$$\boldsymbol{v}^-(x,z) = \mathbf{G}(x,z)\boldsymbol{v}^+(x,z), \qquad x \in \mathbb{R}, \tag{1.11.11a}$$

with

$$\mathbf{G}(x,z) = \mathbf{I} + \boldsymbol{\sigma}_3 z + \mathbf{U}(x), \tag{1.11.11b}$$

where $\mathbf{I}$ is the identity matrix, $\boldsymbol{\sigma}_3 = \mathrm{diag}(1,-1)$, z plays the role of the spectral parameter and $\mathbf{U}(x)$ is a complex z-independent potential function (see also Degasperis, Santini and Ablowitz [1985]; Santini [1986]). (1.11.11) is the scattering problem associated with the Sine-Hilbert equation

$$\mathrm{H}u_t = \sin u, \tag{1.11.12}$$

where H is the Hilbert transform. The I.S.T. scheme for a class of integro-differential equations related to (1.11.11), including the Sine-Hilbert equation (1.11.12), has recently been developed by Santini, Ablowitz and Fokas [1987]. The scheme in this case has a number of novel features (see §4.4).

1.11.5 Partial Differential Equations in Two Spatial Dimensions. The I.S.T. method applies to certain $2+1$-dimensional equations; the methodology which differs significantly from that in $1+1$-dimensions, is reviewed in Chapter 5. The scattering problem associated with the Kadomtsev-Petviashvili (KP) equation

$$(u_t + 6uu_x + u_{xxx})_x = -3\sigma^2 u_{yy}, \qquad \sigma^2 = \pm 1, \tag{1.11.13}$$

is

$$\sigma v_y + v_{xx} + (u + \lambda)v = 0, \tag{1.11.14}$$

which may be thought of as a two-dimensional generalization of the Schrödinger scattering problem used to solve the KdV equation. Recall that in the physical derivation of the KP equation (1.11.14) in §1.2, the change in sign of σ^2 is related to the magnitudes of gravity and surface tension. The parameter λ no longer has the importance it had for the KdV equation; in fact, without loss of generality we may set $\lambda = 0$, since it can be scaled out. In this case a parameter is inserted via a suitable "boundary condition". There exist two important cases of the KP equation, $\sigma = \mathrm{i}$ (KPI), and $\sigma = -1$ (KPII). For $\sigma = \mathrm{i}$, (1.11.14) is the "time-dependent" Schrödinger equation and the I.S.T. scheme for KPI was developed by Manakov [1981], Fokas and Ablowitz [1983c]; Boiti, Leon and Pempinelli [1989a]. This inverse problem is formulated in terms of a nonlocal Riemann-Hilbert boundary value problem, as was the case for the BO equation. However for KPII ($\sigma = -1$), the inverse problem can not be formulated as a Riemann-Hilbert boundary value problem since it turns out that the eigenfunctions are nowhere analytic. In this case a generalization of the Riemann-Hilbert problem is required, the $\bar{\partial}$ ("DBAR") problem. The $\bar{\partial}$ problem concept was introduced by Beals and Coifman [1981, 1982] in connection with the I.S.T. scheme for certain first order one-dimensional systems of differential equations (see also Beals and Coifman [1984, 1985a,b]). However, in these problems this concept is not crucial since the inverse problem for such ordinary differential equations can always be reduced to a Riemann-Hilbert problem. On the other hand, as shown by Ablowitz, BarYaacov and Fokas [1983], the $\bar{\partial}$ approach is essential in the development of the I.S.T. scheme for KPII. It is very significant since it is the first case where the Riemann-Hilbert problem formulation of the I.S.T. scheme is inadequate.

Ablowitz and Haberman [1975a] proposed the two-dimensional generalization of the scattering problem (1.9.6) in the form

$$\boldsymbol{v}_x = \mathbf{Q}\boldsymbol{v} - \sigma \mathbf{J} \boldsymbol{v}_y, \tag{1.11.15}$$

where $\boldsymbol{v}$ is an $N \times 1$ vector, $\mathbf{J} = \mathrm{diag}(J_1, J_2, \ldots, J_N)$ is a real $N \times N$ constant diagonal matrix, and $\mathbf{Q}(x, y, t)$ is an $N \times N$ off-diagonal matrix which contains the potentials. As for the KP equation, there exist two important cases, the hyperbolic case, $\sigma = -1$, and the elliptic case, $\sigma = \mathrm{i}$. The I.S.T. scheme for the hyperbolic case can be formulated in terms of a nonlocal Riemann-Hilbert boundary value problem and in the elliptic case in terms of a $\bar{\partial}$ problem (Fokas [1983]; Fokas and Ablowitz [1983e,f, 1984]). Suppose the time dependence of the eigenfunction is given by

$$\boldsymbol{v}_t = \mathbf{A}\boldsymbol{v} + \mathbf{B}\boldsymbol{v}_y + \mathbf{C}\boldsymbol{v}_{yy}. \tag{1.11.16}$$

Equations (1.11.15,16) are compatible (i.e., $\boldsymbol{v}_{xt} = \boldsymbol{v}_{tx}$) provided that

$$[\mathbf{J}, \mathbf{C}] = \mathbf{0}, \tag{1.11.17a}$$
$$[\mathbf{Q}, \mathbf{C}] = \sigma[\mathbf{J}, \mathbf{B}], \tag{1.11.17b}$$
$$[\mathbf{Q}, \mathbf{B}] = \sigma[\mathbf{J}, \mathbf{A}] + \mathbf{B}_x + \sigma\mathbf{J}\mathbf{B}_y + 2\mathbf{C}\mathbf{Q}_y, \tag{1.11.17c}$$
$$[\mathbf{Q}, \mathbf{A}] = \mathbf{A}_x + \sigma\mathbf{J}\mathbf{A}_y - \mathbf{Q}_t + \mathbf{B}\mathbf{Q}_y + \mathbf{C}\mathbf{Q}_{yy}, \tag{1.11.17d}$$

where it is assumed that $\mathbf{C}$ is a constant matrix (these equations are obtained by equating coefficients of y derivatives of $\boldsymbol{v}$).

For the 2×2 case, suppose

$$\mathbf{J} = \begin{pmatrix} 1 & 0 \\ 0 & -1 \end{pmatrix}, \qquad \mathbf{Q} = \begin{pmatrix} 0 & q \\ r & 0 \end{pmatrix}, \tag{1.11.18}$$

in the scattering problem (1.11.15). For consistency, equations (1.11.17) imply that

$$\mathbf{B} = -(\mathrm{i}/\sigma)\mathbf{Q}, \qquad \mathbf{C} = \mathrm{i}\mathbf{J}, \tag{1.11.19a}$$
$$\mathbf{A} = \begin{pmatrix} A_1 & \frac{1}{2}\mathrm{i}\sigma^2(q_x - \sigma q_y) \\ -\frac{1}{2}\mathrm{i}\sigma^2(r_x + \sigma r_y) & A_2 \end{pmatrix}; \tag{1.11.19b}$$

provided

$$\mathrm{i}q_t + \tfrac{1}{2}\sigma^2(q_{xx} + \sigma^2 q_{yy}) = \mathrm{i}q(A_1 - A_2), \tag{1.11.20a}$$
$$\mathrm{i}r_t - \tfrac{1}{2}\sigma^2(r_{xx} + \sigma^2 r_{yy}) = -\mathrm{i}r(A_1 - A_2), \tag{1.11.20b}$$
$$(\partial_x + \sigma\partial_y)A_1 = -\tfrac{1}{2}\mathrm{i}\sigma^2(\partial_x - \sigma\partial_y)(qr), \tag{1.11.20c}$$
$$(\partial_x - \sigma\partial_y)A_2 = \tfrac{1}{2}\mathrm{i}\sigma^2(\partial_x + \sigma\partial_y)(qr), \tag{1.11.20d}$$

where $\partial_x \equiv \partial/\partial_x$ and $\partial_y \equiv \partial/\partial_y$, which are a generalization of the nonlinear Schrödinger equation to 2+1-dimensions. If $r = \alpha q^*$, with $\alpha^2 = \pm 1$, and $\phi = \mathrm{i}(A_1 - A_2) + \alpha|q|^2$, then (1.11.20) reduce to the physically interesting Davey-Stewartson (DS) equations

$$\mathrm{i}q_t + \tfrac{1}{2}\sigma^2(q_{xx} + \sigma^2 q_{yy}) = -\alpha|q|^2 q + q\phi, \tag{1.11.21a}$$
$$\phi_{xx} - \sigma^2\phi_{yy} = 2\alpha(|q|^2)_{xx}, \tag{1.11.21b}$$

where $\sigma^2 = \pm 1$ (Davey and Stewartson [1974]). $\sigma = -1$ (DSI) is the hyperbolic case and $\sigma = \mathrm{i}$ (DSII) is the elliptic case. The hyperbolic case of the scattering problem (1.11.15) can also be used to develop the I.S.T. scheme for the DSI equation, and the elliptic case for the DSII equation. The issue of boundary conditions is important. Although $q \to 0$ as $(x^2+y^2)^{1/2} \to \infty$, we can take $\phi \to 0$ as $(x^2+y^2)^{1/2} \to \infty$ only for $\sigma = \mathrm{i}$ (DSII), the elliptic case. For DSI, ϕ need not vanish at infinity. However in considering the DS equations as embedded in a physical problem (e.g., water waves; see Ablowitz, Manakov and Schultz [1990]), with localized initial data, then the boundary value problem of DSI that is naturally picked out has ϕ vanishing at one end (but not the other). This boundary value problem was solved by Fokas and Ablowitz [1983e,f, 1984]. When one allows homogeneous solutions to be added at infinity, then interesting solutions to DSI can be found. These solutions decay exponentially fast as $(x^2 + y^2)^{1/2} \to \infty$, and have many properties in common (but not all) with solitons. Motivated in part by these observations, there have been many studies of the I.S.T. solution of the Davey-Stewartson equations (Arkadiev, Pogrebkov and Polivanov [1989b]; Boiti, Leon, Martina and Pempinelli [1988b, 1989]; Boiti, Leon and Pempinelli [1989b, 1990]; Fokas [1989a,b]; Fokas and Ablowitz [1983e,f, 1984]; Fokas and Santini [1989, 1990]; Konopelchenko and Matkarimov [1990a,b]; Santini [1990]). Along other lines the DS system has been proposed as a 2 + 1-dimensional model for quantum field theory (Schultz, Ablowitz and BarYaacov [1987]; Kulish and Lipovsky [1988]).

Analogous to the one-dimensional case, it is possible to generate a class of nonlinear evolution equations associated with the scattering problem (1.11.15) by an appropriate choice for the time evolution of the eigenfunction

$$\boldsymbol{v}_t = \mathbf{T}\boldsymbol{v}, \tag{1.11.22a}$$

for an appropriate $N \times N$ matrix $\mathbf{T}$. Recall that in §1.9, nonlinear evolution equations were generated by expanding $\mathbf{T}$ in powers of the eigenvalue k, in the two-dimensional case we suppose the expansion takes the form

$$\mathbf{T}\boldsymbol{v} = \sum_{j=0}^{n} \mathbf{A}_j \frac{\partial^j \boldsymbol{v}}{\partial y^j}, \tag{1.11.22b}$$

i.e., expand in powers of ∂_y. Letting $v(x,y) = \hat{v}(x)\mathrm{e}^{\mathrm{i}\lambda y}$ yields the 1 + 1-dimensional results. (We remark that there is a similar analogy for the scalar scattering operator (1.11.14) associated with the KP equation and the Schrödinger scattering operator associated with the KdV equation — see §5.2.2.)

In the case when $n = 1$ in (1.11.22) and $\sigma = -1$ in (1.11.15) (i.e., the hyperbolic case), then an associated nonlinear evolution equation is the N-wave interaction equations

$$\frac{\partial Q^{ij}}{\partial t} = \alpha_{ij}\frac{\partial Q^{ij}}{\partial x} + \beta_{ij}\frac{\partial Q^{ij}}{\partial y} + \sum_{k=1}^{n}(\alpha_{ik} - \alpha_{kj})Q^{ik}Q^{kj}, \tag{1.11.23}$$

where $Q^{ij} = \sigma_{ij}Q^{ji*}$, $Q^{ii} = 0$, with $\sigma_{ik}\sigma_{kj} = -\sigma_{ij}$ for $j < k < i$ and α_{ij}, β_{ij} are constants, which are $\frac{1}{2}n(n-1)$ equations (Ablowitz and Haberman [1975a]; Fokas and Ablowitz [1984]). When we take $N = 3$ with $Q^{12} = iu_3$, $Q^{23} = iu_1$, $Q^{31} = iu_2$, and suitable scaling, we obtain the three-wave interaction equations in $2+1$-dimensions [see also equations (1.11.3,4)]

$$\frac{\partial u_i}{\partial t} + a_i\frac{\partial u_i}{\partial x} + b_i\frac{\partial u_i}{\partial y} = c_i u_j^* u_k^*,$$

with a_i, b_i, c_i constants and $i, j, k = 1, 2, 3$ cyclically permuted, where $\boldsymbol{v}_i = (a_i, b_i)$ are the three group velocities associated with the waves (see also Cornille [1979]; Kaup [1980a, 1981a,b]; Fokas and Ablowitz [1984]).

In the case $N = 2$ and $n = 3$ in equation (1.11.22), then an associated nonlinear evolution equation is the following $2+1$-dimensional MKdV equation

$$u_t = \tfrac{1}{4}u_{xxx} + \tfrac{3}{4}u_{xyy} - \tfrac{3}{2}\alpha u^2 u_x + \tfrac{3}{4}(uw)_x - \tfrac{3}{4}u_y v, \tag{1.11.24a}$$

$$w_{xx} - w_{yy} = -2\alpha(u^2)_{yy}, \tag{1.11.24b}$$

$$v_{xx} - v_{yy} = 2\alpha(u^2)_{xy}, \tag{1.11.24c}$$

with α a constant (Boiti, Leon, Martina and Pempinelli [1988b]). We remark that this equation is the "Modified KP" equation which Fokas and Ablowitz [1983e,f, 1984] mention (but do not write down explicitly); equation (1.11.24) is *not* to be confused with the Modified KP equation introduced by Konopelchenko and Dubrovsky [1984] which we mention below — see equation (1.11.28). Therefore the N-wave interaction equations (1.11.23), the Davey-Stewartson equations (1.11.21) and the $2+1$-dimensional MKdV equation (1.11.24) can be thought of as the first three members of a hierarchy of evolution equations associated with the scattering problem (1.11.15) and the time evolution (1.11.22). (Recall that in the one-dimensional case, the nonlinear Schrödinger and MKdV equations arise from taking $n = 2$ and $n = 3$, respectively in (1.9.15) — see Examples 1.9.1 and 1.9.2.. Hence equation (1.11.24) is on the same hierarchy as the DS system, in the analogous way as the $1+1$ MKdV equations is in the same hierarchy as the NLS equation.)

Another example of a nonlinear partial differential equation solvable by an I.S.T. scheme for which (1.11.15), with $N = 2$, is the associated scattering operator, is the following vector generalization of the Davey-Stewartson equation

$$\boldsymbol{S}_t + \boldsymbol{S} \wedge (\boldsymbol{S}_{xx} + \sigma^2 \boldsymbol{S}_{yy}) + \phi_x \boldsymbol{S}_y + \phi_y \boldsymbol{S}_x = 0, \tag{1.11.25a}$$

$$\phi_{xx} - \sigma^2\phi_{yy} + 2\sigma^2 \boldsymbol{S}\bullet(\boldsymbol{S}_x \wedge \boldsymbol{S}_y) = 0, \tag{1.11.25b}$$

where $|\boldsymbol{S}(x,y,t)|^2 = S_1^2 + S_2^2 + S_3^2 = 1$, $\phi(x,y,t)$ is a scalar function (Ishimori [1984]; Beals and Coifman [1989]; Konopelchenko and Matkarimov [1989, 1990a,b]). The scattering operator for equations (1.11.25) is

$$\mathrm{L} = \sigma\partial_y + \mathbf{P}\partial_x, \tag{1.11.26a}$$

where $\mathbf{P} := S_1\boldsymbol{\sigma}_1 + S_2\boldsymbol{\sigma}_2 + S_3\boldsymbol{\sigma}_3$, with $\boldsymbol{\sigma}_j$, $j = 1,2,3$ the Pauli spin matrices given by

$$\boldsymbol{\sigma}_1 = \begin{pmatrix} 0 & 1 \\ 1 & 0 \end{pmatrix}, \qquad \boldsymbol{\sigma}_2 = \begin{pmatrix} 0 & -\mathrm{i} \\ \mathrm{i} & 0 \end{pmatrix}, \qquad \boldsymbol{\sigma}_3 = \begin{pmatrix} 1 & 0 \\ 0 & -1 \end{pmatrix}. \tag{1.11.26b}$$

However, we remark that equations (1.11.26) and the Davey-Stewartson equations (1.11.20 are gauge equivalent (see §5.6.1). They give the M operator associated with (1.11.25,26)

$$\mathrm{M}\boldsymbol{v} = \boldsymbol{v}_t - 2\mathrm{i}\mathbf{P}\boldsymbol{v}_{xx} - (\mathrm{i}\mathbf{P}_x + \mathrm{i}\sigma\mathbf{P}_y\mathbf{P} + \sigma^2\mathbf{P}\phi_x - \phi_y)\boldsymbol{v}_x.$$

Konopelchenko and Dubrovsky [1984] have associated several nonlinear evolution equations in two spatial dimensions (x, y) and one temporal dimension t, with the scalar scattering operator

$$\mathrm{L} = \sum_{j=0}^{N} U_j(x,y,t)\frac{\partial^j}{\partial x^j} + \sigma\frac{\partial}{\partial y}, \qquad \sigma^2 = 1, \tag{1.11.27}$$

where $U_j(x,y,t)$, $j = 0,1,\ldots,N$ are scalar functions (usually chosen in terms of the solution u of the evolution equation and its spatial derivatives). Examples of nonlinear evolution equations obtained by Konopelchenko and Dubrovsky [1984] (see also Konopelchenko [1982]; Konopelchenko and Dubrovsky [1991]), for which (1.11.27) is the associated scattering operator include:
(a), a Modified Kadomtsev-Petviashvili equation

$$u_t = u_{xxx} - 6u^2u_x - 6\sigma u_x(\partial_x^{-1}u_y) + 3\sigma^2(\partial_x^{-1}u_{yy}), \tag{1.11.28}$$

where $(\partial_x^{-1}f)(x) := \int_{-\infty}^{x} f(x')\,\mathrm{d}x'$;
(b), $2+1$-dimensional version of fifth order equations discussed in §1.11.2

$$\begin{aligned} u_t = u_{xxxxx} &+ 10u_xu_{xx} + 10uu_{xxx} + 20u^2u_x + 5\sigma u_{xxy} \\ &- 5\sigma^2(\partial_x^{-1}u_{yy}) + 10\sigma uu_y + 10\sigma u_x(\partial_x^{-1}u_y), \\ u_t = u_{xxxxx} &+ 10u_xu_{xx} + 25uu_{xxx} + 20u^2u_x + 5\sigma u_{xxy} \\ &- 5\sigma^2(\partial_x^{-1}u_{yy}) + 10\sigma uu_y + 10\sigma u_x(\partial_x^{-1}u_y); \end{aligned}$$

(c), a two-dimensional generalization of the Harry-Dym equation (1.9.37),

$$u_t = u^3u_{xxx} + 3\sigma^2u^{-1}[u^2\partial_x^{-1}(u^{-2}u_y)]_y.$$

We note that Jiang and Bullough [1987] and Jiang [1989] have discussed the direct and inverse scattering problems for the scattering operator (1.11.27) using $\bar{\partial}$ and nonlocal Riemann-Hilbert methods.

Examples of other equations in 2+1 dimensions which are solvable by an appropriate I.S.T. scheme include:
(a), a cylindrical KP equation

$$(u_t + 6uu_x + u_{xxx})_x + \frac{1}{2t}u_x + \frac{3\sigma^2}{t^2}u_{yy} = 0, \qquad \sigma^2 = \pm 1$$

(Johnson [1980]; Dryuma [1983]);

(b)
$$(u_t + 6uu_x + u_{xxx})_x - 3u_{yy} + \kappa(|\Phi|^2)_{xx} = 0, \qquad \kappa = \pm,$$
$$\mathrm{i}\Phi_y = u\Phi + \Phi_{xx}$$

(Mel'nikov [1983, 1986, 1987a,b]);
(c), Two equations which arise as special reductions of the Self-dual Yang-Mills equations in 2 + 2-dimensions (see §1.11.8)

(i)
$$\left(\mathbf{U}^\dagger\mathbf{U}_t\right)_t + \left(\mathbf{U}^\dagger\mathbf{U}_z\right)_{\bar{z}} = \mathbf{0},$$

where $z = x + \mathrm{i}y$, $\bar{z} = x - \mathrm{i}y$ and $\mathbf{U}$ is a unitary matrix ($\mathbf{U}^\dagger\mathbf{U} = \mathbf{I}$), in SU($n$) (Manakov and Zakharov [1981]; Villarroel [1989]);

(ii)
$$\begin{aligned} &-\left(\mathbf{J}^{-1}\mathbf{J}_t\right)_t + \left(\mathbf{J}^{-1}\mathbf{J}_x\right)_x + \left(\mathbf{J}^{-1}\mathbf{J}_y\right)_y + n_0\left\{\left(\mathbf{J}^{-1}\mathbf{J}_y\right)_x - \left(\mathbf{J}^{-1}\mathbf{J}_x\right)_y\right\} \\ &+ n_1\left\{\left(\mathbf{J}^{-1}\mathbf{J}_t\right)_y - \left(\mathbf{J}^{-1}\mathbf{J}_y\right)_t\right\} + n_2\left\{\left(\mathbf{J}^{-1}\mathbf{J}_x\right)_t - \left(\mathbf{J}^{-1}\mathbf{J}_t\right)_x\right\} = \mathbf{0}, \end{aligned}$$

where $\boldsymbol{n} = (n_0, n_1, n_2)$ is a constant unit vector (Ward [1988a,b]; Villarroel [1990a]), which is a 2 + 1-dimensional Chiral Field equation.

Boiti, Leon, Manna and Pempinelli (Boiti, Leon, Manna and Pempinelli [1986, 1987]; Boiti, Leon and Pempinelli [1987a,b]) have discussed inverse scattering for 2+1-dimensional equations involving a "weak Lax pair." The standard inverse scattering method involves writing an nonlinear evolution equations as the compatibility condition of two operators L and M, i.e., they commute [L, M] = 0, if and only if a potential(s) satisfies a nonlinear equation. However, in some cases, this requirement maybe too strong. The Lax pair condition can be relaxed by only requiring that the operator M commutes with L on the subspace of the solutions of the spectral equation $\mathrm{L}v = 0$, i.e., $[\mathrm{L}, \mathrm{M}]v = 0$ (see, for example, Vesselov and Novikov [1984a,b]; Boiti, Leon, Manna and Pempinelli [1986, 1987]; Boiti, Leon and Pempinelli [1987a,b]). We discuss this further in §5.6.2. Here we list some examples of solvable evolution equations in two spatial dimensions obtained using this construction:

(1)
$$u_t + u_{zzz} + u_{\bar{z}\bar{z}\bar{z}} = 3(uv_z)_z + 3(uw_{\bar{z}})_{\bar{z}}, \qquad u = v_{\bar{z}} = w_z,$$

where $z = x + iy$, $\bar{z} = x - iy$ which is associated with the spectral problem

$$\mathrm{L}_2 v \equiv v_{xx} + v_{yy} + u(x,y,t)v = 0$$

(see Nizhnik [1980]; Vesselov and Novikov [1984a,b]; Grinevich and Novikov [1986]; Boiti, Leon, Manna and Pempinelli [1986]; Bogdanov and Manakov [1988]).

(2) $$u_t + u_{\xi\xi\xi} = 3(uv)_\xi, \qquad v_\eta = u_\xi,$$

where $\xi = \frac{1}{2}(x+y)$, $\eta = \frac{1}{2}(x-y)$ which is associated with the spectral problem

$$\mathrm{L}_1 v \equiv v_{xx} - v_{yy} - u(x,y,t)v = 0,$$

and reduces to the KdV equation if $\xi = \eta$. Of course L_1 and L_2 are related formally by $y \to iy$, but the I.S.T. schemes have significant differences analogous to KPI and KPII.

(3) $$\left[\mathrm{e}^{-\theta}\left\{\mathrm{e}^{\theta}(\theta_{xt} + \sinh\theta)\right\}_x\right]_x = \sigma^2\left[\mathrm{e}^{-\theta}\partial_x^{-1}(\mathrm{e}^{\theta})_y\right]_{yt} - \tfrac{1}{2}\sigma^2\left[\left(\mathrm{e}^{-\theta}\partial_x^{-1}(\mathrm{e}^{\theta})_y\right)^2\right]_{xt}$$

(Boiti, Leon and Pempinelli [1987a,b]) which is a generalized Sinh-Gordon equation and is associated with the spectral problem

$$\mathrm{L}_3 v \equiv \sigma v_y + v_{xx} - u(x,y,t)v, \qquad \sigma^2 = \pm 1.$$

Similarly,

(4) $$\begin{aligned} u_{ty} &= -(\eta_x + uu_y)_x, \\ \eta_t &= -(u\eta + u + u_{xy})_x \end{aligned}$$

(Boiti, Leon and Pempinelli [1987b]) is associated with the spectral problem

$$\mathrm{L}_4 \equiv v_{xy} - r(x,y,t)v_y - q(x,y,t)v.$$

1.11.6 Multidimensional Scattering Equations. The following higher dimensional generalizations of the scattering problems (1.11.14) and (1.11.15) have also been studied

$$-\sum_{j=1}^{n} \frac{\partial^2 v}{\partial x_j^2} + u(\boldsymbol{x},y)v = \lambda v, \tag{1.11.29}$$

$$\sigma\frac{\partial v}{\partial y} + \sum_{j=1}^{n} \frac{\partial^2 v}{\partial x_j^2} + u(\boldsymbol{x},y)v = 0, \tag{1.11.30}$$

$$\sigma\frac{\partial \boldsymbol{v}}{\partial y} + \sigma\sum_{p=1}^{n} \mathbf{J}_p\frac{\partial \boldsymbol{v}}{\partial x_p} = \mathbf{Q}(x,y)\boldsymbol{v}, \tag{1.11.31}$$

with $\boldsymbol{x} = (x_1, x_2, \ldots, x_n)$, $\boldsymbol{v} = (v_1, v_2, \ldots, v_N)$ and where $\sigma = \sigma_R + i\sigma_I \in \mathbb{C}$, $x \in \mathbb{R}^n$, $y \in \mathbb{R}$, $\mathbf{Q} \in \mathrm{M}_N(\mathbb{C})$ (the space of $N \times N$ matrices over $\mathbb{C}$) and $\mathbf{J}_p = \mathrm{diag}(J_p^1, J_p^2, \ldots, J_p^N)$, with $J_p^i \neq J_p^j$, for $i \neq j$, $i, j = 1, 2, \ldots, N$ and $p = 1, 2, \ldots, n$.

The inverse scattering problem associated with the n-dimensional time independent Schrödinger operator (1.11.29) has been studied by, amongst others, Faddeev [1966, 1976]; Newton [1974, 1980, 1981, 1982a,b,c, 1983]; Nachman and Ablowitz [1984a]; Beals [1985]; Lavine and Nachman [1987]; Novikov and Khenkin [1987a,b]; Aktosun and van der Mee [1990, 1991]; Weder [1990, 1991]; and that with the n-dimensional time dependent Schrödinger operator (1.11.30) has been studied by Nachman and Ablowitz [1984a]. However, at present there are no known nonlinear evolution equations associated with these generalizations in greater than two spatial dimensions (other than the N-wave interaction equations, which we briefly discuss below and in Chapter 6). The prime reason for the difficulty finding nonlinear evolution equations in greater then two spatial dimensions which are associated with an appropriate scattering problem is that it turns out that in general the scattering data must satisfy a nonlinear constraint. This poses an essential difficulty is in characterizing suitable scattering data. We shall discuss this restriction further in Chapter 6.

The inverse scattering problem associated with multidimensional first order systems has been studied by Nachman and Ablowitz [1984b]; Fokas [1986a,b]; Nachman, Fokas and Ablowitz [1986]; Sung and Fokas [1991b]. As before, there are two important cases, that is the hyperbolic case $\sigma = -1$ and the elliptic case $\sigma = \mathrm{i}$. Nachman and Ablowitz [1984b] showed that the scattering operator (1.11.31) in the hyperbolic case is associated with the N-wave interaction equations in $n+1$-spatial dimensions and one-temporal dimension (Ablowitz and Haberman [1975a])

$$\frac{\partial Q^{ij}}{\partial t} = \alpha_{ij}\frac{\partial Q^{ij}}{\partial y} - \sum_{p=1}^{n}(\alpha_{ij}J_p^i - B_p^i)\frac{\partial Q^{ij}}{\partial x_p} + \sum_{p=1}^{n}(\alpha_{ip} - \alpha_{pj})Q^{ip}Q^{pj}, \tag{1.11.32a}$$

with

$$\alpha_{ij} = \frac{B_p^j - B_p^i}{J_p^j - J_p^i}, \qquad 1 \le p \le n, \tag{1.11.32b}$$

for some real B_p^i, $1 \le p \le n$, $1 \le i \le N$, provided that the J_p, $p = 1, 2, \ldots, n$, satisfy the constraint

$$\frac{J_p^k - J_p^j}{J_q^k - J_q^j} = \frac{J_p^i - J_p^j}{J_q^i - J_q^j}, \tag{1.11.32c}$$

for $p, q = 1, 2, \ldots, n$; $i, j, k = 1, 2, \ldots, N$. The N-wave interaction equations appear to be the only known nonlinear evolution system related to a multidimensional scattering problem in greater than two spatial dimensions. However Fokas [1986b] has shown that the constraint (1.11.32c) allows the N-wave interaction equations to be reduced from $n+1$ spatial dimensions to two spatial dimensions (earlier results of Nachman and Ablowitz [1984b] demonstrated reductions to at most three dimensions).

1.11.7 Multidimensional Differential Geometric Equations. The Sine-Gordon equation

$$u_{tt} - u_{xx} = \kappa \sin u, \tag{1.11.33}$$

is a well-known example of an equation solvable by I.S.T. (Ablowitz, Kaup, Newell and Segur [1973a]). It also arises in many physical applications and in the study of surfaces with constant negative curvature in differential geometry. Classical work by Bäcklund [1905]; Bianchi [1922] developed special solutions as well as transformations between solutions.

Natural geometric generalizations of these classical results have been obtained by Tenenblat and Terng [1980] and Terng [1980], in which a multidimensional version of the Sine-Gordon equation (1.11.33), called the generalized Sine-Gordon (GSG) equation, and related transformations were found. The GSG equation has the form

$$\frac{\partial}{\partial x_i}\left(\frac{1}{a_{1i}}\frac{\partial a_{1j}}{\partial x_i}\right) + \frac{\partial}{\partial x_j}\left(\frac{1}{a_{1j}}\frac{\partial a_{1i}}{\partial x_j}\right) + \sum_{k\neq i,j}\frac{1}{a_{ij}^2}\frac{\partial a_{1i}}{\partial x_k}\frac{\partial a_{1k}}{\partial x_j} = \varepsilon a_{1i}a_{1j}, \qquad i \neq j, \tag{1.11.34a}$$

$$\frac{\partial}{\partial x_k}\left(\frac{1}{a_{1j}}\frac{\partial a_{1i}}{\partial x_j}\right) = \frac{1}{a_{1k}a_{1j}}\frac{\partial a_{1i}}{\partial x_k}\frac{\partial a_{1k}}{\partial x_j}, \qquad i,j,k \ \text{ distinct}, \tag{1.11.34b}$$

$$\frac{\partial a_{jk}}{\partial x_i} = \frac{a_{ji}}{a_{1j}}\frac{\partial a_{1k}}{\partial x_i}, \qquad i \neq k. \tag{1.11.34c}$$

where $\varepsilon = 1$. Similar results were obtained by Tenenblat [1985] for the generalized Wave equation (GW), which is given by (1.11.34) with $\varepsilon = 0$. We remark that the seemingly natural generalization of the Sine-Gordon equation to n-spatial dimensions namely

$$\frac{\partial^2 u}{\partial t^2} - \sum_{j=1}^{n}\frac{\partial^2 u}{\partial x_j^2} = \kappa \sin u, \tag{1.11.35}$$

is thought not to be integrable for $n \geq 2$ (cf. Clarkson [1986]).

Ablowitz, Beals and Tenenblat [1986] developed the associated linear scattering equations and the I.S.T. schemes for the GSG and GW equations. They showed that the linear problems for both equations are given by systems of ordinary differential equations and that the direct and inverse problems can be solved for appropriately decaying data on suitable lines. The solutions of the generalized equations were obtained via factorization of a certain Riemann-Hilbert boundary value problem. We discuss the solution of the GSG and GW equations via I.S.T. in §6.4.

Motivated by the results of Ablowitz, Beals and Tenenblat [1986] on the GW and GSG equations, Ablowitz, Costa and Tenenblat [1987] developed a method for obtaining solutions of multidimensional generalizations of other well-known nonlinear systems, in particular for the multidimensional extension of the anti-self-dual Yang-Mills equation for gauge fields in gl$(n, \mathbb{C})$,

$$S_j \frac{\partial}{\partial \bar{z}_{j+1}} \left(\boldsymbol{\Omega}^{-1} \frac{\partial \boldsymbol{\Omega}}{\partial z_i} \right) - S_i \frac{\partial}{\partial \bar{z}_{i+1}} \left(\boldsymbol{\Omega}^{-1} \frac{\partial \boldsymbol{\Omega}}{\partial z_j} \right) = 0, \tag{1.11.36a}$$

$$\sum_{\substack{l=1\\l\neq j}}^{n} S_{jl} \frac{\partial}{\partial \bar{z}_l} \left(\boldsymbol{\Omega}^{-1} \frac{\partial \boldsymbol{\Omega}}{\partial z_i} \right) - \sum_{\substack{k=1\\k\neq i}}^{n} S_{ik} \frac{\partial}{\partial \bar{z}_k} \left(\boldsymbol{\Omega}^{-1} \frac{\partial \boldsymbol{\Omega}}{\partial z_j} \right) = 0, \tag{1.11.36b}$$

$$r_j \frac{\partial}{\partial \bar{z}_{j-r_j(n/2)}} \left(\boldsymbol{\Omega}^{-1} \frac{\partial \boldsymbol{\Omega}}{\partial z_i} \right) - r_i \frac{\partial}{\partial \bar{z}_{i-r_i(n/2)}} \left(\boldsymbol{\Omega}^{-1} \frac{\partial \boldsymbol{\Omega}}{\partial z_j} \right) = 0, \tag{1.11.36c}$$

where n is an even integer

$$S_j = (-1)^j, \qquad S_{jk} = \begin{cases} 1, & \text{if } j > k, \\ -1, & \text{if } j < k \end{cases}, \qquad r_j = \begin{cases} -1, & \text{if } j \leq n/2, \\ 1, & \text{if } j > n/2 \end{cases}$$

and $\boldsymbol{\Omega}(\boldsymbol{z})$ is an $N \times N$ matrix valued function of $\boldsymbol{z} = (z_1, \ldots, z_n) \in \mathbb{C}^n$. Note that when $n = 2$, these equations reduces the the anti-self-dual Yang-Mills equation (cf. Pohlymeyer [1980])

$$\frac{\partial}{\partial \bar{z}_1} \left(\boldsymbol{\Omega}^{-1} \frac{\partial \boldsymbol{\Omega}}{\partial z_1} \right) + \frac{\partial}{\partial \bar{z}_2} \left(\boldsymbol{\Omega}^{-1} \frac{\partial \boldsymbol{\Omega}}{\partial z_2} \right) = 0, \tag{1.11.37}$$

where $\boldsymbol{\Omega}(\boldsymbol{z})$ is an $N \times N$ matrix valued function of $\boldsymbol{z} = (z_1, z_2) \in \mathbb{C}^2$.

1.11.8 The Self-dual Yang-Mills Equations. One of the most interesting and important multidimensional integrable equations is the self-dual Yang-Mills (SDYM) equations (Ward [1977]; Belavin and Zakharov [1978]). These four-dimensional equations arise in the study of field theory (see, for example, Chau [1984]) and relativity (Woodhouse and Mason [1988]; Mason and Newman [1989]). The SDYM equations are regarded as being completely integrable as a consequence of the so-called "twistor" construction, which related solutions of the equations to certain holomorphic vector bundles (cf. Atiyah and Ward [1977]; Corrigan, Fairlie, Yates and Goddard [1978]; Ward [1981]).

It is standard to define the Yang-Mills field as

$$F_{\mu\nu} = \frac{\partial \mathcal{A}_\nu}{\partial x_\mu} - \frac{\partial \mathcal{A}_\mu}{\partial x_\nu} - [\mathcal{A}_\mu, \mathcal{A}_\nu], \tag{1.11.38}$$

where x_μ, $\mu = 0, 1, 2, 3$ are the co-ordinates in Euclidian space E^4. $\mathcal{A}_\mu$ and $\mathcal{A}_\nu$ are the Yang-Mills potentials and take values in some generally Lie algebra, but also algebras related to diffeomorphisms, with [,] being the associated Lie bracket. In terms of standard null co-ordinates $x_\alpha = t + \mathrm{i}z$, $x_{\bar{\alpha}} = t - \mathrm{i}z$, $x_\beta = x + \mathrm{i}y$ and $x_{\bar{\beta}} = x - \mathrm{i}y$, the metric being $\mathrm{d}s^2 = \mathrm{d}x_\alpha \, \mathrm{d}x_{\bar{\alpha}} + \mathrm{d}x_\beta \, \mathrm{d}x_{\bar{\beta}}$, the SDYM field equations are given by

$$F_{\alpha\beta} = 0, \qquad F_{\bar{\alpha}\bar{\beta}} = 0, \qquad F_{\alpha\bar{\alpha}} + F_{\beta\bar{\beta}} = 0. \tag{1.11.39}$$

This is a system of three partial differential equations for $\mathcal{A}_\alpha$, $\mathcal{A}_{\bar{\alpha}}$, $\mathcal{A}_\beta$ and $\mathcal{A}_{\bar{\beta}}$. We note that the SDYM equations are invariant under the gauge transformation

$$\mathcal{A}_\mu \to \mathcal{F}\mathcal{A}_\mu\mathcal{F}^{-1} - \partial_\mu,$$

with $\partial_\mu = \partial/\partial x_\mu$,for any $\mathcal{F}(x_\mu)$. Frequently a gauge condition such as $\mathcal{A}_\alpha = 0$ is imposed. Equations (1.11.39) can also be obtained as the compatibility condition of the Lax pair

$$(\partial_\alpha + \zeta\partial_{\bar{\beta}})\Psi = (\mathcal{A}_\alpha + \zeta\mathcal{A}_{\bar{\beta}})\Psi, \tag{1.11.40a}$$

$$(\partial_\beta - \zeta\partial_{\bar{\alpha}})\Psi = (\mathcal{A}_\beta - \zeta\mathcal{A}_{\bar{\alpha}})\Psi, \tag{1.11.40b}$$

where ζ is the spectral parameter. The compatibility condition is expressed as a polynomial in ζ

$$(\partial_\alpha + \zeta\partial_{\bar{\beta}})(\mathcal{A}_\beta - \zeta\mathcal{A}_{\bar{\alpha}}) - (\partial_\beta - \zeta\partial_{\bar{\alpha}})(\mathcal{A}_\alpha + \zeta\mathcal{A}_{\bar{\beta}}) = [\mathcal{A}_\alpha + \zeta\mathcal{A}_{\bar{\beta}}, \mathcal{A}_\beta - \zeta\mathcal{A}_{\bar{\alpha}}]. \tag{1.11.41}$$

Equating coefficients of powers of ζ yields (1.11.39).

Recently it has been shown that these equations admit reductions to well known 'soliton' equations in one space and one time $(1+1)$ dimension, e.g., the Sine-Gordon, nonlinear Schrödinger, Korteweg-de Vries and Toda Lattice equations (Ward [1984, 1985, 1986]; Mason and Sparling [1989]); additionally it has also been shown that several classical systems of ordinary differential equations including the Euler-Arnold equations for free motion of an n-dimensional rigid body about a fixed point (Arnold [1974]; Manakov [1976]), the Kovalevskaya Top (Kovalevski [1889a,b]), and a generalization of the Nahm equation which is related to a classical third order differential equation (due to Chazy [1910, 1911]) possessing a movable natural boundary in the complex plane, arise as one-dimensional reductions of the SDYM equations (Chakravarty, Ablowitz and Clarkson [1990]). We discuss reductions of the SDYM equations in §6.5.

The self-dual Yang Mills equations are often written in the form due to Pohlmeyer [1976, 1980]

$$\partial_{\bar{\alpha}}(\mathbf{J}^{-1}\partial_\alpha\mathbf{J}) + \partial_{\bar{\beta}}(\mathbf{J}^{-1}\partial_\beta\mathbf{J}) = 0, \tag{1.11.42}$$

where $\mathbf{J} \in \mathrm{GL}(N, \mathbb{C})$ (the Lie algebra of $N \times N$ matrices over the complex field). However as remarked by Ward [1985], whereas this is a neat form of the equations, it is not as general as the original form (1.11.39) since the SO(4)-invariance, which was present in (1.11.39), and the geometrical interpretation (in terms of connections and curvatures) have been lost. Solutions using inverse scattering formalism (direct methods) of (1.11.42) have been studied by Beals and Coifman [1985b, 1986] and Villarroel [1990b]. The associated linear system is given by

$$(\partial_\alpha + \zeta\partial_{\bar{\beta}})\Psi = \mathbf{J}^{-1}(\partial_\alpha\mathbf{J})\Psi, \tag{1.11.43a}$$

$$(\partial_\beta + \zeta\partial_{\bar{\alpha}})\Psi = \mathbf{J}^{-1}(\partial_\beta\mathbf{J})\Psi, \tag{1.11.43b}$$

Ψ is a matrix-valued function.

Chapter Two

Scattering, Inverse Scattering and the Inverse Scattering Transform Associated with the Korteweg-de Vries Equation.

2.1 Introduction.

As a prototype example for the inverse scattering transform (I.S.T.) in one spatial dimension, we consider the Korteweg-de Vries (KdV) equation

$$u_t + 6uu_x + u_{xxx} = 0. \tag{2.1.1}$$

(In the next chapter we discuss certain other nonlinear equations in one spatial dimension which are solvable by I.S.T..) Recall that the Lax pair for the KdV equation is given by

$$\mathrm{L}v = v_{xx} + u(x,t)v = \lambda v, \tag{2.1.2a}$$

$$v_t = \mathrm{M}v = (u_x + \gamma)v - (2u + 4\lambda)v_x, \tag{2.1.2b}$$

where γ is an arbitrary constant parameter and λ is the spectral parameter. These equations are compatible (i.e., $v_{xxt} = v_{txx}$) and satisfy

$$\mathrm{L}_t + [\mathrm{L}, \mathrm{M}] = 0, \tag{2.1.3}$$

provided that u satisfies the KdV equation (2.1.1).

2.2 The Direct Scattering Problem.

Suppose that $\lambda = -k^2$, then (2.1.2) becomes

$$v_{xx} + \left\{u(x) + k^2\right\}v = 0, \tag{2.2.1a}$$

$$v_t = (u_x + \gamma)v + (4k^2 - 2u)v_x, \tag{2.2.1b}$$

where we have suppressed the time dependence in u for convenience. We shall further assume that $u(x)$ lies in the space P_μ, where

$$P_\mu = \left\{u(x) : \int_{-\infty}^{\infty} (1 + |x|^\mu)|u(x)|\,\mathrm{d}x < \infty\right\},$$

for either $\mu = 1$ or $\mu = 2$ (see Deift and Trubowitz [1979] for the case $\mu = 2$; we remark that P_μ is sometimes refered to as L^1_μ, e.g., Deift and Trubowitz [1979]). The case $\mu = 1$ was the original condition given by Faddeev [1963] and has recently been shown to be valid by Marchenko [1986] (see also Melin [1985]). Associated with (2.2.1a) are two complete sets of eigenfunctions which are bounded for all values of x, and which have appropriate analytic extensions. These eigenfunctions are defined by the equation and boundary conditions, that is the four eigenfunctions defined by the following asymptotic behaviors

$$\phi(x;k) \sim \mathrm{e}^{-ikx}, \qquad \bar{\phi}(x;k) \sim \mathrm{e}^{ikx}, \qquad \text{as} \quad x \to -\infty, \tag{2.2.2a}$$

$$\psi(x;k) \sim \mathrm{e}^{ikx}, \qquad \bar{\psi}(x;k) \sim \mathrm{e}^{-ikx}, \qquad \text{as} \quad x \to \infty. \tag{2.2.2b}$$

Therefore $\phi(x;k)$, for example, is that solution of (2.2.1a) which tends to e^{-ikx} as $x \to -\infty$ (it transpires that we only require three of these eigenfunctions). Clearly from (2.2.1a) and the boundary conditions (2.2.2), we see that

$$\phi(x;k) = \bar{\phi}(x;-k), \tag{2.2.3a}$$

$$\psi(x;k) = \bar{\psi}(x;-k). \tag{2.2.3b}$$

Equation (2.2.1a) is a linear second order ordinary differential equation, and therefore by the linear independence of its solutions we obtain the following relationships between the eigenfunctions

$$\phi(x;k) = a(k)\bar{\psi}(x;k) + b(k)\psi(x;k), \tag{2.2.4a}$$

$$\bar{\phi}(x;k) = -\bar{a}(k)\psi(x;k) + \bar{b}(k)\bar{\psi}(x;k), \tag{2.2.4b}$$

for some functions $a(k)$, $\bar{a}(k)$, $b(k)$, $\bar{b}(k)$, which satisfy the symmetry conditions

$$\bar{a}(k) = -a(-k) = -a^*(k^*), \tag{2.2.5a}$$

$$\bar{b}(k) = b(-k) = b^*(k^*), \tag{2.2.5b}$$

where $a^*(k^*)$, $b^*(k^*)$ are the complex conjugates of $a(k)$, $b(k)$, respectively, as is easily verified using equations (2.2.1-4) and their complex conjugates. Additionally the functions $a(k)$, $\bar{a}(k)$, $b(k)$, $\bar{b}(k)$ satisfy the condition

$$a(k)\bar{a}(k) + b(k)\bar{b}(k) = -1. \tag{2.2.5c}$$

This follows since

$$\begin{aligned} W(\phi(x;k),\bar{\phi}(x;k)) &= W(a(k)\bar{\psi}(x;k) + b(k)\psi(x;k), -\bar{a}(k)\psi(x;k) + \bar{b}\bar{\psi}(x;k)) \\ &= \left\{a(k)\bar{a}(k) + b(k)\bar{b}(k)\right\} W(\psi(x;k),\bar{\psi}(x;k)), \end{aligned} \tag{2.2.6}$$

where

$$W(\phi(x;k),\psi(x;k)) := \phi(x;k)\psi_x(x;k) - \phi_x(x;k)\psi(x;k),$$

is the *Wronskian* which has the following properties

$$W(\phi(x;k),\psi(x;k)) = -W(\psi(x;k),\phi(x;k)), \tag{2.2.7a}$$

$$W(c_1\phi(x;k),c_2\psi(x;k)) = c_1c_2\,W(\phi(x;k),\psi(x;k)), \tag{2.2.7b}$$

with c_1, c_2 arbitrary constants. From the asymptotic behaviors (2.2.2), it is easily seen that

$$W(\phi(x;k),\bar{\phi}(x;k)) = 2\mathrm{i}k = -W(\psi(x;k),\bar{\psi}(x;k)),$$

and substituting these into equation (2.2.6) yields (2.2.5c). Furthermore, it is easily seen that

$$a(k) = \frac{W(\phi(x;k),\psi(x;k))}{2\mathrm{i}k}, \qquad b(k) = -\frac{W(\phi(x;k),\bar{\psi}(x;k))}{2\mathrm{i}k}.$$

Rather than work with the eigenfunctions $\phi(x;k)$, $\bar{\phi}(x;k)$, $\psi(x;k)$ and $\bar{\psi}(x;k)$, it turns out that it is more convenient to work with the (modified) eigenfunctions $M(x;k)$, $\bar{M}(x;k)$, $N(x;k)$ and $\bar{N}(x;k)$, defined by

$$M(x;k) := \phi(x;k)\,\mathrm{e}^{\mathrm{i}kx}, \qquad \bar{M}(x;k) := \bar{\phi}(x;k)\,\mathrm{e}^{\mathrm{i}kx}, \tag{2.2.8a}$$

$$N(x;k) := \psi(x;k)\,\mathrm{e}^{\mathrm{i}kx}, \qquad \bar{N}(x;k) := \bar{\psi}(x;k)\,\mathrm{e}^{\mathrm{i}kx}, \tag{2.2.8b}$$

then

$$M(x;k) \sim 1, \qquad \bar{M}(x;k) \sim \mathrm{e}^{2\mathrm{i}kx}, \qquad \text{as} \quad x \to -\infty, \tag{2.2.9a}$$

$$N(x;k) \sim \mathrm{e}^{2\mathrm{i}kx}, \qquad \bar{N}(x;k) \sim 1, \qquad \text{as} \quad x \to \infty. \tag{2.2.9b}$$

However, we shall only require three of these eigenfunctions. Equations (2.2.4a) and (2.2.8) imply that

$$\frac{M(x;k)}{a(k)} = \bar{N}(x;k) + \rho(k)N(x;k), \tag{2.2.10a}$$

$$\frac{\bar{M}(x;k)}{\bar{a}(k)} = -N(x;k) + \bar{\rho}(k)\bar{N}(x;k), \tag{2.2.10b}$$

where

$$\rho(k) := \frac{b(k)}{a(k)}, \qquad \bar{\rho}(k) := \frac{\bar{b}(k)}{\bar{a}(k)}, \tag{2.2.10c}$$

$t(k) := 1/a(k)$ and $\rho(k)$ are called the *transmission* and *reflection* coefficients respectively. From equations (2.2.5,6), we see that $t(k)$ and $\rho(k)$ satisfy

$$|\rho(k)|^2 + |t(k)|^2 = 1.$$

Equations (2.2.3b,8b) imply that

$$N(x;k) = \bar{N}(x;-k)\,\mathrm{e}^{2\mathrm{i}kx}. \tag{2.2.11}$$

Therefore from (2.2.10,11) we obtain

$$\frac{M(x;k)}{a(k)} = \bar{N}(x;k) + \rho(k)\,\mathrm{e}^{2\mathrm{i}kx}\bar{N}(x;-k), \tag{2.2.12}$$

which is the fundamental equation, and we claim that (2.2.12) is equivalent to a *Riemann-Hilbert boundary value problem.* This is a consequence of the analyticity properties for $M(x;k)$, $\bar{N}(x;k)$ and $a(k)$ which are established in the following lemma.

LEMMA 2.2.1

(i), *$M(x;k)$ and $a(k)$ can be analytically extended to the upper half k-plane and tend to unity as $|k| \to \infty$ (for $\operatorname{Im} k > 0$);*

(ii), *$\bar{N}(x;k)$ can be analytically extended to the lower half k-plane and tends to unity as $|k| \to \infty$ (for $\operatorname{Im} k < 0$).*

Proof. These analyticity properties are established by characterizing $M(x;k)$ and $\bar{N}(x;k)$ in terms of linear integral equations.

First, we convert the equation (2.2.1a), that is the Schrödinger equation, into an integral equation by making the transformation $v(x;k) = m(x;k)\mathrm{e}^{-\mathrm{i}kx}$; then $m(x;k)$ satisfies

$$m_{xx}(x;k) - 2\mathrm{i}km_x(x;k) = -u(x)m(x;k),$$

or equivalently (assuming that $m \to 1$, either as $x \to \infty$, or as $x \to -\infty$)

$$m(x;k) = 1 + \int_{-\infty}^{\infty} G(x-\xi;k)\,u(\xi)\,m(\xi;k)\,\mathrm{d}\xi, \tag{2.2.13a}$$

where $G(x;k)$ is the Green's function which solves

$$G_{xx} - 2\mathrm{i}kG_x = -\delta(x). \tag{2.2.13b}$$

Hence, using the Fourier transform method,

$$G(x;k) = \frac{1}{2\pi}\int_C \frac{\mathrm{e}^{\mathrm{i}px}}{p(p-2k)}\,\mathrm{d}p, \tag{2.2.14}$$

where C is an appropriate contour. Therefore consider $G_\pm(x;k)$ defined by

$$G_\pm(x;k) = \frac{1}{2\pi}\int_{C_\pm} \frac{\mathrm{e}^{\mathrm{i}px}}{p(p-2k)}\,\mathrm{d}p,$$

where C_+ and C_- are the contours from $-\infty$ to ∞ which respectively pass below and above both the singularities at $p = 0$ and $p = 2k$ (see Figure 2.2.1).

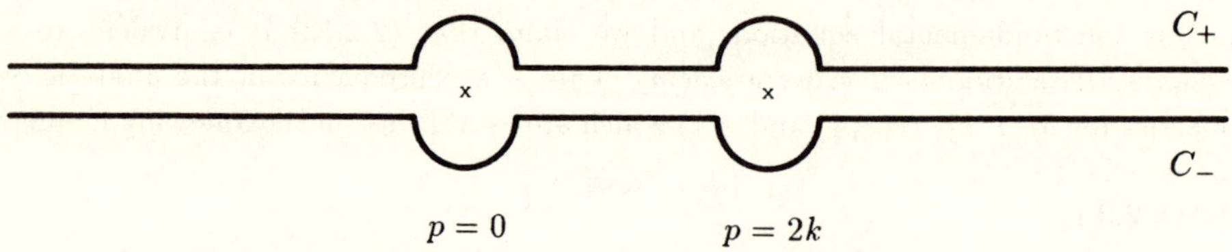

FIGURE 2.2.1 *The contours C_+ and C_-.*

Therefore

$$G_+(x;k) = \frac{1}{2ik}(1 - e^{2ikx})\theta(x) = \begin{cases} \dfrac{1}{2ik}(1 - e^{2ikx}), & \text{if } x > 0, \\ 0, & \text{if } x < 0, \end{cases}$$
$$G_-(x;k) = \frac{-1}{2ik}(1 - e^{2ikx})\theta(-x) = \begin{cases} 0, & \text{if } x > 0, \\ \dfrac{-1}{2ik}(1 - e^{2ikx}), & \text{if } x < 0, \end{cases}$$

where $\theta(x)$ is the Heaviside function given by

$$\theta(x) = \begin{cases} 1, & \text{if } x > 0, \\ 0, & \text{if } x < 0. \end{cases}$$

We require that $M(x;k)$ and $\bar{N}(x;k)$ satisfy the boundary conditions (2.2.9). Since from equation (2.2.9a), $M(x;k) \to 1$ as $x \to -\infty$, then we associate $M(x;k)$ with the Green's function $G_+(x;k)$ (so the integral is from $-\infty$ to x), and analogously since $\bar{N}(x;k) \to 1$ as $x \to \infty$, then we associate $\bar{N}(x;k)$ with the Green's function $G_-(x;k)$. Therefore it follows that $M(x;k)$ and $\bar{N}(x;k)$ are the solutions of the following integral equations

$$\begin{aligned} M(x;k) &= 1 + \int_{-\infty}^{\infty} G_+(x-\xi;k)\, u(\xi)\, M(\xi;k)\, d\xi \\ &= 1 + \frac{1}{2ik}\int_{-\infty}^{x} \left\{1 - e^{2ik(x-\xi)}\right\} u(\xi)\, M(\xi;k)\, d\xi, \end{aligned} \tag{2.2.15}$$

$$\begin{aligned} \bar{N}(x;k) &= 1 + \int_{-\infty}^{\infty} G_-(x-\xi;k)\, u(\xi)\, \bar{N}(\xi;k)\, d\xi \\ &= 1 - \frac{1}{2ik}\int_{x}^{\infty} \left\{1 - e^{2ik(x-\xi)}\right\} u(\xi)\, \bar{N}(\xi;k)\, d\xi. \end{aligned} \tag{2.2.16}$$

These are Volterra integral equations which have unique solutions for each k with $\operatorname{Im} k \geq 0$, for $u \in P_1$ (cf. Lemma 1, §2.1, Deift and Trubowitz [1979]). It is straightforward to show that their Neumann series converge uniformly for u in the same function space. $G_\pm(x;k)$ is analytic for $\operatorname{Im} k \gtrless 0$ and vanishes as $|k| \to \infty$, therefore $M(x;k)$ and $\bar{N}(x;k)$ are analytic for $\operatorname{Im} k > 0$ and $\operatorname{Im} k < 0$ respectively, and tend to unity as $|k| \to \infty$.

The analytic properties of $a(k)$ may be easily established by using an integral representation for $a(k)$. To derive this integral representation we first define

$$\Delta(x;k) := M(x;k) - a(k)\bar{N}(x;k), \tag{2.2.17}$$

and then from (2.2.15) and (2.2.16) it follows that

$$\begin{aligned}\Delta(x;k) = 1 - a(k) &+ \frac{1}{2ik}\int_{-\infty}^{\infty}\left\{1 - e^{2ik(x-\xi)}\right\}u(\xi)\,M(\xi;k)\,d\xi \\ &- \frac{1}{2ik}\int_{x}^{\infty}\left\{1 - e^{2ik(x-\xi)}\right\}u(\xi)\,\Delta(\xi;k)\,d\xi.\end{aligned} \tag{2.2.18}$$

Also, from (2.2.12) we have

$$\Delta(x;k) = b(k)\,e^{2ikx}\bar{N}(x;-k), \tag{2.2.19}$$

therefore from (2.2.16)

$$\Delta(x;k) = b(k)\,e^{2ikx} - \frac{1}{2ik}\int_{x}^{\infty}\left\{1 - e^{2ik(x-\xi)}\right\}u(\xi)\,\Delta(\xi;k)\,d\xi. \tag{2.2.20}$$

Hence by comparing equations (2.2.18) and (2.2.20), and equating coefficients of 1 and e^{2ikx}, we obtain the following integral representations for $a(k)$ and $b(k)$;

$$a(k) = 1 + \frac{1}{2ik}\int_{-\infty}^{\infty} u(\xi)\,M(\xi;k)\,d\xi, \tag{2.2.21}$$

$$b(k) = -\frac{1}{2ik}\int_{-\infty}^{\infty} u(\xi)\,M(\xi;k)\,e^{-2ik\xi}\,d\xi. \tag{2.2.22}$$

From these integral representations it follows that $a(k)$ is analytic for $\operatorname{Im} k > 0$, and tends to unity as $|k| \to \infty$, while $b(k)$ cannot, in general, be continued analytically off the real k-axis. Furthermore, since $M(x;k) \to 1$ as $|k| \to \infty$, then it follows from equation (2.2.21) that $a(k) \to 1$ as $|k| \to \infty$ (for $\operatorname{Im} k > 0$).

Therefore, these analytic properties derived in Lemma 2.2.1 show that the equation

$$\frac{M(x;k)}{a(k)} = \bar{N}(x;k) + \rho(k)\,e^{2ikx}\bar{N}(x;-k), \qquad \rho(k) = \frac{b(k)}{a(k)}, \tag{2.2.12}$$

may be regarded as a Riemann-Hilbert boundary value problem.

Now we discuss the zeros of $a(k)$ by means of following lemma.

LEMMA 2.2.2

The function $a(k)$ can have a finite number of simple zeros at $k_1, k_2, \ldots, k_N$, where $k_j = \mathrm{i}\kappa_j$, $\kappa_j \in \mathbb{R}$, $j = 1, 2, \ldots, N$ (i.e., they all lie on the imaginary axis), in the upper half k-plane.

Proof. First we note that from equations (2.2.5) we have

$$|a(k)|^2 - |b(k)|^2 = 1,$$

for $k \in \mathbb{R}$ and hence $a(k) \neq 0$ for $k \in \mathbb{R}$ [in fact, Deift and Trubowitz [1979] show that $a(k)$ is continuous for $\mathrm{Im}(k) \geq 0$].

Recall that $\phi(x;k)$, $\psi(x;k)$ and $\bar{\psi}(x;k)$ are solutions of the Schrödinger equation (2.2.1a) satisfying the boundary conditions

$$\phi(x;k) \sim \mathrm{e}^{-\mathrm{i}kx}, \qquad \text{as} \qquad x \to -\infty, \tag{2.2.23a}$$

$$\psi(x;k) \sim \mathrm{e}^{\mathrm{i}kx}, \qquad \text{as} \qquad x \to \infty. \tag{2.2.23b}$$

$$\bar{\psi}(x;k) \sim \mathrm{e}^{-\mathrm{i}kx}, \qquad \text{as} \qquad x \to \infty. \tag{2.2.23c}$$

together with the relationship

$$\phi(x;k) = a(k)\bar{\psi}(x;k) + b(k)\psi(x;k). \tag{2.2.24}$$

Now consider the Wronskian $W(\phi(x;k), \psi(x;k))$, then using equations (2.2.23-24), it is easily shown that

$$W(\phi(x;k), \psi(x;k)) := \phi(x;k)\psi_x(x;k) - \phi_x(x;k)\psi(x;k) = 2\mathrm{i}ka(k). \tag{2.2.25}$$

In particular, if $a(k_0) = 0$ and $k_0 = \xi_0 + \mathrm{i}\kappa_0$, then $\phi(x;k_0) = \beta_0\psi(x;k_0)$ for some (nonzero) constant β_0, and so

$$\phi(x;k_0) \sim \mathrm{e}^{-\mathrm{i}\xi_0 x}\mathrm{e}^{\kappa_0 x}, \qquad \text{as} \quad x \to -\infty, \tag{2.2.26a}$$

$$\phi(x;k_0) \sim \beta_0\, \mathrm{e}^{\mathrm{i}\xi_0 x}\mathrm{e}^{-\kappa_0 x}, \qquad \text{as} \quad x \to \infty. \tag{2.2.26b}$$

Assuming that $\phi(x;k_0)$ is in $L^2(\mathbb{R})$, then necessarily $\kappa_0 > 0$ [since $\phi(x;k_0)$ must decay to zero as $|x| \to \infty$].

$\phi(x;k)$ and its complex conjugate $\phi^*(x;k^*)$ satisfy the equations

$$\phi_{xx} + \left\{u(x) + k^2\right\}\phi = 0, \tag{2.2.27a}$$

$$\phi^*_{xx} + \left\{u(x) + k^{*2}\right\}\phi^* = 0, \tag{2.2.27b}$$

respectively. Hence

$$\frac{\partial}{\partial x}W(\phi, \phi^*) + (k^{*2} - k^2)\phi\phi^* = 0, \tag{2.2.28}$$

and so since $\phi \to 0$, $\phi_x \to 0$, $\phi^* \to 0$, $\phi^*_x \to 0$, as $x \to \pm\infty$ then from equation (2.2.28)

$$(k^{*2} - k^2) \int_{-\infty}^{\infty} |\phi(x;k)|^2 \, \mathrm{d}x = 0.$$

Hence if $a(k_0) = 0$, where $k_0 = \xi_0 + \mathrm{i}\kappa_0$, then $\xi_0 \kappa_0 \|\phi(\,\bullet\,; k_0)\|_2^2 = 0$, where $\|\bullet\|_2$ is the $L^2(\mathbb{R})$ norm; since $\kappa_0 > 0$, and so necessarily $\xi_0 = 0$. Therefore we conclude that if $a(k_0) = 0$, where $k_0 = \xi_0 + \mathrm{i}\kappa_0$, then $\xi_0 = 0$ and $\kappa_0 > 0$, i.e., the zeros of $a(k)$ lie on the positive imaginary k-axis.

To show that the zeros of $a(k)$ are simple, differentiate (2.2.27a) with respect to k, then

$$\phi_{xxk} + 2k\phi + \{u(x) + k^2\}\phi_k = 0. \tag{2.2.30}$$

Multiplying this equation by $\phi(x;k)$ and equation (2.2.27a) by $\phi_k(x;k)$, and substracting the resulting equations yields

$$(\phi\phi_{xk} - \phi_x\phi_k)_x + 2k\phi^2 = 0,$$

i.e.,

$$\begin{aligned} W(\phi(\infty;k),&\phi_k(\infty;k)) - W(\phi(-\infty;k), \phi_k(-\infty;k)) \\ &+ 2k \int_{-\infty}^{\infty} \phi^2(x;k) \, \mathrm{d}x = 0. \end{aligned} \tag{2.2.31}$$

Also, differentiating (2.2.27a) with respect to k yields

$$W(\phi_k(x;k), \psi(x;k)) + W(\phi(x;k), \psi_k(x;k)) = 2\mathrm{i}a(k) + 2\mathrm{i}k\frac{\mathrm{d}a}{\mathrm{d}k}. \tag{2.2.32}$$

Now suppose that $a(\mathrm{i}\kappa_0) = 0$, then from (2.2.24,25), we have $\phi(x;\mathrm{i}\kappa_0) = \beta_0\psi(x;\mathrm{i}\kappa_0)$. Hence setting $k = \mathrm{i}\kappa_0$ in (2.2.32) yields

$$\beta_0^2 W(\psi(x;\mathrm{i}\kappa_0), \psi_k(x;\mathrm{i}\kappa_0)) - W(\phi(x;\mathrm{i}\kappa_0), \phi_k(x;\mathrm{i}\kappa_0)) = -2\beta_0\kappa_0\frac{\mathrm{d}a}{\mathrm{d}k}(\mathrm{i}\kappa_0), \tag{2.2.33}$$

where we have used equations (2.2.7). Evaluating (2.2.33) at $x = -\infty$ and adding it to equation (2.2.31) evaluated at $k = \mathrm{i}\kappa_0$ yields

$$\begin{aligned} \beta_0^2 W(\psi(\infty;\mathrm{i}\kappa_0),&\psi_k(\infty;\mathrm{i}\kappa_0)) - W(\phi(-\infty;\mathrm{i}\kappa_0), \phi_k(-\infty;\mathrm{i}\kappa_0)) \\ &= -2\mathrm{i}\kappa_0 \int_{-\infty}^{\infty} \phi^2(x;\mathrm{i}\kappa_0) \, \mathrm{d}x - 2\kappa_0\beta_0\frac{\mathrm{d}a}{\mathrm{d}k}(\mathrm{i}\kappa_0), \end{aligned} \tag{2.2.34}$$

Using the known asymptotics of $\phi(x;k)$ and $\psi(x;k)$ as $x \to -\infty$ and $x \to \infty$ respectively [i.e., equations (2.2.23)], then it follows that the left-hand side of equation (2.2.34) is zero. Therefore if $a(\mathrm{i}\kappa_0) = 0$, then

$$\frac{\mathrm{d}a}{\mathrm{d}k}(\mathrm{i}\kappa_0) = -\mathrm{i}\beta_0^{-1} \int_{-\infty}^{\infty} \phi^2(x;\mathrm{i}\kappa_0) \, \mathrm{d}x. \tag{2.2.35}$$

Setting $k = \mathrm{i}\kappa_0$ in the integral equation (2.2.15) shows that $M(x;\mathrm{i}\kappa_0)$ is real, so $\phi(x;\mathrm{i}\kappa_0)$ is also real, hence from equation (2.2.35)

$$\frac{\mathrm{d}a}{\mathrm{d}k}(\mathrm{i}\kappa_0) \neq 0,$$

and so the zeros of $a(k)$ are simple.

Finally, to prove that $a(k)$ has only a finite number of zeros, we recall that, as shown in Lemma 2.2.1, $a \to 1$ as $|k| \to \infty$ (for $\operatorname{Im} k > 0$); furthermore, as shown by Deift and Trubowitz [1979], $a(k)$ is continuous for $\operatorname{Im} k \geq 0$. Hence there can be no cluster points of zeros of $a(k)$ either in the upper half k-plane or on the real k-axis and so necessarily $a(k)$ can possess only a finite number of zeros. In fact, the number of zeros, N, of $a(k)$ satisfies the inequality

$$N \leq 1 + \int_{-\infty}^{\infty} |x||u(x,0)|\,\mathrm{d}x$$

(cf. Deift and Trubowitz [1979]; Segur [1973]).

Hence, in general, $M(x;k)/a(k)$ is a meromorphic function in the upper half k-plane with a finite number of poles at $k = \mathrm{i}\kappa_1, \mathrm{i}\kappa_2, \ldots, \mathrm{i}\kappa_N$. Then set

$$\frac{M(x;k)}{a(k)} = \mu_+(x;k) + \sum_{j=1}^{N} \frac{A_j(x)}{k - \mathrm{i}\kappa_j}, \tag{2.2.36}$$

where $\mu_+(x;k)$ is analytic in the upper half k-plane. Integrating equation (2.2.36) around $\mathrm{i}\kappa_j$ and using equation (2.2.12) shows that

$$A_j(x) = C_j \bar{N}(x;-\mathrm{i}\kappa_j)\exp(-2\kappa_j x),$$

where C_j, are the *normalization constants*; hence from equation (2.2.12) we obtain

$$\begin{aligned}\mu_+(x;k) = \bar{N}(x;k) - \sum_{j=1}^{N} \frac{C_j}{k - \mathrm{i}\kappa_j}\exp(-2\kappa_j x)\bar{N}(x;-\mathrm{i}\kappa_j)\\ + \rho(k)\exp(2\mathrm{i}kx)\bar{N}(x;-k).\end{aligned} \tag{2.2.37}$$

Furthermore, $\mu_+(x;k) \to 1$ as $|k| \to \infty$ (for $\operatorname{Im} k > 0$).

In general, all the scattering problems arising in the study of one-dimensional partial differential equations, differential-difference equations and partial difference equations which we discuss are of the form

$$(\mathbf{m}_+ - \mathbf{m}_-)(x;k) = \mathbf{m}_-(x;\alpha(k))\,\mathbf{V}(x;k), \qquad \text{on} \quad \Sigma, \tag{2.2.38a}$$

with

$$\mathbf{m}(x;k) \to \mathbf{I}, \qquad \text{as} \quad |k| \to \infty, \tag{2.2.38b}$$

where Σ is an appropriate contour in the complex k-plane, $\alpha(k)$ and $\mathbf{V}$ are given on Σ, V depends explicitly on the scattering data, $\mathbf{m}_{\pm}(x;k)$ are meromorphic $n \times n$ matrix functions of $k \in \mathbb{C}\backslash\Sigma$, for $\operatorname{Im} k \gtrless 0$, and $m_{\pm}(x;k)$ have a finite number of poles at specified locations, $k_1, k_2, \ldots, k_N$, with specified residues. For equation (2.2.12)

$$m_{+}(x;k) = \frac{M(x;k)}{a(k)}, \quad m_{-}(x;k) = \bar{N}(x;k), \tag{2.2.39a}$$

$$\alpha(k) = -k, \quad V(x;k) = \rho(k)\,\mathrm{e}^{2\mathrm{i}kx}. \tag{2.2.39b}$$

2.3 The Inverse Problem.

Equation (2.2.37) defines a Riemann-Hilbert problem in terms of the scattering data $S(\lambda) = \{(\kappa_j, C_j)_{j=1}^{N}, \rho(k), a(k)\}$. For convenience, we shall assume that there are no poles, that is $a(k) \neq 0$, and so we consider equation (2.2.12) rather than equation (2.2.37).

Now consider the $\mathcal{P}^{\pm}$ projection operator defined by

$$(\mathcal{P}^{\pm} f)(k) = \frac{1}{2\pi\mathrm{i}} \int_{-\infty}^{\infty} \frac{f(\zeta)}{\zeta - (k \pm \mathrm{i}0)}\, \mathrm{d}\zeta. \tag{2.3.1}$$

Suppose that $f_{\pm}(k)$ is analytic in the upper/lower half k-plane and $f_{\pm}(k) \to 0$ as $|k| \to \infty$ (for $\operatorname{Im} k \gtrless 0$), then it is easily shown that

$$(\mathcal{P}^{\pm} f_{\mp})(k) = 0, \tag{2.3.2a}$$

$$(\mathcal{P}^{\pm} f_{\pm})(k) = \pm f_{\pm}(k), \tag{2.3.2b}$$

Operating on (2.2.12) with $\mathcal{P}^{-}$ and using the analyticity properties of M, $\bar{N}$ and a yields

$$\begin{aligned} &\mathcal{P}^{-}\left[\frac{M(x;k)}{a(k)} - 1 - (\bar{N}(x;k) - 1) - \rho(k)\mathrm{e}^{2\mathrm{i}kx}\bar{N}(x;-k)\right] \\ &\quad = 0 + \bar{N}(x;k) - 1 - \frac{1}{2\pi\mathrm{i}} \int_{-\infty}^{\infty} \frac{\rho(\zeta)\,\mathrm{e}^{2\mathrm{i}\zeta x}\,\bar{N}(x;-\zeta)}{\zeta - (k - \mathrm{i}0)}\, \mathrm{d}\zeta = 0. \end{aligned}$$

Hence, since $N(x;k) = \mathrm{e}^{2\mathrm{i}kx}\bar{N}(x;-k)$, we have

$$\bar{N}(x;k) = 1 + \frac{1}{2\pi\mathrm{i}} \int_{-\infty}^{\infty} \frac{\rho(\zeta)\,N(x;\zeta)}{\zeta - (k - \mathrm{i}0)}\, \mathrm{d}\zeta, \tag{2.3.3a}$$

$$N(x;k) = \mathrm{e}^{2\mathrm{i}kx}\left\{1 + \frac{1}{2\pi\mathrm{i}} \int_{-\infty}^{\infty} \frac{\rho(\zeta)\,N(x;\zeta)}{\zeta + k + \mathrm{i}0}\, \mathrm{d}\zeta\right\}. \tag{2.3.3b}$$

As $|k| \to \infty$, equation (2.3.3a) shows that

$$\bar{N}(x;k) \sim 1 - \frac{1}{2\pi\mathrm{i}}\,\frac{1}{k} \int_{-\infty}^{\infty} \rho(\zeta)\,N(x;\zeta)\,\mathrm{d}\zeta, \tag{2.3.4}$$

From the integral equation for $\bar{N}(x;k)$, equation (2.2.16), using the Riemann-Lebesgue Lemma we see that as $|k| \to \infty$

$$\bar{N}(x;k) \sim 1 - \frac{1}{2\mathrm{i}k}\int_x^\infty u(\xi)\,\mathrm{d}\xi. \tag{2.3.5}$$

Therefore, by comparing equations (2.3.4) and (2.3.5), we see that the potential can be reconstructed by the expression

$$u(x) = -\frac{\partial}{\partial x}\left\{\frac{1}{\pi}\int_{-\infty}^{\infty}\rho(\zeta)\,N(x;\zeta)\,\mathrm{d}\zeta\right\}. \tag{2.3.6}$$

It is worthwhile to note that the Riemann-Hilbert boundary value problem (2.2.10) or (2.2.12) can be written as the vector Riemann-Hilbert boundary value problem

$$\begin{aligned}\boldsymbol{\phi}_+(x;k) &= \begin{pmatrix} 1+\rho(k)\bar{\rho}(k) & \rho(k)\mathrm{e}^{2\mathrm{i}kx} \\ \bar{\rho}(k)\mathrm{e}^{-2\mathrm{i}kx} & 1\end{pmatrix}\boldsymbol{\phi}_-(x;k) \\ &\equiv \mathbf{G}(k)\boldsymbol{\phi}_-(x;k),\end{aligned}$$

where

$$\boldsymbol{\phi}_+(x;k) = \begin{pmatrix} M(x;k)/a(k) \\ N(x;k)\mathrm{e}^{-2\mathrm{i}kx}\end{pmatrix}, \qquad \boldsymbol{\phi}_-(x;k) = \begin{pmatrix} \bar{N}(x;k) \\ -\bar{M}(x;k)\mathrm{e}^{-2\mathrm{i}kx}/\bar{a}(k)\end{pmatrix}, \tag{2.3.7}$$

and note from (2.2.5) that $\bar{\rho}(k) = -r^*(k)$. Since $\det\mathbf{G} = +1$ and both the eigenvalues of $\mathbf{G}$ are positive then it follows from the work of Gohberg and Krein [1958] that there exists a unique solution to the vector Riemann-Hilbert boundary value problem (2.3.7). This is in a generic sense for the general problem, but explicitly in the case when $a(k)$ has no zeros (which corresponds to no eigenvalues or bound states).

For the case when $a(k)$ has zeros, it is straightforward to modify the above and give a compact expression provided that u decays sufficiently rapidly as $|x| \to \infty$. In either cases the integration from $-\infty$ to ∞ in (2.3.3) and (2.3.6) along the real axis is replaced by a contour $\mathcal{L}$ from $-\infty$ to ∞ which passes above all the zeros of $a(k)$. Suppose that

$$a(\mathrm{i}\kappa_j) = 0, \qquad \kappa_j \in \mathbb{R}, \qquad j = 1,2,\ldots,N,$$

then define

$$M_j(x) := M(x;\mathrm{i}\kappa_j), \tag{2.3.8a}$$
$$N_j(x) := \exp(-2\kappa_j x)\bar{N}(x;-\mathrm{i}\kappa_j), \tag{2.3.8b}$$

for $j = 1,2,\ldots,N$. Applying the $\mathcal{P}^-$ projection operator (2.3.1) to equation (2.2.37) and recalling that $N(x;k) = \mathrm{e}^{2\mathrm{i}kx}\bar{N}(x;-k)$, then the integral equations for the eigenfunctions are given by

$$N(x;k) = \mathrm{e}^{2\mathrm{i}kx}\left\{1 - \sum_{j=1}^{N}\frac{C_jN_j(x)}{k+\mathrm{i}\kappa_j} + \frac{1}{2\pi\mathrm{i}}\int_{-\infty}^{\infty}\frac{\rho(\zeta)\,N(x;\zeta)}{\zeta+k+\mathrm{i}0}\,\mathrm{d}\zeta\right\}, \tag{2.3.9}$$

$$N_p(x) = \exp(-2\kappa_p x)\left\{1 + \mathrm{i}\sum_{j=1}^{N}\frac{C_jN_j(x)}{\kappa_p+\kappa_j} + \frac{1}{2\pi\mathrm{i}}\int_{-\infty}^{\infty}\frac{\rho(\zeta)\,N(x;\zeta)}{\zeta+\mathrm{i}\kappa_p}\,\mathrm{d}\zeta\right\}, \tag{2.3.10}$$

and the potential is reconstructed by the following expression

$$u(x) = \frac{\partial}{\partial x}\left\{ 2\mathrm{i}\sum_{j=1}^{N} C_j N_j(x) - \frac{1}{\pi}\int_{-\infty}^{\infty} \rho(k)\, N(x;k)\,\mathrm{d}k \right\}. \tag{2.3.11}$$

As shown in §2.5.2 below, the integral equations (2.3.9,10) are equivalent to the *Gel'-fand-Levitan-Marchenko equation*, for which we can deduce the existence and uniqueness of solutions using the Fredholm alternative (cf. Dodd, Eilbeck, Gibbon and Morris [1982, §4.1]). We remark that that the existence and uniequeness of solutions to equations (2.3.9-11) is given by Beals, Deift and Tomei [1988, Chapter 2] (after writing the scattering problem (2.1.2a) as a first order system).

With sufficient decay equations (2.3.9-11) can be written as

$$N(x;k) = \mathrm{e}^{2\mathrm{i}kx}\left\{1 + \frac{1}{2\pi\mathrm{i}}\int_{\mathcal{L}_0} \frac{\rho(\zeta)\,N(x;\zeta)}{\zeta + k + \mathrm{i}0}\,\mathrm{d}\zeta\right\}, \tag{2.3.12a}$$

$$N_p(x) = \exp(-2\kappa_p x)\left\{1 + \frac{1}{2\pi\mathrm{i}}\int_{\mathcal{L}_0} \frac{\rho(\zeta)\,N(x;\zeta)}{\zeta + \mathrm{i}\kappa_p}\,\mathrm{d}\zeta\right\}, \tag{2.3.12b}$$

$$u(x) = -\frac{\partial}{\partial x}\left\{\frac{1}{\pi}\int_{\mathcal{L}_0} \rho(k)\,N(x;k)\,\mathrm{d}k\right\}, \tag{2.3.12c}$$

where $\mathcal{L}_0$ is the contour from $-\infty$ to ∞ which passes above all the zeros of $a(k)$.

2.4 The Time Dependence of the Scattering Data.

The time evolution of the scattering data may be obtained by analysing the asymptotic behavior of the associated time evolution operator, which for the KdV equation is

$$v_t = (u_x + \gamma)v + (4k^2 - 2u)v_x, \tag{2.4.1}$$

with γ a constant. Now we make the transformation

$$v(x;k;t) = \phi(x;k;t) = M(x;k;t)\,\mathrm{e}^{-\mathrm{i}kx}, \tag{2.4.2}$$

then M satisfies

$$M_t = (\gamma - 4\mathrm{i}k^3 + u_x + 2\mathrm{i}ku)M + (4k^2 - 2u)M_x. \tag{2.4.3}$$

From (2.2.8) the asymptotic behavior of $M(x;k;t)$ is given by

$$M(x;k;t) \to 1, \qquad \text{as} \quad x \to -\infty, \tag{2.4.4a}$$

$$M(x;k;t) \to a(k;t) + b(k;t)\,\mathrm{e}^{2\mathrm{i}kx}, \qquad \text{as} \quad x \to \infty, \tag{2.4.4b}$$

[recall that from (2.2.10)

$$M(x;k;t) = a(k;t)\bar{N}(x;k;t) + b(k;t)N(x;-k;t), \tag{2.4.5}$$

where $\rho(k;t) = b(k;t)/a(k;t)$]. By considering (2.4.3) as $x \to \pm\infty$ and using the fact that $u \to 0$ as $x \to \pm\infty$ [since $(1+|x|)u(x) \in L^1(\mathbb{R})$], it follows that

$$\gamma - 4\mathrm{i}k^3 = 0, \tag{2.4.6a}$$

$$a_t + b_t e^{2\mathrm{i}kx} = 8\mathrm{i}k^3 b e^{2\mathrm{i}kx}. \tag{2.4.6b}$$

Therefore

$$a_t = 0, \qquad b_t = 8\mathrm{i}k^3 b. \tag{2.4.7}$$

Hence upon solving (2.4.7)

$$a(k;t) = a(k;0), \qquad b(k;t) = b(k;0)\exp(8\mathrm{i}k^3 t), \tag{2.4.8a}$$

and so

$$\rho(k;t) \equiv \frac{b(k;t)}{a(k;t)} = \rho(k;0)\exp(8\mathrm{i}k^3 t). \tag{2.4.8b}$$

Also the discrete eigenvalues, which are the zeros of $a(k)$ and are finite in number, satisfy

$$k_j = \mathrm{i}\kappa_j = \text{constant}, \qquad j = 1, 2, \ldots, N, \tag{2.4.8c}$$

and so we have an isospectral problem. Similarly we find that the time dependence of the normalization constants is given by

$$C_j(t) = C_j(0)\exp(8\mathrm{i}k_j^3 t) = C_j(0)\exp(8\kappa_j^3 t). \tag{2.4.8d}$$

Using (2.4.8) we may now solve the initial value problem for the KdV equation:

1. Given $u(x,0)$, solve equation (2.2.15) for $M(x;k;0)$ and hence, using equations (2.2.21-22), determine the initial scattering data κ_j, $C_j(0)$ and $\rho(k;0)$.
2. Use (2.4.8) to compute $C_j(t)$ and $\rho(k;t)$.
3. Solve (2.3.9) to obtain $N(x;k;t)$ and then construct $u(x,t)$ from (2.3.11).

Schematically this is given by

$$u(x,0) \xrightarrow{\text{direct scattering}} S(\lambda,0) = \left(\{\kappa_n, C_j(0)\}_{j=1}^N, \rho(k,0), a(k,0)\right)$$

$$\Big\downarrow \ \omega(k)\text{: dispersion relation}$$

$$u(x,t) \xleftarrow{\text{inverse scattering}} S(\lambda,t) = \left(\{\kappa_n, C_j(t)\}_{j=1}^N, \rho(k;t), a(k,t)\right)$$

Note the analogy of this method to the Fourier transform method for linear partial differential equations (cf. §1.6).

2.5 Further Remarks.

2.5.1 Soliton Solutions. Pure soliton solutions may be obtained by assuming a special form for $\rho(k;t)$, namely

$$\rho(k;t) = \begin{cases} 0, & \text{if } k \text{ real;} \\ \displaystyle\sum_{j=1}^{N} \frac{C_j(0)}{k - \mathrm{i}\kappa_j} \exp(8\kappa_j^3 t), & \text{if } \operatorname{Im} k > 0. \end{cases}$$

Then we obtain a linear algebraic system for the $N_j(x,t)$

$$N_j(x,t) + \sum_{p=1}^{N} \frac{C_p(0)}{\kappa_p + \kappa_j} \exp(-2\kappa_p x + 8\kappa_p^3 t) N_p(x,t) = \exp(-2\kappa_j x), \tag{2.5.1a}$$

where $N_j(x,t) := N(x; k = \mathrm{i}\kappa_j; t)$, and the potential $u(x,t)$ is given by

$$u(x,t) = -2\frac{\partial}{\partial x}\left\{\sum_{j=1}^{N} \exp(8\kappa_j^3 t) C_j(0) N_j(x,t)\right\}. \tag{2.5.1b}$$

The formula for a one-soliton solution is

$$N_1(x,t) = \frac{2\kappa_1 \exp(-2\kappa_1 x)}{2\kappa_1 + C_1(0)\exp(-2\kappa_1 x + 8\kappa_1^3 t)} \tag{2.5.2a}$$

$$u(x,t) = 2\kappa_1^2 \operatorname{sech}^2\left\{\kappa_1(x - 4\kappa_1^2 t - x_1)\right\} \tag{2.5.2b}$$

where $C_1(0) = 2\kappa_1 \exp(2\kappa_1 x_1)$.

The two-soliton solution of the KdV equation is

$$u(x,t) = \frac{4(\kappa_2^2 - \kappa_1^2)\left[(\kappa_2^2 - \kappa_1^2) + \kappa_1^2\cosh(2\kappa_2\xi_2) + \kappa_2^2\cosh(2\kappa_1\xi_1)\right]}{\left[(\kappa_2 - \kappa_1)\cosh(\kappa_1\xi_1 + \kappa_2\xi_2) + (\kappa_2 + \kappa_1)\cosh(\kappa_2\xi_2 - \kappa_1\xi_1)\right]^2},$$

where $\xi_i = x - 4\kappa_i^2 t - x_i$, and $C_i(0) = 2\kappa_i \exp(2\kappa_i x_i)$, for $i = 1, 2$.

2.5.2 Delta-function Initial Profile. In the previous section we obtained pure soliton solutions for the KdV equation corresponding to relectionless potentials (also recall §1.7.2 above). However, more general choices of the initial data $u(x,0)$ give rise to nonzero reflection coefficients; unfortunately, in this case it generally not possible to solve equations (2.3.9-11) in closed form.

As an example we consider the delta-function initial profile given by

$$u(x;0) = \alpha\delta(x), \tag{2.5.3}$$

where α is a positive constant and $\delta(x)$ is the Dirac delta function. In this case there is a single discrete eigenvalue $\lambda = -\kappa_1^2$, where $\kappa_1 = \alpha$, and normalization constant $C_1(0) = \kappa_1$; the transmission coefficient is nonzero and given by

$$\rho(k;0) = -\frac{\kappa_1}{\kappa_1 + \mathrm{i}k}.$$

Hence the time evolution of the scattering data is given by

$$C_1(t) = \kappa_1 \exp(8\kappa_1^3 t), \qquad \rho(k;t) = -\frac{\kappa_1 \exp(8\mathrm{i}k^3 t)}{\kappa_1 + \mathrm{i}k}.$$

Substituting these into equations (2.3.9-10) yields

$$N(x,t;k) = \mathrm{e}^{2\mathrm{i}kx}\left\{1 - \frac{\kappa_1 N_1(x,t)\exp(8\kappa_1^3 t)}{k + \mathrm{i}\kappa_1} + \frac{\kappa_1}{2\pi\mathrm{i}}\int_{-\infty}^{\infty} \frac{N(x,t;\zeta)\exp(8\mathrm{i}\zeta^3 t)}{(\zeta + k + \mathrm{i}0)(\kappa_1 + \mathrm{i}k)}\,\mathrm{d}\zeta\right\}, \tag{2.5.4a}$$

$$N_1(x,t) = \exp(-2\kappa_1 x)\left\{1 + \tfrac{1}{2}\mathrm{i}N_1(x,t)\exp(8\kappa_1^3 t) + \frac{\kappa_1}{2\pi}\int_{-\infty}^{\infty} \frac{N(x,t;\zeta)\exp(8\mathrm{i}\zeta^3 t)}{\zeta^2 + \kappa_1^2}\,\mathrm{d}\zeta\right\}, \tag{2.5.4b}$$

and the potential $u(x,t)$ is given by

$$u(x,t) = \frac{\partial}{\partial x}\left\{2\mathrm{i}\kappa_1 N_1(x,t)\exp(8\kappa_1^3 t) + \frac{\kappa_1}{\pi}\int_{-\infty}^{\infty} \frac{N(x,t;k)\exp(8\mathrm{i}k^3 t)}{\kappa_1 + \mathrm{i}k}\,\mathrm{d}k\right\}. \tag{2.5.4c}$$

In spite of the fact that it is not possible to solve equations (2.5.4) in closed form, it is possible to obtain useful information through asymptotic analysis. For example, it can be shown that the solution $u(x,t)$ incorporates the single soliton associated with the single discrete eigenvalue $\kappa_1 = \alpha$, and given by

$$u(x,t) \sim 2\kappa_1 \operatorname{sech}^2\{\kappa_1(x - 4\kappa_1^2 t - x_1)\},$$

as $t \to +\infty$, with $x - 4\kappa_1^2 t$ fixed, and where $\exp(2\kappa_1 x_1) = \frac{1}{2}$. By considering the limit as $t \to +\infty$ for $x < 0$ (in this region of (x,t)-space the soliton is exponentially small), it can be shown that the solution is an oscialltory dispersive wave propagating to the left with an amplitude which decays like $t^{-1/3}$. Finally, we remark that if $\alpha < 0$ in (2.5.3), then there is no discrete eigenvalue. Consequently there is no soliton and only the dispersive-wave train arises in the solution for $t > 0$.

2.5.3 Uniqueness. In the above we have not discussed the uniqueness of the map between the potential $u(x)$ and scattering data $S(k)$.

For the direct scattering problem (i.e., the forward problem), this follows from the integral representations of $a(k)$ and $b(k)$ in terms of the $u(x)$ and the eigenfunction $M(x;k)$. (For further details, see, for example, Deift and Trubowitz [1979, p.146]; Dodd, Eilbeck, Gibbon and Morris [1982, p.149].)

For the inverse scattering problem, the uniquesness of the mapping follows from the integral equations (2.3.9-11). (For further details, see, for example, Deift and Trubowitz [1979, p.173] who prove

THEOREM 2.5.1

A potential $u \in P_1$ is determined by its reflection coefficient $\rho(k)$, bound states $\{\kappa_n\}_1^N$, and norming constants $\{C_n\}_1^N$,

which is a generalization of the following theorem due to Levinson [1949]

THEOREM 2.5.2

A potential $u(x) \in P_1$ without bound states is determined by its reflection coefficient $\rho(k)$;

Dodd, Eilbeck, Gibbon and Morris [1982, p.188].)

2.5.4 The Gel'fand-Levitan-Marchenko Integral Equation. The Gel'fand-Levitan-Marchenko equation may be derived from this approach as follows: given the integral equation (2.3.3) for $N(x;k;t)$

$$N(x;k;t) = \mathrm{e}^{2\mathrm{i}kx}\left\{1 + \frac{1}{2\pi\mathrm{i}}\int_{\mathcal{L}} \frac{\rho(\zeta;t)\,N(x;\zeta;t)}{\zeta + k + \mathrm{i}0}\,\mathrm{d}\zeta\right\}, \tag{2.5.5}$$

where $\mathcal{L}$ passes above all the poles of $\rho(k;t)$ (hence we need to assume that $u \to 0$ sufficiently rapidly as $|x| \to \infty$); we seek a "triangular" kernel $K(x,s;t)$ such that

$$N(x;k;t) = \mathrm{e}^{2\mathrm{i}kx}\left\{1 + \int_x^\infty K(x,s;t)\,\mathrm{e}^{\mathrm{i}k(s-x)}\,\mathrm{d}s\right\}. \tag{2.5.6}$$

Substituting (2.5.6) into (2.5.5) and operating with

$$\frac{1}{2\pi}\int_{-\infty}^{\infty} \mathrm{d}k\,\mathrm{e}^{\mathrm{i}k(x-y)}, \qquad \text{for} \quad y > x \tag{2.5.7}$$

(i.e., take the Fourier transform), we see that

$$K(x,y;t) + F(x+y;t) + \int_x^\infty K(x,s;t)\,F(s+y;t)\,\mathrm{d}s = 0, \qquad y > x, \tag{2.5.8}$$

where

$$F(x;t) = \frac{1}{2\pi}\int_{\mathcal{L}} \rho(k;t)\,\mathrm{e}^{\mathrm{i}kx}\,\mathrm{d}k,$$
$$= \sum_{j=1}^{N} C_j(0)\exp\left(8\kappa_j^3 t - \kappa_j x\right) + \frac{1}{2\pi}\int_{-\infty}^{\infty} \rho(k;t)\,\mathrm{e}^{\mathrm{i}kx}\,\mathrm{d}k, \qquad (2.5.9)$$

and $\mathcal{L}$ is a contour which passes above all the poles of $\rho(k)$. The existence and uniqueness of solutions of linear integral equations such as (2.5.8) is usually proven using the Fredholm alternative (cf. Pogorzelski [1966]; Porter and Stirling [1990]); for details see, for example, Dodd, Eilbeck, Gibbon and Morris [1982, §4.1]. Substituting equation (2.5.6) into (2.3.12c) we have

$$u(x,t) = 2\frac{\partial}{\partial x}\left\{K(x,x;t)\right\}. \qquad (2.5.10)$$

This clearly shows that the Gel'fand-Levitan-Marchenko equation (2.5.8) arises as a direct consequence of the Riemann-Hilbert boundary value problem. Whereas there are not necessarily any disadvantages in using the Gel'fand-Levitan-Marchenko equation, it is not as fundamental as the Riemann-Hilbert boundary value problem (since some problems do not have an associated Gel'fand-Levitan-Marchenko equation). We remark that the relationship between Hirota's bilinear method (§2.6.5 below) and solutions obtained through the Gel'fand-Levitan-Marchenko integral equation using Fredholm determinants is discussed by Segur [1974] and Oishi [1979].

2.5.5 A General Class of Solutions of the Korteweg-de Vries Equation.

LEMMA 2.5.3 (Fokas and Ablowitz [1981]; Rosales [1978a]).

Let $\phi(x,t;k)$ be a solution of the linear integral equation

$$\phi(x,t;k) + \mathrm{i}\exp\left\{\mathrm{i}(kx+k^3t)\right\}\int_{\mathcal{L}} \frac{\phi(x,t;p)}{p+k}\,\mathrm{d}\lambda(p) = \exp\left\{\mathrm{i}(kx+k^3t)\right\}, \qquad (2.5.11)$$

where the contour $\mathcal{L}$ and the measure $\mathrm{d}\lambda(p)$ are essentially arbitrary. Assume that the homogeneous integral equation corresponding to (2.5.11) has only the trivial solution. Then

$$u(x,t) = -\frac{\partial}{\partial x}\left\{\int_{\mathcal{L}} \phi(x,t;k)\,\mathrm{d}\lambda(k)\right\}, \qquad (2.5.12)$$

is a solution of the KdV equation

$$u_t + 6uu_x + u_{xxx} = 0. \qquad (2.5.13)$$

This result may be motivated in two different ways. The first follows Rosales [1978a] by seeking a perturbation solution of (2.5.13) about $u = 0$, and then summing the

disturbance is elastic (that is the solitary wave asymptotically preserves its shape and velocity). By definition these solitary waves are *solitons* (see §1.4).

2.6.2 Infinite Number of Conservation Laws. The partial differential equations possess an infinite number of *conservation laws* (see §1.6). For the KdV equation, the first three conservation laws are:

$$\begin{gathered}(u)_t + (3u^2 + u_{xx})_x = 0,\\ (u^2)_t + (4u^3 + 2uu_{xx} - u_x^2)_x = 0,\\ \left(u^3 - \tfrac{1}{2}u_x^2\right)_t + \left(\tfrac{9}{2}u^4 + 3u^2u_{xx} - 6uu_x^2 - u_xu_{xxx} + \tfrac{1}{2}u_{xx}^2\right)_x = 0.\end{gathered}$$

2.6.3 Compatibility of Linear Operators. The partial differential equations in one spatial dimension arise as the compatibility condition of two linear equations

$$\boldsymbol{v}_x = \mathbf{X}\boldsymbol{v}, \qquad \boldsymbol{v}_t = \mathbf{T}\boldsymbol{v}, \tag{2.6.2}$$

where $\mathbf{X}$ and $\mathbf{T}$ are $n \times n$ matrices ($n \geq 2$) and $\boldsymbol{v}$ is an n-component vector. Equations (2.6.2) are compatible, i.e., $\boldsymbol{v}_{xt} = \boldsymbol{v}_{tx}$, if and only if

$$\mathbf{X}_t - \mathbf{T}_x + [\mathbf{X}, \mathbf{T}] = 0, \tag{2.6.3}$$

for suitably chosen $\mathbf{X}$ and $\mathbf{T}$. For this to be effective, the operator $\mathbf{X}$ should depend upon a parameter k, which essentially play the role of an eigenvalue and obeys $\mathrm{d}k/\mathrm{d}t = 0$. An example where $\mathbf{X}$ does not depend on k is the Cole-Hopf transformation (Hopf [1950]; Cole [1951]), $v_x = uv$, which serves to directly linearize the Burgers equation $u_t - 2uu_x - u_{xx} = 0$, to the heat equation $v_t = v_{xx}$. The Cole-Hopf transformation is local and does not require inverse scattering.

For the KdV equation (2.6.1), $n = 2$ and if we take

$$\mathbf{X} = \begin{pmatrix} -\mathrm{i}k & u \\ -1 & \mathrm{i}k \end{pmatrix}, \qquad \mathbf{T} = \begin{pmatrix} A & B \\ C & -A \end{pmatrix},$$

with

$$\begin{aligned} A &= -4\mathrm{i}k^3 + 2\mathrm{i}ku - u_x,\\ B &= 4uk^2 + 2\mathrm{i}u_xk - (2u^2 + u_{xx}),\\ C &= -4k^2 + 2u, \end{aligned}$$

then $\mathbf{X}$ and $\mathbf{T}$ satisfy (2.6.3) if and only if u satisfies the KdV equation (2.6.1). We remark that this is more general than this approach is somewhat more general than that of Lax [1968], which we discussed in §1.8, since here the dependence on the scattering parameter can be rather complicated. If an equation is solvable by an

I.S.T. scheme, then there exists an *associated linear scattering problem*, which is given by (2.6.2).

2.6.4 Completely Integrable Hamiltonian System and Action-Angle Variables. One of the fundamental concepts which underlies the I.S.T. method of solution is the interpretation that nonlinear evolution equations which are solvable by an I.S.T. scheme are (infinite dimensional) *completely integrable Hamiltonian systems* and I.S.T. can be thought of as a nonlinear transformation from physical variables to (an infinite set of) action-angle variables (for an introduction to the Hamiltonian formalism as applied to the KdV equation see, for example, Lax [1978]). The description of the I.S.T. method as a canonical transformation to action-angle variables was first developed for the KdV equation by Zakharov and Faddeev [1971] (following preliminary work by Gardner [1971]); Zakharov and Manakov [1974] and Flaschka and Newell [1975] considered the schemes for the nonlinear Schrödinger and related equations.

A dynamical system is said to be *Hamiltonian* if it is possible to identify generalized coordinates $q(x,t)$, generalized momenta $p(x,t)$ and a Hamiltonian $H(p,q,t)$ such that the equations of motion can be written in the form

$$\frac{\partial q}{\partial t}=\frac{\delta H}{\delta p},\qquad \frac{\partial p}{\partial t}=-\frac{\delta H}{\delta q},\tag{2.6.4}$$

where the Hamiltonian H is a complex-valued functional of (p,q) and their x-derivatives of the form

$$H(p,q,t)=\int_{-\infty}^{\infty}h(p,q,t)\,\mathrm{d}x\tag{2.6.5a}$$

and assuming that H is a functional of q, p and their derivatives, then the functional (or Frechet, or variational) derivatives $\delta H/\delta p$ and $\delta H/\delta q$ are defined by

$$\frac{\delta H}{\delta p}=\sum_{n=0}^{\infty}(-1)^n\frac{\partial^n}{\partial x^n}\left(\frac{\partial h}{\partial p_{nx}}\right),\qquad \frac{\delta H}{\delta q}=\sum_{n=0}^{\infty}(-1)^n\frac{\partial^n}{\partial x^n}\left(\frac{\partial h}{\partial q_{nx}}\right),\tag{2.6.5b}$$

with $p_{nx}:=\partial^n p/\partial x^n$ and $q_{nx}:=\partial^n q/\partial x^n$ [the variables (p,q) are said to be *conjugate*].

The associated Hamiltonian for the KdV equation (2.6.1) is given by

$$H=-\int_{-\infty}^{\infty}(pq_x^2+p^2q_x-p_xq_{xx})\,\mathrm{d}x,\tag{2.6.6}$$

and the associated dynamical equations are

$$q_t=-q_x^2-2pq_x-q_{xxx},\tag{2.6.7a}$$

$$p_t=-(2pq_x)_x-(p^2)_x-p_{xxx}.\tag{2.6.7b}$$

If at $t=0$, $p(x,0)=q_x(x,0)$, then $p(x,t)=q_x(x,t)$ for all t, therefore the evolution equations for p and q_x are identical and moreover they are both the KdV equation for $p(x,t)$.

Since the Hamiltonian (2.6.6) for the KdV equation does not involve q explicitly (though it does depend on the x-derivatives of q), then the dynamical system may be written in the form

$$\frac{\partial u}{\partial t} = \frac{\partial}{\partial x}\left(\frac{\delta \hat{H}}{\delta u}\right), \tag{2.6.8}$$

where

$$\hat{H}(u,t) = -\int_{-\infty}^{\infty} \left(u^3 - \tfrac{1}{2}u_x^2\right) \mathrm{d}x.$$

(Note that $\hat{H}(u,t) \neq H(u,u_x,t)$.)

Magri [1978] (see also Magri [1980]; Gel'fand and Dikii [1977]) showed that the KdV equation actually has a *bi-Hamiltonian* structure since it has the following two Hamiltonian decompositions

$$u_t + \partial_x(3u^2 + u_{xx}) = 0, \tag{2.6.9}$$

$$u_t + (\partial_x^3 + 4u\partial_x + 2u_x\mathrm{I})u = 0, \tag{2.6.10}$$

where $\partial_x \equiv \partial/\partial x$ and I is the identity operator (note that (2.6.9) is just the decomposition discussed above).

Suppose we have a dynamical system with conjugate variables (p,q), then in order to change to another set of conjugate variables (P,Q), we define the *Poisson brackets*:

$$\{A(\alpha), B(\beta)\} \equiv \int_{-\infty}^{\infty}\left\{\frac{\delta A(\alpha)}{\delta q(x)}\frac{\delta B(\beta)}{\delta p(x)} - \frac{\delta A(\beta)}{\delta p(x)}\frac{\delta B(\alpha)}{\delta q(x)}\right\}\mathrm{d}x. \tag{2.6.11}$$

Alternatively, if only the derivatives of q and not q itself arise in the Hamiltonian, then we may replace (2.6.11) by

$$\{A(\alpha), B(\beta)\} = \int_{-\infty}^{\infty}\left\{\frac{\delta A(\alpha)}{\delta u(x)}\frac{\partial}{\partial x}\frac{\delta B(\beta)}{\delta u(x)}\right\}\mathrm{d}x, \tag{2.6.12}$$

where $u = p = q_x$. A transformation from (p,q) to (P,Q) is said to be *canonical* if

$$\{Q(x), Q(y)\} = 0, \tag{2.6.13a}$$

$$\{P(x), P(y)\} = 0, \tag{2.6.13b}$$

$$\{Q(x), P(y)\} = \delta(x-y), \tag{2.6.13c}$$

where $\delta(x)$ is the Dirac delta function. The evolution of a quantity u satisfying a Hamiltonian equation obeys

$$u_t = \{u, H\},$$

where H is the Hamiltonian. The main points are the following:

1. Suppose a system of evolution equations is Hamiltonian, in which the dependent variables (p,q) play the roles of the conjugate variables.

2. There is a subset of the scattering data from which the rest of the scattering data can be constructed.
3. The mapping $(p, q) \to (P, Q)$ is a canonical transformation.
4. The conjugate variables in (P, Q) are of action-angle type; that is $H = H(P)$, so that (2.6.4) become

$$\frac{\partial P}{\partial t} = 0, \qquad \frac{\partial Q}{\partial t} = \frac{\delta H}{\delta P} = \text{constant}. \tag{2.6.14}$$

Suppose that $\phi(x; k)$ is an eigenfunction of the linear scattering problem associated with the KdV equation, i.e.,

$$\phi_{xx} + [u(x) + k^2]\phi = 0, \tag{2.6.15a}$$

with boundary condition $\phi(x; k) \sim \mathrm{e}^{-\mathrm{i}kx}$ as $x \to -\infty$, which also is the solution of the integral equation

$$\phi(x; k) = \mathrm{e}^{-\mathrm{i}kx}\left\{1 + \frac{1}{2\mathrm{i}k}\int_{-\infty}^{x}\left[1 - \mathrm{e}^{2\mathrm{i}k(x-\xi)}\right] u(\xi)\,\phi(\xi; k)\,\mathrm{e}^{\mathrm{i}k\xi}\,\mathrm{d}\xi\right\} \tag{2.6.15b}$$

[recall equations (2.2.1a,2a,15)], where k is the spectral parameter and u satisfies the KdV equation. Further suppose that $a(k)$ and $b(k)$ are given by the integral representations

$$a(k) = 1 + \frac{1}{2\mathrm{i}k}\int_{-\infty}^{\infty} u(\xi)\,\phi(\xi; k)\,\mathrm{e}^{\mathrm{i}k\xi}\,\mathrm{d}\xi, \tag{2.6.16a}$$

$$b(k) = -\frac{1}{2\mathrm{i}k}\int_{-\infty}^{\infty} u(\xi)\,\phi(\xi; k)\,\mathrm{e}^{2\mathrm{i}k\xi}\,\mathrm{d}\xi \tag{2.6.16b}$$

[recall equations (2.2.21,22)]. Now suppose we wish to calculate the Poisson bracket of $a(k)$ and $b(k)$, i.e.,

$$\begin{aligned}\{a(k), b(\ell)\} &= \int_{-\infty}^{\infty}\left\{\frac{\delta a(k)}{\delta u(x)}\frac{\partial}{\partial x}\frac{\delta b(\ell)}{\delta u(x)}\right\}\mathrm{d}x,\\ &= \tfrac{1}{2}\int_{-\infty}^{\infty}\left\{\frac{\delta a(k)}{\delta u(x)}\frac{\partial}{\partial x}\frac{\delta b(\ell)}{\delta u(x)} - \frac{\delta b(\ell)}{\delta u(x)}\frac{\partial}{\partial x}\frac{\delta a(k)}{\delta u(x)}\right\}\mathrm{d}x\end{aligned} \tag{2.6.17}$$

(since the Poisson bracket is anti-symmetric). From equations (2.6.16), using equation (2.6.15), it can be shown that

$$\frac{\delta a(k)}{\delta u(x)} = -\frac{\bar{\psi}(x; k)\phi(x; k)}{2\mathrm{i}k}, \tag{2.6.18a}$$

$$\frac{\delta b(k)}{\delta u(x)} = \frac{\psi(x; k)\phi(x; k)}{2\mathrm{i}k}, \tag{2.6.18b}$$

where $\psi(x;k)$ and $\bar{\psi}(x;k)$ are solutions of (2.6.15a) satisfying the boundary conditions $\psi(x;k) \sim e^{ikx}$, $\bar{\psi}(x;k) \sim e^{-ikx}$, as $x \to \infty$. Substituting these, using the fact that $\phi(x;k)$, $\psi(x;k)$ and $\bar{\psi}(x;k)$ satisfy the Schrödinger scattering equation (2.2.1a), together with the boundary conditions (2.2.2) and the symmetry relation (2.2.4a), one finds that

$$\{a(k), b(\ell)\} = \frac{a(k)\,b(\ell)}{2(k^2-\ell^2)} - \frac{\pi\mathrm{i}\,a(k)\,b(k)\,\delta(k-\ell)}{4k} \tag{2.6.19}$$

(see Novikov, Manakov, Pitaevskii and Zakharov [1984] for further details).

For the KdV equation, we define for real k,

$$P(k) = k\pi^{-1}\ln|a(k)|^2, \tag{2.6.20a}$$

$$Q(k) = -\tfrac{1}{2}\mathrm{i}\ln\left[b(k)/\bar{b}(k)\right], \tag{2.6.20b}$$

with $a(k)$ and $b(k)$ as defined in equations (2.6.16). Then using equation (2.6.19) it can be shown that $P(k)$ and $Q(k)$ are canonical [i.e., satisfy equations (2.6.13)].

In the general case, the scattering data also contains eigenvalues $\mathrm{i}\kappa_j$ and normalization constants C_j, $j = 1, 2, \ldots, n$, which are associated with the discrete spectrum (recall §2.2 above). Here consider

$$P_j = \kappa_j^2, \qquad Q_j = -2\ln|C_j|. \tag{2.6.21}$$

Multiplying (2.6.15) by $\phi(x;\mathrm{i}\kappa_j)$, integrating with respect to x, and taking the functional derivative yields

$$\frac{\delta P_j}{\delta u(x)} = \frac{\phi^2(x;\mathrm{i}\kappa_j)}{\displaystyle\int_{-\infty}^{\infty}\phi^2(x;\mathrm{i}\kappa_j)\,\mathrm{d}x} = \frac{\phi^2(x;\mathrm{i}\kappa_j)}{\mathrm{i}\beta_j a'(\mathrm{i}\kappa_j)}, \tag{2.6.22}$$

where we have used equation (2.2.35). From equation (2.2.38), we have $\beta_j = C_j a'(\mathrm{i}\kappa_j)$; furthermore $\beta_j = \tilde{b}(\mathrm{i}\kappa_j)$, where $\tilde{b}(k)$ is the analytic continuation of $b(k)$ to the upper half k-plane. Thus we can calculate the variational derivative of β_j (and hence also C_j and Q_j) using (2.6.18b). It can then be shown that P_j, Q_j, $j = 1, 2, 3\ldots, n$, satisfy the commutation relations

$$\{P_i, Q_j\} = \delta_{ij}, \qquad \{P_i, P_j\} = 0, \qquad \{Q_i, Q_j\} = 0. \tag{2.6.23}$$

Furthermore, they commute with $P(k)$ and $Q(k)$ as defined in equation (2.6.20).

The set of variables $\{P(k),\ Q(k),\ P_j,\ Q_j,\ j = 1, 2, \ldots, n\}$, is evidently a complete set since all the scattering data is completely determined by them. In these variables the Hamiltonian $\hat{H}$ is given by

$$\hat{H} = -\tfrac{32}{5}\sum_{j=1}^{n} P_j^{5/2} + 8\int_{-\infty}^{\infty} k^3 P(k)\,\mathrm{d}k. \tag{2.6.24}$$

The associated equations of motion are

$$\frac{\partial P}{\partial t} = 0, \qquad \frac{\partial Q}{\partial t} = 8k^3, \tag{2.6.25a}$$
$$\frac{\mathrm{d}P_j}{\mathrm{d}t} = 0, \qquad \frac{\mathrm{d}Q_j}{\mathrm{d}t} = -16P_j^{3/2}, \qquad j = 1, 2, \ldots, n, \tag{2.6.25b}$$

which exactly reproduce the time evolution of the scattering data given in §2.4 above, and are trivially solvable

$$P(k;t) = P(k;0), \qquad Q(k;t) = Q(k;0) + 8k^3t,$$
$$P_j(t) = P_j(0), \qquad Q_j(t) = Q_j(0) - 16P_j^{3/2}, \qquad j = 1, 2, \ldots, n$$

(For further details see Novikov, Manakov, Pitaevskii and Zakharov [1984].)

However even for the KdV equation, several paradoxes arise in the Hamiltonian formulation corresponding to the underlying boundary conditions and properties of the scattering data (cf. Buslaev, Faddeev and Takhtajan [1986]). Since for generic initial data $u(x,0)$ which decay sufficiently rapidly $[(1+x^2)u(x,0) \in L^1(\mathbb{R})]$, then the reflection coefficient $r(k) = b(k)/a(k)$ at $k = 0$ satisfies $r(0) = -1$ (Ablowitz, Kruskal and Segur [1979]; Deift and Trubowitz [1979]), then generically $a(0)$ is undefined. It turns out that the Poisson bracket expression needs to be modified for this case. The above description holds for the nongeneric case $r(0) = -1$. Nevertheless, the basic ideas ae illustrated by the above analysis. This delicate issue underlies the difficulty inherent in the action-angle formulation of the KdV equation. For a more detailed discussion of Hamiltonian theory and action-angle variables see Faddeev and Takhtajan [1987].

2.6.5 Bilinear Representation. Hirota [1971] developed a direct method for finding N-soliton solutions of nonlinear evolution equations (see also Hirota [1980]; Hietarinta [1987b,c,d, 1988] and the references therein). Recall that in §1.7, it was shown that N-soliton solutions of the KdV equation may be written in the form

$$u(x,t) = 2\frac{\partial^2}{\partial x^2}\left\{\ln[\det(\mathbf{I} + \mathbf{C})]\right\}, \tag{2.6.26}$$

where $\mathbf{I}$ is the $N \times N$ identity matrix and $\mathbf{C}$ is an $N \times N$ matrix. Motivated by (2.6.26), in the KdV equation we make the transformation

$$u(x,t) = 2\frac{\partial^2}{\partial x^2}\left\{\ln F(x,t)\right\}. \tag{2.6.27}$$

Then $F(x,t)$ satisfies the bilinear equation

$$F_{xt}F - F_xF_t + F_{xxxx}F - 4F_{xxx}F_x + 3F_{xx}^2 = 0, \tag{2.6.28}$$

which Hirota writes in the elegant form

$$(\mathrm{D}_x\mathrm{D}_t + \mathrm{D}_x^4)F(x,t)F(x',t')\big|_{x=x',t=t'} = 0, \tag{2.6.29}$$

where

$$\mathrm{D}_x := \frac{\partial}{\partial x} - \frac{\partial}{\partial x'}, \qquad \mathrm{D}_t := \frac{\partial}{\partial t} - \frac{\partial}{\partial t'}. \tag{2.6.30}$$

Hirota's method for obtaining soliton solutions proceeds by seeking a formal solution of (2.6.29) in the form

$$F(x,t) = 1 + \epsilon f^{(1)} + \epsilon^2 f^{(2)} + \dots . \tag{2.6.31}$$

Substituting (2.6.31) into (2.6.29) and equating coefficients of powers of ϵ gives the following recursion relations for the $f^{(n)}$

$$\begin{aligned}
\epsilon &: \quad f^{(1)}_{xxxx} + f^{(1)}_{xt} = 0, && (2.6.32a)\\
\epsilon^2 &: \quad f^{(2)}_{xxxx} + f^{(2)}_{xt} = -\tfrac{1}{2}(\mathrm{D}_x\mathrm{D}_t + \mathrm{D}_x^4)f^{(1)}f^{(1)}, && (2.6.32b)\\
\epsilon^3 &: \quad f^{(3)}_{xxxx} + f^{(3)}_{xt} = -(\mathrm{D}_x\mathrm{D}_t + \mathrm{D}_x^4)f^{(1)}f^{(2)}, && (2.6.32c)
\end{aligned}$$

and so on. N-soliton solutions for the KdV equation are found by assuming that $f^{(1)}$ has the form

$$f^{(1)} = \sum_{j=1}^{N} \exp(\eta_j), \qquad \eta_j = \kappa_j x - \omega_j t + x_{j0}, \tag{2.6.33}$$

and κ_j, ω_j and x_{j0} are constants, provided that the series (2.6.31) truncates.

For $N = 1$, we take

$$f^{(1)} = \exp(\eta_1),$$

and by solving (2.6.32) we find that

$$f^{(n)} = 0, \qquad \text{for} \qquad n \geq 2.$$

Therefore we have

$$F_1 = 1 + \exp(\eta_1), \qquad \omega_1 = -\kappa_1^3,$$

and

$$u(x,t) = \tfrac{1}{2}\kappa_1^2 \operatorname{sech}^2\left\{\tfrac{1}{2}(\kappa_1 x - \kappa_1^3 t + x_{10})\right\}.$$

For $N = 2$, if we take

$$f^{(1)} = \exp(\eta_1) + \exp(\eta_2),$$

then

$$f^{(2)} = \exp(\eta_1 + \eta_2 + A_{12}),$$

where

$$\exp(A_{ij}) = \left(\frac{\kappa_i - \kappa_j}{\kappa_i + \kappa_j}\right)^2, \tag{2.6.34}$$

and

$$f^{(n)} = 0 \qquad \text{for} \qquad n \geq 3.$$

Therefore the two-soliton solution for the KdV equation is obtained from

$$u(x,t) = 2\frac{\partial^2}{\partial x^2}(\ln F_2),$$

where

$$F_2 = 1 + \exp(\eta_1) + \exp(\eta_2) + \exp(\eta_1 + \eta_2 + A_{12}).$$

For $N = 3$, a similar calculation shows that the three-soliton solution of the KdV equation has the form

$$u(x,t) = 2\frac{\partial^2}{\partial x^2}(\ln F_3),$$

where

$$\begin{aligned} F_3 = 1 &+ \exp(\eta_1) + \exp(\eta_2) + \exp(\eta_3) + \exp(\eta_1 + \eta_2 + A_{12}) \\ &+ \exp(\eta_2 + \eta_3 + A_{23}) + \exp(\eta_3 + \eta_1 + A_{31}) \\ &+ \exp(\eta_1 + \eta_2 + \eta_3 + A_{12} + A_{23} + A_{31}), \end{aligned}$$

where A_{ij} is as defined in (2.6.34). This process can be continued to higher orders, though the analysis becomes somewhat tedious. Frequently the N-soliton solution is obtained by "guessing" its form and then proving this by induction (see Ablowitz and Segur [1981, p.174] for details). We further remark that it is possible to express these N-soliton solutions in terms of Wronskian determinants (see Freeman [1984] and the references therein).

2.6.6 Bäcklund Transformations. A *Bäcklund transformation* is essentially defined as being a system of equations relating the solution of a given equation either to another solution of the same equation, or to a solution of another equation (see Ablowitz and Segur [1981, p.154] for a precise definition).

In order to give the Bäcklund transformation for the KdV equation (Wahlquist and Estabrook [1973]), we first make the change of variable $u = w_x$, whence w satisfies the Potential KdV (PKdV) equation

$$w_t + 3w_x^2 + w_{xxx} = 0, \tag{2.6.35}$$

where the integration "constant" has been taken to be zero without loss of generality. Then the Bäcklund transformation for (2.6.35) is given by

$$w_x + \tilde{w}_x + 2\kappa^2 + \tfrac{1}{2}(w - \tilde{w})^2 = 0, \tag{2.6.36a}$$

$$w_t - \tilde{w}_t + 3(w_x - \tilde{w}_x)(w_x + \tilde{w}_x) + w_{xxx} - \tilde{w}_{xxx} = 0, \tag{2.6.36b}$$

where κ is a real parameter (the scattering parameter). Equations (2.6.36a,b) are compatible, and moreover it is easily verified that if w is a solution of the PKdV equation then so is $\tilde{w}$, and the solution of the KdV equation is obtained from $\tilde{u} = \tilde{w}_x$. The effect of a Bäcklund transformation on a solution of the KdV equation on the infinite interval is to add or subtract one soliton [or another zero to the scattering for $a(k)$].

Suppose we consider the trivial solution $w = 0$ of the PKdV equation (corresponding to the trivial solution $u = 0$ of the KdV equation). Then solving (2.6.36) gives the solution

$$\tilde{w}(x,t) = 2\kappa \tanh\{\kappa(x - 4\kappa^2 t - x_0)\},$$

provided that $|\tilde{w}| < 2\kappa$ which leads to the one-soliton solution

$$\tilde{u}(x,t) = 2\kappa^2 \operatorname{sech}^2\{\kappa(x - 4\kappa^2 t - x_0)\},$$

of the KdV equation. If $|\tilde{w}| > 2\kappa$ then

$$\tilde{w}(x,t) = 2\kappa \coth\{\kappa(x - 4\kappa^2 t - x_0)\},$$

which leads to the *singular* solution

$$\tilde{u}(x,t) = 2\kappa^2 \operatorname{cosech}^2\{\kappa(x - 4\kappa^2 t - x_0)\},$$

of the KdV equation. This process can be continued to generate N-soliton solutions. Consider two distinct solutions $w_1 = w_1(x,t;\kappa_1)$ and $w_2 = w_2(x,t;\kappa_2)$ which are generated from any given solution w_0 by choosing two different values of the parameter κ_1 and κ_2. Therefore

$$(w_1 + w_0)_x - 2\kappa_1^2 + \tfrac{1}{2}(w_0 - w_1)^2 = 0, \qquad (2.6.37a)$$
$$(w_2 + w_0)_x - 2\kappa_2^2 + \tfrac{1}{2}(w_0 - w_2)^2 = 0. \qquad (2.6.37b)$$

Now generate a solution $w_{12} = w_{12}(x,t;\kappa_1,\kappa_2)$ from w_1 using the parameter κ_2 and $w_{21} = w_{21}(x,t;\kappa_1,\kappa_2)$ from w_2 using the parameter κ_1. Thus we have

$$(w_{12} + w_1)_x - 2\kappa_2^2 + \tfrac{1}{2}(w_1 - w_{12})^2 = 0, \qquad (2.6.38a)$$
$$(w_{21} + w_2)_x + 2\kappa_1^2 - \tfrac{1}{2}(w_2 - w_{21})^2 = 0. \qquad (2.6.38b)$$

By virtue of Bianchi's theorem of permutability (cf. Lamb [1980, p.247]), it follows that $w_{12} = w_{21}$. Eliminating derivatives between equations (2.6.37) and (2.6.38) we find the algebraic relation

$$w_{12} = w_0 + \frac{4(\kappa_2^2 - \kappa_1^2)}{w_2 - w_1}, \qquad (2.6.39)$$

which gives a purely algebraic description of w_{12}! Continuing this process we obtain a three-soliton solution

$$\begin{aligned} w_{123} &= w_1 + \frac{4(\kappa_3^2 - \kappa_2^2)}{w_{13} - w_{12}} \\ &= \frac{\kappa_1^2 w_1(w_2 - w_3) + \kappa_2^2 w_2(w_3 - w_1) + \kappa_3^2 w_3(w_1 - w_2)}{\kappa_1^2(w_2 - w_3) + \kappa_2^2(w_3 - w_1) + \kappa_3^2(w_1 - w_2)}. \end{aligned} \tag{2.6.40}$$

Multi-soliton solutions can be generated by continuing this process and taking

$$w_j(x,t;\kappa_j) = 2\kappa_j^2 \operatorname{sech}^2\{\kappa_j(x - 4\kappa_j^2 t - x_j)\},$$

with x_j a constant. Thus we are able to construct solutions of the KdV equation in a straightforward manner (however the associated algebraic manipulations, although simple, become very labor intensive!). We remark that equations (2.6.39,40) constitute *nonlinear superposition principles* for the KdV equation.

As an example, if in (2.6.39) we take

$$w_0 = 0, \qquad w_1 = 2\kappa_1 \tanh(\kappa_1\xi_1), \qquad w_2 = 2\kappa_2 \coth(\kappa_2\xi_2),$$

with $\xi_i = x - 4\kappa_i^2 t + x_i$, for $i = 1, 2$, we obtain the two-soliton solution

$$\begin{aligned} u(x,t) &= \frac{2(\kappa_2^2 - \kappa_1^2)\left[\kappa_1^2 \operatorname{sech}(2\kappa_1\xi_1) + \kappa_2^2 \operatorname{cosech}(2\kappa_2\xi_2)\right]}{\left[\kappa_1 \tanh(\kappa_1\xi_1) - \kappa_2 \coth(\kappa_2\xi_2)\right]^2} \\ &= \frac{4(\kappa_2^2 - \kappa_1^2)\left[(\kappa_2^2 - \kappa_1^2) + \kappa_1^2 \cosh(2\kappa_2\xi_2) + \kappa_2^2 \cosh(2\kappa_1\xi_1)\right]}{\left[(\kappa_2 - \kappa_1)\cosh(\kappa_1\xi_1 + \kappa_2\xi_2) + (\kappa_2 + \kappa_1)\cosh(\kappa_2\xi_2 - \kappa_1\xi_1)\right]^2}. \end{aligned}$$

2.6.7 Painlevé Property. Another test which in practice has provided a useful criterion for the identification of nonlinear evolution equations which are solvable by an I.S.T. scheme is the *Painlevé ODE test*, which we shall discuss in more detail in §7.2. The *Painlevé conjecture* as formulated by Ablowitz, Ramani and Segur [1978, 1980a] asserts that every ordinary differential equation which arises as a similarity reduction of a nonlinear evolution equation which is solvable by an I.S.T. scheme is of *Painlevé type*, that is it has no movable singularities other than poles, perhaps after a transformation of variables. In practice one discards equations (for the purposes of I.S.T.) when the test produces solutions in which movable branch points cannot be transformed into a single-valued function. Recent work indicates that for I.S.T. equations, single-valuedness of the solution is more essential than merely being only poles (Chakravarty, Ablowitz and Clarkson [1990]; Kruskal [1991]; Kruskal and Clarkson [1991] — see also Painlevé [1902]); the test concentrates on pole solutions.

For example, three similarity reductions of the KdV equation are:

$$u(x,t) = w_1(z), \qquad z = x - \lambda_1 t, \tag{2.6.41a}$$

$$u(x,t) = w_2(z) + \lambda_2 t, \qquad z = x - 3\lambda_2 t^2, \tag{2.6.41b}$$

$$u(x,t) = (3t)^{-2/3} w_3(z), \qquad z = x/(3t)^{1/3}. \tag{2.6.41c}$$

with λ_1, λ_2 arbitrary constants and where $w_1(z)$, $w_2(z)$ and $w_3(z)$ satisfy

$$\frac{d^3 w_1}{dz^3} + 6w_1 \frac{dw_1}{dz} - \lambda_1 \frac{dw_1}{dz} = 0, \tag{2.6.42a}$$

$$\frac{d^3 w_2}{dz^3} + 6w_2 \frac{dw_2}{dz} + \lambda_2 = 0, \tag{2.6.42b}$$

$$\frac{d^3 w_3}{dz^3} + 6w_3 \frac{dw_3}{dz} - 2w_3 - z\frac{dw_3}{dz} = 0 \tag{2.6.42c}$$

(see, for example, Olver [1986] for a derivation of these similarity reductions). As we discussed in §1.2, equation (2.6.42a) is solvable either in terms of elementary or elliptic functions, and in both cases $w_1(z)$ is a meromorphic function. Integrating equation (2.6.42b) yields

$$\frac{d^2 w_2}{dz^2} + 3w_2^2 + \lambda_2 z = c_2,$$

with c_2 an arbitrary constant, which is equivalent to the first Painlevé equation $\mathrm{P_I}$

$$\frac{d^2 y}{dx^2} = 6y^2 + x,$$

after an appropriate rescaling and translation of the variables. Equation (2.6.42c) is solvable in terms of the second Painlevé equation $\mathrm{P_{II}}$

$$\frac{d^2 y}{dz^2} = 2y^3 + zy + \alpha,$$

with α a constant, since there is a one-to-one relationship between solutions of (2.6.42c) and the solution of $\mathrm{P_{II}}$ given by

$$w_3(z) = -y'(z) - y^2(z), \qquad y(z) = \frac{w_3'(z) + \alpha}{2w_3 - z},$$

with $' = d/dz$ (see Fokas and Ablowitz [1982] for details).

Recently Weiss, Tabor and Carnevale [1983] have proposed the Painlevé property for partial differential equations, or *Painlevé PDE test*, as a method of applying the Painlevé conjecture directly to a partial differential equation without having to reduce it to an ordinary differential equation. A partial differential equation is said to *possess the Painlevé property* if its solutions are "single-valued" in the neighborhood of noncharacteristic movable singularity manifolds. The method introduced by Weiss, Tabor and Carnevale [1983], involves seeking a solution of the KdV equation in the form

$$u(x,t) = \phi^p(x,t) \sum_{j=0}^{\infty} u_j(t)\phi^j(x,t), \tag{2.6.43}$$

where $u_j(t)$, $j = 0, 1, 2, \ldots$, are analytic functions of t such that $u_0 \not\equiv 0$, in the neighborhood of a noncharacteristic movable singularity manifold defined by $\phi(x,t) = 0$.

Kruskal (private communication) has shown that the test may be simplified considerably (without loss of generality) by taking

$$\phi(x,t) = x + \psi(t), \tag{2.6.44}$$

where $\psi(t)$ is an arbitrary analytic function. Leading order analysis shows that the u_{xxx} and $6uu_x$ terms balance if $p = -2$ and $u_0 = -2$. Substituting (2.6.43) into the KdV equation and equating the coefficient of ϕ^{j-5} defines the recursion relation

$$(j+1)(j-4)(j-6)u_j = F_j(u_0, u_1, \ldots, u_{j-1}, \psi), \tag{2.6.45a}$$

where

$$F_j = -3(j-4)\sum_{k=1}^{j-1} u_k u_{j-k} - \frac{\mathrm{d}u_{j-3}}{\mathrm{d}t} - (j-4)\frac{\mathrm{d}\psi}{\mathrm{d}t}u_{j-2}, \tag{2.6.45b}$$

for $j \geq 1$ (define $u_j = 0$ for $j < 0$). (2.6.45) defines u_j for $j \geq 1$ unless $j = 4$ or $j = 6$, which are called *resonances* (and so u_4 and u_6 are arbitrary functions). In order that the KdV equation has solutions of the form (2.6.43) then it is necessary that $F_4 \equiv 0$ and $F_6 \equiv 0$, which are called *compatibility conditions*. [If either of these compatibility conditions is not identically satisfied then it is necessary to introduce logarithmic terms into the expansion (2.6.43) which will then not be single-valued.] For $j = 1, 2, 3$, we find that

$$u_1 = 0, \qquad u_2 = -\frac{1}{6}\frac{\mathrm{d}\psi}{\mathrm{d}t}, \qquad u_3 = 0,$$

respectively. For $j = 4$,

$$F_4 = -\frac{\mathrm{d}u_1}{\mathrm{d}t},$$

and so $F_4 \equiv 0$. For $j = 5$,

$$u_5 = -\frac{1}{36}\frac{\mathrm{d}^2\psi}{\mathrm{d}t^2},$$

and for $j = 6$

$$F_6 = -6(2u_1u_5 + 2u_2u_4 + u_3^2) - \frac{\mathrm{d}u_3}{\mathrm{d}t} - 2\frac{\mathrm{d}\psi}{\mathrm{d}t}u_4.$$

It is easily shown that $F_6 \equiv 0$. The remaining functions F_j, $j \geq 7$ are straightforward to compute. (For further details on the application of the Painlevé tests see §7.2.)

2.6.8 Prolongation Structure. Given a nonlinear evolution equation, at present there is no completely systematic method of determining the linear scattering problem associated with the equation (assuming one exists). One relatively systematic method is the prolongation method, which was introduced by Wahlquist and Estabrook [1975, 1976] using the language of differential geometry.

In order to derive the linear scattering problem for the KdV equation using this method we suppose that it has the form

$$\boldsymbol{v}_x = \mathbf{X}(u;k)\boldsymbol{v}, \tag{2.6.46a}$$

$$\boldsymbol{v}_t = \mathbf{T}(u, u_x, u_{xx})\boldsymbol{v}, \tag{2.6.46b}$$

where $\mathbf{X}$ and $\mathbf{T}$ are $n \times n$ matrices and $\boldsymbol{v}$ is a n-component vector (it will turn out that for the KdV equation $n = 2$). Note that $\mathbf{X}$ is assumed to be function of u and a parameter k only, whilst $\mathbf{T}$ is assumed to be a function of u, u_x and u_{xx} (since the KdV equation is a third order equation). Equations (2.6.46) are compatible provided that

$$\begin{aligned}\mathbf{X}_t &= \mathbf{T}_x + [\mathbf{T}, \mathbf{X}] \\ &= \mathbf{T}_x + (\mathbf{TX} - \mathbf{XT}).\end{aligned} \tag{2.6.47}$$

Since $\mathbf{X}$ is a function of u only then using the KdV equation

$$-(6uu_x + u_{xxx})\frac{\partial \mathbf{X}}{\partial u} = u_x\frac{\partial \mathbf{T}}{\partial u} + u_{xx}\frac{\partial \mathbf{T}}{\partial u_x} + u_{xxx}\frac{\partial \mathbf{T}}{\partial u_{xx}} + [\mathbf{T}, \mathbf{X}], \tag{2.6.48}$$

where $[\mathbf{T}, \mathbf{X}] = \mathbf{TX} - \mathbf{XT}$. By equating coefficients of u_{xxx} we see that

$$\frac{\partial \mathbf{T}}{\partial u_{xx}} = -\frac{\partial \mathbf{X}}{\partial u}$$

so that

$$\mathbf{T} = -u_{xx}\frac{\partial \mathbf{X}}{\partial u} + \mathbf{F}(u, u_x),$$

for some $\mathbf{F}(u, u_x)$ to be determined. Hence from (2.6.48)

$$-6uu_x\frac{\partial \mathbf{X}}{\partial u} = [\mathbf{F}, \mathbf{X}] + u_{xx}\left(\frac{\partial \mathbf{F}}{\partial u_x} - \left[\frac{\partial \mathbf{X}}{\partial u}, \mathbf{X}\right] - u_x\frac{\partial^2 \mathbf{X}}{\partial u^2}\right) + u_x\frac{\partial \mathbf{F}}{\partial u}.$$

Since neither $\mathbf{X}$ nor $\mathbf{F}$ contain u_{xx} then

$$\mathbf{F}(u, u_x) = \tfrac{1}{2}u_x^2\frac{\partial^2 \mathbf{X}}{\partial u^2} + u_x\left[\frac{\partial \mathbf{X}}{\partial u}, \mathbf{X}\right] + \mathbf{G}(u),$$

for some $\mathbf{G}(u)$ to be determined. By continuing this process (equating coefficients of powers of u_x etc.), we find that

$$\frac{\partial^3 \mathbf{X}}{\partial u^3} = \mathbf{0}, \tag{2.6.49a}$$

$$\frac{\partial}{\partial u}\left[\frac{\partial \mathbf{X}}{\partial u}, \mathbf{X}\right] + \frac{1}{2}\left[\frac{\partial^2 \mathbf{X}}{\partial u^2}, \mathbf{X}\right] = \mathbf{0}, \tag{2.6.49b}$$

$$\frac{\mathrm{d}\mathbf{G}}{\mathrm{d}u} - 6u\frac{\partial \mathbf{X}}{\partial u} + \left[\left[\frac{\partial \mathbf{X}}{\partial u}, \mathbf{X}\right], \mathbf{X}\right] = \mathbf{0}, \tag{2.6.49c}$$

$$[\mathbf{G}, \mathbf{X}] = \mathbf{0}. \tag{2.6.49d}$$

Therefore on solving (2.6.49a,b) we find that

$$\mathbf{X} = \mathbf{X}_1 + u\mathbf{X}_2 + u^2\mathbf{X}_3, \tag{2.6.50a}$$

$$[\mathbf{X}_1, \mathbf{X}_3] = \mathbf{0}, \tag{2.6.50b}$$

$$[\mathbf{X}_2, \mathbf{X}_3] = \mathbf{0}, \tag{2.6.50c}$$

with $\mathbf{X}_1$, $\mathbf{X}_2$, $\mathbf{X}_3$, constant matrices of, as yet, unknown dimension. Solving (2.6.49c) then yields

$$\mathbf{G} = 4u^3\mathbf{X}_3 + 3u^2\mathbf{X}_2 - \tfrac{1}{2}u^2[\mathbf{X}_2, [\mathbf{X}_1, \mathbf{X}_2]] - u[\mathbf{X}_1, [\mathbf{X}_1, \mathbf{X}_2]] + \mathbf{X}_4, \tag{2.6.51}$$

with $\mathbf{X}_4$ a constant matrix. If we define

$$[\mathbf{X}_1, \mathbf{X}_2] = -\mathbf{X}_7, \tag{2.6.52a}$$

$$[\mathbf{X}_1, \mathbf{X}_7] = \mathbf{X}_5, \tag{2.6.52b}$$

$$[\mathbf{X}_2, \mathbf{X}_7] = \mathbf{X}_6, \tag{2.6.52c}$$

then $[\mathbf{G}, \mathbf{X}] = \mathbf{0}$ provided that

$$[\mathbf{X}_1, \mathbf{X}_4] = \mathbf{0}, \tag{2.6.53a}$$

$$[\mathbf{X}_3, \mathbf{X}_6] = \mathbf{0}, \tag{2.6.53b}$$

$$[\mathbf{X}_3, \mathbf{X}_4] + \tfrac{1}{2}[\mathbf{X}_1, \mathbf{X}_6] + [\mathbf{X}_2, \mathbf{X}_5] = 3\mathbf{X}_7, \tag{2.6.53c}$$

$$[\mathbf{X}_2, \mathbf{X}_4] + [\mathbf{X}_1, \mathbf{X}_5] = \mathbf{0}, \tag{2.6.53d}$$

$$\tfrac{1}{2}[\mathbf{X}_2, \mathbf{X}_6] + [\mathbf{X}_3, \mathbf{X}_5] = \mathbf{0}. \tag{2.6.53e}$$

(These conditions are obtained by equating coefficients of powers of u.)

Equations (2.6.52) and (2.6.53) constitute an incomplete Lie algebra (the *prolongation algebra* — it is incomplete since we have not determined all products). The second step in the prolongation method is to complete the algebra, i.e., by adding additional matrices $\mathbf{X}_j$, if necessary, such all the matrix brackets are expressible in the form

$$[\mathbf{X}_i, \mathbf{X}_j] = \sum_{j=1}^{n} \alpha_{ijk}\mathbf{X}_k, \qquad 1 \le i, j, k \le n. \tag{2.6.54}$$

If this is possible then the Lie algebra will be finite dimensional (since there are only a finite number of generators).

Further relations are derived using the Jacobi identity

$$[\mathbf{X}_i, [\mathbf{X}_j, \mathbf{X}_k]] + [\mathbf{X}_k, [\mathbf{X}_i, \mathbf{X}_j]] + [\mathbf{X}_j, [\mathbf{X}_k, \mathbf{X}_i]] = \mathbf{0}. \tag{2.6.55}$$

For example, by taking the bracket of (2.6.52c) with $\mathbf{X}_1$, using the Jacobi identity (2.6.55) and equations (2.6.52a,b) we obtain

$$[\mathbf{X}_2, \mathbf{X}_5] = [\mathbf{X}_1, \mathbf{X}_6], \tag{2.6.56}$$

Proceeding in a similar manner yields the remaining relations and we obtain a seven-dimensional algebra.

The final step is to attempt to find a representation for this algebra (that is to seek matrices $\mathbf{X}_i$, $i = 1, 2, \ldots, 7$, of some dimension which satisfy the commutation relations above), then we will have determined $\mathbf{X}$ and $\mathbf{T}$, and thus the associated linear scattering problem. Therefore the problem of finding the linear spectral problem for the KdV equation is reduced to finding a set of matrices satisfying the commutation relations. We can consider these to be a set of generators and relations used to define a Lie algebra. In such circumstances there is an algebraic procedure for determining a matrix representation of the elements $\mathbf{X}_i$ and hence the spectral problem (this procedure is described in detail by Dodd and Fordy [1983, 1984] — see also Fordy [1990b]). For those whose primary interest is the determination of the spectral problem, then the prolongation method can be very useful; one feature is that it often fails quite spectacularly when the equation is not integrable. For example, for the equation

$$u_t + u^3 u_x + u_{xxx}, \tag{2.6.57}$$

the algebra is empty and no spectral problem exists (cf. Dodd and Fordy [1983]).

The classification of the algebra is important, often it can be associated with the smallest simple Lie algebra, sl$(2, \mathbb{R})$ (i.e., 2×2 traceless matrices). The standard basis for sl$(2, \mathbb{R})$ is

$$\boldsymbol{\sigma}_3 = \begin{pmatrix} 1 & 0 \\ 0 & -1 \end{pmatrix}, \qquad \boldsymbol{\sigma}_+ = \begin{pmatrix} 0 & 1 \\ 0 & 0 \end{pmatrix}, \qquad \boldsymbol{\sigma}_- = \begin{pmatrix} 0 & 0 \\ 1 & 0 \end{pmatrix}, \tag{2.6.58}$$

which have the following commutation relations

$$[\boldsymbol{\sigma}_3, \boldsymbol{\sigma}_\pm] = \pm 2\boldsymbol{\sigma}_\pm, \qquad [\boldsymbol{\sigma}_+, \boldsymbol{\sigma}_-] = \boldsymbol{\sigma}_3. \tag{2.6.59}$$

For the KdV equation it is found that (Wahlquist and Estabrook [1975])

$$\mathbf{X}_1 = \begin{pmatrix} 0 & 1 \\ -k^2 & 0 \end{pmatrix}, \qquad \mathbf{X}_2 = \begin{pmatrix} 0 & 0 \\ 1 & 0 \end{pmatrix}, \tag{2.6.60a}$$

$$\mathbf{X}_3 = \begin{pmatrix} 0 & 0 \\ 0 & 0 \end{pmatrix}, \qquad \mathbf{X}_4 = \begin{pmatrix} 0 & 4k^2 \\ -4k^4 & 0 \end{pmatrix}, \tag{2.6.60b}$$

and so

$$\mathbf{X} = \boldsymbol{\sigma}_+ - (u + k^2)\boldsymbol{\sigma}_-, \tag{2.6.61a}$$

$$\mathbf{T} = (4k^2 - 2u)\boldsymbol{\sigma}_+ + [u_{xx} - (4k^2 - 2u)(u + k^2)]\boldsymbol{\sigma}_- - u_x\boldsymbol{\sigma}_3. \tag{2.6.61b}$$

Note that (2.6.61a) is the usual Schrödinger scattering problem

$$v_{xx} + (u + k^2)v = 0. \tag{2.6.62}$$

We remark that the solution (2.6.60) is not unique, since any similarity transformation will yield another solution. For example, since the KdV equations possesses the scaling symmetry

$$(x, t, u) \rightarrow (\lambda x, \lambda^3 t, \lambda^{-2} u),$$

then the algebra can be rescaled in an analogous manner, $\hat{\mathbf{X}}_1 \rightarrow \lambda^{-1}\mathbf{X}_1$, $\hat{\mathbf{X}}_2 \rightarrow \lambda \mathbf{X}_2$, etc.. However, all 2×2 matrix solutions will lead to the same scalar equation (2.6.62). A solution such as (2.6.60), in canonical form, is a feature of associated the algebraic machinery; the matrices are elements of $\mathrm{sl}(2, \mathbb{R})$ written in fundamental representation. In any other representation, such as the 3×3 adjoint, will also yield a solution. However, this would lead to a different scalar equation from (2.6.62).

The prolongation method has been successfully used to identify the previously unknown scattering problem for several nonlinear evolution equations (cf. Dodd and Gibbon [1977]). Recently, Dodd and Fordy [1983, 1984] (see also Kaup [1980c]; Fordy [1990b]) have cast the method into the form of a (potential) algorithm for finding the scattering problem of a given nonlinear evolution equation which appears to be a useful criterion for the identification of completely integrable equations.

Chapter Three

General Inverse Scattering in One Dimension

3.1 Inverse Scattering and Riemann-Hilbert Problems for $N \times N$ systems.

3.1.1 The Direct and Inverse Scattering Problem: 2nd Order Case. In this section we discuss the Riemann-Hilbert boundary value problems associated with the 2×2 scattering problem

$$v_{1,x} = -\mathrm{i}kv_1 + qv_2, \tag{3.1.1a}$$
$$v_{2,x} = \mathrm{i}kv_2 + rv_1. \tag{3.1.1b}$$

and its $N \times N$ generalization

$$\boldsymbol{v}_x = \mathrm{i}k\mathbf{J}\boldsymbol{v} + \mathbf{Q}\boldsymbol{v}, \tag{3.1.2}$$

where $\boldsymbol{v} = (v_1, v_2, \ldots, v_N)$ is an $N \times 1$ vector and $\mathbf{J}$ and $\mathbf{Q}$ are $N \times N$ matrices such that $\mathbf{J} = \mathrm{diag}(J^1, J^2, ..., J^N)$, with $J^i \neq J^j$ for $i \neq j$, and $Q^{ii} = 0$.

The 2×2 problem provides a valuable starting point, and sets the stage for explaining the essential difference between 2×2 matrix problems and $N \times N$ problems.

For the scattering problem (3.1.1), the standard approach is to define two complete sets of eigenfunctions by the asymptotic conditions

$$\boldsymbol{\phi}(x;k) \sim \begin{pmatrix}1\\0\end{pmatrix}\mathrm{e}^{-\mathrm{i}kx}, \quad \bar{\boldsymbol{\phi}}(x;k) \sim \begin{pmatrix}0\\-1\end{pmatrix}\mathrm{e}^{\mathrm{i}kx}, \quad \text{as} \quad x \to -\infty, \tag{3.1.3a}$$
$$\boldsymbol{\psi}(x;k) \sim \begin{pmatrix}0\\1\end{pmatrix}\mathrm{e}^{\mathrm{i}kx}, \quad \bar{\boldsymbol{\psi}}(x;k) \sim \begin{pmatrix}1\\0\end{pmatrix}\mathrm{e}^{-\mathrm{i}kx}, \quad \text{as} \quad x \to \infty. \tag{3.1.3b}$$

The completeness relationship of the eigenfunctions is given by

$$\boldsymbol{\phi}(x;k) = a(k)\bar{\boldsymbol{\psi}}(x;k) + b(k)\boldsymbol{\psi}(x;k), \tag{3.1.4a}$$
$$\bar{\boldsymbol{\phi}}(x;k) = -\bar{a}(k)\boldsymbol{\psi}(x;k) + \bar{b}(k)\bar{\boldsymbol{\psi}}(x;k). \tag{3.1.4b}$$

Now define

$$\boldsymbol{M}(x;k) := \boldsymbol{\phi}(x;k)\,\mathrm{e}^{\mathrm{i}kx}, \qquad \bar{\boldsymbol{M}}(x;k) := \bar{\boldsymbol{\phi}}(x;k)\,\mathrm{e}^{-\mathrm{i}kx}, \tag{3.1.5a}$$
$$\boldsymbol{N}(x;k) := \boldsymbol{\psi}(x;k)\,\mathrm{e}^{-\mathrm{i}kx}, \qquad \bar{\boldsymbol{N}}(x;k) := \bar{\boldsymbol{\psi}}(x;k)\,\mathrm{e}^{\mathrm{i}kx}, \tag{3.1.5b}$$

then as $x \to -\infty$,

$$\boldsymbol{M}(x;k) \sim \begin{pmatrix}1\\0\end{pmatrix}, \qquad \bar{\boldsymbol{M}}(x;k) \sim \begin{pmatrix}0\\-1\end{pmatrix},$$

and also as $x \to \infty$,

$$\boldsymbol{N}(x;k) \sim \begin{pmatrix}0\\1\end{pmatrix}, \qquad \bar{\boldsymbol{N}}(x;k) \sim \begin{pmatrix}1\\0\end{pmatrix}.$$

Integral equations governing $\boldsymbol{M}$, $\bar{\boldsymbol{M}}$, $\boldsymbol{N}$, $\bar{\boldsymbol{N}}$ can be readily deduced. Writing (3.1.1) in the form

$$\frac{\partial}{\partial x}\left(v_1 \mathrm{e}^{\mathrm{i}kx}\right) = q v_2 \mathrm{e}^{\mathrm{i}kx}, \tag{3.1.6a}$$

$$\frac{\partial}{\partial x}\left(v_2 \mathrm{e}^{-\mathrm{i}kx}\right) = r v_1 \mathrm{e}^{-\mathrm{i}kx}, \tag{3.1.6b}$$

Integrating and using the relevant boundary conditions yields

$$\boldsymbol{M}(x;k) = \begin{pmatrix}1\\0\end{pmatrix} + \int_{-\infty}^{\infty} \mathbf{G}_+(x-\xi;k)\,\mathbf{Q}(\xi)\,\boldsymbol{M}(\xi;k)\,\mathrm{d}\xi, \tag{3.1.7a}$$

$$\boldsymbol{N}(x;k) = \begin{pmatrix}0\\1\end{pmatrix} + \int_{-\infty}^{\infty} \hat{\mathbf{G}}_+(x-\xi;k)\,\mathbf{Q}(\xi)\,\boldsymbol{N}(\xi;k)\,\mathrm{d}\xi, \tag{3.1.7b}$$

$$\bar{\boldsymbol{M}}(x;k) = \begin{pmatrix}0\\-1\end{pmatrix} + \int_{-\infty}^{\infty} \mathbf{G}_-(x-\xi;k)\,\mathbf{Q}(\xi)\,\bar{\boldsymbol{M}}(\xi;k)\,\mathrm{d}\xi, \tag{3.1.7c}$$

$$\bar{\boldsymbol{N}}(x;k) = \begin{pmatrix}1\\0\end{pmatrix} + \int_{-\infty}^{\infty} \hat{\mathbf{G}}_-(x-\xi;k)\,\mathbf{Q}(\xi)\,\bar{\boldsymbol{N}}(\xi;k)\,\mathrm{d}\xi, \tag{3.1.7d}$$

where

$$\mathbf{Q}(x) = \begin{pmatrix}0 & q(x)\\ r(x) & 0\end{pmatrix}, \tag{3.1.8a}$$

and

$$\mathbf{G}_+(x;k) = \begin{pmatrix}1 & 0\\ 0 & \mathrm{e}^{2\mathrm{i}kx}\end{pmatrix}\theta(x), \tag{3.1.8b}$$

$$\hat{\mathbf{G}}_+(x;k) = -\begin{pmatrix}\mathrm{e}^{-2\mathrm{i}kx} & 0\\ 0 & 1\end{pmatrix}\theta(-x), \tag{3.1.8c}$$

$$\mathbf{G}_-(x;k) = \begin{pmatrix}\mathrm{e}^{-2\mathrm{i}kx} & 0\\ 0 & 1\end{pmatrix}\theta(x), \tag{3.1.8d}$$

$$\hat{\mathbf{G}}_+(x;k) = -\begin{pmatrix}1 & 0\\ 0 & \mathrm{e}^{2\mathrm{i}kx}\end{pmatrix}\theta(-x), \tag{3.1.8e}$$

where $\theta(x)$ is the Heaviside function ($\theta(x) = 1$ if $x > 0$, $\theta(x) = 0$ if $x < 0$). The kernels $\mathbf{G}_+$, $\hat{\mathbf{G}}_+$ ($\mathbf{G}_-$, $\hat{\mathbf{G}}_-$) are analytic in the upper (lower) half k-planes, and in fact the integral equations (3.1.7) are Volterra integral equations. Their Neumann series are absolutely convergent for $\mathbf{Q}(x) \in L^1(\mathbb{R})$ (see, for example, Ablowitz and Segur [1981]), in the relevant half planes. In fact if

$$\int_{-\infty}^{\infty} |x|^n |q(x)|\,\mathrm{d}x < \infty, \qquad \int_{-\infty}^{\infty} |x|^n |r(x)|\,\mathrm{d}x < \infty, \tag{3.1.9}$$

for all n, then analyticity continues to the real k-axis. The scattering coefficients $a(k)$, $\bar{a}(k)$, $b(k)$, $\bar{b}(k)$ can be expressed in terms of integrals over the potentials $q(x)$ and $r(x)$ and the eigenfunctions. The procedure used for the Schrödinger scattering problem (see equations (2.2.17-22)) is also useful here; namely we call

$$\mathbf{\Delta}(x;k) = \boldsymbol{M}(x;k) - a(k)\bar{\boldsymbol{N}}(x;k), \tag{3.1.10}$$

substitute the integral equations (3.1.7) into (3.1.10), and use the "symmetry" condition

$$(\mathbf{G}_+ - \hat{\mathbf{G}}_-)(x;k) = -\begin{pmatrix} 1 & 0 \\ 0 & \mathrm{e}^{2\mathrm{i}kx} \end{pmatrix}, \tag{3.1.11}$$

to yield

$$\begin{aligned} \mathbf{\Delta}(x;k) &- \int_{-\infty}^{\infty} \hat{\mathbf{G}}_-(x-\xi;k)\,\mathbf{Q}(\xi)\,\mathbf{\Delta}(\xi;k)\,\mathrm{d}\xi \\ &= [1-a(k)]\begin{pmatrix} 1 \\ 0 \end{pmatrix} - \int_{-\infty}^{\infty} \begin{pmatrix} 1 & 0 \\ 0 & \mathrm{e}^{2\mathrm{i}k(x-\xi)} \end{pmatrix} \mathbf{Q}(\xi)\,\boldsymbol{M}(\xi;k)\,\mathrm{d}\xi. \end{aligned} \tag{3.1.12}$$

Alternatively

$$\mathbf{\Delta}(x;k) = b(k)\mathrm{e}^{2\mathrm{i}kx}\boldsymbol{N}(x;k), \tag{3.1.13}$$

whereupon substituting the integral equation for $\boldsymbol{N}(x;k)$ and using

$$\mathrm{e}^{2\mathrm{i}kx}\hat{\mathbf{G}}_+(x;k) = \hat{\mathbf{G}}_-(x;k), \tag{3.1.14}$$

gives

$$\mathbf{\Delta}(x;k) - \int_{-\infty}^{\infty} \hat{\mathbf{G}}_-(x-\xi;k)\,\mathbf{Q}(\xi)\,\mathbf{\Delta}(\xi;k)\,\mathrm{d}\xi = b(k)\mathrm{e}^{2\mathrm{i}kx}\begin{pmatrix} 0 \\ 1 \end{pmatrix}. \tag{3.1.15}$$

Comparing equations (3.1.12) and (3.1.15) results in the formulae

$$a(k) = 1 + \int_{-\infty}^{\infty} q(x)\,M_2(x;k)\,\mathrm{d}x, \tag{3.1.16a}$$

$$b(k) = -\int_{-\infty}^{\infty} \mathrm{e}^{-2\mathrm{i}kx} r(x)\,M_1(x;k)\,\mathrm{d}x, \tag{3.1.16b}$$

where $M_1(x;k)$ and $M_2(x;k)$ are the components of $\boldsymbol{M}(x;k)$. Similar formulae for $\bar{a}(k)$ and $\bar{b}(k)$ may be found in an analogous manner; these are

$$\bar{a}(k) = 1 + \int_{-\infty}^{\infty} r(x)\,\bar{M}_1(x;k)\,\mathrm{d}x, \tag{3.1.16c}$$

$$\bar{b}(k) = -\int_{-\infty}^{\infty} \mathrm{e}^{2\mathrm{i}kx} q(x)\,\bar{M}_2(x;k)\,\mathrm{d}x. \tag{3.1.16d}$$

Alternatively $a(k)$, $\bar{a}(k)$, $b(k)$, $\bar{b}(k)$ can be expressed in terms of Wronskian relationships. Namely if we call $W(\boldsymbol{\phi},\boldsymbol{\psi}) = \phi_1\psi_2 - \psi_1\phi_2$, etc., then straightforward calculations of (3.1.4) give

$$a(k) = W(\boldsymbol{\phi},\boldsymbol{\psi}), \qquad \bar{a}(k) = W(\bar{\boldsymbol{\phi}},\bar{\boldsymbol{\psi}}), \qquad b(k) = -W(\boldsymbol{\phi},\bar{\boldsymbol{\psi}}), \qquad \bar{b}(k) = W(\bar{\boldsymbol{\phi}},\boldsymbol{\psi}).$$

From (3.1.16) or the Wronskian relationship it is clear that $a(k)$ ($\bar{a}(k)$) are analytic in the upper (lower) half k-planes, since $\boldsymbol{M}(x;k)$ ($\bar{\boldsymbol{M}}(x;k)$) have these properties. In fact, if the potentials have sufficient decay (i.e., satisfy (3.1.7) for all n), then $a(k)$ ($\bar{a}(k)$) are analytic in the half k-plane $\operatorname{Im} k \geq 0$ ($\operatorname{Im} k \leq 0$); there are no cluster points and consequently $a(k) = 0$ ($\bar{a}(k) = 0$) have only a finite number of zeros.

The Riemann-Hilbert problem is derived from (3.1.4) or in terms of $\boldsymbol{M}$, $\bar{\boldsymbol{M}}$, $\boldsymbol{N}$, $\bar{\boldsymbol{N}}$

$$\frac{\boldsymbol{M}(x;k)}{a(k)} = \bar{\boldsymbol{N}}(x;k) + \rho(k)\,\mathrm{e}^{2\mathrm{i}kx}\boldsymbol{N}(x;k), \tag{3.1.17a}$$

$$\frac{\bar{\boldsymbol{M}}(x;k)}{\bar{a}(k)} = -\boldsymbol{N}(x;k) + \bar{\rho}(k)\,\mathrm{e}^{-2\mathrm{i}kx}\bar{\boldsymbol{N}}(x;k), \tag{3.1.17b}$$

where

$$\rho(k) = \frac{b(k)}{a(k)}, \qquad \bar{\rho}(k) = \frac{\bar{b}(k)}{\bar{a}(k)} \tag{3.1.17c}$$

(note the analogy with equation (2.2.10)). We can write (3.1.17) as a Riemann-Hilbert boundary value problem. Calling

$$\mathbf{m}_+(x;k) = \left(\frac{\boldsymbol{M}(x;k)}{a(k)}, \boldsymbol{N}(x;k)\right), \tag{3.1.18a}$$

$$\mathbf{m}_-(x;k) = \left(\bar{\boldsymbol{N}}(x;k), -\frac{\bar{\boldsymbol{M}}(x;k)}{\bar{a}(k)}\right), \tag{3.1.18b}$$

we have

$$(\mathbf{m}_+ - \mathbf{m}_-)(x;k) = \mathbf{m}_-(x;k)\mathbf{V}(x;k), \tag{3.1.19a}$$

where

$$\mathbf{V}(x;k) = \begin{pmatrix} \rho(k)\bar{\rho}(k) & \bar{\rho}(k)\mathrm{e}^{-2\mathrm{i}kx} \\ \rho(k)\mathrm{e}^{2\mathrm{i}kx} & 0 \end{pmatrix}, \tag{3.1.19b}$$

and $\mathbf{m}_\pm(x;k) \to \mathbf{I}$ as $|k| \to \infty$. The integral equations (3.1.7) give the asymptotic formulae as $k \to \infty$

$$\mathbf{m}_\pm(x;k) \sim \begin{pmatrix} 1 + \dfrac{1}{2\mathrm{i}k}\displaystyle\int_x^\infty q(\xi)r(\xi)\,\mathrm{d}\xi & \dfrac{q(x)}{2\mathrm{i}k} \\ -\dfrac{r(x)}{2\mathrm{i}k} & 1 - \dfrac{1}{2\mathrm{i}k}\displaystyle\int_{-\infty}^x q(\xi)r(\xi)\,\mathrm{d}\xi \end{pmatrix}. \tag{3.1.20}$$

If $r(x) = \pm q^*(x)$ then the following symmetry conditions hold (for k real)

$$\bar{a}(k) = a^*(k), \qquad \bar{b}(k) = \mp b^*(k),$$
$$\bar{\rho}(k) = \frac{\bar{b}(k)}{\bar{a}(k)} = \mp\rho^*(k),$$
$$\bar{\boldsymbol{N}}(x;k) = \begin{pmatrix}\bar{N}_1\\ \bar{N}_2\end{pmatrix}(x;k) = \begin{pmatrix}N_2^*\\ \pm N_1^*\end{pmatrix}(x;k),$$
$$\bar{\boldsymbol{M}}(x;k) = \begin{pmatrix}\bar{M}_1\\ \bar{M}_2\end{pmatrix}(x;k) = \begin{pmatrix}\mp M_2^*\\ -M_1^*\end{pmatrix}(x;k).$$

As in the Schrödinger scattering problem (see equation (2.3.7)), when $\bar{\rho}(k) = -\rho^*(k)$, then so long as $a(k) \neq 0$, the results of Gohberg and Krein [1958] show that these exists a unique solution to (3.1.19). For other cases, existence is guaranteed in the generic sense. Similarly, if there are no zeros of $a(k)$ and $\bar{a}(k)$, then taking the minus projection of (3.1.19a) yields

$$\mathbf{m}_-(x;k) = \mathbf{I} + \frac{1}{2\pi \mathrm{i}}\int_{-\infty}^{\infty} \frac{\mathbf{m}_-(x;\zeta)\,\mathbf{v}(x;\zeta)}{\zeta - (k - \mathrm{i}0)}\,\mathrm{d}\zeta, \tag{3.1.21}$$

which yields the formal solution to the problem. If $a(k) = 0$ then we must add into (3.1.21) the pole contributions — which contain the soliton solutions. Comparing asymptotic formulae (as $|k| \to \infty$), yields the potentials, a symmetric form is given by (again $a(k) \neq 0$, $\bar{a}(k) \neq 0$)

$$\begin{pmatrix} \displaystyle\int_x^{\infty} q(\xi)r(\xi)\,\mathrm{d}\xi & q(x) \\ -r(x) & \displaystyle -\int_x^{\infty} q(\xi)r(\xi)\,\mathrm{d}\xi \end{pmatrix}$$
$$= -\frac{1}{\pi}\begin{pmatrix} \displaystyle\int_{-\infty}^{\infty} \rho(\zeta)\mathrm{e}^{2\mathrm{i}\zeta x} N_1(x;\zeta)\,\mathrm{d}\zeta & \displaystyle\int_{-\infty}^{\infty} \bar{\rho}(\zeta)\mathrm{e}^{-2\mathrm{i}\zeta x}\bar{N}_1(x;\zeta)\,\mathrm{d}\zeta \\ \displaystyle\int_{-\infty}^{\infty} \rho(\zeta)\mathrm{e}^{2\mathrm{i}\zeta x} N_2(x,\zeta)\,\mathrm{d}\zeta & \displaystyle\int_{-\infty}^{\infty} \bar{\rho}(\zeta)\mathrm{e}^{-2\mathrm{i}\zeta x}\bar{N}_2(x;\zeta)\,\mathrm{d}\zeta \end{pmatrix}. \tag{3.1.22}$$

An alternative procedure is to solve the inverse problem by Gel'fand-Levitan-Marchenko integral equations (see Ablowitz and Segur [1981]). Assuming the triangular forms

$$\boldsymbol{N}(x;k) = \begin{pmatrix}0\\1\end{pmatrix} + \int_x^{\infty} \boldsymbol{K}(x,s)\mathrm{e}^{\mathrm{i}k(x-s)}\,\mathrm{d}s, \tag{3.1.23a}$$
$$\bar{\boldsymbol{N}}(x;k) = \begin{pmatrix}1\\0\end{pmatrix} + \int_x^{\infty} \bar{\boldsymbol{K}}(x,s)\mathrm{e}^{-\mathrm{i}k(x-s)}\,\mathrm{d}s, \tag{3.1.23b}$$

substituting these into equation (3.1.21) with (3.1.17b) and taking the Fourier transform (see also §2.5.2) we find

$$\boldsymbol{K}(x,y) - \begin{pmatrix}1\\0\end{pmatrix}\bar{F}(x+y) - \int_x^{\infty} \bar{\boldsymbol{K}}(x,s)\bar{F}(s+y)\,\mathrm{d}s = 0, \tag{3.1.24a}$$
$$\bar{\boldsymbol{K}}(x,y) + \begin{pmatrix}0\\1\end{pmatrix}F(x+y) + \int_x^{\infty} \boldsymbol{K}(x,s)F(s+y)\,\mathrm{d}s = 0, \tag{3.1.24b}$$

where

$$F(x) = \frac{1}{2\pi}\int_{-\infty}^{\infty} \rho(k)\mathrm{e}^{\mathrm{i}kx}\,\mathrm{d}k, \qquad \bar{F}(x) = \frac{1}{2\pi}\int_{-\infty}^{\infty} \bar{\rho}(k)\mathrm{e}^{-\mathrm{i}kx}\,\mathrm{d}k, \tag{3.1.24c}$$

and the connection to the potentials is via

$$q(x) = -2K_1(x,x), \qquad r(x) = -2\bar{K}_2(x,x). \tag{3.1.24d}$$

We note that if $a(k) = 0$ ($\bar{a}(k) = 0$) in the upper (lower) half k-plane then pole contributions modify $F(x)$ ($\bar{F}(x)$), i.e.

$$F(x) = \frac{1}{2\pi}\int_{-\infty}^{\infty} \rho(k)\mathrm{e}^{\mathrm{i}kx}\,\mathrm{d}k - \mathrm{i}\sum_{j=1}^{N} C_j \exp(\mathrm{i}k_j x), \tag{3.1.25a}$$

$$\bar{F}(x) = \frac{1}{2\pi}\int_{-\infty}^{\infty} \bar{\rho}(k)\mathrm{e}^{-\mathrm{i}kx}\,\mathrm{d}k + \mathrm{i}\sum_{j=1}^{N} \bar{C}_j \exp(-\mathrm{i}k_j x), \tag{3.1.25b}$$

where C_j ($\bar{C}_j$) are the contributions due to $a(k) = 0$ ($\bar{a}(k) = 0$), at a finite number of locations. We note that, unlike the Schrödinger scattering problem, $a(k)$ ($\bar{a}(k)$) can vanish on the real axis $\operatorname{Im} k = 0$ and $a(k)$ ($\bar{a}(k)$) can have multiple roots (we treat multiple roots as a coalescence of single roots). When $\rho(k) = \bar{\rho}(k) = 0$, we obtain pure soliton soliton solutions; e.g., when $N = 1$, $F(x) = -\mathrm{i}C\exp(\mathrm{i}kx)$, $k = \xi + \mathrm{i}\eta$, $\eta > 0$, we have the one-soliton solution

$$\begin{aligned} q(x,t) = 2\eta \exp(-2\mathrm{i}\xi x)\exp\{2\mathrm{i}[\operatorname{Im} A_-(k)]t - \mathrm{i}(\psi_0 + \tfrac{1}{2}\pi)\} \\ \times \operatorname{sech}(2\eta x + [\operatorname{Re} A_-(k)]t - \chi_0), \end{aligned} \tag{3.1.26a}$$

with

$$C = |C_0|\exp(\mathrm{i}\chi_0)\exp\{-2A_-(k)t\}, \tag{3.1.26b}$$

$$\chi_0 = \ln(|C_0|/2\eta), \tag{3.1.26c}$$

$$A_-(k) = \lim_{x\to-\infty} A(x;k), \tag{3.1.26d}$$

where $A(x;k)$ is the coefficient in the associated time dependence of the eigenfunction (see equation (1.9.9)). In the case of the nonlinear Schrödinger equation

$$\mathrm{i}q_t = q_{xx} \pm 2|q|^2 q, \tag{3.1.27}$$

$A_-(k) = 2\mathrm{i}k^2$ and we have the one-soliton solution

$$q(x,t) = 2\eta\exp(-2\mathrm{i}\xi x)\exp\{4\mathrm{i}(\xi^2 - \eta^2)t - \mathrm{i}(\phi_0 + \tfrac{1}{2}\pi)\}\operatorname{sech}(2\eta x - 8\xi\eta t - \chi_0).$$

Note that the velocity of this solution is given by 4ξ and its amplitude by η. We remark that the nonlinear Schrödinger equation (3.1.27) arises in several physical applications including: the evolution of an envelope of weakly nonlinear dispersive water waves (cf., Ablowitz and Segur [1979, 1981]; Hasimoto and Ono [1972]; Infeld and Rowlands [1990]; Yuen and Ferguson [1978]; Yuen and Lake [1975]; Zakharov [1968]); to model the coupling of Langmuir oscillations to ion acoustic waves in a plasma (Zakharov [1972]; Nicholson and Goldman [1978]); the propagation of pulses in optical fibers (cf., Hasegawa [1990]; Hasegawa and Kodama [1981]; Kodama [1985]; Kodama and Hasegawa [1987]; Mollenauer [1985]; Mollenauer and Stolen [1982]; Mollenauer, Stolen and Gordon [1983]); in the description of energy transport along alpha-helix proteins (cf., Davydov [1981, 1985]; Hyman, McLaughlin and Scott [1981]); and quantum field theories (cf., Jackiw [1977]; Rajaraman [1982]).

3.1.2 The Direct and Inverse Scattering Problem: *N*th Order Case. It is natural to ask how one can generalize the scattering analysis for the 2×2 system to the $N \times N$ case, i.e.

$$\frac{\mathrm{d}\mathbf{\Psi}}{\mathrm{d}x}(x;k) = \mathrm{i}k\mathbf{J}\mathbf{\Psi}(x;k) + \mathbf{Q}\mathbf{\Psi}(x;k). \tag{3.1.28}$$

where $\mathbf{J} = \mathrm{diag}(J^1, J^2, \ldots, J^N)$, $Q^{ii} = 0$, $\mathbf{Q} \in \mathbb{C}^{N\times N}$ (see equation (1.10.22)). In order to do this we shall find that the integral equations governing the eigenfunctions with nice analytic properties are now of Fredholm type. This problem has been studied by a number of authors (see Beals and Coifman [1981, 1982, 1984, 1987, 1989]; Caudrey [1982]; Mikhailov [1979]; Sattinger [1989, 1990]; Zhou [1989]). We follow the work of Beals and Coifman in our brief discussion.

It is convenient to assume that J^j, $j = 1, 2, \ldots, N$, are real and $J^1 > J^2 > \ldots > J^N$. Otherwise the analysis is more complicated and the resulting Riemann-Hilbert boundary value problem is in multisectors as opposed to the upper/lower half planes. We assume that $\mathbf{Q}(x) \in L^1(\mathbb{R})$ (i.e., $Q^{jk}(x) \in L^1(\mathbb{R})$). It is useful to transform (3.1.28) by setting

$$\mathbf{\Psi}(x;k) = \mathbf{m}(x;k)\exp\{\mathrm{i}xk\mathbf{J}\},$$

whereupon $\mathbf{m}(x;k)$ satisfies

$$\begin{aligned}\frac{\mathrm{d}\mathbf{m}}{\mathrm{d}x} &= \mathrm{i}k[\mathbf{J},\mathbf{m}] + \mathbf{Q}\mathbf{m},\\ &\equiv \mathrm{i}k(\mathrm{ad}\mathbf{J} + \mathbf{Q})\mathbf{m},\end{aligned} \tag{3.1.29}$$

where $[\mathbf{J},\mathbf{m}] \equiv (\mathrm{ad}\mathbf{J})\mathbf{m} = \mathbf{J}\mathbf{m} - \mathbf{m}\mathbf{J}$. Beals and Coifman [1984] prove that there exists a unique solution $\mathbf{m}(x;k)$ of (3.1.29) bounded for all x, meromorphic in k off the real k-axis and satisfying

$$\lim_{k\to\infty} \mathbf{m}(x;k) = \mathbf{I}. \tag{3.1.30}$$

To understand why there exist meromorphic eigenfunctions for $\operatorname{Im} k \gtrless 0$, we consider an integral equation

$$\begin{aligned}\mathbf{m}(x;k) &= \mathbf{I} + \int_{-\infty}^{\infty} \mathbf{G}(x-\xi,k)\,\mathbf{Q}(\xi)\,\mathbf{m}(\xi;k)\,\mathrm{d}\xi,\\ &= \mathbf{I} + \tilde{\mathbf{G}}(\mathbf{Qm}),\end{aligned} \tag{3.1.31}$$

or in its component form

$$m^{ij}(x;k) = \delta^{ij} + \int_{-\infty}^{\infty} G^{ij}(x-\xi;k)(\mathbf{Q}(\xi)\,\mathbf{m}(\xi;k))^{ij}\,\mathrm{d}\xi,$$

where $\mathbf{G}$ is the matrix Green's function whose rsth component G^{rs} satisfies the equation

$$\frac{\partial G^{rs}}{\partial x} - \mathrm{i}k(J^r - J^s)G^{rs} = \delta(x). \tag{3.1.32}$$

Meromorphic eigenfunctions in the upper/lower half k-planes $\mathbf{m}_\pm(x;k)$ are induced by Greens functions which are analytic in the upper/lower half k-planes for $r \neq s$

$$G_\pm^{rs}(x;k) = \frac{1}{2\pi\mathrm{i}}\int_{-\infty}^{\infty} \frac{\mathrm{e}^{\mathrm{i}px}}{p - k(J^r - J^s) \mp \operatorname{sgn}(J^r - J^s)\mathrm{i}0}\,\mathrm{d}p. \tag{3.1.33}$$

When $r = s$ there is freedom of normalization. We take, for example

$$G_\pm^{rr}(x;k) = \frac{1}{2\pi\mathrm{i}}\int_{-\infty}^{\infty} \frac{\mathrm{e}^{\mathrm{i}px}}{p+\mathrm{i}0}\,\mathrm{d}p = -\theta(x). \tag{3.1.34}$$

Evaluating $G_\pm^{rs}(x;k)$

$$G_\pm^{rs}(x;k) = \begin{cases} \pm\exp\{\mathrm{i}kx(J^r - J^s)\}\,\theta(\pm x). & \text{if } r > s,\\ \mp\exp\{\mathrm{i}kx(J^r - J^s)\}\,\theta(\mp x), & \text{if } r < s,\end{cases} \tag{3.1.35}$$

and so (3.1.31) becomes

$$\begin{aligned}\mathbf{m}_\pm(x;k) = \mathbf{I} &+ \int_{-\infty}^{x} \exp\{\mathrm{i}k(x-\xi)\hat{\mathbf{J}}\}(\boldsymbol{\pi}_0 + \boldsymbol{\pi}_\mp)(\mathbf{Qm}_\pm)(\xi;k)\,\mathrm{d}\xi\\ &- \int_{x}^{\infty} \exp\{\mathrm{i}k(x-\xi)\hat{\mathbf{J}}\}\boldsymbol{\pi}_\pm(\mathbf{Qm}_\pm)(\xi;k)\,\mathrm{d}\xi,\end{aligned} \tag{3.1.36}$$

where $\boldsymbol{\pi}_0\mathbf{M}$, $\boldsymbol{\pi}_+\mathbf{M}$ and $\boldsymbol{\pi}_-\mathbf{M}$ denote the diagonal, strictly upper and strictly lower diagonal parts of the matrix $\mathbf{M}$ and

$$\exp\{\mathrm{i}kx\hat{\mathbf{J}}\}\mathbf{F} = \exp\{\mathrm{i}kx\mathbf{J}\}\mathbf{F}\exp\{-\mathrm{i}kx\mathbf{J}\}.$$

The 2×2 case above differs from the 2×2 case discussed earlier (§3.1.1) in that the normalization is different. Namely

$$\mathbf{m}(x;k) \equiv \begin{pmatrix} m^{11}(x;k) & m^{12}(x;k) \\ m^{21}(x;k) & m^{22}(x;k) \end{pmatrix} = \begin{pmatrix} 1 & 0 \\ 0 & 1 \end{pmatrix}$$
$$+ \int_{-\infty}^{x} \begin{pmatrix} q(\xi)m_{+}^{21}(\xi;k) & 0 \\ \exp\{ik(J^2 - J^1)(x-\xi)\}r(\xi)m_{+}^{11}(\xi;k) & r(\xi)m_{+}^{12}(\xi;k) \end{pmatrix} d\xi$$
$$- \int_{x}^{\infty} \begin{pmatrix} 0 & q(\xi)m_{+}^{22}(\xi;k)\exp\{ik(J^1 - J^2)(x-\xi)\} \\ 0 & 0 \end{pmatrix} d\xi. \tag{3.1.37}$$

So columnwise if $J^1 = -1$, $J^2 = 1$

$$\begin{pmatrix} m_{+}^{11}(x;k) \\ m_{+}^{21}(x;k) \end{pmatrix} = \begin{pmatrix} 1 \\ 0 \end{pmatrix} + \int_{-\infty}^{x} \begin{pmatrix} q(\xi)m_{+}^{21}(\xi;k) \\ \exp\{-2ik(x-\xi)\}r(\xi)m_{+}^{11}(\xi;k) \end{pmatrix} d\xi, \tag{3.1.38a}$$

$$\begin{pmatrix} m_{+}^{12}(x;k) \\ m_{+}^{22}(x;k) \end{pmatrix} = \begin{pmatrix} 0 \\ 1 + \int_{-\infty}^{\infty} r(\xi)m_{+}^{12}(\xi;k)\, d\xi \end{pmatrix}$$
$$- \int_{x}^{\infty} \begin{pmatrix} q(\xi)m_{+}^{22}(\xi;k)\exp\{2ik(x-\xi)\} \\ r(\xi)m_{+}^{12}(\xi;k) \end{pmatrix} d\xi, \tag{3.1.38b}$$

Clearly (3.1.38a) is equivalent to $\boldsymbol{M}(x;k)$ given by (3.1.7a), but (3.1.38b) is not the same as $\boldsymbol{N}(x;k)$ given by (3.1.7b) — the difference due to the normalization chosen here. Indeed the equations (3.1.31) with (3.1.33-35) or (3,1.36) are Fredholm integral equations as is (3.1.38b); whereas equations (3.1.7a-d) are all Volterra type. The mechanism for introducing bound states is now via poles due to the Fredholm nature of $\mathbf{m}_{\pm}(x;k)$ (i.e., the Fredholm determinant vanishes) and not due to scattering coefficients vanishing per se. In the $N \times N$ case, Beals and Coifman [1984] show how scattering eigenfunctions (satisfying Volterra equations and consequently not having nice analyticity properties) are related to analytic eigenfunctions.

The integral equations governing the eigenfunctions $\mathbf{m}_{\pm}(x;k)$ produce two fundamental matrix solutions to equation (3.1.29). Consequently we have that they are linearly dependent upon each other, hence we have

$$(\mathbf{m}_{+} - \mathbf{m}_{-})(x;k) = \mathbf{m}_{-}(x;k)\,\mathbf{V}(x;k), \tag{3.1.39a}$$

where

$$\mathbf{V}(x;k) = \exp(ikx\mathbf{J})\mathbf{v}(k)\exp(-ikx\mathbf{J}) \tag{3.1.39b}$$

We note any independent solution $\tilde{\mathbf{m}}(x;k)$, given another solution $\mathbf{m}(x;k)$ of (3.1.4), is given by

$$\tilde{\mathbf{m}}(x;k) = \mathbf{m}(x;k)\exp(ikx\mathbf{J})\mathbf{A}(k)\exp(-ikx\mathbf{J}),$$

for an arbitrary matrix $\mathbf{A}(k)$. (3.1.39) generalizes to the $N \times N$ case the results of the 2×2 case (3.1.19). We can derive integral expressions for $\mathbf{V}(x;k)$ in a manner similar to that described earlier. Since the calculation is straightforward we will not do this here. However it should be noted that there is less compelling need to do this in the $N \times N$ case since unlike $a(k)$, $\bar{a}(k)$, which in the 2×2 case give information regarding the location of the bound states, here $\mathbf{V}(x;k)$ plays the role of the scattering data — but does not immediately yield this information. Rather the location of the poles is dictated by the locations of the homogeneous solutions of (3.1.31). It is worth noting that the kernels of the Fredholm integral equations are analytic and are compact operators (see Beals and Coifman [1984]). Consequently the solutions $\mathbf{m}_\pm(x;k)$ are meromorphic in the upper/lower half k-planes. Indeed equation (3.1.39) defines $\mathbf{V}(x;k)$ if we know $\mathbf{m}_\pm(x;k)$, which is in fact known at $t=0$. The evolution of $\mathbf{V}(x;k)$ as it relates to a nonlinear evolution equation is elementary. Once we have equation (3.1.39) then $\mathbf{m}_\pm$ can be computed by taking projections of (3.1.39) after subtracting out any contributions due to poles. (We use $\mathbf{m}_\pm \to \mathbf{I}$ as $|k| \to \infty$.)

An alternative method to develop an integral equation for the inverse problem governing the eigenfunctions is to compute the $\bar{\partial}$ derivative of $\mathbf{m}(x;k)$ in the entire k-plane. The eigenfunction $\mathbf{m}(x;k)$ has the following properties: (i), it is meromorphic in the upper/lower half k-planes possessing a jump across $\operatorname{Im} k = 0$; (ii), it has poles at $k = k_j$ with residue $\mathbf{m}_j(x)$. We take these poles as simple — obtaining multiple poles as a limit of the coalescence of simple poles. Consequently the $\bar{\partial}$ derivative of $\mathbf{m}(x;k)$ satisfies

$$\frac{\partial \mathbf{m}}{\partial \bar{k}} = (\mathbf{m}_+ - \mathbf{m}_-)(x;k)\,\mu(k,\xi) + \sum_{j=1}^{N} \pi \mathbf{m}_j(x)\delta(k-k_j), \qquad \text{for} \qquad \operatorname{Im} k = 0, \tag{3.1.40}$$

where

$$\mu(k,\xi) = \begin{cases} 1 & \text{if } k=\xi, \text{ real}, \\ 0 & \text{otherwise}, \end{cases} \qquad \mathbf{m}_j(x) = \mathbf{m}(x;k_j), \qquad k = k_R + \mathrm{i}k_I,$$

$$\frac{\partial}{\partial \bar{k}} = \frac{1}{2}\left(\frac{\partial}{\partial k_R} + \mathrm{i}\frac{\partial}{\partial k_I}\right), \tag{3.1.41a}$$

and the relationship (proven by integration over a small circle about $k = k_j$ against an analytic test function in the neighborhood of k_j)

$$\frac{\partial}{\partial \bar{k}}\left(\frac{1}{k-k_j}\right) = \pi\delta(k-k_j). \tag{3.1.41b}$$

On the real k axis we have (from equations (3.1.37-39)) that

$$(\mathbf{m}_+ - \mathbf{m}_-)(x;k) = \mathbf{m}_-(x;k)\,\mathbf{V}(x;k), \tag{3.1.42a}$$

where

$$\mathbf{V}(x;k) = \exp(\mathrm{i}kx\mathbf{J})\mathbf{v}(k)\exp(-\mathrm{i}kx\mathbf{J}). \tag{3.1.42b}$$

$\mathbf{v}(k)$ plays the role of the scattering data, i.e., it is analogous to the reflection coefficient — but in this case is not obtained from physical scattering (Volterra equations). The inverse problem is governed by an integral equation derived from the well known generalized Cauchy integral formula (cf. Hörmander [1973])

$$\mathbf{m}(x;k) = \frac{1}{2\pi \mathrm{i}} \iint_R \frac{\partial \mathbf{m}}{\partial \bar{z}}(x;z) \frac{\mathrm{d}z \wedge \mathrm{d}\bar{z}}{z-k} + \frac{1}{2\pi \mathrm{i}} \int_C \frac{\mathbf{m}(x;z)}{z-k} \mathrm{d}z, \tag{3.1.43}$$

where R and C are an appropriate region and contour in the z-plane, $z = z_R + \mathrm{i}z_I$ and $\mathrm{d}z \wedge \mathrm{d}\bar{z} = -2\mathrm{i}\mathrm{d}z_R \mathrm{d}z_I$. In this problem we shall take R to be the entire complex plane R_∞ and C an "infinite" large circle in the plane. The normalization $\mathbf{m} \to \mathbf{I}$ as $k \to \infty$ is consistent with the solutions of $\mathbf{m}_\pm$ via the integral equations (3.1.31). Using equations (3.1.40,42) in (3.1.43) we have

$$\mathbf{m}(x;k) = \mathbf{I} + \sum_{j=1}^{N} \frac{\mathbf{m}_j(x)}{k-k_j} + \frac{1}{2\pi \mathrm{i}} \int_{-\infty}^{\infty} \frac{\mathbf{m}_-(x;z)\,\mathbf{v}(x;z)}{z-k} \mathrm{d}z. \tag{3.1.44}$$

The relevant Riemann-Hilbert boundary value problem is obtained from equation (3.1.44) by taking $\operatorname{Im} k \to \pm 0$ namely

$$\mathbf{m}_\pm(x;\xi) = \mathbf{I} + \sum_{j=1}^{N} \frac{\mathbf{m}_j(x)}{\xi-k_j} + \frac{1}{2\pi \mathrm{i}} \int_{-\infty}^{\infty} \frac{\mathbf{m}_-(x;z)\,\mathbf{v}(x;z)}{z-(\xi \pm \mathrm{i}0)} \mathrm{d}z. \tag{3.1.45}$$

The system is closed by evaluating (3.1.44) at the location of the poles $k = k_j$

$$\mathbf{0} = \mathbf{I} + \sum_{\substack{j=1 \\ j\neq \ell}}^{n} \frac{\mathbf{m}_j(x)}{k_\ell-k_j} + \frac{1}{2\pi \mathrm{i}} \int_{-\infty}^{\infty} \frac{\mathbf{m}_-(x;z)\,\mathbf{v}(x;z)}{z-k_\ell} \mathrm{d}z. \tag{3.1.46}$$

Finally, the potential $\mathbf{Q}(x)$ is reconstructed by comparing the asymptotics of the integral equation (3.1.31) as $|k| \to \infty$ (or the differential equation) with that of (3.1.44) to obtain

$$\mathbf{Q}(x) = \left[\mathbf{J}, \frac{1}{2\pi \mathrm{i}} \int_{-\infty}^{\infty} \mathbf{m}_-(x;k)\,\mathbf{V}(x;k)\,\mathrm{d}k - \sum_{j=1}^{N} \mathbf{m}_j(x)\right]. \tag{3.1.47}$$

3.1.3 The time dependence. In order to solve the initial value problem for an associated nonlinear evolution equation, we need to determine the time dependence of the scattering data. This is usually straightforward since the scattering data evolves simply.

Suppose that the associated time-dependent part of the Lax pair is $\boldsymbol{v}_t = \mathbf{M}\boldsymbol{v}$, and we assume that $\mathbf{M} \to \mathbf{M}_0$, as $|x| \to \infty$, where $\mathbf{M}_0$ is a constant real matrix and $\mathbf{M}_0\mathbf{J} = \mathbf{J}\mathbf{M}_0$. Now let $\boldsymbol{v} = \exp\{\mathrm{i}kx\mathbf{J} - \mathrm{i}kt\mathbf{M}_0\}\boldsymbol{w}$, then $\boldsymbol{w}$ satisfies

$$\boldsymbol{w}_t = (\mathbf{M} - \mathrm{i}k\mathbf{M}_0)\boldsymbol{w}. \tag{3.1.48}$$

Applying the operator $\mathcal{M} := \partial/\partial t - \mathbf{M}$ to the scattering equation (3.1.1) and using (3.1.48), it follows that

$$\mathbf{v}_t(k;t) = \mathrm{i}k(\mathbf{M}_0\mathbf{v} - \mathbf{v}\mathbf{M}_0),$$

and so the evolution of the scattering data is given by

$$\mathbf{v}(k;t) = \exp\{\mathrm{i}kt\mathbf{M}_0\}\mathbf{v}(k;0)\exp\{-\mathrm{i}kt\mathbf{M}_0\}. \tag{3.1.49}$$

The 2×2 scattering problem (3.1.1) is of course a special case of these more general result, but we again remark on the difference (due to normalization) between $\mathbf{m}_\pm$ satisfying Fredholm integral equations as defined here and the ones used standardly in the 2×2 problem.

The following general Riemann-Hilbert boundary value problem is the inverse problem associated with not only (3.1.2), but also for Nth order Sturm Liouville problems (see below), a special case of which is the second order Schrödinger operators discussed earlier;

$$(\mathbf{m}_+ - \mathbf{m}_-)(x,t;k) = \mathbf{m}_-(x,t;\alpha(k))\mathbf{V}(x,t;k), \qquad \text{on} \qquad \Sigma, \tag{3.1.50a}$$

$$\mathbf{m}(x,t;k) \to \mathbf{I}, \qquad \text{as} \quad |k| \to \infty, \tag{3.1.50b}$$

where $\alpha(k)$ and $\mathbf{V}(x,t;k)$ given on Σ in terms of the scattering data. The formal I.S.T. scheme is:

(1), Direct problem, given $\boldsymbol{u}(x,0) \in \mathcal{P}$, determine $\mathbf{m}_\pm(x,0;k)$ by solving the integral equation governing the direct problem (e.g. (3.1.36) solves (3.1.2)). The normalization $\mathbf{m}_\pm \to \mathbf{I}$, as $|k| \to \infty$ is useful.

(2), Using equation (3.1.50) determine $\mathbf{V}(x,0;k) \in S$.

(3), Time evolution, determine the time evolution of $\mathbf{V}(x,t;k)$ from the time part of the Lax pair.

(4), Inverse problem, given $\mathbf{V}(x,t;k) \in S$, the space of scattering functions, solve (3.1.50) for $\mathbf{m}_\pm(x,t;k)$.

(5), Determine $\boldsymbol{u}(x,t)$ from $\mathbf{m}_\pm(x,t;k)$.

This may be written schematically as follows:

$$\begin{array}{ccccc}
\boldsymbol{u}(x,0) & \xrightarrow[\text{From L operator}]{\text{Direct Problem}} & \mathbf{m}_\pm(x,0;k) & \longrightarrow & \mathbf{V}(x,0;k) \\
 & & & & \Big\downarrow \text{From } \mathcal{M} \text{ operator} \\
\boldsymbol{u}(x,t) & \longleftarrow & \mathbf{m}_\pm(x,t;k) & \xleftarrow[\text{via RHBVP or } \bar\partial]{\text{Inverse Problem}} & \mathbf{V}(x,t;k)
\end{array}$$

We shall discuss the $\bar\partial$ problem further in Chapters 5 and 6 in connection with the Kadomtsev-Petviashvili equation, and inverse scattering in general.

3.1.4 Hamiltonian System and Action-Angle Variables for the Nonlinear Schrödinger Equation. In §2.6.4 above we discussed the Hamiltonian structure and action-angle variables for the KdV equation. In this section we discuss the Hamiltonian structure and action-angle variables for the nonlinear Schrödinger equation

$$\mathrm{i}q_t + q_{xx} \pm 2|q|^2 q = 0, \tag{3.1.51}$$

which for convenience, we write as the system

$$\mathrm{i}q_t + q_{xx} - 2q^2 r = 0, \tag{3.1.52a}$$

$$\mathrm{i}r_t - r_{xx} + 2qr^2 = 0; \tag{3.1.52b}$$

we shall assume that $r(x,0) = \mp q^*(x,0) \in L^1(\mathbb{R})$. The system (3.1.52) is Hamiltonian, as may be seen by the identification

$$\text{coordinates } (q): \quad q(x,t), \tag{3.1.53a}$$

$$\text{momenta } (p): \quad r(x,t), \tag{3.1.53b}$$

$$\text{Hamiltonian } (H): \quad -\mathrm{i}\int_{-\infty}^{\infty} \{q_x r_x + (qr)^2\}\,\mathrm{d}x, \tag{3.1.53c}$$

together with the Poisson brackets

$$\{A(\alpha), B(\beta)\} \equiv \int_{-\infty}^{\infty} \left\{ \frac{\delta A(\alpha)}{\delta q(x)} \frac{\delta B(\beta)}{\delta p(x)} - \frac{\delta A(\beta)}{\delta p(x)} \frac{\delta B(\alpha)}{\delta q(x)} \right\} \mathrm{d}x. \tag{3.1.54}$$

Alternatively, we may assert that (3.1.51) is Hamiltonian, and identify coordinates $[q(x,t)]$, momenta $[q^*(x,t)]$ and a Hamiltonian $[-\mathrm{i}\int_{-\infty}^{\infty}\{\pm|q_x|^2 + |q|^4\}\,\mathrm{d}x]$ (provided that $q(x,t)$ and $q^*(x,t)$ are regarded as being independent functions).

Suppose that $\boldsymbol{\phi}(x;k)$ and $\boldsymbol{\psi}(x;k)$ are the eigenfunction of the 2×2 scattering problem (3.1.1), which tend to $\mathrm{e}^{-\mathrm{i}kx}\begin{pmatrix}1\\0\end{pmatrix}$ as $x \to -\infty$ and $\mathrm{e}^{\mathrm{i}kx}\begin{pmatrix}0\\1\end{pmatrix}$ as $x \to \infty$, respectively. Then we have seen that their components $\phi_1(x;k)$, $\phi_2(x;k)$ and $\psi_1(x;k)$, $\psi_2(x;k)$ satisfy the integral equations

$$\phi_1(x;k) = \mathrm{e}^{-\mathrm{i}kx}\left\{1 + \int_{-\infty}^{x} q(\xi)\phi_2(\xi;k)\mathrm{e}^{\mathrm{i}k\xi}\,\mathrm{d}\xi\right\}, \tag{3.1.55a}$$

$$\phi_2(x;k) = \mathrm{e}^{\mathrm{i}kx}\int_{-\infty}^{x} r(\xi)\phi_1(\xi;k)\mathrm{e}^{-\mathrm{i}k\xi}\,\mathrm{d}\xi, \tag{3.1.55b}$$

$$\psi_1(x;k) = -\mathrm{e}^{-\mathrm{i}kx}\int_{x}^{\infty} q(\xi)\psi_2(\xi;k)\mathrm{e}^{\mathrm{i}k\xi}\,\mathrm{d}\xi, \tag{3.1.56a}$$

$$\psi_2(x;k) = \mathrm{e}^{\mathrm{i}kx}\left\{1 - \int_{x}^{\infty} r(\xi)\psi_1(\xi;k)\mathrm{e}^{\mathrm{i}k\xi}\,\mathrm{d}\xi\right\} \tag{3.1.56b}$$

(recall equations (3.1.7)).

Further using the results of §3.1.1 that the scattering data $a(k)$, $b(k)$ are given by the integral representations

$$a(k) = 1 + \int_{-\infty}^{\infty} q(\xi)\phi_2(\xi;k)e^{ik\xi}\,d\xi, \tag{3.1.57a}$$

$$b(k) = -\int_{-\infty}^{\infty} r(\xi)\phi_1(\xi;k)e^{-ik\xi}\,d\xi \tag{3.1.57b}$$

(recall equations (3.1.16)); $a(k)$ and $b(k)$ also possess the Wronskian representation

$$a(k) = W(\boldsymbol{\phi}, \boldsymbol{\psi}), \qquad b(k) = -W(\boldsymbol{\phi}, \bar{\boldsymbol{\psi}}), \tag{3.1.58}$$

where $W(\boldsymbol{\phi}, \boldsymbol{\psi}) = \phi_1\psi_2 - \phi_2\psi_1$. For the sake of simplicity, we shall only consider the continuous scattering data, i.e., we assume that there is no contribution from the discrete scattering data.

From equation (3.1.55a) we have

$$\frac{\delta\phi_1(x;k)}{\delta q(x)} = \theta(x-\xi)\phi_2(x;k)e^{ik(\xi-x)},$$

where $\theta(x)$ is the Heaviside step function. For and function $f(x)$ that is differentiable except at finitely many points, we define

$$\frac{\delta f(x)}{\delta q(x)} = \lim_{\xi\uparrow x}\frac{\delta f(y)}{\delta q(x)},$$

so that

$$\frac{\delta\phi_1(x;k)}{\delta q(x)} = \phi_2(x;k). \tag{3.1.59a}$$

Similarly,

$$\frac{\delta\phi_2(x;k)}{\delta r(x)} = \phi_1(x;k), \qquad \frac{\delta\phi_1(x;k)}{\delta r(x)} = 0, \qquad \frac{\delta\phi_2(x;k)}{\delta q(x)} = 0, \tag{3.1.59b}$$

and

$$\frac{\delta\psi_1(x;k)}{\delta q(x)} = 0, \qquad \frac{\delta\psi_1(x;k)}{\delta r(x)} = 0, \qquad \frac{\delta\psi_2(x;k)}{\delta q(x)} = 0, \qquad \frac{\delta\psi_2(x;k)}{\delta r(x)} = 0. \tag{3.1.60}$$

Hence from equations (3.1.58) (or from (3.1.57) with some further calculations), we see that

$$\frac{\delta a(k)}{\delta q(x)} = \frac{\delta}{\delta r(x)}\{\phi_1\psi_2 - \phi_2\psi_1\} = \phi_2(x;k)\psi_2(x;k), \tag{3.1.61a}$$

$$\frac{\delta a(k)}{\delta r(x)} = -\phi_1(x;k)\psi_1(x;k). \tag{3.1.61b}$$

Similarly

$$\frac{\delta b(k)}{\delta q(x)} = -\phi_2(x;k)\bar{\psi}_2(x;k), \qquad \frac{\delta b(k)}{\delta r(x)} = \phi_1(x;k)\bar{\psi}_1(x;k). \tag{3.1.62}$$

Hence from equations (3.1.61,61), using the equations satisfied by the eigenfunctions, $\phi_j(x;k)$, $\psi_j(x;k)$, $\bar{\psi}_j(x;k)$, $j=1,2$ together with their asymptotics, we find that

$$\{a(k), b(\ell)\} = \frac{a(k)b(\ell)}{2(k-\ell)} - \tfrac{1}{2}\pi \mathrm{i} a(k)b(\ell)\delta(k-\ell). \tag{3.1.63}$$

For the system (3.1.52), we define

$$P(k) = \frac{1}{\pi}\ln|a(k)|^2, \qquad Q(k) = \tfrac{1}{2}\mathrm{i}\ln\left(\frac{b(k)}{\bar{b}(k)}\right), \tag{3.1.64}$$

with $a(k)$, $b(k)$ as defined in equations (3.1.57). Then using equation (3.1.63), it can be shown that $P(k)$ and $Q(k)$ are canonical, i.e., satisfy

$$\{P(k), P(\ell)\} = 0, \qquad \{Q(k), Q(\ell)\} = 0, \quad \{Q(k), P(\ell)\} = \delta(k-\ell), \tag{3.1.65}$$

and so $P(k)$ and $Q(k)$ are action-angle coordinates. Furthermore, in terms of these coordinates, the Hamiltonian is given by

$$H = 4\int_{-\infty}^{\infty} k^2 P(k)\,\mathrm{d}k. \tag{3.1.66}$$

Therefore we have

$$\frac{\partial P}{\partial t} = \{P, H\} = \frac{\delta H}{\delta Q} = 0, \tag{3.1.67a}$$

$$\frac{\partial Q}{\partial t} = \{Q, H\} = \frac{\delta H}{\delta P} = 4k^2. \tag{3.1.67b}$$

(For further details, including the contribution from the discrete scattering data, see Ablowitz and Segur [1981]; Novikov, Manakov, Pitaevskii and Zakharov [1984].)

3.1.5 Riemann-Hilbert Problems for Nth order Sturm-Liouville Scattering Problems. We remark that the Nth order Sturm-Liouville scattering problem

$$\frac{\mathrm{d}^N v}{\mathrm{d}x^N} + \sum_{j=2}^{N} u_j(x)\frac{\mathrm{d}^{N-j}v}{\mathrm{d}x^{N-J}} = z^n v, \tag{3.1.71}$$

has been discussed in detail by Beals [1985] and in the monograph by Beals, Deift and Tomei [1988]. Included amongst the evolution equations associated scattering problem

(3.1.71), for $N = 3$, are:
(a) The Boussinesq equation

$$u_{tt} = u_{xx} + 3(u^2)_{xx} + u_{xxxx}, \tag{3.1.72}$$

(Zakharov [1974]; Ablowitz and Haberman [1976b]; Caudrey [1980, 1982]; Deift, Tomei and Trubowitz [1982]); the associated scattering problem is

$$\mathrm{L}_1 v = 4v_{xxx} + (1 + 6u)v_x + (3u_x - \mathrm{i}\sqrt{3}w_t)v = \lambda v, \tag{3.1.73a}$$

$$v_t = \mathrm{M}_1 = \mathrm{i}\sqrt{3}(v_{xx} + uv), \tag{3.1.73b}$$

with $u + w_x$. We note that there is another version of the Boussinesq equation which is well posed; it is obtained from (3.1.72) by changing th sign of the last term. However in the physical problem arising from a long wave approximation, (3.1.72) is the one that arises.
(b) Two fifth order KdV type equations:

$$u_t + u_{xxxxx} + 30uu_{xxx} + 30u_x u_{xx} + 180u^2 u_x = 0 \tag{3.1.74}$$

(Sawada and Kotera [1974]; Caudrey, Dodd and Gibbon [1976] — see also Dodd and Gibbon [1977]; Satsuma and Kaup [1978]) and

$$u_t = u_{xxxxx} + 30uu_{xxx} + 45u_x u_{xx} + 180u^2 u_x. \tag{3.1.75}$$

(Kaup [1980b] — see also Gibbon [1985]); the associated scattering problems are

$$\mathrm{L}_2 v = v_{xxx} + 6uv_x = \lambda v, \tag{3.1.76a}$$

$$v_t = \mathrm{M}_2 = (9\lambda - 18u_x)v_{xx} + 6(u_{xx} - 6u^2)v_x + 36\lambda uv, \tag{3.1.76b}$$

and

$$\mathrm{L}_3 v = v_{xxx} + 6uv_x + 3u_x v = \lambda v, \tag{3.1.77a}$$

$$v_t = \mathrm{M}_3 = 9\lambda v_{xx} - 3(u_{xx} + 12u^2)v_x + 3(u_{xxx} + 12u\lambda + 24uu_x)v, \tag{3.1.77b}$$

respectively.

We remark that equations (3.1.74) and (3.1.75) are related by a Miura type transformation (Fordy and Gibbons [1980a]; Hirota and Ramani [1980]).

The study of the scattering problem (3.1.71) can be reduced to the study $N \times N$ systems. Beals [1985] and Beals, Deift and Tomei [1988] show that one can construct suitable sectionally meromorphic functions and hence obtain a Riemann-Hilbert boundary value problem. Several of the results for $N \times N$ systems (Beals and Coifman [1981, 1982, 1984]) are applicable, however the point $z = 0$ poses some extra technical difficulties in the determination of the required analyticity properties.

3.2 Riemann-Hilbert Problems for Discrete Scattering Problems.

3.2.1 Differential-Difference Equations: Discrete Schrödinger Scattering Problem. In this section we briefly discuss the Riemann-Hilbert boundary value problem associated with the discrete Schrödinger scattering problem

$$\alpha_n v_{n+1} + \alpha_{n-1} v_{n-1} + \beta_n v_n = k v_n, \tag{3.2.1}$$

which is the scattering operator associated with the Toda Lattice (Toda [1967, 1975, 1976])

$$\frac{\partial^2 u_n}{\partial t^2} = \exp\{-(u_n - u_{n-1})\} - \exp\{-(u_{n+1} - u_n)\}, \tag{3.2.2}$$

with

$$\alpha_n = \tfrac{1}{2}\exp\left\{-\tfrac{1}{2}(u_n - u_{n-1})\right\}, \qquad \beta_n = -\tfrac{1}{2}\frac{\partial u_{n-1}}{\partial t}. \tag{3.2.3}$$

The I.S.T. scheme for the Toda Lattice equation was developed by Flaschka [1974a,b] and Manakov [1975].

For the scattering problem (3.2.1), define the eigenfunctions with the following asymptotic behaviors

$$\phi_n(z) \sim z^{-n}, \qquad\qquad \text{as} \quad n \to -\infty, \tag{3.2.4a}$$

$$\psi_n(z) \sim z^n, \qquad \bar{\psi}_n(z) \sim z^{-n}, \qquad \text{as} \quad n \to \infty, \tag{3.2.4b}$$

where we have assumed that $\alpha_n \to \frac{1}{2}$ and $\beta_n \to 0$ sufficiently rapidly as $|n| \to \infty$, and $k = \frac{1}{2}(z + z^{-1})$. The completeness relation of the eigenfunctions is

$$\phi_n(z) = a(z)\bar{\psi}_n(z) + b(z)\psi_n(z), \qquad \text{on} \qquad |z| = 1. \tag{3.2.5}$$

Now define the eigenfunctions

$$M_n(z) := \phi_n(z) z^n, \qquad N_n(z) := \psi_n(z) z^{-n}, \qquad \bar{N}_n(z) := \bar{\psi}_n(z) z^n.$$

Then by analysis which is similar to that for the KdV equation outlined in Chapter 2 above, it can be shown that:

(i), $M_n(z)$, $N_n(z)$ and $a(z)$ are analytic for $|z| > 1$;

(ii), $\bar{N}_n(z)$ is analytic for $|z| < 1$.

The symmetry condition is

$$\psi_n(z) = \bar{\psi}_n(z^{-1}),$$

that is

$$N_n(z) = \bar{N}_n(z^{-1}) z^{-2n},$$

and hence the Riemann-Hilbert boundary value problem associated with the scattering operator (3.2.1) is

$$\frac{M_n(z)}{a(z)} = \bar{N}_n(z) + \rho(z) z^{-2n} \bar{N}_n(z^{-1}), \qquad \text{on} \qquad |z| = 1, \tag{3.2.6}$$

where

$$\rho(z) := \frac{b(z)}{a(z)},$$

and we have the following asymptotic relations

$$\lim_{z \to \infty} \left(\frac{M_n(z)}{a(z)} \right) = 1, \qquad \lim_{z \to 0} \left(\bar{N}_n(z) \right) = 1.$$

Nonlinear wave equations are solved by considering appropriate associated time dependence for the eigenfunctions which induces time dependence for $\rho(z)$. For example, the time dependent scattering problem associated to the Toda Lattice (3.2.1) is given by

$$\frac{\partial v_n}{\partial t} = 2\alpha_n v_{n+1} + \left(\sum_{k=-\infty}^{n} (\ln \alpha_{k-1})_t + \gamma \right) v_n, \tag{3.2.7}$$

where γ is to be determined. Substituting equation (3.2.6a) into (3.2.7) and taking the limit as $n \to -\infty$ (using the fact that $\alpha_n \to \frac{1}{2}$ as $|n| \to \infty$), shows that

$$z^{-n-1} + \gamma z^{-n} \sim 0, \qquad \text{on} \quad |z| = 1. \tag{3.2.8}$$

Therefore for consistency we choose $\gamma = -z^{-1}$. From equations (3.2.4,5) we see that

$$\phi_n(z) \sim a(z) z^{-n} + b(z) z^n, \qquad \text{on} \quad |z| = 1. \tag{3.2.9}$$

Hence substituting (3.2.9) into (3.2.7) with $\gamma = -z^{-1}$ and taking the limit as $n \to +\infty$ yields

$$\frac{\partial a}{\partial t} z^{-n} + \frac{\partial b}{\partial t} z^n \sim b(z - z^{-1}) z^n,$$

i.e.,

$$\frac{\partial a}{\partial t} = 0, \qquad \frac{\partial b}{\partial t} = (z - z^{-1}) b, \tag{3.2.10a}$$

and so

$$a(z;t) = a(z;0), \qquad b(z;t) = b(z;0) \exp\{(z - z^{-1})t\}.$$

Therefore the appropriate time dependence for the Toda Lattice is given by

$$\rho(z;t) = \rho(z;0) \exp\{(z - z^{-1})t\}, \tag{3.2.10b}$$

whereas other time dependencies give different nonlinear evolution equations. In a similar manner to that for the KdV equation, the zeros of $a(z)$, or the poles of the meromorphic function $M_n(z)/a(z)$, give rise to the soliton solutions (pure soliton solutions correspond to the case where $\rho(z;t) = 0$). A one-soliton solution of the Toda Lattice equation (3.2.1), corresponding to an eigenvalue $z = \sigma e^{-W}$, with $\sigma = \pm 1$, is given by

$$\exp\{-(u_n - u_{n-1})\} = 1 + \sinh^2 W \operatorname{sech}^2(nW + \sigma t \sinh W + W_0), \tag{3.2.11}$$

with W_0 a constant.

3.2.2 Differential-Difference Equations: Discrete 2 × 2 Scattering Problem. In this section we briefly discuss the Riemann-Hilbert boundary value problem associated with the discrete 2×2 scattering problem

$$v_{1,n+1} = zv_{1,n} - \sigma u_n^* v_{2,n}, \tag{3.2.12a}$$
$$v_{2,n+1} = z^{-1} v_{2,n} + u_n v_{1,n}. \tag{3.2.12b}$$

which is the scattering operators associated with the discrete nonlinear Schrödinger equation

$$\mathrm{i}\frac{\partial u_n}{\partial t} = (u_{n+1} + u_{n-1} - 2u_n) + \sigma u_n u_n^*(u_{n+1} + u_{n-1}), \tag{3.2.13}$$

where $\sigma = \pm 1$. The I.S.T. scheme for the discrete nonlinear Schrödinger equation was developed Ablowitz and Ladik [1975, 1976a].

The analysis for the scattering problem (3.2.12) is similar to that for the discrete Schrödinger scattering problem (3.2.1). If we assume that $\lim_{|n|\to\infty} u_n = 0$ and $\lim_{|n|\to\infty} u_n^* = 0$ sufficiently rapidly, then the eigenfunctions are defined by the asymptotic conditions,

$$\boldsymbol{\phi}_n(z) \sim \begin{pmatrix}1\\0\end{pmatrix} z^n, \qquad \bar{\boldsymbol{\phi}}_n(z) \sim \begin{pmatrix}0\\-1\end{pmatrix} z^{-n}, \qquad \text{as} \quad n \to -\infty, \tag{3.2.14a}$$
$$\boldsymbol{\psi}_n(z) \sim \begin{pmatrix}0\\1\end{pmatrix} z^{-n}, \qquad \bar{\boldsymbol{\psi}}_n(z) \sim \begin{pmatrix}1\\0\end{pmatrix} z^{n}, \qquad \text{as} \quad n \to \infty. \tag{3.2.14b}$$

The completeness of these eigenfunctions is given by

$$\boldsymbol{\phi}_n(z) = a(z)\bar{\boldsymbol{\psi}}_n(z) + b(z)\boldsymbol{\psi}_n(z), \tag{3.2.15a}$$
$$\bar{\boldsymbol{\phi}}_n(z) = \bar{a}(z)\boldsymbol{\psi}_n(z) + \bar{b}(z)\bar{\boldsymbol{\psi}}_n(z). \tag{3.2.15b}$$

Now define

$$\boldsymbol{M}_n(z) := \boldsymbol{\phi}_n(z) z^{-n}, \qquad \bar{\boldsymbol{M}}_n(z) := \bar{\boldsymbol{\phi}}_n(z) z^{n}, \tag{3.2.16a}$$
$$\boldsymbol{N}_n(z) := \boldsymbol{\psi}_n(z) z^{n}, \qquad \bar{\boldsymbol{N}}_n(z) := \bar{\boldsymbol{\psi}}_n(z) z^{-n}. \tag{3.2.16b}$$

The analyticity properties are:

(i), $\boldsymbol{M}_n(z), \boldsymbol{N}_n(z)$ and $a(z)$ are analytic for $|z| > 1$;

(ii), $\bar{\boldsymbol{M}}_n(z), \bar{\boldsymbol{N}}_n(z)$ and $\bar{a}(z)$ are analytic for $|z| < 1$.

The Riemann-Hilbert boundary value problem associated with the scattering operator (3.2.12) is on $|z| = 1$

$$\frac{\boldsymbol{M}_n(z)}{a(z)} = \bar{\boldsymbol{N}}_n(z) + \rho(z) z^{-2n} \boldsymbol{N}_n(z), \tag{3.2.17a}$$

$$\frac{\bar{\boldsymbol{M}}_n(z)}{\bar{a}(z)} = -\boldsymbol{N}_n(z) + \bar{\rho}(z) z^{2n} \bar{\boldsymbol{N}}_n(z), \tag{3.2.17b}$$

where $\rho(z) := b(z)/a(z)$, $\bar{\rho}(z) := \bar{b}(z)/\bar{a}(z)$ and $\bar{\boldsymbol{M}}_n(z)$, $\bar{\boldsymbol{N}}_n(z)$, $\bar{a}(z)$ all have constant asymptotic values (with respect to z) as $|z| \to \infty$, and similarly for $\boldsymbol{M}_n(z)$, $\boldsymbol{N}_n(z)$, $a(z)$ as $z \to 0$.

Nonlinear wave equations are solved by considering associated time dependence of the eigenfunctions which induces time dependence for $\rho(z)$ and $\bar{\rho}(z)$. The time dependent scattering problem associated with the discrete nonlinear Schrödinger equation (3.2.13), is given by

$$\frac{\partial v_{1,n}}{\partial t} = A_n v_{1,n} + B_n v_{2,n}, \tag{3.2.18a}$$

$$\frac{\partial v_{2,n}}{\partial t} = C_n v_{1,n} + D_n v_{2,n}, \tag{3.2.18d}$$

where

$$A_n = \mathrm{i}(1 - z^2 \mp u_n u_{n-1}^*), \tag{3.2.19a}$$

$$B_n = \mathrm{i}(-z u_n + z^{-1} u_{n-1}^*), \tag{3.2.19b}$$

$$C_n = \pm\mathrm{i}(z u_{n-1}^* - z^{-1} u_n^*), \tag{3.2.19c}$$

$$D_n = -\mathrm{i}(1 - z^{-2} \mp u_n^* u_{n-1}). \tag{3.2.19d}$$

Since u_n and u_n^* tend to zero sufficiently rapidly as $|n| \to \infty$, then on $|z| = 1$

$$\frac{\partial v_{1,n}}{\partial t} \sim \mathrm{i}(1 - z^2) v_{1,n}, \qquad \frac{\partial v_{2,n}}{\partial t} \sim -\mathrm{i}(1 - z^{-2}) v_{2,n}, \tag{3.2.20}$$

as $n \to \pm\infty$, and so the time dependent eigenfunctions are chosen to satisfy

$$\boldsymbol{\phi}_n(z;t) = \boldsymbol{\phi}_n(z;0) \exp\{\mathrm{i}(1 - z^2)t\}, \tag{3.2.21a}$$

$$\bar{\boldsymbol{\phi}}_n(z;t) = \bar{\boldsymbol{\phi}}_n(z;0) \exp\{-\mathrm{i}(1 - z^{-2})t\}, \tag{3.2.21b}$$

$$\boldsymbol{\psi}_n(z;t) = \boldsymbol{\psi}_n(z;0) \exp\{-\mathrm{i}(1 - z^{-2})t\}, \tag{3.2.21c}$$

$$\bar{\boldsymbol{\psi}}_n(z;t) = \bar{\boldsymbol{\psi}}_n(z;0) \exp\{\mathrm{i}(1 - z^2)t\}. \tag{3.2.21d}$$

Then from equations (3.2.15), it follows that

$$a(z;t) = a(z;0), \qquad b(z;t) = b(z;0)\exp\{-\mathrm{i}(z^2 + z^{-2} - 2)t\}, \tag{3.2.22a}$$

$$\bar{a}(z;t) = \bar{a}(z;0), \qquad \bar{b}(z;t) = \bar{b}(z;0)\exp\{\mathrm{i}(z^2 + z^{-2} - 2)t\}, \tag{3.2.22b}$$

and so

$$\rho(z;t) = \rho(z;0)\exp\{-\mathrm{i}(z^2 + z^{-2} - 2)\}, \tag{3.2.22c}$$

$$\bar{\rho}(z;t) = \bar{\rho}(z;0)\exp\{-\mathrm{i}(z^2 + z^{-2} - 2)\}. \tag{3.2.22d}$$

Similarly, the zeros of $a(z)$, or the poles of the meromorphic function $\boldsymbol{M}_n(z)/a(z)$, give rise to the soliton solutions. A one-soliton solution of the discrete nonlinear Schrödinger equation (3.2.13), corresponding to an eigenvalue $z = \mathrm{e}^{-W+\mathrm{i}\theta}$, with is given by

$$u_n = \sinh(2W)\exp\{-2\mathrm{i}[n\theta + (\cosh(2W)\cos(2W) - 1)t]\} \\ \times \operatorname{sech}[2nW - 2t\sinh(2W)\sin(2\theta) + \phi_0], \tag{3.2.23}$$

with ϕ_0 a constant.

3.2.3 Partial-Difference Equations. Here we note that similar ideas apply to fully discrete problems (that is nonlinear evolution equations which are discrete in space and discrete in time). Corresponding to every partial differential equation there is a class of differential-difference equations; similarly, for every differential-difference equations there is a class of partial difference equations (the discretized equations can be made to relax to their continuous counterparts as the mesh is refined). The sense in which these statements are meant is how well the linearized version of the nonlinear equation is approximated by its discrete analogue. For example, consider the linear Schrödinger equation

$$\mathrm{i}u_t = u_{xx}. \tag{3.2.24}$$

A linearized differential-difference version, accurate to $O((\Delta x)^2)$ in the continuum limit, is

$$\mathrm{i}\frac{\partial u_n}{\partial t} = \frac{1}{(\Delta x)^2}(\delta^2 u_n), \tag{3.2.25a}$$

where

$$\delta^2 u_n = u_{n+1} + u_{n-1} - 2u_n, \tag{3.2.25b}$$

is the second central difference. A linearized partial difference version of these equations, accurate to $O((\Delta x)^2)$ and $O((\Delta t)^2)$ for the partial differential equation (3.2.24) or to order $O((\Delta t)^2)$ for its differential-difference analogue (3.2.25), is

$$\frac{\mathrm{i}\Delta^m u_n^m}{\Delta t} = \frac{1}{2(\Delta x)^2}(\delta^2 u_n^m + \delta^2 u_n^{m+1}), \tag{3.2.26}$$

where $\Delta^m u_n^m = u_n^{m+1} - u_n^m$, and $\delta^2 u_n^m$ is the second central difference evaluated at time step $m\Delta t$ and space step $n\Delta x$. A partial difference equation which is consistent with the nonlinear Schrödinger equation

$$\mathrm{i}u_t = u_{xx} + 2u^2u^*, \tag{3.2.27}$$

is given by

$$\begin{aligned}\frac{\mathrm{i}\Delta^m u_n^m}{\Delta t} = \frac{1}{2(\Delta x)^2}&\left[\left(u_{n+1}^m - 2u_n^m + u_{n-1}^m \prod_{k=-\infty}^{n-1} \Lambda_k^m\right)\right.\\
&\left.+\left(u_{n+1}^{m+1} \prod_{k=-\infty}^{n} \Lambda_k^m - 2u_n^{m+1} + u_{n-1}^{m+1}\right)\right]\\
+\frac{1}{4}&\left[u_n^m\left(u_n^{m*}u_{n+1}^m + u_n^{m+1*}u_{n+1}^{m+1}\right) + u_n^{m+1}\left(u_{n-1}^m u_n^{m*} + u_{n-1}^{m+1}u_n^{m+1*}\right)\right.\\
&\left.+ 2u_n^m u_n^{m*} u_{n+1}^{m+1} \prod_{k=-\infty}^{n} \Lambda_k^m + 2u_n^{m+1}u_n^{m+1*}u_{n-1}^m \prod_{k=-\infty}^{n-1} \Lambda_k^m\right]\\
-u_n^m &\sum_{k=-\infty}^{n} \Delta^m S_k^m - u_n^{m+1} \sum_{k=-\infty}^{n-1} \Delta^m S_k^{m*},\end{aligned} \tag{3.2.28a}$$

with

$$\Lambda_k^m = \frac{1 \pm u_k^{m+1}u_k^{m+1*}}{1 \pm u_k^m u_k^{m*}}, \tag{3.2.28b}$$

$$S_n^m = u_k^m u_{k-1}^{m*} + u_{k+1}^m u_k^{m*}, \tag{3.2.28c}$$

$$\Delta^m u_n^m = u_n^{m+1} - u_n^m, \tag{3.2.28d}$$

which is a nonlinear Crank-Nicholson scheme. The truncation error is accurate to $O((\Delta t)^2)$, and $O((\Delta x)^2)$; in the linear limit the scheme reduces to the standard Crank-Nicholson scheme

$$\frac{\mathrm{i}\Delta^m u_n^m}{\Delta t} := \frac{1}{2(\Delta x)^2}\left(u_{n+1}^{m+1} - 2u_n^{m+1} + u_{n-1}^{m+1} + u_{n+1}^m - 2u_n^m + u_{n-1}^m\right).$$

Nonlinear partial difference equations such as (3.2.28) provide an excellent basis for numerical simulations for the associated nonlinear evolution equation. In particular, for an integrable equation such as the nonlinear Schrödinger equation (3.2.24), the associated partial difference equations are usually constructed using methods related to the I.S.T. method of solution (for further details see Ablowitz and Ladik [1977]; §2.2 of Ablowitz and Segur [1981]; Taha and Ablowitz [1984a,b,c]). A difference schemes such as (3.2.28) have the following advantages:

(i), the scheme maintains many of the important properties of the original equation. For example, one can associate with the scheme an infinite of conservation laws and has travelling wave solutions, which have special properties; these are the solitons;

(ii), the associated linear scheme is always neutrally stable;

(iii), the scheme maintains a certain joint x, t symmetry of the original equation;

(iv), the order of accuracy is the same for both the linear and nonlinear schemes;

(v), the schemes depend globally on the mesh points

(see Ablowitz and Ladik [1977]; Taha and Ablowitz [1984a]).

With regard to the I.S.T. for partial difference equations, the procedure is similar to that for differential-difference equations outlined above except for the time dependence which for equation (3.2.28) is given by

$$\rho(z;m) = \rho(z;0)\left(\frac{2 - \mathrm{i}\sigma(z^2 + z^{-2} - 2)}{2 + \mathrm{i}\sigma(z^2 + z^{-2} - 2)}\right)^m, \tag{3.2.29}$$

with $\sigma = \Delta t/(\Delta x)^2$. The one-soliton solution corresponding the eigenvalue $z = \mathrm{e}^{-W+\mathrm{i}\theta}$ is given by (write $Q_n^m = q_n^m \Delta x$)

$$\begin{aligned} Q_n^m = \sinh(2W)\exp\{\mathrm{i}[2n\theta + 2m(\arg\omega) + \theta_0]\} \\ \times \operatorname{sech}(2nW - mn|\omega| - \phi_0), \end{aligned} \tag{3.2.30}$$

with θ_0, ϕ_0 constants and where $\omega = |\omega|\exp(\mathrm{i}\arg\omega)$.

We note that recent work has shown that I.S.T. based schemes are extremely valuable when integrating for long periods of time. Indeed non-I.S.T. based schemes can lead to chaotic behaviour (Herbst and Ablowitz [1989a,b]; Ablowitz and Herbst [1990]) — see also the discussion in the following section.

3.3 Homoclinic Struture and Numerically Induced Chaos for the Nonlinear Schrödinger Equation.

3.3.1 Introduction. In these notes we have been primarily concerned with completely integrable equations, the I.S.T. schemes for their solutions and soliton solutions. Whereas the soliton is a remarkable stable entity (recall Definition 1.4.1), chaos, in a sense, represents the opposite end of the spectrum — solutions that are extremely sensitive to small disturbances. The "modern" history of deterministic chaos began with the seminal paper of Lorenz [1963] who investigated a three mode Galerkin truncation of the Oberbeck-Boussinesq equations for two dimensional Rayleigh-Bénard convection (it must be pointed out that Poincare at the turn of the century discovered many aspects of what we now consider chaos in Hamiltonian dynamical systems). The phase plane of the Lorentz system contains a "strange attractor", i.e., a lower dimensional manifold where the orbits of the solution reside. Subsequently, chaos has been observed in many different physical situations and nonlinear dynamical systems with chaotic solutions have been studied extensively and is frequently associated with homoclinic orbits in deterministic nonlinear dynamics. An understanding of the mechanism of chaos in these situations depends on a detailed knowledge of the geometric structure of the finite dimensional phase space (see, for example, Guckenheimer and Holmes [1983]; Wiggens [1988]; Tabor [1989]). For partial differential equations such detailed information about infinite dimensional phase space is not yet available. Only in the case of the completely integrable soliton equations has it been possible to make some progress. The spectral theory and associated geometric structure of the sine-Gordon equation

$$u_{tt} - u_{xx} + \sin u = 0, \tag{3.3.1}$$

have been worked out in detail (Ercolani, Forest and McLaughlin [1990a,b,c]; Flesch, Forest and Sinha [1991]). The homoclinic structure underlying the sine-Gordon equation (3.3.1) has profound implications on the stability of the equation and analytical and numerical studies demonstrate how homoclinic crossings in the weakly driven, damped sine-Gordon equation

$$u_{tt} - u_{xx} + \sin u = \epsilon\left[-\alpha u_t + \Gamma \sin(\omega t)\right],$$

with ϵ, α, Γ and ω constants, can lead to temporal chaos (see, for example, Bishop, Forest, McLaughlin and Overman [1986, 1988]; Taki, Spatschek, Fernandez, Grauer and Reinisch [1989]).

The nonlinear Schrödinger (NLS) equation,

$$\mathrm{i}u_t + u_{xx} + 2|u|^2 u = 0, \tag{3.3.2}$$

plays a ubiquitous role in physics. It arises as an asymptotic limit of a slowly varying dispersive wave envelope in a nonlinear medium and as such has significant applications; e.g. nonlinear optics, water waves, plasma physics etc.. Moreover, as we discussed in §3.2, it is solvable through the I.S.T.. As such we are ensured that the NLS equation does not possess chaotic behavior for the standard initial value problem. Recently there has also been significant interest in particular forced versions of the NLS equation and approximate solutions via low mode truncations (Bishop, Flesch, Forest, McLaughlin and Overman [1990]; Forest [1988] — see equation (3.3.18) below). On the other hand, there are numerous popular discretizations of the NLS equation (3.3.2) which provide a vehicle for numerical solutions. Some of these discretizations are physically important in their own right, with applications to nonlinear dimers, self trapping phenomena, biological systems etc. (cf. Kenkre and Campbell [1986]; Kenkre and Tsironis [1987]; Eilbeck, Lomdahl and Scott [1985]; Davydov [1982]). A numerical scheme can be thought of as a small perturbation of the governing equation.

In this section we are concerned with a novel numerical instability involving the NLS equation (3.3.2) with periodic boundary conditions. It turns out that the instability is manifest in the 'standard' numerical schemes employed to solve the NLS equation (3.3.2). When operative, the instability is deeply connected with the underlying homoclinic orbits of the NLS equation (3.3.2). In a series of papers, Herbst and Ablowitz have referred to this instability as Numerical Homoclinic Instability (NHI) (Herbst and Ablowitz [1989a,b]; Ablowitz and Herbst [1990a]). The instability disappears when the discretization mesh is sufficiently refined. In fact, it does not exist at all when the integrable difference scheme of Ablowitz and Ladik [1976b] is employed! The studies of Herbst and Ablowitz show that this instability is in a sense generic and will apply to standard numerical schemes approximating partial differential equations which have underlying homoclinic structures.

This instability explains the irregular 'stochastic-like' numerical solutions which have been observed by many authors (see, for example, Caponi, Saffman and Yuen [1982]; Weideman [1986]; Shen and Nicolson [1987]). Herbst and Ablowitz [1989b] show that NHI may disappear only at levels of refinement much higher than one would normally expect; e.g., up to 64 modes may be required to solve the NLS equation for suitable initial values by the exponentially convergent Fourier Spectral Method, whereas the solution has most of its energy contained in the first 4–8 modes.

Like the sine-Gordon equation (3.3.1), the NLS equation (3.3.2) with periodic boundary conditions has been studied from the point of view of inverse spectral theory. Relatively complicated formulae describing a broad class of periodic solutions in terms of Riemann Theta functions of N phases can be computed (see, for example, Ma and Ablowitz [1981]; Tracy [1984]; Ercolani, Forest and McLaughlin [1990c]) and via limits, Bäcklund Transformations, etc., elementary formulae describing homoclinic orbits can be found Ercolani, Forest and McLaughlin [1990c] (see also Akhmediev, Korneev and Mitskevich [1985, 1988]). Here we show how the homoclinic structure of the NLS equation (3.3.2) can be readily ascertained by employing the formula for N-soliton solutions of Hirota. Hence, the combination homoclinic orbits of the focusing NLS equation (3.3.2) are elementary functions and are ultimately related to the N-'dark hole' soliton solutions of the defocusing NLS equation (Hirota [1980]).

In the §3.3.2 we give a linearized stability analysis of the plane-wave solution of the NLS equation (3.3.2). The analytical description of the homoclinic orbits obtained from Hirota's method in §3.3.3 show how the unstable eigendirections associated with the unstable modes of the linearized analysis, can be related to the homoclinic orbits. Combination homoclinic orbits arise when the linearized analysis admits more than one unstable mode and these are discussed in §3.3.4. In general, there are $2^N - 1$ single- and combination homoclinic orbits in the case of N unstable modes. Some of the consequences of this relatively complicated homoclinic structure for numerical studies of the NLS equation (3.3.2) are discussed in §3.3.5.

3.3.2 A Linearized Stability Analysis. It is convenient for our purposes to study the NLS equation (3.3.2) with periodic boundary conditions, $u(x+L,t) = u(x,t)$, initial condition, $u(x,0) = f(x)$.

The starting point for developing the homoclinic structure is to find a suitable fixed point, in this case given by

$$\tilde{u}(x,t) = a\exp(2\mathrm{i}|a|^2 t),$$

where a is any complex constant. Note that it is possible to eliminate the time dependence by a simple transformation, justifying our use of the term 'fixed point'. In fact, the fixed point may also be written as

$$\tilde{u}(x,t) = a\exp\{\mathrm{i}(2a^2 t + \tilde{\phi})\}, \tag{3.3.3}$$

where a in this case is any *real* constant and $\tilde{\phi} \in [0, 2\pi)$. In this sense there is a ring of fixed points characterized by the different values of the phase, $\tilde{\phi}$. Next we investigate the stability of the fixed point by considering small perturbations of the form (see Stuart and DiPrima [1978]),

$$u(x,t) = \tilde{u}(x,t)[1 + \epsilon(x,t)], \tag{3.3.4}$$

where $|\epsilon| \ll 1$. Substituting (3.3.4) into (3.3.2) and keeping linear terms in ϵ yields

$$\epsilon_t = \mathrm{i}\epsilon_{xx} + 2\mathrm{i}|a|^2(\epsilon + \epsilon^*).$$

Assuming

$$\epsilon(x,t) = \hat{\epsilon}_{-n}(t)\exp(-\mathrm{i}\mu_n x) + \hat{\epsilon}_n(t)\exp(\mathrm{i}\mu_n x),$$

where $\mu_n = 2\pi n/L$, it follows that the growth rate, σ_n, of the nth mode, $\hat{\epsilon}_n(t) = \hat{\epsilon}_n(0)\exp(\sigma_n t)$, is given by

$$\sigma_{n\pm} = \pm\mu_n(4|a|^2 - \mu_n^2)^{1/2}.$$

Hence, the fixed point is hyperbolic provided,

$$0 < \mu_n^2 < 4|a|^2. \tag{3.3.5}$$

The number of unstable modes, which also determines the complexity of the homoclinic structure, is given by the largest integer N, satisfying,

$$0 < N < |a|L/\pi.$$

The directions of the stable and unstable eigenspaces of the n-th mode at the fixed point are

$$\begin{pmatrix} \hat{\epsilon}_n(0) \\ \hat{\epsilon}_{-n}(0) \end{pmatrix} = (\mathrm{i}\mu_n^2 - \sigma_{n\pm}) \begin{pmatrix} \exp(\mathrm{i}\phi) \\ -\exp(-\mathrm{i}\phi) \end{pmatrix},$$

where ϕ is an arbitrary real constant. This translates into an initial condition,

$$u(x,0) = a + \epsilon_0(\mu_n^2 + \mathrm{i}\sigma_{n\pm})\sin(\mu_n x + \phi), \tag{3.3.6}$$

where $0 < |\epsilon_0| \ll 1$. Since the NLS equation (3.3.2) admits the symmetry, $(x,t,u) \to (x + x_0, t, u)$, it is sufficient to consider the initial condition,

$$u(x,0) = a + \epsilon(\mu_n^2 + \mathrm{i}\sigma_{n\pm})\cos(\mu_n x). \tag{3.3.7}$$

The additional symmetry, $(x,t,u) \to (-x,t,u)$, ensures that even initial conditions, $u(x,0) = u(-x,0)$, evolve in time as even solutions. This fact is exploited in our numerical calculations, allowing us to reduce the amount of computation by a factor of two, roughly. Herbst and Ablowitz [1989b] illustrate that this spatial symmetry is sometimes lost during numerical computations due to rounding error and the extreme sensitivity of the problem in the neighborhood of the homoclinic orbit.

We remark that it is well-known (see Stuart and DiPrima [1978]) that the NLS equation (3.3.2) allows steady wave-like solutions that are unstable under small perturbations (the so-called Benjamin-Feir instability — Benjamin and Feir [1967]). This means that if (3.3.2) is solved with the initial values

$$u(x,0) = a[1 + \epsilon\cos(\mu_n x)], \qquad |\epsilon| \ll 1,$$

as the initial condition, where $\mu_n = 2\pi n/L$, a, ϵ are constants and L is the period, then all wave numbers n satisfying equation (3.3.5) will at first grow exponentially according to linear analysis and then saturate due to the nonlinear term. These instabilities display the phenomenon known as recurrence which has been observed both experimentally and numerically (Yuen and Lake [1975]; Yuen and Ferguson [1978]).

3.3.3 Hirota's Method for the Single Homoclinic Orbit. The starting point of Hirota's method is to assume a solution of (3.3.2) of the form, $u(x,t) = G(x,t)/F(x,t)$, where without loss of generality we assume that F is real. Substituting this expression into (3.3.2) the resulting equation is split into,

$$\mathrm{i}\mathrm{D}_t G \bullet F + \mathrm{D}_x^2 G \bullet F = \lambda G \bullet F, \tag{3.3.8a}$$

$$\mathrm{D}_x^2 F \bullet F - 2G \bullet G^* = \lambda F \bullet F, \tag{3.3.8b}$$

where

$$\mathrm{D}_t G \bullet F = G_t F - G F_t,$$

$$\mathrm{D}_x^2 G \bullet F = G_{xx} F - 2G_x F_x + G F_{xx},$$

(see, for example, Hirota [1980]; Ablowitz and Segur [1981], for a more detailed discussion). Since we are interested in a homoclinic orbit to the fixed point, (3.3.3), a solution of the form,

$$\begin{aligned} G &= a\exp(\mathrm{i}a^2t)[1 + b_1\exp(\mathrm{i}px + \Omega t + \gamma) \\ &\qquad + b_2\exp(-\mathrm{i}px + \Omega t + \gamma) + b_3\exp(2\Omega t + 2\gamma)], \\ F &= 1 + b_4[\exp(\mathrm{i}px + \Omega t + \gamma) + \exp(-\mathrm{i}px + \Omega t + \gamma)] + b_5\exp(2\Omega t + 2\gamma), \end{aligned}$$

where γ is an arbitrary phase and a is assumed to be real, is assumed. The values of the coefficients, b_j, are determined by substituting G and F into (3.3.8a) and (3.3.8b) and collecting coefficients of all the independent functions. For instance, the coefficients of the constant term of (3.3.8a) yields

$$\lambda = -2a^2,$$

and the coefficient of $\exp(\mathrm{i}px + \Omega t)$ yields

$$b_1(\mathrm{i}\Omega - p^2) = b_4(\mathrm{i}\Omega + p^2),$$

etc.. Proceeding in this manner we may solve for the b_j to find that,

$$\begin{aligned} &b_1 = b_2 = \exp(2\mathrm{i}\phi), \qquad b_3 = A_{12}b_1^2, \qquad b_4 = 1, \qquad b_5 = A_{12}, \\ &\Omega = \pm p(4a^2 - p^2)^{1/2}, \end{aligned}$$

where $p = 2a\sin\phi$, $A_{12} = \sec^2\phi$. Hence, we obtain the solution,

$$\begin{aligned} u(x,t) &= a\exp(2\mathrm{i}a^2t) \\ &\times \frac{1 + 2\cos(px)\exp(\Omega t + 2\mathrm{i}\phi + \gamma) + A_{12}\exp(2\Omega t + 4\mathrm{i}\phi + 2\gamma)}{1 + 2\cos(px)\exp(\Omega t + \gamma) + A_{12}\exp(2\Omega t + 2\gamma)}. \end{aligned} \tag{3.3.9}$$

(Note that the value $\phi = \pi/2$ is excluded from (3.3.9) by the condition, $0 < p < 2a$, to be discussed below.) For given (spatial) period L, it follows that

$$p = \mu_n := 2\pi n/L,$$

for any integer value n. Hence, provided that $p = \mu_n$, satisfies the linearized instability condition (3.3.5), the solution (3.3.9) has the characteristics of a homoclinic orbit, leaving the ring of fixed points at $a\exp(2\mathrm{i}a^2t)$ as $t \to -\infty$ and returning at $a\exp(2\mathrm{i}a^2t)\exp(4\mathrm{i}\phi)$ as $t \to \infty$ (note the phase shift from $t \to -\infty$ to $t \to \infty$). Thus, with *each* unstable mode (3.3.5), there is an associated homoclinic orbit, with the wavenumber associated with each homoclinic orbit restricted by (3.3.5).

Note the only difference between equations (3.3.9) in the linear limit and (3.3.7) is a multiplicative phase factor. Hence, (3.3.9) describes the temporal evolution of the linearly unstable modes given by (3.3.7) for a particular choice of phase. This follows immediately from a linearization of (3.3.9) where $\epsilon_0 = \exp(\Omega t + \gamma)$ is chosen to be small and second order terms in ϵ_0 neglected. Accordingly, taking the initial values (3.3.6), with $\epsilon_0 \to 0$, a sequence of periodic orbits is obtained which approach the homoclinic orbit described by (3.3.9). Figure 3.3.1 illustrates the homoclinic orbit corresponding to the initial condition,

$$u(x,0) = \tfrac{1}{2} + \epsilon_0(1+\mathrm{i})\cos(px),$$

with $L = 2\sqrt{2}\,\pi$ and $\epsilon_0 = 10^{-5}$. Notice that for these parameter values, (3.3.5) admits only one unstable mode, $p = 2\pi/L$.

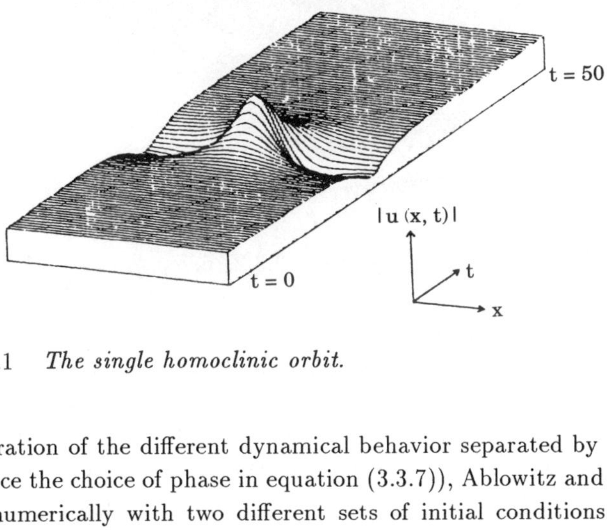

FIGURE 3.3.1 *The single homoclinic orbit.*

As an illustration of the different dynamical behavior separated by the homoclinic orbit (and hence the choice of phase in equation (3.3.7)), Ablowitz and Herbst [1990a] solve (3.3.2) numerically with two different sets of initial conditions, on 'opposite' sides of the homoclinic orbit. Figure 3.3.2a shows the solution obtained from the initial condition,

$$u(x,0) = 0.5 + 0.1\cos(px),$$

and Figure 3.3.2b the solution obtained from,

$$u(x,0) = 0.5 + 0.1\mathrm{i}\cos(px),$$

where $p = 2\pi/L$ and $L = 2\sqrt{2}\,\pi$. Figure 3.3.2a shows $\cos(px)$ appearing periodically whereas interchanges between $\cos(px)$ and $\cos\{p(x+\frac{1}{2}L)\}$ take place in Figure 3.3.2b.

(a)

(b)

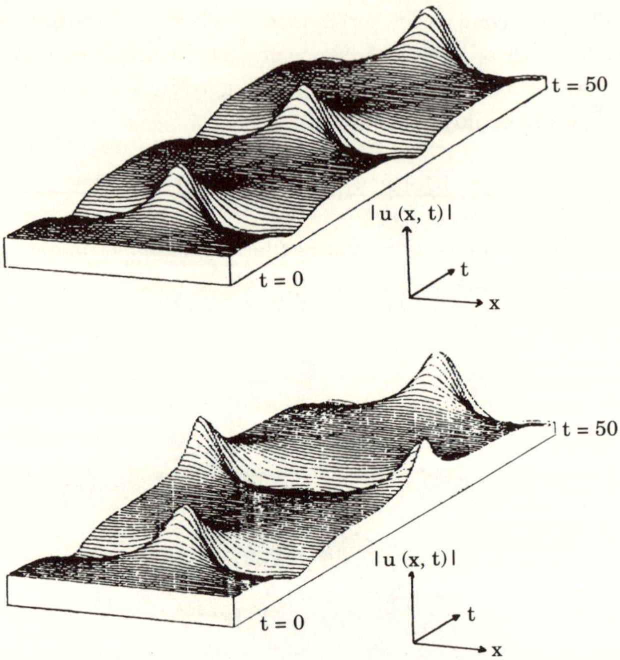

FIGURE 3.3.2 *The solution of equation* (3.3.2) *using the initial conditions:* (a), $u(x,0) = 0.5 + 0.1\cos(px)$; (b), $u(x,0) = 0.5 + 0.1\mathrm{i}\cos(px)$.

Each homoclinic orbit described above is characterized by a specific mode, $p = \mu_n$, satisfying (3.3.5). However, these modes may also exist in combination, giving rise to a more complicated homoclinic structure; a possibility that is not obvious from linearized analysis. In the next section we describe these combination homoclinic orbits.

3.3.4 Combination Homoclinic Orbits. Instead of generalizing the solution obtained in the previous section to include more modes, Ablowitz and Herbst [1990a] show how the homoclinic structure of the NLS equation (3.3.2) is associated with the dark-hole soliton solutions of the defocusing NLS equation,

$$\mathrm{i}u_t + u_{xx} - 2|u|^2u = 0. \tag{3.3.10}$$

The N-dark-hole soliton solution may be obtained, using Hirota's method as outlined in the previous section and is given by Hirota [1980],

$$u(x,t) = a\exp(-2\mathrm{i}a^2t)g(x,t)/f(x,t) \tag{3.3.11}$$

where

$$f(x,t) = \sum_{\tilde{\beta}=0,1} \exp\left\{\sum_{j>k}^{(N)} \alpha_{jk}\beta_j\beta_k + \sum_{j=1}^{N} \beta_j\eta_j\right\},$$

$$g(x,t) = \sum_{\tilde{\beta}=0,1} \exp\left\{\sum_{j>k}^{(N)} \alpha_{jk}\beta_j\beta_k + \sum_{j=1}^{N} \beta_j(\eta_j + 2\mathrm{i}\phi_j)\right\},$$

and

$$\begin{aligned} \exp(\alpha_{jk}) &= \left\{\frac{\sin[\frac{1}{2}(\phi_j - \phi_k)]}{\sin[\frac{1}{2}(\phi_j + \phi_k)]}\right\}^2, \\ \eta_j &= p_j x - \Omega_j t + \gamma_j, \\ p_j &= 2a\sin\phi_j, \\ \Omega_j &= \pm p_j(4a^2 - p_j^2)^{1/2}, \end{aligned}$$

where ϕ_j are distinct real constants, $\sum_{\tilde{\beta}=0,1}$ is the summation over all possible combinations of $\beta_j = 0, 1$, $j = 1, \ldots, N$, $\sum_{j>k}^{(N)}$ indicates the summation over all possible pairs α_{jk} $(j > k)$ chosen from N elements and γ_j are arbitrary phases.

It is easily verified that solutions of (3.3.10), analytic on the real axis will be solutions of (3.3.2) under the transformation,

$$x \to \mathrm{i}x, \qquad t \to -t, \tag{3.3.12}$$

provided the evenness condition, $u(x,t) = u(-x,t)$, is satisfied (this condition is required by the complex conjugate term in the NLS equation). The dark-hole solitons of the defocusing NLS equation (3.3.10), satisfying the evenness condition become the homoclinic structure associated with (3.3.2) under this transformation.

Note that the evenness condition cannot be satisfied for $N = 1$ and is not a solution of (3.3.2) under the transformation (3.3.12). The single homoclinic orbit (3.3.9) is obtained from (3.3.11) for $N = 2$ choosing

$$p_1 = -p_2 = p, \qquad \gamma_1 = \gamma_2 = \gamma, \qquad \Omega_1 = \Omega_2 = \Omega,$$

which implies that

$$\phi_2 = \phi_1 + \pi.$$

The simplest example of a combination mode arises from the case $N = 4$. Proceeding as above, let $p_2 = -p_1$, $p_4 = -p_3$, $\gamma_2 = \gamma_1$, $\gamma_4 = \gamma_3$, etc., and using the transformation, (3.3.12), a combination homoclinic orbit of the form,

$$u(x,t) = a\exp(2\mathrm{i}a^2 t)[g(x,t)/f(x,t)] \tag{3.3.13}$$

is obtained, where

$$
\begin{aligned}
f(x,t) = 1 &+ 2\cos(p_1 x)\exp(\Omega_1 t + \gamma_1) + A_{12}\exp(2\Omega_1 t + 2\gamma_1) \\
&+ 2\cos(p_3 x)\exp(\Omega_3 t + \gamma_3) + A_{34}\exp(2\Omega_3 t + 2\gamma_3) \\
&+ 2\left[A_{13}\cos\{(p_1 + p_3)x\} + A_{23}\cos\{(p_3 - p_1)x\}\right] \\
&\qquad \times \exp\{(\Omega_1 + \Omega_3)t + (\gamma_1 + \gamma_3)\} \\
&+ 2A_{13}A_{23}A_{34}\cos(p_1 x)\exp\{(\Omega_1 + 2\Omega_3)t + (\gamma_1 + 2\gamma_3)\} \\
&+ 2A_{12}A_{13}A_{23}\cos(p_3 x)\exp\{(2\Omega_1 + \Omega_3)t + (2\gamma_1 + \gamma_3)\} \\
&+ A_{12}A_{13}^2A_{23}^2A_{34}\exp\{2(\Omega_1 + \Omega_3)t + 2(\gamma_1 + \gamma_3)\}, \\
g(x,t) = 1 &+ 2\exp(2\mathrm{i}\phi_1)\cos(p_1 x)\exp(\Omega_1 t + \gamma_1) + A_{12}\exp(4\mathrm{i}\phi_1)\exp(2\Omega_1 t + 2\gamma_1) \\
&+ 2\exp(2\mathrm{i}\phi_3)\cos(p_3 x)\exp(\Omega_3 t + \gamma_3) + A_{34}\exp(4\mathrm{i}\phi_3)\exp(2\Omega_3 t + 2\gamma_3) \\
&+ 2\exp\{2\mathrm{i}(\phi_1 + \phi_3)\}\left[A_{13}\cos\{(p_1 + p_3)x\} + A_{23}\cos\{(p_3 - p_1)x\}\right] \\
&\qquad \times \exp\{(\Omega_1 + \Omega_3)t + (\gamma_1 + \gamma_3)\} \\
&+ 2A_{13}A_{23}A_{34}\exp\{\mathrm{i}(2\phi_1 + 4\phi_3)\cos(p_1 x)\exp\{(\Omega_1 + 2\Omega_3)t + (\gamma_1 + 2\gamma_3)\} \\
&+ 2A_{12}A_{13}A_{23}\exp\{\mathrm{i}(4\phi_1 + 2\phi_3)\cos(p_3 x)\exp\{(2\Omega_1 + \Omega_3)t + (2\gamma_1 + \gamma_3)\} \\
&+ A_{12}A_{13}^2A_{23}^2A_{34}\exp\{4\mathrm{i}(\phi_1 + \phi_3)\}\exp\{2(\Omega_1 + \Omega_3)t + 2(\gamma_1 + \gamma_3)\},
\end{aligned}
$$

where $A_{jk} = \exp(\alpha_{jk})$, $\Omega_j = \pm p_j(4a^2 - p_j^2)^{1/2}$, $j, k = 1, 3$, etc..

Again this is a homoclinic orbit to the ring of spatially periodic fixed points with period L, provided $p_1 = \mu_n$, $p_3 = \mu_m$ satisfy the instability condition, (3.3.5), for some integer values m and n. In this case the homoclinic orbit leaves the ring of fixed points at

$$u = a\exp(2\mathrm{i}a^2 t), \qquad \text{as} \quad t \to -\infty,$$

and returns to the ring at

$$u = a\exp(2\mathrm{i}a^2 t)\exp\{4\mathrm{i}(\phi_1 + \phi_3)\} \qquad \text{as} \quad t \to \infty.$$

In contrast with the previous homoclinic orbit this one is characterized by the two modes, $\cos p_1 x$ and $\cos p_3 x$, temporally separated by the parameters γ_1 and γ_3.

A suitable initial condition for the calculation of this combination homoclinic orbit is obtained by choosing t and the phases γ_1 and γ_3 such that $\epsilon_1 = \exp(\Omega_1 t + \gamma_1)$, $\epsilon_3 = \exp(\Omega_3 t + \gamma_3)$ are small. Neglecting second order terms in ϵ_1 and ϵ_3 lead to the linearized initial condition,

$$u(x,0) = a[1 + 4\mathrm{i}\epsilon_1 \sin\phi_1 \exp(\mathrm{i}\phi_1)\cos(p_1 x) + 4\mathrm{i}\epsilon_3 \sin\phi_3 \exp(\mathrm{i}\phi_3)\cos(p_3 x)]. \quad (3.3.14)$$

Note that ϵ_1 and ϵ_3 depend on γ_1 and γ_3 their relative values determine the temporal separation between the two modes.

Figure 3.3.3 illustrates the combination homoclinic orbit numerically. Ablowitz and Herbst [1990a] solve (3.3.2) numerically using (3.3.14) (and the numerical scheme given by (3.3.17) below) as the initial condition with $L = 4\sqrt{2}\,\pi$, $a = \frac{1}{2}$, $p_3 = 2p_1 = 2\pi/L$ and the value of ϕ_j given by $p_j = 2a\sin\phi_j$. In Figure 3.3.3a, $\epsilon_1 = \epsilon_3 = 10^{-4}$ and for Figure 3.3.3b, $\epsilon_1 = 10^{-4}$, $\epsilon_3 = 10^{-5}$. Note how the different initial conditions change the temporal separation between the two modes. (Incidentally, Ablowitz and Herbst [1990a] checked their numerical results against the analytical solution!)

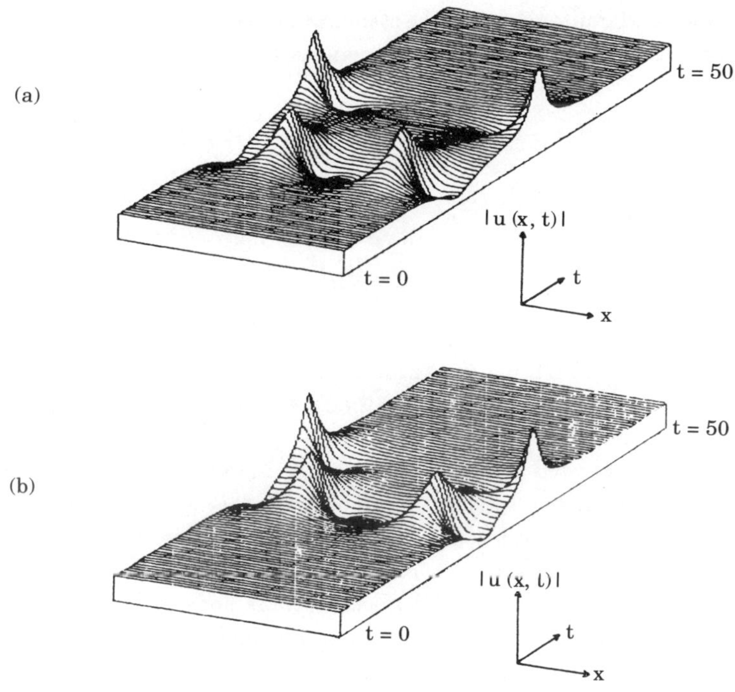

FIGURE 3.3.3 *The combination homoclinic orbit:* (a), $\epsilon_1 = \epsilon_3 = 10^{-4}$; (b), $\epsilon_1 = 10^{-4}$, $\epsilon_3 = 10^{-5}$.

Although we have given an example of a combination mode consisting of only two modes, it is clear that more complicated structures can easily be constructed via (3.3.11). Again we note that for a solution with N unstable modes, $2^N - 1$ single and combination homoclinic orbits exist.

3.3.5 Numerical Homoclinic Instability. In order to illustrate some of the issues involved in numerical simulations of the homoclinic orbits and the sensitivity caused by the homoclinic structure, Ablowitz and Herbst [1990a] considered various numerical schemes. Here we discuss only finite difference simulations, but Herbst and Ablowitz [1989b] show that exponentially convergent Fourier Spectral schemes have similar difficulties as do the "standard" finite difference schemes discussed below. The first one is the standard second order scheme,

$$\mathrm{i}\frac{\mathrm{d}U_j}{\mathrm{d}t} + (U_{j-1} - 2U_j + U_{j+1})/h^2 + 2|U_j|^2 U_j = 0, \tag{3.3.15}$$

where $h = L/N_e$ is the grid spacing; periodic boundary conditions are used, $U_{j+N_e} = U_j$. This scheme is Hamiltonian with the standard Poisson brackets and two constants of motions are easily obtained, the Hamiltonian itself

$$H = -\mathrm{i}\sum_{j=0}^{N_e-1} \left(|U_{j+1} - U_j|^2/h^2 - |U_j|^4\right),$$

and the L^2 norm,

$$I = \sum_{j=0}^{N_e-1} |U_j|^2.$$

This implies that the system is integrable when $N_e = 2$ and it has been shown by Weideman and Herbst [1987] that this crude discretization captures certain qualitative properties of (3.3.2) under appropriate conditions. However, if N_e is increased further, integrability is lost and the scheme loses even the qualitative properties of solutions of (3.3.2). This is illustrated in Figure 3.3.4a where the same parameter values as in Figure 3.3.3a were used, but solving the problem with (3.3.15) and $N_e = 32$. The time integration was performed with the Runge-Kutta-Merson routine, D02BBF, in the NAG library with the relative error specified as 10^{-6} (experiments with smaller values have ensured that the effects described here, are not consequences of the time discretization). It is important to note that the correct qualitative behavior is obtained for a while before the solution becomes chaotic (which occurs rather abruptly, as indicated in the figure). This implies that we are dealing with an instability and not merely the effect of truncation error. If the level of discretization is refined to $N_e = 64$, all irregularity disappears, over the time interval under consideration (up to $t = 50$), as illustrated in Figure 3.3.4b.

Next, another aspect of the instability is illustrated. For this example $L = 2\sqrt{2}\,\pi$, $a = \frac{1}{2}$ are chosen and according to (3.3.5) only one unstable mode is admitted. Because of the simpler homoclinic structure, only a single homoclinic orbit, this problem is considerably less sensitive than the previous one. Nevertheless, an instability is observed. Using the initial condition

$$u(x, 0) = 0.5 + 0.05\cos(\mu x) + 10^{-5}\mathrm{i}\sin(\mu x), \tag{3.3.16}$$

(a)

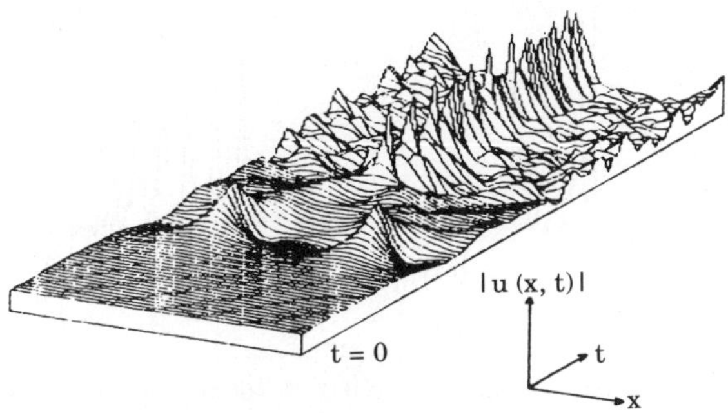

(b)

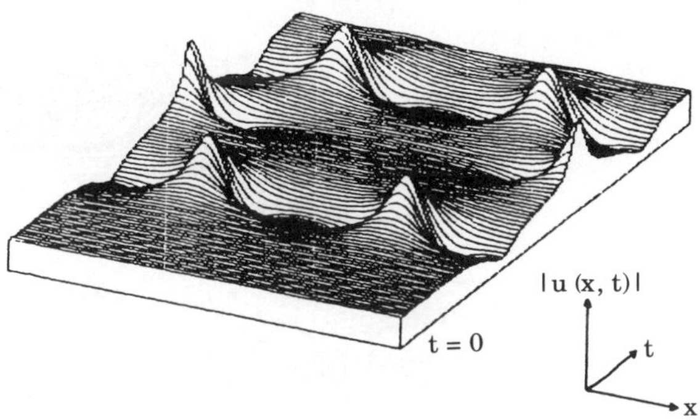

FIGURE 3.3.4 *The standard finite difference solution of equation* (3.3.15)*:* (a), $N_e = 32$; (b), $N_e = 64$.

with $\mu = 2\pi/L$, Figure 3.3.5a shows the time evolution of the modulus of the solution at $x = 0$, obtained from (3.3.15) with $N = 20$. The time evolution of the numerical solution develops irregularities after about 40 time units. Figure 3.3.5b shows the time evolution of the spatial structure of the same problem over the first 100 time units. A comparison of Figures 3.3.5a and 3.3.5b shows that the temporal irregularity coincides with a loss of spatial symmetry. Although the spatial structure is retained (additional numerical experiments indicate to a very high degree), it suffers from a significant phase shift, resulting in increasingly irregular temporal behavior. Figure 3.3.6 shows how the instability disappears as the spatial resolution is refined to $N_e \geq 24$. It should be pointed out that the rather sudden improvement as N_e is increased, appears to be typical.

The numerical experiments of Ablowitz and Herbst [1990a] show that destabilization of the phase in the proximity of a homoclinic orbit may also be triggered by rounding error. It is surprising that this effect of rounding error should disappear as the grid is

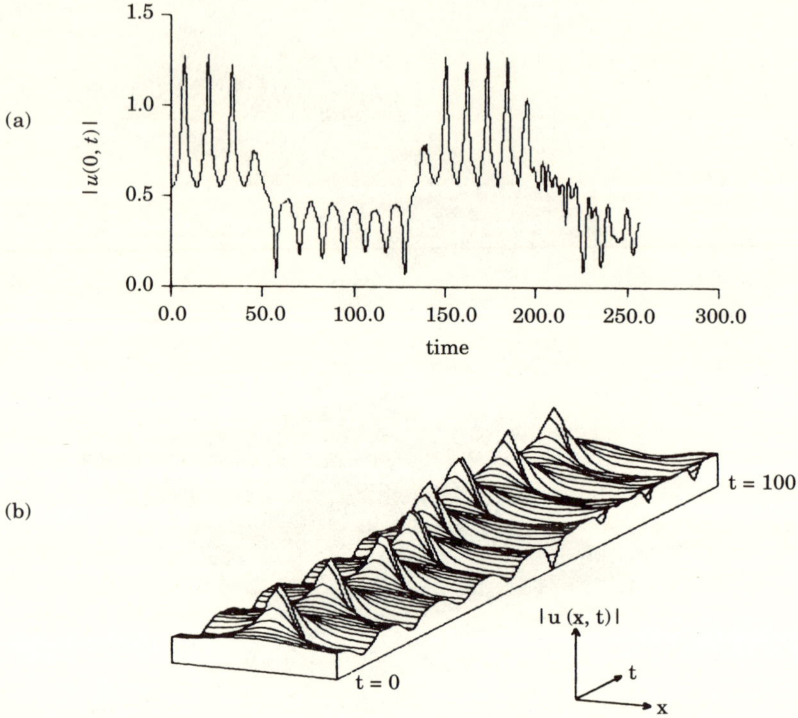

FIGURE 3.3.5 *The finite difference solution of equation* (3.3.17) *using* $N_e = 20$*:* (a), *The modulus of the solution at* $x = 0$*;* (b), *The time evolution of the spatial structure over the first* 100 *time units.*

refined, i.e., when more calculations are performed. Again this is due to the sensitivity of the nonintegrable scheme in the vicinity of a homoclinic orbit.

The numerical scheme which follows has been specially designed for solving the NLS equation (3.3.2) and it turns out to be eminently suitable for capturing the salient features of the NLS equation; in particular, it is designed to be integrable. It is given by Ablowitz and Ladik [1976b] (see the discussion in Ablowitz and Segur [1981] and here §1.11.3 and §3.3.2)

$$\mathrm{i}\frac{\mathrm{d}U_j}{\mathrm{d}t} + (U_{j-1} - 2U_j + U_{j+1})/h^2 + |U_j|^2(U_{j-1} + U_{j+1}) = 0. \tag{3.3.17}$$

Note that it is also of second order accuracy and that the differences with the standard finite difference scheme appear at second order and only in the *nonlinear* term.

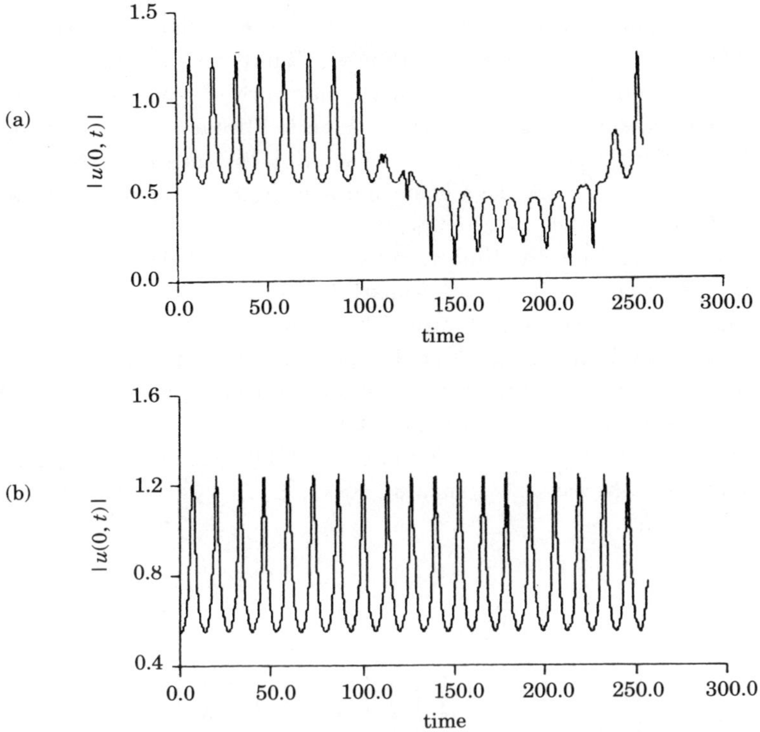

FIGURE 3.3.6 *The finite difference solution of equation* (3.3.17): (a), $N_e = 22$; (b), $N_e = 24$.

The Hamiltonian structure of (3.3.17) is given (for $h = 1$) by the Hamiltonian

$$H = -\mathrm{i} \sum_{j=0}^{N_e - 1} \left[U_j^* (U_{j-1} + U_{j+1}) - 2 \ln(1 + U_j U_j^*) \right],$$

together with the non-standard Poisson brackets,

$$\{q_m, p_n\} = (1 + q_n p_n) \delta_{m,n}, \qquad \{q_m, q_n\} = 0 = \{p_m, p_n\}$$

where $q_n = N_n$, $p_n = (U_{n+1} - U_N)/h$ Kulish [1981]. In this case a second constant of integration is given by

$$I = \sum_{j=0}^{N_e - 1} \left[U_j^* (U_{j-1} + U_{j+1}) \right].$$

Of course there are infinitely many for this scheme Ablowitz and Ladik [1976b] (on the infinite interval — or N independent ones on the periodic interval).

The numerical results shown in Figures 3.3.1–3 were produced by Ablowitz and Herbst [1990a] using this scheme. In particular, Figure 3.3.3a was produced using $N_e = 32$, the same value as for standard finite difference scheme, (3.3.15), in Figure 3.3.4a, yet, the integrable scheme displays the expected (from an integrable equation) behavior. In addition, Figure 3.3.7 shows the solution obtained from the initial condition (3.3.16) using the integrable scheme and only $N_e = 8$. Again a quasi-periodic solution is obtained and there is no indication of any phase instability. In fact, the integrable scheme does not suffer from any of the maladies of NHI described above for the standard difference scheme! This is remarkable considering the fact that the two schemes differ only in the nonlinear term. The crucial difference being that (3.3.17) is *integrable* at all levels of discretization and possesses its own homoclinic structure which is a small perturbation of the homoclinic structure of the NLS equation. Herbst and Ablowitz [1990] show that homoclinic orbits correspond to multiple eigenvalues in the associated spectral problem. The integrable difference scheme preserves the multiple eigenvalues whereas the "standard" does not.

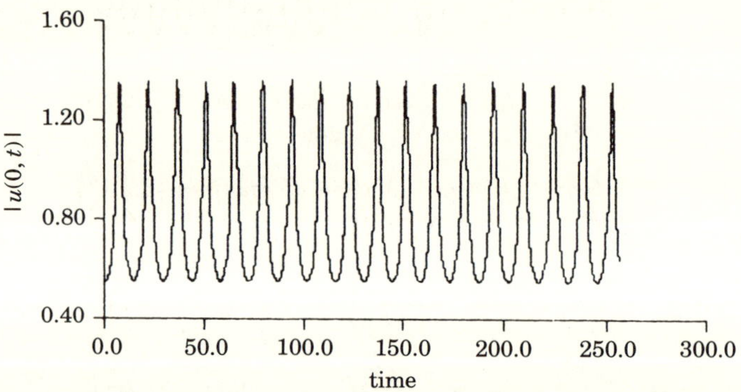

FIGURE 3.3.7 *The solution obtained from the integrable scheme using the initial condition* (3.3.16) *and* $N_e = 8$.

The phase instability shown by Ablowitz and Herbst [1990a], which we have just discussed, can be avoided by exploiting the symmetry $u(x,t) = u(-x,t)$; this alternative approach has been used by Ablowitz, Herbst and Keiser [1990] and we shall now briefly discuss their results. Ablowitz, Herbst and Keiser [1990] illustrated the instability by considering the initial condition

$$u(x,0) = 0.5 + 0.05\cos(\mu x),$$

where $\mu = 2\pi/L$, with increasing values of the spatial period, $L = 2\sqrt{2}\,m\pi$, $m = 1, 2, 3$. For the growth rates of the unstable modes, these choices correspond to maximal growth for the mth mode.

For $m = 1$ (i.e., only one unstable mode and a single homoclinic orbit), both schemes perform satisfactorily. This is illustrated in Figure 3.3.8 showing the solutions obtained from the standard scheme (3.3.15), using $N = 32$. The time evolution of the modulus and the time evoltuion of the spectral decomposition of the spatial structure are shown. It is clear that the spatial structure consists of only a few low order modes.

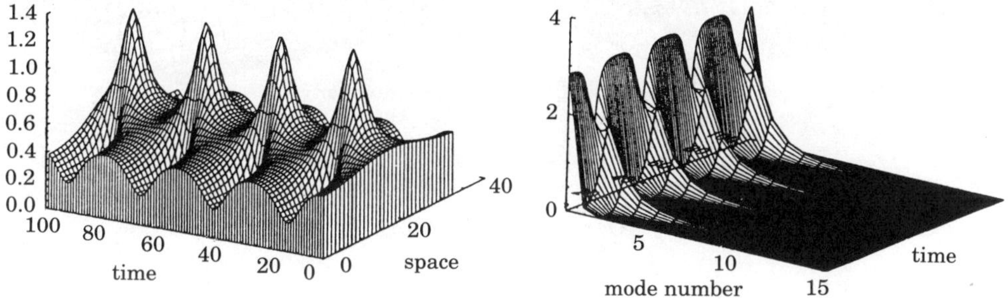

(a) (b)

FIGURE 3.3.8 *Standard scheme: $m = 1$, $N = 32$.*

If the value of m is increased to two, two modes are unstable according to the linearized instability condition (3.3.5). This implies a three homoclinic orbit structure — two single ones (each corresponding to one of the unstable modes), and a combination orbit. Because of the complicated homoclinic structure, this example is not easily solved by the standard finite difference scheme ($N = 32$), as shown in Figure 3.3.9. The time evolution has becme completely irregular. This solution should be compared with the solution obtained from the integrable scheme, using the same parameter values, as shown in Figure 3.3.10 (We remark that this situation differs slightly from the experiments by Herbst and Ablowitz [1989a] where the second mode was just on the edge of the instability region.) If the number of elements of the standard scheme is increased to $N = 64$, the irregularity in the time evolution improves dramatically (see Figure 3.3.11). It is of interest to note that the higher modes in the Fourier decomposition of the spatial structure are apparently negligible according to Figure 3.3.11b. Yet, the time evolution still becomes increasingly irregular with increasing time. Figure 3.3.12a show the time evolution of the modulus of the solution ar $x = 0$ over 256 time units, using $N = 64$. After about 150 time units, the time evolution becomes quite irregular. The solution obtained from the integrable scheme, also for $N = 64$, is shown in Figure 3.3.12b. This solution is well-behaved over the entire interval of intergration.

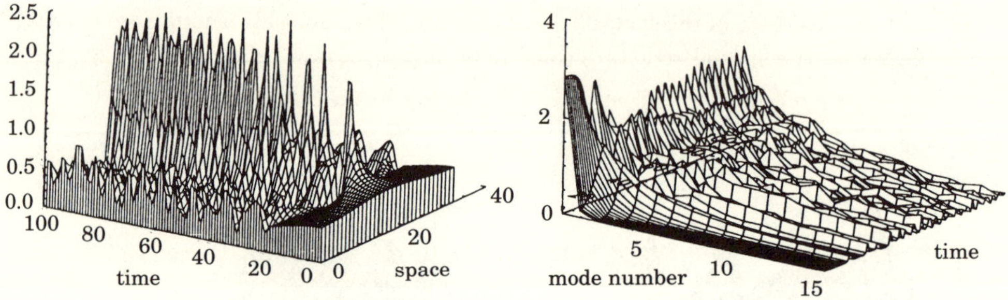

(a) (b)

FIGURE 3.3.9 *Standard scheme:* $m = 2$, $N = 32$.

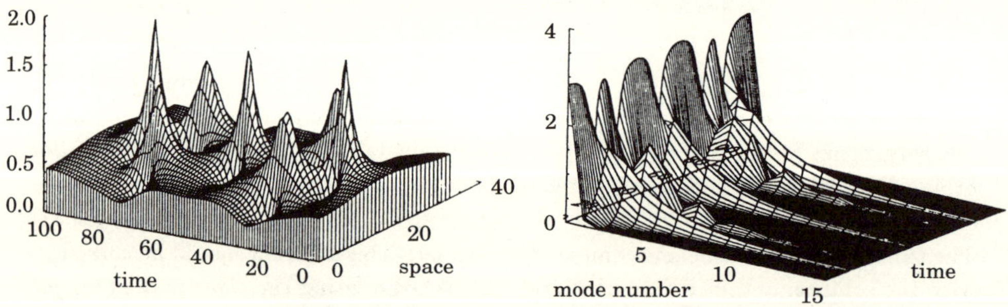

(a) (b)

FIGURE 3.3.10 *Integrable scheme:* $m = 2$, $N = 32$.

Finally, m is increased to three, resulting in four unstable modes according to the instability condition (3.3.5). Now the time evolution becomes very complicated, but not chaotic. Figure 3.3.13 shows the results from the standard scheme, using $n = 64$. Although the time evoltuion is apparently completely irregular, the higher order

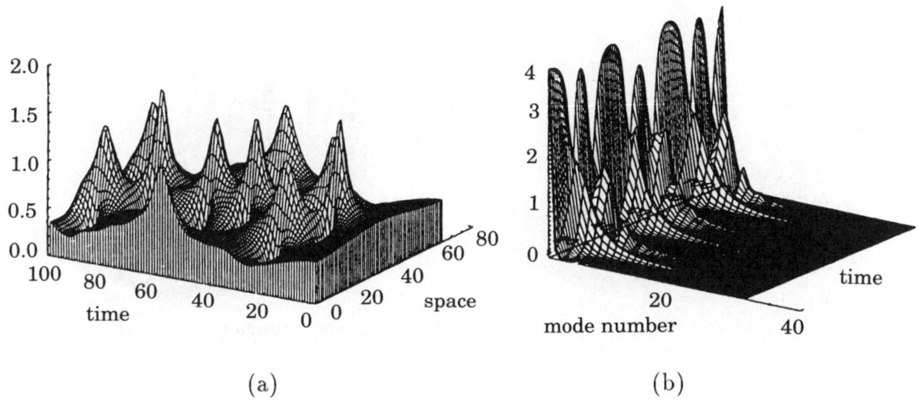

FIGURE 3.3.11 *Standard scheme: $m = 2$, $N = 64$.*

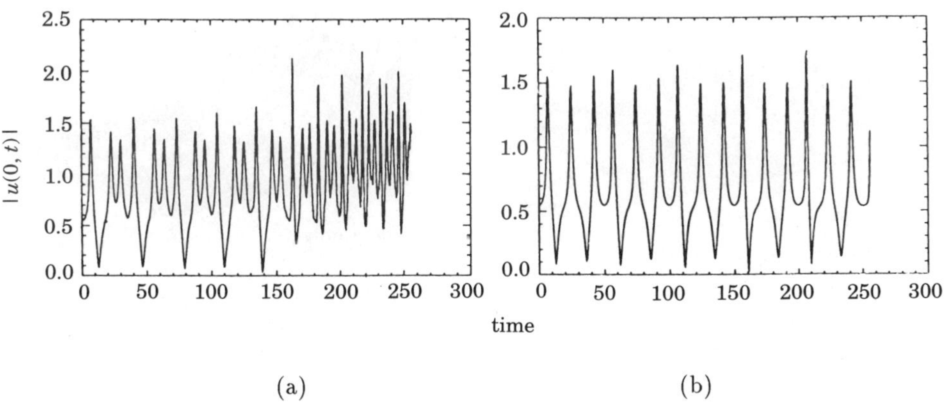

FIGURE 3.3.12 (a), *Standard scheme,* (b), *Integrable scheme: $m = 2$, $N = 64$.*

modes remain relatively unaffected, as seen in Figure 3.3.13b. Again a remarkable improvement is obtained when we resort to the integrable scheme, as shown in Figure 3.3.14 for $N = 32$. From this Figure it is clear that the time evolution is not simple!

When the NLS equation is perturbed its integrability is lost, in general. For instance, Bishop, Flesch, Forest, McLaughlin and Overman [1990] and Forest [1988] study the

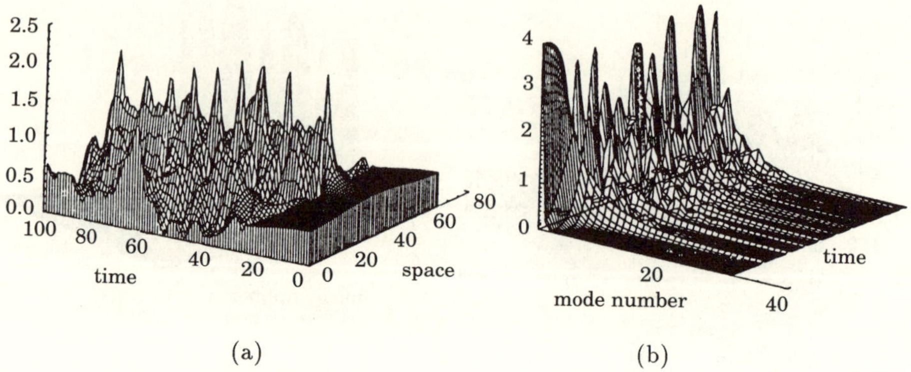

FIGURE 3.3.13 *Standard scheme:* $m = 3$, $N = 64$.

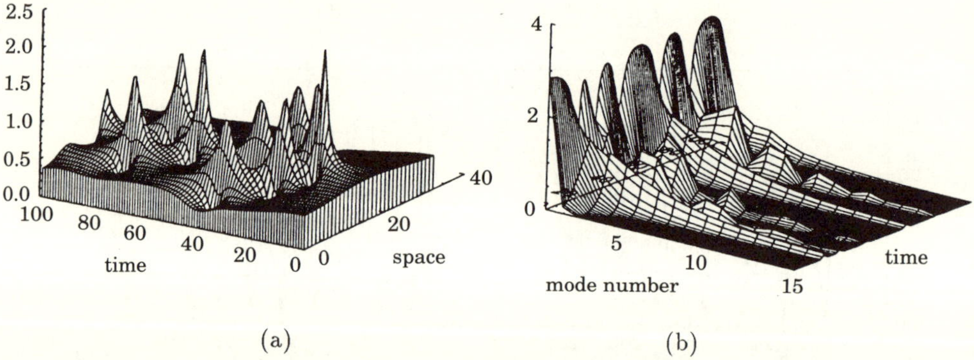

FIGURE 3.3.14 *Integrable scheme:* $m = 3$, $N = 32$.

following perturbed problem (apart from an elementary transformation),

$$-\mathrm{i}u_t + u_{xx} + |u|^2 u = \mathrm{i}\epsilon\left[\alpha u + \Gamma \exp(-\mathrm{i}t)\right], \tag{3.3.18}$$

where α, Γ are real constants, $\epsilon = 1$ and with the initial condition given by, $u(x, 0) = c + b\cos(\mu x)$ where $\mu = 2\pi/L$, $L = 12\sqrt{0.26}$ and c and b are complex constants. Fixing $\alpha = 0.155$, the equation is destabilized by increasing the magnitude of the driver, Γ. Chaos is found for $\Gamma = 0.275$, using $c = -0.4641 + 0.6103\mathrm{i}$, $b = 0.3883 - 0.5206\mathrm{i}$. Herbst and Ablowitz [1989a] considered the qualitative differences between solutions of the perturbed NLS equation using the finite difference methods (3.3.15,17). Solutions are compared as the value of ϵ is increased, with the values of the remaining parameters fixed.

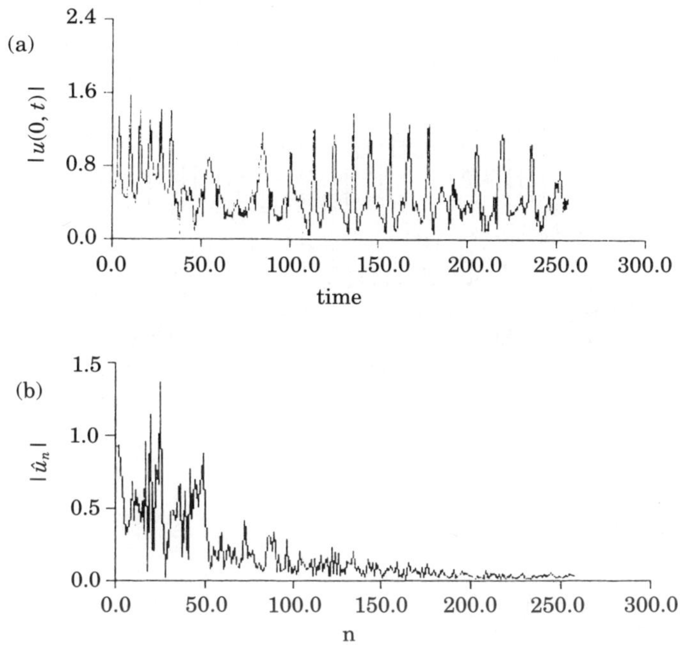

FIGURE 3.3.15 *The NLS equation, standard differences, $N = 32$:*
(a), *the time evolution of the modulus of the solution;*
(b), *the Fourier spectrum of the time evolution of the modulus.*

For $\epsilon = 0$, the solutions from both methods for all N tried, show quasi-periodic behavior, illustrating that only certain regions of the phase-space of scheme (3.3.15) contain chaotic solutions. For small positive values of ϵ, the damping dominates and the solutions from both methods quickly settle down to a rest state. However, as the value of ϵ increases into the chaotic region, the solutions from the two methods again differ markedly. The modulus of the solution at $x = 0$ up to a time $T = 256$, using $\epsilon = 1$ and $N = 8$ for the two schemes corresponding to (3.3.15,17) are shown in Figures 3.3.15a and 3.3.16a. Although both solutions display irregular behavior, the qualitative features of the solutions are quite different. This is supported by the Fourier spectra of the two solutions, which are shown in Figures 3.3.15b and 3.3.16b.

The solution obtained from the integrable method (3.3.17) is dominated by a few low frequencies, superimposed on a continuous background. This is in contrast with the solution of the nonintegrable method (3.3.15), where the Fourier spectrum is more evenly spread over the whole domain. (For further details see Herbst and Ablowitz [1989a].)

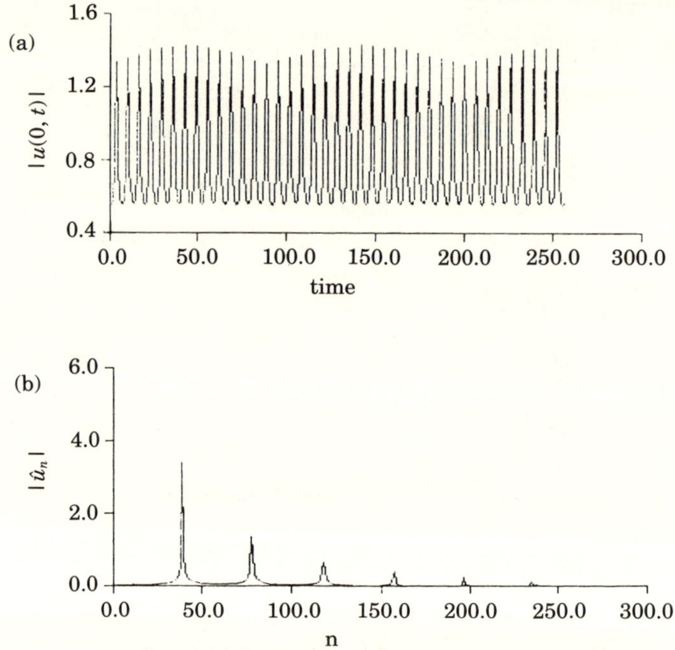

FIGURE 3.3.16 *The NLS equation, the integrable difference scheme, $N = 32$: (a), the time evolution of the modulus of the solution; (b), the Fourier spectrum of the time evolution of the modulus.*

Therefore the homoclinic structure associated with the NLS equation (3.3.2) may be obtained from the N-soliton solution of the defocusing NLS equation (3.3.10). As in the case of nonlinear ordinary differential equations, proximity to homoclinic orbits can act as a source of chaos in weakly perturbed problems. This provides an explanation for the irregular behavior observed with nonintegrable discretizations of the NLS equation (3.3.2) at intermediate levels of discretization. For instance, the nonintegrable scheme (3.3.15) may be viewed as a perturbation of the integrable scheme (3.3.17), i.e. (3.3.15), may be written as

$$i\frac{dU_j}{dt} + (U_{j-1} - 2U_j + U_{j+1})/h^2 + |U_j|^2(U_{j-1} + U_{j+1}) \\ = |U_j|^2(U_{j-1} - 2U_j + U_{j+1}). \qquad (3.3.19)$$

This perturbation, of order $h^2|u|^2u_{xx}$, is sufficient to allow orbits to become unstable in the proximity of the homoclinic structure associated with the integrable scheme, which, in turn, is an approximation of the homoclinic structure described above for the analytical equation. Proximity to or frequent homoclinic crossings leads to chaotic temporal behavior. We also note that McLaughlin and Schober [1990] analize the chaotic structure by I.S.T. spectral decomposition.

As we remarked above, the numerical results of Herbst and Ablowitz [1989a,b] and Ablowitz and Herbst [1990a] illustrate the marked differences in the solutions of two finite difference methods (3.3.15) and (3.3.17); identical, except for the small change in the discretization of the nonlinear term. Both are second order accurate and the usual comparisons based on convergence estimates predict similar results for the two methods. The fact that this is not the case in practice, stems from the crucial difference between the two methods — that of integrability. The situation for the Fourier methods is similar to the one described above. For $N = 2$ the methods are integrable and reasonable qualitative agreement with the analytical solution is obtained (Weideman [1986]). This qualitative agreement is lost when N is increased and the solution becomes chaotic as well. The exponential convergence only comes into play if N is increased even further; considerably beyond the number of modes one would expect intuitively. The situation is analogous to the time step restrictions imposed on numerical methods to prevent numerical instabilities; the nonintegrable scheme (3.3.15) being analogous to conditionally stable methods (N needs to be large enough) and the integrable scheme (3.3.17) analogous to the unconditionally stable schemes.

NHI is similar to the chaos observed by Bishop, Flesch, Forest, McLaughlin and Overman [1990] and Forest [1988] in their studies of equation (3.3.18). They find good qualitative description of the analytical situation by means of a two-mode Fourier truncation that is known to be integrable for the nearby unperturbed problem. It is unclear whether the inclusion of an intermediate number of modes, will still give a reasonable qualitative description of the analytical situation. Their experiments are based on a forced/damped perturbation. If we consider the Hamiltonian situation, low mode truncations are not likely to be effective.

Admittedly, integrability is a special property. Numerically viable integrable discretizations may not always be available and many problems are not even close to integrable ones. The danger of confusing numerically induced chaos with physical chaos in these cases, emphasizes the importance of current attempts at estimating the

Fourier dimension of inertial manifolds (see, for example, Doering, Gibbon, Holm and Nicolaenco [1987, 1988]). Knowledge of the finite dimensionality of an attractor is of considerable value and estimates of the dimension might provide information regarding the resolution required for eliminating spurious numerical effects and the ability of the numerical scheme to capture the intrinsically important properties of the nonlinear evolution equation at intermediate levels of discretization.

These results indicate that NHI is in a sense a generic phenomenon. We expect to find the instability with wide classes of discretizations of problems possessing a homoclinic structure. In addition, because of the physical significance of equations such as NLS and sine-Gordon, one has to be careful not to confuse NHI with true physical behavior, especially in more complicated problems.

3.3.6 Duffing's Equation and Mel'nikov Analysis. In order to investigate further some of the issues related to NHI, Herbst and Ablowitz [1990a] considered Duffing equation in the form

$$\frac{\mathrm{d}^2 w}{\mathrm{d}x^2} - w + 2w^3 = 0 \tag{3.3.20}$$

which can be obtained from the NLS equation (3.3.1) by the reduction $u(x,t) = \exp(it)w(x)$, where $w(x)$ is a real function. Employing similar reductions to the discretizations (3.3.15,17), one obtains an integrable discretization (arising from (3.3.17))

$$W_{n+1} = V_n, \qquad V_{n+1} = -W_n + \frac{2\mu V_n}{1 + h^2 V_n^2}, \tag{3.3.21}$$

and a nonintegrable discretization of (3.3.20) (arising from (3.3.15))

$$W_{n+1} = V_n, \qquad V_{n+1} = -W_n + \frac{2\mu V_n}{1 + h^2 V_n^2} + h^4 V_n^3 \left(\frac{1 - 2V_n^2}{1 + h^2 v_n^2} \right), \tag{3.3.22}$$

with $\mu := 1 + \frac{1}{2}h^2$, where both discretizations have been written as first order systems. Both these schemes are second order accurate with leading order truncation errors of the form

$$T_I = (\tfrac{1}{12}u_{xxxx} + \tfrac{1}{2}u^2 u_{xx})h^2,$$
$$T_{II} = \tfrac{1}{12}u_{xxxx}h^2,$$

and being derived from (3.3.17,15), inherit many of the properties of the schemes. Furthermore both reductions are symplectic (i.e., area preserving) and (3.3.21) inherits the integrability properties of (3.3.17). One constant of the motion

$$H(V,W) := \tfrac{1}{2}\left[V^2W^2 + (V^2 + W^2 - 2\mu VW)/h^2\right] = E, \tag{3.3.23}$$

where E is a constant, is easily obtained.

For intermediate values of h the solutions of (3.3.21) are constrained to the curve (3.3.23) whereas the solutions of (3.3.22) display chaotic behavior, which is analogous to the situation for the schemes (3.3.17,15). This is illustrated in Figure 3.3.17 for $h = 0.2$ (note the typical features of nonintegrable symplectic mappings — chaotic regions interspersed with islands of stability, resonant islands, etc.). From Figure 3.3.17, it is observed that the chaotic behavior is confined to a thin region of order 10^{-7} in the vicinity of the homoclinic orbit of the unperturbed problem for a relatively large value of the discretization parameter h. This appears to be much smaller than the second order accuracy might suggest (since chaos exists, one would expect it to be present in a region of order $O(h^2)$).

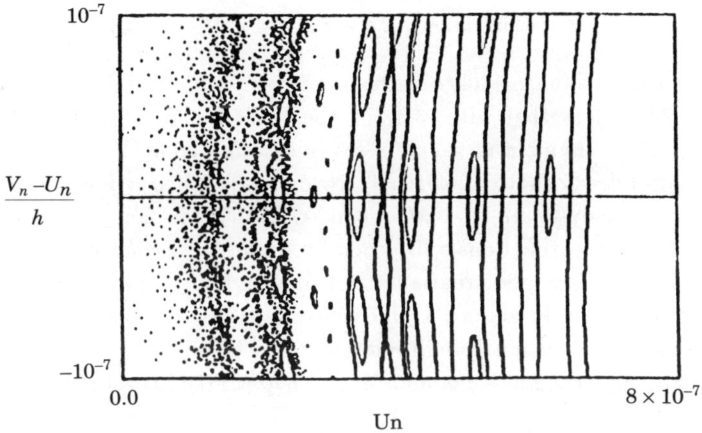

FIGURE 3.3.17 *Chaotic behavior associated with the standard discretization* (3.3.32) *in the vicinity of the homoclinic orbit with* $h = 0.2$.

Herbst and Ablowitz [1990a] investigated this further by studying the mapping

$$W_{n+1} = V_n, \qquad V_{n+1} = -W_n + \frac{2\mu V_n}{1 + h^2 V_n^2} + \epsilon h^2 V_n^3 \left(\frac{1 - 2V_n^2}{1 + h^2 V_n^2} \right), \tag{3.3.24}$$

and its associated Mel'nikov function

$$M(\xi; h) = -2h^2 \sum_{n=-\infty}^{\infty} W_v^2 \left(\frac{\frac{1}{2} - W_n^2}{1 + h^2 V_n^2} \right) \left[W_n^2 V_n + (V_n - \mu W_n)/h^2 \right]. \tag{3.3.25}$$

Although (3.3.24) is different to (3.3.22), it is still a second order symplectic map (assuming $\epsilon = O(h^2)$, or smaller) that can be used to numerically integrate Duffing's equation. A Mel'nikov function $M(\xi; h)$ provides a reliable estimate of the splitting distance between the stable and unstable manifolds at a phase point ξ for a mapping

$$Q_\epsilon : \boldsymbol{v} \rightarrow \boldsymbol{F}(\boldsymbol{x}) + \epsilon \boldsymbol{G}(\boldsymbol{x}, \epsilon), \tag{3.3.26}$$

where $\boldsymbol{x} \in \mathbb{R}^2$ and $\boldsymbol{F}, \boldsymbol{G} : \mathbb{R}^2 \to \mathbb{R}^2$ and is given by

$$M(\xi) = \sum_{n=-\infty}^{\infty} \boldsymbol{G}(\hat{\boldsymbol{x}}_{n-1}(\xi), \epsilon) \wedge \hat{\boldsymbol{v}}_n(\xi), \tag{3.3.26}$$

where $\hat{\boldsymbol{x}}_n(\xi)$ is the homoclinic orbit of the unperturbed, $\epsilon = 0$, problem and $\hat{\boldsymbol{v}}_n(\xi)$ is the nth iterate of the tangent vector to the unperturbed homoclinic orbit (cf. Easton [1984]; Papageorgiou [1988]).

The homoclinic orbit for the unpertubed problem (i.e., (3.3.21)) is given by

$$W_n(\xi) = \sinh \omega \operatorname{sech}(\tilde{\omega} n - \xi), \qquad V_n(\xi) = W_{n+1}(\xi), \tag{3.3.27}$$

with $\mu = \cosh \tilde{\omega}$, $\sinh \omega = (1/h) \sinh \tilde{\omega}$ and where $\hat{\boldsymbol{x}}_n = (W_n, V_n)^t$. Herbst and Ablowitz [1990a] show that with (3.3.27), the Mel'nikov function (3.3.25) is periodic in ξ, has an infinite number of simple zeros (indicating homoclinic tangles and chaos) and exponential decay (both analytically, by expressing it in a Fourier series, and numerically, where they found that $M \propto \exp(-7.57/h)$). In fact, exponential decay of the related Mel'nikov functions will generally occur for second order perturbed systems with Hamiltonians perturbations (and smooth homoclinic orbits). These results suggest why there are fundamental differences between symplectic and nonsymplectic discretizations of certain Hamiltonian systems (symplectic integrators have been studied by amongst others, Kang and Meng-Zhao [1987]; Scovel [1990]).

3.4 Cellular Automata.

We have seen that both solitons and chaos are associated with solutions of partial or ordinary differential equations. However they also exist for totally discrete systems — a cellular automaton may be viewed as an extreme form of discretization where the solution is not only defined at discrete locations in space and time, but the solution itself exists in a small number of states, e.g. only two. During the past few years there has been a significant increase in interest in the study of Cellular Automata which have a wide applicability in computer science, physics, information processing, biology, etc. (see Wolfram [1986] for a collection of papers discussing various aspects of Cellular Automata). Physical applications of Cellular Automata include reaction-diffusion systems, crystal growth and fluid dynamics (in particular, we note the application of numerical simulation of the Navier-Stokes equation through Cellular Automata — Frisch, Hasslacher and Pomeau [1986]). The study of Cellular Automatas in general has been primarily limited to extensive computer experimentation. A major reason for this is the inherent difficulty in giving a representation of the Cellular Automata in such a way that analytical progress can be made.

In this section we turn to the question of finding soliton interactions and developing some analytical insight in Cellular Automata. It turns out that there is a class of

Cellular Automata that admits soliton interactions, bearing a rather close resemblance to KdV solitons. Recently Park, Steiglitz and Thurston [1986] introduced their so-called Parity Rule Filter Automata (PRFA) and numerical studies indicate that the PRFA admit *soliton* solutions. Considering the history of soliton solutions of partial differential equations in multi-dimensions, it is remarkable that the PRFA rule may easily be generalized to yield solitary waves and sometimes even soliton interactions in $2+1$ *and* $3+1$ dimensions (Ablowitz, Herbst and Keiser [1990]).

The prototype model which shall be discussed here referred to as the PRFA (Park, Steiglitz and Thurston [1986]), is given by the following implicit rule. Define the sum

$$S_i^{t+1} = S(a_i^{t+1}) = \sum_{j=1}^{r} a_{i-j}^{t+1} + \sum_{j=0}^{r} a_{i+j}^{t}, \tag{3.4.1}$$

on the interval $-\infty < i < \infty$ where a_i^t denotes values at site i, time t with a_i^t taking on values 0, 1 only. The new state at level $t+1$ is given by

$$a_i^{t+1} = \begin{cases} 0 & \text{if } S(a_i^{t+1}) \text{ is odd or zero,} \\ 1 & \text{if } S(a_i^{t+1}) \text{ is even, nonzero.} \end{cases} \tag{3.4.2}$$

The computation is carried out by sweeping from left to right, assuming that at the initial time there is a finite number of nonzero sites, a_i^t, and that to the left we always have an infinite number of zeros.

An alternative formulation of the PRFA, referred to as the Fast Rule Theorem (FRT), is given by Papatheodorou, Ablowitz and Saridakis [1988],

$$a_{i-r}^{t+1} = \begin{cases} a_i^t, & \text{if } i \notin B(t), \\ \bar{a}_i^t, & \text{if } i \in B(t), \end{cases} \tag{3.4.3}$$

where $\bar{a}_i^t = 1 - a_i^t$, i.e., the complement of a_i^t, and $B(t)$ is a certain set described as follows:

(a) The site of the first nonzero value is an element of $B(t)$.

(b) We place all subsequent sites in steps of $r+1$ bits in $B(t)$ so long as there is at least one nonzero value in any of the intervening $r+1$ bits.

(c) If there is at least $r+1$ zeros after a site in $B(t)$, then we go to the next available nonzero value and repeat step (b).

Steps (a) to (c) are repeated until all nonzero values are exhausted. The FRT theorem is proved by Papatheodorou, Ablowitz and Saridakis [1988] and has been used to study many of the analytical features of the aforementioned Cellular Automata, including, stability, soliton interactions and periodic particles (see also Papatheodorou, Ablowitz and Saridakis [1988]; Fokas, Papadopoulou, Saridakis and Ablowitz [1989]; Keiser [1989]; Fokas, Papadopoulou and Saridakis [1990]).

As an example of the value of the FRT Theorem, we prove the following stability theorem for the PRFA.

THEOREM 3.4.1

The PRFA is a <u>*stable automata;*</u> *namely a finite number of ones at any given time can generate only a finite number of ones at a subsequent time*

Proof. This follows from the fact that if the box set at time t, $B(t)$, is finite (i.e., there is a finite number of ones at time t), then then number of ones at time $t+1$ is finite since the state at time $t+1$ differences from the state at time t only over the set $B(t)$.

We shall consider some examples. In the following the underlined cells are in $B(t)$.

EXAMPLE 3.4.2

Take $r = 3$, basic fast rule concept:

$$\begin{array}{ll} a^t: & \ldots 0\ 0\ 0\ \underline{1}\ 0\ 1\ 1\ \underline{1}\ 1\ 0\ 1\ \underline{0}\ 0\ 0\ 0\ 0\ 0\ \underline{1}\ 1\ 1\ 0\ \underline{0}\ 0\ 0\ \ldots \\ a^{t+1}: & \quad 0\ 0\ \underline{1}\ 1\ 0\ 1\ \underline{0}\ 1\ 1\ 0\ \underline{0}\ 0\ 0\ 0\ 0\ \underline{1}\ 1\ 0\ 1\ \underline{0}\ 0\ 0\ \ldots \end{array}$$

EXAMPLE 3.4.3

Take general r. This is a solitary wave with maximum speed

$$\begin{array}{ll} a^t: & \ldots \overbrace{0\ 0\ 0\ \ldots\ 0}^{r}\ \underline{1}\ \overbrace{1\ 1\ 1\ \ldots\ 1}^{r}\ \underline{0}\ 0\ \ldots \\ a^{t+1}: & \quad 0\ \underbrace{\underline{1}\ 1\ 1\ \ldots\ 1}_{r+1}\ \underline{0}\ \ldots \end{array}$$

This solitary wave has speed $c = r - 1$, i.e., in one time unit the left most q is displaced $r - 1$ units to the left.

EXAMPLE 3.4.4

Take general r. This is general one periodic solitary wave.

$$a^t: \qquad 0\ \ldots\ 0\ \overbrace{S\ S\ S\ \ldots\ S}^{r+1}\ 0\ \ldots,$$

where S is the substring with one nonzero value and z zeros; $S = 1\ \underbrace{0\ 0\ \ldots\ 0}_{z+1}$. The FRT yields

$$a^{t+1}: \qquad 0\ \ldots\ 0\ \overbrace{S\ S\ S\ \ldots\ S}^{r+1}\ 0\ \ldots,$$

which has the same string as a_i^t (hence it is a solitary wave) except that the first one is displaced $r - (1 + z)$ units to th left. If we denote the total number of ones in a_i^t to be ℓ which equals the number of substrings S then $\ell(1 + z) = r + 1$ and the speed is given by

$$c = r - \frac{r+1}{\ell} = \frac{(\ell - 1)r - 1}{\ell}. \qquad (3.4.4)$$

EXAMPLE 3.4.5 Nonreversibility.

Let us call the null string

$$a^t = N_b = \underbrace{0\ 0\ \ldots\ 0}_{r},$$

then by the FRT

$$a^{t+1} = N_b = \underbrace{0\ 0\ \ldots\ 0}_{r+1}.$$

If we call the prenull string PN_b

$$a^t = PN_b = \underbrace{1\ 0\ \ldots\ 0}_{r},$$

then by the FRT

$$a^{t+1} = N_b = \underbrace{0\ 0\ \ldots\ 0}_{r+1}.$$

Clearly the Cellular Automata (3.4.2) is not reversible (i.e., a state has many predecessors); information is lost in the evolution of the Cellular Automata.

The irreversibility of this Cellular Automata makes the existence of solitons even more interesting, since for well-known differential and difference equations mentioned elsewhere in these notes, this is *not* the case.

Periodic particles have the following property

$$a_{i-d}^{t+p} = a_i^t, \tag{3.4.5}$$

where p is the period and d the displacement. Numerically speaking such particles are plentiful and frequently behave like solitons upon interaction with other particles (Park, Steiglitz and Thurston [1986]; Steiglitz, Kamal and Watson [1988]).

A ***basic string*** is a special particle consisting of $r+1$ bits being neither a null nor a prenull string. A particle is thought of as a collection of basic strings. Separate particles at any given time have at least one null string separating them. Two particles which are distinct at a given time (i.e., separated by at least a null string), can join at a late time.

A large class of periodic particles may be constructed by a *linear* difference scheme (Keiser [1989]; Ablowitz and Keiser [1990]). The following parameters need to specified: (a), radius r; (b), period p; (c), a set of displacements $d_1, d_2, \ldots, d_p$ $(0 \le d_j \le r-1)$, The constructed particle evolves through p orbital phases with subsequent leading ones located at positions $d_1, d_2, \ldots, d_p$, for time steps $t = 0, 1, 2, \ldots, p-1$. The formula for the periodic particle is given by

$$\chi_i = \chi_{i-\rho} + (1 - 2\chi_{i-\rho}) g_i^{(p)}, \tag{3.4.6a}$$

$$\chi_i = 0, \qquad i = -1, -2, \ldots, -\rho \qquad (\chi_0 = 1), \tag{3.4.6b}$$

with $g_i^{(p)}$ defined by (3.4.7) below, unless a condition "D" is satisfied, where $\rho = pr - \sum_{m=1}^{p} d_m$. Condition "D" is a set of linear diophantine equations which, if satisfied, result in an invalid particle (too many zeros separate nontrivial basic strings). $g_i^{(p)}$ is given by

$$g_i^{(p)} = \sum_{j=1}^{p} A_j b_j, \qquad a_j = (-1)^{j+1} j, \qquad b_j = \sum_{j_1, j_2, \ldots, j_k}^{p} \delta_{j_1} \delta_{j_2} \ldots \delta_{j_p}, \tag{3.4.7}$$

where $\sum_{j_1, j_2, \ldots, j_k}^{p}$ is the sum of all unique products of delta functions

$$\begin{aligned} \delta_j &= \delta\left(\frac{i - \rho_{j-1}}{r+1}\right), \\ \rho_j &= \rho_{j-1} + r d_j, \qquad \rho_0 = 0, \\ \delta(k) &= \begin{cases} 1, & \text{if } k \text{ is a nonnegative integer,} \\ 0, & \text{otherwise.} \end{cases} \end{aligned}$$

So the first three forcing functions $g_i^{(p)}$ for $p = 1, 2, 3$ are

$$g_i^{(1)} = \delta_1 = \delta\left(\frac{i - \rho_0}{r+1}\right) = \delta\left(\frac{i}{r+1}\right), \tag{3.4.8a}$$

$$g_i^{(2)} = \delta_1 + \delta_2 - 2\delta_1\delta_2, \tag{3.4.8b}$$

$$g_i^{(3)} = \delta_1 + \delta_2 + \delta_3 - 2(\delta_1\delta_2 + \delta_2\delta_3 + \delta_3\delta_1) + 3\delta_1\delta_2\delta_3. \tag{3.4.8c}$$

All the above periodic particles satisfy a reversability criterion. Namely if at any time we run the particles backwards in time, the subsequent evolution is obtained by doing the FRT from the right or simply sweeping from right to left in the standard PRFA rule. Indeed Ablowitz and Keiser [1990] show that a state can be reversed in general, so long as no prenull basic string is created, left to right.

A particularly interesting group of periodic particles are *basic strings* Write the string in the form

$$B^t = \underbrace{S\,S\,\ldots\,S}_{\alpha},$$

where α is the number of substrings S. If z_s denotes the number of zeros in S then $\alpha(\ell_s + z_s) = r + 1$ where ℓ_s denotes the number of ones in S and $\ell = \alpha\ell_s$ is the total number of ones in B^t. Then the FRT shows that B^t evolves as a solitary wave with period $p = \ell_s = \ell/\alpha$ and speed $c = [(\ell - 1)r - 1]/\ell$. (Note that the speed is the same as in (3.4.4).) The number of ones characterize the *speed* and *amplitude* of the basic string B^t. The period is the number of *orbital phases inherent* in the solitary wave.

Based upon experience with solitons and solitary wave interactions it is natural to consider what happens when one solitary wave meets another. In particular, is the interaction solitonic? Indeed for basic strings described the answer is affirmative (cf.

Fokas, Papadopoulou, Saridakis and Ablowitz [1989]). Namely if at $t = 0$ we take a state

$$B_1 Z_\alpha B_2, \tag{3.4.9}$$

where B_1, B_2 are two basic strings and Z_α is a collection of zeros with the number of bits being of width $r + 1 + z_\alpha$. If the number of ones in B_i is ℓ_i $(i = 1, 2)$, then the speeds c_i satisfy $c_1 > c_2$ if $\ell_1 > \ell_2$ and $c_1 < c_2$ if $\ell_1 < \ell_2$. If $c_1 > c_2$, the basic strings never interact and $z_\alpha \to \infty$ as $t \to \infty$. If $c_1 < c_2$, the basic strings will interact. Depending on the period, orbital phase and distribution of the nonzero bits in he basic strings, a variety of interactions can take plane (they can "oscillate" a finite number of times), but as $t \to \infty$, the basic strings interchange their positions and the final state is found to be solitonic

$$B_2 Z_\beta B_1, \tag{3.4.10}$$

where $Z_\beta = r + 1 + z_\beta$ and $z_\beta \to \infty$ as $t \to \infty$. The phase shift due the interaction can be determined explicitly. It is found the the fast soliton is pushed forward (with phase shift δ_+) and the the slow soliton is pushed back (with phase shift δ_-). The phase shift formula for $p = 1$ waves is given by

$$\delta_+ = 2\ell_1(r - c_2), \qquad \delta_2 = -2(r + 1), \tag{3.4.11}$$

and the total phase shift is $\Delta = \delta_+ - \delta_-$. It should be noted that for general p, the phase shift of the slow soliton is the same whereas for the fast soliton, in general, it depends on the relative distribution of the nonzero bits and the number of times the particles oscillate. However, with an interpretation of a uniformly moving *envelope*, the phase shift formula (3.4.11) can be employed if the particles do not oscillate.

It is interesting to note that interactions between general particles can produce a variety of solitonic/nonsolitonic collisions. Goldberg [1988] gives conditions under which a two particle collision will be locally solitonic (i.e., they separate for a while but could interact again). He also gives a notion of energy $D^t = \sum_{i=-\infty}^{\infty}(a_i - a_{i-t-1})$ and proves that E^t is nonincreasing, i.e., $E^{t+1} \leq E^t$. Periodic particle conserve energy. Ablowitz and Keiser [1990] have investigated a wide variety of two-particle collisions. They find that sometimes particles may interact solitonically and be reversible, solitonically and be nonreversible, and nonsolitonically and be nonreversible.

It is interesting to note that the Cellular Automata can also be written as a local difference scheme (Ablowitz, Herbst and Keiser [1990]). For reference we give the formula below. The Cellular Automata in equation (3.4.2) is a special case of

$$a_i^{t+1} = f(a_i^{t+1}, a_{i+1}^{t+1}, \ldots, a_{i+r}^{t+1}; a_{i-r}^{t+1}, \ldots, a_{i-1}^{t+1}). \tag{3.4.12}$$

We rewrite the rule as

$$a_i^{t+1} = f(a_1, a_2, \ldots, a_n), \tag{3.4.13}$$

and then the Cellular Automata takes the form

$$a_i^{t+1} = \sum_{k=2}^{n} (-1)^k \beta_k \sum_{j_1, j_2, \ldots, j_k} a_{j_1} a_{j_2} \ldots a_{j_k}, \tag{3.4.14}$$

where $\beta_k = 2^{k-1} - 1$ and $\sum_{j_1, j_2, \ldots, j_k}$ denotes the summands over all distinct combinations of products of k different forms.

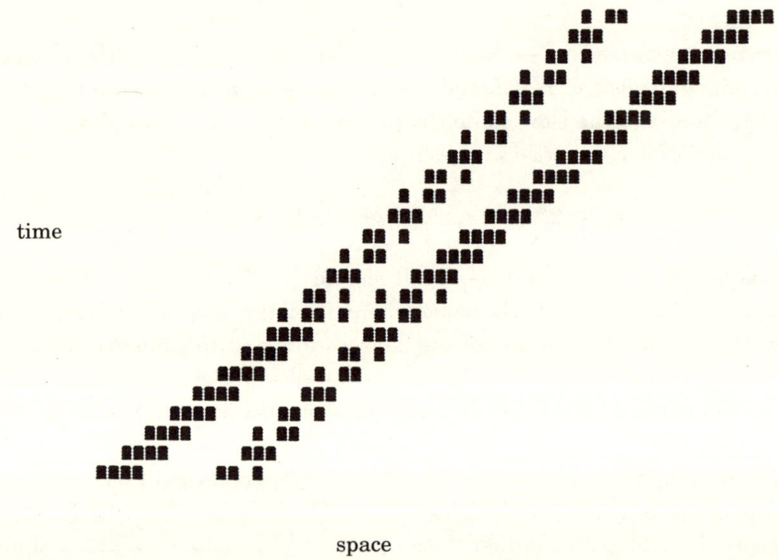

FIGURE 3.4.1 *Soliton interaction of two particles* ($r = 3$)*: the two particles.*

Now we shall discuss examples of soliton interactions in 1 + 1-, 2 + 1- and 3 + 1-dimensions. Figure 3.4.1 shows two particles, initially completely separated by a large number of zeros, emerging intact after an interaction. Note that a one is represented by a black box and a zero by a blank space. The resemblance with the KdV soliton interaction is striking. Note, for instance, a similar phase shift. Changing the representation of the Cellular Automata to plot the 'sum profile' (i.e., S_i^{t+1} from equation (3.4.1)), allows the resemblance to be even more pronounced. In Figure 3.4.2 the particles of Figure 3.4.1 are represented by their sum profiles (note that the Cellular

Automata rule is not changed). The sum profile at level $t+1$ is simply given by S_i^{t+1}, $-\infty < i < \infty$, in (3.4.1). It should be noted that the value of $S(a_i^{t+1})$ is always in the range zero to $2r+1$.

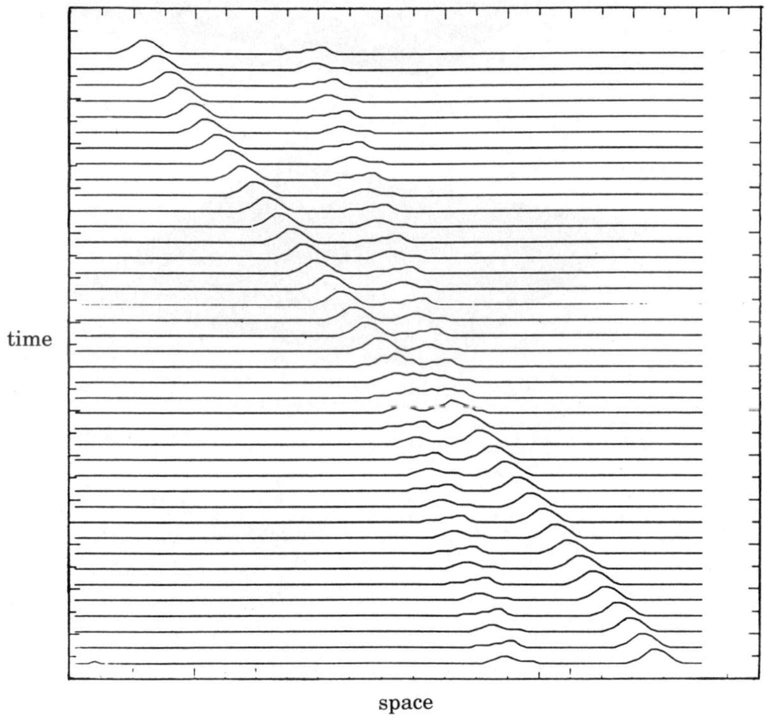

FIGURE 3.4.2 *Soliton interaction of two particles: representation by their sum profiles.*

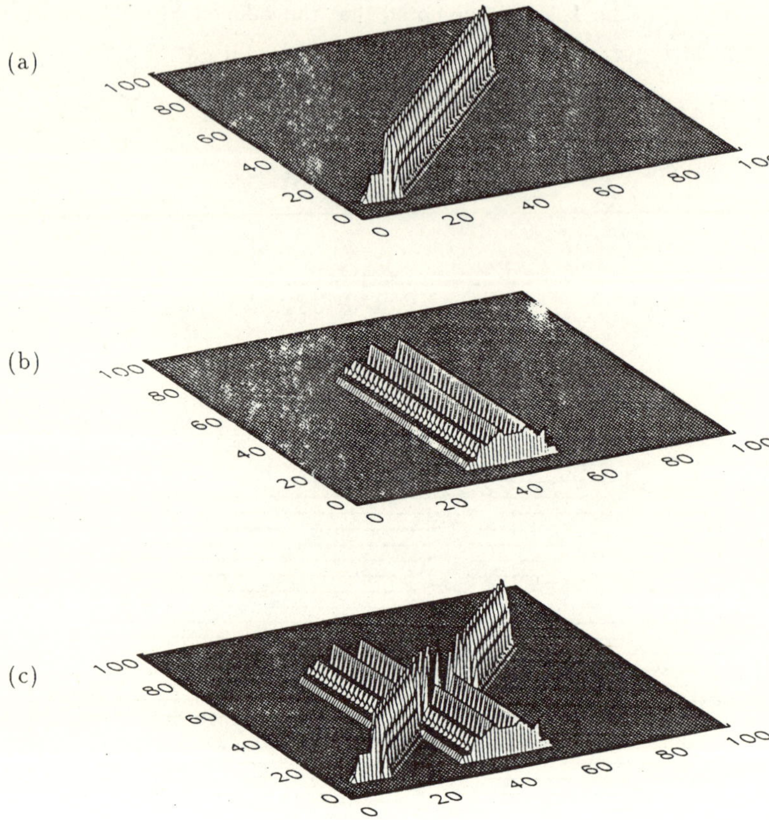

FIGURE 3.4.3 *Soliton interaction in* $2+1$*-dimensions.*

These ideas can be easily extended to higher dimensions. Consider a two-dimensional grid and let a_{ij}^t denotes the value at site (i, j) at time level t. In order to advance to the next time level the one-dimensional PRFA is applied to each of the successive horizontal levels, sweeping from left to right. Next the PRFA is successively applied to each of the vertical levels, sweeping from top to bottom (these sweeping directions are clearly not important). The two-dimensional automaton is represented by its sum profile by adding the horizontal and vertical sums. The main observation is that we find that in multi-dimensions it is easy to construct localized solitary waves. Indeed, interactions of these waves can be solitonic. Figures 3.4.3 and 3.4.4 show the interaction of two different pairs of particles in $2+1$-dimensions ($r = 3$). We start initially with two localized particles. The time evolutions of the two individual

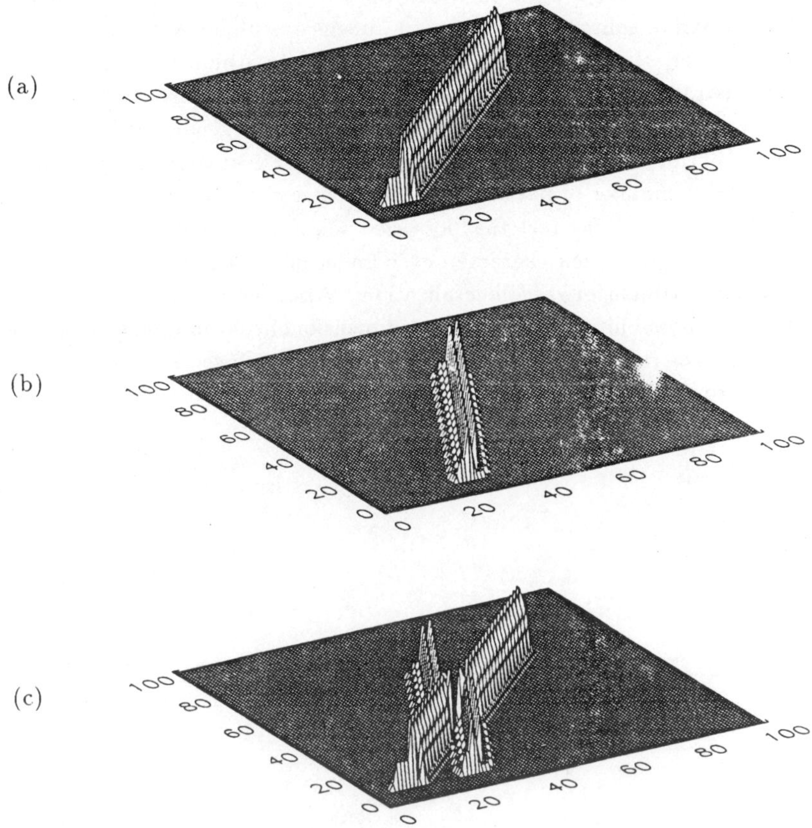

FIGURE 3.4.4 *Soliton interaction in 2 + 1-dimensions.*

particles, without any interaction, are shown in Figures 3.4.3a,b and 3.4.4a,b. As the time evolves, the images of the particles at the old positions are frozen. In Figures 3.4.3c and 3.4.4c the particles are allows to interact. Again the particles emerge virtually intact from the interaction, apart from a phase shift — the particles behave like solitons.

Recently, Ablowitz, Keiser and Takhtajan [1991a,b] have found modifications to the PRFA and FRT which are stable, reversible and have an arbitrary number of states (i.e., a "q-state" Cellular Automata).

These ideas are generalized in a straightforward manner to three spatial dimensions, or, for that matter, to $n+1$-dimensions, for any $n \geq 1$. We apply the one-dimensional PRFA successively in each of the three spatial directions in order to advance to the next time level. Again the order in which the rule is applied is not important. It is difficult to illustrate soliton interactions in dimensions higher than $2+1$ in a visally attractive manner; the interested reader is referred to Ablowitz, Herbst and Keiser [1990] for illustrations of soliton interactions in $3+1$-dimensions.

Finally we note that the way in which the PRFA is generalized here, increases dissipation. By this we mean that the 'energy' decrease is greater in higher dimensions and that a larger number of particles will loose their character under propagation by this rule. Nevertheless, the fact that localized solitary waves exist and the fact that some particles maintain their character, even under interactions as we have seen, merit interest in these straightforward generalizations. Also, these generalizations are by no means the only possibilities. In fact, $2+1$-dimensional soliton interactions have been observed for other rules based on the FRT mentioned above. Although not quite as straightforward as the ones described above, preliminary studies show that it may be less dissipative.

Chapter Four

Inverse Scattering for Integro-Differential Equations

4.1 Introduction.

In this Chapter we discuss I.S.T. for integro-differential equations, primarily considering as prototype examples, the Intermediate Long Wave (ILW) equation

$$u_t + \delta^{-1} u_x + 2uu_x + \mathrm{T}u_{xx} = 0, \tag{4.1.1a}$$

where δ is a constant and $\mathrm{T}u$ is the singular integral operator

$$(\mathrm{T}f)(x) := \frac{1}{2\delta} \fint_{-\infty}^{\infty} \coth\left\{\frac{\pi}{2\delta}(y-x)\right\} f(y)\,\mathrm{d}y, \tag{4.1.1b}$$

and the Benjamin-Ono (BO) equation

$$u_t + 2uu_x + \mathrm{H}u_{xx} = 0, \tag{4.1.2a}$$

where $\mathrm{H}u$ is the Hilbert transform [formally letting $\delta \to \infty$ in (4.1.1)]

$$(\mathrm{H}f)(x) := \frac{1}{\pi} \fint_{-\infty}^{\infty} \frac{f(y)}{y-x}\,\mathrm{d}y \tag{4.1.2b}$$

($\fint_{-\infty}^{\infty}$ represents the Cauchy principal value integral). In Chapter 1 we discussed how the ILW equation arises in the context of long internal gravity waves in a stratified fluid with finite depth, characterized by the parameter δ. We note that the ILW equation (4.1.1) reduces to the KdV equation

$$u_t + 2uu_x + \tfrac{1}{3}\delta u_{xxx} = 0, \tag{4.1.3}$$

as $\delta \to 0$ (shallow water limit), and to the BO equation (4.1.2), as $\delta \to \infty$ (deep water limit). The BO equation physically arises in the context of long internal gravity waves in a stratified fluid (Benjamin [1967]; Davies and Acrivos [1967]; Ono [1975]). In addition to the significance of the ILW and BO equations as models of physically interesting phenomena, the I.S.T. schemes associated with these two equations are interesting from a mathematical point of view.

The limit of the I.S.T. scheme for the ILW equation as $\delta \to 0$ reduces to that for the KdV equation (Satsuma, Ablowitz and Kodama [1979]; Kodama, Ablowitz and Satsuma [1982]), and the limit is rather straightforward. The limit of the I.S.T. scheme for the ILW equation as $\delta \to \infty$ to that for the BO equation was established by Santini, Ablowitz and Fokas [1984a], though it is subtle and presents some difficulties. There are certain significant differences between the I.S.T. schemes for the ILW equation (for finite δ) and the BO equation. The I.S.T. scheme for the ILW (for finite δ) equation is conceptually similar to that for the KdV equation; however, the I.S.T. scheme for the BO equation is similar in many respects to that of some 2 + 1-dimensional equations (for example the Kadomtsev-Petviashvili equation — which we shall discuss in Chapter 5). Therefore the limit $\delta \to \infty$ in some sense provides a bridge between the I.S.T. schemes for 1 + 1- and 2 + 1-dimensional problems.

In §4.4 we discuss other integro-differential evolution equations which are solvable by an I.S.T. scheme.

4.2 The Intermediate Long Wave Equation.

It has been established that the ILW equation (4.1.1) possesses the following properties: a Lax pair (Satsuma, Ablowitz and Kodama [1979]); N-soliton solutions (Joseph and Egri [1978]; Chen and Lee [1979]; Matsuno [1979b]); an infinite number of conserved quantities; a Bäcklund transformation (Satsuma, Ablowitz and Kodama [1979]; Kodama, Satsuma and Ablowitz [1981]); Hamiltonian formalism and action-angle variables (Radul [1985]); a bi-Hamiltonian structure and a recursion operator (Santini [1989b]). The I.S.T. scheme for the ILW equation was developed by Kodama, Satsuma and Ablowitz [1981] (see also Kodama, Ablowitz and Satsuma [1982]), and which we outline below. The periodic case of the ILW equation has been considered by Ablowitz, Fokas, Satsuma and Segur [1982].

4.2.1 The Direct Scattering Problem. The Lax pair for the ILW equation (4.1.1) is

$$\mathrm{i}v_x^+ + (u-\lambda)v^+ = \mu v^-, \tag{4.2.1a}$$

$$\mathrm{i}v_t^\pm + \mathrm{i}\left(2\lambda+\delta^{-1}\right)v_x^\pm + v_{xx}^\pm + (\pm\mathrm{i}u_x - \mathrm{T}(u_x)+\nu)v^\pm = 0, \tag{4.2.1b}$$

where it is convenient to parametrize λ, μ as:

$$\lambda(k) = -\tfrac{1}{2}k\coth(k\delta) \qquad \text{and} \qquad \mu(k) = \tfrac{1}{2}k\,\mathrm{cosech}(k\delta),$$

k is a constant which is interpreted as a spectral parameter and ν is an arbitrary constant. Given u, equation (4.2.1a) defines a Riemann-Hilbert problem in a strip of the complex z-plane (where z is the complex extension of x), $v^{\pm}(x)$ represent the boundary values of functions which are analytic in the horizontal strips between $\mathrm{Im}(z) = 0$ and $\mathrm{Im}(z) = 2\delta$ ($z = x + \mathrm{i}y$), and periodically extended vertically (that is $v^{\pm}(x) := \lim_{y\downarrow 0} v(x \pm \mathrm{i}y)$; $v^{-}(x) = v^{+}(x + 2\mathrm{i}\delta)$).

By using the operator $\mathrm{T}(\bullet)$, $v^{\pm}(x)$ may be written in the form

$$v^{+}(x) = \lim_{y\downarrow 0} v(x) = \tfrac{1}{2}(\mathrm{I} - \mathrm{i}\mathrm{T})\psi(x), \tag{4.2.2a}$$

$$v^{-}(x) = \lim_{y\uparrow 0} v(x) = \tfrac{1}{2}(\mathrm{I} + \mathrm{i}\mathrm{T})\psi(x), \tag{4.2.2b}$$

where $\psi(x)$ defined on the real axis is a proper function for the operator $\mathrm{T}(\bullet)$ (i.e., $\psi(x) \in L^1(\mathbb{R})$), and $\psi(x)$ satisfies the Hölder condition on the real axis, i.e., there exist constants C and α such that $|v(x) - v(y)| < |x - y|^{\alpha}$, for some $0 < \alpha < 1$, and for $\mathrm{Im}\, z \neq 0 \pmod{2\delta}$, $v(z)$ is given by

$$v(z) = \frac{1}{4\mathrm{i}\delta} \int_{-\infty}^{\infty} \coth\left\{\frac{\pi}{2\delta}(y - z)\right\} \psi(x)\, \mathrm{d}y. \tag{4.2.2c}$$

Hence from equations (4.2.2) we have the relation

$$\mathrm{T}(v^{+} - v^{-}) = \mathrm{i}(v^{+} + v^{-}).$$

From the periodicity (period 4δ) of (4.2.2c), we have the relation between $v^{\pm}(x)$ in the form

$$v^{-}(x) = v^{+}(x + 2\mathrm{i}\delta). \tag{4.2.3}$$

We remark that the linear equation (4.2.1a) is solvable without the use of Riemann-Hilbert theory since it may be viewed as a differential-difference equation; this is a consequence of the periodicity condition (4.2.3).

It is convenient to define

$$W^{\pm}(x;k) := v^{\pm}(x;k) \exp\left\{\tfrac{1}{2}\mathrm{i}k(x \mp \mathrm{i}\delta)\right\}.$$

Hence (4.2.1) becomes

$$\mathrm{i}W_x^{+} + [\zeta_{+} + 1/(2\delta)]\,(W^{+} - W^{-}) = -uW^{+}, \tag{4.2.4a}$$

$$\mathrm{i}W_t^{\pm} - 2\mathrm{i}\zeta_{+}W_x^{\pm} + W_{xx}^{\pm} + [\pm\mathrm{i}u_x - \mathrm{T}(u_x) + \rho]\,W^{\pm} = 0, \tag{4.2.4b}$$

where

$$\zeta_{\pm}(k) : = \tfrac{1}{2}k \pm \tfrac{1}{2}k\coth(k\delta) \mp 1/(2\delta), \tag{4.2.4c}$$

$$\rho = -k\zeta_{+} + \tfrac{1}{4}k^2 + \nu \tag{4.2.4d}$$

($\zeta_-(k)$ will be required shortly). The solution of (4.2.4a) which we are interested in is given by the integral equation

$$W^+(x;k) = W_0(x;k) + \int_{-\infty}^{\infty} G(x,y;k)\, u(y)\, W^+(y;k)\, \mathrm{d}y,$$

where $W_0(x;k)$ is the solution of the homogeneous version of (4.2.4a) (that is $u(x) = 0$), and $G(x,y;k)$ is the Green's function satisfying

$$\mathrm{i}\frac{\partial}{\partial x}\{G^+(x,y;k)\} + [\zeta_+ + 1/(2\delta)]\,[G^+(x,y;k) - G^-(x,y;k)] = -\delta(x-y), \tag{4.2.5}$$

where $G^\pm(x,y;k) = G(x \mp \mathrm{i}\delta, y;k)$. Note that the Green's functions $G^+(x,y;k)$ and $G^-(x,y;k)$ are analytic functions in the upper (+) and lower (−) strips respectively, defined with respect to the real axis of the complex z-plane, where z is the complex extension of x. From (4.2.5) we obtain the Fourier representation of the Green's function $G(x,y;k)$

$$G(x,y;k) = \frac{1}{2\pi}\int_C \hat{G}(p;k)\, \exp\{\mathrm{i}p(x-y)\}\, \mathrm{d}p, \tag{4.2.6a}$$

where

$$\begin{aligned}\hat{G}(p;k) &= \frac{1}{2}\,\frac{\operatorname{cosech}(p\delta)}{\zeta_+(p) - \zeta_+(k)} \\ &= \frac{\exp(-p\delta)}{p - [\zeta_+(k) + 1/(2\delta)][1 - \exp(-2p\delta)]},\end{aligned} \tag{4.2.6b}$$

and the contour C is taken to be a contour from $-\infty$ to ∞, suitably chosen. From (4.2.6b) we see that $\hat{G}(p;k)$ has poles at $p = 0$, $p = \zeta_+^{-1}(\zeta_+(k))$. Since $\zeta_+^{-1}(\bullet)$ is a multivalued function, there are an infinite number of poles. Therefore from equation (4.2.6) we have

$$G_\pm(x,y;k) = \frac{1}{2\pi}\int_{C_\pm} \frac{\exp\{\mathrm{i}p(x-y) - p\delta\}}{p - [\zeta_+(k) + 1/(2\delta)][1 - \exp(-2p\delta)]}\, \mathrm{d}p, \tag{4.2.7}$$

where the contours $C_\pm$ for $G_\pm(x,y;k)$ are indented below (+), above (−) the singularities on $\mathrm{Re}\,p = 0$. Note that the Green's functions $G_+(x,y;k)$ and $G_-(x,y;k)$ are the boundary values of analytic functions in the upper (+) and and lower (−) half-planes respectively, defined with respect to the real axis of the complex κ-plane, where κ is the complex extension of k. We distinguish these $G_\pm$ from $G^\pm$ by now using subscripts as opposed to superscripts used previously.

In order to develop the scattering theory, it turns out that we only need the eigenfunction $W^+(x;k)$. (The eigenfunction $W^-(x;k)$ can be obtained from the relation $W^-(x;k) = W^+(x+2\mathrm{i}\delta;k)$.) In analogy with the Schrödinger scattering problem which we discussed in Chapter 2, we denote M, $\bar{M}$ to be the "left" eigenfunctions, and N, $\bar{N}$ the "right" eigenfunctions. These eigenfunctions have the following asymptotic behavior

$$M(x;k) \sim 1, \qquad \bar{M}(x;k) \sim \mathrm{e}^{ikx+k\delta}, \qquad \text{as} \quad x \to -\infty, \tag{4.2.8a}$$

$$N(x;k) \sim \mathrm{e}^{ikx+k\delta}, \qquad \bar{N}(x;k) \sim 1, \qquad \text{as} \quad x \to \infty. \tag{4.2.8b}$$

The eigenfunctions M, $\bar{M}$, N and $\bar{N}$ satisfy the following Fredholm integral equations

$$\begin{pmatrix} M(x;k) \\ \bar{M}(x;k) \end{pmatrix} = \begin{pmatrix} 1 \\ \mathrm{e}^{ikx+k\delta} \end{pmatrix} + \int_{-\infty}^{\infty} G_{+}(x,y;k)u(y) \begin{pmatrix} M(y;k) \\ \bar{M}(y;k) \end{pmatrix} \mathrm{d}y, \tag{4.2.9a}$$

$$\begin{pmatrix} N(x;k) \\ \bar{N}(x;k) \end{pmatrix} = \begin{pmatrix} \mathrm{e}^{ikx+k\delta} \\ 1 \end{pmatrix} + \int_{-\infty}^{\infty} G_{-}(x,y;k)u(y) \begin{pmatrix} N(y;k) \\ \bar{N}(y;k) \end{pmatrix} \mathrm{d}y, \tag{4.2.9b}$$

where $G_{\pm}(x,y;k)$ is as defined equation (4.2.7). It is important to remark that (4.2.9) are *Fredholm* integral equations, whereas for the Schrödinger scattering problem associated with the KdV equation, the eigenfunctions M, $\bar{M}$, N and $\bar{N}$ satisfy the *Volterra* integral equations [see equations (2.2.15,16)].

The eigenfunctions M, N, $\bar{N}$ are related through the completeness relation

$$M(x;k) = a(k)\bar{N}(x;k) + b(k)N(x;k). \tag{4.2.10}$$

In order to derive integral representations for $a(k)$ and $b(k)$, we require a symmetry condition (or analytic connection) between N and $\bar{N}$. From (4.2.6-7) we have

$$G_{\pm}(x,y;k) = G_{\pm}(x,y;-k)\exp\{ik(x-y)\}, \tag{4.2.11}$$

hence from (4.2.9b),

$$N(x;k) = \bar{N}(x;-k)\,\mathrm{e}^{ikx+k\delta}. \tag{4.2.12}$$

The scattering data is obtained as follows. Define

$$\Delta(x;k) := M(x;k) - a(k)\bar{N}(x;k), \tag{4.2.13}$$

then from (4.2.9) we have

$$\begin{aligned} \Delta(x;k) = 1 - a(k) &+ \int_{-\infty}^{\infty} G_{-}(x,y;k)\,u(y)\,\Delta(y;k)\,\mathrm{d}y \\ &+ \int_{-\infty}^{\infty} [G_{+}(x,y;k) - G_{-}(x,y;k)]\,u(y)\,M(x,y;k)\,\mathrm{d}y. \end{aligned} \tag{4.2.14}$$

Using (4.2.10) and the symmetry condition (4.2.12) we have by the definition (4.2.13)

$$\Delta(x;k) = b(k)\bar{N}(x;-k)\,\mathrm{e}^{ikx+k\delta}, \tag{4.2.15}$$

therefore from (4.2.9b) and (4.2.11)

$$\Delta(x;k) = b(k)\,\mathrm{e}^{\mathrm{i}kx+k\delta} + \int_{-\infty}^{\infty} G_-(x,y;k)\,u(y)\,\Delta(y;k)\,\mathrm{d}y. \tag{4.2.16}$$

By comparing (4.2.14) and (4.2.16) we see that

$$\begin{aligned} 1 - a(k) + \int_{-\infty}^{\infty} [G_+(x,y;k) - G_-(x,y;k)]\,u(y)\,M(x,y;k)\,\mathrm{d}y \\ = b(k)\,\mathrm{e}^{\mathrm{i}kx+k\delta}. \end{aligned} \tag{4.2.17}$$

By virtue of the fact that the jump in the Green's function (4.2.6-7) satisfies

$$G_+(x,y;k) - G_-(x,y;k) = \frac{1}{2\delta\zeta_+} - \frac{1}{2\delta\zeta_-}\theta\left(\zeta_+ + 1/(2\delta)\right)\mathrm{e}^{\mathrm{i}k(x-y)}, \tag{4.2.18}$$

where

$$\theta(k) = \begin{cases} 1, & \text{if } k > 0, \\ 0, & \text{if } k < 0 \end{cases}$$

(that is $\theta(k)$ is the Heaviside step function), we obtain the following integral representations for $a(k)$ and $b(k)$

$$a(k) := 1 + \frac{1}{2\mathrm{i}\zeta_+(k)} \int_{-\infty}^{\infty} u(y)\,M(y;k)\,\mathrm{d}y, \tag{4.2.19a}$$

$$b(k) := -\frac{1}{2\mathrm{i}\zeta_+(k)}\,\theta\left(\zeta_+ + 1/(2\delta)\right) \int_{-\infty}^{\infty} u(y)\,M(y;k)\,\mathrm{e}^{-\mathrm{i}ky-k\delta}\,\mathrm{d}y \tag{4.2.19b}$$

[(4.2.18) is derived by studying the poles of $G(p;k)$, as defined by equation (4.2.6b)].

The "bound states" defined as $M(x;k) \to 0$ as $x \to \infty$ and $N(x;k) \to 0$ as $x \to -\infty$ are given by those k_j for which

$$a_j := a(k_j) = 0, \qquad M_j(x) = b_j N_j(x), \tag{4.2.20}$$

where $b_j := b(k_j)$, $M_j(x) := M(x;k_j)$ and $N_j(x) := N(x;k_j)$, for $j = 1, 2, \ldots, n$. As for the KdV equation, $a(k)$ has only simple zeros which lie on the imaginary k-axis, i.e., $k_j = \mathrm{i}\kappa_j$, with $0 < \kappa_j < \pi/(2\delta)$ (for further details see Kodama, Ablowitz and Satsuma [1982, Appendix D]). We shall assume that these zeros (eigenvalues) are distinct and finite in number.

4.2.2 The Inverse Scattering Problem. The inverse scattering problem is based on equation (4.2.10). Given $a(k)$, $b(k)$ and appropriate information about the bound states, we wish to reconstruct M, N, $\bar{N}$. In order to regard (4.2.10) as a Riemann-Hilbert problem in the complex $\zeta_+(k)$-plane, it is necessary to establish analyticity properties for the eigenfunctions M, N, $\bar{N}$ in $\zeta_+(k)$. From (4.2.7), it is clear that the kernels $G_\pm(x,y;k)$ of the integral equations for the eigenfunctions M and $\bar{N}$ [equations (4.2.9)] are (+) and (−) functions in $\zeta_+(k)$, respectively, that is they may be analytically extended in appropriate regions of the $\zeta_+(k)$-plane. Since the inhomogeneous term in (4.2.9) for M and $\bar{N}$ is unity, then Fredholm theory implies that M and $\bar{N}$ are also (+) and (−) functions in $\zeta_+(k)$ respectively, provided that there are no non-trivial solutions to the homogeneous versions of equations (4.2.9) (when δ is finite, it can be shown that for suitable potentials this is actually the case). Furthermore, equation (4.2.19a) implies that a is a (+) function in ζ_+ as well.

Equation (4.2.10), together with the analytic properties of M, $\bar{N}$ and a and equation (4.2.12) defines a Riemann-Hilbert problem (with a shift) in $\zeta_+(k)$. Since we are considering the $\zeta_+(k)$ plane, we define $M(x;\zeta_+) \equiv M(x;\zeta_+(k)) := M(x;k)$, $N(x;\zeta_+) \equiv N(x;\zeta_+(k)) := N(x;k)$, $\bar{N}(x;\zeta_+) \equiv \bar{N}(x;\zeta_+(k)) := \bar{N}(x;k)$ and $a(\zeta_+) \equiv a(\zeta_+(k)) := a(k)$. The function $M(x;\zeta_+)/a(\zeta_+)$ is meromorphic in the upper half ζ_+ plane (it has poles at the zeros $\zeta_{+j} := \zeta_+(k_j)$ of $a(\zeta_+)$, where, from equation (4.2.20a) the bound states k_j are defined by $a(k_j) = 0$). Hence

$$\frac{M(x;\zeta_+)}{a(\zeta_+)} = 1 + \mu_+(x;\zeta_+) + \mathrm{i}\sum_{j=1}^{n} \frac{C_j N_j(x)}{\zeta_+ - \zeta_{+j}}, \tag{4.2.21a}$$

with

$$C_j := -\mathrm{i}\rho(k_j), \qquad N_j(x) := N_j(x;k_j), \tag{4.2.21b}$$

where

$$\rho(k) = \frac{b(k)}{a(k)} = \frac{b(k)}{a(\zeta_+)}, \tag{4.2.21c}$$

and $\mu_+(x;\zeta_+)$ is analytic in the upper half ζ_+-plane vanishing as $\zeta_+ \to \infty$.

Expressing $\theta\left(\zeta_+ + 1/(2\delta)\right)\rho(\zeta_+)N(x;\zeta_+)$ in terms of its (+) and (−) parts,

$$\theta\left(\zeta_+ + 1/(2\delta)\right)\rho(\zeta_+)\,N(x;\zeta_+) = U^+(x;\zeta_+) - U^-(x;\zeta_+), \tag{4.2.22a}$$

where

$$U^\pm(x;\zeta_+) = \frac{1}{2\pi\mathrm{i}}\int_{-1/(2\delta)}^{\infty} \frac{\rho(\lambda_+)\,N(x;\lambda_+)}{\lambda_+ - (\zeta_+ \pm \mathrm{i}0)}\,\mathrm{d}\lambda_+, \tag{4.2.22b}$$

then from equation (4.2.10) we find that $\mu_+(x;\zeta_+) = U^+(x;\zeta_+)$ and we obtain the following integral equation

$$\bar{N}(x;\zeta_+) = 1 + \mathrm{i}\sum_{j=1}^{n} \frac{C_j N_j(x)}{\zeta_+ - \zeta_{+j}} + \frac{1}{2\pi\mathrm{i}}\int_{-1/(2\delta)}^{\infty} \frac{\rho(\lambda_+)\,N(x;\lambda_+)}{\lambda_+ - (\zeta_+ - \mathrm{i}0)}\,\mathrm{d}\lambda_+.$$

Multiplying this equation by $\exp(\mathrm{i}kx + k\delta)$, letting $k \to -k$, and using the symmetry condition (4.2.12) yields

$$N(x;\zeta_-) = \mathrm{e}^{\mathrm{i}kx+k\delta}\left\{1 - \mathrm{i}\sum_{j=1}^{n} \frac{C_j N_j(x)}{\zeta_- + \zeta_{+j}} \right.$$
$$\left. + \frac{1}{2\pi\mathrm{i}}\int_{-1/(2\delta)}^{\infty} \frac{\rho(\lambda_-)\,N(x;\lambda_-)}{\lambda_- + \zeta_- + \mathrm{i}0}\,\mathrm{d}\lambda_-\right\}. \tag{4.2.23}$$

(The Gel'fand-Levitan-Marchenko equation derived by Kodama, Ablowitz and Satsuma [1982] is obtained by taking an appropriate Fourier transform of the resulting integral equation for $N(x;k)$ — see §4.2.4 below.)

Now consider (4.2.23) for large ζ_+, so we are required to evaluate $N(x;\zeta_+)$ for large ζ_+. By studying the operators E and T_+ defined by

$$(\mathrm{E}f)(x) := f(x+2\mathrm{i}\delta), \tag{4.2.24a}$$

$$(\mathrm{T}_+f)(x) := \frac{1}{4\delta}\int_{-\infty}^{\infty} \coth\left\{\frac{\pi}{2\delta}(y-x-\mathrm{i}0)\right\} f(y)\,\mathrm{d}y, \tag{4.2.24b}$$

then it is easily shown that

$$-\mathrm{i}\mathrm{T}_+(\mathrm{I}-\mathrm{E}) = \mathrm{I}, \tag{4.2.25}$$

where I is the identity operator. Hence from equation (4.2.7)

$$\begin{aligned} G_-(x,y;k) &= \frac{1}{4\mathrm{i}\delta}\int_{-\infty}^{\infty} \coth\left\{\frac{\pi}{2\delta}(\eta-x-\mathrm{i}0)\right\} \times \\ &\qquad \left(\frac{1}{2\pi}\int_{C_-} \frac{\exp\{\mathrm{i}p(x-y)\}[1-\exp(-2p\delta)]}{p-[\zeta_+(k)+1/(2\delta)][1-\exp(-2p\delta)]}\,\mathrm{d}p\right)\mathrm{d}\eta \\ &= -\frac{1}{4\mathrm{i}\delta\zeta_+}\coth\left\{\frac{\pi}{2\delta}(y-x-\mathrm{i}0)\right\} + O(\zeta_+^{-2}), \end{aligned} \tag{4.2.26}$$

as $\zeta_+ \to \infty$. Therefore from the integral equation (4.2.9b), as $\zeta_+ \to \infty$,

$$N(x;k) \sim 1 - \frac{1}{4\mathrm{i}\delta\zeta_+}\int_{-\infty}^{\infty} \coth\left\{\frac{\pi}{2\delta}(y-x-\mathrm{i}0)\right\} u(y)\,\mathrm{d}y, \tag{4.2.27}$$

Then from equations (4.2.23) and (4.2.27) we obtain the following relationship

$$u^+(x) = \frac{1}{2\pi\mathrm{i}}\int_{-1/(2\delta)}^{\infty} \rho(\zeta_+)\,N(x;\zeta_+)\,\mathrm{d}\zeta_+ - \mathrm{i}\sum_{j=1}^{n} C_j N_j(x), \tag{4.2.28a}$$

where

$$u^+(x) := \frac{1}{4\mathrm{i}\delta}\int_{-\infty}^{\infty} \cosh\left\{\frac{\pi}{2\delta}(y-x-\mathrm{i}0)\right\} u(y)\,\mathrm{d}y. \tag{4.2.28b}$$

Finally, the reality of the potential u, implies that u is given by

$$u(x) = u^+(x) + u^-(x) = u^+(x) + \overline{u^+(x)}. \tag{4.2.29}$$

Equation (4.2.23) defines N in terms of ρ, C_j, k_j, and (4.2.28) defines u^+ in terms of ρ, C_j, N. Hence equations (4.2.23) and (4.2.28) define u^+ in terms of the scattering data, whose time evolution is given below.

4.2.3 The Time Dependence. The time evolution of the scattering data is obtained from the t-part of the Lax pair, equation (4.2.1b), and the analysis is similar to that for the KdV equation (§2.4). $M(x,t;k) = v^+(x,t;k)\exp\left\{\frac{1}{2}\mathrm{i}k(x-\mathrm{i}\delta)\right\}$ satisfies

$$\mathrm{i}M_t - 2\mathrm{i}\zeta_+ M_x + M_{xx} + (\mathrm{i}u_x \mathrm{T} u_x + \rho)M. \tag{4.2.30}$$

Since $u(x,t)\to 0$ as $|x|\to\infty$ then the boundary condition (4.2.8a) shows that the constant ρ must be zero. Also from the boundary condition (4.2.8b) and the completeness relation (4.2.10) we see that

$$a_t = 0, \qquad b_t = \mathrm{i}k[\coth(k\delta) - 1/\delta]b.$$

Hence

$$a(k,t) = a(k,0), \qquad b(k,t) = b(k,0)\exp\left\{\mathrm{i}k\left[k\coth(k\delta) - 1/\delta\right]t\right\},$$

and so

$$\rho(k;t) = \rho(k;0)\exp\left\{\mathrm{i}k\left[k\coth(k\delta) - 1/\delta\right]t\right\}. \tag{4.2.31}$$

Similarly we find that

$$k_j(t) = k_j(0), \tag{4.2.32a}$$
$$C_j(t) = C_j(0)\exp\left\{\mathrm{i}k_j\left[k_j\coth(k_j\delta) - 1/\delta\right]t\right\}, \tag{4.2.32b}$$

for $j = 1,2,\ldots,n$.

4.2.4 Further Remarks.

1. Soliton solutions.

Pure soliton solutions are obtained by taking $\rho(k;t) = 0$, hence

$$N(x,t;k) = \exp(\mathrm{i}kx + k\delta)\left(1 + \mathrm{i}\sum_{j=1}^{n}\frac{C_j(t)N_j(x,t)}{\zeta_-(k)+\zeta_{+j}}\right),$$

and so setting $k = \mathrm{i}\kappa_\ell$ yields

$$N_\ell(x,t) = \exp(-\kappa_\ell x + \mathrm{i}\kappa_\ell\delta)\left(1 + \mathrm{i}\sum_{j=1}^{n}\frac{C_j(t)N_j(x,t)}{\zeta_{-\ell}+\zeta_{+j}}\right),$$

with $\zeta_{-\ell} = \zeta_-(\mathrm{i}\kappa_\ell)$. Hence from (4.2.28a,29,32b)

$$u(x) = \mathrm{i}\sum_{j=1}^{n}[\bar{N}_j(x,t) - N_j(x,t)]C_j(0)\exp(\kappa\gamma t),$$

where $\gamma = \delta^{-1} - \kappa\cot(\kappa\delta)$. In particular, if $n = 1$ then

$$N_1(x,t) = 1 + \mathrm{i}\frac{C_1(t)N_1(x,t)}{\zeta_+ - \zeta_{+1}},$$

and so the one-soliton solution of the ILW equation is given by

$$u(x,t) = \frac{\kappa\sin(\kappa\delta)}{\cos(\kappa\delta) + \cosh\{\kappa(x - \gamma t) + x_0\}}, \tag{4.2.33}$$

where x_0 is a constant. As $\delta \to 0$

$$u(x,t) \sim \tfrac{1}{2}\delta^2\kappa\,\mathrm{sech}^2\left\{\tfrac{1}{2}\kappa\left(x - \tfrac{1}{3}\delta\kappa^2 t\right) + x_0\right\},$$

which is the one-soliton solution of the KdV equation (4.1.3). For N-soliton solutions of the ILW equation (which reduce to the N-soliton solutions of the KdV equation as $\delta \to 0$), see Joseph and Egri [1978]; Chen and Lee [1979]; Matsuno [1979b]; Satsuma and Ablowitz [1980].

2. The Gel'fand-Levitan-Marchenko Equation.

The Gel'fand-Levitan-Marchenko integral equation given by Kodama, Ablowitz and Satsuma [1982] may be derived as follows. For $N(x,t;k)$, from the analyticity and the asymptotic behavior (4.2.27), the following integral representation is suggested:

$$N(x,t;k) = 1 + \int_x^\infty K(x,s;t)\exp[\mathrm{i}\zeta_+(x-s)]\,\mathrm{d}s \quad \text{for} \quad \mathrm{Im}\,\zeta_+ < 0. \tag{4.2.34}$$

Hence from (4.2.24) and the symmetry condition (4.2.12), we also have

$$N(x,t;k) = \exp(\mathrm{i}kx + k\delta)\left\{1 + \int_x^\infty K(x,s;t)\exp[-\mathrm{i}\zeta_-(x-s)]\,\mathrm{d}s\right\} \tag{4.2.35}$$

(note that $\zeta_+(-k) = -\zeta_-(k)$ and $\zeta_+(k) + \zeta_-(k) = k$). Recall from (4.2.20b) that $N_j(x,t) = N(x,t;k_j)$, therefore from (4.2.35)

$$N_j(x,t) = \exp(\mathrm{i}k_jx + k_j\delta) + \int_x^\infty K(x,s;t)\exp\{\mathrm{i}\zeta_{+j}x + \mathrm{i}\zeta_{-j}s + k_j\delta)\}\,\mathrm{d}s, \tag{4.2.36}$$

where $\zeta_{-j} := \zeta_-(k_j)$. Applying the operator

$$\frac{1}{2\pi}\int_{-\infty}^\infty \mathrm{d}\zeta_+\exp\{\mathrm{i}\zeta_+(y-x)\} \tag{4.2.37}$$

(i.e., the Fourier transform) to equation (4.2.23), for $y > x$, we obtain

$$\begin{aligned} &\frac{1}{2\pi}\int_{-\infty}^\infty (N(x,t;\zeta_+) - 1)\exp\{\mathrm{i}\zeta_+(y-x)\}\,\mathrm{d}\zeta_+ \\ &\qquad + \frac{1}{2\pi}\int_{-1/(2\delta)}^\infty \rho(\zeta_+,t)\,N(x,t;\zeta_+)\exp[i\zeta_+(y-x)]\,\mathrm{d}\zeta_+ \\ &\qquad\qquad = -\sum_{j=1}^n C_j(t)N_j(x,t)\exp\{\mathrm{i}\zeta_{+j}(y-x)\}. \end{aligned} \tag{4.2.38}$$

Hence, using (4.2.34-36), we obtain the Gel'fand-Levitan-Marchenko integral equation

$$K(x,y;t)+F(x,y;t)+\int_x^\infty K(x,s;t)\,F(s,y;t)\,\mathrm{d}s=0 \quad \text{for} \quad y>x, \tag{4.2.39}$$

with

$$\begin{aligned} F(x,y;t)=\frac{1}{2\pi}\int_{-\infty}^{\infty}\rho(k;t)\exp(\mathrm{i}\zeta_+ y+\mathrm{i}\zeta_- x+k\delta)\,\frac{\mathrm{d}\zeta_+}{\mathrm{d}k}\,\mathrm{d}k \\ +\sum_{j=1}^{n}C_j(t)\exp(\mathrm{i}\zeta_{+j}y+\mathrm{i}\zeta_{-j}x+k_j\delta), \end{aligned} \tag{4.2.40}$$

where we have used the fact that

$$\int_{-1/(2\delta)}^{\infty} f(\zeta_+)\,\mathrm{d}\zeta_+=\int_{-\infty}^{\infty} f(k)\,\frac{\mathrm{d}\zeta_+}{\mathrm{d}k}\,\mathrm{d}k.$$

4.3 The Benjamin Ono Equation.

It has been established that the BO equation (4.1.2) possesses the following properties: N-soliton solutions (Case [1978, 1979]; Chen, Lee and Pereira [1979]; Matsuno [1979a, 1980, 1984]; Satsuma and Ishimori [1979]); Bäcklund transformations and a novel Lax pair (Nakamura [1979]; Bock and Kruskal [1979]); infinitely many commuting symmetries and constants of the motion in involution (Bock and Kruskal [1979]; Fokas and Fuchssteiner [1981]); a bi-Hamiltonian structure and a recursion operator (Fokas and Santini [1988b]; Santini [1989b]). The I.S.T. scheme for the BO equation related to the scattering problem (4.3.1a) below was developed by Fokas and Ablowitz [1983b] (see also Ablowitz, Fokas and Anderson [1983]), which we outline below. Mathematically, whilst being interesting in its own right, this provides a pivot to the development of I.S.T. for multidimensional equations. Although essentially a 1 + 1-dimensional equation, the BO equation has many features similar to those of multidimensional equations (this is a consequence of its nonlocal character). We note that an alternative scattering problem has been derived by Anderson and Taflin [1985], and rigorously studied by Coifman and Wickerhauser [1990]. At present, it is unclear whether there is an intermediate version of this scattering problem. As with the scattering problem (4.3.1a) below, the scattering data is generically singular at the origin in scattering space.

4.3.1 The Direct Scattering Problem. The Lax pair for the BO equation can be written in the form

$$\mathrm{i}w_x^+ + k(w^+ - w^-) = -uw^+, \tag{4.3.1a}$$

$$\mathrm{i}w_t^\pm - 2\mathrm{i}kw_x^\pm + w_{xx}^\pm - 2\mathrm{i}[u]_x^\pm w^\pm = -\rho w^\pm, \tag{4.3.1b}$$

with

$$[u]^{\pm} := \pm\tfrac{1}{2}u - \tfrac{1}{2}\mathrm{i}\mathrm{H}(u), \tag{4.3.1c}$$

where k is a constant, which is interpreted as the spectral parameter and ρ is an arbitrary constant. $w^{\pm}(x)$ represent the boundary values of functions which are analytic in the upper $(+)$ and lower $(-)$ half x-plane, that is $w^{\pm}(x) = \lim_{y\downarrow 0} w(x \pm \mathrm{i}y)$; similarly, $[u]^{+}$ and $[u]^{-}$ are analytic in the upper and lower half z-plane respectively (where z is the complex extension of x). Given u, equation (4.3.1a) defines a differential Riemann-Hilbert problem in the complex x-plane. Naturally we note the similarity of (4.3.1) with (4.2.3). Indeed Santini, Ablowitz and Fokas [1984a] show how the limit as $\delta \to \infty$ of the results for the ILW equation tend to those for the BO equation.

As for the ILW equation above, it is sufficient to consider only the $(+)$ eigenfunctions. Let M, $\bar{M}$ denote the "left" eigenfunctions and N, $\bar{N}$ denote the "right" eigenfunctions. These are specified by the following asymptotic behavior

$$M(x;k) \sim 1, \qquad \bar{M}(x;k) \sim \mathrm{e}^{\mathrm{i}kx}, \qquad \text{as} \quad x \to -\infty, \tag{4.3.2a}$$

$$N(x;k) \sim \mathrm{e}^{\mathrm{i}kx}, \qquad \bar{N}(x;k) \sim 1, \qquad \text{as} \quad x \to \infty. \tag{4.3.2b}$$

The eigenfunctions M, $\bar{M}$, N, $\bar{N}$ satisfy the following Fredholm integral equations

$$\begin{pmatrix} M(x;k) \\ \bar{M}(x;k) \end{pmatrix} = \begin{pmatrix} 1 \\ \mathrm{e}^{\mathrm{i}kx} \end{pmatrix} + \int_{-\infty}^{\infty} G_{+}(x,y;k)\, u(y) \begin{pmatrix} M(y;k) \\ \bar{M}(y;k) \end{pmatrix} \mathrm{d}y, \tag{4.3.3a}$$

$$\begin{pmatrix} N(x;k) \\ \bar{N}(x;k) \end{pmatrix} = \begin{pmatrix} \mathrm{e}^{\mathrm{i}kx} \\ 1 \end{pmatrix} + \int_{-\infty}^{\infty} G_{-}(x,y;k)\, u(y) \begin{pmatrix} N(y;k) \\ \bar{N}(y;k) \end{pmatrix} \mathrm{d}y, \tag{4.3.3b}$$

where G_{+}, G_{-} are the $(+)$ and $(-)$ parts of the sectionally holomorphic function

$$G(x,y;k) = \frac{1}{2\pi} \int_0^{\infty} \frac{\mathrm{e}^{\mathrm{i}(x-y)p}}{p-\kappa}\, \mathrm{d}p, \tag{4.3.4a}$$

where κ denotes the complex extension of k, that is,

$$G_{\pm}(x,y;k) = \frac{1}{2\pi} \int_0^{\infty} \frac{\mathrm{e}^{\mathrm{i}(x-y)p}}{p-(k \pm \mathrm{i}0)}\, \mathrm{d}p. \tag{4.3.4b}$$

The positive k-axis is a line of discontinuity of $G(x,y;k)$, hence using the Plemelj formulae (cf. Muskheilishvili [1953])

$$G_{+}(x,y;k) - G_{-}(x,y;k) = \mathrm{i}\theta(k) \exp\{\mathrm{i}(x-y)k\}, \tag{4.3.5}$$

where $\theta(k)$ is the Heaviside function.

The eigenfunctions M, N, $\bar{N}$ are related through the scattering equation

$$M(x;k) = \bar{N}(x;k) + \rho(k) N(x;k), \tag{4.3.6a}$$

where

$$\rho(k) = \mathrm{i}\theta(k) \int_{-\infty}^{\infty} u(y) M(y;k) \,\mathrm{e}^{-\mathrm{i}ky}\,\mathrm{d}y, \tag{4.3.6b}$$

In order to derive these equations we define

$$\Delta(x;k) := M(x;k) - \bar{N}(x;k),$$

and then from equations (4.3.3) (i.e., subtracting the relevant equations)

$$\begin{aligned}\Delta(x;k) &= \int_{-\infty}^{\infty} G_+(x,y;k)\,u(y)\,M(y;k)\,\mathrm{d}y - \int_{-\infty}^{\infty} G_-(x,y;k)\,u(y)\,\bar{N}(y;k)\,\mathrm{d}y \\ &= \int_{-\infty}^{\infty} G_-(x,y;k)\,u(y)\,\Delta(y;k)\,\mathrm{d}y \\ &\qquad - \int_{-\infty}^{\infty} [G_+(x,y;k) - G_-(x,y;k)]\,u(y)\,M(y;k)\,\mathrm{d}y. \end{aligned} \tag{4.3.7}$$

Hence from (4.3.5) and (4.3.6)

$$\begin{aligned}\Delta(x;k) = \int_{-\infty}^{\infty} & G_-(x,y;k)\,u(y)\,\Delta(y;k)\,\mathrm{d}y \\ & + \mathrm{i}\mathrm{e}^{\mathrm{i}kx} \int_{-\infty}^{\infty} \mathrm{e}^{-\mathrm{i}ky}\,u(y)\,M(y;k)\,\mathrm{d}y, \end{aligned} \tag{4.3.8}$$

which implies that

$$\Delta(x;k) = \rho(k) N(x;k). \tag{4.3.9}$$

4.3.2 The Inverse Scattering Problem. The solitons of the ILW equation correspond to the "bound states" which are generated from the zeros of $a(k)$. However in the scattering equation (4.3.6), the coefficient of $\bar{N}$ is 1, that is $a(k) = 1$ for the BO equation. Therefore the solitons for the BO equation must arise in a different manner. The solitons arise as poles of the eigenfunctions M and $\bar{N}$ which are a consequence of the fact that the homogeneous versions of the integral equations (4.3.3) for the eigenfunctions, in contrast to the homogeneous versions of the integral equations (4.2.9) for the eigenfunctions of the ILW equation, have nontrivial solutions $\Phi_j(x)$, for some k_j, where $k_j < 0$. Hence

$$\Phi_j(x) = \int_{-\infty}^{\infty} G(x,y;k_j)\,u(y)\,\Phi_j(y)\,\mathrm{d}y, \qquad k_j < 0, \tag{4.3.10a}$$

where

$$G(x,y;k_j) = \frac{1}{2\pi} \int_0^{\infty} \frac{\mathrm{e}^{\mathrm{i}(x-y)p}}{p - k_j}\,\mathrm{d}p. \tag{4.3.10b}$$

The kernels of the integral equations (4.3.3) for the eigenfunctions M and $\bar{N}$ are (+) and (−) functions in k respectively. Hence

$$M(x;k) = 1 + \sum_{j=1}^{n} \frac{C_j \Phi_j(x)}{k - k_j} + m_+(x;k), \tag{4.3.11a}$$

$$\bar{N}(x;k) = 1 + \mathrm{i} \sum_{j=1}^{n} \frac{C_j \Phi_j(x)}{k - k_j} + \bar{n}_-(x;k). \tag{4.3.11b}$$

where m_+, $\bar{n}_-$ are (+) and (−) functions in k, respectively. The constants C_j, $j = 1, 2, \ldots, n$, are determined by the normalization $x\Phi_j(x) \to 1$ as $x \to \infty$; then we find that $C_j = -\mathrm{i}$, $j = 1, 2, \ldots, n$.

In order to view (4.3.6) as a Riemann-Hilbert problem in the k-plane, we need to establish a symmetry relation between N and $\bar{N}$ and appropriate analytic properties of the eigenfunctions. By differentiating equation (4.3.4b) with respect to k and then integrating by parts with respect to p, we find

$$\frac{\partial}{\partial k}\left(G_\pm(x,y;k)\right) = -\frac{1}{2\pi k} + \mathrm{i}(x-y)G_\pm(x,y;k). \tag{4.3.12}$$

Also, differentiating equation (4.3.3b) with respect to k implies that

$$\begin{aligned}\frac{\partial}{\partial k}\left(N(x;k)\,\mathrm{e}^{-\mathrm{i}kx}\right) &= \int_{-\infty}^{\infty} \frac{\partial}{\partial k}\left(G_-(x,y;k)\,\mathrm{e}^{-\mathrm{i}k(x-y)}\right) u(y)\, N(y;k)\,\mathrm{e}^{-\mathrm{i}ky}\,\mathrm{d}y \\ &+ \int_{-\infty}^{\infty} G_-(x,y;k)\,\mathrm{e}^{-\mathrm{i}k(x-y)}\, u(y) \frac{\partial}{\partial k}\left(N(y;k)\,\mathrm{e}^{-\mathrm{i}ky}\right)\,\mathrm{d}y.\end{aligned} \tag{4.3.13}$$

Hence

$$\begin{aligned}\frac{\partial}{\partial k}\left(N(x;k)\,\mathrm{e}^{-\mathrm{i}kx}\right) &= \mathrm{e}^{-\mathrm{i}kx} f(k) \\ &+ \int_{-\infty}^{\infty} G_-(x,y;k)\,\mathrm{e}^{-\mathrm{i}k(x-y)}\, u(y) \frac{\partial}{\partial k}\left(N(y;k)\,\mathrm{e}^{-\mathrm{i}ky}\right)\,\mathrm{d}y.\end{aligned} \tag{4.3.14a}$$

where

$$f(k) := -\frac{1}{2\pi k} \int_{-\infty}^{\infty} u(y)\, N(y;k)\,\mathrm{d}y. \tag{4.3.14b}$$

By comparing equations (4.3.3b) and (4.3.14), we obtain the symmetry condition

$$\frac{\partial}{\partial k}\left(N(x;k)\,\mathrm{e}^{-\mathrm{i}kx}\right) = f(k)\bar{N}(x;k)\,\mathrm{e}^{-\mathrm{i}kx}, \tag{4.3.15}$$

Therefore from (4.3.6) and (4.3.15)

$$M(x;k) = \bar{N}(x;k) + \rho(k)\,\mathrm{e}^{\mathrm{i}xk} \int_0^k f(\lambda)\,\mathrm{e}^{-\mathrm{i}x\lambda} \bar{N}(x;\lambda)\,\mathrm{d}\lambda. \tag{4.3.16}$$

Analytically continuing equation (4.3.3b) for $k<0$ implies

$$\bar{N}^{(j)}(x;k) - \int_{-\infty}^{\infty} G(x,y;k)\,u(y)\,\bar{N}^{(j)}(y;k)\,\mathrm{d}y$$
$$= 1 + \frac{\mathrm{i}}{k-k_j}\left[\Phi_j(x) - \int_{-\infty}^{\infty} G(x,y;k)\,u(y)\,\Phi_j(y)\,\mathrm{d}y\right], \quad (4.3.17a)$$

where

$$\bar{N}^{(j)}(x;k) := \bar{N}(x;k) + \frac{\mathrm{i}\Phi_j(x)}{k-k_j}. \quad (4.3.17b)$$

Hence, letting $k \to k_j$ and using (4.3.12) (with $\epsilon = 0$ and $k<0$), it follows that

$$\bar{N}^{(j)}(x;k_j) - \int_{-\infty}^{\infty} G(x,y;k_j)\,u(y)\,\bar{N}^{(j)}(y;k_j)\,\mathrm{d}y$$
$$= a_j + \int_{-\infty}^{\infty} (x-y)\,G(x,y;k_j)\,u(y)\,\Phi_j(y)\,\mathrm{d}y, \quad (4.3.18a)$$

where

$$a_j := 1 + \frac{\mathrm{i}}{2\pi k_j}\int_{-\infty}^{\infty} \Phi_j(y)\,u(y)\,\mathrm{d}y. \quad (4.3.18b)$$

The asymptotics of equation (4.3.4a) as $x \to \infty$ shows that

$$G(x,y;k_j) = \frac{1}{2\pi \mathrm{i} x k_j} + O(x^{-2}).$$

$\Phi_j^{\pm}(x)$ satisfy

$$\mathrm{i}\frac{\partial \Phi_j^+}{\partial x} + k_j(\Phi_j^+ - \Phi_j^-) = -u(x)\Phi_j^+.$$

Furthermore $\Phi_j^{\pm} \sim 1/x$ as $x \to \infty$ and then we consider

$$-\int_{-\infty}^{\infty} u(y)\Phi_j^+(y)\,\mathrm{d}y = k_j\left\{\int_{-\infty}^{\infty}\left(\Phi_j^+(y) - \frac{1}{y+\mathrm{i}}\right)\mathrm{d}y - \int_{-\infty}^{\infty}\left(\Phi_j^-(y) - \frac{1}{y-\mathrm{i}}\right)\mathrm{d}y\right.$$
$$\left. - \int_{-\infty}^{\infty}\left(\frac{1}{y+\mathrm{i}} - \frac{1}{y-\mathrm{i}}\right)\mathrm{d}y\right\}.$$

By integrating round a semi-circle in the upper/lower half plane it is shown that

$$\int_{-\infty}^{\infty}\left(\Phi_j^+(y) - \frac{1}{y+\mathrm{i}}\right)\mathrm{d}y = 0 = \int_{-\infty}^{\infty}\left(\Phi_j^-(y) - \frac{1}{y-\mathrm{i}}\right)\mathrm{d}y,$$

and so we have

$$k_j = \frac{1}{2\pi\mathrm{i}}\int_{-\infty}^{\infty} u(y)\Phi_j^+(y)\,\mathrm{d}y. \quad (4.3.19)$$

Therefore $a_j \equiv 0$ and a particular solution of (4.3.18) is given by $\bar{N}^{(j)}(x;k_j) = x\Phi_j(x)$. Hence applying Fredholm's alternative theorem, we find that

$$\lim_{k\to k_j}\left\{N(x;k)+\frac{\mathrm{i}\Phi_j(x)}{(k-k_j)}\right\}=[x+\gamma_j(t)]\,\Phi_j(x). \qquad (4.3.20)$$

Note that the symmetry condition (4.3.15) is *continuous* and not *discrete* as it was the case for the KdV and ILW scattering problems. Equations (4.3.6,16-17,20) define a *nonlocal Riemann-Hilbert problem in the complex k-plane* (the fact that the the Riemann-Hilbert problem is nonlocal is a consequence of the fact that the symmetry condition is continuous). This is the first time that nonlocal Riemann-Hilbert appear in the inverse problems in one spatial dimension. Nonlocal Riemann-Hilbert problems also appear to be features of scattering problems associated with evolution equations in two spatial dimensions (see Chapter 5, especially §5.2 in which KPI is discussed). We note that knowledge of the scattering equation (4.3.6) and the symmetry condition (4.3.15) are sufficient in order to solve the BO equation by the I.S.T. method. The Riemann-Hilbert problem is equivalent to the integral equation

$$\begin{aligned}N(x,t;k)-\frac{1}{2\pi}\int_0^\infty w(x,t;k,\lambda)\,\rho(\lambda,t)\,N(x,t;\lambda)\,\mathrm{d}\lambda&\\ +\sum_{j=1}^{n}\Phi_j(x,t)\,w(x,t;k,k_j)&=v(x,t;k),\end{aligned} \qquad (4.3.21)$$

and

$$[x+\gamma_j(t)]\,\Phi_j(x,t)-\frac{1}{2\pi\mathrm{i}}\int_0^\infty\frac{\rho(\lambda,t)\,N(x,t;\lambda)}{\lambda-k_j}\,\mathrm{d}\lambda+\mathrm{i}\sum_{\substack{\ell=1\\ \ell\neq j}}^{n}\frac{\Phi_\ell(x,t)}{(k_j-k_\ell)}=1, \qquad (4.3.22)$$

where

$$v(x,t;k)=\int_0^k\left[f(\lambda,t)\,\mathrm{e}^{\mathrm{i}x(k-\lambda)}+g(\lambda)\,\mathrm{e}^{\mathrm{i}\lambda x}\right]\mathrm{d}\lambda, \qquad (4.3.23\text{a})$$

$$g(\lambda)=\frac{1}{\lambda\ln\lambda}, \qquad (4.3.23\text{b})$$

$$w(x,t;k,\lambda)=\mathrm{e}^{\mathrm{i}(k-\lambda)x}\int_{-\infty}^{x}v(s;k)\,\mathrm{e}^{-\mathrm{i}(k-\lambda)s}\,\mathrm{d}s,\quad \lambda>0, \qquad (4.3.23\text{c})$$

$$\begin{aligned}w(x,t;k,k_j)&=\mathrm{e}^{\mathrm{i}(k-\lambda)x}\int_{\alpha}^{x}v(s;k)\,\exp\{-\mathrm{i}(k-k_j)s\}\,\mathrm{d}s\\ &\quad+\exp\{\mathrm{i}(k-k_j)(x-\alpha)\}\int_0^k\left(\frac{f(\lambda,t)\,\mathrm{e}^{\mathrm{i}\alpha(k-\lambda)}}{k_j-\lambda}+\frac{g(\lambda)}{k_j}\right)\mathrm{d}\lambda.\end{aligned} \qquad (4.3.23\text{d})$$

Equations (4.3.21-23) may be derived as follows: substitute equations (4.3.11) into equation (4.3.6) and then consider the $(-)$ part of (4.3.6), hence

$$N(x,t;k)-\frac{1}{2\pi\mathrm{i}}\int_0^\infty\frac{\rho(\lambda,t)\,N(x,t;\lambda)}{\lambda-(k-\mathrm{i}0)}\,\mathrm{d}\lambda+\mathrm{i}\sum_{j=1}^{n}\frac{\Phi_j(x,t)}{k-k_j}=1. \qquad (4.3.24)$$

Equation (4.3.21) follows by considering equation (4.3.24) with $k = 0$, equation (4.3.15) and the formulae

$$\frac{e^{ix(\lambda-k)}}{\lambda-(k-i0)} = i\int_{\infty}^{x} e^{is(\lambda-k)}\,ds,$$
$$\frac{e^{ix(\lambda-k)}}{\lambda-k} = i\int_{\alpha}^{x} e^{is(\lambda-k)}\,ds + \frac{e^{i\alpha(\lambda-k)}}{\lambda-k}.$$

Equation (4.3.22) may be derived by analytically continuing (4.3.24) for $k < 0$, taking its limit as $k \to k_j$ and then using equation (4.3.20). It follows from the integral equation (4.3.3b) that

$$\bar{N} \sim 1 - \frac{u^+}{k}, \qquad \text{as} \quad k \to \infty. \tag{4.3.25}$$

Therefore by considering the asymptotics of equation (4.3.24) as $k \to \infty$, and using (4.3.25) we find that

$$u^+(x,t) = \frac{1}{2\pi i}\int_0^{\infty} \rho(k;t)\,N(x,t;k)\,dk + i\sum_{j=1}^{n} \Phi_j(x,t). \tag{4.3.26}$$

Finally assuming that u is real, the potential is given by

$$u(x,t) = u^+(x,t) + \overline{u^+(x,t)}. \tag{4.3.27}$$

Equations (4.3.21-23,26-27) define the potential $u(x,t)$ in terms of the scattering data k_j, $\gamma_j(t)$, $\rho(k;t)$, $f(k,t)$ whose time evolution is given below.

4.3.3 The Time Dependence. The time dependence of the scattering data is determined from the time dependent part of the Lax pair, equation (4.3.1b). $M(x,t;k)$ satisfies

$$iM_t - 2ikM_x + M_{xx} - 2i[u]_x^+ M + \rho M = 0. \tag{4.3.28}$$

Since $|u| \to 0$ as $|x| \to \infty$ then from the boundary condition (4.3.2a) we see that $\rho = 0$ and so

$$iM_t - 2ikM_x + M_{xx} \sim 0, \tag{4.3.29}$$

as $|x| \to \infty$. Substituting (4.3.6a) in (4.3.29) and using (4.3.2b) yields

$$\rho_t(k,t) = ik^2\rho(k;t),$$

and so

$$\rho(k;t) = \rho(k;0)\exp(ik^2 t). \tag{4.3.30a}$$

Similarly substituting (4.3.16) in (4.3.29) yields

$$f_t(k,t) = -ik^2 f(k,t),$$

and so

$$f(k,t) = f(k,0)\exp(-\mathrm{i}k^2 t). \tag{4.3.30b}$$

Finally from (4.3.20) we have

$$\gamma_{j,t}(t) = 2k_j,$$

and so

$$\gamma_j(t) = 2k_j t + \mathrm{i}\gamma_{j,0} \qquad (k_j \text{ constant}), \tag{4.3.30c}$$

for $j = 1, 2, \dots, n$.

4.3.4 Further Remarks.

1. Soliton solutions.

Pure soliton solutions are obtained by taking

$$\rho(k;t) = 0, \qquad N(x,t;k) = 1 - \mathrm{i}\sum_{j=1}^{n} \frac{\Phi_j}{k - k_j},$$

hence from (4.3.16)

$$(x + 2k_j t + \mathrm{i}\gamma_{j,0})\Phi_j(x) = 1 - \mathrm{i}\sum_{\substack{p=1\\ p\neq j}}^{n} \frac{\Phi_j(x)}{k_j - k_p}. \tag{4.3.31}$$

This equation is a linear system (analogous to equation (2.5.2) for the KdV equation), however the coefficients of the system in this case are *algebraic* rather than exponential. In particular, the one-soliton eigenfunctions and solution of the BO equation are given by

$$\Phi_1(x) = \frac{C}{C(x - Ct) + \mathrm{i}}, \tag{4.3.32a}$$

$$u(x,t) = \frac{2C}{C^2(x - Ct)^2 + 1}, \tag{4.3.32b}$$

where $C = 1/\gamma_{1,0} = -2k_1$. Note that if we consider the limit as $\delta \to \infty$ of equation (4.2.33), that is the one-soliton solution of the ILW equation, then there is no proper limit unless $\kappa \to 0$ such that $\kappa\delta$ is finite. Now if we suppose that $\delta\kappa = \pi - \kappa/C$ and set $x_0 = 0$, where C is a real positive constant, then taking the limit of (4.2.33) yields (4.3.32), the one-soliton solution of the BO equation.

N-soliton solutions of the BO equation (which also may be derived by taking the limit as $\delta \to \infty$ of the N-soliton solution of the ILW equation in an appropriate manner), can also be derived as the motion of poles. If we seek a solution of the BO equation in the form

$$u(x,t) = \sum_{j=1}^{N} \frac{\mathrm{i}}{x - x(t)} - \sum_{j=1}^{N} \frac{\mathrm{i}}{x - \overline{x(t)}}, \tag{4.3.33}$$

then it is easily shown that the $x_j(t)$ satisfy

$$\frac{\mathrm{d}^2 x_j}{\mathrm{d}t^2} = 8 \sum_{\substack{k=1 \\ k \neq j}}^{N} \frac{1}{(x_j - x_k)^3}, \qquad j = 1, 2, \ldots, N \tag{4.3.34}$$

(see Case [1978, 1979]; Chen, Lee and Pereira [1979]). Equation (4.3.34) is the well-known Calogero-Moser N-body problem, which is an integrable Hamiltonian system (see Calogero [1971]; Moser [1975]). It might be remarked that the dynamical system of the motion of the poles for the ILW equation is considerably more complicated than (4.3.34) and requires consistency conditions (see Ablowitz and Segur [1981, p.213]).

A periodic solution of the BO equation is given by

$$u(x,t) = \frac{\kappa \tanh \phi_0}{1 + \operatorname{sech} \phi_0 \cos[\kappa(x - ct) + \xi_0]},$$

where κ, ϕ_0, ξ_0 and c are arbitrary constants (Satsuma and Ishimori [1979]).

2. The Gel'fand-Levitan-Marchenko Equation.

If $\Phi_j = 0$, $j = 1, 2, \ldots, n$ (that is if there are *no* solitons), then equation (4.3.21) can be formally reduced to the Gel'fand-Levitan-Marchenko equation. Hence we obtain

$$\begin{aligned} N(x,t;k) &- v(x,t;k) \\ &= \frac{1}{2\pi} \int_0^\infty \mathrm{e}^{\mathrm{i}(k-p)x} \left(\int_\infty^x v(s;k)\, \mathrm{e}^{-\mathrm{i}(k-p)s}\, \mathrm{d}s \right) \rho(p,t)\, N(x,t;p)\, \mathrm{d}p, \end{aligned} \tag{4.3.35}$$

Now multiply equation (4.3.35) by $\frac{1}{2\pi}\rho(k;t)\,\mathrm{e}^{\mathrm{i}(y-x)k}$ and integrate with respect to k from 0 to ∞. Then it is easily shown that

$$K(x,y;t) + \int_x^\infty K(x,s;t)\, F(s,y;t)\, \mathrm{d}s = F(x,y;t), \qquad \text{for} \qquad y > x, \tag{4.3.36}$$

where

$$K(x,y;t) = \frac{1}{2\pi} \int_0^\infty \rho(k;t)\, N(x,t;k)\, \mathrm{e}^{\mathrm{i}(y-x)k}\, \mathrm{d}k, \tag{4.3.37a}$$

and

$$F(x,y;t) = \frac{1}{2\pi} \int_0^\infty \rho(k;t)\, v(x,t;k)\, \mathrm{e}^{\mathrm{i}(y-x)k}\, \mathrm{d}k, \tag{4.3.37b}$$

Also from equation (4.3.26)

$$u^+(x,t) = -\mathrm{i}K(x,x;t). \tag{4.3.38}$$

4.4 Classes of Integro-Differential Evolution Equations.

4.4.1 Introduction. In this section we discuss some other integro-differential equations which are solvable by inverse scattering. In Chapter 1 we saw that Miura's transformation maps solutions of the mKdV equation into solutions of the KdV equation. Analogously a modified Intermediate Long Wave (MILW) equation

$$\tfrac{1}{3}\delta v_t + \mathrm{T}(v_{xx}) + v_x\left[\delta^{-1}\mathrm{e}^v + \mathrm{T}(v_x)\right] = 0, \tag{4.4.1a}$$

where

$$(\mathrm{T}f)(x) := \frac{1}{2\delta}\,⨍_{-\infty}^{\infty} \coth\left\{\frac{\pi}{2\delta}(y-x)\right\} f(y)\,\mathrm{d}y, \tag{4.4.1b}$$

has been introduced and a "Miura transformation" derived which maps solutions of the MILW equation (4.4.1) into solutions of the ILW equation (Gibbons and Kupershmidt [1980]; Satsuma and Ablowitz [1980]). Satsuma, Taha and Ablowitz [1984] have studied the Bäcklund transformation, an associated linear scattering problem and conservation laws for the MILW equation. Further generalizations in this direction have been made by Lebedev and Radul [1983] who constructed a general class of ILW-type evolution equations and Bobenko, Matveev and Sall [1982] who introduced an intermediate Kadomtsev-Petviashvili equation, which naturally reduces to the Kadomtsev-Petviashvili equation as $\delta \to 0$. Recently, Degasperis, Lebedev, Olshanetsky, Pakuliak, Perelomov and Santini [1991a,b] have discussed hierarchies of integro-differential equations.

Degasperis and Santini [1983] generated a class of matrix nonlinear integro-differential evolution equations using the following 2×2 matrix spectral problem:

$$\mathbf{V}^-(x,t;z) = \mathbf{G}(x,t;z)\mathbf{V}^+(x,t;z), \qquad x \in \mathbb{R}, \tag{4.4.2a}$$

$$\mathbf{G}(x,t;z) \equiv \mathbf{I} + z\boldsymbol{\sigma}_3 + \mathbf{U}(x,t), \tag{4.4.2b}$$

where $\mathbf{I}$ is the identity matrix, $\boldsymbol{\sigma}_3 = \mathrm{diag}(1,-1)$, and z plays the role of the spectral parameter and the (complex) z-independent potential matrix $\mathbf{U}(x,t)$ and $\det\mathbf{U} = 1$ is imposed. The matrices $\mathbf{V}^{\pm}(x,t)$ are the boundary values of a function a function $\boldsymbol{\Psi}(x,t;z)$, holomorphic in the horizontal strip between $\mathrm{Im}(x) = 0$ and $\mathrm{Im}(x) = \delta$:

$$\mathbf{V}^+(x,t;z) = \lim_{y\downarrow 0} \boldsymbol{\Psi}(x+\mathrm{i}y,t;z), \qquad x \in \mathbb{R},$$

$$\mathbf{V}^-(x,t;z) = \lim_{y\uparrow \delta} \boldsymbol{\Psi}(x+\mathrm{i}y,t;z), \qquad x \in \mathbb{R},$$

In fact $\boldsymbol{\Psi}(x;z)$ can be written as

$$\boldsymbol{\Psi}(x;z) = \exp\{-\mathrm{i}\boldsymbol{\Omega}(z)x\}\left[\frac{1}{2\mathrm{i}\delta}\int_{-\infty}^{\infty} \coth\left\{\frac{\pi}{\delta}(y-x)\right\}\mathbf{H}(y;z)\mathrm{d}y + \alpha\right],$$

for $0 < \operatorname{Im} x < \delta$, where α is a constant,

$$\exp\{\delta \boldsymbol{\Omega}(z)\} = \mathbf{I} + z\boldsymbol{\sigma}_3,$$
$$\mathbf{H}(x;z) = (\mathbf{I} + z\boldsymbol{\sigma}_3)^{-1} \exp\{\mathrm{i}\boldsymbol{\Omega}(z)x\}\mathbf{U}(x)\mathbf{V}^{+}(x;z),$$

such that $\mathbf{H}(x;z)$ is Hölder continuous on $x \in \mathbb{R}$ and satisfies the condition

$$\left|\int_{-\infty}^{\infty} \mathbf{H}(x;z)\,\mathrm{d}x\right| < \infty.$$

It should be noted that (4.4.2) is a pure Riemann-Hilbert problem (in x), and hence differs from that of the ILW equation which is a *differential* Riemann-Hilbert problem. Degasperis, Santini and Ablowitz [1985] and Santini [1986] show that unique solutions of (4.4.2) will exist if $\mathbf{U}(x,t)$ has the form

$$\mathbf{U}(x,t) = \sqrt{\mathbf{I} + \mathbf{Q}^2(x,t)} - \mathbf{I} + \mathbf{Q}(x,t) \tag{4.4.3a}$$
$$= \begin{pmatrix} \sqrt{1 + Q^{12}Q^{21}} - 1 & Q^{12} \\ Q^{21} & \sqrt{1 + Q^{12}Q^{21}} - 1 \end{pmatrix},$$
$$\mathbf{Q} = \begin{pmatrix} 0 & Q^{12} \\ Q^{21} & 0 \end{pmatrix}, \tag{4.4.3b}$$

where $\mathbf{I}$ is the identity matrix and $\mathbf{Q}(x,t)$ is an off-diagonal matrix, which is just the off-diagonal part of $\mathbf{U}(x,t)$. Given the matrix function $\mathbf{U}(x,t)$, equations (4.4.2-3) define a homogeneous Riemann-Hilbert boundary value problem on a strip of the complex plane. The associated class of nonlinear evolution equations may be written in the following form

$$\mathbf{Q}_t = \boldsymbol{\sigma}_3 p(L_\delta)\mathbf{Q}, \tag{4.4.4a}$$

where

$$L_\delta \mathbf{F} \equiv \mathrm{i}\boldsymbol{\sigma}_3 \left(\sqrt{\mathbf{I} + \mathbf{Q}^2}\,\mathrm{D}_\delta \mathbf{F} + \tfrac{1}{2}\mathbf{Q}\mathrm{D}_\delta^{-1} \frac{[\mathbf{Q},\mathbf{F}]}{\sqrt{1 + Q^2}} \right), \tag{4.4.4b}$$
$$\sqrt{1 + Q^2} = \sqrt{1 + Q^{12}Q^{21}}, \tag{4.4.4c}$$

with $\mathbf{F}$ an off-diagonal 2×2 matrix, $p(y)$ is an arbitrary polynomial in y, $[\ ,\ ]$ the usual commutator and the integral operator D_δ and inverse D_δ^{-1} are defined by

$$(\mathrm{D}_\delta f)(x) = \frac{1}{\delta} \int_{-\infty}^{\infty} f(y) \operatorname{cosech}\left\{\frac{\pi}{\delta}(y - x)\right\} \mathrm{d}y, \tag{4.4.4d}$$
$$(\mathrm{D}_\delta^{-1} f)(x) = -\frac{1}{\delta} \int_{-\infty}^{\infty} f(y) \coth\left\{\frac{\pi}{\delta}(y - x)\right\} \mathrm{d}y. \tag{4.4.4e}$$

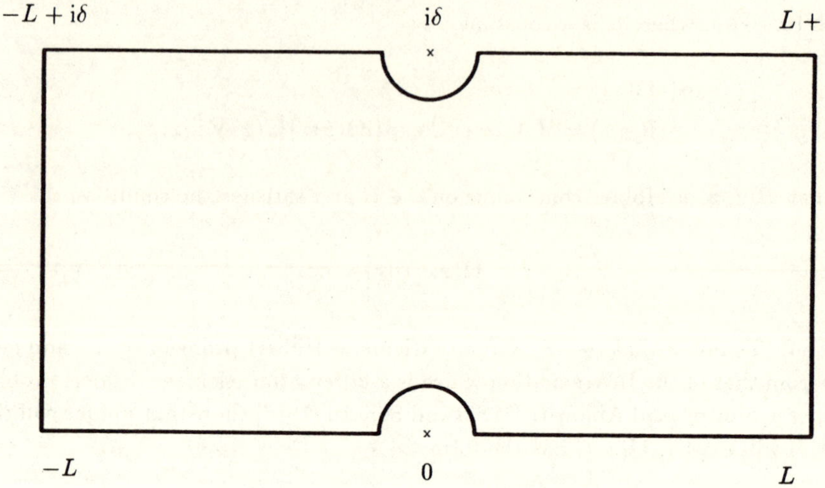

FIGURE 4.4.1 *The contour C_L.*

It can be seen that (4.4.4d,e) are the inverses of each other by studying their Fourier transforms

$$
\begin{aligned}
(\mathrm{D}_\delta f)\hat{}(k) &= \int_{-\infty}^{\infty} (\mathrm{D}_\delta f)(x)\,\mathrm{e}^{-\mathrm{i}kx}\,\mathrm{d}x \\
&= \frac{1}{\delta}\left(\fint_{-\infty}^{\infty} \operatorname{cosech}(\pi y/\delta)\,\mathrm{e}^{-\mathrm{i}ky}\,\mathrm{d}y\right)\left(\int_{-\infty}^{\infty} f(x)\mathrm{e}^{-\mathrm{i}kx}\,\mathrm{d}x\right), \\
(\mathrm{D}_\delta^{-1} f)\hat{}(k) &= \int_{-\infty}^{\infty} (\mathrm{D}_\delta^{-1} f)(x)\,\mathrm{e}^{-\mathrm{i}kx}\,\mathrm{d}x \\
&= -\frac{1}{\delta}\left(\fint_{-\infty}^{\infty} \coth(\pi y/\delta)\,\mathrm{e}^{-\mathrm{i}ky}\,\mathrm{d}y\right)\left(\int_{-\infty}^{\infty} f(x)\mathrm{e}^{-\mathrm{i}kx}\,\mathrm{d}x\right).
\end{aligned}
$$

Since $\coth(\pi y/\delta) \notin L^1(\mathbb{R})$, then we have to consider its Fourier transform in a generalized sense. Suppose C_L is the contour given in Figure 4.4.1

Consider

$$
\begin{aligned}
I_L &= \int_{C_L} \coth(\pi y/\delta)\,\mathrm{e}^{-\mathrm{i}ky}\,\mathrm{d}y \\
&= \fint_{-L}^{L} \coth(\pi y/\delta)\,\mathrm{e}^{-\mathrm{i}ky}\,\mathrm{d}y - \frac{\mathrm{i}\pi}{\pi/\delta} + I_{\delta 1} \\
&\quad - \fint_{-L}^{L} \coth(\mathrm{i}\pi + \pi y/\delta)\exp\{-\mathrm{i}k(y+\mathrm{i}\delta)\}\,\mathrm{d}y - \frac{\mathrm{i}\pi\mathrm{e}^{k\delta}}{\pi/\delta} - I_{\delta 2},
\end{aligned}
$$

where

$$
I_{\delta 1} = \mathrm{i}\int_0^{\delta} \coth[\pi(L+\mathrm{i}s)/\delta]\exp\{-\mathrm{i}k(L+\mathrm{i}s)\}\,\mathrm{d}s,
$$

$$I_{\delta 2} = \mathrm{i} \int_0^{\delta} \coth\left[\pi(-L + \mathrm{i}s)/\delta\right] \exp\{-\mathrm{i}k(-L + \mathrm{i}s)\}\, \mathrm{d}s.$$

It is easily shown that as $L \to \infty$

$$I_{\delta 1} \to \frac{\mathrm{i}\mathrm{e}^{-\mathrm{i}kL}(\mathrm{e}^{k\delta} - 1)}{k},$$
$$I_{\delta 2} \to -\frac{\mathrm{i}\mathrm{e}^{\mathrm{i}kL}(\mathrm{e}^{k\delta} - 1)}{k}.$$

In the sense of generalized functions

$$\frac{\mathrm{e}^{\pm \mathrm{i}kL}}{\mathrm{i}k} \to \pm\pi\delta(k), \qquad \text{and} \quad L \to \infty,$$

and so $(I_{\delta 1} - I_{\delta 2}) \to 0$ as $L \to \infty$ and so

$$\lim_{L\to\infty} \fint_{-L}^{L} \coth\left(\pi y/\delta\right) \mathrm{e}^{-\mathrm{i}ky}\, \mathrm{d}y = \mathrm{i}\delta \coth\left(\tfrac{1}{2}k\delta\right).$$

To evaluate the Fourier transform of $\operatorname{cosech}(\pi y/\delta)$ (as a Cauchy principal valued integral), we use the same contour, except this is simpler since as $L \to \infty$

$$\int_0^{\delta} \operatorname{cosech}\left[\pi(L + \mathrm{i}s)/\delta\right] \exp\{-\mathrm{i}k(L + \mathrm{i}s)\}\, \mathrm{d}s \to 0,$$
$$\int_0^{\delta} \operatorname{cosech}\left[\pi(-L + \mathrm{i}s)/\delta\right] \exp\{-\mathrm{i}k(-L + \mathrm{i}s)\}\, \mathrm{d}s \to 0.$$

Therefore

$$\fint_{-\infty}^{\infty} \operatorname{cosech}\left(\pi y/\delta\right) \mathrm{e}^{-\mathrm{i}ky}\, \mathrm{d}y = \mathrm{i}\delta \tanh\left(\tfrac{1}{2}k\delta\right).$$

Hence we conclude that D_δ and D_δ^{-1} are inverses of each other.

Recall from §1.10 that for general evolution equation in the AKNS method may be written in the form

$$\boldsymbol{\sigma}_3 \boldsymbol{u}_t + A_0(\mathrm{L})\boldsymbol{u} = \mathbf{o}, \tag{4.4.5}$$

where $A_0(\mathrm{L})$ is an entire function, L an integro-differential operator, defined by equation (1.10.2), and $\boldsymbol{u} = (r, q)$. Equation (4.4.5) may equivalently be written in the form

$$\mathbf{Q}_t = -\boldsymbol{\sigma}_3 A_0(\mathrm{L})\mathbf{Q}, \tag{4.4.6a}$$

where

$$\mathrm{L}\mathbf{F} := \tfrac{1}{2}\mathrm{i}\boldsymbol{\sigma}_3(\mathbf{F}_x - 2\mathbf{Q}\mathrm{I}_-[\mathbf{Q}, \mathbf{F}]), \tag{4.4.6b}$$

with $\mathbf{F}$ an arbitrary off-diagonal matrix and $(\mathrm{I}_- f)(x) := \int_{-\infty}^{x} f(y)\, \mathrm{d}y$.

The first three evolution equations in this class (4.4.4) are the following, using the reduction of Degasperis and Santini [1983] and Degasperis, Santini and Ablowitz [1985], i.e., (4.4.3) with $Q^{12} = Q^{21} = q$,
(i) an intermediate wave equation

$$p(y) = -\mathrm{i}cy, \qquad \mathbf{Q} = \begin{pmatrix} 0 & q \\ q & 0 \end{pmatrix}, \qquad \mathbf{U} = \begin{pmatrix} u-1 & q \\ q & u-1 \end{pmatrix},$$
$$q_t = cu\mathrm{D}_\delta q, \qquad u = \sqrt{1+q^2}; \tag{4.4.7}$$

(ii) an intermediate nonlinear Schrödinger equation

$$p(y) = \mathrm{i}cy^2, \qquad \mathbf{Q} = \begin{pmatrix} 0 & \mathrm{i}q^* \\ \mathrm{i}q & 0 \end{pmatrix}, \qquad \mathbf{U} = \begin{pmatrix} u-1 & \mathrm{i}q^* \\ \mathrm{i}q & u-1 \end{pmatrix},$$
$$\mathrm{i}q_t + c\left[u\mathrm{D}_\delta(u\mathrm{D}_\delta q) - q\mathrm{D}_\delta^{-1}\mathrm{Re}(q^*\mathrm{D}_\delta q)\right] = 0, \qquad u = \sqrt{1-qq^*}; \tag{4.4.8}$$

(iii) an intermediate modified KdV equation (distinct from (4.4.1) which arises from a differential Riemann-Hilbert scattering problem)

$$p(y) = \mathrm{i}cy^3, \qquad \mathbf{Q} = \begin{pmatrix} 0 & q \\ q & 0 \end{pmatrix}, \qquad \mathbf{U} = \begin{pmatrix} u-1 & q \\ q & u-1 \end{pmatrix},$$
$$q_t = cu\mathrm{D}_\delta(u\mathrm{D}_\delta(u\mathrm{D}_\delta q)\mathrm{D}_\delta^{-1}q\mathrm{D}_\delta q), \qquad u = \sqrt{1+q^2}. \tag{4.4.9}$$

Taking the limit as $\delta \to 0$ of equations (4.4.7-9), we obtain the linear wave equation, the nonlinear Schrödinger equation and the Modified KdV equation, respectively (see Degasperis and Santini [1983]). Note that the first equation (4.4.7) in this class can be written in the simple form

$$\mathrm{D}_\delta^{-1}\theta_t = c\sin\theta, \qquad \theta = \theta(x,t), \tag{4.4.10}$$

where $q(x,t) = \mathrm{i}\sin\theta(x,t)$.

If in equation (4.4.4) we take the limit $\delta \to \infty$, that is replacing D_δ by H and D_δ^{-1} by $-\mathrm{H}$, where

$$(\mathrm{H}f)(x) = \frac{1}{\pi} \int_{-\infty}^{\infty} \frac{f(y)}{y-x}\,\mathrm{d}y,$$

is the Hilbert transform, then we obtain the class of matrix evolution equations given by

$$\mathbf{Q}_t = \sigma_3 p(\mathrm{L})\mathbf{Q}, \tag{4.4.11a}$$

where

$$\mathbf{LF} \equiv \mathrm{i}\sigma_3\left(\sqrt{1+Q^2}\,\mathrm{H}\mathbf{F} - \tfrac{1}{2}\mathbf{Q}\mathrm{H}\frac{[\mathbf{Q},\mathbf{F}]}{\sqrt{1+Q^2}}\right), \tag{4.4.11b}$$
$$\sqrt{1+Q^2} = \sqrt{1+Q^{12}Q^{21}}. \tag{4.4.11c}$$

Santini, Ablowitz and Fokas [1987] have developed the I.S.T. scheme associated with this class of evolution equations (see also Santini [1986]).

4.4.2 The Sine-Hilbert Equation. The first (and simplest) equation in this class is obtained by letting $\delta \to \infty$ in (4.4.10), then by making use of the fact that $\mathrm{H}^2 = -\mathrm{I}$, where I is the identity operator, we obtain the equation

$$\mathrm{H}u_t = -c \sin u, \qquad q(x,t) = \mathrm{i} \sin u(x,t). \tag{4.4.12}$$

Due to the analogy between equation (4.4.12) and the Sine-Gordon equation

$$u_{xt} = \sin u,$$

Degasperis, Santini and Ablowitz [1985] named it the *Sine-Hilbert equation*. (Note also the compelling analogy with the KdV equation and the BO equation, whereby one goes from the KdV equation

$$u_t + uu_x + u_{xxx} = 0,$$

to the BO equation

$$u_t + uu_x + \mathrm{H}u_{xx} = 0,$$

by simply replacing one of the x-derivatives by the Hilbert transform H.)

We now sketch the main results of Santini, Ablowitz and Fokas [1987]. For convenience we take $p(y) = a_n y^n$ in (4.4.11). The scattering problem is (4.4.2-3), where $v^{\pm}(x)$ are the boundary values of $v(x)$ in the upper $(+)$/lower $(-)$ half x-plane. The associated linear time equation is

$$\frac{\partial \mathbf{V}^{\pm}}{\partial t}(x,t,z) = \tfrac{1}{2} a_n \left[z^n \boldsymbol{\sigma}_3 + \sum_{j=0}^{n-1} z^j \left(\mathcal{P}^{\pm} \mathbf{W}^{\pm}_{n,j}\right)(x,t) \right] \mathbf{V}^{\pm}(x,t,z), \tag{4.4.13}$$

where

$$\mathbf{W}^{\pm}_{n,j} \equiv 2\mathrm{L}^{n-j-1}\mathbf{Q} \mp \frac{1}{\sqrt{1+Q^2}} \left[\mathbf{Q}, \mathrm{L}^{n-j-1}\mathbf{Q}\right]. \tag{4.4.14a}$$

LF is given by (4.4.11b) and $\mathcal{P}^{\pm}$ are the usual projection operators

$$(\mathcal{P}^{\pm} f)(x) \equiv \pm \frac{1}{2\pi \mathrm{i}} \int_{-\infty}^{\infty} \frac{f(y)}{y - (x \pm \mathrm{i}0)}\, \mathrm{d}y. \tag{4.4.14b}$$

Equation (4.4.3) implies that

$$\det[\mathbf{I} + z\boldsymbol{\sigma}_3 + \mathbf{U}(x,t)] = 1 - z^2, \tag{4.4.15}$$

with the following important consequences (Degasperis, Santini and Ablowitz [1985]): (i), the matrix $\mathbf{I}+z\boldsymbol{\sigma}_3+\mathbf{U}(x,t)$ is invertible for every $x\in\mathbb{R}$; this is a necessary condition for the solvability of (4.4.2); and (ii), the total index κ of the matrix Riemann-Hilbert problem (4.4.2) is zero since

$$\kappa=\frac{1}{2\pi}\left[\arg(\det(\mathbf{I}+z\boldsymbol{\sigma}_3+\mathbf{U}(x,t)))\right]_{-\infty}^{\infty}, \tag{4.4.16}$$

where $[f(x)]_{-\infty}^{\infty}=f(\infty)-f(-\infty)$. Then an important theorem due to Gohberg and Krein [1958] shows that "generically" the two partial indices κ_1, κ_2 $(\kappa=\kappa_1+\kappa_2)$ are both zero.

This fact guarantees the existence and uniqueness of the fundamental matrix solutions $\boldsymbol{\Psi}^{\pm}(x,z)$ of (4.4.2) (here and in the following we omit the time dependence when it is not needed), satisfying the following boundary conditions

$$\lim_{|x|\to\infty}\boldsymbol{\Psi}^{+}(x,z)=\mathbf{I},\qquad \lim_{|x|\to\infty}\boldsymbol{\Psi}^{-}(x,z)=\mathbf{I}+z\boldsymbol{\sigma}_3, \tag{4.4.17}$$

which are a consequence of the requirement that $\mathbf{Q}(x)$ vanishes as $|x|\to\infty$ sufficiently rapidly. Equation (4.4.17) suggests the introduction of functions $\mathbf{m}^{\pm}(x,z)$ defined by

$$\mathbf{m}^{+}(x,z):=\boldsymbol{\Psi}^{+}(x,z),\qquad \mathbf{m}^{-}(x,z):=\boldsymbol{\Psi}^{-}(x,z)(\mathbf{I}+z\boldsymbol{\sigma}_3)^{-1}.$$

They obviously satisfy the Riemann-Hilbert boundary value problem

$$\mathbf{m}^{+}(x,z)-\mathbf{m}^{-}(x,z)+z\left(\boldsymbol{\sigma}_3\mathbf{m}^{+}(x,z)-\mathbf{m}^{-}(x,z)\boldsymbol{\sigma}_3\right)$$
$$+\,\mathbf{U}(x)\mathbf{m}^{+}(x,z)=\mathbf{0}, \tag{4.4.18a}$$
$$\lim_{|x|\to\infty}\mathbf{m}^{\pm}(x,z)=\mathbf{I}, \tag{4.4.18b}$$

and have the property

$$\det\mathbf{m}^{\pm}(x,z)=1, \tag{4.4.19}$$

which is a direct consequence of equation (4.4.16).

The unique solutions of this problem can be given in terms of the following matrix integral equations

$$\mathbf{m}^{+}(x,z)+(\mathbf{I}+z\boldsymbol{\sigma}_3)^{-1}\mathcal{P}^{+}(\mathbf{U}\mathbf{m}^{+})(x,z)=\mathbf{I}, \tag{4.4.20a}$$
$$\mathbf{m}^{-}(x,z)-\mathcal{P}^{-}(\mathbf{U}\mathbf{m}^{+})(x,z)(\mathbf{I}+z\boldsymbol{\sigma}_3)^{-1}=\mathbf{I}. \tag{4.4.20b}$$

These equations can have homogeneous matrix solutions $\boldsymbol{\Phi}^{\pm(j)}(x)$, $j=1,2,\ldots$ (corresponding to the eigenvalues $z=z_j$) that satisfy the following equations:

$$\boldsymbol{\Phi}^{+(j)}(x)+(\mathbf{I}+z_j\boldsymbol{\sigma}_3)^{-1}\mathcal{P}^{+}(\mathbf{U}\boldsymbol{\Phi}^{+(j)})(x)=0, \tag{4.4.21a}$$
$$\boldsymbol{\Phi}^{-(j)}(x)-\mathcal{P}^{-}(\mathbf{U}\boldsymbol{\Phi}^{+(j)})(x)(\mathbf{I}+z_j\boldsymbol{\sigma}_3)^{-1}=0, \tag{4.4.21b}$$

together with the boundary conditions

$$\Phi^{\pm(j)}(x) = O(x^{-1}), \qquad |x| \gg 1. \tag{4.4.22}$$

Because of (4.4.22), the eigenvalues z_j are bound states of (4.4.18); they correspond to the nongeneric case in which the partial indices κ_1 and κ_2 are different from zero ($\kappa_1 = -\kappa_2 \neq 0$), when there does not exist a unique solution of (4.4.18).

For suitable potentials $\mathbf{Q}(x)$ the structure of (4.4.20) together with Fredholm theory imply that the matrices $\mathbf{m}^{\pm}(x,z)$ are holomorphic in the complex z plane except for possible poles that generically cluster at ± 1. (The proof in given in Degasperis, Santini and Ablowitz [1985].) The poles of $\mathbf{m}^{\pm}(x,z)$ correspond to the homogeneous solutions of equations (4.4.20); where we assume that then are simple (with a genericity argument) and that the following representation formula holds

$$\mathbf{m}^{\pm}(x,z) = \mathbf{I} + \sum_{j=-\infty}^{\infty} \frac{\Phi^{\pm(j)}(x)}{z - z_j}, \qquad \lim_{j\to\pm\infty} z_j = \pm 1, \qquad j \in \mathbb{Z}. \tag{4.4.23}$$

The first consequence of equations (4.4.21) is that the 2×2 matrices $\Phi^{\pm(\ell)}(x)$ are singular; precisely

$$\Phi^{\pm(\ell)}(x) = \begin{pmatrix} \Phi_{11}^{\pm(\ell)} & \Phi_{12}^{\pm(\ell)} \\ \Phi_{21}^{\pm(\ell)} & \Phi_{22}^{\pm(\ell)} \end{pmatrix}(x) = (\psi^{\pm(\ell)}, \alpha_\ell^{\pm}\psi^{\pm(\ell)}), \tag{4.4.24a}$$

where $\psi^{\pm(\ell)}$ indicate the first column vectors of the matrices $\Phi^{\pm(\ell)}$, i.e.

$$\psi^{\pm(\ell)}(x) = \begin{pmatrix} \psi_1^{\pm(\ell)} \\ \psi_2^{\pm(\ell)} \end{pmatrix} \equiv \begin{pmatrix} \Phi_{11}^{\pm(\ell)} \\ \Phi_{21}^{\pm(\ell)} \end{pmatrix}, \tag{4.4.24b}$$

and from (4.4.21)

$$\alpha_\ell^{-} - \frac{1 + z_\ell}{1 - z_\ell}\alpha_\ell^{+}. \tag{4.4.24c}$$

Equations (4.4.21) also imply that

$$\lim_{|x|\to\infty} x\psi_1^{+(\ell)}(x) = \lim_{|x|\to\infty} x\psi_1^{-(\ell)}(x) \equiv c_\ell, \tag{4.4.25a}$$

$$(1 - z_\ell)\lim_{|x|\to\infty} x\psi_2^{+(\ell)}(x) = (1 + z_\ell)\lim_{|x|\to\infty} x\psi_2^{-(\ell)}(x), \tag{4.4.25b}$$

while equations (4.4.19,23-25) imply that

$$\lim_{|x|\to\infty} x\psi_2^{+(\ell)}(x) = -c_\ell/\alpha_\ell^{\pm}. \tag{4.4.26}$$

Finally, a representation for $\mathbf{U}(x)$ is obtained by expanding equation (4.4.18a) for large z and using equation (4.4.23);

$$\mathbf{U}(x) = \sum_{j=-\infty}^{\infty} \left[\Phi^{-(j)}(x)\sigma_3 - \sigma_3\Phi^{+(j)}(x)\right]. \tag{4.4.27}$$

Significantly it turns out that the vector solutions $\psi^{\pm(\ell)}(x)$, $\ell \in \mathbb{Z}$, of equations (4.4.21) satisfy the following infinite dimensional algebraic system

$$(x + \gamma_\ell^\pm)\psi^{\pm(\ell)}(x) = c_\ell \left[\begin{pmatrix} 1 \\ -1/\alpha_\ell^\pm \end{pmatrix} + \sum_{\substack{j=-\infty \\ j\neq \ell}}^{\infty} \frac{\alpha_\ell^\pm - \alpha_j^\pm}{\alpha_\ell^\pm (z_\ell - z_j)} \psi^{\pm(\ell)}(x) \right], \tag{4.4.28a}$$

with

$$\gamma_\ell^+ - \gamma_\ell^- = \frac{2c_\ell}{1 - z_\ell^2}. \tag{4.4.28b}$$

The direct problem is a linear mapping from the potential $\mathbf{Q}(x)$ to the scattering data:

$$S \equiv \left\{ z_\ell, c_\ell, \alpha_\ell^\pm, \gamma_\ell^\pm; \ell \in \mathbb{Z} \right\}. \tag{4.4.29}$$

More precisely, given $\mathbf{Q}(x)$ [and then $\mathbf{U}(x)$ through equation (4.4.3)], $\mathbf{m}^\pm(x,z)$ and the bound states z_ℓ are in principle given by solving equations (4.4.20), and then the $\mathbf{\Phi}^{\pm(\ell)}(x)$ are obtained by taking the limit

$$\mathbf{\Phi}^{\pm(\ell)}(x) = \lim_{z \to z_\ell} (z - z_\ell)\mathbf{m}^\pm(x,z).$$

All the scattering data can be constructed through the following sequence of steps: c_ℓ via equation (4.4.25a); α_ℓ^+ via $\alpha_\ell^+ = -c_\ell / \lim_{|x|\to\infty} x\psi_2^{+(\ell)}(x)$ [a consequence of equation (4.4.26)]; γ_ℓ^+ via equation (4.4.28) and the α^- and γ_ℓ^-, which are not independent data, can obtained using equations (4.4.24c) and (4.4.28), respectively.

The inverse scattering problem, a linear mapping from the scattering data S to the potential $\mathbf{Q}$, is formally performed by solving the infinite dimensional algebraic system (4.4.28) and then by making use of equations (4.4.27).

The time evolution of the scattering data is obtained by observing first that the solutions $\mathbf{m}^\pm$ of equation (4.4.18) evolve in time according to the following equations

$$\frac{\partial \mathbf{m}^\pm}{\partial t}(x,t,z) = \tfrac{1}{2} a_n \left\{ z^n [\boldsymbol{\sigma}_3, \mathbf{m}^\pm(x,t,z)] + \sum_{j=0}^{n-1} z^j (\mathcal{P}^\pm \mathbf{W}_{n,j}^\pm)(x,t) \right\} \mathbf{m}^\pm(x,t,z). \tag{4.4.30}$$

From this equation the data is found to evolve as follows:

$$z_\ell(t) = z_\ell(0) = z_\ell, \tag{4.4.31a}$$

$$c_\ell(t) = c_\ell(0) = c_\ell, \tag{4.4.31b}$$

$$\alpha_\ell^\pm(t) = \alpha_\ell^\pm(0)\exp\{a_n z_\ell^n t\}, \tag{4.4.31c}$$

$$\gamma_\ell^\pm(t) = \gamma_\ell^\pm(0) - n a_n c_\ell z_\ell^{n-1} t. \tag{4.4.31d}$$

In order to obtain the above equations for the time evolution of the scattering data we first substitute equations (4.4.23,27) into equation (4.4.30). This yields $\dot{z}_\ell = 0$.

The other relationships can be found by taking the time derivative of (4.4.28), using (4.4.30) at $z = z_\ell$, with $\mathbf{m}^+(x,t,z_\ell)$ replaced by $\mathbf{\Phi}^+(\ell)$ when z is replaced by z_ℓ and evaluating the results as $x \to \infty$.

These results complete the I.S.T. scheme:

$$\begin{array}{ccc} u(x,0) & \xrightarrow{\text{direct scattering}} & S(0) = \{z_\ell(0), c_\ell(0), \alpha_\ell^\pm(0), \gamma_\ell^\pm(0); \ell \in \mathbb{Z}\} \\ & & \Big\downarrow \text{ evolution of scattering data} \\ u(x,t) & \xleftarrow{\text{inverse scattering}} & S(t) = \{z_\ell(t), c_\ell(t), \alpha_\ell^\pm(t), \gamma_\ell^\pm(t); \ell \in \mathbb{Z}\} \end{array}$$

which reduces the solution of the Cauchy problem for equations (4.4.11) to a sequence of linear steps.

The pure soliton solutions associated with the class of equations (4.4.11) correspond to a finite number N of poles in equation (4.4.23). If $N = 1$ ($c_j = 0, j \geq 1$), equations (4.4.27-28,31) yield the following one-soliton solution

$$Q^{12}(x,t) = -\frac{2c_0\alpha_0^+(t)}{1-z_0}\left\{\frac{1}{x+\gamma_0^-(t)} - \frac{c_0}{(1+z_0)[x+\gamma_0^+(t)]}\right\},$$

$$Q^{21}(x,t) = -\frac{2c_0}{(1+z_0)\,\alpha_0^+(t)}\left\{\frac{1}{x+\gamma_0^-(t)} - \frac{c_0}{(1-z_0)[x+\gamma_0^+(t)]}\right\}.$$

It should be mentioned that in the reduction case $Q^{12} = Q^{21}$ (which is allowed for all the evolution equations (4.4.11) corresponding to n odd), the singularities of $\mathbf{m}^\pm(x,t,z)$ come in pairs $z_{-\ell} = -z_\ell$. Correspondingly it can be shown that

$$\alpha_\ell^+(0)\alpha_{-\ell}^+(0) = 1, \qquad c_{-\ell} = c_\ell, \qquad \gamma_{-\ell}^+ = \gamma_\ell^+,$$

if $\ell = 0$, $z_0 = 0$ and $|\alpha_0^+(0)|^2 = 1$.

The Sine-Hilbert equation (4.4.12) is the $n = 1$ case of the above results. In particular it has the one-soliton solution

$$u(x,t) = -2\tan^{-1}\left(\frac{a}{x-at-b}\right). \tag{4.4.32}$$

Differentiating this with respect to x yields

$$u_x(x,t) = \frac{2a}{(x-at-b)^2+a^2},$$

which is the one-soliton solution of the Benjamin-Ono equation. A periodic solution of the Sine-Hilbert equation, with period π/β, is given by

$$u(x,t) = -2\tan^{-1}\left(\frac{2\beta a\cot[\beta(x-at-b)]}{1+\sqrt{1+4\beta^2a^2}}\right),$$

which reduces to the one-soliton solution (4.4.32) in the limit as $\beta \to 0$ (Matsuno [1987a]). N-soliton solutions of the Sine-Hilbert equation (4.4.12) have been constructed by Matsuno [1986, 1987b].

The inverse problem associated with (4.4.11) is novel. Specifically (4.4.16) admits only bound states; there is *no* continuous spectrum. The inverse scattering is carried out via a solution of an infinite dimensional algebraic system, i.e., equation (4.4.28). This is in contrast to all previous studies where there is both a continuous spectrum and a linear integral equation governing the inverse scattering.

We remark that it is expected that many of the results presented in this section should be extendable to 2 + 1-dimensional systems. Indeed Bobenko, Matveev and Sall [1982] have shown that one can add the operator $\partial/\partial y$ to the linear differential Riemann-Hilbert operator governing the ILW problem (4.2.1). The following intermediate KP equation is a result of compatibility

$$(u_t + (\mathrm{T}u)_{xx} - 6uu_x)_x = (3\delta^{-2}u_y + (\mathrm{T}u)_{xx} - 6uu_x)_y, \tag{4.4.33}$$

which in the limit as $\delta \to \infty$ becomes

$$(u_t + (\mathrm{H}u)_{xx} - 6uu_x)_x = ((\mathrm{H}u)_{xx} - 6uu_x)_y. \tag{4.4.34}$$

4.4.3 Further Examples. We next show that there are other nonlinear singular-differential equations which are exactly solvable by using a certain reduction process. These ideas are given by Ablowitz, Fokas and Kruskal [1987], motivated by some recent results of Constantin, Lax and Majda [1985], who proposed the following equation as a model for the motion of vorticity for an inviscid incompressible fluid flow

$$u_t = u\mathrm{H}u, \tag{4.4.35}$$

where $\mathrm{H}u$ is the Hilbert transform of u. Introduce the transformation

$$w = u + \mathrm{i}v, \qquad v = \mathrm{H}u, \tag{4.4.36}$$

then it follows that w satisfies the ordinary differential equation

$$\frac{\mathrm{d}w}{\mathrm{d}t} = \tfrac{1}{2}w^2. \tag{4.4.37}$$

It should be noted that the above result can be obtained as follows: operate on equation (4.4.35) with $(\mathrm{I} + \mathrm{iH})$ and use

$$\mathrm{H}(u\mathrm{H}u) = \tfrac{1}{2}((\mathrm{H}u)^2 - u^2), \tag{4.4.38}$$

which is a special case of the known formula

$$\mathrm{H}(f\mathrm{H}g) + \mathrm{H}(g\mathrm{H}f) = (\mathrm{H}f)(\mathrm{H}g) - fg. \tag{4.4.39}$$

Equation (4.4.37) can easily be extended. Since, as is well known, $\mathrm{H}^2 = -\mathrm{I}$, we have that $\mathrm{H}w = -\mathrm{i}w$. Thus w is the boundary value of a function analytic in the lower half plane (a "lower function"), vanishing at infinity. Since $\mathrm{H}w = -\mathrm{i}w$, then more generally

$$\mathrm{H}w^n = -\mathrm{i}w^n, \qquad (n > 0, \quad \text{integer}), \tag{4.4.40a}$$

$$\mathrm{He}^w = -\mathrm{ie}^w, \tag{4.4.40b}$$

etc. where $w := u + \mathrm{iH}u$. This enables us to construct arbitrarily many reducible equations. For example, associated with the following ordinary differential equations

$$\frac{\mathrm{d}w}{\mathrm{d}t} = -\tfrac{1}{2}\mathrm{i}w^2, \tag{4.4.41a}$$

$$\frac{\mathrm{d}w}{\mathrm{d}t} = w^3, \tag{4.4.42a}$$

$$\frac{\mathrm{d}w}{\mathrm{d}t} = \mathrm{ie}^{-\mathrm{i}w}, \tag{4.4.43a}$$

are the following singular integro-differential equations

$$u_t = u\mathrm{H}u, \tag{4.4.41b}$$

$$u_t = u^3 - 3u(\mathrm{H}u)^2, \tag{4.4.42b}$$

$$u_t = \mathrm{e}^{\mathrm{H}u}\sin u, \tag{4.4.43b}$$

respectively.

The extension of the above results to partial differential equations is straightforward once it is noted that the above considerations go through even if a linear operator is substituted for the time derivative in the above equations. For example, associated with the following partial differential equations

$$w_t = w_{xx} - \mathrm{i}(w^2)_x, \tag{4.4.44a}$$

$$w_t + w_{xxx} - \mathrm{i}\alpha(w^2)_x + \beta(w^3)_x = 0, \tag{4.4.45a}$$

$$w_{xt} = \mathrm{ie}^{-\mathrm{i}w}, \tag{4.4.46a}$$

$$w_t + \mathrm{i}(w_{xx} + 2ww_x) = 0, \tag{4.4.47a}$$

are the singular integro-differential equations

$$u_t = u_{xx} + 2(u\mathrm{H}u)_x, \tag{4.4.44b}$$

$$u_t + u_{xxx} + 2\alpha(u\mathrm{H}u)_x + \beta(u^3 - 3u(\mathrm{H}u)^2)_x = 0, \tag{4.4.45b}$$

$$u_{xt} = \mathrm{e}^{\mathrm{H}u}\sin u, \tag{4.4.46b}$$

$$u_t + (\mathrm{H}u)_{xx} + 2(u\mathrm{H}u)_x, \tag{4.4.47b}$$

respectively, where α, β are arbitrary real constants. Equation (4.4.44a) is essentially Burgers' equation which can be linearized via the Cole-Hopf transformation $w = -\mathrm{i}(\ln f)_x$. Equation (4.4.44b) arises in various population ecological models and to our knowledge, was first considered and solved via a dependent variable transformation and splitting into upper and lower functions by Satsuma [1981]. Equation (4.4.45b) is the analog of the Gardner equation (a combination of the KdV and mKdV equations). Equation (4.4.46b) is related to the Liouville equation (4.4.46a) which is known to be linearizable.

Let us consider the initial value problem for each of the above equations with u real. Given $u(x,0)$, initial values for $w(x,t)$ are obtained from $w(x,0) = u(x,0) + \mathrm{i}\mathrm{H}u(x,0)$, and the solutions $u(x,t)$ is recovered from $u(x,t) = \mathrm{Re}\, w(x,t)$.

Multidimensional analogs can also be readily constructed. For example, a 2 + 1-dimensional analog of the KP equation is

$$(u_t + u_{xxx} + 2(u\mathrm{H}u)_x)_x + 3\sigma^2 u_{yy} = 0, \qquad \sigma^2 = \pm 1, \tag{4.4.48}$$

and is linearized via the KP equation

$$(w_t + w_{xxx} - \mathrm{i}(w^2)_x)_x + 3\sigma^2 w_{yy} = 0, \qquad \sigma^2 = \pm 1. \tag{4.4.49}$$

A 3 + 1-dimensional equation can also be linearized via the KP equation (4.4.49). Namely let $\mathrm{H}_z u$ denote the Hilbert transform of $u(x,y,z,t)$ with respect to the variable z, i.e.,

$$\mathrm{H}_z u = \frac{1}{\pi} \int_{-\infty}^{\infty} \frac{u(x,y,\xi,t)}{\xi - z}\, \mathrm{d}\xi. \tag{4.4.50}$$

Then consider a 3 + 1-dimensional analog of (4.4.48)

$$(u_t + u_{xxx} + 2(u\mathrm{H}_z u)_x)_x + 3\sigma^2 u_{yy} = 0, \qquad \sigma^2 = \pm 1. \tag{4.4.51}$$

It is also mapped to the KP equation (4.4.48) via $w = u + \mathrm{i}\mathrm{H}_z u$.

Let us now consider complex u. For example

$$\mathrm{i}u_t = u_{xx} + 2(u\mathrm{H}u)_x. \tag{4.4.52}$$

In association with (4.4.52) define $w_\pm := u \pm \mathrm{i}\mathrm{H}u$, then

$$\mathrm{i}(w_\pm)_t = (w_\pm)_{xx} \mp \mathrm{i}(w_\pm^2)_x, \tag{4.4.53}$$

which is linearized via $w_\pm = \mp\mathrm{i}(\ln f_\pm)_x$. The initial values are obtained as before but the complex solution $u(x,t)$ is now recovered from $u(x,t) = \frac{1}{2}(w_+ + w_-)$. The above approach can also be used for dealing with complex initial values of the equations considered earlier.

As discussed above we deal with $\mathrm{H}u$ by extending the function u to its upper and lower functions. Similarly we can deal with $\mathrm{T}u$ by extending u to a function sectionally holomorphic in horizontal strips of thickness δ (cf. Satsuma, Ablowitz and Kodama [1979]; Kodama, Ablowitz and Satsuma [1982]). Operators associated with certain other geometries can also be considered (cf. Ablowitz, Fokas, Satsuma and Segur [1982]).

Chapter Five

Inverse Scattering in Two Dimensions

5.1 Introduction.

In this Chapter we discuss I.S.T. schemes for 2+1-dimensional equations, primarily considering the Kadomtsev-Petviashvili (KP) equation (Kadomtsev and Petviashvili [1970])

$$(u_t + 6uu_x + u_{xxx})_x + 3\sigma^2 u_{yy} = 0, \tag{5.1.1}$$

where $\sigma^2 = \pm 1$, as the prototype example. The KP equation (5.1.1), which may be thought of as a two spatial dimensional analog of the KdV equation (2.1.1), and plays the role in 2 + 1 dimensions (two space, one time) that the KdV equation plays in 1 + 1 dimensions. It is one of the classical prototype problems in the field of exactly solvable equations and arises generically in physical contexts with both choices of the sign of σ^2, e.g. plasma physics (Kadomtsev and Petviashvili [1970]) and surface water waves (Ablowitz and Segur [1979]). (Recall that in §1.2 we showed that for surface water waves, the sign of σ^2 depends upon the relevant magnitude of gravity and surface tension.) The choice of sign is critical with respect to the stability characteristics of one dimensional (i.e., y independent) line solitons subject to slow transverse perturbations (in the y-direction). For $\sigma^2 = -1$, the solitons are unstable, whereas for $\sigma^2 = +1$, they are stable. As we show below, the sign of σ^2 is also critical in the development of the I.S.T. scheme for solving the initial value problem for the KP equation. In particular, for $\sigma^2 = -1$, there exist lump type solitons, which decay as $r = \sqrt{x^2+y^2} \to \infty$; whilst for $\sigma^2 = +1$, it is known that for appropriately decaying and sufficiently small initial data, lump type solitons do not exist (it seems unlikely that lump type solitons will exist even for large initial data since they have neither been observed nor is there any mathematical evidence suggesting their existence).

The linear scattering problem associated with the KP equation was derived by Dryuma [1974] and is given by

$$\sigma v_y + v_{xx} + (u+\lambda)v = 0, \tag{5.1.2}$$

$$v_t + 4v_{xxx} + 6uv_x + 3u_x v - 3\sigma(\partial_x^{-1} u_y)v + \alpha v = 0, \tag{5.1.3}$$

where α is an arbitrary constant and $(\partial_x^{-1} f)(x) := \frac{1}{2}\left(\int_{-\infty}^{x} - \int_{x}^{\infty}\right) f(x')\,\mathrm{d}x'$. (Note that: (i), in the case when $\sigma = \mathrm{i}$, equation (5.1.2) becomes the "time"-dependent

Schrödinger equation; (ii) the definition of ∂_x^{-1} is not the usual one (i.e., $\int_{-\infty}^{x}$) — we discuss this further in §5.3.5 below). Equation (5.1.2) may be thought of as a two-dimensional generalization of the linear scattering problem associated with the KdV equation, equation (2.1.2a). The KP equation also possesses N-"line"-soliton solutions (Satsuma [1976]; Satsuma and Ablowitz [1979]; Nakamura [1981a,b,c,d, 1982]); periodic solutions (Bobenko and Borgag [1989]) an infinite number of symmetries and conservation laws (Oevel and Fuchssteiner [1982]; Chen, Lee and Liu [1983]; Fokas [1987]); an infinite dimensional symmetry group, in fact this involves an infinite dimensional Lie algebra with a Kac-Moody-Virasoro loop structure (Schwarz [1982]; David, Kamran, Levi and Winternitz [1986]); similarity reductions to the first, second and fourth Painlevé equations (Tajiri, Nishitani and Kawamoto [1982]; David, Kamran, Levi and Winternitz [1986]; Clarkson and Winternitz [1991]); possesses the Painlevé property (Weiss, Tabor and Carnevale [1983]); a Bäcklund transformation (Chen [1975]; David, Levi and Winternitz [1987b]); a recursion operator and bi-Hamiltonian structure (Fokas and Santini [1986, 1988a,b]; Boiti, Leon, Martina and Pempinelli [1987]; Santini and Fokas [1988]).

The I.S.T. scheme for (5.1.1) with $\sigma^2 = -1$, which we call KPI, was developed by Manakov [1981]; Segur [1982]; Fokas and Ablowitz [1983c,d]. However there are various significant differences in the solution of KPI by an I.S.T. scheme in comparison with the standard one-dimensional equations we discussed in Chapters 2 and 3: (i), the inverse problem is formulated in terms of a nonlocal (as opposed to local) Riemann-Hilbert boundary value problem; (ii), algebraically decaying solutions, the so-called lump solutions, arise in a rather novel way. (Recall that the BO equation, which we discussed in §4.3, possesses all the above features.) In Manakov's treatment of KPI (Manakov [1981]), the initial value problem was solved for a restricted class of initial data in terms of a Gel'fand-Levitan-Marchenko (GLM) equation. However this GLM equation is defined in terms of a certain function $f(k,l)$. The main difficulty in the solution of the inverse problem for KPI is to relate $f(k,l)$ to the scattering data; this relationship is established by use of another GLM-type equation. The lump solutions were excluded in this formulation. Segur [1982] obtains the scattering data in terms of Volterra integral equations, and proved the existence of solutions to the direct problem; otherwise the development parallels that of Manakov [1981]. Fokas and Ablowitz [1983c,d] obtain explicit expressions for the scattering data in terms of a newly introduced eigenfunction which is neither analytic in the upper nor the lower half spectral parameter plane (the development is analogous to that for the Benjamin-Ono equation). Their analysis incorporated the lump solutions naturally. In §5.2 we discuss the I.S.T. scheme for KPI, following Fokas and Ablowitz [1983c,d]. Boiti, Leon and Pempinelli [1989a] show how to compute $f(k,l)$ directly in terms of analytic eigenfunctions; otherwise the point of view is similar. We mention their results in §5.2.

However for (5.1.1) with $\sigma^2 = +1$, which we call KPII, the previously used ideas in scattering theory are no longer sufficient. The inverse problem cannot be formulated as a Riemann-Hilbert boundary value problem since there exist eigenfunctions for the associated spectral problem which, although bounded, are nowhere analytic in the spectral parameter. In this case a generalization of the Riemann-Hilbert problem is required (the $\bar{\partial}$ "DBAR" problem) and was developed for KPII by Ablowitz, BarYaacov and Fokas [1983]. The $\bar{\partial}$ problem concept was introduced by Beals and Coifman [1981, 1982] in connection with the I.S.T. scheme for certain first order one-dimensional systems of differential equations (see also Beals and Coifman [1984, 1985a,b]). However, in these latter problems the $\bar{\partial}$ method is not crucial since the inverse problem for ordinary differential equations can always be reduced to a Riemann-Hilbert problem. On the other hand, Ablowitz, Bar Yaacov and Fokas [1983] showed that the $\bar{\partial}$ approach is essential in the development of the I.S.T. scheme for KPII. This is significant since it is the first case where the Riemann-Hilbert problem formulation of the I.S.T. scheme is inadequate. We discuss the I.S.T. scheme for KPII in §5.3, following Ablowitz, BarYaacov and Fokas [1983]. Rigorous results for the direct and inverse problems for small norms were obtained by Wickerhauser [1987].

The second important scattering problem in one spatial dimension is the $N \times N$ scattering problem, which we discussed in §3.2. There exist two-dimensional generalizations of this scattering problem analogous to the two-dimensional generalizations of the Schrödinger scattering problem discussed above. The two-dimensional generalization of the $N \times N$ scattering problem proposed by Ablowitz and Haberman [1975b] is

$$\boldsymbol{v}_x = \mathbf{Q}\boldsymbol{v} + \sigma \mathbf{J}\boldsymbol{v}_y, \tag{5.1.4}$$

where $\mathbf{J}$ is a constant real diagonal matrix (with diagonal entries $J^1, J^2, \dots, J^N$ such that $J^i \neq J^j \neq 0$, for $i \neq j$, $i,j = 1,2,\dots,N$); $\mathbf{Q}(x,y) = (Q^{ij}) \in \mathbb{C}^{N\times N}$ is an off-diagonal matrix containing the potentials and $\sigma = \sigma_R + \mathrm{i}\sigma_I$ is a complex parameter. As for the two-dimensional generalization of the Schrödinger equation, i.e. equation (5.1.2), there exist two important cases of equation (5.1.4), the hyperbolic ($\sigma = -1$) and elliptic ($\sigma = \mathrm{i}$) cases. As for the KP equation, the I.S.T. schemes for these two cases are quite different; in the hyperbolic case, an I.S.T. scheme was formulated in terms of a Riemann-Hilbert boundary value problem and in the elliptic case, an I.S.T. scheme was formulated in terms of a $\bar{\partial}$ problem (Fokas [1983]; Fokas and Ablowitz [1983e,f, 1984]). In particular, the hyperbolic case of the scattering problem (5.1.4) has been used to develop an I.S.T. scheme for the Davey-Stewartson I, n-wave interactions and Modified KPI equations and the elliptic case for the Davey-Stewartson II and Modified KPII equations. We briefly discuss I.S.T. schemes for both the scattering problem (5.1.4) in §5.4 (for both hyperbolic and elliptic versions). The issue of boundary conditions is quite important for the Davey-Stewartson equations, especially DSI. We mention this in §5.5.5.

We introduced the concept of the $\bar{\partial}$ method in §3.2 (in connection with $N \times N$

scattering problems) and the main steps in this method can be summarized as follows:

(i) Define an eigenfunction $m(x,y;k)$ which is bounded for all complex values of the spectral parameter k and is suitably normalized (i.e., $m \to 1$ as $|k| \to \infty$). This eigenfunction is usually defined in terms of a Fredholm linear integral equation of the second kind, and it may have different representations in different sections of the complex k-plane. To derive this integral equation for $m(x,y,k)$, one regards terms involving the potential $u(x,y,t)$ as "forcing terms" and uses a Fourier transform or Green's function formulation as appropriate. This integral equation may have homogeneous solutions which are rather important since they give rise to the lump solutions. When a Riemann-Hilbert formulation is used (e.g., for equations such as the BO and KPI equations), $m(x,y;k)$ is a *sectionally meromorphic* function of k, i.e., it is meromorphic in regions of the complex k-plane separated by contours Σ, and it has a jump(s) across these contour(s) [e.g., the real k-axis $\operatorname{Im} k = 0$ separates m^+ and m^-, where $m^\pm := \lim_{\varepsilon\to 0^+} m(k \pm \mathrm{i}\varepsilon)$].

(ii) Compute $\partial m/\partial\bar{k}$ $[= \frac{1}{2}(\partial m/\partial\bar{k}_R + \mathrm{i}\partial m/\partial\bar{k}_I)]$ by taking the $\partial/\partial\bar{k}$ derivative of the integral equation. This is in general expressed in terms of some other bounded eigenfunction, which we call $N(x,y;k)$, and appropriate scattering data. In the Riemann-Hilbert formulation, $\partial m/\partial\bar{k}$ effectively means these jump(s), i.e., it measures the "departure of $m(x,y;k)$ from holomorphicity", and will be zero everywhere except on pole locations and on the above contour(s).

(iii) If necessary, establish an appropriate "symmetry" relationship between the eigenfunctions $N(x,y;k)$ and $m(x,y;k)$ to express $\partial m/\partial\bar{k}$ in terms of m and appropriate scattering data. By using the properties of integral equations (and Green's functions), this is shown to be a linear relationship of the form

$$\frac{\partial m}{\partial\bar{k}} = \tilde{\mathrm{T}}_{x,y} m,$$

where $\tilde{\mathrm{T}}_{x,y}$ is a linear operator which explicitly contains the scattering data. If m has homogeneous solutions, then one needs additionally to establish a relationship between m and these homogeneous modes. This symmetry condition may be either "discrete" (e.g. KPII) or "differential" (e.g. KPI). The relationship between $\partial m/\partial\bar{k}$, m and the scattering data, which we call the scattering equation, is the central equation associated with the inverse problem for a given equation. This scattering equation defines, in general, the $\bar{\partial}$ problem, i.e., given $\partial m/\partial\bar{k}$, determine m. In the case that m is sectionally meromorphic this $\bar{\partial}$ problem degenerates into a Riemann-Hilbert problem.

(iv) Use the following generalized Cauchy formula

$$m(x,y;k) = \frac{1}{2\pi\mathrm{i}} \iint_R \frac{\partial m}{\partial\bar{z}}(x,y;z) \frac{\mathrm{d}z \wedge \mathrm{d}\bar{z}}{z-k} + \frac{1}{2\pi\mathrm{i}} \int_C \frac{m(x,y;z)}{z-k}\,\mathrm{d}z, \qquad (5.1.5)$$

where R and C are an appropriate domain and contour respectively in the complex z-plane (typically $R = R_\infty$ the entire complex plane, and $C = C_\infty$ an arbitrarily large circle in the plane), $\partial m/\partial \bar{k}$ exists at least in the sense of distributions, $z = z_R + \mathrm{i} z_I$, and $\mathrm{d}z \wedge \mathrm{d}\bar{z} = -2\mathrm{i}\, \mathrm{d}z_R\, \mathrm{d}z_I$. Its solution is found, in general, in terms of a linear integral equation for $m(x, y; k)$. Equation (5.1.5) is uniquely defined in terms of the above mentioned scattering data. If m has a homogeneous solution at k_j, then $\partial m/\partial \bar{k}$ has a delta function at $k = k_j$; thus pure lump solutions are always found in closed form as the solution of a linear system of algebraic equations.

(v) Having solved (5.1.5) for $m(x, y; k)$, then calculate the potential $u(x, y, t)$ directly from the solution of the inverse problem. Although we can recover the potential from the governing differential equation, we typically derive it from large k asymptotics and is given in terms of integrals over $m(x, y; k)$ and the scattering data.

(vi) In order to solve the initial value problem of some related nonlinear evolution equations, it is sufficient to determine the time evolution of the scattering data. This is straightforward and can be achieved by using the time-dependent part of the Lax pair for the given equation. Furthermore, the initial scattering data can always be expressed in terms of the initial data $u(x, y, 0)$. Thus equation (5.1.5) [and hence the formula for the potential $u(x, y, t)$] is uniquely defined in terms of the scattering data.

We note again (1), the close analogy to Fourier analysis. Not only are the steps similar, but moreover the duality between the scattering data and potentials is analogous to the direct and inverse Fourier transforms. (2) a new feature of solving multi-dimensional nonlinear evolution equations by I.S.T. in $2 + 1$ dimensions is the role of the boundary conditions. We discuss this further in §§5.3.5 and 5.5.5.

5.2 The Kadomtsev-Petviashvili I Equation.

5.2.1 The Direct Scattering Problem. The KPI equation

$$(u_t + 6uu_x + u_{xxx})_x - 3u_{yy} = 0, \tag{5.2.1}$$

can be written as the compatibility condition of the following Lax pair

$$\mathrm{i}v_y + v_{xx} + (u + \lambda)v = 0, \tag{5.2.2}$$

$$v_t + 4v_{xxx} + 6uv_x + 3u_x v - 3\mathrm{i}(\partial_x^{-1} u_y)v + \alpha(k)v = 0, \tag{5.2.3}$$

where $(\partial_x^{-1} f)(x) := \frac{1}{2}\left(\int_{-\infty}^{x} - \int_{x}^{\infty}\right) f(x')\, \mathrm{d}x'$. In equation (5.2.2) we may assume without loss of generality that $\lambda = 0$ (by the scaling property of v). It should be noted that (5.2.3) can also be written in the form

$$v_t + (-4\mathrm{i}\partial_y + 2u)v_x + \left(\alpha(k) - 3\mathrm{i}(\partial_x^{-1} u_y) - u_x\right) v = 0, \tag{5.2.4}$$

where $\partial_y := \partial/\partial y$. Furthermore, if in (5.2.2,4) we set $v(x,y) = \tilde{v}(x)\exp(\mathrm{i}\lambda y)$, then $\tilde{v}(x)$ satisfies

$$\tilde{v}_{xx} + u\tilde{v} = \lambda\tilde{v},$$
$$\tilde{v}_t + (4\lambda + 2u)\tilde{v}_x + (\alpha - u_x)\tilde{v} = 0,$$

which is just the Lax pair for the KdV equation [recall equation (2.1.2)]. We seek a solution in the form

$$v(x,y;k) = m(x,y;k)\exp\{\mathrm{i}(kx - k^2y)\},$$

hence the eigenfunction $m(x,y;k)$ satisfies

$$\mathrm{i}m_y + m_{xx} + 2\mathrm{i}km_x = -um, \tag{5.2.5}$$

$$m_t + 4m_{xxx} + 12\mathrm{i}km_{xx} - 12k^2m_x + 6um_x + 6\mathrm{i}kum + 3u_xm - 3\mathrm{i}(\partial_x^{-1}u_y)m + (\alpha(k) - 4\mathrm{i}k^3)m = 0. \tag{5.2.6}$$

We seek a eigenfunction $m(x,y;k)$ of (5.2.5) which is bounded for all x, y, for all complex values of $k = k_R + \mathrm{i}k_I$ and such that $m \to 1$ as $|k| \to \infty$. (This last condition is a convenient normalization.) Two such solutions of (5.2.5) may be written in the form

$$m_\pm = 1 + \tilde{G}_\pm(um_\pm), \tag{5.2.7±}$$

where

$$\tilde{G}_\pm f \equiv G_\pm * f \equiv \int_{-\infty}^{\infty}\int_{-\infty}^{\infty} G_\pm(x-x', y-y'; k)\, f(x',y')\,\mathrm{d}x'\,\mathrm{d}y', \tag{5.2.8±}$$

and the Green's function satisfies

$$\mathrm{i}G_y + G_{xx} + 2\mathrm{i}kG_x = -\delta(x-x')\,\delta(y-y'). \tag{5.2.9}$$

We seek a bounded Green's function, so taking the Fourier transform of (5.2.9) in both x and y we obtain

$$G(x,y;k) = \frac{\mathrm{i}}{2\pi}\int_{-\infty}^{\infty} \mathrm{e}^{\mathrm{i}x\xi} g(y;k,\xi)\,\mathrm{d}\xi, \tag{5.2.10a}$$

with

$$g(y;k,\xi) = \frac{1}{2\pi\mathrm{i}}\int_{-\infty}^{\infty} \frac{\mathrm{e}^{\mathrm{i}y\eta}}{\eta + \xi(\xi + 2k)}\,\mathrm{d}\eta. \tag{5.2.10b}$$

It is clear that if $k = k_R + \mathrm{i}k_I$, then g will have a jump across $k_I = 0$. Indeed the integral (5.2.10b) is not defined for $k_I = 0$ until we decide how to avoid the singularity $\eta = -\xi(\xi + 2k_R)$. Let g_+, g_- denote g for $k_I = 0\pm$ respectively (i.e., we take the contour below/above $k_I = 0+$, $k_I = 0-$ respectively). These functions can be analytically continued for $k_I > 0$ and $k_I < 0$ respectively. Then

$$g_\pm(y;k,\xi) = \mathrm{e}^{-\mathrm{i}\xi(\xi+2k)y}\left\{\theta(y)\,\theta(\mp\xi) - \theta(-y)\,\theta(\pm\xi)\right\}, \tag{5.2.11}$$

where $\theta(x)$ is the Heaviside step function. Hence

$$G_{\pm}(x,y,k)=\frac{\mathrm{i}}{2\pi}\int_{-\infty}^{\infty}\mathrm{e}^{\mathrm{i}\xi x-\mathrm{i}\xi(\xi+2k)y}\left\{\theta(y)\,\theta(\mp\xi)-\theta(-y)\,\theta(\pm\xi)\right\}\mathrm{d}\xi. \qquad (5.2.12)$$

For potential $u(x,y)$ decaying sufficiently fast at ∞, Fredholm theory implies that equation (5.2.7) has solutions which are meromorphic in k. Hence $m_{\pm}(x,y;k)$ is a ***sectionally meromorphic function*** of k with some jump across $k_I=0$.

Next consider the eigenfunction $N(x,y;k,l)$ which solves equation (5.2.5) and which also satisfies the integral equation

$$\begin{aligned}N(x,y;k,l)&=\exp\{\beta(x,y;k,l)\}\\&+\int_{-\infty}^{\infty}\int_{-\infty}^{\infty}G_{-}(x-x',y-y';k)\,u(x',y')\,N(x',y';k,l)\,\mathrm{d}x'\,\mathrm{d}y',\end{aligned} \qquad (5.2.13)$$

where $\beta(x,y;k,l):=\mathrm{i}(l-k)x-\mathrm{i}(l^2-k^2)y$. (Note that $N(x,y;k,k)=m_{-}(x,y;k)$.) The eigenfunction $N(x,y;k,k)$ can be naturally motivated by relating $m_{+}(x,y;k)$ and $m_{-}(x,y;k)$. Define

$$\Delta(x,y;k)=m_{+}(x,y;k)-m_{-}(x,y;k), \qquad (5.2.14)$$

then from equation (5.2.7)

$$\begin{aligned}\Delta(x,y;k)&=\int_{-\infty}^{\infty}\int_{-\infty}^{\infty}[G_{+}-G_{-}](x-x',y-y';k)\,u(x',y')\,m_{+}(x',y';k)\,\mathrm{d}x'\,\mathrm{d}y'\\&+\int_{-\infty}^{\infty}\int_{-\infty}^{\infty}G_{-}(x-x',y-y';k)\,u(x',y')\,\Delta(x',y';k)\,\mathrm{d}x'\,\mathrm{d}y'.\end{aligned}$$

From equations (5.2.12)

$$[G_{+}-G_{-}](x,y;k)=\frac{1}{2\pi\mathrm{i}}\int_{-\infty}^{\infty}\mathrm{sgn}(\xi)\exp\{\mathrm{i}\xi x-\mathrm{i}\xi(\xi+2k)y\}\,\mathrm{d}\xi,$$

hence

$$\begin{aligned}\Delta(x,y;k)&=\int_{-\infty}^{\infty}T(k,l)\,\mathrm{e}^{\beta(x,y;k,l)}\,\mathrm{d}l\\&+\int_{-\infty}^{\infty}\int_{-\infty}^{\infty}G_{-}(x-x',y-y';k)\,u(x',y')\,\Delta(x',y';k)\,\mathrm{d}x'\,\mathrm{d}y',\end{aligned} \qquad (5.2.15)$$

where

$$T(k,l)=\frac{\mathrm{i}}{2\pi}\,\mathrm{sgn}(k-l)\int_{-\infty}^{\infty}\int_{-\infty}^{\infty}u(x',y')\,m_{+}(x',y';k)\,\mathrm{e}^{\beta(x',y';k,l)}\,\mathrm{d}x'\,\mathrm{d}y'. \qquad (5.2.16)$$

Then, by multiplying equation (5.2.13) by $T(k,l)$ and integrating with respect to l, and comparing with equation (5.2.15) we obtain

$$m_+(x,y;k) - m_-(x,y;k) = \int_{-\infty}^{\infty} T(k,l)\, N(x,y;k,l)\, \mathrm{d}l. \tag{5.2.17}$$

In order to view equation (5.2.17) as a Riemann-Hilbert boundary value problem, we need to relate $N(x,y;k,l)$ and $m_-(x,y;k)$. This crucial relationship may be derived as follows: multiply equation (5.2.13) by $\exp\{\mathrm{i}(kx - k^2y)\}$ then

$$[\mathrm{L}_{k,u}\hat{N}(\,\bullet\,,\,\bullet\,;k,l)](x,y) = \exp\{\mathrm{i}(lx - l^2y)\}, \tag{5.2.18}$$

where the operator $\mathrm{L}_{k,u}$ is defined by

$$[\mathrm{L}_{k,u}f](x,y) = f(x,y) - \int_{-\infty}^{\infty}\int_{-\infty}^{\infty} \hat{G}_-(x-x',y-y';k)\, u(x',y')\, f(x',y')\, \mathrm{d}x'\, \mathrm{d}y', \tag{5.2.19}$$

with

$$\hat{N}(x,y;k,l) := N(x,y;k,l)\exp\{\mathrm{i}(kx - k^2y)\}, \tag{5.2.20a}$$

$$\hat{G}_-(x,y;k) := G_-(x,y;k)\exp\{\mathrm{i}(kx - k^2y)\}. \tag{5.2.20b}$$

Differentiating equation (5.2.18) with respect to k implies that

$$[\mathrm{L}_{k,u}\frac{\partial \hat{N}}{\partial k}(\,\bullet\,,\,\bullet\,;k,l)](x,y) = -F(k,l)\exp\{\mathrm{i}(kx - k^2y)\}, \tag{5.2.21}$$

where

$$F(k,l) = \frac{\mathrm{i}}{2\pi}\int_{-\infty}^{\infty}\int_{-\infty}^{\infty} u(x',y')\, N(x',y';k,l)\, \mathrm{d}x'\, \mathrm{d}y'. \tag{5.2.22}$$

Also, equation (5.2.7) implies that

$$[\mathrm{L}_{k,u}\mu_-(\,\bullet\,,\,\bullet\,;k)](x,y) = \exp\{\mathrm{i}(kx - k^2y)\}, \tag{5.2.23}$$

where

$$\mu_-(x,y;k) := m_-(x,y;k)\exp\{\mathrm{i}(kx - k^2y)\}. \tag{5.2.24}$$

Therefore by comparing equations (5.2.21) and (5.2.24), it follows that

$$\frac{\partial}{\partial k}\left(N(x,y;k,l)\exp\{\mathrm{i}(kx - k^2y)\}\right) = -F(k,l)m_-(x,y;k)\exp\{\mathrm{i}(kx - k^2y)\}. \tag{5.2.25}$$

Finally, by integrating (5.2.25) with respect to k and using the fact that $N(x,y;k,k) = m_-(x,y;k)$, we obtain the required relationship

$$N(x,y;k,l) = m_-(x,y;l)\mathrm{e}^{\beta(x,y;k,l)} - \int_l^k F(p,l)\, m_-(x,y;p)\mathrm{e}^{\beta(x,y;k,p)}\, \mathrm{d}p. \tag{5.2.26}$$

Using equations (5.2.16,17,26), we shall now derive the scattering equation

$$m_+(x,y;k) - m_-(x,y;k) = \int_{-\infty}^{\infty} f(k,l)\exp\{\beta(x,y;k,l)\}\, m_-(x,y;l)\,\mathrm{d}l, \tag{5.2.27a}$$

where

$$f(k,l) = \frac{\mathrm{i}\,\mathrm{sgn}(k-l)}{2\pi}\int_{-\infty}^{\infty}\int_{-\infty}^{\infty} u(x',y')\, N(x',y',k,l)\,\mathrm{d}x\,\mathrm{d}y. \tag{5.2.27b}$$

Substituting $N(x,y;k,l)$ from equation (5.2.26) into equation (5.2.17) yields

$$\begin{aligned} m_+(x,y;k) - m_-(x,y;k) &= \int_{-\infty}^{\infty} \hat{T}(k,l) m_-(x,y;l)\,\mathrm{d}l \\ &- \int_{-\infty}^{k}\left(\int_{-\infty}^{l} \hat{A}(k,p,l)\,\mathrm{d}p\right) m_-(x,y;l)\,\mathrm{d}l \\ &+ \int_{k}^{\infty}\left(\int_{l}^{\infty} \hat{A}(k,p,l)\,\mathrm{d}p\right) m_-(x,y;l)\,\mathrm{d}l, \end{aligned} \tag{5.2.28a}$$

where

$$\hat{T}(k,l) := T(k,l)\exp\{\beta(x,y;k,l)\}, \tag{5.2.28b}$$

$$\hat{A}(k,p,l) := T(k,p)\, F(p,l)\exp\{\beta(x,y;k,l)\}. \tag{5.2.28c}$$

Hence we obtain the scattering equation (5.2.27a) with

$$\begin{aligned} f(k,l) = T(k,l) - \theta(k-l)&\int_{-\infty}^{l} T(k,p)F(p,l)\,\mathrm{d}p \\ + \theta(l-k)&\int_{l}^{\infty} T(k,p)F(p,l)\,\mathrm{d}p. \end{aligned}$$

It turns out that there exists a more convenient characterization of the scattering data. Equation (5.2.28) can also be written in the form

$$\begin{aligned} m_+(x,y;k) - m_-(x,y;k) &= \int_{-\infty}^{\infty} \hat{T}(k,l) m_-(x,y;l)\,\mathrm{d}l \\ &- \int_{-\infty}^{k}\left[\left(\int_{-\infty}^{l} + \int_{k}^{\infty}\right) \hat{A}(k,p,l)\,\mathrm{d}p\right] m_-(x,y;l)\,\mathrm{d}l \\ &+ \int_{k}^{\infty}\left[\left(\int_{l}^{\infty} + \int_{-\infty}^{k}\right) \hat{A}(k,p,l)\,\mathrm{d}p\right] m_-(x,y;l)\,\mathrm{d}l, \end{aligned}$$

which implies (5.2.27a) with

$$\begin{aligned} f(k,l) &= T(k,l) - \mathrm{sgn}(k-l)\int_{-\infty}^{\infty} T(k,p)\, F(p,l)\,\mathrm{d}p \\ &+\theta(k-l)\int_{l}^{k} T(k,p)\, F(p,l)\,\mathrm{d}p - \theta(l-k)\int_{k}^{l} T(k,p)\, F(p,l)\,\mathrm{d}p. \end{aligned} \tag{5.2.29}$$

In order to simplify the expression for $f(k,l)$, suppose that

$$f(k,l) = \frac{\mathrm{i}\,\mathrm{sgn}(k-l)}{2\pi} \int_{-\infty}^{\infty}\int_{-\infty}^{\infty} u(x',y')\,\tilde{N}(x',y';l,k)\,\mathrm{d}x'\,\mathrm{d}y',$$

then from equation (5.2.29)

$$\begin{aligned}\tilde{N}(x,y;l,k) = m_+(x,y;k)\exp\{\beta(x,y;k,l)\} &- \int_{-\infty}^{\infty} T(k,p)\,N(x,y;p,l)\,\mathrm{d}p \\ &+ \theta(k-l)\int_l^k T(k,p)\,N(x,y;p,l)\,\mathrm{d}p \\ &+ \theta(l-k)\int_k^l T(k,p)\,N(x,y;p,l)\,\mathrm{d}p. \end{aligned} \tag{5.2.30}$$

From equation (5.2.29,30) it follows that

$$\begin{aligned}\tilde{N}(x,y;k,k) &= m_+(x,y;k) - \int_{-\infty}^{\infty} T(k,p)N(x,y;p,k)\,\mathrm{d}p \\ &= m_-(x,y;k) \\ &= N(x,y;k,k), \end{aligned} \tag{5.2.31a}$$

and also

$$\begin{aligned}\frac{\partial}{\partial l}\Big(\tilde{N}(x,y;l,k)\exp\{\mathrm{i}(lx-l^2y)\}\Big) &= m_-(x,y;l)\exp\{\mathrm{i}(lx-l^2y)\} \\ &\times \frac{1}{2\pi\mathrm{i}}\int_{-\infty}^{\infty}\int_{-\infty}^{\infty} u(x',y')\tilde{N}(x',y';l,k)\,\mathrm{d}x'\,\mathrm{d}y'. \end{aligned} \tag{5.2.31b}$$

Equations (5.2.25) and (5.2.31) imply that $\tilde{N}(x,y;l,k) = N(x,y;k,l)$ and hence in the scattering equation (5.2.28),

$$\begin{aligned} f(k,l) &= \mathrm{sgn}(k-l)F(k,l) \\ &= \frac{\mathrm{i}\,\mathrm{sgn}(k-l)}{2\pi}\int_{-\infty}^{\infty}\int_{-\infty}^{\infty} u(x',y')\,N(x',y';k,l)\,\mathrm{d}x'\,\mathrm{d}y'. \end{aligned} \tag{5.2.32}$$

This relates the scattering data $f(k,l)$ to the initial data $u(x,y)$ by the introduction of an additional (nonanalytic) eigenfunction $N(x,y;k,l)$. It is useful to note that the spectral transform can be affected without recourse to the extra eigenfunction N. Recently Boiti, Leon and Pempinelli [1989a] show how to compute $f(k,l)$ in terms of integrals over $m_\pm$. They defined a new spectral transform $r(k,l)$ of $u(x,y)$ by

$$r_\pm(k,l) = \frac{1}{2\pi\mathrm{i}}\int_{-\infty}^{\infty}\int_{-\infty}^{\infty} \exp\{\mathrm{i}[(k-l)\xi - (k^2-l^2)\eta]\}m_\pm(\xi,\eta;k)u(\xi,\eta)\,\mathrm{d}\xi\,\mathrm{d}\eta.$$

$r_\pm(k,l)$ satisfy the following characterization equations

$$r_\pm(k,l) + \bar{r}_\pm(l,k) = \pm\left(\int_k^\infty - \int_{-\infty}^l\right) r_\pm(k,p)\bar{r}_\pm(p,l)\,\mathrm{d}p,$$

$$r_+(k,l) + \bar{r}_-(l,k) = -\int_l^k r_+(k,p)\bar{r}_-(p,l)\,\mathrm{d}p,$$

where $\bar{r}_\pm$ is the complex conjugate of $r_\pm$, and $f(k,l)$ can be explicitly expresses in terms of $r_+(k,l)$

$$f(k,l) = \theta(k-l)r_+(k,l) + \theta(l-k)r_+(l,k) + \int_{-\infty}^k (l-p)r_+(k,p)\bar{r}_+(l,p)\,\mathrm{d}p.$$

This is all that is necessary if there are no homogeneous solutions. However, if there exist homogeneous solutions of equation (5.2.8) then we need to establish a relationship between the bounded and discrete eigenfunctions. These homogeneous solutions are important since they give rise to the slowly decaying lump solutions.

Equations (5.2.8) are Fredholm equations of the second type which we assume to be regular (actually, by defining $M_\pm := m_\pm\sqrt{|u|}$ it follows that $M_\pm$ satisfy integral equations with weakly singular kernels and hence Fredholm theory applies). Let $\phi_{j+}(x,y)$ denote the homogeneous solutions corresponding to the eigenvalues k_{j+}, where $\mathrm{Im}(k_{j+}) > 0$ and $\mathrm{Im}(k_{j-}) < 0$. We assume that there exist a finite number of such eigenvalues and that they are all simple. (Actually, if u is real, then one can show that $k_{j-} = \overline{k_{j+}}$, where $\bar{k}$ denotes the complex conjugate of k.) Fredholm theory implies that $m_\pm$ have the following representations

$$m_\pm(x,y;k) = 1 + \sum_{j=1}^n \frac{C_{j\pm}\phi_{j\pm}(x,y)}{k-k_{j\pm}} + \mu_\pm(x,y,k), \qquad (5.2.33\pm)$$

where $\mu_\pm$ are $\pm$ functions with respect to k respectively ($C_{j\pm}$ are introduced only for normalization purposes). We have the following important relations

$$\lim_{k\to k_{j\pm}}\left(m_\pm(x,y;k) - \frac{C_{j\pm}\phi_{j\pm}(x,y)}{k-k_{j\pm}}\right) = (x - 2k_{j\pm}y + \gamma_{j\pm})\phi_{j\pm}(x,y), \qquad (5.2.34\pm)$$

where $\gamma_{j\pm}$ are constants and $C_{j\pm} = \mathrm{i}$ if $\phi_{j\pm}(x,y)$ are normalized by the requirement that $(x - 2k_{j\pm}y)\phi_{j\pm}(x,y)_{j\pm} \to 1$ as $(x^2+y^2)^{1/2} \to \infty$. Equation (5.2.34) can be proven as follows: define

$$\mu_{j-}(x,y;k) = \left(m_- - \frac{C_{j-}\phi_{j-}(x,y)}{k-k_{j-}}\right)\exp\{\mathrm{i}(kx-k^2y)\}, \qquad (5.2.35\mathrm{a})$$

$$\hat{\phi}_{j-}(x,y;k) = \phi_{j-}(x,y)\exp\{\mathrm{i}(kx-k^2y)\}. \qquad (5.2.35\mathrm{b})$$

Then from equation (5.2.7) [using the definition of $\mathrm{L}_{k,u}$ in (5.2.19)]

$$[\mathrm{L}_{k,u}\mu_{j-}(\bullet,\bullet;k)](x,y) = \exp\{\mathrm{i}(kx-k^2y)\} - \frac{C_{j-}}{k-k_{j-}}[\mathrm{L}_{k,u}\hat{\phi}_{j-}(\bullet,\bullet;k)](x,y). \tag{5.2.36}$$

Taking the limit of equation (5.2.36) as $k \to k_{j-}$ and using the fact that

$$[\mathrm{L}_{k_{j-},u}\hat{\phi}_{j-}(\bullet,\bullet;k_{j-})](x,y) = 0,$$

it follows that

$$[\mathrm{L}_{k_{j-},u}\mu_{j-}(\bullet,\bullet;k_{j-})](x,y) = -C_{j-}[\mathrm{L}_{k_{j-},u}\frac{\partial\phi_{j-}}{\partial k}(\bullet,\bullet;k_{j-})](x,y)$$
$$+\left(1-\frac{\mathrm{i}C_{j-}}{2\pi}\int_{-\infty}^{\infty}\int_{-\infty}^{\infty}u(x',y')\,\phi_{j-}(x',y')\,\mathrm{d}x'\,\mathrm{d}y'\right)\exp\{\mathrm{i}(k_{j-}x-k_{j-}^2y)\}.$$

Then Fredholm's alternative theorem implies that

$$\mu_{j-}(x,y;k_{j-}) = -C_{j-}\frac{\partial\phi_{j-}}{\partial k}(x,y;k_{j-}) + \gamma_{j-}\hat{\phi}_{j-}(x,y;k_{j-}),$$

provided that ($C_{j-}=\mathrm{i}$)

$$1+\frac{1}{2\pi}\int_{-\infty}^{\infty}\int_{-\infty}^{\infty}u(x',y')\,\phi_{j-}(x',y')\,\mathrm{d}x'\,\mathrm{d}y' = 0.$$

Equation (5.2.34+) is proven in a similar manner. Hence equations (5.2.34±) are valid provided that (see also equation (4.3.19) for the Benjamin-Ono equation)

$$1\mp\frac{1}{2\pi}\int_{-\infty}^{\infty}\int_{-\infty}^{\infty}u(x',y')\,\phi_{j\pm}(x',y')\,\mathrm{d}x'\,\mathrm{d}y' = 0. \tag{5.2.37}$$

(Note that the constants $\gamma_{j\pm}$ are determined by the asymptotics of equations (5.2.33±) in the neighborhood of $k=k_{j\pm}$.)

5.2.2 The Inverse Scattering Problem. Equation (5.2.28), with $f(k,l)$ given by equation (5.2.32) is the central equation associated with the solution of the inverse problem for KPI. Equation (5.2.28), together with equations (5.2.33±) and (5.2.34±), define a Riemann-Hilbert boundary value problem for the sectionally meromorphic function $m(x,y;k)$. By splitting equation (5.2.28) into its (+) and (−) parts, say (5.2.28±), and using equation (5.2.28−) together with equation (5.2.33±) (with $C_{j\pm}=\mathrm{i}$), we obtain the linear integral equation

$$1 = m_-(x,y;k) - \mathrm{i}\sum_{l=1}^{n}\left(\frac{\phi_{l+}(x,y)}{k-k_{l+}}+\frac{\phi_{l-}(x,y)}{k-k_{l-}}\right)$$
$$-\frac{1}{2\pi\mathrm{i}}\int_{-\infty}^{\infty}\int_{-\infty}^{\infty}f(p,l)\exp\{\beta(x,y;l,p)\}\,m_-(x,y;l)\frac{\mathrm{d}l\,\mathrm{d}p}{p-k+\mathrm{i}0}. \tag{5.2.38}$$

Now, by analytically continuing (in k) equation (5.2.38) [and the analogous equation for $m_+(x,y;k)$], and taking the limit as $k \to k_{j-}$, we obtain [using equation (5.2.34−)] another linear integral equation

$$1 = [x - 2k_{j\pm}y + \gamma_{j\pm}]\phi_{j\pm}(x,y) - \mathrm{i}\sum_{l=1}^{n}{}' \left(\frac{\phi_{l+}(x,y)}{k_{j\pm} - k_{l+}} + \frac{\phi_{l-}(x,y)}{k_{j\pm} - k_{l-}} \right) - \frac{1}{2\pi\mathrm{i}} \int_{-\infty}^{\infty}\int_{-\infty}^{\infty} f(p,l) \exp\{\beta(x,y;l,p)\}\, m_-(x,y;l) \frac{\mathrm{d}l\,\mathrm{d}p}{p - k_{j\pm}}, \qquad (5.2.39\pm)$$

where $\sum'$ means summation unless any of the denominators vanishes. Equations (5.2.38) and (5.2.39) define $m_-(x,y;k)$ and $\{\phi_{j\pm}(x,y)\}_{j=1}^{n}$ in terms of the scattering data

$$\{k_{j\pm}, \gamma_{j\pm}\}_{j=1}^{n}, \qquad f(k,l). \qquad (5.2.40)$$

From equation (5.2.38) it follows that asymptotically as $k \to \infty$

$$m_-(x,y;k) \sim 1 + \mathrm{i}\sum_{j=1}^{n} [\phi_{j+}(x,y) + \phi_{j-}(x,y)] - \frac{1}{2\pi\mathrm{i}} \int_{-\infty}^{\infty}\int_{-\infty}^{\infty} f(p,l) \mathrm{e}^{\beta(x,y;l,p)} m_-(x,y;l)\,\mathrm{d}l\,\mathrm{d}p, \qquad (5.2.41)$$

and therefore using equation (5.2.5) it is seen that the solutions $u(x,y,t)$ of KPI (for initial data which decay sufficiently rapidly) is obtained from

$$u(x,y,t) = \frac{\partial}{\partial x}\left(2\sum_{j=1}^{n} [\phi_{j+}(x,y,t) + \phi_{j-}(x,y,t)] + \frac{1}{\pi} \int_{-\infty}^{\infty}\int_{-\infty}^{\infty} f(k,l,t) \mathrm{e}^{\beta(x,y;k,l)} m_-(x,y,t;l)\,\mathrm{d}l\,\mathrm{d}k \right). \qquad (5.2.42)$$

5.2.3 The Time Dependence. In order to derive the time evolution of the scattering data (5.2.40) it is more convenient to consider the eigenfunction $v(x,y,t;k)$, which satisfies equations (5.2.2) and (5.2.3), rather then $m(x,y,t;k)$. The time evolution of $v(x,y,t;k)$ is given by

$$\mathrm{M}_k v := [\partial_t + 4\partial_x^3 + 6u\partial_x + 3u_x - 3\mathrm{i}(\partial_x^{-1}u_y) + \alpha(k)]v = 0, \qquad (5.2.43)$$

where $\alpha(k)$ is some function of k. Recall that $m \to 1$ as $k \to \infty$, hence $v \sim \exp\{\mathrm{i}(kx - k^2y)\}$ as $k \to \infty$. Therefore equation (5.2.43) implies that $\alpha(k) = 4\mathrm{i}k^3$. Also, equation (5.2.28) implies that

$$v_+(x,y,t;k) = v_-(x,y,t;k) + \int_{-\infty}^{\infty} f(k,l,t) v_-(x,y,t;l)\,\mathrm{d}l. \qquad (5.2.44)$$

Applying the operator M_k with $\alpha(k) = 4ik^3$ to equation (5.2.44) and using

$$M_k = M_l + 4i(k^3 - l^3),$$

it follows that

$$\frac{\partial f}{\partial t}(k,l,t) = 4i(l^3 - k^3)f(k,l,t), \tag{5.2.45}$$

and hence

$$f(k,l,t) = f(k,l,0)\exp\{4i(l^3 - k^3)t\}. \tag{5.2.46}$$

In a similar manner it may be shown that

$$\frac{\partial k_{j\pm}}{\partial t} = 0, \qquad \gamma_{j\pm}(t) = 12(k_{j\pm})^2 t + \gamma_{j\pm}(0). \tag{5.2.47}$$

5.2.4 Further Remarks.

1. Lump solutions. The pure lump solutions correspond to $f(k,l,0) = 0$ and are then characterized by

$$u(x,y,t) = 2\frac{\partial}{\partial x}\left\{\sum_{j=1}^{n}[\phi_{j+}(x,y,t) + \phi_{j-}(x,y,t)]\right\}, \tag{5.2.48}$$

where $\phi_{j\pm}(x,y,t)$ satisfy the following system of *linear algebraic* equations

$$(x - 2k_{j\pm}y + \gamma_{j\pm})\,\phi_{j\pm}(x,y,t) - i\sum_{l=1}^{n}\left(\frac{\phi_{l+}(x,y,t)}{k_{j\pm} - k_{l+}} + \frac{\phi_{l-}(x,y,t)}{k_{j\pm} - k_{l-}}\right) = 1. \tag{5.2.49}$$

In particular, if $n = 1$ then

$$\left[x - 2k_+y + 12k_+^2t + \gamma_+(0)\right]\phi_+(x,y,t) = 1 + i\frac{\phi_-(x,y,t)}{k_+ - k_-},$$

$$\left[x - 2k_-y + 12k_-^2t + \gamma_-(0)\right]\phi_-(x,y,t) = 1 - i\frac{\phi_+(x,y,t)}{k_+ - k_-}.$$

Hence if $\gamma_\pm(0) = \gamma_R + i\gamma_I(0)$, then

$$\phi_\pm(x,y,t) = \frac{1}{\Delta}\left\{\left[x - 2k_\mp y + 12k_\mp^2 t + \gamma_\mp(0)\right] \pm \frac{i}{k_+ - k_-}\right\},$$

where

$$\Delta = (\xi - 2k_R\eta)^2 + 4k_I^2\eta^2 + \frac{1}{4k_I^2},$$

with

$$\xi = x - 12(k_R^2 + k_I^2)t + \xi_0, \qquad \eta = y - 12k_Rt + \eta_0,$$

$$\xi_0 = \gamma_R + \gamma_I, \qquad \eta_0 = \frac{\gamma_I}{2k_I}, \qquad k_\pm = k_R \pm ik_I.$$

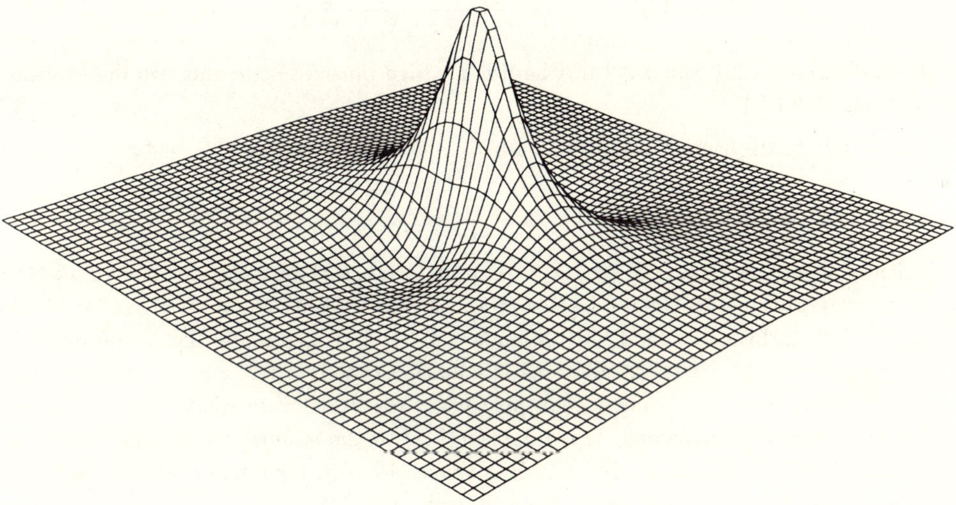

FIGURE 5.2.1 *Lump solution of KPI.*

Therefore we obtain the one-lump solution

$$\begin{aligned} u(x,y,t) &= 2\frac{\partial^2}{\partial x^2}\left\{\ln\left[(\xi - 2k_R\eta)^2 + 4k_I^2\eta^2 + \frac{1}{4k_I^2}\right]\right\} \\ &= 16\left\{\frac{-4(\xi - 2k_R\eta)^2 + 16k_I^2\eta^2 + 1/k_I^2}{[4(\xi - 2k_R\eta)^2 + 16k_I^2\eta^2 + 1/k_I^2]^2}\right\} \end{aligned}$$

(see Figure 5.2.1).

2. Line-soliton solutions. A one line-soliton solution of KPI is

$$u(x,y,t) = 2\kappa^2 \operatorname{sech}^2 \left\{\kappa(x + \lambda y - (4\kappa^2 - 3\lambda^2)t + \delta_0)\right\}.$$

A two line-soliton (written in Hirota's notation — recall §2.6.5) is

$$u(x,y,t) = 2\frac{\partial^2}{\partial x^2} \ln F(x,y,t),$$

with

$$F(x,y,t) = 1 + \exp(\eta_1) + \exp(\eta_2) + \exp(\eta_1 + \eta_2 + A_{12}),$$

where

$$\eta_i = 2\kappa_i[x + \lambda_i y - (4\kappa_i^2 + 3\sigma^2\lambda_i^2)t] + \delta_i,$$

$$\exp(A_{12}) = \frac{4(\kappa_1 - \kappa_2)^2 - \sigma^2(\lambda_1 - \lambda_2)^2}{4(\kappa_1 + \kappa_2)^2 - \sigma^2(\lambda_1 - \lambda_2)^2}.$$

(Recall Figures 1.2.1 and 1.2.2 in Chapter 1 which illustrate one and two line-soliton solutions of KPI.)

3. Gel'fand-Levitan-Marchenko equation. In the case when $\phi_{j\pm} = 0$, using

$$\mathrm{i}\int_\infty^x \exp\{\mathrm{i}\xi(k - p - \mathrm{i}\varepsilon)\}\,\mathrm{d}\xi = \frac{\exp\{\mathrm{i}x(k-p)\}}{k - p - \mathrm{i}\varepsilon},$$

multiplying equation (5.2.38) by $f(k,k')\exp\{\beta(x,y;k,k')\}$ and integrating over $\mathrm{d}k\,\mathrm{d}k'$, we directly obtain the Gel'fand-Levitan-Marchenko (GLM) equation given by Manakov [1981]. In his solution of KPI, Manakov [1981] first introduces the eigenfunctions m_+ and m_- which are those solutions of equation (5.2.5) which also satisfy the integral equation (5.2.7). Manakov then postulates the scattering equation (5.2.27) and through a series of ingenious steps relates $f(k,l)$ to the standard scattering data via certain GLM-type equations in scattering space. Finally, assuming that m_+ and m_- are (+) and (−) functions respectively, he solves (5.2.27) through a GLM equation in physical space. In this approach, lump solutions were excluded. Segur [1982] goes somewhat further by expressing $f(k,l)$ via Volterra integral equations and proves the existence of solutions to the direct problem.

4. Reduction to KdV equation. We remark that the above analysis for the KPI equation (formally) reduces to that for the KdV equation, assuming that there are no line solitons, if u is a function of x (i.e., $u_y \equiv 0$). Formally, the Riemann-Hilbert formulation for the KPI equation reduces easily to that for the KdV equation [take $f(k,l) = r(k)\delta(k-l)$ in equation (5.2.27)].

5. Rigorous Analysis. Zhou [1990] has considered rigorous aspects of the direct and inverse scattering problems associated with the time dependent Schödinger equation

$$\mathrm{i}v_y + v_{xx} + u(x,y)v = 0.$$

In particular he proves, assuming that $\int_{-\infty}^{\infty} u(x,y)\,\mathrm{d}x = 0$ (see §5.3.5 below), that if $\|\hat{u}\|_1 < \sqrt{2\pi}$, where $\hat{u}(p,y,0)$ is the Fourier transform of $u(x,y,0)$ with respect to x and $\|\bullet\|_1$ is the $L^1(\mathbb{R}^2)$ norm, together with certain conditions on the y derivatives of $\hat{u}(p,y,0)$, then KPI has a unique real solution $u(x,y,t)$ for all real t.

6. Direct Linearization of KPI. In this section we give a rather general linear integral equation by which one can obtain solutions of KPI (Fokas and Ablowitz [1983c]). Let $m(x,y,t;k)$ be a solution of the linear integral equation

$$m(x,y,t;k) + \iint_{\mathcal{L}} w(x,y,t;k,l,p)\, m(x,y,t;l)\,\mathrm{d}\zeta(l,p) = v(x,y,t;k), \tag{5.2.50}$$

where the contour $\mathcal{L}$ and measure $\mathrm{d}\zeta(l,p)$ are essentially arbitrary; where v is any solution of

$$\mathrm{i}v_y + v_{xx} + 2\mathrm{i}kv_x = 0, \tag{5.2.51a}$$

$$v_t + 4v_{xxx} + 12\mathrm{i}kv_{xx} - 12k^2 v_x = 0, \tag{5.2.51b}$$

and where

$$\begin{aligned} w(x,y,t;k,l,p) &= \exp\{\hat{\beta}(x,y,t;k,l,p)\} \\ &\times \left(\int_a^x v(\xi,y,t;k)\mathrm{e}^{\mathrm{i}(k-p)\xi}\,\mathrm{d}\xi + B(y,t;k,l,p)\mathrm{e}^{\mathrm{i}a(k-p)} \right), \end{aligned} \tag{5.2.52}$$

in which

$$\hat{\beta}(x,y,t;k,l,p) := \mathrm{i}(l-k)x - \mathrm{i}(l^2-p^2)y + 4\mathrm{i}(l^3-p^3)t,$$

and $B = B(y,t;k,l,p)$ is defined in terms of v through

$$B_y + \mathrm{i}(p^2-k^2)B - \mathrm{i}v_x(a) - (k+p)v(a), \tag{5.2.53a}$$

$$B_t + 4\mathrm{i}(k^3-p^3)B = -4\left[v_{xx}(a) + \mathrm{i}(p+2k)v_x(a) - (k^2+p^2+pk)v(a)\right], \tag{5.2.54b}$$

with $v(a) := v(a,y,t;k)$. Then m solves both equations (5.2.5) and (5.2.6), and u satisfies KPI where

$$u(x,y,t) = 2\frac{\partial}{\partial x}\left[\iint_{\mathcal{L}} \exp\{\beta(x,y,t;l,p,p)\}\, m(x,y,t;l)\,\mathrm{d}\zeta(l,p)\right]. \tag{5.2.55}$$

(i) It is easily shown that equations (5.2.38) and (5.2.39) are special cases of equation (5.2.50). Consider equation (5.2.50) with the following particular choice of measure $\mathrm{d}\zeta(l,p)$ and of the function $v(x,y,t;k)$:

$$\mathrm{d}\zeta(l,p) = \begin{cases} \tfrac{1}{2} f(p,l,0)\,\mathrm{d}l\,\mathrm{d}p, & l,p \in \mathbb{R}, \\ -\mathrm{i}[\delta(l-k_{l+})\,\delta(p-k_{l+}) + \delta(l-k_{l-})\,\delta(p-k_{l-})], & l,p \notin \mathbb{R}, \end{cases}$$

$$v(x,y,t;k) = 1, \qquad k \in \mathbb{R} \quad \text{and/or} \quad k = k_{j\pm}.$$

Then equation (5.2.50) reduces to equations (5.2.38), (5.2.39±), where

$$m(x,y,t;k) = \begin{cases} m_-(x,y,t;k), & k \in \mathbb{R}, \\ \phi_{j\pm}(x,y,t), & k = k_{j\pm}. \end{cases}$$

Similarly, equation (5.2.55) reduces to equation (5.2.42) (for details see Fokas and Ablowitz [1983d]).

(ii) If $\partial_t = 0, l = p$, the linearization given by equation (5.2.50) reduces to the linearization of the Benjamin-Ono equation

$$u_y + 2uu_x + \frac{1}{\pi}\frac{\partial^2}{\partial x^2}\left(\fint_{-\infty}^{\infty} u(\xi,y)\frac{d\xi}{\xi - x}\right) = 0,$$

given by Ablowitz, Fokas and Anderson [1983]. This is consistent with the fact that equation (5.2.5) is related to the y-part of the Lax pair for the Benjamin-Ono equation.

(iii) If $\partial_y = 0$, then equation (5.2.50) reduces to the direct linearization of the KdV equation due to Fokas and Ablowitz [1981], which we discussed in §2.5.3 above.

This method can be extended to give more general solutions and contain the $\sigma^2 = 1$ KPII equation as well as AKNS type systems (see Santini, Ablowitz and Fokas [1984b], and the references therein, for further details on the direct linearization of nonlinear evolution equations which is analogous to the Dressing method of Zakharov and Shabat [1974, 1979], which was based on Volterra-type linear integral equations, and Zakharov and Manakov [1985], which used a $\bar{\partial}$ problem on the complex plane — see also Zakharov [1990]; Fokas and Zakharov [1990].)

5.3 The Kadomtsev-Petviashvili II Equation.

5.3.1 The Direct Scattering Problem. The KPII equation

$$(u_t + 6uu_x + u_{xxx})_x + 3u_{yy} = 0, \tag{5.3.1}$$

can be written as the compatibility condition of the following Lax pair (we take, $\sigma = -1$)

$$-v_y + v_{xx} + uv = 0, \tag{5.3.2}$$

$$v_t + 4v_{xxx} + 6uv_x + 3u_x v + 3(\partial_x^{-1}u_y)v + \alpha(k)v = 0, \tag{5.3.3}$$

where $(\partial_x^{-1}f)(x) := \frac{1}{2}\left(\int_{-\infty}^{x} - \int_{x}^{\infty}\right) f(\xi)\, d\xi$. Consider the eigenfunction m defined by

$$v(x,y;k) = m(x,y;k)\exp\{ikx - k^2y\}, \tag{5.3.4}$$

so m satisfies

$$-m_y + m_{xx} + 2\mathrm{i}km_x = -um, \tag{5.3.5}$$

$$m_t + 4m_{xxx} + 12\mathrm{i}km_{xx} - 12k^2m_x + 6um_x + 6\mathrm{i}kum \\ + 3u_xm + 3(\partial_x^{-1}u_y)m + (\alpha(k) - 4\mathrm{i}k^3)m = 0. \tag{5.3.6}$$

As for KPI, we seek a eigenfunction $m(x, y; k)$ of (5.3.5) which is bounded for all x, y and such that $m \to 1$ as $|k| \to \infty$. Notice that the dominant operator in (5.3.5) is the heat operator which is ill-posed as an initial value problem. Even though we pose a boundary value problem, immediately it is suggested that in this case there will be some type of unusual behavior. Solutions to (5.3.5) which are bounded for all complex k may be written in the form

$$m = 1 + \tilde{G}(um), \tag{5.3.7}$$

where

$$\tilde{G}f \equiv G * f \equiv \int_{-\infty}^{\infty}\int_{-\infty}^{\infty} G(x - x', y - y'; k)\, f(x', y')\, \mathrm{d}x'\, \mathrm{d}y' \tag{5.3.8}$$

and the Green's function $G(x, y; k)$ satisfies

$$-G_y + G_{xx} + 2\mathrm{i}kG_x = -\delta(x)\,\delta(y). \tag{5.3.9}$$

We seek a bounded Green's function, so taking the Fourier transform of (5.3.9) in both x and y we obtain

$$G(x, y; k) = \frac{1}{4\pi^2}\int_{-\infty}^{\infty}\int_{-\infty}^{\infty} \frac{\exp\{\mathrm{i}(\xi x + \eta y)\}}{\xi^2 + 2\xi k + \mathrm{i}\eta}\, \mathrm{d}\xi\, \mathrm{d}\eta, \tag{5.3.10}$$

where k is the complex number $k = k_R + \mathrm{i}k_I$. Note that the kernel in (5.3.10) has singularities at the points $\xi = -2k_R$, $\eta = 4k_Rk_I$ and $\xi = \eta = 0$. These singularities are integrable in the ξ, η plane. This is different from the case for KPI, where the singularities are dense for $k_I = 0$ (along, say, the real η line for any given real ξ). Furthermore, we can evaluate $G(x, y; k)$

$$\begin{aligned} G(x, y; k) &= \frac{\operatorname{sgn}(-y)}{2\pi}\int_{-\infty}^{\infty} \theta(y(\xi^2 + 2k_R\xi)) \exp\{\mathrm{i}\xi x - \xi(\xi + 2k)y\}\, \mathrm{d}\xi \\ &= \left\{\theta(k_R)\left[-\theta(-y)\int_{-2k_R}^{0} \mathrm{d}\xi + \theta(y)\left(\int_0^{\infty} \mathrm{d}\xi + \int_{-\infty}^{2k_R} \mathrm{d}\xi\right)\right]\right. \\ &\quad \left. + \theta(-k_R)\left[-\theta(-y)\int_0^{-2k_R} \mathrm{d}\xi + \theta(y)\left(\int_{-\infty}^{0} \mathrm{d}\xi + \int_{-2k_R}^{\infty} \mathrm{d}\xi\right)\right]\right\} \\ &\quad \times \frac{1}{2\pi}\exp\{\mathrm{i}\xi x - \xi(\xi + 2k)y\}. \end{aligned} \tag{5.3.11}$$

Notice that the splitting for the scattering function $m(x,y;k)$ depends not only on whether $k_R < 0$ or $k_R > 0$ but also k_R enters explicitly in the integral equation. For example, for $k_R > 0$

$$\begin{aligned} m(x,y;k) = 1 + \frac{1}{2\pi}\Bigg\{ &- \int_y^\infty \mathrm{d}y' \int_{-\infty}^\infty \mathrm{d}x' \int_{-2k_R}^0 \mathrm{d}\xi \\ &+ \int_{-\infty}^y \mathrm{d}y' \int_{-\infty}^\infty \mathrm{d}x' \left(\int_{-\infty}^{-2k_R} \mathrm{d}\xi + \int_0^\infty \mathrm{d}\xi \right) \Bigg\} \\ &\times \exp\{\mathrm{i}\xi(x-x') - \xi(\xi+2k)(y-y')\}\, u(x',y')\, m(x',y';k). \end{aligned} \tag{5.3.12}$$

(Compare (5.3.12) with (5.2.7).) Since we have $k_R = \frac{1}{2}(k+\bar{k}), k_I = -\frac{1}{2}\mathrm{i}(k-\bar{k})$, where $\bar{k}$ is the complex conjugate of k, it is clear that both $G(x,y;k)$ and $m(x,y;k)$ are analytic nowhere in the complex k-plane. Despite this apparent difficulty, an inverse problem may be constructed.

We now compute $\partial m/\partial\bar{k}$ (this is computable even though $m(x,y;k)$ is nowhere analytic; recall that $\partial/\partial\bar{k} = \frac{1}{2}(\partial/\partial k_R + \mathrm{i}\partial/\partial k_I)$. Hence

$$\begin{aligned} \frac{\partial m}{\partial \bar{k}}(x,y;k) = &\int_{-\infty}^\infty \int_{-\infty}^\infty \frac{\partial G}{\partial \bar{k}}(x-x',y-y';k)\, u(x',y')\, m(x',y';k)\, \mathrm{d}x'\, \mathrm{d}y' \\ &+ \int_{-\infty}^\infty \int_{-\infty}^\infty G(x-x',y-y';k)\, u(x',y')\, \frac{\partial m}{\partial \bar{k}}(x',y';k)\, \mathrm{d}x'\, \mathrm{d}y'. \end{aligned} \tag{5.3.13}$$

Using the relation

$$\frac{\partial}{\partial \bar{k}}\left(\frac{1}{k-k_0}\right) = \pi\delta(k-k_0), \tag{5.3.14}$$

then from equation (5.3.10)

$$\frac{\partial G}{\partial \bar{k}}(x,y;k) = \frac{1}{2\pi}\operatorname{sgn}(-k_R)\exp\{\mathrm{i}(px+qy)\}, \tag{5.3.15a}$$

where

$$p = -2k_R = -(k+\bar{k}), \tag{5.3.15b}$$

$$q = 4k_Rk_I = -\mathrm{i}(k+\bar{k})(k-\bar{k}), \tag{5.3.15c}$$

$$k = -\tfrac{1}{2}\left(p + \mathrm{i}q/p\right). \tag{5.3.15d}$$

Equations (5.3.13-15) imply [assuming that there are no homogeneous solutions to equation (5.3.7)]

$$\frac{\partial m}{\partial \bar{k}}(x,y;k) = F(k_R,k_I)\, N(x,y;k_R,k_I), \tag{5.3.16}$$

where

$$F(k_R,k_I) = \frac{\operatorname{sgn}(-k_R)}{2\pi} \int_{-\infty}^\infty \int_{-\infty}^\infty u(x',y')\, m(x',y';k)\, \mathrm{e}^{-\mathrm{i}(px'+qy')}\, \mathrm{d}x'\, \mathrm{d}y', \tag{5.3.17}$$

and $N(x, y; k_R, k_I)$ satisfies

$$N(x, y; k_R, k_I) = \exp\{i(px + qy)\} + \int_{-\infty}^{\infty}\int_{-\infty}^{\infty} G(x - x', y - y'; k)\, u(x', y')\, N(x', y'; k_R, k_I)\, dx'\, dy'. \quad (5.3.18)$$

To derive equation (5.3.16), substitute (5.3.15) into (5.3.13), which using (5.3.17) yields

$$\frac{\partial m}{\partial \bar{k}}(x, y; k) = F(k_R, k_I) \exp\{i(px + qy)\} + \int_{-\infty}^{\infty}\int_{-\infty}^{\infty} G(x - x', y - y'; k)\, u(x', y')\, \frac{\partial m}{\partial \bar{k}}(x', y'; k)\, dx'\, dy'.$$

Hence if $N(x, y; k_R, k_I)$ satisfies (5.3.18), then (5.3.16) follows. Note that equations (5.3.13,15-18) could have been derived without the use of equation (5.3.14). One can simply take $\partial/\partial\bar{k}$ of either equation (5.3.11) or (5.3.12), where $k_R = \frac{1}{2}(k + \bar{k})$ appears explicitly.)

In order to view equation (5.3.16) as a $\bar{\partial}$ problem, we require a "symmetry" relationship between N and m. To obtain this, observe that

$$G(x, y; -\bar{k}) = G(x, y; k) \exp\{-i(px + qy)\}. \quad (5.3.19)$$

Hence equations (5.3.7) and (5.3.18) imply that this important relationship is given by

$$N(x, y; k_R, k_I) = m(x, y; -\bar{k}) \exp\{i(px + qy)\}, \quad (5.3.20)$$

which is a discrete "symmetry" relationship. It follows from equations (5.3.16) and (5.3.20) that

$$\frac{\partial m}{\partial \bar{k}}(x, y; k) = F(k_R, k_I)\, m(x, y; -\bar{k}) \exp\{-i(2k_R x - 4k_R k_I y)\}. \quad (5.3.21)$$

This defines a $\bar{\partial}$ problem in k for the bounded eigenfunction m, in terms of the scattering data $F(k_R, k_I)$ defined by equation (5.3.17).

5.3.2 The Inverse Scattering Problem. The inverse problem is developed using the generalized Cauchy integral formula

$$m(x, y; k) = \frac{1}{2\pi i} \iint_R \frac{\partial m}{\partial \bar{z}}(x, y; z)\, \frac{dz \wedge d\bar{z}}{z - k} + \frac{1}{2\pi i} \int_C \frac{m(x, y; z)}{z - k}\, dz, \quad (5.3.22)$$

where R and C are an appropriate domain and contour respectively in the complex z-plane, $\partial m/\partial\bar{k}$ exists at least in the sense of distributions, $z = z_R + iz_I$, and $dz \wedge d\bar{z} = -2i\, dz_R\, dz_I$. In this case we take R to be the entire complex plane R_∞ and C to be the contour C_∞ at infinity (i.e., an arbitrarily large circle). Since $m \sim 1$ as $|k| \to \infty$, then

$$\int_{C_\infty} \frac{m(x,y;z)}{z-k}\,\mathrm{d}z = 2\pi\mathrm{i}.$$

Hence from equations (5.3.21-22)

$$m(x,y;k) = 1 + \frac{1}{2\pi\mathrm{i}} \int_{-\infty}^{\infty}\int_{-\infty}^{\infty} F(z_R, z_I)\, m(x,y;-\bar{z}) \exp\{\mathrm{i}(\hat{p}x + \hat{q}y)\} \frac{\mathrm{d}z \wedge \mathrm{d}\bar{z}}{z-k}, \quad (5.3.23)$$

where $\hat{p} = -2z_R$, $\hat{q} = 4z_Rz_I$. Equation (5.3.23) characterizes the solution of the inverse problem. (Note that for appropriately decaying $u(x,y)$, (5.3.23) is again a linear Fredholm integral equation.) Once $m(x,y;k)$ has been found, the potential $u(x,y)$ is reconstructed from the formula

$$u(x,y) = \frac{\partial}{\partial x}\left(\frac{1}{\pi}\int_{-\infty}^{\infty}\int_{-\infty}^{\infty} F(z_R, z_I)\, m(x,y;-\bar{z}) \exp\{\mathrm{i}(\hat{p}x + \hat{q}y)\}\, \mathrm{d}z \wedge \mathrm{d}\bar{z}\right). \quad (5.3.24)$$

This is derived by considering the limit as $k \to \infty$ in equations (5.3.7) and (5.3.21) and then comparing terms of $O(k^{-1})$ as follows: from equations (5.3.7) and (5.3.21) we have

$$\begin{aligned}\int_{-\infty}^{\infty}\int_{-\infty}^{\infty} & G(x-x', y-y'; k)\, u(x',y')\, m(x',y';k)\, \mathrm{d}x'\, \mathrm{d}y' \\ &= \frac{1}{2\pi\mathrm{i}}\int_{-\infty}^{\infty}\int_{-\infty}^{\infty} F(z_R, z_I)\, m(x,y;-\bar{z}) \exp\{\mathrm{i}(\hat{p}x + \hat{q}y)\} \frac{\mathrm{d}z \wedge \mathrm{d}\bar{z}}{z-k}, \quad (5.3.25)\end{aligned}$$

where

$$G(x,y;k) = \frac{1}{4\pi^2}\int_{-\infty}^{\infty}\int_{-\infty}^{\infty} \frac{\exp\{\mathrm{i}(\xi x + \eta y)\}}{\xi^2 + 2\xi k + \mathrm{i}\eta}\, \mathrm{d}\xi\, \mathrm{d}\eta. \quad (5.3.26)$$

As $k \to \infty$, from equations (5.3.23) and (5.3.26), we see that

$$m(x,y;k) = 1 + O(k^{-1}), \quad (5.3.27)$$

$$\begin{aligned} G(x,y;k) &= \frac{1}{8\pi^2 k}\int_{-\infty}^{\infty}\left\{ ⨍_{-\infty}^{\infty} \exp\{\mathrm{i}(\xi x + \eta y)\}\xi^{-1}\, \mathrm{d}\xi \right\} \mathrm{d}\eta + O(k^{-2}), \\ &= \frac{\mathrm{i}}{4\pi}\, \mathrm{sgn}(x)\, \delta(y), \end{aligned} \quad (5.3.28)$$

respectively. Now define

$$\hat{G}(x,x',y,y') := \frac{1}{4\pi^2}\int_{-\infty}^{\infty}\int_{-\infty}^{\infty} \exp\{\mathrm{i}[\xi(x-x') + \eta(y-y')]\} \frac{\mathrm{d}\xi\, \mathrm{d}\eta}{\xi},$$

$$U(x,y) = \tfrac{1}{2}\left(\int_{-\infty}^{x} - \int_{x}^{\infty}\right) u(\xi,y)\, \mathrm{d}\xi,$$

and hence

$$\int_{-\infty}^{\infty}\int_{-\infty}^{\infty} G(x-x',y-y';k)\,u(x',y')\,m(x',y';k)\,\mathrm{d}x'\,\mathrm{d}y' \\ = \tfrac{1}{2}\mathrm{i}k^{-1}U(x,y)+O(k^{-2}). \qquad (5.3.29)$$

Also

$$\frac{1}{2\pi\mathrm{i}}\int_{-\infty}^{\infty}\int_{-\infty}^{\infty} F(z_R,z_I)\,m(x,y;-\bar{z})\exp\{\mathrm{i}(\hat{p}x+\hat{q}y)\}\,\frac{\mathrm{d}z\wedge\mathrm{d}\bar{z}}{z-k} \\ = \frac{\mathrm{i}}{2\pi k}\int_{-\infty}^{\infty}\int_{-\infty}^{\infty} F(z_R,z_I)\,m(x,y;-\bar{z})\exp\{\mathrm{i}(\hat{p}x+\hat{q}y)\}\,\mathrm{d}z\wedge\mathrm{d}\bar{z} \\ +O(k^{-2}). \qquad (5.3.30)$$

Therefore by comparing terms of $O(k^{-1})$ in equations (5.3.29,30) we obtain

$$U(x,y)=\frac{1}{\pi}\int_{-\infty}^{\infty}\int_{-\infty}^{\infty} F(z_R,z_I)\,m(x,y;-\bar{z})\exp\{\mathrm{i}(\hat{p}x+\hat{q}y)\}\,\mathrm{d}z\wedge\mathrm{d}\bar{z},$$

and so (5.3.24) follows.

5.3.3 The Time Dependence. The time dependence of the scattering data $F(k_R,k_I;t)$ may be determined using the time dependent part of the Lax pair, i.e., equation (5.3.3). It is more convenient to use the eigenfunction $v(x,y,t;k)$ as opposed to $m(x,y,t;k)$, where v and m are related by equation (5.3.4). From equation (5.3.3), the time evolution of v is given by

$$\mathrm{M}_k v := [\partial_t+4\partial_x^3+6u\partial_x+3u_x+3(\partial_x^{-1}u_y)+\alpha(k)]v=0, \qquad (5.3.31)$$

where $\alpha(k)$ is a function of k to be determined. Since

$$v(x,y,t;k)=m(x,y,t;k)\exp\{\mathrm{i}kx-k^2y\},$$

and $m\sim 1$ for large k, then

$$v\sim\exp\{\mathrm{i}kx-k^2y\}, \qquad \text{as}\quad |k|\to\infty,$$

and hence equation (5.3.31) implies that $\alpha(k)=4\mathrm{i}k^3$. Equations (5.3.4) and (5.3.21) imply that

$$\frac{\partial v}{\partial\bar{k}}(x,y,t;k)=v(x,y,t;-\bar{k})\,F(k_R,k_I,t). \qquad (5.3.32)$$

Note that if $v(x,y,t;k)$ satisfies equation (5.3.31) then $\partial v/\partial\bar{k}$ also satisfies equation (5.3.31) (since $\mathrm{d}f/\mathrm{d}\bar{k}=0$ if $f(k)$ is analytic). Therefore, applying the operator M_k to equation (5.3.32) and using the fact that

$$\mathrm{M}_k=\mathrm{M}_l+4\mathrm{i}(k^3-l^3),$$

with $l = -\bar{k}$, it follows that $F(k_R, k_I; t)$ satisfies

$$\frac{\partial F}{\partial t}(k_R, k_I; t) = -4\mathrm{i}\left(k^3 + \bar{k}^3\right) F(k_R, k_I; t).$$

Hence

$$F(k_R, k_I; t) = F(k_R, k_I; 0) \exp\left\{-4\mathrm{i}\left(k^3 + \bar{k}^3\right) t\right\}. \tag{5.3.33}$$

5.3.4 Comments on Rigorous Analysis. We first make the following remarks.

1. We note that if $u(x, y) \in L^1(\mathbb{R}^2) \cap L^\infty(\mathbb{R}^2)$, then the kernel

$$K(x, y, x', y'; k) = \mathrm{sgn}[u(x', y')] G(x - x', y - y'; k) \sqrt{|u(x', y')\, u(x, y)|},$$

is compact. This kernel is obtained by multiplying equation (5.3.7) by $\sqrt{|u(x, y|}$ and considering the integral equation for the function

$$\mu(x, y; k) = \sqrt{|u(x, y)|}\, m(x, y; k).$$

Compactness follows from the fact that the Green's function can be split as $G = G_1 + G_\infty$, where $G_1 \in L^1(\mathbb{R}^2)$ and $G_\infty \in L^\infty(\mathbb{R}^2)$. Then we are guaranteed the existence, for all $k = k_R + \mathrm{i}k_I$, of m and hence μ, for u sufficiently small.

2. We expect, in fact, that there will be no points in the complex k-plane where homogeneous solutions to equation (5.3.7) will occur. If they did, then this would suggest the existence of lump type solutions (a situation which has never been demonstrated for KPII).

We now discuss the results on the analytic treatment of KPII for "small" initial data due to Wickerhauser [1987], who investigated the inverse scattering problem associated with

$$\mathrm{L}v \equiv -v_y + v_{xx} + u(x, y)v = 0. \tag{5.3.34}$$

Wickerhauser proves two main results.

First, if $u(x, y) \in L^1(\mathbb{R}^2) \cap L^2(\mathbb{R}^2)$ and is small in the sense that $\|u\|_1 + \|u\|_2 < \varepsilon$, for some ε sufficiently small (where $\|\bullet\|_p$ is the $L^p(\mathbb{R}^2)$ norm), the operator is determined by the leading coefficients of asymptotically exponential functions in its kernel. This theorem has several parts.

(a). Let $m(x, y; k)$ be a bounded function such that

$$v(x, y; k) = m(x, y; k) \exp\left\{\mathrm{i}kx - k^2 y\right\},$$

is a solution of (5.3.34) (i.e., lies in the kernel of L), and $m(x, y; k)$ satisfies

$$-m_y + m_{xx} + 2\mathrm{i}k m_x + u(x, y)m = 0. \tag{5.3.35}$$

Then for $u(x,y)$ small in $L^1(\mathbb{R}^2) \cap L^2(\mathbb{R}^2)$, there exists a unique solution $m(x,y;k) \in L^\infty(\mathbb{R}^2 \times \mathbb{C})$ satisfying (5.3.35) together with the boundary condition

$$m(x,y;k) \to 1, \qquad \text{as} \qquad \sqrt{x^2+y^2} \to \infty; \tag{5.3.36}$$

hence

$$v(x,y;k) \sim \exp\left\{ikx - k^2 y\right\}, \qquad \text{as} \qquad \sqrt{x^2+y^2} \to \infty.$$

(b). Suppose that $m(x,y;k) \in L^\infty(\mathbb{R}^2 \times \mathbb{C})$, is the solution of the integral equation

$$m(x,y;k) = 1 + \int_{-\infty}^{\infty}\int_{-\infty}^{\infty} G(x-x', y-y'; k)\, u(x',y')\, m(x',y';k)\, \mathrm{d}x'\, \mathrm{d}y', \tag{5.3.37a}$$

where the Green's function is given by

$$G(x,y;k) = \frac{1}{(2\pi)^2}\int_{-\infty}^{\infty}\int_{-\infty}^{\infty} \frac{\exp\{\mathrm{i}(\xi x + \eta y)\}}{\xi^2 + 2\xi k + \mathrm{i}\eta}\, \mathrm{d}\xi\, \mathrm{d}\eta, \tag{5.3.37b}$$

and that

$$\left(1 + \sqrt{x^2+y^2}\right) |u(x,y)| \in L^1(\mathbb{R}^2) \cap L^2(\mathbb{R}^2).$$

Then for $k_R \neq 0$, $m(x,y;k)$ has the asymptotic expansion

$$\begin{aligned} m(x,y;k) = 1 &+ \frac{\mathrm{i}\,\mathrm{sgn}(k_R) A(k_R,k_I)}{x + 2\mathrm{i}ky} - \frac{\mathrm{i}F(k_R,k_I)}{x - 2\mathrm{i}\bar{k}y} \exp\left\{-\mathrm{i}(k+\bar{k})x + (k^2 - \bar{k}^2)y\right\} \\ &+ o\left((x^2+y^2)^{-1/2}\right), \qquad \text{as} \qquad \sqrt{x^2+y^2} \to \infty, \end{aligned} \tag{5.3.38}$$

where $F(k_R,k_I)$ and $A(k_R,k_I)$ are bounded functions given by (5.3.17) and

$$A(k_R,k_I) = \frac{1}{2\pi}\int_{-\infty}^{\infty}\int_{-\infty}^{\infty} u(x,y)\, m(x,y,k)\, \mathrm{d}x\, \mathrm{d}y, \tag{5.3.39}$$

respectively, and $\bar{k}$ is the complex conjugate of k.

(c). The coefficient $F(k_R,k_I)$ suffices to determine the $\bar{\partial}$ data of the operator; that is, if $\|u\|_1 + \|u\|_2$ is small so that $m(x,y;k)$ is the unique solution in $L^\infty(\mathbb{R}^2 \times \mathbb{C})$ of (5.3.35,36), then $\partial m/\partial \bar{k}$ exists, is bounded on $\mathbb{R}^2 \times \mathbb{C}$, and in fact may be written as

$$\begin{aligned} \frac{\partial m}{\partial \bar{k}}(x,y;k) &= -\mathrm{i}F(k_R,k_I) \exp\left\{-\mathrm{i}(k+\bar{k})x + (k^2-\bar{k}^2)y\right\} m(x,y;-\bar{k}) \\ &\equiv (Tm)(x,y;k) \end{aligned} \tag{5.3.40}$$

[i.e., this is really equation (5.3.21)]. Furthermore, if $\|u\|_1 + \|u\|_2$ is small, then there exists a bounded function $g(k_R,k_I)$ with $\lim_{k_R\to\infty} g(k_R,k_I) = 0$, such that

$$|F(k_R,k_I) \leq \frac{\|u\|_1}{1 - g(k_R,k_I)\left[\|u\|_1 + \|u\|_2\right]}. \tag{5.3.41}$$

(d). Suppose that $u(x,y)$ is small in $L^1(\mathbb{R}^2) \cap L^2(\mathbb{R}^2)$ and that u_x, u_y, u_{xx}, u_{xy} $u_{yy} \in L^1(\mathbb{R}^2) \cap L^2(\mathbb{R}^2)$ then the unique solution $m(x,y;k)$ to (5.3.35,36) is also a unique solution to the equation

$$m(x,y;k) = 1 + \frac{1}{2\pi \mathrm{i}} \int_{-\infty}^{\infty}\int_{-\infty}^{\infty} (Tm)(x,y;z)\, \frac{\mathrm{d}z \wedge \mathrm{d}\bar{z}}{z-k}, \tag{5.3.42}$$

where $(Tm)(x,y;k)$ is given by equation (5.3.40) and is the scattering data associated to u by the heat operator [this is equivalent of equation (5.3.23)]. Hence both $m(x,y;k)$ and $\partial m/\partial \bar{k}(= (Tm)(x,y;k)$ are determined by $F(k_R,k_I)$ whilst $u(x,y)$ is determined by

$$\begin{aligned} u(x,y) &= -\frac{1}{\pi \mathrm{i}} \frac{\partial}{\partial x}\left(\int_{-\infty}^{\infty}\int_{-\infty}^{\infty} (Tm)(x,y;z)\,\mathrm{d}z \wedge \mathrm{d}\bar{z}\right) \\ &= \frac{1}{\pi}\frac{\partial}{\partial x}\left(\int_{-\infty}^{\infty}\int_{-\infty}^{\infty} \exp\{-\mathrm{i}(z+\bar{z})x + (z^2-\bar{z}^2)y\} \right. \\ &\qquad \left. \times F(z_R,z_I)\, m(x,y;-\bar{z})\,\mathrm{d}z\wedge\mathrm{d}\bar{z}\right) \end{aligned} \tag{5.3.43}$$

[i.e., this is really equation (5.3.24)]. The maps $u(x,y) \to F(k_R,k_I)$ and $F(k_R,k_I) \to u(x,y)$ are the forward and inverse scattering transforms, respectively (behaving much like the Fourier transform and its inverse). Furthermore, if $\|u\|_1 + \|u\|_2$ is small and $u(x,y)$ has two continuous derivatives in $L^1(\mathbb{R}^2) \cap L^2(\mathbb{R}^2)$, then $F(k_R,k_I)$ is bounded and $|k_R|^{1/2}F(k_R,k_I) \in L^2(\mathbb{R}^2)$.

(e). If $u(x,y)$ has n derivatives in $L^1(\mathbb{R}^2) \cap L^2(\mathbb{R}^2)$ (i.e., $\partial^{p+q}u/\partial x^p \partial y^q \in L^1(\mathbb{R}^2) \cap L^2(\mathbb{R}^2)$ for all p, q, such that $p+q \le n$), then

$$F(k_R,k_I) = O\left(\left(1+\sqrt{|k|^2+|k|^4}\right)^{-n}\right), \qquad \text{as} \qquad |k| \to \infty. \tag{5.3.44a}$$

Conversely, if

$$F(k_R,k_I) = O\left(\left(1+\sqrt{|k|^2+|k|^4}\right)^{-n-2}\right), \qquad \text{as} \qquad |k| \to \infty. \tag{5.3.44b}$$

then $u(x,y)$ has n derivatives in $L^2(\mathbb{R}^2)$ with bounded Fourier transform.

Secondly Wickerhauser [1987] finds the time evolution of $F(k_R,k_I;t)$ and $A(k_R,k_I;t)$, which are needed for $u(x,y,t)$ to satisfy KPII

$$(u_t + 6uu_x + u_{xxx})_x + 3u_{yy} = 0.$$

Let $t \in \mathbb{R}^+$ be regarded as time then

$$\begin{aligned} &\frac{\partial F}{\partial t}(k_R,k_I;t) = -4\mathrm{i}(k^3+\bar{k}^3)\,F(k_R,k_I;t), \\ &\frac{\partial A}{\partial t}(k_R,k_I;t) = 0, \end{aligned}$$

i.e.,

$$F(k_R, k_I; t) = \exp\left\{-4\mathrm{i}(k^3 + \bar{k}^3)t\right\} F(k_R, k_I; 0). \tag{5.3.45a}$$

$$A(k_R, k_I; t) = A(k_R, k_I; 0) \tag{5.3.45b}$$

[equation (5.3.45a) is really equation (5.3.33)]. Since $k^3 + \bar{k}^3$ is purely real, then $|F(k_R, k_I; t)| = |F(k_R, k_I; 0)|$, for all $k \in \mathbb{C}$, $t \geq 0$. He then uses result (e) to show that if the initial value $u(x, y, 0)$ has small derivatives up to order 10 in $L^1(\mathbb{R}^2) \cap L^2(\mathbb{R}^2)$, then there is a uniformly bounded solution $u(x, y, t)$ to the initial value problem for KPII, for all t in the space of $L^2(\mathbb{R}^2)$ functions with bounded Fourier transform.

Wickerhauser discusses these results in relation to the Kadomtsev-Petviashvili hierarchy as given by Date, Jimbo, Kashiwara and Miwa [1981a,b,c,d,e 1982].

5.3.5 Boundary Conditions and the Choice of the Operator ∂_x^{-1}. In this section we discuss the role of boundary conditions in the solutions of the KP equation by I.S.T. and how they determine the appropriate choice of ∂_x^{-1} in the scattering problem (5.1.2,3). We note that if we assume that u along with all its derivatives (including u_t) tend to zero as $\sqrt{x^2 + y^2} \to \infty$, then it follows that

$$\chi_1 := \int_{-\infty}^{\infty} u_{yy}(x, y, t)\,\mathrm{d}x = \frac{\partial^2}{\partial y^2} \int_{-\infty}^{\infty} u(x, y, t)\,\mathrm{d}x = 0. \tag{5.3.46a}$$

Since this condition is not preserved in time, we are forced to require that u satisfies an infinite number of constraints χ_n, $n = 1, 2, \ldots$. By differentiating χ_1 with respect to t we find

$$\chi_2 = \frac{\partial^4}{\partial y^4} \int_{-\infty}^{\infty} \left\{ \left(\int_{-\infty}^{x} - \int_{x}^{\infty} \right) u(x', y, t)\,\mathrm{d}x' \right\} \mathrm{d}x = 0, \tag{5.3.46b}$$

and similarly for the higher order constraints. It is important to determine to what extent these constraints are necessary.

First we write the KP equation in evolution form, i.e.

$$u_t + 6uu_x + u_{xxx} + 3\sigma^2 \partial_x^{-1}(u_{yy}) = 0. \tag{5.3.47}$$

This is an infinite family of different equations according to the choice we make for ∂_x^{-1}. However, once the boundary conditions are selected (here implicitly through equations (5.2.7,10) for KPI and through equations (5.3.7,10) for KPII), then these define ∂_x^{-1} uniquely.

We can see how the I.S.T. method defines ∂_x^{-1} by appealing to the asymptotic expansion for $m(x, y; k)$ as $k \to \infty$

$$m \sim 1 + \frac{m_1(x, y)}{k} + \frac{m_2(x, y)}{k^2} + o(k^{-2}). \tag{5.3.48}$$

From the definitions of the Green's functions (5.2.10,3.10) we have

$$G(x,y;k) \sim \frac{\mathrm{i}\,\mathrm{sgn}(x)\delta(y)}{4k} + o(k^{-1}),$$

and so equations (5.2.5,3.5) yield

$$m_1(x,y) = \tfrac{1}{4}\mathrm{i}\left(\int_{-\infty}^{x} - \int_{x}^{\infty}\right) u(\xi,y)\,\mathrm{d}\xi.$$

Further substitution into the associated time evolution equations of the eigenfunctions, equations (5.2.6,3.6) yields

$$\partial_x^{-1} u_y = -2\mathrm{i}m_{1,y} = \tfrac{1}{2}\left(\int_{-\infty}^{x} - \int_{x}^{\infty}\right) u_y(\xi,y)\,\mathrm{d}\xi,$$

and consequently we have

$$\partial_x^{-1} \equiv \tfrac{1}{2}\left(\int_{-\infty}^{x} - \int_{x}^{\infty}\right), \tag{5.3.49}$$

for both KPI associated with the scattering problem (5.2.5,6) and KPII associated with the scattering problem (5.3.5,6).

Next we shall consider the I.S.T. solutions for KPII. This will show how the I.S.T. data leads to the underlying constraints if $\partial_x^{-1} = \int_{-\infty}^{x}$ and also how it relates to the constants of motion. We make the following remarks:

1. We are assuming the $u(x,y,0) \in L^1(\mathbb{R}^2) \cap L^2(\mathbb{R}^2)$, sufficiently small and smooth (which guarantees that the integral equation for $m(x,y,0;k)$ is bounded and $(1+|k_R|)F(k_R,k_I) \in L^1(\mathbb{R}^2) \cap L^2(\mathbb{R}^2)$.
2. The problem is ill-posed for general σ unless $\sigma^2 \in \mathbb{R}$ (i.e., for KPI and KPII). In the case of for general σ, the scattering data $F(k_R,k_I;t)$ satisfies

$$\frac{\partial F}{\partial t} = 8\mathrm{i}\kappa_0(6k\kappa_0 - 4\kappa_0^2 - 3k^2)F, \tag{5.3.50}$$

   where $\kappa_0 = k_R + (\sigma_I/\sigma_R)k_I$, $k = k_R + \mathrm{i}k_I$, $\sigma = \sigma_R + \mathrm{i}\sigma_I$ (with $\sigma_R \neq 0$). Hence there will be exponential growth/decay since

$$\mathrm{Re}[8\mathrm{i}\kappa_0(6k\kappa_0 - 4\kappa_0^2 - 3k^2)] = -\frac{48k_I^2\sigma_I}{\sigma_R^2}(k_R\sigma_R + k_I\sigma_I) \neq 0,$$

   even in the so-called $\bar{\partial}$ limit case where $\sigma \to \varepsilon \pm \mathrm{i}$ (and even if we rescale $\hat{k}_I = k_I/\sigma_R$, $\sigma_R = \varepsilon$, there is exponential growth for $k_R < 0$).
3. We shall assume that

$$(1 + 8|z_R|^3 + 16z_R^2 z_I^2)F(z_R,z_I)m(x,y;-\bar{z}) \in L^1(\mathbb{R}^2) \cap L^2(\mathbb{R}^2),$$

which implies that u_{xxx} and u_{yy} tend to zero as $\sqrt{x^2+y^2} \to \infty$ and that differentiation under the integral sign can be performed in equation (5.3.24).

4. The condition $F(k_R = 0, k_I) = 0$ guarantees that (5.3.24) solves KPII and that differentiation with respect to t can be performed under the integral sign. Consequently $u_t \to 0$ as $\sqrt{x^2+y^2} \to \infty$ which ensures that the infinite number of constraints are satisfied.
5. Some insight into the behavior of (5.3.24) is obtained by making the transformation $\hat{p} = -2z_R$, $\hat{q} = 4z_R z_I$, whence

$$u(x,y) = \frac{1}{2\pi\mathrm{i}}\frac{\partial}{\partial x}\left(\int_{-\infty}^{\infty}\int_{-\infty}^{\infty} m(x,y;\tfrac{1}{2}(\hat{p}-\mathrm{i}\hat{q}/\hat{p}))\,\hat{F}(\hat{p},\hat{q})\exp\{\mathrm{i}(\hat{p}x+\hat{q}y)\}\,\frac{\mathrm{d}\hat{p}\,\mathrm{d}\hat{q}}{\hat{p}}\right)$$
$$= \frac{1}{2\pi\mathrm{i}}\int_{-\infty}^{\infty}\int_{-\infty}^{\infty}(\mathrm{i}m+\Phi)\,\hat{F}(\hat{p},\hat{q})\exp\{\mathrm{i}(\hat{p}x+\hat{q}y)\}\,\mathrm{d}\hat{p}\,\mathrm{d}\hat{q}, \qquad (5.3.51)$$

where

$$\hat{F}(\hat{p},\hat{q}) = F(-\tfrac{1}{2}\hat{p}, -\tfrac{1}{2}\hat{q}/\hat{p}), \qquad \Phi = m_x/\hat{p}.$$

Furthermore

$$\hat{F}(\hat{p},\hat{q};t) = \hat{F}(\hat{p},\hat{q};0)\exp\{\mathrm{i}(\hat{p}^3 - 3\hat{q}^2/\hat{p})t\}, \qquad (5.3.52)$$

and so

$$u(x,y) = \int_{-\infty}^{\infty}\int_{-\infty}^{\infty}\exp\{\mathrm{i}[\hat{p}x+\hat{q}y+(\hat{p}^3-3\hat{q}^2/\hat{p})t]\}\hat{\Phi}(x,y,\hat{p},\hat{q},t)\,\mathrm{d}\hat{p}\,\mathrm{d}\hat{q}, \quad (5.3.53\text{a})$$

where

$$\hat{\Phi}(x,y,\hat{p},\hat{q},t) = \frac{1}{2\pi\mathrm{i}}\hat{F}(\hat{p},\hat{q})(\mathrm{i}m+\Phi)(x,y,t;\tfrac{1}{2}(\hat{p}-\mathrm{i}\hat{q}/\hat{p})). \qquad (5.3.53\text{b})$$

We note that $\hat{p} \to 0$ with $\hat{q}$ fixed corresponds to $k_R \to 0$, $k_I \to \infty$, whereupon $m \to 1$ and

$$m_x \sim \frac{m_{1,x}}{\mathrm{i}k_I} = \frac{2\mathrm{i}\hat{p}}{\hat{q}}\,m_{1,x}(x,y,t),$$

and consequently Φ is regular as $\hat{p} \to 0$. Moreover we can formally evaluate the integral of $u(x,y,t)$ using equation (5.3.53). In particular, using the fact that $m(x,y,t;k) \to 1$ as $x \to \pm\infty$ then

$$\lim_{L\to\infty}\int_{-L}^{L} u(x,y,t)\,\mathrm{d}x = \lim_{L\to\infty}\left\{\frac{1}{2\pi\mathrm{i}}\int_{-\infty}^{\infty}\int_{-\infty}^{\infty}\hat{F}(\hat{p},\hat{q};t)\mathrm{e}^{\mathrm{i}\hat{q}y}\right.$$
$$\left.\times\left[\mathrm{e}^{\mathrm{i}\hat{p}L}m(L,y;\tfrac{1}{2}(\hat{p}-\mathrm{i}\hat{q}/\hat{p})) - \mathrm{e}^{-\mathrm{i}\hat{p}L}m(-L,y;\tfrac{1}{2}(\hat{p}-\mathrm{i}\hat{q}/\hat{p}))\right]\mathrm{d}\hat{p}\,\mathrm{d}\hat{q}\right\}$$
$$= \lim_{L\to\infty}\left\{\frac{1}{2\pi\mathrm{i}}\int_{-\infty}^{\infty}\int_{-\infty}^{\infty}\hat{F}(\hat{p},\hat{q};t)\mathrm{e}^{\mathrm{i}\hat{q}y}\left(\frac{\mathrm{e}^{\mathrm{i}\hat{p}L}-\mathrm{e}^{-\mathrm{i}\hat{p}L}}{\hat{p}}\right)\mathrm{d}\hat{p}\,\mathrm{d}\hat{q}\right\}. \quad (5.3.54)$$

It is shown by Ablowitz and Villarroel [1990b] using the Riemann-Lebesgue Lemma and contour integration that this integral vanishes (nonuniformly in time) as $L \to \infty$. Hence we have

$$\lim_{L\to\infty} \int_{-L}^{L} u(x,y,t)\,\mathrm{d}x = 0.$$

Consequently, even in

$$\int_{-\infty}^{\infty} u(x,y,0)\,\mathrm{d}x \neq 0,$$

initially, eventually the solution "seeks" this natural condition.

6. Next we discuss the conserved quantities associated with KPII and show how the "standard" conserved quantities need to be modified. As mentioned earlier, when $r = \sqrt{x^2+y^2} \to \infty$,

$$m(x,y,t) = 1 + \frac{\mathrm{i}\,\mathrm{sgn}(k_R)A(k_R,k_I)}{x+2\mathrm{i}ky} - \frac{\mathrm{i}F(k_R,k_I)\mathrm{e}^{\mathrm{i}(px+qy)}}{x-2\mathrm{i}\bar{k}y} + o(r^{-1}), \qquad (5.3.55)$$

with $p = -2k_R$, $q = 4k_Rk_I$ and where $A(k_R,k_I)$ and $F(k_R,k_I)$ are given by equations (5.3.39) and (5.3.17) respectively [recall equation (5.3.38)]. From equations (5.3.21) and (5.3.39) we have

$$\frac{\partial A}{\partial \bar{k}} = -\,\mathrm{sgn}(k_R)F(k_R,k_I)F(-k_R,k_I),$$

and consequently

$$A(k_R,k_I) = A_\infty - \frac{1}{2\pi\mathrm{i}}\int_{-\infty}^{\infty}\int_{-\infty}^{\infty} \frac{|F(z_R,z_I,t)|^2\,\mathrm{sgn}(z_R)}{z-k}\,\mathrm{d}z\wedge\mathrm{d}\bar{z}, \qquad (5.3.56)$$

with $A_\infty = A(k_R,k_I)|_{|k|=\infty}$ and where we have used $\bar{F}(z_R,z_I,t) = F(-z_R,z_I,t)$. From equation (5.3.31), (5.3.36) is time independent and hence an asymptotic expansion

$$A(k_R,k_I) - A_\infty = \sum_{n=0}^{\infty} \frac{I_n}{k^n},$$

will yield an infinite number constants of the motion I_n. If $F(0,z_I,0) = 0$, then finding the asymptotic expansion in powers of k is straightforward. Namely expanding equation (5.3.56) as well its symmetrized version ($z_R \to -z_R$) yields the constants of the motion I_n in the form

$$\begin{aligned} I_n &= -\frac{1}{2\pi}\int_{-\infty}^{\infty}\int_{-\infty}^{\infty} \mathrm{sgn}(z_R)|F(z_R,z_I,t)|^2[z^{n-1} - (-\bar{z})^{n-1}]\,\mathrm{d}z_R\,\mathrm{d}z_I \\ &= \frac{1}{8\pi}\int_{-\infty}^{\infty}\int_{-\infty}^{\infty} \frac{|\hat{F}(\hat{p},\hat{q};t)|^2}{(-2)^{n-1}\hat{p}} \\ &\qquad \times \left\{ \left(\hat{p} + \frac{\mathrm{i}\hat{q}}{\hat{p}}\right)^{n-1} - \left(-\hat{p} + \frac{\mathrm{i}\hat{q}}{\hat{p}}\right)^{n-1} \right\} \mathrm{d}\hat{p}\,\mathrm{d}\hat{q}. \end{aligned} \qquad (5.3.57)$$

However if $F(0, z_I, 0) \neq 0$, then equation (5.3.57) does not make sense in general. Nevertheless, these integrals can be regularized when necessary. For example, the first two constraints for $n = 2, 3$ are well defined (and are proportional to the "horizontal momenta" $P_x = -4\pi I_2$, $P_y = -4\pi \mathrm{i} I_3$). These are given by

$$P_x = \frac{1}{2}\int_{-\infty}^{\infty}\int_{-\infty}^{\infty} |\hat{F}(\hat{p}, \hat{q}; t)|^2 \, \mathrm{d}\hat{p}\, \mathrm{d}\hat{q}, \tag{5.3.58a}$$

$$P_y = \frac{1}{2}\int_{-\infty}^{\infty}\left\{\int_{-\infty}^{\infty} \frac{\hat{q}}{\hat{p}} |\hat{F}(\hat{p}, \hat{q}; t)|^2 \, \mathrm{d}\hat{q}\right\} \mathrm{d}\hat{p}. \tag{5.3.58b}$$

The next quantity I_4, which is proportional to the Hamiltonian ($H = -16\pi I_4$), is divergent as it stands. It is regularized as follows:

$$H = \frac{1}{2} \,{\textstyle\int\!\!\!\!\!-}_{-\infty}^{\infty}\left\{\int_{-\infty}^{\infty}\left[\hat{p}^2 |\hat{F}(\hat{p}, \hat{q}; t)|^2 - \frac{3\hat{q}^2}{\hat{p}^2}\left(|\hat{F}(\hat{p}, \hat{q}; t)|^2 - |\hat{F}(0, \hat{q}, t)|^2\right)\right] \mathrm{d}\hat{q}\right\} \mathrm{d}\hat{p}. \tag{5.3.58c}$$

H is conserved and reduces to the standard Hamiltonian (cf. Lipovskii [1986]; also see §5.3.6 below), if $F(0, \hat{q}, 0) = 0$. In order to check that it is the Hamiltonian of the problem it suffices to check that the equation of motions follow from

$$u_t = \frac{\partial}{\partial x}\frac{\delta H}{\delta u},$$

which it does — note that

$$\frac{\partial}{\partial x}\frac{\delta F(0, q, t)}{\delta u(x, y, t)} = 0.$$

5.3.6 Hamiltonian Formalism and Action-Angle Variables. Using the $\bar{\partial}$ results and assuming the constraint condition

$$\frac{\partial^2}{\partial y^2}\int_{-\infty}^{\infty} u(x, y, t)\, \mathrm{d}x = 0$$

(i.e., $F(0, k_I, 0) = 0$) and in this case

$$\partial_x^{-1} \equiv \int_{-\infty}^{x} \equiv \tfrac{1}{2}\left(\int_{-\infty}^{x} - \int_{x}^{\infty}\right),$$

Lipovskii [1986] developed the Hamiltonian formalism for KPII and found the action-angle variables (see also Wickerhauser [1988]). The Hamiltonian and Poisson brackets are given by

$$H = \int_{-\infty}^{\infty}\int_{-\infty}^{\infty}\left\{\tfrac{1}{2}u_x^2 - u^3 - \tfrac{3}{2}\left(\partial_x^{-1} u_y\right)^2\right\} \mathrm{d}x\, \mathrm{d}y, \tag{5.3.59a}$$

$$\{F, G\} = \tfrac{1}{2}\int_{-\infty}^{\infty}\int_{-\infty}^{\infty}\left[\frac{\delta F}{\delta u}\frac{\partial}{\partial x}\frac{\delta G}{\delta u} - \frac{\delta G}{\delta u}\frac{\partial}{\partial x}\frac{\delta F}{\delta u}\right] \mathrm{d}x\, \mathrm{d}y. \tag{5.3.59b}$$

(If the constraints (5.3.46) do not hold then, as alluded to in the previous section, it is necessary to correct them by adding an extra term to H. However, we shall consider the simpler case when the constraints are satisfied.)

Suppose that $\phi(x,y;k)$ is the solution of the eigenfunction equation

$$-\phi_y + \phi_{xx} + u(x,y)\phi = 0 \tag{5.3.60}$$

[recall equation (5.3.2)], with boundary condition

$$\phi(x,y;k) \sim \exp\{\mathrm{i}kx - k^2 y\}, \qquad \text{as} \quad |k| \to \infty;$$

$\phi(x;y;k)$ is also a solution of the integral equation

$$\begin{aligned}\phi(x;y;k) = \exp\{\mathrm{i}kx - k^2y\}\Bigg\{1 + \int_{-\infty}^{\infty} G(x-x', y-y'; k)\\ \times u(x',y')\,\phi(x';y';k)\exp\{-\mathrm{i}kx' + k^2y'\}\,\mathrm{d}x'\,\mathrm{d}y'\Bigg\},\end{aligned} \tag{5.3.61}$$

where the Green's function $G(x,y;k)$ is given by equations (5.3.10,11). Suppose also that $\tilde{\phi}(x;y;k)$ is the solution of

$$\tilde{\phi}_y + \tilde{\phi}_{xx} + u(x,y)\tilde{\phi} = 0 \tag{5.3.62}$$

[which is formally adjoint to equation (5.3.60)], with boundary condition

$$\tilde{\phi}(x;y;k) \sim \exp\{-\mathrm{i}kx + k^2y\}, \qquad \text{as} \quad |k| \to \infty.$$

Furthermore, the scattering data $F(k_R,k_I)$ is given by

$$F(k_R,k_I) = \frac{\operatorname{sgn}(-k_R)}{2\pi}\int_{-\infty}^{\infty}\int_{-\infty}^{\infty} u(x,y)\phi(x,y;k)\exp\{\mathrm{i}\bar{k}x + \bar{k}^2y\}\,\mathrm{d}x\,\mathrm{d}y \tag{5.3.63}$$

[recall equation (5.3.17)].

It follows then from equations (5.3.60-64) that

$$\frac{\delta F(k_R,k_I)}{\delta u(x,y)} = \phi(x,y;k)\tilde{\phi}(x,y;k)\exp\{\mathrm{i}(k+\bar{k})x - (k^2-\bar{k}^2)y\}, \tag{5.3.64a}$$

$$\frac{\delta \bar{F}(k_R,k_I)}{\delta u(x,y)} = \phi(x,y;-\bar{k})\tilde{\phi}(x,y;-\bar{k})\exp\{-\mathrm{i}(k+\bar{k})x + (k^2-\bar{k}^2)y\}. \tag{5.3.64b}$$

Hence we have

$$\{F(k_R,k_I), F(\ell_R,\ell_I)\} = 0, \tag{5.3.65a}$$

$$\{F(k_R,k_I), \bar{F}(\ell_R,\ell_I)\} = \tfrac{1}{4}\mathrm{i}\delta(k_R-\ell_R)\delta(k_I-\ell_I). \tag{5.3.65b}$$

Now define the action-angle coordinates as follows

$$P(k_R, k_I) = -4F(k_R, k_I)\,\bar{F}(k_R, k_I), \tag{5.3.66a}$$

$$Q(k_R, k_I) = \frac{1}{2\mathrm{i}} \ln \left[\frac{F(k_R, k_I)}{\bar{F}(k_R, k_I)} \right]. \tag{5.3.66b}$$

Using equations (5.3.65), it is easily shown that

$$\{P(k_R, k_I), P(\ell_R, \ell_I)\} = 0, \qquad \{Q(k_R, k_I), Q(\ell_R, \ell_I)\} = 0, \tag{5.3.67a}$$

$$\{Q(k_R, k_I), P(\ell_R, \ell_I)\} = \delta(k_R - \ell_R)\delta(k_I - \ell_I), \tag{5.3.67b}$$

and so $P(k_R, k_I)$ and $Q(k_R, k_I)$ are canonical variables.

The Hamiltonian can be written as

$$\begin{aligned} H &= 16 \int_{-\infty}^{\infty} \left\{ \int_{-\infty}^{\infty} \operatorname{sgn}(z_R)\, F(z_R, z_I)\, \bar{F}(z_R, z_I)\, (z^3 + \bar{z}^3)\, \mathrm{d}z_I \right\} \mathrm{d}z_R \\ &= -4 \int_{-\infty}^{\infty} \left\{ \int_{-\infty}^{\infty} \operatorname{sgn}(z_R)\, P(z_R, z_I)\, (z^3 + \bar{z}^3)\, \mathrm{d}z_I \right\} \mathrm{d}z_R, \\ &= \tfrac{1}{8} \int_{-\infty}^{\infty} \left\{ \int_{-\infty}^{\infty} \hat{P}(\hat{p}, \hat{q})\, (\hat{p}^2 - 3\hat{q}^2/\hat{p}^2)\, \mathrm{d}\hat{q} \right\} \mathrm{d}\hat{p}, \end{aligned}$$

where $\hat{p} = -2z_R$, $\hat{q} = 4z_R z_I$ and $\hat{P}(\hat{p}, \hat{q}) = P(z_R, z_I)$, which is in agreement with (5.3.58c).

5.4 Hyperbolic and Elliptic Systems in the Plane in 2+1 Dimensions.

In this section we briefly discuss the two-dimensional generalization of the $N \times N$ scattering problem proposed by Ablowitz and Haberman [1975b], namely

$$\boldsymbol{v}_x = \lambda \mathbf{B} \boldsymbol{v} + \mathbf{Q} \boldsymbol{v} + \sigma \mathbf{J} \boldsymbol{v}_y, \tag{5.4.1}$$

where $\mathbf{B}$ is a constant diagonal matrix and $\mathbf{J}$ is a real constant diagonal matrix with diagonal entries $J^1, J^2, \ldots, J^N$ such that $J^i \neq J^j$, for $i \neq j$, $i,j = 1, 2, \ldots, N$ (for simplicity we take $J^1 > J^2 > \ldots > J^N$); $\mathbf{Q}(x,y) = (Q^{ij}) \in \mathbb{C}^{N \times N}$ is an off-diagonal matrix containing the potentials and $\sigma = \sigma_R + \mathrm{i}\sigma_I$ is a complex parameter. We assume that $Q^{ij}(x,y) \to 0$ sufficiently rapidly for large x, y and without loss of generality we take $\mathbf{B} = 0$. We shall discuss an I.S.T. scheme for the two important cases of equation (4.1.6), namely the hyperbolic ($\sigma = 1$) and elliptic ($\sigma = \mathrm{i}$) cases. The analysis is very similar to that for KPI (§5.2) and KPII (§5.3), respectively (for further details see Fokas [1983]; Fokas and Ablowitz [1983e,f]).

5.4.1 Hyperbolic Systems. In order to study inverse scattering associated with (5.4.1) in the hyperbolic case ($\sigma = 1$), we seek a fundamental matrix solution $\mathbf{\Psi}(x, y; k)$ (an $N \times N$ matrix valued function) in the form

$$\mathbf{\Psi}(x, y; k) = \mathbf{m}(x, y; k) \exp\{ik(x\mathbf{J} + y\mathbf{I})\}, \tag{5.4.2}$$

where $\mathbf{I}$ is the identity matrix. Hence the $N \times N$ matrix valued function $\mathbf{m}(x, y; k)$ satisfies

$$\mathbf{m}_x = \mathrm{i}k[\mathbf{J}, \mathbf{m}] + \mathbf{Q}\mathbf{m} + \mathbf{J}\mathbf{m}_y, \qquad [\mathbf{J}, \mathbf{f}] := \mathbf{J}\mathbf{f} - \mathbf{f}\mathbf{J}. \tag{5.4.3}$$

Let $\boldsymbol{\pi}_0\mathbf{m}$, $\boldsymbol{\pi}_+\mathbf{m}$, $\boldsymbol{\pi}_-\mathbf{m}$ denote the diagonal, strictly upper diagonal and strictly lower diagonal parts of the matrix $\mathbf{m}$, respectively. A solution of (5.4.3), bounded for all values of $k = k_R + \mathrm{i}k_I$ and tending to the identity matrix $\mathbf{I}$ as $k \to \infty$, is given by $\mathbf{m}(x, y; k) = \mathbf{m}_+(x, y; k)$ for $k_I > 0$, $\mathbf{m}(x, y; k) = \mathbf{m}_-(x, y; k)$ for $k_I < 0$, where $\mathbf{m}_\pm(x, y; k)$ satisfy the following linear integral equations:

$$\begin{aligned}\mathbf{m}_\pm(x, y; k) = \mathbf{I} + \frac{1}{2\pi}\int_{-\infty}^{x} &\mathrm{E}(x - x'; \mathbf{J}) \exp\{\mathrm{i}k(x - x')\hat{\mathbf{J}}\} \\ &\times (\boldsymbol{\pi}_0 + \boldsymbol{\pi}_\pm)[\mathbf{Q}(x', y)\, \mathbf{m}_\pm(x', y; k)]\, \mathrm{d}x' \\ -\frac{1}{2\pi}\int_{x}^{\infty} \mathrm{E}(x - x'; \mathbf{J}) &\exp\{\mathrm{i}k(x - x')\hat{\mathbf{J}}\}\boldsymbol{\pi}_\pm[\mathbf{Q}(x', y)\, \mathbf{m}_\mp(x', y; k)]\, \mathrm{d}x', \end{aligned} \tag{5.4.4$\pm$}$$

where the linear operator $\mathrm{E}(x; \mathbf{J})$ is defined by

$$\begin{aligned}[\mathrm{E}(x; \mathbf{J})\mathbf{F}(\bullet)](x, y) :&= \int_{-\infty}^{\infty}\int_{-\infty}^{\infty} \exp\{\mathrm{i}\mu x\mathbf{J} + \mathrm{i}\mu(y - y')\}\, \mathbf{F}(y')\, \mathrm{d}\mu\, \mathrm{d}y' \\ &= \mathbf{F}(y + x\mathbf{J}),\end{aligned} \tag{5.4.5}$$

in which $\mathbf{F}(y + x\mathbf{J})$ denotes the matrix obtained from $\mathbf{F}(x)$ by evaluating the ℓth row of $\mathbf{F}$ at $y + xJ^\ell$ (for $\ell = 1, 2, \ldots, N$), and

$$\exp\{\mathrm{i}a\hat{\mathbf{J}}\}\mathbf{F} := \exp\{\mathrm{i}a\mathbf{J}\}\, \mathbf{F} \exp\{-\mathrm{i}a\mathbf{J}\}. \tag{5.4.6}$$

The integral equations (5.4.4) can be derived by first applying a Fourier transform in the y direction which reduces (5.4.3) to a one-dimensional scattering problem (we outline this below), or alternatively using a Green's function approach similar to that used for the KP equation in §§5.2, 5.3.

Taking a Fourier transform of equation (5.4.3) yields

$$\hat{\mathbf{m}}_x - \mathrm{i}\lambda\mathbf{J}\hat{\mathbf{m}} = \mathrm{i}k[\mathbf{J}, \hat{\mathbf{m}}] + (\widehat{\mathbf{Q}\mathbf{m}}), \tag{5.4.7a}$$

where

$$\hat{\mathbf{m}}(x, \lambda; k) = \int_{-\infty}^{\infty} \mathbf{m}(x, y; k) \mathrm{e}^{-\mathrm{i}\lambda y}\, \mathrm{d}y, \tag{5.4.7b}$$

which can be written in the form

$$\tilde{\mathbf{m}}_x - \mathrm{i}k[\mathbf{J}, \tilde{\mathbf{m}}] = \tilde{\mathbf{Q}}, \tag{5.4.8a}$$

where

$$\tilde{\mathbf{m}}(x, \lambda; k) = \hat{\mathbf{m}}(x, y; k) \exp\{-\mathrm{i}\lambda x \mathbf{J}\}, \tag{5.4.8b}$$

$$\tilde{\mathbf{Q}}(x, \lambda; k) = (\widehat{\mathbf{Qm}})(x, y; k) \exp\{-\mathrm{i}\lambda x \mathbf{J}\}. \tag{5.4.8c}$$

Equations (5.4.8) can be solved (is nothing but the $N \times N$ system, see Chapter 3) and by taking the inverse Fourier transform we can obtain equation (5.4.4). Namely by taking the Fourier transform (in x) of the equation which the Green's function $\tilde{G}^{ij}(x; k)$ associated with equation (5.4.8) satisfies

$$\frac{\partial \tilde{G}^{ij}}{\partial x} - \mathrm{i}k(J^i - J^j)\tilde{G}^{ij} = \delta(x),$$

we find that for $i \neq j$ (recall that $J^i \neq J^j$ if $i \neq j$)

$$\begin{aligned} \tilde{G}^{ij} &= \frac{1}{2\pi\mathrm{i}} \int_{-\infty}^{\infty} \frac{\mathrm{e}^{\mathrm{i}px}}{p - k(J^i - J^j)}\, \mathrm{d}p \\ &= \begin{cases} \exp\{\mathrm{i}k(J^i - J^j)x\} & \text{if } x > 0,\ \mathrm{Im}[k(J^i - J^j)] > 0, \\ 0, & \text{if } x < 0,\ \mathrm{Im}[k(J^i - J^j)] > 0, \\ 0, & \text{if } x > 0,\ \mathrm{Im}[k(J^i - J^j)] < 0, \\ -\exp\{\mathrm{i}k(J^i - J^j)x\} & \text{if } x < 0,\ \mathrm{Im}[k(J^i - J^j)] < 0. \end{cases} \end{aligned}$$

In the case $i = j$, we require (normalization) that $\mathbf{m} \to \mathbf{I}$ as $x \to -\infty$, so we take a contour C_+ from $-\infty$ to $+\infty$ which passes below the origin so that

$$\tilde{G}^{ii}(x; k) - \frac{1}{2\pi\mathrm{i}} \int_{C_+} \frac{\mathrm{e}^{\mathrm{i}px}}{p}\, \mathrm{d}p = \begin{cases} 1, & \text{if } x > 0, \\ 0, & \text{if } x < 0. \end{cases}$$

Therefore

$$\begin{aligned} \tilde{m}^{ij}(x, \lambda; k) &= \delta^{ij} + \int_{-\infty}^{\infty} \tilde{G}^{ij}(x - \xi; k)\, \tilde{Q}^{ij}(\xi, \lambda; k)\, \mathrm{d}\xi \\ &= \begin{cases} \displaystyle\int_{-\infty}^{x} \exp\{\mathrm{i}k(J^i - J^j)(x - \xi)\}\, \tilde{Q}^{ij}(\xi, \lambda; k)\, \mathrm{d}\xi, & \text{if } i > j, \\ \displaystyle 1 + \int_{-\infty}^{x} \tilde{Q}^{ij}(\xi, \lambda; k)\, \mathrm{d}\xi, & \text{if } i = j, \\ \displaystyle -\int_{x}^{\infty} \exp\{\mathrm{i}k(J^i - J^j)(x - \xi)\}\, \tilde{Q}^{ij}(\xi, \lambda; k)\, \mathrm{d}\xi, & \text{if } i < j. \end{cases} \end{aligned}$$

Since $\mathbf{J}$ is a diagonal matrix then

$$\exp\{\mathrm{i}k(J^i - J^j)(x - \xi)\}\, \tilde{\mathbf{Q}}(\xi, \lambda; k) = \Big[\exp\{\mathrm{i}k(x - \xi)\hat{\mathbf{J}}\}\, \tilde{Q}^{ij}(\xi, \lambda; k)\Big]^{ij}$$

[recall equation (5.4.6)], and thus: (i), if $k_I > 0$,

$$\tilde{\mathbf{m}}_+(x,\lambda;k) = \mathbf{I} + \int_{-\infty}^{x} \exp\{ik(x-\xi)\hat{\mathbf{J}}\}(\boldsymbol{\pi}_0 + \boldsymbol{\pi}_+)\tilde{\mathbf{Q}}_+(\xi,\lambda;k)\,\mathrm{d}\xi - \int_{-\infty}^{x} \exp\{ik(x-\xi)\hat{\mathbf{J}}\}\boldsymbol{\pi}_-\tilde{\mathbf{Q}}_+(\xi,\lambda;k)\,\mathrm{d}\xi; \quad (5.4.9a)$$

(ii), if $k_I < 0$,

$$\tilde{\mathbf{m}}_-(x,\lambda;k) = \mathbf{I} + \int_{-\infty}^{x} \exp\{ik(x-\xi)\hat{\mathbf{J}}\}(\boldsymbol{\pi}_0 + \boldsymbol{\pi}_-)\tilde{\mathbf{Q}}_-(\xi,\lambda;k)\,\mathrm{d}\xi - \int_{-\infty}^{x} \exp\{ik(x-\xi)\hat{\mathbf{J}}\}\boldsymbol{\pi}_+\tilde{\mathbf{Q}}_-(\xi,\lambda;k)\,\mathrm{d}\xi; \quad (5.4.9b)$$

(iii), if $k_I = 0$,

$$\tilde{\mathbf{m}}(x,\lambda;k) = \mathbf{I} + \int_{-\infty}^{x} \exp\{ik(x-\xi)\hat{\mathbf{J}}\}\tilde{\mathbf{Q}}(\xi,\lambda;k)\,\mathrm{d}\xi. \quad (5.4.9c)$$

Combining (5.4.9a,b) yields

$$\tilde{\mathbf{m}}_\pm(x,\lambda;k) = \mathbf{I} + \int_{-\infty}^{x} \exp\{ik(x-\xi)\hat{\mathbf{J}}\}(\boldsymbol{\pi}_0 + \boldsymbol{\pi}_\pm)\tilde{\mathbf{Q}}_\pm(\xi,\lambda;k)\,\mathrm{d}\xi - \int_{-\infty}^{x} \exp\{ik(x-\xi)\hat{\mathbf{J}}\}\boldsymbol{\pi}_\mp\tilde{\mathbf{Q}}_\pm(\xi,\lambda;k)\,\mathrm{d}\xi,$$

and hence

$$\hat{\mathbf{m}}_\pm(x,\lambda;k) = \exp(i\lambda x\mathbf{J}) + \int_{-\infty}^{x} \exp\{ik(x-\xi)\hat{\mathbf{J}}\}(\boldsymbol{\pi}_0 + \boldsymbol{\pi}_\pm)(\widehat{\mathbf{Q}\mathbf{m}_\pm})(\xi,\lambda;k)\,\mathrm{d}\xi - \int_{-\infty}^{x} \exp\{ik(x-\xi)\hat{\mathbf{J}}\}\boldsymbol{\pi}_\mp(\widehat{\mathbf{Q}\mathbf{m}_\pm})(\xi,\lambda;k)\,\mathrm{d}\xi.$$

Therefore, by taking the inverse Fourier transform (in y) and noting that

$$\frac{1}{2\pi}\int_{-\infty}^{\infty} \exp\{i\lambda(x\mathbf{J}+y)\}\,\mathbf{F}(\lambda)\,\mathrm{d}\lambda = \mathbf{F}(x\mathbf{J}+y),$$

we obtain the integral equations (5.4.4).

With the assumption that the linear integral equations have no homogeneous solutions it follows that $\mathbf{m}_+$ and $\mathbf{m}_-$ are holomorphic functions of k, for $k_I > 0$ and $k_I < 0$, respectively. Hence the function $\mathbf{m}(x,y;k)$ is a *sectionally holomorphic function of k having a jump across the real axis $k_I = 0$*. Thus $\partial\mathbf{m}/\partial\bar{k} = 0$ for all k such that $k_I \neq 0$ and $\partial\mathbf{m}/\partial\bar{k} = \mathbf{m}_+(x,y;k) - \mathbf{m}_-(x,y;k)$ if $k_I = 0$ (i.e., for $k = k_R$).

By formulating an integral equation for the function

$$\boldsymbol{\Delta}(x,y;k) := \mathbf{m}_+(x,y;k) - \mathbf{m}_-(x,y;k), \qquad \text{for} \qquad k \in \mathbb{R}, \quad (5.4.10)$$

and using equation (5.4.4), the following scattering equation may be derived:

$$\mathbf{m}_+(x,y;k) - \mathbf{m}_-(x,y;k) = \int_{-\infty}^{\infty} \mathbf{m}_-(x,y;l) \exp\{\mathrm{i}l(x\mathbf{J}+y)\} \, \mathbf{f}(l,k) \exp\{-\mathrm{i}k(x\mathbf{J}+y)\} \, \mathrm{d}l, \tag{5.4.11a}$$

on $k = k_R$, where the scattering data $\mathbf{f}(l,k)$, for l, k real satisfies

$$\mathbf{f}(l,k) - \int_{-\infty}^{\infty} \mathbf{T}_+(l,p) \, \mathbf{f}(p,k) \, \mathrm{d}p = \mathbf{T}_+(l,k) - \mathbf{T}_-(l,k), \tag{5.4.11b}$$

where

$$\mathbf{T}_\pm(l,k) := \frac{1}{2\pi} \int_{-\infty}^{\infty}\int_{-\infty}^{\infty} \exp\{-\mathrm{i}l(x'\mathbf{J}+y')\} \times \boldsymbol{\pi}_\pm[\mathbf{Q}(x',y') \, \mathbf{m}_\mp(x',y';k)] \exp\{\mathrm{i}k(x'\mathbf{J}+y')\} \, \mathrm{d}x' \, \mathrm{d}y'. \tag{5.4.11c}$$

Prior to outlining the derivation of these equations, we note the remarkable fact that since the kernel of equation (5.4.11b) is strictly upper triangular, then the equation can be solved in closed form in terms of $\mathbf{T}_+$ and $\mathbf{T}_-$ (which are strictly upper and lower triangular matrices, respectively). For example, if $N = 2$, then

$$f^{11}(l,k) = -\int_{-\infty}^{\infty} T_+^{12}(l,p) \, T_-^{21}(p,k) \, \mathrm{d}p, \tag{5.4.12a}$$

$$f^{21}(l,k) = -T_+^{12}(l,k), \tag{5.4.12b}$$

$$f^{12}(l,k) = T_-^{21}(l,k), \tag{5.4.12c}$$

$$f^{22}(l,k) = 0, \tag{5.4.12d}$$

where T_+^{12} and T_-^{21} are the (only) nonzero elements in $\mathbf{T}_+$ and $\mathbf{T}_-$ respectively.

To derive the scattering equation (5.4.11), consider (5.4.4+) − (5.4.4−), which yields the following integral equation for $\boldsymbol{\Delta}(x,y;k)$

$$\boldsymbol{\Delta}(x,y;k) = \frac{1}{2\pi} \int_{-\infty}^{x} \mathrm{E}(x-\xi;\mathbf{J}) \exp\{\mathrm{i}k(x-\xi)\hat{\mathbf{J}}\} \boldsymbol{\Delta}(\xi,y;k) \, \mathrm{d}\xi + \frac{1}{2\pi} \int_{-\infty}^{\infty} \mathrm{E}(x-\xi;\mathbf{J}) \exp\{\mathrm{i}k(x-\xi)\hat{\mathbf{J}}\} \, [\boldsymbol{\pi}_+(\mathbf{Q}\mathbf{m}_-) - \boldsymbol{\pi}_-(\mathbf{Q}\mathbf{m}_+)] \, \mathrm{d}\xi.$$

Postulating that $\boldsymbol{\Delta}(x,y;k)$ is given by equations (5.4.10,11a), i.e.,

$$\boldsymbol{\Delta}(x,y;k) = \int_{-\infty}^{\infty} \mathbf{m}_-(x,y;l) \exp\{\mathrm{i}l(x\mathbf{J}+y)\} \, \mathbf{f}(l,k) \exp\{-\mathrm{i}k(x\mathbf{J}+y)\} \, \mathrm{d}l,$$

yields

$$\int_{-\infty}^{\infty} \mathbf{m}_-(x,y;l) \exp\{\mathrm{i}l(x\mathbf{J}+y)\} \, \mathbf{f}(l,k) \exp\{-\mathrm{i}k(x\mathbf{J}+y)\} \, \mathrm{d}l = \frac{1}{2\pi} \int_{-\infty}^{x} \mathrm{E}(x-\xi;\mathbf{J}) \exp\{\mathrm{i}k(x-\xi)\hat{\mathbf{J}}\} \times \left(\int_{-\infty}^{\infty} \mathbf{m}_-(\xi,y;l) \exp\{\mathrm{i}l(\xi\mathbf{J}+y)\} \, \mathbf{f}(l,k) \exp\{-\mathrm{i}k(\xi\mathbf{J}+y)\} \, \mathrm{d}l \right) \mathrm{d}\xi + \frac{1}{2\pi} \int_{-\infty}^{\infty} \mathrm{E}(x-\xi;\mathbf{J}) \exp\{\mathrm{i}k(x-\xi)\hat{\mathbf{J}}\} \, [\boldsymbol{\pi}_+(\mathbf{Q}\mathbf{m}_-) - \boldsymbol{\pi}_-(\mathbf{Q}\mathbf{m}_+)] \, \mathrm{d}\xi. \tag{5.4.13}$$

Henceforth we shall assume that (i), all integrals are well defined (so, in particular, $\mathbf{f}(l,k) \to 0$ as $l, k \to \infty$) and (ii), all interchanges or orders of integration are valid. Next multiply the equation for $\mathbf{m}_-$, i.e., equation (5.4.4−), by

$$\exp\{\mathrm{i}l(x\mathbf{J}+y)\}\,\mathbf{f}(l,k)\,\exp\{-\mathrm{i}k(x\mathbf{J}+y)\},$$

and integrate with respect to l from $-\infty$ to ∞; substituting the resulting expression into equation (5.4.13) yields

$$\begin{aligned}
&\int_{-\infty}^{\infty} \exp\{\mathrm{i}l(x\mathbf{J}+y)\}\,\mathbf{f}(l,k)\,\exp\{-\mathrm{i}k(x\mathbf{J}+y)\}\,\mathrm{d}l \\
&\qquad + \frac{1}{2\pi}\int_{-\infty}^{x}\int_{-\infty}^{\infty} \mathrm{E}(x-\xi;\mathbf{J})\exp\{\mathrm{i}l(x-\xi)\hat{\mathbf{J}}\}(\boldsymbol{\pi}_0+\boldsymbol{\pi}_-)[\mathbf{Q}(\xi,y)\mathbf{m}_-(\xi,y;l)] \\
&\qquad\qquad \times \exp\{\mathrm{i}l(x\mathbf{J}+y)\}\,\mathbf{f}(l,k)\,\exp\{-\mathrm{i}k(x\mathbf{J}+y)\}\,\mathrm{d}l\,\mathrm{d}\xi \\
&\qquad - \frac{1}{2\pi}\int_{x}^{\infty}\int_{-\infty}^{\infty} \mathrm{E}(x-\xi;\mathbf{J})\exp\{\mathrm{i}l(x-\xi)\hat{\mathbf{J}}\}\boldsymbol{\pi}_+[\mathbf{Q}(\xi,y)\mathbf{m}_-(\xi,y;l)] \\
&\qquad\qquad \times \exp\{\mathrm{i}l(x\mathbf{J}+y)\}\,\mathbf{f}(l,k)\,\exp\{-\mathrm{i}k(x\mathbf{J}+y)\}\,\mathrm{d}l\,\mathrm{d}\xi \\
&\quad = \frac{1}{2\pi}\int_{-\infty}^{x}\int_{-\infty}^{\infty} \mathrm{E}(x-\xi;\mathbf{J})\exp\{\mathrm{i}k(x-\xi)\hat{\mathbf{J}}\}\mathbf{Q}(\xi,y)\,\mathbf{m}_-(\xi,y;l) \\
&\qquad\qquad \times \exp\{\mathrm{i}l(\xi\mathbf{J}+y)\}\,\mathbf{f}(l,k)\,\exp\{-\mathrm{i}k(\xi\mathbf{J}+y)\}\,\mathrm{d}l\,\mathrm{d}\xi \\
&\qquad + \frac{1}{2\pi}\int_{-\infty}^{\infty} \mathrm{E}(x-\xi;\mathbf{J})\exp\{\mathrm{i}k(x-\xi)\hat{\mathbf{J}}\}\,[\boldsymbol{\pi}_+(\mathbf{Q}\mathbf{m}_-)-\boldsymbol{\pi}_-(\mathbf{Q}\mathbf{m}_+)]\,\mathrm{d}\xi. \qquad (5.4.14)
\end{aligned}$$

First, it is easily shown that

$$\begin{aligned}
&\frac{1}{2\pi}\int_{-\infty}^{x}\int_{-\infty}^{\infty} \mathrm{E}(x-\xi;\mathbf{J})\exp\{\mathrm{i}l(x-\xi)\hat{\mathbf{J}}\}\mathbf{Q}(\xi,y)\mathbf{m}_-(\xi,y;l) \\
&\qquad\qquad \times \exp\{\mathrm{i}l(x\mathbf{J}+y)\}\,\mathbf{f}(l,k)\,\exp\{-\mathrm{i}k(x\mathbf{J}+y)\}\,\mathrm{d}l\,\mathrm{d}\xi \\
&\quad = \frac{1}{2\pi}\int_{-\infty}^{x}\int_{-\infty}^{\infty} \mathrm{E}(x-\xi;\mathbf{J})\exp\{\mathrm{i}k(x-\xi)\hat{\mathbf{J}}\}\mathbf{Q}(\xi,y)\,\mathbf{m}_-(\xi,y;l) \\
&\qquad\qquad \times \exp\{\mathrm{i}l(\xi\mathbf{J}+y)\}\,\mathbf{f}(l,k)\,\exp\{-\mathrm{i}k(\xi\mathbf{J}+y)\}\,\mathrm{d}l\,\mathrm{d}\xi. \qquad (5.4.15)
\end{aligned}$$

Then, after a little manipulation, we obtain

$$\begin{aligned}
&\frac{1}{2\pi}\int_{-\infty}^{\infty}\int_{-\infty}^{\infty} \mathrm{E}(x-\xi;\mathbf{J})\exp\{\mathrm{i}l(x-\xi)\hat{\mathbf{J}}\}\boldsymbol{\pi}_+[\mathbf{Q}(\xi,y)\mathbf{m}_-(\xi,y;l)] \\
&\qquad\qquad \times \exp\{\mathrm{i}l(x\mathbf{J}+y)\}\,\mathbf{f}(l,k)\,\exp\{-\mathrm{i}k(x\mathbf{J}+y)\}\,\mathrm{d}l\,\mathrm{d}\xi \\
&\quad = \int_{-\infty}^{\infty}\int_{-\infty}^{\infty} \exp\{\mathrm{i}l(x\mathbf{J}+y)\}\,\mathbf{G}(l,p)\,\mathbf{F}(p,k)\exp\{-\mathrm{i}k(x\mathbf{J}+y)\}\,\mathrm{d}l\,\mathrm{d}p, \qquad (5.4.16a)
\end{aligned}$$

where

$$\begin{aligned}
\mathbf{G}(l,p) = -\frac{1}{2\pi}&\int_{-\infty}^{\infty}\int_{-\infty}^{\infty} \exp\{-\mathrm{i}l(x\mathbf{J}+y)\} \\
&\times \boldsymbol{\pi}_+[\mathbf{Q}(\xi,y)\mathbf{m}_-(\xi,y;p)]\exp\{\mathrm{i}k(x\mathbf{J}+y)\}\,\mathrm{d}l\,\mathrm{d}p, \qquad (5.4.16b)
\end{aligned}$$

and

$$\frac{1}{2\pi}\int_{-\infty}^{\infty} \mathrm{E}(x-\xi;\mathbf{J}) \exp\{\mathrm{i}k(x-\xi)\hat{\mathbf{J}}\} \left[\boldsymbol{\pi}_+(\mathbf{Q}\mathbf{m}_-) - \boldsymbol{\pi}_-(\mathbf{Q}\mathbf{m}_+)\right](\xi,y;k)\,\mathrm{d}\xi$$
$$= \int_{-\infty}^{\infty} \exp\{\mathrm{i}p(x\mathbf{J}+y)\}\,\mathbf{A}(p,k)$$
$$\times \exp\{-\mathrm{i}k(x\mathbf{J}+y)\}\,\mathrm{d}p, \tag{5.4.17a}$$

where

$$\mathbf{A}(p,k) = \frac{1}{2\pi}\int_{-\infty}^{\infty}\int_{-\infty}^{\infty} \exp\{-\mathrm{i}p(\xi\mathbf{J}+y)\} \left[\boldsymbol{\pi}_+(\mathbf{Q}\mathbf{m}_-) - \boldsymbol{\pi}_-(\mathbf{Q}\mathbf{m}_+)\right](\xi,y;k)$$
$$\times \exp\{\mathrm{i}k(\xi\mathbf{J}+y)\}\,\mathrm{d}\xi\,\mathrm{d}y. \tag{5.4.17b}$$

Hence substituting (5.4.15-17) in (5.4.14) yields (after a Fourier transform),

$$\mathbf{f}(l,k) + \int_{-\infty}^{\infty} \mathbf{G}(l,p)\,\mathbf{f}(p,k)\,\mathrm{d}p = \mathbf{A}(l,k).$$

Finally, we obtain the scattering equation (5.4.11) by observing that

$$\mathbf{G}(l,k) = -\mathbf{T}_+(l,k), \qquad \mathbf{A}(l,k) = \mathbf{T}_+(l,k) - \mathbf{T}_-(l,k),$$

with $\mathbf{T}_\pm(l,k)$ as defined in equation (5.4.11c).

Equation (5.4.11a) defines a Riemann-Hilbert problem for $\mathbf{m}(x,y;k)$ in terms of $\mathbf{f}(l,k)$. Its solution is given by the following linear integral equation [which can be obtained by taking the "minus" projection of equation (5.4.11a)]:

$$\mathbf{m}_-(x,y;k) = \mathbf{I} - \frac{1}{2\pi\mathrm{i}}\int_{-\infty}^{\infty}\int_{-\infty}^{\infty} \frac{\mathbf{m}_-(x,y;l)\,\mathbf{F}(x,y,l,p;\mathbf{J})}{p-(k-\mathrm{i}0)}\,\mathrm{d}l\,\mathrm{d}p, \tag{5.4.18a}$$

where

$$\mathbf{F}(x,y,l,p;\mathbf{J}) := \exp\{\mathrm{i}l(x\mathbf{J}+y)\}\,\mathbf{f}(l,p)\,\exp\{-\mathrm{i}p(x\mathbf{J}+y)\}, \tag{5.4.18b}$$

By taking the large k limit of equation (5.4.18) and comparing with that of equation (5.4.4) we obtain the reconstruction formula

$$\mathbf{Q}(x,y) = -\left[\mathbf{J}, \frac{1}{2\pi}\int_{-\infty}^{\infty}\int_{-\infty}^{\infty} \mathbf{m}_-(x,y;l)\,\mathbf{F}(x,y,l,p;\mathbf{J})\,\mathrm{d}l\,\mathrm{d}p\right]. \tag{5.4.19}$$

Hence, the formal solution of the inverse problem associated with equation (5.4.4) is given by equation (5.4.19), where $\mathbf{m}_-(x,y;k)$ is obtained from equation (5.4.18) and the inverse data $\mathbf{f}(l,k)$ can be found in terms of $\mathbf{T}_\pm(l,k)$ from equation (5.4.11c).

5.4.2 Elliptic Systems. We now discuss inverse scattering associated with (5.4.1), in the elliptic case ($\sigma = -\mathrm{i}$). We seek a fundamental matrix solution $\mathbf{\Psi}(x, y; k)$ (an $N \times N$ matrix valued function) in the form

$$\mathbf{\Psi}(x, y; k) = \mathbf{m}(x, y; k) \exp\{\mathrm{i}kx\mathbf{J} - ky\mathbf{I}\},$$

where $\mathbf{I}$ is the identity matrix. Hence the $N \times N$ matrix valued function $\mathbf{m}(x, y; k)$ satisfies

$$\mathbf{m}_x = \mathrm{i}k[\mathbf{J}, \mathbf{m}] + \mathbf{Q}\mathbf{m} - \mathrm{i}\mathbf{J}\mathbf{m}_y, \qquad [\mathbf{J}, \mathbf{f}] := \mathbf{J}\mathbf{f} - \mathbf{f}\mathbf{J}, \tag{5.4.20}$$

where $\mathbf{m}(x, y; k)$ is an $N \times N$ matrix, $\mathbf{J}$ is a constant real diagonal matrix with elements $J^1 > J^2 > \ldots > J^N$, and $\mathbf{Q}(x, y)$ is an $N \times N$ off-diagonal matrix containing the potentials $Q^{ij}(x, y)$. (Equation (5.4.20) is obtained from equation (5.4.3) by replacing $\mathbf{m}_y$ by $-\mathrm{i}\mathbf{m}_y$.) A matrix eigenfunction $\mathbf{m}(x, y; k) = [m^{ij}(x, y; k)]$ which is a solution of (5.4.20), bounded for all values of $k = k_R + \mathrm{i}k_I$ and tending to the identity matrix as $k \to \infty$, is given by the solution of the integral equation

$$\mathbf{m} = \mathbf{I} + \tilde{\mathbf{G}}(\mathbf{Q}\mathbf{m}), \tag{5.4.21a}$$

that is

$$m^{ij} = \delta^{ij} + \int_{-\infty}^{\infty}\int_{-\infty}^{\infty} G^{ij}(x - x', y - y'; k) \sum_{p=1}^{N} Q^{ip}(x', y')\, m^{pj}(x', y'; k)\, \mathrm{d}x'\, \mathrm{d}y', \tag{5.4.21b}$$

where the Green's function $G^{ij}(x, y; k)$ satisfies the equation

$$\frac{\partial G^{ij}}{\partial x} - \mathrm{i}k(J^i - J^j)G^{ij} + \mathrm{i}J^i\frac{\partial G^{ij}}{\partial y} = \delta(x)\delta(y).$$

By taking the Fourier transform of this we find that

$$G^{ij}(x, y; k) = \frac{1}{4\pi^2\mathrm{i}}\int_{-\infty}^{\infty}\int_{-\infty}^{\infty} \frac{\exp\{\mathrm{i}(x\xi + y\eta)\}}{\xi - k(J^i - J^j) + \mathrm{i}\eta J^i}\, \mathrm{d}\xi\, \mathrm{d}\eta, \tag{5.4.21c}$$

which can be evaluated explicitly by contour integration using the integrals

$$\int_{-\infty}^{\infty} \frac{\mathrm{e}^{\mathrm{i}\xi x}}{\xi + a + \mathrm{i}b}\, \mathrm{d}\xi = 2\pi\mathrm{i}\, \mathrm{sgn}(x) \exp\{-\mathrm{i}(a + \mathrm{i}b)x\}\Theta(-bx),$$

$$\int_{-\infty}^{\infty} \mathrm{e}^{(c+\mathrm{i}d)\eta}\Theta(\alpha - c\eta)\, \mathrm{d}\eta = \frac{\mathrm{sgn}(c)}{c + \mathrm{i}d} \exp\{(c + \mathrm{i}d)\alpha/c\},$$

where $\Theta(x)$ is the Heaviside step function. Hence

$$G^{ij}(x, y; k) = \frac{\mathrm{sgn}(J^i)}{2\pi(xJ^i + \mathrm{i}y)} \exp\{-\theta^{ij}(x, y; k_R, k_I)\}, \tag{5.4.22a}$$

where

$$\theta^{ij}(x,y;k_R,k_I) = \mathrm{i}C^{ij}(J^i k_R x + k_I y), \tag{5.4.22b}$$

$$C^{ij} = (J^i - J^j)/J^i; \tag{5.4.22c}$$

furthermore we have

$$\frac{\partial G^{ij}}{\partial \bar{k}}(x,y;k) = -\tfrac{1}{2}C^{ij}(xJ^i + \mathrm{i}y)G^{ij}. \tag{5.4.22d}$$

Comparing equation (5.4.22a) with (5.4.4) it follows that (a), equation (5.4.21), in contrast to equation (5.4.4), has no jump across $k_I = 0$, and (b), equation (5.4.21) depends explicitly on k_I. Hence the solution $\mathbf{m}(x,y;k)$, although bounded for all complex values of k, is nowhere analytic with respect to k, since $\partial\mathbf{m}/\partial\bar{k} \neq 0$. Assuming that equation (5.4.21) has no homogeneous solutions, it follows by differentiation that $\partial\mathbf{m}/\partial\bar{k}$ satisfies an equation obtained from (5.4.21) by replacing $\mathbf{I}$ by $\mathbf{\Omega}(x,y;k_R,k_I)$, where $\Omega^{ii} = 0$ and

$$\Omega^{ij}(x,y;k_R,k_I) = T^{ij}(k_R,k_I)\exp\{\theta^{ij}(x,y;k_R,k_I)\}, \tag{5.4.23a}$$

where

$$T^{ij}(k_R,k_I) = \frac{\mathrm{i}}{4\pi}\,\mathrm{sgn}(J^i)C^{ij}\int_{-\infty}^{\infty}\int_{-\infty}^{\infty}[\mathbf{Q}(x',y'),\mathbf{m}(x',y';k)]^{ij} \times \exp\{-\theta^{ij}(x',y';k_R,k_I)\}\,\mathrm{d}x'\,\mathrm{d}y'. \tag{5.4.23b}$$

Specifically we may express $\partial\mathbf{m}/\partial\bar{k}$ in terms of $\mathbf{m}$. Explicitly, carrying out this procedure we find

$$\frac{\partial\mathbf{m}}{\partial\bar{k}}(x,y;k) = \sum_{\substack{i,j=1\\ i\neq j}}^{N} \mathbf{m}\left(x,y;k_R + \mathrm{i}(J^j/J^i)k_I\right) T^{ij}(k_R,k_I)\,\mathbf{w}^{ij}(x,y;k_R,k_I), \tag{5.4.24}$$

where $\mathbf{w}^{ij}$ is a matrix with zeros everywhere except at its ijth entry which equals $\exp\{\theta^{ij}(x,y;k_R,k_I)\}$. Introduce an eigenfunction $\mathbf{N}^{ij}(x,y;k_R,k_I)$ which satisfies an equation obtained from equation (5.4.21) by replacing $\mathbf{I}$ by $\mathbf{w}^{ij}$, and then by using the important "symmetry" condition

$$\mathbf{N}^{ij}(x,y;k_R,k_I) = \mathbf{m}\left(x,y;k_R + \mathrm{i}(J^j/J^i)k_I\right)\mathbf{w}^{ij}(x,y;k_R,k_I).$$

Using (5.4.24) and the generalized Cauchy formula

$$m(x,y;k) = \frac{1}{2\pi\mathrm{i}}\iint_R \frac{\partial m}{\partial\bar{z}}(x,y;z)\,\frac{\mathrm{d}z\wedge\mathrm{d}\bar{z}}{z-k} + \frac{1}{2\pi\mathrm{i}}\int_C \frac{m(x,y;z)}{z-k}\,\mathrm{d}z,$$

where R and C are an appropriate domain and contour respectively in the complex z-plane, $\partial\mathbf{m}/\partial\bar{k}$ exists at least in the sense of distributions, $z = z_R + \mathrm{i}z_I$, and $\mathrm{d}z \wedge \mathrm{d}\bar{z} = -2\mathrm{i}\,\mathrm{d}z_R\,\mathrm{d}z_I$, one obtains the following linear integral equation for $m(x,y;k)$:

$$\mathbf{m}(x,y;k) = \mathbf{I} + \frac{1}{2\pi\mathrm{i}} \int_{-\infty}^{\infty}\int_{-\infty}^{\infty} \sum_{\substack{i,j=1\\ i\neq j}}^{N} \mathbf{m}\left(x,y;z_R + \mathrm{i}(J^j/J^i)z_I\right) \times T^{ij}(z_R,z_I)\,\mathbf{w}^{ij}(x,y;z_R,z_I)\,\frac{\mathrm{d}z\wedge\mathrm{d}\bar{z}}{z-k}. \tag{5.4.25}$$

Once $\mathbf{m}$ is found, $\mathbf{Q}$ can be reconstructed using the formula

$$\mathbf{Q}(x,y) = \frac{1}{2\pi}\left[\mathbf{J}, \frac{1}{2\pi\mathrm{i}} \int_{-\infty}^{\infty}\int_{-\infty}^{\infty} \sum_{\substack{i,j=1\\ i\neq j}}^{N} \left\{\mathbf{m}\left(x,y;z_R + \mathrm{i}(J^j/J^i)z_I\right) \times T^{ij}(z_R,z_I)\,\mathbf{w}^{ij}(x,y;z_R,z_I)\right\}\mathrm{d}z\wedge\mathrm{d}\bar{z}\right]. \tag{5.4.26}$$

Hence the solution of the inverse problem associated with the elliptic scattering problem is given by equation (5.4.26), where $m(x,y;k)$ is obtained from equation (5.4.25) and the scattering data $T^{ij}(k_R,k_I)$ are defined by equation (5.4.23b). One can incorporate slowly decaying lump-type soliton solutions into this formalism (see Fokas and Ablowitz [1983e,f]), and which we shall not do here.

We shall now consider some physically interesting evolution equations which can be written as the compatibility condition of a Lax pair, the time-independent part of which is of the form (5.4.1). We briefly show how the above results can be used to solve the initial-value problem for these evolution equations. We consider three examples: the n-wave interaction and Davey-Stewartson I equations which are hyperbolic systems and for one boundary value problem is solvable using the method outlined in §5.4.1 and §5.5.2, respectively; and the Davey-Stewartson II equations which are an elliptic system and solvable using the (essential) $\bar{\partial}$ approach, this is described in §5.5.3.

5.4.3 The n-Wave Interaction Equations. The first $2+1$-dimensional equation to be solved through an I.S.T. scheme was three-wave interaction equations

$$\frac{\partial u_j}{\partial t} + a_j\frac{\partial u_j}{\partial x} + b_j\frac{\partial u_j}{\partial y} = \mathrm{i}u_k^* u_l^*, \tag{5.4.27}$$

with $j,k,l = 1,2,3$ cyclically permutated [Kaup [1980a], with a different approach to that presented here — see also Zakharov and Shabat [1974, 1979] who obtain a wide class of solutions to (5.4.27)]. These equations arise in several physical applications including nonlinear optics, fluid mechanics and plasma physics (cf. Ablowitz and Segur [1981]). The Lax pair for equations (5.4.27) was derived by Ablowitz and Haberman

[1975a]. Additionally the equations are known to possess an infinite set of conserved quantities (Kaup [1981a,b]); a Bäcklund transformation (Kaup [1981a,b]; Levi, Pilloni and Santini [1981]); and an infinite dimensional symmetry group and reductions to the third, fourth, fifth and sixth Painlevé equations (Leo, Martina and Soliani [1986]; Leo, Martina, Soliani and Tondo [1986]; Martina and Winternitz [1989]; Kitaev [1990]).

For the n-wave interaction equations, consider the Lax pair has the form

$$\boldsymbol{v}_x = \mathbf{J}\boldsymbol{v}_y + \mathbf{Q}\boldsymbol{v}, \tag{5.4.28a}$$

$$\boldsymbol{v}_t = \mathbf{A}_1\boldsymbol{v} + \mathbf{A}_2\boldsymbol{v}_y, \tag{5.4.28b}$$

and we assume that

$$\mathbf{A}_1 \to 0, \qquad \mathbf{A}_2 \to \mathbf{A}_{20} \qquad \text{as} \quad \sqrt{x^2+y^2} \to \infty, \tag{5.4.28c}$$

where $\mathbf{A}_{20}$ is a constant real diagonal matrix ($\mathbf{A}_1$ and $\mathbf{A}$ are defined in equations (5.4.29) below). Hence for large x and y the eigenfunction $\boldsymbol{v} \sim \exp\{ik(x\mathbf{J}+y\mathbf{I}+t\mathbf{A}_{20})\}$. Since the I.S.T. scheme is based on an eigenfunction normalized to $\mathbf{I}$ for large x and y, we seek a fundamental matrix solution $\mathbf{\Phi}(x,y,t;k)$ in the form

$$\mathbf{\Phi}(x,y,t;k) = \mathbf{m}(x,y,t;k)\exp\{ik(x\mathbf{J}+y+\mathbf{A}_{20}t)\}, \tag{5.4.29a}$$

then $\mathbf{m}$ satisfies

$$\mathbf{m}_x = ik[\mathbf{J},\mathbf{m}] + \mathbf{Q}\mathbf{m} + \mathbf{J}\mathbf{m}_y, \tag{5.4.29b}$$

$$\mathbf{m}_t = ik(\mathbf{A}_2\mathbf{m} - \mathbf{m}\mathbf{A}_{20}) + \mathbf{A}_1\mathbf{m} + \mathbf{A}_2\mathbf{m}_y. \tag{5.4.29c}$$

The n-wave interaction equations are

$$\frac{\partial Q^{ij}}{\partial t} = a^{ij}\frac{\partial Q^{ij}}{\partial x} + b^{ij}\frac{\partial Q^{ij}}{\partial y} + \sum_{k=1}^{n}(a^{ik}-a^{kj})Q^{ik}Q^{kj}, \tag{5.4.30a}$$

for $i,j,k = 1,2,\ldots,n$, where a^{ij} and b^{ij} are real constants related to the x and y components of the underlying group velocities. Equations (5.4.30a) are the compatibility conditions of the Lax pair (5.4.28) with $\mathbf{A}_1$ and $\mathbf{A}_2$ defined by

$$(\mathbf{A}_1)^{ij} = a^{ij}Q^{ij}, \qquad (\mathbf{A}_1)^{ii} = 0, \qquad \mathbf{A}_2 = \operatorname{diag}(C^1,\ldots,C^n), \tag{5.4.30b}$$

where J^p, C^p are given in terms of a^{ij}, b^{ij} by

$$a^{ij} = \frac{C^i - C^j}{J^i - J^j}, \qquad b^{ij} = C^i - J^i a^{ij}. \tag{5.4.30c}$$

The initial-value problem for the n-wave interaction equations can be solved through equations (5.4.18) and (5.4.19), where $\mathbf{f}(l,k,0)$ is determined from equation (5.4.11b) and the time evolution of $\mathbf{f}(l,k,t)$ is given by

$$\mathbf{f}(l,k,t)=\exp\{-l^2t\mathbf{A}_{20}\}\mathbf{f}(l,k,0)\exp\{k^2t\mathbf{A}_{20}\}$$

where $\mathbf{A}_{20}=\mathrm{diag}(C^1,\ldots,C^n)$. (This may be obtained by applying the operator

$$\mathrm{M}_k:=\frac{\partial}{\partial t}-\mathrm{i}k\mathbf{A}_2-\mathbf{A}_1-\mathbf{A}_2\frac{\partial}{\partial y}$$

to the scattering equation (5.4.18) and using equation (5.4.29c).) We note that the n-wave interaction equations in n-dimensions can be reduced to a $2+1$-dimensional system (see Nachman and Ablowitz [1984b] and Fokas [1986a]).

5.4.4 Comments on Rigorous Analysis for the Elliptic Scattering Problem. In this section we sketch some of the results of Beals and Coifman [1985b], on the rigorous analysis of the scattering and inverse scattering for the 2×2 elliptic spectral problem. In the 2×2 case, (5.4.20) reduces to

$$\mathrm{D}\boldsymbol{v}\equiv\begin{pmatrix}\partial/\partial\bar{z} & 0\\ 0 & \partial/\partial z\end{pmatrix}\boldsymbol{v}=\begin{pmatrix}0 & q\\ r & 0\end{pmatrix}\boldsymbol{v}, \tag{5.4.31}$$

where $z=x+\mathrm{i}y\in\mathbb{C}$ [we have set $\sigma=-\mathrm{i}$, $J^1=1$, $J^2=-1$ and $\mathbf{B}=0$ in (5.4.1)]. We seek a fundamental matrix solution of (5.4.31) in the form

$$\boldsymbol{\Psi}(x,y;k)=\mathbf{m}(x,y;k)\begin{pmatrix}\mathrm{e}^{\mathrm{i}(x+\mathrm{i}y)k} & 0\\ 0 & \mathrm{e}^{-\mathrm{i}(x-\mathrm{i}y)k}\end{pmatrix}, \tag{5.4.32a}$$

where $k\in\mathbb{C}$, together with the boundary condition

$$\mathbf{m}(x,y;k)\to\mathbf{I},\qquad \text{as}\qquad \sqrt{x^2+y^2}\to\infty. \tag{5.4.32b}$$

The equation for $\mathbf{m}(x,y;k)$ is then

$$(\mathrm{D}_k\mathbf{m})(x,y;k)\equiv[\mathbf{E}^{-1}(z;k)\mathrm{D}\mathbf{E}(z;k)\mathbf{m}](x,y;k)=(\mathbf{Q}\mathbf{m})(x,y;k), \tag{5.4.33}$$

with $z=x+\mathrm{i}y$ and the matrices $\mathbf{E}(z;k)$ and $\mathbf{E}^{-1}(z;k)$ operate on matrix functions $\mathbf{M}(z)$ and $\mathbf{EM}$, $\mathbf{E}^{-1}\mathbf{M}$ are given by

$$\mathbf{E}(z;k)\mathbf{M}(z)=\begin{pmatrix}M^{11}(z) & \exp\{-\mathrm{i}(\bar{z}k+z\bar{k})\}M^{21}(z)\\ \exp\{\mathrm{i}(zk+\bar{z}\bar{k})\}M^{12}(z) & M^{22}(z)\end{pmatrix},$$
$$\mathbf{E}^{-1}(z;k)\mathbf{M}(z)=\begin{pmatrix}M^{11}(z) & \exp\{-\mathrm{i}(zk+\bar{z}\bar{k})\}M^{21}(z)\\ \exp\{\mathrm{i}(\bar{z}k+z\bar{k})\}M^{12}(z) & M^{22}(z)\end{pmatrix},$$

with $\bar{z} = x - iy$. Equation (5.4.33) is equivalent to the integral equation

$$\mathbf{m} = \mathbf{I} + \mathrm{D}_k^{-1}(\mathbf{Qm}) \equiv \mathbf{I} + \tilde{G}_k(\mathbf{Qm}), \tag{5.4.34}$$

(we note that (5.4.34) results from (5.4.21) in the 2×2 case), where

$$\mathrm{D}_k^{-1} = \mathbf{E}^{-1}(z;k)\mathrm{D}^{-1}\mathbf{E}(z;k),$$

with D^{-1} the fundamental solution of the operator D given by

$$(\mathrm{D}^{-1}\mathbf{M})(z) = \frac{1}{\pi}\int_{\mathbb{C}} \begin{pmatrix} (z-\xi)^{-1} & 0 \\ 0 & (\bar{z}-\bar{\xi})^{-1} \end{pmatrix} \mathbf{M}(\xi)\,\mathrm{d}\xi.$$

Hence

$$\begin{aligned}
(\mathrm{D}^{-1}\mathbf{M})^{11}(z) &= \frac{1}{\pi}\int_{\mathbb{C}} \frac{M^{11}(\xi)}{z-\xi}\,\mathrm{d}\xi, \\
(\mathrm{D}^{-1}\mathbf{M})^{12}(z) &= \frac{1}{\pi}\int_{\mathbb{C}} \exp\{-\mathrm{i}[(\bar{\xi}-\bar{z})k + (\xi - z)\bar{k}]\}\,\frac{M^{12}(\xi)}{\bar{z}-\bar{\xi}}\,\mathrm{d}\xi, \\
(\mathrm{D}^{-1}\mathbf{M})^{21}(z) &= \frac{1}{\pi}\int_{\mathbb{C}} \exp\{-\mathrm{i}[(\xi - z)k + (\bar{z}-\bar{\xi})\bar{k}]\}\,\frac{M^{21}(\xi)}{z-\xi}\,\mathrm{d}\xi, \\
(\mathrm{D}^{-1}\mathbf{M})^{22}(z) &= \frac{1}{\pi}\int_{\mathbb{C}} \frac{M^{22}(\xi)}{\bar{z}-\bar{\xi}}\,\mathrm{d}\xi.
\end{aligned}$$

Beals and Coifman [1985b] prove two main results. First, if $\mathbf{Q} \in L^{2-\varepsilon}(\mathbb{R}^2) \cap L^{2+\varepsilon}(\mathbb{R}^2)$, for some $\varepsilon > 0$ then $\tilde{G}_k\mathbf{Q}$ is a compact map in the space of continuous matrix-valued functions. Also if the norms $\|\mathbf{Q}\|_{2-\varepsilon}$ and $\|\mathbf{Q}\|_{2+\varepsilon}$ are sufficiently small, then $\|\tilde{G}_k\mathbf{Q}\| < 1$, for all k, and so there is a unique bounded solution to the integral equation (5.4.34). Furthermore if $z\mathbf{Q} \in L^{2-\varepsilon}(\mathbb{R}^2) \cap L^{2+\varepsilon}(\mathbb{R}^2)$ then (i), as $\sqrt{x^2+y^2} \to \infty$

$$\mathbf{m}(x,y;k) = \mathbf{I} + \mathbf{E}^{-1}(z;k)\begin{pmatrix} z^{-1} & 0 \\ 0 & \bar{z}^{-1} \end{pmatrix} \frac{\hat{\mathbf{S}}(k)}{\pi} + o\left((x^2+y^2)^{-1/2}\right), \tag{5.4.35a}$$

where

$$S^{jj}(k) = \int_{-\infty}^{\infty}\int_{-\infty}^{\infty} (\mathbf{Qm})^{jj}(x,y;k)\,\mathrm{d}x\,\mathrm{d}y, \qquad j = 1,2, \tag{5.4.35b}$$

$$S^{12}(k) = \int_{-\infty}^{\infty}\int_{-\infty}^{\infty} \exp\{-\mathrm{i}(\bar{z}k + z\bar{k})\}(\mathbf{Qm})^{12}(x,y;k)\,\mathrm{d}x\,\mathrm{d}y, \tag{5.4.35c}$$

$$S^{21}(k) = \int_{-\infty}^{\infty}\int_{-\infty}^{\infty} \exp\{\mathrm{i}(zk + \bar{z}\bar{k})\}(\mathbf{Qm})^{21}(x,y;k)\,\mathrm{d}x\,\mathrm{d}y, \tag{5.4.35d}$$

and (ii), $\mathbf{m}(x,y;k)$ also satisfies the equation

$$\begin{aligned}
\frac{\partial \mathbf{m}}{\partial \bar{k}}(x,y;k) &= (\mathbf{Tm})(x,y;k) \qquad\qquad (5.4.36)\\
&\equiv \mathbf{m}(x,y;-\bar{k})\begin{pmatrix} 0 & S^{12}(k)\exp\{\mathrm{i}(\bar{z}k + z\bar{k})\} \\ S^{21}(k)\exp\{-\mathrm{i}(zk + \bar{z}\bar{k})\} & 0 \end{pmatrix}.
\end{aligned}$$

Conversely, the second main result is that if $S^{ij}(k) \in L^{2-\varepsilon}(\mathbb{C}) \cap L^{2+\varepsilon}(\mathbb{C})$ for some ε, and the norms $\|S^{ij}\|_{2-\varepsilon}$ and $\|S^{ij}\|_{2+\varepsilon}$ are sufficiently small, for $i \neq j$, then there exists a unique solution to the equation

$$\mathbf{m}(x,y;k) = \mathbf{I} + \frac{1}{2\pi\mathrm{i}} \int_{-\infty}^{\infty}\int_{-\infty}^{\infty} (\mathbf{Tm})(x,y;\zeta)\, \frac{\mathrm{d}\zeta \wedge \mathrm{d}\bar{\zeta}}{\zeta - k}.$$

Additionally, if $kS^{ij}(k) \in L^{2-\varepsilon}(\mathbb{C}) \cap L^{2+\varepsilon}(\mathbb{C})$, for $i \neq j$, then $\mathbf{m}(x,y;k)$ satisfies (5.4.33) with

$$q(x,y) = -\frac{1}{2\pi\mathrm{i}} \int_{-\infty}^{\infty}\int_{-\infty}^{\infty} (\mathbf{Tm})^{12}(x,y;\zeta)\, \mathrm{d}\zeta \wedge \mathrm{d}\bar{\zeta}, \tag{5.4.37a}$$

$$r(x,y) = -\frac{1}{2\pi\mathrm{i}} \int_{-\infty}^{\infty}\int_{-\infty}^{\infty} (\mathbf{Tm})^{21}(x,y;\zeta)\, \mathrm{d}\zeta \wedge \mathrm{d}\bar{\zeta}, \tag{5.4.37b}$$

where $q(x,y)$, $r(x,y)$ are the (off-diagonal) elements of the matrix $\mathbf{Q}$.

We note that Sung and Fokas [1991b] discuss rigorous aspects of the direct and inverse scattering problems associated with the $N \times N$ hyperbolic system (5.4.3) for $N \geq 2$.

5.5 The Davey-Stewartson equations.

5.5.1 Introduction. The general Davey-Stewartson (DS) equations are given by

$$\mathrm{i}q_t + \tfrac{1}{2}(\sigma^2 q_{xx} + q_{yy}) = (\phi - qr)q, \tag{5.5.1a}$$

$$-\mathrm{i}r_t + \tfrac{1}{2}(\sigma^2 r_{xx} + r_{yy}) = (\phi - qr)r, \tag{5.5.1b}$$

$$\phi_{xx} + \frac{1}{a}\phi_{yy} = -\frac{b}{a}(qr)_{xx}, \tag{5.5.1c}$$

where the physical water wave problem is obtained by taking $r = \pm q^*$, with q^* the complex conjugate of q. The parameters (σ^2, a, b) are functions of two dimensionless parameters $\tilde{h} \equiv kh$, $\tilde{T} \equiv k^2T/g$, where k is the wave number in the x-direction of the underlying wave train, h the depth, T the surface tension and g the gravity. In the general case a singularity can develop in finite time (see Ablowitz and Segur [1979]; Ghidaglia and Saut [1990]). The standard DS equations (i.e., the ones amenable to I.S.T.) result from a suitable limit ($kh \to 0$) of equations (5.5.1) in which case it is found that

$$\sigma^2 = -\frac{1}{a} = \pm 1, \qquad b = 2.$$

Hence we obtain the DS equations

$$\mathrm{i}q_t + \tfrac{1}{2}(\sigma^2 q_{xx} + q_{yy}) = (\phi - qr)q, \tag{5.5.2a}$$

$$-\mathrm{i}r_t + \tfrac{1}{2}(\sigma^2 r_{xx} + r_{yy}) = (\phi - qr)r, \tag{5.5.2b}$$

$$\phi_{xx} - \sigma^2\phi_{yy} = 2(qr)_{xx}, \tag{5.5.2c}$$

For DSI, $\sigma^2 = +1$ and for DSII, $\sigma^2 = -1$. The DS equations (5.5.2) also possess N-line soliton solutions (Anker and Freeman [1978]; Arkadiev, Pogrebkov and Polivanov [1989a,b]; Nakamura [1982, 1983]; Satsuma and Ablowitz [1979]); localized exponentially decaying solitons (Boiti, Leon, Martina and Pempinelli [1988b]); an infinite-dimensional symmetry group, in fact this involves an infinite dimensional Lie algebra with a Kac-Moody-Virasoro loop structure (Champagne and Winternitz [1988]; Omote [1988]); similarity reductions to the second and fourth Painlevé equations (Tajiri and Hagiwari [1983]); a Bäcklund transformation and the Painlevé property (Ganesan and Lakshmanan [1987]; Leo, Mancarella, Soliani and Solombrino [1988]); an infinite number of commuting symmetries and conservation quantities, a recursion operator and bi-Hamiltonian structure (Fokas and Santini [1988a]; Santini and Fokas [1988]).

The Lax pair for the DS equations (5.5.2) is

$$\boldsymbol{v}_x = -\sigma \mathbf{J} \boldsymbol{v}_y + \mathbf{Q}\boldsymbol{v}, \tag{5.5.3a}$$

$$\boldsymbol{v}_t = \mathbf{A}\boldsymbol{v} - (\mathrm{i}/\sigma)\mathbf{Q}\boldsymbol{v}_y + \mathrm{i}\mathbf{J}\boldsymbol{v}_{yy}, \tag{5.5.3b}$$

with

$$\mathbf{J} = \mathrm{diag}(1,-1), \qquad \mathbf{Q} = \begin{pmatrix} 0 & q(x,y) \\ r(x,y) & 0 \end{pmatrix}, \qquad \mathbf{A} = \begin{pmatrix} A^{11}(x,y) & A^{12}(x,y) \\ A^{21}(x,y) & A^{22}(x,y) \end{pmatrix},$$

and where the entries of $\mathbf{A}$ satisfy

$$A^{12} = \tfrac{1}{2}\mathrm{i}\sigma^2(\partial_x - \sigma\partial_y)q, \tag{5.5.4a}$$

$$A^{21} = -\tfrac{1}{2}\mathrm{i}\sigma^2(\partial_x + \sigma\partial_y)r, \tag{5.5.4b}$$

$$(\partial_x + \sigma\partial_y)A^{11} = -\tfrac{1}{2}\mathrm{i}\sigma^2(\partial_x - \sigma\partial_y)(qr), \tag{5.5.4c}$$

$$(\partial_x - \sigma\partial_y)A^{22} = \tfrac{1}{2}\mathrm{i}\sigma^2(\partial_x + \sigma\partial_y)(qr), \tag{5.5.4d}$$

with $\partial_x \equiv \partial/\partial x$, $\partial_y \equiv \partial/\partial y$ and the mean field ϕ in the DS equations (5.5.2) is determined from

$$\phi = \mathrm{i}(A^{11} - A^{22}) + qr. \tag{5.5.4e}$$

Seeking a fundamental matrix solution $\mathbf{\Phi}(x,y,t;k)$ of equations (5.5.3) in the form

$$\mathbf{\Phi}(x,y,t;k) = \mathbf{m}(x,y,t;k)\exp\{\mathrm{i}k(-x\sigma\mathbf{J} + y - kt\mathbf{J})\},$$

then $\mathbf{m}(x,y,t;k)$ satisfies

$$\mathbf{m}_x = \mathbf{Q}\mathbf{m} - \sigma\mathbf{J}\mathbf{m}_y + \mathrm{i}k\sigma[\mathbf{m},\mathbf{J}], \tag{5.5.5a}$$

$$\mathbf{m}_t = \mathrm{i}\mathbf{J}\mathbf{m}_{yy} - (\mathrm{i}/\sigma)\mathbf{Q}\mathbf{m}_y + \mathbf{A}\mathbf{m} - \mathrm{i}k^2[\mathbf{J},\mathbf{m}] - 2k\mathbf{J}\mathbf{m}_y + (k/\sigma)\mathbf{Q}\mathbf{m}. \tag{5.5.5b}$$

We remark that the two-dimensional nonlinear Schrödinger equation

$$\mathrm{i}q_t + \tfrac{1}{2}(q_{xx} + q_{yy}) \pm |q|^2 q = 0,$$

does *not* appear to be solvable by inverse scattering (cf., Ablowitz and Segur [1979]). Furthermore, the relationship between the number of spatial dimensions (2) is critical since the solution may develop a singularity at some finite time $t = t^*$ (cf., Landman, Le Mesurier., Papanicolaou, Sulem and Sulem [1989]; Landman, Papanicolaou, Sulem and Sulem [1988]; Le Mesurier, Papanicolaou, Sulem and Sulem [19488a,b]; Rasmussen and Rypdal [1986]; Rypdal and Rasmussen [1986]; Wood [1984]).

5.5.2 Inverse Scattering for the DSI Equations. The DSI equations are

$$\mathrm{i}q_t + \tfrac{1}{2}(q_{xx} + q_{yy}) = (\phi - qr)q, \tag{5.5.6a}$$
$$-\mathrm{i}r_t + \tfrac{1}{2}(r_{xx} + r_{yy}) = (\phi - qr)r, \tag{5.5.6b}$$
$$\phi_{xx} - \phi_{yy} = 2(qr)_{xx}, \tag{5.5.6c}$$

with $r = \pm q^*$, which are the compatibility conditions of equations (5.5.3) with $\sigma = -1$. Since we are dealing with a 2×2 system, the results of §5.4.1 may be simplified. We give the results in the following. Equation (5.5.5a) with $\sigma = -1$ becomes

$$m_x^{ij} - J^i m_y^{ij} - \mathrm{i}k(J^i - J^j)m^{ij} = \sum_{l=1}^{2} Q^{il} m^{lj}, \qquad 1 \le i, j \le 2,$$

with $J^1 = 1$, $J^2 = -1$, $Q^{12} = q$, $Q^{21} = r$, $Q^{11} = Q^{22} = 0$. To solve this it is convenient to make a change of variables to characteristic coordinates, $\xi = x + y$, $\eta = x - y$ (so $\partial_x = \partial_\xi + \partial_\eta$, $\partial_y = \partial_\xi - \partial_\eta$, with $\partial_\xi \equiv \partial/\partial\xi$ and $\partial_\eta \equiv \partial/\partial\eta$), and $m^{ij}(x, y; k) = \mu^{ij}(\xi, \eta; k)$ which yields

$$\partial_\eta \mu^{11}(\xi, \eta; k) = \tfrac{1}{2} q(\xi, \eta)\mu^{21}(\xi, \eta; k), \tag{5.5.7a}$$
$$(\partial_\eta - \mathrm{i}k)\mu^{12}(\xi, \eta; k) = \tfrac{1}{2} q(\xi, \eta)\mu^{22}(\xi, \eta; k), \tag{5.5.7b}$$
$$(\partial_\xi + \mathrm{i}k)\mu^{21}(\xi, \eta; k) = \tfrac{1}{2} r(\xi, \eta)\mu^{11}(\xi, \eta; k), \tag{5.5.7c}$$
$$\partial_\xi \mu^{22}(\xi, \eta; k) = \tfrac{1}{2} r(\xi, \eta)\mu^{12}(\xi, \eta; k). \tag{5.5.7d}$$

Furthermore, in characteristic coordinates, the DSI equations (5.5.6) become

$$\mathrm{i}q_t + q_{\xi\xi} + q_{\eta\eta} = (\phi - qr)q, \tag{5.5.8a}$$
$$-\mathrm{i}r_t + r_{\xi\xi} + r_{\eta\eta} = (\phi - qr)r, \tag{5.5.8b}$$
$$2\phi_{\xi\eta} = (qr)_{\xi\xi} + 2(qr)_{\xi\eta} + (qr)_{\eta\eta}, \tag{5.5.8c}$$

We require solutions of equations (5.5.7) which are analytic in the complex k-plane; in fact the eigenfunctions $\mu_\pm^{ij}(\xi, \eta; k)$ satisfy the Volterra integral equations

$$\mu_\pm^{11}(\xi, \eta; k) = 1 + \tfrac{1}{2}\int_{-\infty}^{\eta} q(\xi, \eta')\mu_\pm^{21}(\xi, \eta'; k)\,\mathrm{d}\eta',$$

$$\mu_\pm^{12}(\xi,\eta;k) = \tfrac{1}{2}\int_{\mp\infty}^{\eta} q(\xi,\eta')\mu_\pm^{22}(\xi,\eta';k)\exp\{ik(\eta-\eta')\}\,d\eta',$$
$$\mu_\pm^{21}(\xi,\eta;k) = -\tfrac{1}{2}\int_{\xi}^{\pm\infty} r(\xi',\eta)\mu_\pm^{11}(\xi',\eta;k)\exp\{-ik(\xi-\xi')\}\,d\xi',$$
$$\mu_\pm^{22}(\xi,\eta;k) = 1 + \tfrac{1}{2}\int_{-\infty}^{\xi} r(\xi',\eta)\mu_\pm^{12}(\xi',\eta;k)\,d\xi'.$$

The eigenfunctions $\mu_+^{ij}(\xi,\eta;k)$ and $\mu_-^{ij}(\xi,\eta;k)$ are related through the scattering equations

$$(\boldsymbol{\mu}_+^1 - \boldsymbol{\mu}_-^1)(\xi,\eta;k) = \int_{-\infty}^{\infty} S_2(k,l)\exp\{-i(l\xi+k\eta)\}\boldsymbol{\mu}_+^2(\xi,\eta;l)\,dl, \qquad (5.5.9a)$$
$$(\boldsymbol{\mu}_+^2 - \boldsymbol{\mu}_-^2)(\xi,\eta;k) = -\int_{-\infty}^{\infty} S_1(k,l)\exp\{i(l\xi+k\eta)\}\boldsymbol{\mu}_-^1(\xi,\eta;l)\,dl, \qquad (5.5.9b)$$

where $\boldsymbol{\mu}_\pm^j := (\mu_\pm^{1j},\mu_\pm^{2j})$, for $j=1,2$, and the scattering data is defined by

$$S_2(k,l) = -\frac{1}{4\pi}\int_{-\infty}^{\infty}\int_{-\infty}^{\infty} r(\xi,\eta)\mu_+^{11}(\xi,\eta;k)\exp\{i(k\eta+l\xi)\}\,d\xi\,d\eta,$$
$$S_1(k,l) = \frac{1}{4\pi}\int_{-\infty}^{\infty}\int_{-\infty}^{\infty} q(\xi,\eta)\mu_-^{22}(\xi,\eta;k)\exp\{-i(k\eta+l\xi)\}\,d\xi\,d\eta.$$

It is easily shown that

$$S_1(k,l) = -T_+^{12}(k,l) = f^{21}(k,l), \qquad S_2(k,l) = T_-^{21}(k,l) = f^{12}(k,l)$$

(recall equations (5.4.11c,12) above). Equations (5.5.9) define a Riemann-Hilbert problem, from whose solution it can be shown that

$$q(\xi,\eta) = \frac{1}{\pi}\int_{-\infty}^{\infty}\int_{-\infty}^{\infty} S_1(k,l)\exp\{i(k\xi+l\eta)\}\mu_-^{11}(\xi,\eta;l)\,dk\,dl,$$
$$r(\xi,\eta) = \frac{1}{\pi}\int_{-\infty}^{\infty}\int_{-\infty}^{\infty} S_2(k,l)\exp\{-i(k\xi+l\eta)\}\mu_+^{22}(\xi,\eta;l)\,dk\,dl.$$

Hence, assuming that $A^{11}(x,y)$ and $A^{22}(x,y)$ tend to zero as $R=\sqrt{x^2+y^2}\to\infty$, the initial value problem for the DSI equations can be solved through the methodology of §5.4 above, where $\mathbf{f}(k,l)$ is determined from equation (5.4.12) and the time evolution of $\mathbf{f}(k,l,t)$ is given by

$$\mathbf{f}(k,l,t) = \exp\{-ik^2t\mathbf{J}\}\mathbf{f}(k,l,0)\exp\{il^2t\mathbf{J}\}, \qquad (5.5.10a)$$

with $\mathbf{J} = \mathrm{diag}(1,-1)$. Therefore the time evolution of the scattering data is given by

$$S_1(k,l;t) \equiv f^{21}(k,l;t) = S_1(k,l;0)\exp\{i(k^2+l^2)t\}, \qquad (5.5.10b)$$
$$S_2(k,l;t) \equiv f^{12}(k,l;t) = S_2(k,l;0)\exp\{-i(k^2+l^2)t\}. \qquad (5.5.10c)$$

Equation (5.5.10a) may be derived by applying the operator

$$\mathrm{M}_k := \frac{\partial}{\partial t} - \mathbf{A} + k\mathbf{Q} + \mathrm{i}k^2\mathbf{J} + (2k\mathbf{J} - \mathrm{i}\mathbf{Q})\frac{\partial}{\partial y} - \mathrm{i}\mathbf{J}\frac{\partial^2}{\partial y^2},$$

to the scattering equation (5.4.11) and using equation (5.5.5b) together with

$$\mathrm{M}_k\mathbf{m} = \mathrm{i}k^2\mathbf{m}\mathbf{J}, \qquad \mathrm{M}_k = \mathrm{M}_l + \mathrm{i}(k^2 - l^2) + (k - l)\mathbf{Q} + 2(k - l)\mathbf{J}\frac{\partial}{\partial y}.$$

Schultz and Ablowitz [1989a] show that each of the solutions of DSI that has been found using the I.S.T. formalism due to Fokas and Ablowitz [1983e] corresponds to a different choice of boundary conditions for the quantities $A^{11}(x,y)$ and $A^{22}(x,y)$ which appear in the matrix $\mathbf{A}$ [recall equation (5.5.3b)]. If $\sigma = -1$ (or more generally, if σ is real), there are, strictly speaking, an infinite number of choices which Schultz and Ablowitz [1989a] show correspond to the boundary conditions for the solution of the underlying linear problem. In the case when $r = q^*$ (that is the DS case), the naive choice $A^{11}(x,y) \to 0$ and $A^{22}(x,y) \to 0$ as $R = \sqrt{x^2 + y^2} \to \infty$, corresponds to the trivial solution $|q(x,y)| = 0$. Schultz and Ablowitz [1989a] also show that the general solution for DSI (5.5.6) for arbitrary boundary conditions for $A^{11}(x,y)$ and $A^{22}(x,y)$ remains an open problem. We shall discuss this further in §5.5.5 below.

An I.S.T. scheme for DSI has been developed in the $\bar{\partial}$ limit case by Nachman and Ablowitz [1984b] (see also Schultz and Ablowitz [1989a] and §5.5.5 below). I.S.T. schemes for DSI are also discussed by Boiti, Leon and Pempinelli [1989b, 1990a]; Fokas and Santini [1989, 1990]; Santini [1990].

5.5.3 Inverse Scattering for the DSII Equations. The DSII equations are

$$\mathrm{i}q_t + \tfrac{1}{2}(-q_{xx} + q_{yy}) = (\phi + qr)q, \tag{5.5.11a}$$
$$-\mathrm{i}r_t + \tfrac{1}{2}(-r_{xx} + r_{yy}) = (\phi + qr)r, \tag{5.5.11b}$$
$$\phi_{xx} + \phi_{yy} = -2(qr)_{xx}, \tag{5.5.11c}$$

with $r = \pm q^*$, which are the compatibility conditions of equations (5.5.3) with $\sigma = \mathrm{i}$. Equation (5.5.5a) with $\sigma = \mathrm{i}$ becomes

$$m_x^{ij} + \mathrm{i}J^i m_y^{ij} + k(J^i - J^j)m^{ij} = \sum_{l=1}^{2} Q^{il}m^{lj}, \qquad 1 \le i,j \le 2,$$

with $J^1 = 1$, $J^2 = -1$, $Q^{12} = q$, $Q^{21} = r$, $Q^{11} = Q^{22} = 0$.

Since the associated scattering problem for DSII is elliptic, it is natural to make the change of variable $z = x + \mathrm{i}y$, $\bar{z} = x - \mathrm{i}y$ so that $\partial_x = \partial_z + \bar{\partial}_z$, $\partial_y = \mathrm{i}(\partial_z - \bar{\partial}_z)$, hence

$m^{ij}(x,y;k) = \mu^{ij}(z,\bar{z};k)$ satisfy

$$\begin{aligned}\partial_{\bar{z}}\mu^{11}(z,\bar{z};k) &= \tfrac{1}{2}q(z,\bar{z})\mu^{21}(z,\bar{z};k),\\ (\partial_{\bar{z}} - k)\mu^{12}(z,\bar{z};k) &= \tfrac{1}{2}q(z,\bar{z})\mu^{22}(z,\bar{z};k),\\ (\partial_z + k)\mu^{21}(z,\bar{z};k) &= \tfrac{1}{2}r(z,\bar{z})\mu^{11}(z,\bar{z};k),\\ \partial_z\mu^{22}(z,\bar{z};k) &= \tfrac{1}{2}r(z,\bar{z})\mu^{12}(z,\bar{z};k),\end{aligned}$$

and the DSII equations (5.5.11) take the form

$$\mathrm{i}q_t = q_{zz} + q_{\bar{z}\bar{z}} + (\phi + qr)q, \tag{5.5.12a}$$

$$-\mathrm{i}r_t = r_{zz} + r_{\bar{z}\bar{z}} + (\phi + qr)r, \tag{5.5.12b}$$

$$2\phi_{z\bar{z}} + (qr)_{zz} + 2(qr)_{z\bar{z}} + (qr)_{\bar{z}\bar{z}} = 0. \tag{5.5.12c}$$

The I.S.T. formalism for DSII is completely analogous to that for KPII discussed in §5.3 above and is a special case of the analysis in §5.4.2.

Hence, assuming that $A^{11}(x,y)$ and $A^{22}(x,y)$ tend to zero $R = \sqrt{x^2+y^2} \to \infty$, the initial value problem for the DSII equations can be solved through the methodology of §5.4.2. The time evolution of the scattering data $\mathbf{w}^{ij}(x,y,t;k_R,k_I)$ [recall equations (5.4.23,24)] is given by

$$\mathbf{w}^{ij}(x,y,t;k_R,k_I) = \exp\{-\mathrm{i}\kappa_{ij}^2 t\mathbf{J}\}\mathbf{w}^{ij}(x,y,0;k_R,k_I)\exp\{\mathrm{i}k^2 t\mathbf{J}\}, \tag{5.5.13}$$

where $\kappa_{ij} := k_R + \mathrm{i}J^iJ^jk_I$. To derive this, define the linear operator

$$\mathrm{M}_k := \frac{\partial}{\partial t} - (\mathbf{A} + k\mathbf{Q} + \mathrm{i}k^2\mathbf{J}) + (\mathbf{Q} + 2\mathrm{i}k\mathbf{J})\frac{\partial}{\partial y} - \mathrm{i}\mathbf{J}\frac{\partial^2}{\partial y^2}, \tag{5.5.14a}$$

then (5.5.5b), with $\sigma = \mathrm{i}$, yields

$$\mathrm{M}_k\mathbf{m} + \mathrm{i}k^2\mathbf{mJ} = 0, \qquad \mathrm{M}_k\frac{\partial \mathbf{m}}{\partial \bar{k}} + \mathrm{i}k^2\frac{\partial \mathbf{m}}{\partial \bar{k}}\mathbf{J} = 0. \tag{5.5.14b}$$

Applying the operator M_k to equation (5.4.24) and using (5.5.14b), gives (5.5.13)

I.S.T. schemes for DSII are also discussed by Arkadiev, Pogrebkov and Polivanov [1989b]; Beals and Coifman [1985, 1989]; Konopelchenko and Matkarimov [1990a].

A one-soliton solution of the DSII equations was derived by Arkadiev, Pogrebkov and Polivanov [1989b] from the inverse scattering formalism and is given by

$$q(x,y,t) = \frac{2\bar{\nu}\exp\{\lambda(x+\mathrm{i}y) - \bar{\lambda}(x-\mathrm{i}y) - \mathrm{i}(\lambda^2+\bar{\lambda}^2)t\}}{|(x+\mathrm{i}y) + \mu - 2\mathrm{i}\lambda t|^2 - |\nu|^2},$$

where with λ, μ and ν arbitrary (complex) constants. We note that this solution decays like $(x^2+y^2)^{-1}$ as $x,y \to \infty$ whereas the solition solution derived by Fokas and Ablowitz [1984] decays like $(x^2+y^2)^{-1/2}$. (Arkadiev, Pogrebkov and Polivanov [1989a] derived rational solutions of the DSII equations, which are special cases of the above solution.)

5.5.4 The Strong Coupling Limit. The strong coupling limit of the general DS equations (5.5.1) is obtained by making the transformation

$$q \to q/\varepsilon, \qquad r \to r/\varepsilon, \qquad \phi \to \phi/\varepsilon^2, \qquad t \to \varepsilon^2 t,$$

and letting $\varepsilon \to 0$. In this limit equations (5.5.1) reduce to

$$\mathrm{i}q_t = (\phi - qr)q, \tag{5.5.15a}$$

$$-\mathrm{i}r_t = (\phi - qr)r, \tag{5.5.15b}$$

$$\phi_{xx} + \frac{1}{a}\phi_{yy} = -\frac{b}{a}(qr)_{xx}. \tag{5.5.15c}$$

These equations shed additional light on the nonlocal nature of the Davey-Stewartson problem. The general solution of equations (5.5.15) is readily obtained. Multiplying (5.5.15a) by r, (5.5.15b) by q and subtracting yields

$$\frac{\partial}{\partial t}\left[q(x,y,t)\,r(x,y,t)\right] = 0,$$

hence

$$q(x,y,t)\,r(x,y,t) = q(x,y,0)\,r(x,y,0) \equiv q_0(x,y)r_0(x,y). \tag{5.5.16}$$

We shall be primarily concerned with the boundary value problem of $q \to 0$, $r \to 0$ as $R = \sqrt{x^2+y^2} \to \infty$. Equation (5.5.16) implies that

$$\phi(x,y,t) = \phi(x,y,0) \equiv \phi_0(x,y).$$

Thus equation (5.5.15c) can be solved by elementary methods, i.e.

$$\phi(x,y,t) = \phi_0(x,y) = -\frac{b}{a}\int_{-\infty}^{\infty}\int_{-\infty}^{\infty} G(x-\xi, y-\eta)\left[q_0(\xi,\eta)r_0(\xi,\eta)\right]_{\xi\xi}\,\mathrm{d}\xi\,\mathrm{d}\eta, \tag{5.5.17}$$

where the Green's function $G(x,y)$ satisfies

$$G_{xx} + \sigma' G_{y'y'} = \delta(x)\,\delta(y'), \tag{5.5.18}$$

with $y' = \sqrt{|a|}\,y$ and $\sigma' = \operatorname{sgn}(a)$. For $a > 0$, $\sigma' = +1$ (relevant for the strong coupling limit of DSII), the usual Green's function for the Laplacian is satisfactory; i.e.,

$$G(x,y) = \frac{1}{4\pi}\ln(x^2+y^2),$$

corresponding to the boundary value $\phi \to 0$ as $R \to \infty$. However for $a < 0$, $\sigma' = -1$ (relevant for the strong coupling limit of DSI), there are other natural Green's functions which arise:

$$G(x,y) = \tfrac{1}{4}\theta(x-y)\,\theta(x+y), \tag{5.5.19a}$$

$$G(x,y) = \frac{1}{4\pi i}\ln(x^2-y^2) + \tfrac{1}{4}\left[\theta(x+y)-\theta(-x-y)\right]\left[\theta(x-y)-\theta(-x+y)\right], \tag{5.5.19b}$$

$$G(x,y) = \tfrac{1}{4}\left[\theta(x+y)-\theta(-x-y)\right]\left[\theta(x-y)-\theta(-x+y)\right], \tag{5.5.19c}$$

where $\theta(\bullet)$ is the Heaviside function. Case (a) corresponds to the Riemann-Hilbert formulation of the I.S.T. solution of equation (5.5.1), as discussed by Fokas and Ablowitz [1983e] and Kulish and Lipovsky [1988]; case (b) to the "$\bar\partial$ limit" approach of Nachman and Ablowitz [1984b] and Schultz and Ablowitz [1989a]; and case (c) corresponds to a "symmetric" case of (5.5.1).

The general solution of (5.5.15) is written in the following form

$$q(x,y,t) = q(x,y,0)\exp\left\{-i\psi_0(x,y)\,t\right\}, \tag{5.5.20a}$$

$$r(x,y,t) = r(x,y,0)\exp\left\{i\psi_0(x,y)\,t\right\}, \tag{5.5.20b}$$

where

$$\psi_0(x,y) = \phi_0(x,y) - q_0(x,y)r_0(x,y), \tag{5.5.20c}$$

and $\phi_0(x,y)$ is given by (5.5.17).

The strong coupling limit solution is both illuminating and potentially very useful for a numerical implementation of a split step Fourier scheme for the DS equations.

The Hamiltonian structure of the more general evolution equation (5.5.1), is only completely understood in case (b) (Schultz and Ablowitz [1989a] — see also §5.5.5 below), though here the Green's function $G(x,y)$ is seen to have an imaginary homogeneous part, and we can not in general let $r = q^*$. The Hamiltonian structure in case (a) is not yet well-understood, even in the strong coupling limit. It is straightforward to write down the Hamiltonian generating the equations of motion (5.5.1) as well as its limit (2), corresponding to case (c). Finding an I.S.T. scheme in this "symmetric" case is still an open problem.

Depending on the boundary conditions [i.e., the symmetry properties of the Green's function (5.5.18)], the following conserved quantity may be used as the Hamiltonian for the system (5.5.15):

$$H = -\tfrac{1}{2}i\int_{-\infty}^{\infty}\int_{-\infty}^{\infty} q(x,y,t)\,r(x,y,t)\,\psi_0(x,y)\,dx\,dy, \tag{5.5.21}$$

where $\psi_0(x,y)$ is given by (5.5.20c). Clearly H is constant in time, since qr and ψ_0 are constants of the motion. We employ the natural Poisson brackets

$$\{f,g\} = \int_{-\infty}^{\infty}\int_{-\infty}^{\infty}\left[\frac{\delta f}{\delta q(x,y)}\frac{\delta g}{\delta r(x,y)} - \frac{\delta f}{\delta r(x,y)}\frac{\delta g}{\delta q(x,y)}\right] dx\,dy. \tag{5.5.22}$$

Using the fact that

$$\frac{\delta q(x',y')}{\delta q(x,y)} = \frac{\delta r(x',y')}{\delta r(x,y)} = \delta(x-x')\,\delta(y-y'),$$
$$\frac{\delta q(x',y')}{\delta r(x,y)} = \frac{\delta r(x',y')}{\delta q(x,y)} = 0,$$

we immediately have that (5.5.15) is given by

$$q_t = \{q, H\}, \qquad r_t = \{r, H\}, \tag{5.5.23}$$

whenever the Green's function $G(x,y)$ [i.e., the solution of equation (5.5.18)] has the symmetry property $G(x,y) = G(-x,-y)$.

Equations (5.5.23) demonstrate the Hamiltonian nature of the flow. The above definitions imply the canonical Poisson structure:

$$\{q(x,y), q(x',y')\} = 0,$$
$$\{r(x,y), r(x',y')\} = 0,$$
$$\{q(x,y), r(x',y')\} = \delta(x-x')\,\delta(y-y').$$

These relationships also show that the following quantities are also canonical variables:

$$P(x,y,t) = q(x,y,t)r(x,y,t), \tag{5.5.24a}$$
$$Q(x,y,t) = \tfrac{1}{2}\ln\left[r(x,y,t)/q(x,y,t)\right]. \tag{5.5.24b}$$

Namely we have that

$$\{P(x,y), P(x',y')\} = 0,$$
$$\{Q(x,y), Q(x',y')\} = 0,$$
$$\{P(x,y), Q(x',y')\} = \delta(x-x')\,\delta(y-y').$$

Moreover from the solution to equation (5.5.15) [i.e., equations (5.5.16,19)]

$$P(x,y,t) = q_0(x,y)r_0(x,y), \tag{5.5.25a}$$
$$Q(x,y,t) = \mathrm{i}\psi_0(x,y)\,t + \tfrac{1}{2}\ln\left[r(x,y,t)/q(x,y,t)\right], \tag{5.5.25b}$$

i.e., $P_t = 0$, and $Q(x,y,t)$ flows linearly in time. Hence (5.5.24) are the action-angle variables.

5.5.5 The $\bar{\partial}$-limit Case. In this section, following Schultz and Ablowitz [1989a], we establish the complete integrability (in the Hamiltonian sense) of the DSI equations (5.5.6) for a particular choice of boundary conditions, which we refer to as the $\bar{\partial}$-limit

Case. This case was solved by I.S.T. by Nachman and Ablowitz [1984b] by taking a limit as $\sigma \to -1$ of the I.S.T. equations of the more general $\bar{\partial}$ problem. We also give the canonical action-angle variables in this case.

First we discuss I.S.T. method for DSI (5.5.6) in the $\bar{\partial}$-limit Case. Nachman and Ablowitz [1984b] construct a family of solutions to equations (5.5.3) which arise as a special limiting case of the solution of the more general $\bar{\partial}$ problem given by

$$\Phi^{ij}(\xi,\eta;k) = \mu^{ij}(\xi,\eta;k)\exp\{\mathrm{i}J^j(k_R\zeta_j - k_R^2 t)\}, \tag{5.5.26}$$

where $\mu^{ij}(\xi,\eta;k)$ satisfies the integral equation

$$\mu^{ij}(\xi,\eta;k) = \delta^{ij} + \int_{-\infty}^{\infty}\int_{-\infty}^{\infty} G^{ij}(\xi-\xi',\eta-\eta';k)[\mathbf{Q}(\xi',\eta')\boldsymbol{\mu}(\xi',\eta';k)]^{ij}\,\mathrm{d}\xi'\,\mathrm{d}\eta', \tag{5.5.27}$$

and the differential equations,

$$\frac{\partial\mu^{ij}}{\partial\zeta_{\bar{i}}} - \mathrm{i}k_R(J^i - J^j)\mu^{ij} = [\mathbf{Q}\boldsymbol{\mu}]^{ij}, \tag{5.5.28}$$

where the Green's function is

$$\begin{aligned} G^{ij}(\xi,\eta;k) = \frac{1}{4\pi}\int_{-\infty}^{\infty} &\exp\{\mathrm{i}[\ell + \kappa_R^{ij})J^i\zeta_i - k_R J^j\zeta_j]\} \\ &\times [\theta(\xi+\eta)\theta(-J^i\ell) - \theta(-\xi-\eta)\theta(J^i\ell)]\,\mathrm{d}\ell, \end{aligned} \tag{5.5.29}$$

with

$$\frac{\partial G^{ij}}{\partial\zeta_{\bar{i}}} - \mathrm{i}k_R(J^i - J^j)G^{ij} = \delta(\zeta_i)\delta(\zeta_{\bar{i}}),$$

where

$$\begin{aligned} \zeta_1 &= \zeta_{\bar{2}} = \xi = x + y, \\ \zeta_2 &= \zeta_{\bar{1}} = \eta = x - y, \\ \kappa_R^{ij} &= k_I + J^iJ^j(k_R - k_I), \qquad \kappa_I^{ij} = k_I, \end{aligned}$$

and $\theta(\bullet)$ is the Heaviside function.

Applying $\partial/\partial k_I$ to both sides of equation (5.5.27) we find that $\partial\mu^{ij}/\partial k_I$ can be written as

$$\begin{aligned} \frac{\partial\mu^{ij}}{\partial k_I}(\xi,\eta;k) = \frac{1}{2\pi}\sum_{p=1}^{2}&(J^p - J^j)\mu^{ip}(\xi,\eta;\kappa^{pj}) \\ &\times \exp(\mathrm{i}\kappa^{pj}J^p\zeta_p)T^{pj}(k)\exp(-\mathrm{i}k_R J^j\zeta_j), \end{aligned} \tag{5.5.30}$$

with

$$T^{ij}(k) = \frac{1}{2}\int_{-\infty}^{\infty}\int_{-\infty}^{\infty} \exp(-\mathrm{i}\kappa_R^{ij}J^i\zeta^i)[\mathbf{Q}(\xi,\eta)\boldsymbol{\mu}(\xi,\eta;k)]^{ij}\exp(\mathrm{i}k_R J^j\zeta^j)\,\mathrm{d}\xi\,\mathrm{d}\eta, \tag{5.5.31}$$

which are the scattering data. Then a special limit of the general Cauchy integral formula is applied to equation (5.5.30) to give the following integral equation

$$\mu^{ij}(\xi,\eta;k) = \delta^{ij} + \frac{1}{2\pi \mathrm{i}}\int_{-\infty}^{\infty}\int_{-\infty}^{\infty}\left(\frac{\theta(k_I-\ell_I)}{\ell_R-k_R-\mathrm{i}0}+\frac{\theta(\ell_I-k_I)}{\ell_R-k_R+\mathrm{i}0}\right) \times \frac{\partial \mu^{ij}}{\partial \ell_I}(\xi,\eta;\ell)\,\mathrm{d}\ell_R\,\mathrm{d}\ell_I. \tag{5.5.32}$$

Finally the potential $\mathbf{Q}$ is reconstructed through

$$\mathbf{Q}(\xi,\eta) = \left[\mathbf{J}, \frac{1}{2\pi}\int_{-\infty}^{\infty}\int_{-\infty}^{\infty}\frac{\partial \boldsymbol{\mu}}{\partial \ell_I}(\xi,\eta;\ell)\,\mathrm{d}\ell_R\,\mathrm{d}\ell_I\right]. \tag{5.5.33}$$

The inverse formulae (5.5.32,33) are derived at a fixed time (e.g., $t = 0$), but can easily be generalized to any $t \neq 0$ through the (simple) evolution of the scattering data. Suppose that equation (5.5.27) is valid for all t, then from equations (5.5.26), (5.5.3b) and (5.5.30)

$$T^{ij}(k,t) = T^{ij}(k,0)\exp\{-\mathrm{i}[J^i(\kappa_R^{ij})^2 - J^j k_R^2]t\}. \tag{5.5.34}$$

In the above analysis, it is assumed that the integral equations (5.5.27), satisfied by $\boldsymbol{\mu}$, are valid at an arbitrary time t. We now show that this assumption imposes a restriction on A^{11} and A^{22} in the matrix $\mathbf{A}$ [recall equation (5.5.3b)]. Using (5.5.26), with $\mathbf{m}(x,y;k) = \boldsymbol{\mu}(\xi,\eta;k)$, in (5.5.3b) yields

$$\mathbf{m}_t = \mathrm{i}\mathbf{J}\mathbf{m}_{yy} - (\mathrm{i}/\sigma)\mathbf{Q}\mathbf{m}_y + \mathbf{A}\mathbf{m} - \mathrm{i}k_R^2[\mathbf{J},\mathbf{m}] - 2k_R\mathbf{J}\mathbf{m}_y + (k_R/\sigma)\mathbf{Q}\mathbf{m}. \tag{5.5.35}$$

Of course, the DS equations (5.5.2) are the compatibility condition for equations (5.5.28) and (5.5.35); however, when equation (5.5.28) is written as an integral equation (as is necessary when implementing the I.S.T. formalism), certain boundary conditions are imposed upon $\boldsymbol{\mu}$. Thus we need to insure that the compatibility of the linear integral equation (5.5.27) together with equation (5.5.35) and this imposes constraints on A^{11} and A^{22}. To demonstrate this, expand $\mu^{jj}(\xi,\eta;k)$, for $j = 1,2$, in inverse powers of k_R

$$\mu^{jj}(\xi,\eta;k) = 1 + \sum_{n=1}^{\infty}\frac{\mu_j^{(n)}(\xi,\eta)}{k_R^n}, \qquad j = 1,2. \tag{5.5.36}$$

The off-diagonal elements are given by

$$\mu^{12}(\xi,\eta;k) = \frac{2}{r(\xi,\eta)}\,\partial_\xi \mu^{22}(\xi,\eta;k),$$
$$\mu^{21}(\xi,\eta;k) = \frac{2}{q(\xi,\eta)}\,\partial_\eta \mu^{11}(\xi,\eta;k).$$

By substituting (5.5.36) in equations (5.5.27) and (5.5.28), the coefficients $\mu_j^{(n)}(\xi,\eta)$ can be determined. In particular,

$$\mu_1^{(1)}(\xi,\eta) = \frac{1}{8\mathrm{i}}\int_{-\infty}^{\infty}\int_{-\infty}^{\infty} G^{11}(\xi-\xi',\eta-\eta')q(\xi',\eta')r(\xi',\eta')\,\mathrm{d}\xi'\,\mathrm{d}\eta', \quad (5.5.37\mathrm{a})$$

$$\mu_2^{(1)}(\xi,\eta) = \frac{\mathrm{i}}{8}\int_{-\infty}^{\infty}\int_{-\infty}^{\infty} G^{22}(\xi-\xi',\eta-\eta')q(\xi',\eta')r(\xi',\eta')\,\mathrm{d}\xi'\,\mathrm{d}\eta', \quad (5.5.37\mathrm{b})$$

where the Green's functions $G^{11}(\xi,\eta)$ and $G^{22}(\xi,\eta)$ are given by equation (5.5.29) above (note that these are independent of k). Then substituting equations (5.5.36) in (5.5.35) yields

$$A^{11}(\xi,\eta) = \partial_\xi \mu_1^{(1)}(\xi,\eta), \qquad A^{22}(\xi,\eta) = \partial_\eta \mu_2^{(1)}(\xi,\eta). \quad (5.5.38)$$

Consequently, substituting (5.5.37) in (5.5.38) yields explicit expressions for $A^{11}(\xi,\eta)$ and $A^{22}(\xi,\eta)$. These clearly satisfy the original differential equations for $A^{11}(\xi,\eta)$ and $A^{22}(\xi,\eta)$; furthermore, we see that the choice of boundary conditions for $A^{11}(\xi,\eta)$ and $A^{22}(\xi,\eta)$ is intimately related to the choice of boundary conditions for $\boldsymbol{\mu}(\xi,\eta)$, that is, to the choice of the diagonal elements for the Green functions $G^{11}(\xi,\eta)$ and $G^{22}(\xi,\eta)$ in equation (5.5.27). In the $\bar{\partial}$-limit Case which we are presently interested in, we find that

$$\begin{aligned} A^{11}(\xi,\eta) = \frac{1}{8\pi\mathrm{i}}\int_{-\infty}^{\infty}\int_{-\infty}^{\infty}\int_{-\infty}^{\infty} & \exp\{\mathrm{i}\ell(\xi-\xi')\}[q(\xi',\eta')r(\xi',\eta')]_{\xi'} \\ & \times[\theta(\xi+\eta-\xi'-\eta')\theta(-\ell)-\theta(\xi'+\eta'-\xi-\eta)\theta(\ell)]\,\mathrm{d}\xi'\,\mathrm{d}\eta'\,\mathrm{d}\ell, \end{aligned} \quad (5.5.39\mathrm{a})$$

$$\begin{aligned} A^{22}(\xi,\eta) = \frac{\mathrm{i}}{8\pi}\int_{-\infty}^{\infty}\int_{-\infty}^{\infty}\int_{-\infty}^{\infty} & \exp\{-\mathrm{i}\ell(\eta-\eta')\}[q(\xi',\eta')r(\xi',\eta')]_{\eta'} \\ & \times[\theta(\xi+\eta-\xi'-\eta')\theta(-\ell)-\theta(\xi'+\eta'-\xi-\eta)\theta(\ell)]\,\mathrm{d}\xi'\,\mathrm{d}\eta'\,\mathrm{d}\ell, \end{aligned} \quad (5.5.39\mathrm{b})$$

and also

$$\begin{aligned} \phi(x,y) = \frac{1}{8}\int_{-\infty}^{\infty}\int_{-\infty}^{\infty} & [\theta(x-x'+y-y')-\theta(-x+x'-y+y')][\theta(x-x'-y+y') \\ & -\theta(-x+x'+y-y')]\partial_{x'}^2[q(x',y')r(x',y')]\,\mathrm{d}x'\,\mathrm{d}y' \\ -\frac{\mathrm{i}}{4\pi}\int_{-\infty}^{\infty}\int_{-\infty}^{\infty} & \ln|(x-x')^2-(y-y')^2|\{(\partial_{x'}^2+\partial_{y'}^2)[q(x',y')r(x',y')]\}\,\mathrm{d}x'\,\mathrm{d}y' \end{aligned}$$

[recall equation (5.5.4e)]. We note that because $A^{11}(\xi,\eta)$ and $A^{22}(\xi,\eta)$ are purely imaginary, then, in general, $r \neq q^*$ for the $\bar{\partial}$-limit Case.

The result (5.5.38) for the DS equations (5.5.2), related to the generalized 2×2 scattering problem in $2+1$-dimensions. However, such relationships between nonlocal terms in the nonlinear problem and the solution of the underlying linear problem also appear for other multidimensional nonlinear evolution equations and seem to be a generic and novel feature of integrable systems in higher dimensions (in comparison with the situation in $1+1$-dimensions). It is important to take (5.5.38) into account when considering the Hamiltonian structure of the DS equations.

In the Hamiltonian formulation of the $\bar{\partial}$-limit Case, where A^{11} and A^{22} are given in terms of q and r by equation (5.5.39), equations (5.5.6) can be viewed as Hamilton's equations of motion for q and r

$$q_t = \{q, H\}, \qquad r_t = \{r, H\},$$

where the Poisson bracket structure is given by (5.5.22) and the Hamiltonian is

$$H = \tfrac{1}{2}\mathrm{i}\int_{-\infty}^{\infty}\int_{-\infty}^{\infty} \left[(q_{xx}+q_{yy})r - \mathrm{i}qr(A^{11}-A^{22})\right]\, \mathrm{d}x\, \mathrm{d}y. \tag{5.5.40}$$

In $1+1$-dimensions, equations solvable by an I.S.T. method are found to have an associated infinite number of constants of motion. In fact, the scattering data is found to be closely related to action-angle variables of the theory and can be used to construct the conserved quantities. The relationship between the infinite set of conserved quantities and the scattering data is given by the trace-formula (first found by Zakharov and Shabat [1971] for the KdV equation). Schultz and Ablowitz [1989b] generalized these ideas to higher dimensions, identifying canonical action-angle variables and deriving the trace formula for the $\bar{\partial}$-limit Case of equations (5.5.2).

Schultz, Ablowitz and BarYaacov [1987] show that the Poisson bracket relations for the scattering data were given by

$$\{T^{12}(k), T^{12}(\ell)\} = 0, \tag{5.5.41a}$$
$$\{T^{21}(k), T^{21}(\ell)\} = 0, \tag{5.5.41b}$$
$$\{T^{12}(k), T^{21}(\ell)\} = 4\pi^2\delta(k_R - \ell_R^{ij})\delta(\lambda_R^{ij} - \ell_R), \tag{5.5.41c}$$

with $\lambda_R^{ij} = \ell_I + J^i J^j(\ell_R - \ell_I)$, $\lambda_I^{ij} = \ell_I$. Defining new variables

$$P(k) = \tfrac{1}{4}\pi^{-2}T^{12}(k)T^{21}(\kappa_R^{ij}), \tag{5.5.42a}$$
$$Q(k) = \tfrac{1}{2}\ln[T^{21}(\kappa_R^{ij})/T^{12}(k)], \tag{5.5.42b}$$

then it can easily be checked that $P(k)$ and $Q(k)$ are action-angle variables with canonical Poisson bracket relations

$$\{P(k), P(\ell)\} = 0, \tag{5.5.43a}$$
$$\{Q(k), Q(\ell)\} = 0, \tag{5.5.43b}$$
$$\{P(k), Q(\ell)\} = \delta(k_R - \ell_R)\delta(k_I - \ell_I). \tag{5.5.43c}$$

Furthermore, multiplying equations (5.5.32) from the left by $\mathbf{Q}$ and integrating, we find that

$$T^{11}(k) = 1 + \frac{\mathrm{i}}{2\pi^2}\int_{-\infty}^{\infty}\int_{-\infty}^{\infty}\left\{\frac{\theta(k_I-\ell_I)}{\ell_R-k_R-\mathrm{i}0}+\frac{\theta(\ell_I-k_I)}{\ell_R-k_R+\mathrm{i}0}\right\} \times T^{12}(\kappa_R^{21})T^{21}(\ell)\,\mathrm{d}\ell_R\,\mathrm{d}\ell_I, \tag{5.5.44a}$$

$$T^{22}(k) = 1 - \frac{\mathrm{i}}{2\pi^2}\int_{-\infty}^{\infty}\int_{-\infty}^{\infty}\left\{\frac{\theta(k_I-\ell_I)}{\ell_R-k_R-\mathrm{i}0}+\frac{\theta(\ell_I-k_I)}{\ell_R-k_R+\mathrm{i}0}\right\} \times T^{21}(\kappa_R^{12})T^{12}(\ell)\,\mathrm{d}\ell_R\,\mathrm{d}\ell_I, \tag{5.5.44b}$$

and thus from equations (5.5.41)

$$\{T^{11}(k), T^{11}(\ell)\} = 0, \tag{5.5.45a}$$

$$\{T^{22}(k), T^{22}(\ell)\} = 0, \tag{5.5.45b}$$

$$\{T^{11}(k), T^{22}(\ell)\} = 0. \tag{5.5.45c}$$

Next we obtain the trace formula by using (5.5.44) and expanding $T^{11}(k) \pm T^{22}(k)$ in inverse powers of k_R, noting that from equation (5.5.34) all coefficients in the expansions are time independent

$$\begin{aligned}\tfrac{1}{2}[T^{11}(k) \pm T^{22}(k)] &= \sum_{n=1}^{\infty}(k_R)^{-n}\gamma_n^{\pm} \\ &= \frac{1}{2\pi^2\mathrm{i}}\sum_{n=1}^{\infty}(k_R)^{-n}\int_{-\infty}^{\infty}\int_{-\infty}^{\infty}[(\ell_R)^{n-1}-(\lambda_R^{ij})^{n-1}] \\ &\qquad \times T^{12}(\lambda_R^{ij})T^{21}(\ell)\,\mathrm{d}\ell_R\,\mathrm{d}\ell_I.\end{aligned} \tag{5.5.46}$$

Furthermore, using the expansion of $T^{11}(k) \pm T^{22}(k)$ with equations (5.5.45), one sees that the Poisson bracket relations of all pairs of coefficients are given by

$$\{\gamma_m^{\pm}, \gamma_n^{\pm}\} = 0, \qquad m, n = 1, 2, \ldots\ . \tag{5.5.47}$$

We can write the coefficients in the expansion in terms of the original variables q and r by making use of the expansions (5.5.36) in (5.5.27,28). The first few are

$$\gamma_1^{\pm} = -\tfrac{1}{2}\mathrm{i}\int_{-\infty}^{\infty}\int_{-\infty}^{\infty}[q(\xi,\eta)r(\xi,\eta) \mp q(\xi,\eta)r(\xi,\eta)]\,\mathrm{d}\xi\,\mathrm{d}\eta, \tag{5.5.48a}$$

$$\gamma_2^{\pm} = \tfrac{1}{4}\int_{-\infty}^{\infty}\int_{-\infty}^{\infty} q(\xi,\eta)(\partial_\xi \mp \partial_\eta)r(\xi,\eta)\,\mathrm{d}\xi\,\mathrm{d}\eta, \tag{5.5.48b}$$

$$\gamma_3^{\pm} = \tfrac{1}{8}\mathrm{i}\int_{-\infty}^{\infty}\int_{-\infty}^{\infty} q(\xi,\eta)(\partial_\xi^2 \mp \partial_\eta^2)r(\xi,\eta)\,\mathrm{d}\xi\,\mathrm{d}\eta,$$

$$- \tfrac{1}{8}\mathrm{i} \int_{-\infty}^{\infty}\int_{-\infty}^{\infty} q(\xi,\eta)r(\xi,\eta)\left\{\int_{-\infty}^{\infty}\int_{-\infty}^{\infty} G^{11}(\xi-\xi',\eta-\eta')[q(\xi',\eta')r(\xi',\eta')]_{\xi'}\,\mathrm{d}\xi'\,\mathrm{d}\eta'\right\}\mathrm{d}\xi\,\mathrm{d}\eta,$$

$$\mp \tfrac{1}{8}\mathrm{i} \int_{-\infty}^{\infty}\int_{-\infty}^{\infty} q(\xi,\eta)r(\xi,\eta)\left\{\int_{-\infty}^{\infty}\int_{-\infty}^{\infty} G^{22}(\xi-\xi',\eta-\eta')[q(\xi',\eta')r(\xi',\eta')]_{\eta'}\,\mathrm{d}\xi'\,\mathrm{d}\eta'\right\}\mathrm{d}\xi\,\mathrm{d}\eta, \quad (5.5.48c)$$

where the use of certain symmetry properties of $G^{11}(\xi,\eta)$ and $G^{22}(\xi,\eta)$ have enabled us to simplify the expressions for $\gamma_n^{\pm}$. We recognize γ_3^- to be, up to a constant coefficient, the Hamiltonian H [i.e., equation (5.5.40)] in the $\bar{\partial}$-limit Case.

We again remark that for the DSI equations, there are several choices for the Green's function, depending upon the associated boundary conditions. The situation is completely analogous to that in the strong coupling limit discussed in §5.5.4 above [recall the three Green's functions (5.5.19)]. As we have shown in this section, the Hamiltonian formalism is fully understood in the $\bar{\partial}$-limit Case [corresponding to the Green's function (5.5.19b)]. However, at present, it is unknown for the Riemann-Hilbert formulation of the I.S.T. method for the DSI equations as discussed by Fokas and Ablowitz [1983e] [corresponding to the Green's function (5.5.19a)]. A technical reason is that, in this case, $G(x,y) \neq G(-x,-y)$ (so it is necessary to modify the Poisson brackets); arising from the fact that $\phi \to 0$ as $\xi \to -\infty$ ($\xi = x+y$), whereas $\phi \not\to 0$ as $\eta \to \infty$ ($\eta = x-y$). In the case of the symmetric Green's function (5.5.19c), at present, the I.S.T. formalism is unknown.

5.5.6 Hamiltonian Formalism for the DSII Equations. In this section we discuss the complete integrability (in the Hamiltonian sense) of a generalized version of the DSII equations, following Villarroel and Ablowitz [1991]. In particular, it is shown that this is a completely integrable Hamiltonian system and we give the action-angle variables. For an appropriate choice of σ we obtain, as a special case, the action-angle variables for the DSII equations (5.5.11).

Consider the DS equations (5.5.2) with $\sigma = \alpha + \mathrm{i}\beta$, an arbitrary complex number such that $\beta \neq 0$. In particular, by setting $\alpha = 0$, $\beta = 1$, then we have DSII, whilst letting $\alpha = -1$, $\beta \to 0$ in a certain way, then we obtain the $\bar{\partial}$-limit case discussed in the previous section. The equations (5.5.2) can be viewed of as Hamilton's equations of motion for q and r with Hamiltonian

$$H = \tfrac{1}{2}\mathrm{i} \int_{-\infty}^{\infty}\int_{-\infty}^{\infty} \left[(\sigma^2 q_{xx} + q_{yy})r - q^2r^2 - qr\phi\right]\mathrm{d}x\,\mathrm{d}y, \quad (5.5.49)$$

where ϕ is solved in terms of q and r through

$$\phi(x,y) = -2\int_{-\infty}^{\infty}\int_{-\infty}^{\infty} G(x-x',y-y')[q(x',y')r(x',y')]_{x'x'}\,\mathrm{d}x'\,\mathrm{d}y', \tag{5.5.50a}$$

where the Green's function is given by

$$G(x,y) = \frac{\mathrm{i}}{4\pi\sigma}\ln[(y+\sigma x)(y-\sigma x)]. \tag{5.5.50b}$$

Then a direct calculation shows that

$$q_t = \{q,H\}, \qquad r_t = \{r,H\}, \tag{5.5.51}$$

where the canonical Poisson bracket $\{\ ,\ \}$ is given by equation (5.5.22).

The I.S.T. method for the DSII equations (5.5.11) was discussed in §5.5.3 above, following Fokas and Ablowitz [1983e]. Nachman and Ablowitz [1984b] consider a more general problem in n dimensions (a special case of which is the two dimensional case discussed here, see also Chapter 6) with arbitrary σ. The DS equations (5.5.2) are associated with the scattering problem (5.5.3a), where $\mathbf{\Phi}(x,y;k)$ satisfies

$$\begin{aligned}\Phi^{ij}(x,y;k) &= \delta^{ij}\exp\{k(x\sigma J^j - y)\}\\ &\quad + \int_{-\infty}^{\infty}\int_{-\infty}^{\infty} G^{ij}(x-x',y-y';\kappa_{ij})[\mathbf{Q}(x',y')\mathbf{\Phi}(x',y';k)]^{ij}\,\mathrm{d}x'\,\mathrm{d}y',\end{aligned} \tag{5.5.52a}$$

where the Green's function is given by

$$\begin{aligned}G^{ij}(x,k;\kappa_{ij}) = \frac{1}{2\pi}\int_{-\infty}^{\infty} &[\theta(x)\theta(-J^i\beta\xi) - \theta(-x)\theta(J^i\beta\xi)]\\ &\times\exp\{\mathrm{i}(\xi+\mathrm{i}\kappa_{ij})(y-x\sigma J^i)\}\,\mathrm{d}\xi,\end{aligned} \tag{5.5.52b}$$

with

$$\kappa_{ij} = k_R + \mathrm{i}(\alpha k_R + J^iJ^j[\beta k_I - \alpha k_R])/\beta. \tag{5.5.52c}$$

For convenience, we shall often write $\kappa_{12} = \hat{k}$. Taking the $\partial/\partial\bar{k}$ derivative of $\mathbf{\Phi}$ yields

$$\frac{\partial\Phi^{ij}}{\partial\bar{k}}(x,y;k) = \sum_{m=1}^{2}\frac{\sigma(J^j - J^m)}{\mathrm{sgn}(\beta)}\Phi^{im}(x,y;\kappa_{mj})T^{mj}(k), \tag{5.5.53a}$$

where the scattering data $T^{ij}(k)$ is given by

$$\begin{aligned}T^{ij}(k) = \frac{\mathrm{sgn}(\beta)}{2\pi}\int_{-\infty}^{\infty}\int_{-\infty}^{\infty} &\exp\{-\kappa_{ij}(x\sigma J^i - y)\}\\ &\times[\mathbf{Q}(x',y')\mathbf{\Phi}(x',y';k)]^{ij}\,\mathrm{d}x'\,\mathrm{d}y'.\end{aligned} \tag{5.5.53b}$$

We shall assume that the potentials q and r are rapidly decreasing functions belonging to $L^1(\mathbb{R}^2)$; this excludes the possibility of the weakly decaying lump-type solutions; $\mathbf{\Phi}(x,y;k)$ is bounded for all k. From the asympototic behavior of the Green's function, we have

$$\Phi^{ij}(x,y;k) = \delta^{ij}\exp\{k(x\sigma J^j - y)\} + \frac{T^{ij}(k)\exp\{\kappa_{ij}(x\sigma J_i - y)\}}{\mathrm{i}(yJ^i - x\sigma)} + o(1/R), \quad (5.5.54)$$

with $R = \sqrt{x^2+y^2}$. For convenience we write $\mathbf{\Phi} = \mathbf{\Phi}_H\mathbf{m}$, where $\mathbf{\Phi}_H = \exp[k\sigma(\mathbf{J}x - y0]$ and $\mathbf{m} \to \mathbf{I}$ as $k \to \infty$. Thus we have the asymptotic expansion

$$\mathbf{\Phi} = \mathbf{\Phi}_H\left[\mathbf{I} + (\mathbf{C}_1/k) + (\mathbf{C}_2/k^2) + \ldots\right], \quad (5.5.55)$$

where $\mathbf{C}_n$ are matrices whose elements depend on x and y but not k. The time evolution of the scattering data is given by

$$T^{ij}(k,t) = T^{ij}(k,0)\exp\{\mathrm{i}[J^i(\kappa_{ij})^2 - J^jk^2]t\}. \quad (5.5.56)$$

We note that the problem is ill-posed (i.e., $T^{ij}(k,t)$ has a unbounded growth rate) unless $\alpha = 0$, which corresponds to the usual DSII equations (5.5.11). By using the asymptotic expansion (5.5.55), the boundedness of $\mathbf{\Phi}$ and $\mathbf{\Phi}_H$ and $\mathbf{Q} \in L^1(\mathbb{R})$, we have that $\lim_{k\to\infty} T^{ij}(k) = 0$.

Next we obtain the dispersion relations, i.e., relationships connecting the scattering data. By differentiating $T^{11}(k)$ with respect to $\bar{k}$ and using equations (5.5.53) we have

$$\begin{aligned}\frac{\partial T^{11}}{\partial \bar{k}}(k) &= \frac{\mathrm{sgn}(\beta)}{2\pi}\int_{-\infty}^{\infty}\int_{-\infty}^{\infty} q(x',y')T^{21}(k)\Phi^{22}(x',y';\hat{k})\exp\{-k(x\sigma - y)\}\,\mathrm{d}x'\,\mathrm{d}y', \\ &= \mathrm{sgn}(\beta)T^{21}(k)T^{12}(\hat{k}), \quad (5.5.57\mathrm{a})\\ \frac{\partial T^{22}}{\partial \bar{k}}(k) &= \mathrm{sgn}(\beta)T^{12}(k)T^{21}(\hat{k}), \quad (5.5.57\mathrm{b})\end{aligned}$$

where $\hat{k} = k_R + \mathrm{i}(2\alpha k_R - \beta k_I)/\beta$. From these relationships together with the asymptotic behavior of T^{ij} we have

$$T^{11}(k) = \mathrm{sgn}(\beta)\int_{-\infty}^{\infty}\int_{-\infty}^{\infty} T^{12}(\hat{z})T^{21}(z)\frac{\mathrm{d}z\wedge\mathrm{d}\bar{z}}{z-k}, \quad (5.5.58\mathrm{a})$$

$$T^{22}(k) = \mathrm{sgn}(\beta)\int_{-\infty}^{\infty}\int_{-\infty}^{\infty} T^{12}(z)T^{21}(\hat{z})\frac{\mathrm{d}z\wedge\mathrm{d}\bar{z}}{z-k}, \quad (5.5.58\mathrm{b})$$

with $\hat{z} = z_R + \mathrm{i}(2\alpha z_R - \beta z_I)/\beta$.

Evaluating the Poisson brackets yields

$$\{T^{11}(k), T^{11}(\ell)\} = 0, \quad (5.5.59\mathrm{a})$$

$$\{T^{22}(k), T^{22}(\ell)\} = 0, \quad (5.5.59\mathrm{b})$$

$$\{T^{12}(k), T^{12}(\ell)\} = 0, \tag{5.5.59c}$$

$$\{T^{21}(k), T^{21}(\ell)\} = 0, \tag{5.5.59d}$$

$$\{T^{12}(k), T^{21}(\ell)\} = \frac{4\beta^2\pi^2}{\alpha^2+\beta^2}\delta(k_R-\ell_R)\delta([k_I-\ell_I]\beta-[k_R-\ell_R]\alpha), \tag{5.5.59e}$$

$$\{T^{11}(k), T^{12}(\ell)\} = -\frac{2\,\mathrm{sgn}(\beta)}{\sigma(k-\ell)}T^{12}(\ell), \tag{5.5.59f}$$

together with several other similar relations. From these formulae we deduce that the canonical action-angle variables are

$$P(k) = \frac{\beta}{\sqrt{\alpha^2+\beta^2}}\,T^{12}(k)T^{21}(\hat{k}), \tag{5.5.60a}$$

$$Q(k) = \frac{\beta}{\sqrt{\alpha^2+\beta^2}}\,\ln[T^{21}(\hat{k})/T^{12}(k)], \tag{5.5.60b}$$

which satisfy the canonical commutation relations

$$\{P(k), P(\ell)\} = 0, \qquad \{P(k), Q(\ell)\} = 0, \qquad \{Q(k), Q(\ell)\} = 0. \tag{5.5.61}$$

In particular, by letting $\alpha = 0$, $\beta = \pm 1$, we obtain the action-angle variables for the DSII equations (5.5.11). We remark that the $\bar{\partial}$-limit Case of DSI is obtained by setting $\alpha = -1$, $\beta = 0$; since this is a singular limit (κ_{ij} does not exist for $\beta = 0$), we let $k_R \to 0$, $\beta \to 0$ and $a = -1$, with k_R/β finite. Introduce the modified spectral parameter $K = K_R + \mathrm{i}K_I$ by setting $K_I = k_R/\beta$, $K_R = -k_I$, then the Green's function (5.5.52b) yields the Green's function in the $\bar{\partial}$-limit Case [recall equation (5.5.29)]. Further, equations equations (5.5.59) give

$$\{T^{ij}(K), T^{ij}(K')\} = 0,$$
$$\{T^{12}(K), T^{21}(K')\} = 4\pi^2\delta(K_R+K'_R) - K_I - K'_I)\delta(K_I - K'_I]\alpha),$$
$$\{T^{11}(K), T^{12}(K')\} = \left[\frac{\mathrm{i}}{K_R-K'_R-\mathrm{i}0} + 2\pi\delta(K_R-K'_R)\theta(K_I-K'_I)\right]T^{12}(K').$$

Then, using the well-known distributional formula

$$\frac{1}{x\pm\mathrm{i}\varepsilon} = P(1/x) \mp \mathrm{i}\pi\delta(x),$$

and so we recover the results of §5.5.5 above.

The DSII equations (5.5.11) are an integrable Hamiltonian system with an infinite number of integrals of motion in involution. Expanding $T^{11}(k) \pm T^{22}(k)$ in inverse powers of k

$$T^{11}(k) \pm T^{22}(k) = \sum_{n=1}^{\infty}\gamma_n^{\pm}k^{-n}, \tag{5.5.62}$$

and using equations (5.5.59) yields the Poisson bracket relations

$$\{\gamma_m^{\pm}, \gamma_n^{\pm}\} = 0, \qquad m,n = 1,2,\dots\,, \tag{5.5.63}$$

with the $\gamma_n^{\pm}$ explicitly given by

$$\gamma_n^{\pm} = -\int_{-\infty}^{\infty}\int_{-\infty}^{\infty} (z^{n-1} \mp \hat{z}^{n-1})T^{21}(z)T^{12}(\hat{z})\,\mathrm{d}z \wedge \mathrm{d}\bar{z}. \tag{5.5.64}$$

Hence it follows from equation (5.5.56) that the $\gamma_n^{\pm}$ are time-independent and so are the constants of motion. In particular, up to a constant, γ_3^- is the Hamiltonian for the problem and is related to H given by equation (5.5.49) in physical space. It can be written in action-angle coordinates as follows

$$\gamma_3^- = \frac{\sqrt{\alpha^2+\beta^2}}{\beta}\int_{-\infty}^{\infty}\int_{-\infty}^{\infty}(k^2+\hat{k}^2)P(k)\,\mathrm{d}k\wedge\mathrm{d}\bar{k},$$

and depends only upon the actions $P(k)$, but not the angles $Q(k)$. Thus we can assert that the DSII equations (5.5.11) are indeed a completely integrable Hamiltonian system.

Once again we remark that the boundary conditions for the scattering problem (5.5.3a) and the DSII equations (5.5.11), in particular for ϕ, are intimately related. In order to see this we expand the diagonal elements for $\mathbf{\Phi}$ [i.e., Φ^{11} and Φ^{22} — recall equations (5.5.36)], in inverse powers of k. We note that the off-diagonal elements are given by

$$\Phi^{12}(x,y;k) = \frac{1}{r(x,y)}(\partial_x - \sigma\partial_y)\Phi^{22}(x,y;k),$$
$$\Phi^{21}(x,y;k) = \frac{1}{q(x,y)}(\partial_x + \sigma\partial_y)\Phi^{11}(x,y;k).$$

Substituting equation (5.5.55) into (5.5.52) yields

$$\mathbf{C}_1^{11} = -\tfrac{1}{2}\mathrm{i}\int_{-\infty}^{\infty}\int_{-\infty}^{\infty} G^{11}(x-x',y-y')q(x',y')r(x',y')\,\mathrm{d}x'\,\mathrm{d}y', \tag{5.5.65a}$$

$$\mathbf{C}_1^{22} = \tfrac{1}{2}\mathrm{i}\int_{-\infty}^{\infty}\int_{-\infty}^{\infty} G^{22}(x-x',y-y')q(x',y')r(x',y')\,\mathrm{d}x'\,\mathrm{d}y'. \tag{5.5.65b}$$

Substituting these into equation (5.5.3b) and noting that the temporal part of the Lax pair is the one which contains the function ϕ, we obtain

$$A^{11} = (\partial_x - \sigma\partial_y)\mathbf{C}_1^{11}, \qquad A^{22} = (\partial_x + \sigma\partial_y)\mathbf{C}_1^{22}, \tag{5.5.66}$$

and then using $\phi = \mathrm{i}(A^{11} - A^{22}) + qr$, we obtain equation (5.5.50). Thus we see that once the boundary conditions for the spectral function $\mathbf{\Phi}$ have been fixed, the temporal part of the Lax pair picks a particular form for the function ϕ, i.e., in this case equation (5.5.50).

Similar ideas apply to other nonlocal forms of the equations (and to other nonlocal evolution equations). For example, we apply the above construction to the DSI equation (5.5.4) with the choice of boundary used by Fokas and Ablowitz [1983e].

Suppose we consider an eigenfunction $\mathbf{\Phi}$ satisfying

$$\Phi^{ij}(x,y;k) = \delta^{ij}\exp\{-k(x\sigma J^j + y)\} + \int_{-\infty}^{\infty}\int_{-\infty}^{\infty} \hat{G}^{ij}(x-x', y-y')[\mathbf{Q}(x',y')\mathbf{\Phi}(x',y';k)]^{ij}\,\mathrm{d}x'\,\mathrm{d}y',$$

where the Green's function is given by

$$\hat{G}^{ij}(x,k) = J^i\delta(x + J^i y)\theta(J^i x).$$

In this case we recall that $\mathbf{\Phi}$ satisfies a Riemann-Hilbert boundary value problem. The eigenfunction $\mathbf{\Phi}$ has the asymptotic expansion

$$\mathbf{\Phi}(x,y;k) = \left(\mathbf{I} + \sum_{n=1}^{\infty} k^{-n}\mathbf{C}_n\right)\exp\{-k(x\sigma J^j + y)\},$$

and consequently

$$A^{11}(x,y) = -\tfrac{1}{2}\mathrm{i}(\partial_x + \partial_y)\int_{-\infty}^{\infty}\int_{-\infty}^{\infty}\hat{G}^{11}(x-x',y-y')q(x',y')r(x',y')\,\mathrm{d}x'\,\mathrm{d}y',$$

$$A^{22}(x,y) = \tfrac{1}{2}\mathrm{i}(\partial_x - \partial_y)\int_{-\infty}^{\infty}\int_{-\infty}^{\infty}\hat{G}^{22}(x-x',y-y')q(x',y')r(x',y')\,\mathrm{d}x'\,\mathrm{d}y'.$$

Integrating by parts and using the formula

$$\partial_x[\theta(x-y)\theta(-x-y)] = \theta(x-y)\delta(-x) + \theta(x+y)\delta(x),$$

yields

$$\begin{aligned}\phi(x,y) &\equiv \mathrm{i}[A^{11}(x,y) - A^{22}(x,y)] + q(x,y)r(x,y)\\ &= \tfrac{1}{2}\int_{-\infty}^{\infty}\int_{-\infty}^{\infty}[\theta(x-x'+y'-y)\theta(x'-x+y'-y)]\partial_{x'}^2[q(x',y')r(x',y')]\,\mathrm{d}x'\,\mathrm{d}y',\end{aligned}$$

that is

$$\phi(x,y) = \int_{-\infty}^{\infty}\int_{-\infty}^{\infty}\mathcal{G}(x-x',y-y')\partial_{x'}^2[q(x',y')r(x',y')]\,\mathrm{d}x'\,\mathrm{d}y',$$

where

$$\mathcal{G}(x,y) = \theta(x-y)\theta(-x-y).$$

Thus again the choice of $\mathcal{G}$ is completely determined by the choice of $\hat{G}^{ij}$.

5.5.7 Localized Rapidly Decaying Solutions of DSI Equations. The initial value problem associated with the DSI equations (5.5.6) where $q(x,y,0)$ and $r(x,y,0)$ are given and decay sufficiently rapidly $R=\sqrt{x^2+y^2}\to\infty$, and $\phi\to 0$ as $x\to-\infty$ (say) was solved by Fokas and Ablowitz [1983e]. Due to interesting discoveries by Boiti, Leon, Martina and Pempinelli [1988b], there has been renewed interest in the solution of the initial value problem for the DSI equations (5.5.6). It will turn out that the boundary conditions for ϕ are modified. It is convenient to introduce characteristic coordinates $\xi=x+y$, $\eta=x-y$, then defining

$$U_1(\xi,\eta,t):=-\phi_\eta+\tfrac{1}{2}qr,\qquad U_2(\xi,\eta,t):=-\phi_\xi+\tfrac{1}{2}qr,$$

and integrating equation (5.5.6c) yields

$$\mathrm{i}q_t+\tfrac{1}{2}(q_{\xi\xi}+q_{\eta\eta})+(U_1+U_2)q=0, \tag{5.5.67a}$$

$$-\mathrm{i}r_t+\tfrac{1}{2}(r_{\xi\xi}+r_{\eta\eta})+(U_1+U_2)r=0, \tag{5.5.67b}$$

$$U_1(\xi,\eta,t)=u_1(\eta,t)-\tfrac{1}{2}\int_{-\infty}^{\xi}[q(\xi',\eta,t)r(\xi',\eta,t)]_\eta\,\mathrm{d}\xi', \tag{5.5.67c}$$

$$U_2(\xi,\eta,t)=u_2(\xi,t)-\tfrac{1}{2}\int_{-\infty}^{\eta}[q(\xi,\eta',t)r(\xi,\eta',t)]_\xi\,\mathrm{d}\eta', \tag{5.5.67d}$$

where

$$u_1(\eta,t):=\lim_{\xi\to-\infty}U_1(\xi,\eta,t),\qquad u_2(\xi,t):=\lim_{\eta\to-\infty}U_2(\xi,\eta,t). \tag{5.5.67e}$$

As mentioned in §5.5.5 above, there is considerable freedom in the choice of the boundary conditions for ϕ. Schultz and Ablowitz [1989a] show that the boundary value problem solved by Fokas and Ablowitz [1983e] corresponds to $u_1\equiv 0$, $u_2\equiv 0$ (i.e., $\phi\to 0$ as $x\to-\infty$). In this latter case (which is typical of the situations which have arisen for dispersive multidimensional equations with strong decay at infinity), arbitrary initial data disperse away as $t\to\infty$.

In [1988b] Boiti, Leon, Martina and Pempinelli showed that the DSI equations (5.5.67), with $r=-q^*$, possess a special type of localized soliton, namely a *two-dimensional breather solution*

$$q(\xi,\eta,t)=\frac{4\rho\sqrt{\lambda\mu}\exp\{-\mu(\xi-\xi_0)-\lambda(\eta-\eta_0)+\mathrm{i}(\mu^2+\lambda^2)t\}}{[1+\exp\{-2\mu(\xi-\xi_0)\}]\,[1+\exp\{-2\lambda(\eta-\eta_0)\}]+|\rho|^2}, \tag{5.5.68}$$

where $\lambda\in\mathbb{R}^+$, $\mu\in\mathbb{R}^+$, $\xi_0\in\mathbb{R}$, $\eta_0\in\mathbb{R}$, $\rho\in\mathbb{C}$ are constant parameters, which decays exponentially in both spatial coordinates. The solution (5.5.68) was obtained via a Bäcklund transformation; though the genericity of such solutions and their relationship to an initial boundary problem was not discussed by Boiti, Leon, Martina and Pempinelli [1988b]. This is an important discovery since a disappointing feature of previously studied dispersive, multidimensional equations has been the absence of two-dimensional exponentially decaying solitons. (Recall that the soliton solution (1.2.8) of the KP equation exponentially decays only in one spatial dimension, and the lump solution (1.2.10) of the KP equation only decays algebraically.) Subsequently Boiti, Leon, Martina and Pempinelli [1989a,b] constructed the two-breather (i.e., a two-soliton like mode) solution using the nonlinear superposition formula for Bäcklund transformations and showed that the only effect of the two interacting solutions is a two-dimensional phasé shift. The solutions (5.5.68) can be made to move in time by observing that the DSI equation admits a Galilean transformation. Clearly this procedure can be repeated and the existence of these nodes depends crucially on the boundary conditions at infinity. In this sense they are not like the "free" solitons arising in $1+1$-dimensional theories.

Fokas and Santini [1989, 1990] have studied the DSI equation (5.5.67) with arbitrary time-dependent boundary conditions using an associated I.S.T. formalism (see also Santini [1990]; Boiti, Leon and Pempinelli [1990a,b]). It is conjectured that any arbitrary initial condition will decompose as $t \to \infty$ either (i), for arbitrary time-independent boundary conditions, into a number of breathers (5.5.68) or (ii), for arbitrary time-dependent boundary conditions, into a number of *two-dimensional soliton-like modes*,

$$q(\xi,\eta,t) = \frac{4\rho\sqrt{\lambda_R\mu_R}\exp\left\{-\mu_R(\hat{\xi}+\xi_0)-\lambda_R(\hat{\eta}+\eta_0)\right\}}{\left[1+\exp\{-2\mu_R(\hat{\xi}+\xi_0)\}\right]\left[1+\exp\{-2\lambda_R(\hat{\eta}+\eta_0)\}\right]+|\rho|^2} \times \exp\left\{\mathrm{i}\left[-(\mu_I\hat{\xi}+\lambda_I\hat{\eta})+(|\mu|^2+|\lambda|^2)t+c_0\right]\right\}, \tag{5.5.69a}$$

where

$$\hat{\xi} := \xi - 2\mu_I t, \qquad \xi_0 = \mu_R^{-1}\ln\left(\frac{|\mu|}{\sqrt{2\mu_r}}\right), \tag{5.5.69b}$$

$$\hat{\eta} := \eta - 2\lambda_I t, \qquad \eta_0 = \lambda_R^{-1}\ln\left(\frac{|\lambda|}{\sqrt{2\lambda_r}}\right), \tag{5.5.69c}$$

$c_0 := \arg(c\bar{c})$ with $\lambda_R \in \mathbb{R}^+$, $\mu_R \in \mathbb{R}^+$, $\xi_0 \in \mathbb{R}$ $\eta_0 \in \mathbb{R}$, $\lambda = \lambda_R + \mathrm{i}\lambda_I \in \mathbb{C}$, $\mu = \mu_R + \mathrm{i}\mu_I \in \mathbb{C}$, $\rho \in \mathbb{C}$, $c \in \mathbb{C}$, $\bar{c} \in \mathbb{C}$ are constant parameters. The associated boundary conditions for this solution are

$$u_1(\eta,t) = 2\lambda_R^2\,\mathrm{sech}^2[\lambda_R(\hat{\eta}-\eta_0)], \qquad u_2(\xi,t) = 2\mu_R^2\,\mathrm{sech}^2[\mu_R(\hat{\xi}-\xi_0)]. \tag{5.5.70}$$

Fokas and Santini [1990] termed the solutions (5.5.69), which are localized coherent structures, *dromions*. (We remark that for the nonlinear Schrödinger (NLS) equation

$$\mathrm{i}u_t + u_{xx} + u|u|^2 = 0, \tag{5.5.71}$$

with the boundary condition $u(x,t) \to 0$ as $|x| \to \infty$, an arbitrary initial condition $u(x,0) = u_0(x)$ decomposes into a finite number of solitons.)

The solution of the initial-boundary value problem for DSI (5.5.67) with $r = -q^*$, and with $q(\xi,\eta,0)$, $u_1(\eta,t)$, $u_2(\xi,t)$ given, involves formulating a modified equation for the time dependence of the scattering data, $S^{12}(k,l)$ and $S^{21}(k,l)$, in the presence of nontrivial boundary conditions. The Fourier transforms of S^{12} and S^{21} (i.e., $\hat{S}^{12}$ and $\hat{S}^{21}$, respectively) satisfy nontrivial time-evolution equations

$$\mathrm{i}\frac{\partial \hat{S}^{12}}{\partial t} + \frac{\partial^2 \hat{S}^{12}}{\partial \xi^2} + \frac{\partial^2 \hat{S}^{12}}{\partial \eta^2} + \{u_2(\eta,t) + u_1(\xi,t)\}\,\hat{S}^{12} = 0, \tag{5.5.72a}$$

$$-\mathrm{i}\frac{\partial \hat{S}^{21}}{\partial t} + \frac{\partial^2 \hat{S}^{21}}{\partial \xi^2} + \frac{\partial^2 \hat{S}^{21}}{\partial \eta^2} + \{u_2(\eta,t) + u_1(\xi,t)\}\,\hat{S}^{21} = 0, \tag{5.5.72b}$$

where

$$\hat{S}^{ij}(\xi,\eta) := \int_{-\infty}^{\infty}\int_{-\infty}^{\infty} S^{ij}(k,l)\exp\{\mathrm{i}(k\xi + l\eta)\}\,\mathrm{d}k\,\mathrm{d}l.$$

Thus given $q(\xi,\eta,0)$, $u_1(\eta,t)$, $u_2(\xi,t)$, one computes $\hat{S}^{12}(\xi,\eta,0)$, then (5.5.72) implies $\hat{S}^{12}(\xi,\eta,t)$, and hence one obtains $S^{12}(\xi,\eta,t)$ (see Fokas and Santini [1989, 1990], Santini [1990] for details). We remark that equations (5.5.72) are the linear limit of equations (5.5.67a,b), which is consistent with the fact the I.S.T. method reduces to the Fourier transform method in the linear limit.

Fokas and Santini [1989, 1990] show that dromions arises generically arise generically from the solution of an initial-boundary value problem for the DSI equation (5.5.67) in which the (complex) initial condition $q_0(\xi,\eta) = q(\xi,\eta,0)$ together with the (real) boundary conditions $u_1(\eta,t)$ and $u_2(\xi,t)$ are given, decay sufficiently rapidly as $\xi,\eta \to \infty$ and are bounded in t. For arbitrary nonzero boundary conditions $u_1(\eta,t)$ and $u_2(\xi,t)$ there is a transfer of energy from the mean flow to the surface waves (where focusing effect balancing dispersion can be created), as mentioned above. The basic mechanism for generating dromions is due to the boundary conditions $u_1(\eta,t)$ and $u_2(\xi,t)$, whilst the initial condition $q_0(\xi,\eta)$ affects the amplitudes only and so has a secondary role. During interactions, dromion solutions of the DSI equations (5.5.67) not only exhibit a two-dimensional phase shift, but, generally, also a change of form which is associated with an exchange of energy. (We remark that during interactions, solitons of the NLS equation only exhibit a phase shift.) This is a novelty of such localized coherent structures in $2+1$-dimensions; in $1+1$-dimensions, solitons regain their shape and form after interaction whilst dromions do not regain the same form, though their velocity is conserved. Dromions are intimately related to the boundary conditions $u_1(\eta,t)$ and $u_2(\xi,t)$ and so can be driven anywhere in the (ξ,η) plane for a suitable choice of u_1 and u_2.

Fokas and Santini [1989, 1990] show how to obtain multi-dromion solutions of the DSI equations (5.5.67). It turns out that the solution of equations (5.5.72) is intimately related to the solution of the Schrödinger equation

$$\mathrm{i}\psi_t + \psi_{xx} + u(x,t)\psi = 0. \tag{5.5.73}$$

If the scattering data $\mathbf{S}(\xi,\eta,t)$ is degenerate, then $\hat{\mathbf{S}}(\xi,\eta,t)$ is also degenerate, and so

$$\hat{S}(\xi,\eta,t) = \sum_{n=1}^{N}\sum_{m=1}^{M} \rho^{nm} X^n(\xi,t) Y^m(\eta,t), \qquad \rho^{nm} \in \mathbb{C}, \tag{5.5.74}$$

where $X^n(\xi,t)$ and $Y^m(\eta,t)$ satisfy the systems of equations

$$\mathrm{i}X_t^n + X_{\xi\xi}^n + u_2 X^n = 0, \qquad n = 1,2,\ldots,N, \tag{5.5.75a}$$

$$\mathrm{i}Y_t^m + Y_{\eta\eta}^m + u_2 Y^m = 0, \qquad m = 1,2,\ldots,M. \tag{5.5.75b}$$

The inverse problem reduces to the solution of the algebraic equation

$$(\mathbf{I} - \mathbf{A})\mathbf{Z} = \mathbf{P}, \tag{5.5.76}$$

where

$$\mathbf{A}(\xi,\eta,t) = \mathbf{P}\boldsymbol{\beta}(\eta,t)\mathbf{P}^\dagger \boldsymbol{\alpha}^*(\xi,t), \tag{5.5.77a}$$

$$(\mathbf{P})^{ij} = \rho^{ij}, \qquad (\mathbf{P}^\dagger)^{ij} = \rho^{ji*}, \tag{5.5.77b}$$

$$\alpha^{ij}(\xi,t) = \int_{-\infty}^{\xi} X^i(\xi',t) X^{j*}(\xi',t)\,\mathrm{d}\xi', \tag{5.5.77c}$$

$$\beta^{ij}(\eta,t) = \int_{-\infty}^{\eta} Y^i(\eta',t) Y^{j*}(\eta',t)\,\mathrm{d}\eta'. \tag{5.5.77d}$$

Hence we obtain the closed form solutions of the DSI equations (5.5.67)

$$q(\xi,\eta,t) = 2\sum_{n=1}^{N}\sum_{m=1}^{M} Z^{nm}(\xi,\eta,t) X^n(\xi,t) Y^m(\eta,t), \tag{5.5.78a}$$

$$|q|^2(\xi,\eta,t) = -4\partial_\xi\partial_\eta \ln\det(\mathbf{I} - \mathbf{A}), \tag{5.5.78b}$$

$$U_1(\xi,\eta,t) = u_1(\eta,t) + 2\partial_\eta^2 \ln\det(\mathbf{I} - \mathbf{A}), \tag{5.5.78c}$$

$$U_2(\xi,\eta,t) = u_2(\xi,t) + 2\partial_\xi^2 \ln\det(\mathbf{I} - \mathbf{A}), \tag{5.5.78d}$$

If additionally, the boundary conditions are themselves reflectionless potentials for the Schrödinger equations (5.5.75), then $X^n(\xi,t)$ and $Y^m(\eta,t)$ can be found in closed form through the system

$$\phi_n + \sum_{j=1}^{n} C^{nj}\phi_j = c_n \exp\{-p_n(x - \mathrm{i}p_n t)\}, \qquad \mathrm{Re}(p_n) > 0, \tag{5.5.79a}$$

$$C^{nj}(x,t,\mathbf{c},\mathbf{p}) = \frac{c_n c_j^*}{p_n + p_j^*} \exp\{-(p_n + p_j^*)[x - \mathrm{i}(p_n - p_j^*)t]\}, \tag{5.5.79b}$$

$$u(x,t) = -2\partial_x \left(\sum_{j=1}^{N} c_j^* \exp\{-p_j^*(x + \mathrm{i}p_j^* t)\}\phi_j \right), \tag{5.5.79c}$$

where $\phi_i = X^i$, $u = u_2$, if $x \to \xi$, $p_i \to \mu_i$ and $\phi_i = Y^i$, $u = u_1$, if $x \to \eta$, $p_i \to \lambda_i$, $c_i \to c_i^*$. Then equations (5.5.78,79) yields the (N, M)-dromion solution of the DSI equations. In the special case $\mathrm{Im}(p_j) = 0$ and $c_j \in \mathbb{R}$, then $\phi_j(x,t) = f_j(x)\exp(\mathrm{i}p_j^2 t)$, and

$$f_n + \sum_{j=1}^{n} \frac{c_n c_j}{p_n + p_j} \exp\{-(p_n + p_j)x\} f_j(x) = c_n \exp(-p_n x), \tag{5.5.80a}$$

$$u(x) = -2\partial_x \left(\sum_{j=1}^{n} c_j \exp(-p_j x) f_j \right), \tag{5.5.80b}$$

then the solution $q(x,t)$ oscillates in time and we obtain an (N, M)-breather solution.

5.5.8 On the Physical Derivation of the Boundary Conditions for the Daevy-Stewartson Equations. In this section we are concerned with the question of deducing the proper boundary conditions associated with the DS equations. This issue and analysis was considered by Ablowitz, Manakov and Schultz [1990]. In order to do this we embed the DS equations within the KP equation while maintaining well-posedness in time. Even though the equations are specific for our application, the point of view is quite general and widely applicable. In particular, asymptotic derivations of nonlocal evolution equations with freedom of boundary values will require one to determine a unique specification as for the DS equations and the porcess of embedding will be a valuable tool. For the DS equations, the problem is known to depend crucially on imposed boundary conditions (Schultz and Ablowitz [1989a], Fokas and Ablowitz [1983e]).

It is well known that the KP equation

$$(u_t + u_{xxx} + 6uu_x)_x + 3\sigma^2 u_{yy} = 0. \tag{5.5.81}$$

with $u(x,y,t) \to 0$ as $r = \sqrt{x^2+y^2} \to \infty$, reduces to the DS equations through a multiple-scales reduction. Consider a wave moving in the (negative) x-direction with modulations in x, y slowly varying in t, i.e.,

$$\begin{aligned} u = \varepsilon\{\Psi \exp[\mathrm{i}(kx + k^3 t)] + \Psi^* \exp[-\mathrm{i}(kx + k^3 t)]\} \\ + \varepsilon^2\{u_0 + \Psi_2 \exp[2\mathrm{i}(kx + k^3 t)] + \Psi_2^* \exp[-2\mathrm{i}(kx + k^3 t)]\} + \ldots, \end{aligned} \tag{5.5.82}$$

where Ψ^* denotes the complex conjugate of Ψ, $\Psi = \Psi(X, Y, T)$ with $X = \varepsilon(x + 3k^2 t)$, $Y = \varepsilon y$, $T = \varepsilon^2 t$ as well as for u_0, Ψ_2, etc.. With this choice of scaling Ψ satisfies at order ε^3

$$\Psi_T + 3\mathrm{i}k\Psi_{XX} - \frac{3\mathrm{i}}{k}\sigma^2\Psi_{YY} + 6\mathrm{i}k(u_0\Psi + \Psi_2\Psi^*) = 0, \tag{5.5.83}$$

and at the next order, we find $\Psi_2 = k^{-2}\Psi$. Meanwhile the mean term satisfies, in a proper embedding sense (i.e., a small time derivative term in the equation for the mean), the following equation

$$\varepsilon u_{0,XT} + 3k^2 u_{0,XX} + 3\sigma^2 u_{0,YY} + 6(|\Psi|^2)_{XX} = 0. \tag{5.5.84}$$

For convenience we take $k = 1$, $T \to T/3$ and arrive at an embedding of the DS equations within the KP equation

$$\mathrm{i}\Psi_T = \Psi_{XX} - \sigma^2\Psi_{YY} + 2(|\Psi|^2\Psi + u_0\Psi), \tag{5.5.85a}$$

$$\varepsilon u_{0,XT} + u_{0,XX} + \sigma^2 u_{0,YY} = -2(|\Psi|^2)_{XX}. \tag{5.5.85b}$$

Without the $\varepsilon u_{0,XT}$ term these are the usual DS equations. The small term $\varepsilon u_{0,XT}$ is used in the embedding procedure so as to properly take care of initial values for $u_0(X,Y,T)$. For example, if we take in the mean term $u(x,y,t=0) = 0$, the question arises as to how to make this compatible with the expansion (5.5.82). If the $\varepsilon u_{0,XT}$ is not included then the situation is ambiguous.

We solve the system (5.5.85a) using a Fourier transform and, for convenience, take $u(x,y,t=0) = 0$. Setting $Q := -2(|\Psi|^2)_X$, and

$$u_0(X,Y,T) = \frac{1}{(2\pi)^2}\int_{-\infty}^{\infty}\int_{-\infty}^{\infty} \hat{u}_0(\xi,\eta,T)\exp[\mathrm{i}(\xi X + \eta Y)]\,\mathrm{d}\xi\,\mathrm{d}\eta, \tag{5.5.86}$$

we immediately have

$$\varepsilon\mathrm{i}\xi\hat{u}_{0,T} - (\xi^2 + \sigma^2\eta^2)\hat{u}_0 = \mathrm{i}\xi\hat{Q}, \tag{5.5.87}$$

where $\hat{Q}(\xi,\eta,T)$ is the Fourier transform of $Q(X,Y,T)$.

Equation (5.5.87) can be readily solved and we obtain

$$\hat{u}_0(\xi,\eta,T) = \frac{1}{\varepsilon}\int_0^T \hat{Q}(\xi,\eta,T)\exp\left\{\mathrm{i}\frac{T-T'}{\varepsilon}(\Omega - \mathrm{i}\delta_+)\right\}\mathrm{d}T', \tag{5.5.88}$$

where $\Omega = (\xi^2 + \sigma^2\eta^2)/\xi$; for convenience we insert the $-\mathrm{i}\delta_+$ term in the exponent as though there was some infinitesimal damping ($\delta_+ > 0$) in (5.5.87) — without this term the analysis is somewhat more clumbersome. We also assume that $T > 0$. Taking the limit $\varepsilon \to 0$ we get

$$\hat{u}_0(\xi,\eta,T) \to \frac{Q(\xi,\eta,T)}{\mathrm{i}(\Omega - \mathrm{i}\delta_+)} - \frac{Q(\xi,\eta,0)}{\mathrm{i}(\Omega - \mathrm{i}\delta_+)}\exp[-\mathrm{i}(\Omega - \mathrm{i}\delta_+)T/\varepsilon]. \tag{5.5.89}$$

Using the well known fact that

$$\frac{\exp(-\mathrm{i}\Omega t/\varepsilon)}{\Omega - \mathrm{i}\delta_+} \to 0, \qquad \text{as} \qquad \varepsilon \to 0,$$

it follows that $u_0(X,Y,T)$ is given by

$$u_0(X,Y,T) = -2\int_{-\infty}^{\infty}\int_{-\infty}^{\infty} G(X-x',Y-y')(|\Psi|^2)_{x'}\,dx'\,dy', \tag{5.5.90a}$$

where

$$G(X,Y) = -\frac{i}{(2\pi)^2}\int_{-\infty}^{\infty}\int_{-\infty}^{\infty} \frac{\xi\exp[i(\xi X+\eta Y)]}{\xi^2+\sigma^2\eta^2-i\xi\delta_+}\,d\xi\,d\eta. \tag{5.5.90b}$$

Certainly this procedure uniquely defines the Green's function $G(X,Y)$. For $\sigma^2=1$ (KPII, DSII) the limit $\delta_+\to 0$ is well defined. Without further calculations, it is clear that the integral vanishes as $R=\sqrt{X^2+Y^2}\to\infty$. On the other hand, when $\sigma^2=-1$ (KPI, DSI), the situation is slightly more subtle. Namely, the zeros $\xi=\xi(\eta,\delta_+)$ of the denominator in (5.5.90b) lie in the upper half-plane. Consequently the integral must vanish as $X\to-\infty$, i.e., the Green's function governing $u_0(X,Y,T)$ reads

$$G(X-x',Y-y') \sim \int_{-\infty}^{\infty}\int_{-\infty}^{\infty} \frac{\xi\exp\{i[\xi(X-x')+\eta(Y-y')]\}}{(\xi-\eta-\frac{1}{2}i\xi\delta_+)(\xi+\eta-\frac{1}{2}i\xi\delta_+)}\,d\xi\,d\eta,$$

for $X<x'$, $T>0$. We note that for $T<0$, we must reverse the sign of δ and $u_0(X,Y,T)\to 0$ as $X\to+\infty$; i.e., the Green's function vanishes for $X>x'$. Moreover, if one wishes to include homogeneous solutions in equations (5.5.85b) or (5.5.87) [i.e., to u_0, the mean term in (5.5.82)], the situation is straightforward. We observe that time-independent homogeneous solutions satisfy the wave equation and the inclusion of these terms which do not vanish at infinity (namely they are not of finite energy in an associated physical problem such as a water wave problem), would then allow boundary conditions, as derived above, to be nontrivial and open the way for multidimensional localized waves in $\Psi(X,Y,T)$ to exist (see Boiti, Leon, Martina and Pempinelli [1988b]; Fokas and Santini [1989]).

We conclude by observing that the DS equations are only one of many nonlocal evolution equations derived via the I.S.T. in which the solution method depends greatly on the imposed boundary conditions. In the DS equations, the field variable $\Psi(X,Y,T)$ vanishes as $R=\sqrt{X^2+Y^2}\to\infty$ for both DSI and DSII, however $u_0(X,Y,T)$ only does so for DSII.

The I.S.T. procedure needs to be properly implemented when $u_0(X,Y,T)$ does not vanish everywhere and the form of inverse scattering is deeply connected to the choice of boundary conditions on $u_0(X,Y,T)$. The usual choice, i.e., the standard Riemann-Hilbert boundary value method (see for example Schultz and Ablowitz [1989a], Fokas and Ablowitz [1983e]), is consistent with $u_0(X,Y,T)$ vanishing as either $X\to-\infty$ or $X\to\infty$, but not both. This situation, as shown above, is the physically acceptable one when decaying boundary conditions are assumed in the governing problem.

5.6 Further Examples.

In this section we briefly make some remarks on other nonlinear evolution equation in 2 + 1-dimensions which are solvable by a suitable I.S.T. scheme.

5.6.1 Equations Related to the Davey-Stewartson Equation. Ishimori studied the equation (Ishimori [1984])

$$\boldsymbol{S}_t + \boldsymbol{S} \wedge (\boldsymbol{S}_{xx} + \sigma^2 \boldsymbol{S}_{yy}) + \phi_x \boldsymbol{S}_y + \phi_y \boldsymbol{S}_x = \mathbf{0}, \tag{5.6.1a}$$

$$\phi_{xx} - \sigma^2 \phi_{yy} + 2\sigma^2 \boldsymbol{S} \bullet (\boldsymbol{S}_x \wedge \boldsymbol{S}_y) = \mathbf{0}, \tag{5.6.1b}$$

where

$$|\boldsymbol{S}(x,y,t)|^2 \equiv S_1^2(x,y,t) + S_2^2(x,y,t) + S_3^2(x,y,t) = 1, \tag{5.6.1c}$$

with $\sigma^2 = \pm 1$, which is a 2 + 1-dimensional generalization of the Heisenberg Ferromagnet model

$$\boldsymbol{S}_t = \boldsymbol{S} \wedge \boldsymbol{S}_{xx} \tag{5.6.2a}$$

where

$$|\boldsymbol{S}(x,t)|^2 \equiv S_1^2(x,t) + S_2^2(x,t) + S_3^2(x,t) = 1. \tag{5.6.2b}$$

The I.S.T. scheme in the case $\sigma = \mathrm{i}$, has been discussed by Konopelchenko and Matkarimov [1989, 1990a,b] and Beals and Coifman [1989]. The system is solvable using the $\bar{\partial}$ method, and the method is analogous to that used to solve DSII (§5.5.2). The Lax pair for equations (5.6.1) are:

$$\mathrm{L}\boldsymbol{v} \equiv \sigma \boldsymbol{v}_y + \mathbf{P}\boldsymbol{v}_x = \mathbf{0}, \tag{5.6.3a}$$

$$\mathrm{M}\boldsymbol{v} \equiv \boldsymbol{v}_t - 2\mathrm{i}\mathbf{P}\boldsymbol{v}_{xx} - (\mathrm{i}\mathbf{P}_x + \mathrm{i}\sigma\mathbf{P}_y\mathbf{P} + \sigma^2\mathbf{P}\phi_x - \phi_y)\boldsymbol{v}_x = \mathbf{0}, \tag{5.6.3b}$$

where

$$\mathbf{P} = \sum_{j=1}^{3} S_j \boldsymbol{\sigma}_j \equiv S_1 \begin{pmatrix} 0 & 1 \\ 1 & 0 \end{pmatrix} + S_2 \begin{pmatrix} 0 & -\mathrm{i} \\ \mathrm{i} & 0 \end{pmatrix} + S_3 \begin{pmatrix} 1 & 0 \\ 0 & -1 \end{pmatrix}. \tag{5.6.3c}$$

However equations (5.6.1) and the DS equations are gauge equivalent. Suppose that $\mathbf{g}(x,y,t)$ is a 2×2 matrix such that

$$\mathbf{P} = -\mathbf{g}^{-1}\boldsymbol{\sigma}_3\mathbf{g}, \tag{5.6.4a}$$

$$\mathbf{Q} = \begin{pmatrix} 0 & q \\ r & 0 \end{pmatrix} = \boldsymbol{\sigma}_3\mathbf{g}_x\mathbf{g}^{-1} - \sigma\mathbf{g}_y\mathbf{g}^{-1}. \tag{5.6.4b}$$

Under this transformation, the Lax pair (5.6.3) is mapped into

$$\tilde{\mathrm{L}}\boldsymbol{v} \equiv \sigma\boldsymbol{v}_y - \boldsymbol{\sigma}_3\boldsymbol{v}_x + \mathbf{Q}\boldsymbol{v} = \mathbf{0}, \tag{5.6.5a}$$

$$\tilde{\mathrm{M}}\boldsymbol{v} \equiv \boldsymbol{v}_t - 2\mathrm{i}\boldsymbol{\sigma}_3\boldsymbol{v}_{xx} - 2\mathrm{i}\mathbf{Q}\boldsymbol{v}_x - \mathrm{i}(2\mathbf{A} + \mathbf{Q}_x + \sigma\boldsymbol{\sigma}_3\mathbf{Q}_y)\boldsymbol{v} = \mathbf{0}, \tag{5.6.5b}$$

with $\mathbf{A} = \mathrm{diag}(A_1, A_2)$, which is the Lax pair for the DS equations in the form

$$iq_t - (q_{xx} + \sigma^2 q_{yy}) + 2q\phi = 0, \tag{5.6.6a}$$
$$-ir_t - (r_{xx} + \sigma^2 r_{yy}) + 2r\phi = 0, \tag{5.6.6b}$$
$$\phi_{xx} - \sigma^2\phi_{yy} + (rq)_{xx} + \sigma^2(rq)_{yy} = 0, \tag{5.6.6c}$$

with $\phi = \frac{1}{2}(A_1 - A_2)$.

5.6.2 Multidimensional Isospectral Flows Associated with Second Order Scalar Operators. Recently Boiti, Leon, Manna and Pempinelli (Boiti, Leon, Manna and Pempinelli [1986, 1987]; Boiti, Leon and Pempinelli [1987a,b]) have discussed inverse scattering for 2 + 1-dimensional equations involving a "weak Lax pair." The standard inverse scattering method involves writing an nonlinear evolution equations as the compatibility condition of two operators L and M, i.e., they commute

$$[\mathrm{L}, \mathrm{M}] = 0, \tag{5.6.7}$$

if and only if a potential u satisfies a nonlinear equation. However, in some cases, this requirement maybe too strong. For example, consider the linear spectral problem

$$\mathrm{L}v \equiv v_{xx} - v_{yy} - u(x,y,t)v = 0. \tag{5.6.8a}$$

The time evolution of $u(x,y,t)$ is determined by requiring that an auxiliary operator M given by

$$\mathrm{M}v \equiv v_t + F(\partial_x, \partial_y, u)\, v, \tag{5.6.8b}$$

with $\partial_x \equiv \partial/\partial x$ and $\partial_y \equiv \partial/\partial y$, satisfies (5.6.7). It is known that only trivial local operators M can commute with L (Novikov, Manakov, Pitaevski and Zakharov [1984]). Boiti *et al.* showed that the Lax pair condition can be relaxed by only requiring that the operator M commutes with L on the subspace of the solutions of the spectral equation (5.6.8),

$$[\mathrm{L}, \mathrm{M}]v = 0. \tag{5.6.9}$$

It is convenient to introduce the variables $\xi = \frac{1}{2}(x+y)$, and $\eta = \frac{1}{2}(x-y)$, then

$$\mathrm{L}v \equiv v_{\xi\eta} - u(\xi,\eta,t)v = 0. \tag{5.6.10}$$

A third-order nonlinear partial differential equation can be found which arises as the compatibility condition of equations (5.6.9,10) is given by

$$\begin{aligned} u_t = &-(ua)_{\xi\xi\xi} + (ua_\xi)_{\xi\xi} + (ua_{\xi\xi})_\xi + 3(ua\partial_\eta^{-1}u_\xi)_\xi + (ua_\xi\partial_\eta^{-1}u)_\xi - (uc)_\xi \\ &- (ub)_{\eta\eta\eta} + (ub_\eta)_{\eta\eta} + (ub_{\eta\eta})_\eta + 3(ub\partial_\xi^{-1}u_\eta)_\eta + (ub_\eta\partial_\xi^{-1}u)_\eta - (ud)_\eta, \end{aligned} \tag{5.6.11}$$

where $a(\xi,t)$, $b(\eta,t)$, $c(\xi,t)$, $d(\eta,t)$ are arbitrary functions and

$$(\partial_\eta^{-1} f)(\xi,\eta,t) := \int_{-\infty}^{\eta} f(\xi,\eta',t)\,\mathrm{d}\eta', \qquad (\partial^{-1} I_\xi f)(\xi,\eta,t) := \int_{-\infty}^{\xi} f(\xi',\eta,t)\,\mathrm{d}\xi'.$$

Choosing $a = 1$, $b = c = d = 0$, yields

$$u_t + u_{\xi\xi\xi} = 3(u\partial_\eta^{-1} u_\xi)_\xi, \tag{5.6.12}$$

which reduces to the KdV equation if $\xi = \eta$. Boiti, Leon, Manna and Pempinelli [1986] developed an I.S.T. scheme to solve the Cauchy problem for equation (5.6.11) for initial data decaying sufficiently rapidly at infinity using the spectral equation (5.6.10). This was formulated as a nonlocal Riemann-Hilbert boundary value problem. We remark that equation (5.6.12) arises as a reduction of the equation

$$w_{xxxy} - 3(w_x w_y)_x + w_{yt} - 3w_{xz} = 0, \tag{5.5.13}$$

which was introduced by Jimbo and Miwa [1983] as the second member in a hierarchy of the KP equation; however the equations in this hierarchy such as (5.6.13) are not completely integrable in these used in these notes (see also Dorizzi, Grammaticos, Ramani and Winternitz [1986]).

Similar ideas apply for the spectral problem

$$\mathrm{L}v \equiv v_{xx} + v_{yy} + u(x,y,t)v = 0, \tag{5.6.14}$$

or defining $z = x + \mathrm{i}y$, $\bar{z} = x - \mathrm{i}y$

$$\mathrm{L}v \equiv v_{z\bar{z}} + u(z,\bar{z},t)v = 0. \tag{5.6.15}$$

Dubrovin, Krichever and Novikov [1976] and Vesselov and Novikov [1984a,b] have studied the direct and inverse problem in the periodic case by using algebraic geometric methods (see also Nizhnik [1980]; Grinevich and Novikov [1986]). In this case the most general third-order partial differential equation is given by

$$\begin{aligned} u_t = &-(ua)_{zzz} + (ua_z)_{zz} + (ua_{zz})_z + 3(ua\partial_{\bar{z}}^{-1} u_z)_z + (ua_z\partial_{\bar{z}}^{-1} u)_z - (uc)_z \\ &- (ub)_{\bar{z}\bar{z}\bar{z}} + (ub_{\bar{z}})_{\bar{z}\bar{z}} + (ub_{\bar{z}\bar{z}})_{\bar{z}} + 3(ub\partial_z^{-1} u_{\bar{z}})_{\bar{z}} + (ub_{\bar{z}}\partial_z^{-1} u)_{\bar{z}} - (ud)_{\bar{z}}, \end{aligned} \tag{5.6.16}$$

where $a(z,t)$, $b(\bar{z},t)$, $c(z,t)$, $d(\bar{z},t)$ are arbitrary functions and

$$(\partial_{\bar{z}}^{-1} f)(z,\bar{z},t) := \int_{-\infty}^{\bar{z}} f(z,\bar{z}',t)\,\mathrm{d}\bar{z}', \qquad (\partial_z^{-1} f)(z,\bar{z},t) := \int_{-\infty}^{z} f(z',\bar{z},t)\,\mathrm{d}z'.$$

If it is required that u is real, then $a = \bar{b}$ and $c = \bar{d}$. If $a = b = 1$ and $c = d = 0$, then (5.6.17) simplifies to (Nizhnik [1980]; Vesselov and Novikov [1984a,b]; Grinevich and Novikov [1986])

$$u_t + u_{zzz} + u_{\bar{z}\bar{z}\bar{z}} = 3(u\partial_{\bar{z}}^{-1} u_z)_z + 3(u\partial_z^{-1} u_{\bar{z}})_{\bar{z}}. \tag{5.6.17}$$

Bogdanov and Manakov [1988] also found solutions to this equation through a nonlocal $\bar{\partial}$ problem. Boiti, Leon, Manna and Pempinelli [1987] developed an I.S.T. scheme for equation (5.6.16), using the spectral equation (5.6.15) which they was formulated as a $\bar{\partial}$ problem.

Boiti, Leon and Pempinelli [1987a,b] also used these ideas on Lax pairs

$$\mathrm{L}_1 v = \sigma v_y + v_{xx} - u(x,y,t)v, \qquad \sigma^2 = \pm 1, \tag{5.6.18a}$$
$$\mathrm{M}_1 v = v_{xt} + \alpha(x,y,t)v_t + \beta(x,y,t), \tag{5.6.18b}$$
$$\mathrm{L}_2 v = v_{xy} - r(x,y,t)v_y - q(x,y,t)v, \tag{5.6.19a}$$
$$\mathrm{M}_2 v = v_{xx} + v_t + \gamma(x,y,t), \tag{5.6.19b}$$

together the "weak Lax" compatibility condition

$$[\mathrm{L}_j, \mathrm{M}_j]v = 0, \qquad \mathrm{L}_j v = 0, \qquad j = 1,2. \tag{5.6.20}$$

(We remark that these pairs of operators are related since one can be obtained from the other be the exchange $y \leftrightarrow t$, choosing σ and renaming the potentials.)

The compatibility (5.6.18) yields the following system of evolution equations are

$$u_t = -2\beta_x, \tag{5.6.21a}$$
$$u_x = (\alpha^2 - \alpha_x)_x - \sigma\alpha_y, \tag{5.6.21b}$$
$$\beta_{xx} + 2(\alpha\beta)_x - \sigma\beta_y = 0. \tag{5.6.21c}$$

Defining $\beta = \frac{1}{4}\mathrm{e}^\theta$, then

$$\alpha = -\tfrac{1}{2}\left(\theta_x - \sigma \mathrm{e}^{-\theta}\partial_x^{-1}(\mathrm{e}^\theta)_y\right),$$

and so we obtain the generalized Sinh-Gordon equation

$$\left[\mathrm{e}^{-\theta}\left\{\mathrm{e}^\theta(\theta_{xt}+\sinh\theta)\right\}_x\right]_x = \sigma^2\left[\mathrm{e}^{-\theta}\partial_x^{-1}(\mathrm{e}^\theta)_y\right]_{yt} - \tfrac{1}{2}\sigma^2\left[\left(\mathrm{e}^{-\theta}\partial_x^{-1}(\mathrm{e}^\theta)_y\right)^2\right]_{xt}. \tag{5.6.22}$$

If θ is independent of y and $\mathrm{e}^\theta \to 1$ as $x \to \infty$, then this reduces to the sinh-Gordon equation

$$\theta_{xt} + \sinh\theta = 0,$$

which is known to be completely integrable (Ablowitz, Kaup, Newell and Segur [1974]). Another integrable one-dimensional equation is obtained from (5.6.22) by setting $y = x$, then $\Theta(x,t) := \theta(x,x,t)$ satisfies

$$\left[\mathrm{e}^\Theta(\Theta_{xt} + \sinh\Theta)\right]_x = -\sigma^2(\mathrm{e}^\Theta)_t, \qquad \sigma^2 = \pm 1.$$

The I.S.T. scheme for the generalized Sinh-Gordon equation (5.6.22) it is formulated through a nonlocal Riemann-Hilbert boundary value problem if $\sigma = \mathrm{i}$, and a $\bar{\partial}$ problem if $\sigma = 1$.

The compatibility (5.6.19) yields the following system of evolution equations are

$$r_{yt} = 2q_{xx} - (r_x + r^2)_{xy}, \tag{5.6.23a}$$

$$q_t = q_{xx} - 2(rq)_x, \tag{5.6.23b}$$

together with $\gamma_y = -2q_x$. If we make the change of variables

$$4q = -1 - \eta + u_y, \qquad 2r = u,$$

we obtain the generalized dispersive long wave equation

$$u_{ty} = -(\eta_x + uu_y)_x, \tag{5.6.24a}$$

$$\eta_t = -(u\eta + u + u_{xy})_x, \tag{5.6.24b}$$

In the one-dimensional reduction $x = y$, equations (5.6.24) becomes the dispersive long wave equations

$$u_t = -\eta_x - uu_x, \tag{5.6.25a}$$

$$\eta_t = -(u\eta + u + u_{xx})_x, \tag{5.6.25b}$$

which are known to be completely integrable (Kaup [1975]; Kuperschmidt [1984, 1985]; Antonowicz and Fordy [1987, 1989]). Boiti, Leon and Pempinelli [1987b] derived an I.S.T. scheme for the generalized long wave equations (5.6.24); it is formulated through a nonlocal Riemann-Hilbert boundary value problem.

Chapter Six

Inverse Scattering in Multidimensions

6.1 Introduction.

In this Chapter we discuss I.S.T. schemes for multidimensional equations. The application of the $\bar{\partial}$ method to study nonlinear evolution equations in two spatial dimensions has led to the development of a general formalism to implement inverse scattering in n spatial dimensions (Nachman and Ablowitz [1984a,b]; Ablowitz and Nachman [1986] — see also Beals and Coifman [1985b, 1986]).

In §6.2 we discuss inverse scattering associated with the natural multidimensional generalization of equation (5.1.2), namely

$$\sigma \frac{\partial v}{\partial y} + \sum_{j=1}^{n} \frac{\partial^2 v}{\partial x_j^2} + u(\boldsymbol{x}, y) v = 0, \tag{6.1.1}$$

with $\sigma = \sigma_R + i\sigma_I$, $\boldsymbol{x} = (x_1, x_2, \ldots, x_n) \in \mathbb{R}^n$, $y \in \mathbb{R}$. It turns out that the inverse scattering for (6.1.1) is very similar to that for equation (5.1.2) except that there is a very important constraint which the scattering data must satisfy. The form of this constraint suggests why no local nonlinear multidimensional evolution equations have been found to be associated with equation (6.1.1) (for $n > 2$). We remark that the study of multidimensional inverse scattering associated with the time-independent Schrödinger problem (i.e., (6.1.1) with $\sigma = 0$), has a long history. Important contributions have been made by Faddeev [1966, 1976]; Newton [1970, 1980, 1981, 1982a,b,c, 1983, 1989]; and more recently by Lavine and Nachman [1987]; Beals and Coifman [1985b]; Novikov and Khenkin [1987a,b]; Aktosun and van der Mee [1990, 1991]; Weder [1990, 1991].

Indeed inverse scattering is a rather substantial topic on its own. Here we shall primarily be concerned with the $\bar{\partial}$ especially as it applies to operators which arise as natural generalizations of the scattering problems we studied in Chapter 5.

The second important scattering problem in one spatial dimension is the $N \times N$ scattering problem, which we discussed in §3.2. There exist multidimensional generalizations of this scattering problem analogous to the multidimensional generalizations of the Schrödinger scattering problem discussed above. The multidimensional generalization of the $N \times N$ scattering problem is

$$\frac{\partial \boldsymbol{v}}{\partial y} + \sigma \sum_{p=1}^{n} \mathbf{J}_p \frac{\partial \boldsymbol{v}}{\partial x_p} = \mathbf{Q}(\boldsymbol{x}, y) \boldsymbol{v}, \tag{6.1.2}$$

with $\sigma = \sigma_R + i\sigma_I$, $\boldsymbol{x} = (x_1, x_2, \ldots, x_n) \in \mathbb{R}^n$, $y \in \mathbb{R}$, where the $\mathbf{J}_p$ are constant real diagonal matrices (we denote the diagonal entries of $\mathbf{J}_p$ by $J_p^1, J_p^2, \ldots, J_p^N$ and assume that $J_p^i \neq J_p^j \neq 0$, for $i \neq j$, $i, j = 1, 2, \ldots, N$, $p = 1, 2, \ldots, n$; $\mathbf{Q}(x, y) = (Q^{ij}) \in \mathbb{C}^{N\times N}$ is an off-diagonal matrix containing the potentials. The analysis for the inverse problem associated with equation (6.1.2) is analogous to that for the n-dimensional Schrödinger equation (6.1.1); i.e., the scattering data must satisfy a nonlinear constraint (Nachman and Ablowitz [1984b]; Fokas [1986a,b]). We discuss inverse scattering associated with the operator (6.1.2) in §6.3. Sung and Fokas [1991a] have considered rigorous aspects of the scattering problem (6.1.2) in the case $\sigma_I \neq 0$.

Over one hundred years ago, Bäcklund investigated transformation properties of pseudospherical surfaces, that is surfaces of constant negative curvature (Bäcklund [1876, 1880, 1882, 1905]). (A surface of constant curvature is one with the same total curvature κ, the product of the principal curvatures, at every point and a pseudospherical surface is one with $\kappa < 0$). As a consequence of this classical study of surfaces $M^2 \in \mathbb{R}^3$, Bäcklund derived a transformation which generates a new solution of the Sine-Gordon equation

$$u_{tt} - u_{xx} - \kappa \sin u = 0, \tag{6.1.3}$$

from a given solution. Motivated by the work of Cherg and Terng [1980], Bäcklunds results were generalized to higher dimensions by Terng and Tenenblat (Tenenblat and Terng [1980]; Terng [1980]). In particular, by considering surfaces $M^n \in \mathbb{R}^{2n-1}$, they obtained an n-dimensional generalized Sine-Gordon (GSG) equation

$$\begin{aligned}
&\frac{\partial}{\partial x_i}\left(\frac{1}{a_{1i}}\frac{\partial a_{1j}}{\partial x_i}\right) + \frac{\partial}{\partial x_j}\left(\frac{1}{a_{1j}}\frac{\partial a_{1i}}{\partial x_j}\right) \\
&\qquad + \sum_{k\neq i,j} \frac{1}{a_{ij}^2}\frac{\partial a_{1i}}{\partial x_k}\frac{\partial a_{1k}}{\partial x_j} = \epsilon a_{1i}a_{1j}, \qquad i \neq j, \\
&\frac{\partial}{\partial x_k}\left(\frac{1}{a_{1j}}\frac{\partial a_{1i}}{\partial x_j}\right) = \frac{1}{a_{1k}a_{1j}}\frac{\partial a_{1i}}{\partial x_k}\frac{\partial a_{1k}}{\partial x_j}, \qquad i, j, k \quad \text{distinct} \\
&\qquad \frac{\partial a_{jk}}{\partial x_i} = \frac{a_{ji}}{a_{1i}}\frac{\partial a_{1k}}{\partial x_i}, \qquad i \neq k.
\end{aligned} \tag{6.1.4}$$

where $\epsilon = 1$. Note that the n-dimensional GSG equation is not obtained by simply replacing the u_{xx} term in (6.1.3) by Δu, where Δ is the n-dimensional Laplacian operator. In fact, there is considerable evidence suggesting that the equation

$$\frac{\partial^2 u}{\partial t^2} - \sum_{j=1}^{n} \frac{\partial^2 u}{\partial x_j^2} - \kappa \sin u = 0,$$

is not completely integrable for $n \geq 2$ (see, for example, Clarkson [1986]). Also, by considering appropriate surfaces embedded in spheres: $M^n \in S^{2n-1}$, Tenenblat [1985] derived an n-dimensional generalized Wave (GW) equation, which is exactly (6.1.4) with $\epsilon = 0$. In §6.7 we discuss the scattering and inverse scattering problems associated with the GW and GSG equations.

It turns out that the initial value solution of the n-dimensional GW and GSG equations reduces to the study of the scattering and inverse scattering associated with a coupled system of n one-dimensional ordinary differential equations. This is as opposed to the multidimensional scattering and inverse scattering which we discuss in §§6.2 and 6.3, where, as mentioned above it turns out that there is a serious constraint inherent in the scattering theory. The analysis for the GW and GSG equations does not have such restrictions since the linear problem is simply a compatible set of n linear one-dimensional scattering problems. However, the results show that the initial value problem is posed by giving data along lines and not on $(n-1)$-dimensional manifolds.

6.2 Multidimensional Inverse Scattering Associated with the Time-Dependent and Time-Independent Schrödinger Equation.

In this section we describe the results of Ablowitz and Nachman [1986] (see also Nachman and Ablowitz [1984a]) on the inverse scattering problem associated with a generalization of the n-dimensional time-dependent (and time-independent) Schrödinger equation (6.1.1). As mentioned earlier, equation (6.2.1) is a natural multidimensional generalization of the linear scattering problem associated with the KP equation, which we discussed in detail in §§5.2, 5.3 above.

6.2.1 The Direct Scattering Problem. In order to study the scattering problem (6.1.1), we seek a solution in the form

$$v(\boldsymbol{x}, y; \boldsymbol{k}) = m(\boldsymbol{x}, y; \boldsymbol{k}) \exp\left\{ \mathrm{i}\boldsymbol{k} \bullet \boldsymbol{x} + \boldsymbol{k}^2 y/\sigma \right\}, \tag{6.2.1}$$

where $\boldsymbol{k} = \boldsymbol{k}_R + \mathrm{i}\boldsymbol{k}_I \in \mathbb{C}^n$. Hence the eigenfunction m satisfies

$$\sigma \frac{\partial m}{\partial y} + \sum_{j=1}^{n} \frac{\partial^2 m}{\partial x_j^2} + 2\mathrm{i}\boldsymbol{k} \bullet \nabla m + u(\boldsymbol{x}, y) m = 0. \tag{6.2.2}$$

We shall consider the case $\sigma_R \neq 0$. In a similar manner to that used above in §5.3, we seek a solution m of (6.2.2) which is bounded for all $\boldsymbol{x}$, y and $m \to 1$ as $|\boldsymbol{k}| \to \infty$ (which is a convenient normalization). When we consider (6.2.1) for $\sigma = \pm 1$ (as with the scattering problem for the KPII equation), we immediately notice that the dominant operator is the heat operator which is ill-posed as an initial value problem. Even though we pose a boundary value problem, immediately we suspect that in this case there will be some type of unusual behavior. In fact it will be shown that the bounded function m for $\sigma_R \neq 0$ is nowhere analytic as a function of the spectral parameter $\boldsymbol{k}$; specifically $m = m(\boldsymbol{x}, y, \boldsymbol{k}_R, \boldsymbol{k}_I)$. In particular m is constructed from the following equation. Given that $u(\boldsymbol{x}, y) \to 0$ sufficiently rapidly as $\sqrt{(\boldsymbol{x}^2 + y^2)} \to \infty$, the direct problem is given by

$$m = 1 + \tilde{\mathrm{G}}(um), \tag{6.2.3a}$$

where

$$\tilde{\mathrm{G}}f \equiv G * f \equiv \iint_{\mathbb{R}^{n+1}} G(\boldsymbol{x} - \boldsymbol{x}', y - y'; \boldsymbol{k}_R, \boldsymbol{k}_I)\, f(\boldsymbol{x}', y')\, \mathrm{d}\boldsymbol{x}'\, \mathrm{d}y'. \tag{6.2.3b}$$

We seek a bounded Green's function and take the Fourier transform in both $\boldsymbol{x}$ and y, hence

$$G(\boldsymbol{x}, y; \boldsymbol{k}_R, \boldsymbol{k}_I) = C_{n+1} \iint_{\mathbb{R}^{n+1}} \frac{\exp\{\mathrm{i}(\boldsymbol{x} \bullet \boldsymbol{\xi} + y\eta)\}}{\boldsymbol{\xi}^2 + 2\boldsymbol{k} \bullet \boldsymbol{\xi} - \mathrm{i}\sigma\eta}\, \mathrm{d}\boldsymbol{\xi}\, \mathrm{d}\eta \tag{6.2.4a}$$

$$= \operatorname{sgn}(-y)\sigma^{-1} C_n \int_{\mathbb{R}^n} \exp\left\{y(\boldsymbol{\xi}^2 + 2\boldsymbol{k} \bullet \boldsymbol{\xi})\sigma^{-1} + \mathrm{i}\boldsymbol{x} \bullet \boldsymbol{\xi}\right\} \times \Theta\left(-y\left[\sigma_R \boldsymbol{\xi}^2 + 2\left(\sigma_R \boldsymbol{k}_R + \sigma_I \boldsymbol{k}_I\right) \bullet \boldsymbol{\xi}\right]\right) \mathrm{d}\boldsymbol{\xi}, \tag{6.2.4b}$$

where

$$C_n = (2\pi)^{-n}; \qquad \Theta(x) = \begin{cases} 1, & x > 0, \\ 0, & x < 0. \end{cases}$$

(Note that since $\sigma_R \neq 0$, then the Green's function is nowhere analytic, as was the case for KPII in §5.3.)

Taking the $\bar{\partial}$ derivative of (6.2.3a) with respect to $\bar{k}_j$ we find [recall $\partial/\partial\bar{k}_j = \frac{1}{2}\left(\partial/\partial k_{R_j} + \mathrm{i}\partial/\partial k_{I_j}\right)$]:

$$\frac{\partial m}{\partial \bar{k}_j} = \frac{\partial \tilde{\mathrm{G}}}{\partial \bar{k}_j}(um) + \tilde{\mathrm{G}}\left(u\,\frac{\partial m}{\partial \bar{k}_j}\right). \tag{6.2.5}$$

The first term in (6.2.5) is evaluated directly using the definition of the Green's function (6.2.4)

$$\frac{\partial \tilde{\mathrm{G}}}{\partial \bar{k}_j}(um) = -\frac{C_n}{|\sigma_R|} \int_{\mathbb{R}^n} \exp\{\mathrm{i}\beta(\boldsymbol{x}, y; \boldsymbol{k}_R, \boldsymbol{k}_I, \boldsymbol{\xi})\} \times T(\boldsymbol{k}_R, \boldsymbol{k}_I, \boldsymbol{\xi})\, (\xi_j - k_{R_j})\, \delta\left(s(\boldsymbol{\xi})\right) \mathrm{d}\boldsymbol{\xi}, \tag{6.2.6a}$$

where

$$T(\boldsymbol{k}_R, \boldsymbol{k}_I, \boldsymbol{\xi}) \equiv \iint_{\mathbb{R}^{n+1}} \exp\{-\mathrm{i}\beta(\boldsymbol{x}, y; \boldsymbol{k}_R, \boldsymbol{k}_I, \boldsymbol{\xi})\} \times u(\boldsymbol{x}, y)\, m(\boldsymbol{x}, y; \boldsymbol{k}_R, \boldsymbol{k}_I)\, \mathrm{d}\boldsymbol{x}\, \mathrm{d}y, \tag{6.2.6b}$$

$$\beta(\boldsymbol{x}, y; \boldsymbol{k}_R, \boldsymbol{k}_I, \boldsymbol{\xi}) = \left(\boldsymbol{x} + \frac{2y}{\sigma_R}\boldsymbol{k}_I\right) \bullet (\boldsymbol{\xi} - \boldsymbol{k}_R), \tag{6.2.6c}$$

$$s(\boldsymbol{\xi}) = s(\boldsymbol{\xi}, \boldsymbol{k}_R, \boldsymbol{k}_I) \equiv \left(\boldsymbol{\xi} + \frac{\sigma_I}{\sigma_R}\boldsymbol{k}_I\right)^2 - \left(\boldsymbol{k}_R + \frac{\sigma_I}{\sigma_R}\boldsymbol{k}_I\right)^2, \tag{6.2.6d}$$

and $\delta(\bullet)$ is the usual Dirac delta function. Equation (6.2.6) can be derived either by taking the $\partial/\partial\overline{k}_j$ derivative directly on (6.2.4b) or on (6.2.4a) and using the well-known fact

$$\frac{\partial}{\partial\overline{z}}\left(\frac{1}{z-z_0}\right)=\pi\delta(z-z_0).$$

Using equation (6.2.6), we can readily calculate $\partial m/\partial\overline{k}_j$ (assuming that equation (6.2.3a) has no homogeneous solutions)

$$\frac{\partial m}{\partial\overline{k}_j}=\tilde{\mathrm{T}}_j m=-\frac{C_n}{|\sigma_R|}\int_{\mathbb{R}^n}\exp\{\mathrm{i}\beta(\boldsymbol{x},y;\boldsymbol{k}_R,\boldsymbol{k}_I,\boldsymbol{\xi})\}\,T(\boldsymbol{k}_R,\boldsymbol{k}_I,\boldsymbol{\xi})$$
$$\times(\xi_j-k_{R_j})\,\delta(s(\boldsymbol{\xi}))\,m(\boldsymbol{x},y,\boldsymbol{\xi},\boldsymbol{k}_I)\,\mathrm{d}\boldsymbol{\xi}. \qquad (6.2.7j)$$

This is derived by noting that $\partial m/\partial\overline{k}_j$ is a suitable superposition over a fundamental solution $W(\boldsymbol{x},y;\boldsymbol{k}_R,\boldsymbol{k}_I,\boldsymbol{\xi})$ satisfying

$$W(\boldsymbol{x},y;\boldsymbol{k}_R,\boldsymbol{k}_I,\boldsymbol{\xi})=\exp\{\mathrm{i}\beta(\boldsymbol{x},y;\boldsymbol{k}_R,\boldsymbol{k}_I,\boldsymbol{\xi})\}+\tilde{\mathrm{G}}(uW). \qquad (6.2.8)$$

Using the symmetry condition on the Green's function

$$\exp\{-\mathrm{i}\beta(\boldsymbol{x},y;\boldsymbol{k}_R,\boldsymbol{k}_I,\boldsymbol{\xi})\}\,G(\boldsymbol{x},y;\boldsymbol{k}_R,\boldsymbol{k}_I)=G(\boldsymbol{x},y,\boldsymbol{\xi},\boldsymbol{k}_I), \qquad (6.2.9)$$

on $s(\boldsymbol{\xi})=0$, it follows that

$$W(\boldsymbol{x},y;\boldsymbol{k}_R,\boldsymbol{k}_I)=\exp\{\mathrm{i}\beta(\boldsymbol{x},y;\boldsymbol{k}_R,\boldsymbol{k}_I,\boldsymbol{\xi})\}\,m(\boldsymbol{x},y;\boldsymbol{\xi},\boldsymbol{k}_I), \qquad (6.2.10)$$

on $s(\boldsymbol{\xi})=0$, and then equation (6.2.7j) follows.

For the special case $n=1$ of equation (6.2.7j), $\partial m/\partial\overline{k}_j$ depends locally on m. For $n=1$, let $x_1\equiv x$ and $k_1\equiv k$; then (6.2.7) reduces to

$$\frac{\partial m}{\partial\overline{k}}=\frac{1}{2\pi|\sigma_R|}\,\mathrm{sgn}\left(k_R+\frac{\sigma_I}{\sigma_R}k_I\right)\exp\{\mathrm{i}\beta(x,y;k_R,k_I,\xi_0)\}$$
$$\times\,T(k_R,k_I,\xi_0)\,m(x,y;\xi_0,k_I), \qquad (6.2.11)$$

where $\xi_0=-k_R-(2\sigma_I/\sigma_R)k_I$. Equation (6.2.11) defines the $\overline{\partial}$ derivative of m which can be used to solve the KP equation; for KPII we take $\sigma_I=0$, $\sigma_R=-1$ [in §5.3, compare equation (6.2.11) with equation (5.3.21)] and for KPI let $\sigma_I=1$, take the limit $\sigma_R\to0$ (with $\sigma_R<0$) after making the scaling $\hat{k}_I=k_I/\sigma_R$, in fact this limiting $\overline{\partial}$ procedure gives an alternative I.S.T. solution for KPI (also see the discussion of the limit to the time-dependent Schrödinger equation in §6.2.4), though as we noted in Chapter 5, this solution is ill-posed.

6.2.2 The Inverse Scattering Problem. The discussion in §6.2.2 above has been solely concerned with the direct scattering problem. However, it suggests what the natural data might be for this problem. We shall call $T(\boldsymbol{k}_R,\boldsymbol{k}_I,\boldsymbol{\xi})$ the *inverse data*.

The inverse problem is: given $T(\boldsymbol{k}_R, \boldsymbol{k}_I, \boldsymbol{\xi})$, construct $u(\boldsymbol{x}, y)$. However, it is immediately obvious that there is a serious redundancy question. Namely $T(\boldsymbol{k}_R, \boldsymbol{k}_I, \boldsymbol{\xi})$ is a function of $3n$ parameters with one restriction [due to $\delta(s(\boldsymbol{\xi}))$ in equation (6.2.7)]: i.e., T will be given as a function of $3n-1$ variables and we wish to construct a function $u(\boldsymbol{x}, y)$ depending on $n+1$ variables. (Note that for $n=1$, namely for the problem in two spatial dimensions, this difficulty disappears since $T = T(k_R, k_I, \xi_0(k_R, k_I))$ is a function of two parameters, as is $u(x, y)$.)

Using equation (6.2.7) there are several possible reconstruction formulae for $u(\boldsymbol{x}, y)$. However, crucial restrictions on T are imposed by the requirement that u depends only on $\boldsymbol{x}$, y and decays at ∞. This is part of the characterization question: i.e. which inverse data $T(\boldsymbol{k}_R, \boldsymbol{k}_I, \boldsymbol{\xi})$ are "admissible".

One set of inversion formulae for m may be obtained using the one-dimensional generalized Cauchy formula

$$m(\boldsymbol{x}, y; k) = \frac{1}{2\pi \mathrm{i}} \iint_R \frac{\partial m}{\partial \bar{z}}(\boldsymbol{x}, y; z) \frac{\mathrm{d}z \wedge \mathrm{d}\bar{z}}{z-k} + \frac{1}{2\pi \mathrm{i}} \int_C \frac{m(\boldsymbol{x}, y; z)}{z-k} \mathrm{d}z,$$

where R and C are an appropriate domain and contour respectively, $z = z_R + \mathrm{i}z_I$ and $\mathrm{d}z \wedge \mathrm{d}\bar{z} = -2\mathrm{i}\mathrm{d}z_R\mathrm{d}z_I$. Taking R to be the entire complex plane R_∞ and C to be the contour C_∞ at infinity, then since $m \to 1$ as $|k| \to \infty$, the second integral is unity and so we have

$$m(\boldsymbol{x}, y; \boldsymbol{k}_R, \boldsymbol{k}_I) = 1 + \frac{1}{\pi} \int_{-\infty}^{\infty} \int_{-\infty}^{\infty} \frac{\partial m}{\partial \bar{\kappa}_j}(\boldsymbol{x}, y; \boldsymbol{\kappa}_R^j, \boldsymbol{\kappa}_I^j) \frac{\mathrm{d}\kappa_{R_j} \, \mathrm{d}\kappa_{I_j}}{k_j - \kappa_j}, \tag{6.2.12}$$

where we use the simplified notation $\boldsymbol{\kappa}_R^j = (k_{R_1}, \ldots, k_{R_{j-1}}, \kappa_{R_j}, k_{R_{j+1}}, \ldots, k_{R_n})$ and similarly for $\boldsymbol{\kappa}_I^j$. (6.2.12) is a linear integral equation and by using equation (6.2.7j) it can be shown (see below) that the potential is reconstructed from the formula

$$u(\boldsymbol{x}, y) = \frac{2\mathrm{i}}{\pi} \frac{\partial}{\partial x_j} \left(\int_{-\infty}^{\infty} \int_{-\infty}^{\infty} \frac{\partial m}{\partial \bar{\kappa}_j}(\boldsymbol{x}, y; \boldsymbol{\kappa}_R^j, \boldsymbol{\kappa}_I^j) \, \mathrm{d}\kappa_{R_j} \, \mathrm{d}\kappa_{I_j} \right). \tag{6.2.13}$$

Equation (6.2.13) is derived by considering the limit as $|k_j| \to \infty$ in equations (6.2.3) and (6.2.12) and then comparing the k_j^{-1} terms as follows; from equations (6.2.3) and (6.2.12)

$$\begin{aligned} \iint_{\mathbb{R}^{n+1}} & G(\boldsymbol{x} - \boldsymbol{x}', y - y'; \boldsymbol{k}_R, \boldsymbol{k}_I) \, u(\boldsymbol{x}', y') \, m(\boldsymbol{x}', y'; \boldsymbol{k}_R, \boldsymbol{k}_I) \, \mathrm{d}\boldsymbol{x}' \, \mathrm{d}y' \\ &= \frac{1}{\pi} \int_{-\infty}^{\infty} \int_{-\infty}^{\infty} \frac{\partial m}{\partial \kappa_j}(\boldsymbol{x}, y; \boldsymbol{\kappa}_R^j, \boldsymbol{\kappa}_I^j) \frac{\mathrm{d}\kappa_{R_j} \, \mathrm{d}\kappa_{I_j}}{k_j - \kappa_j}, \end{aligned} \tag{6.2.14}$$

where

$$G(\boldsymbol{x}, y; \boldsymbol{k}_R, \boldsymbol{k}_I) = C_{n+1} \iint_{\mathbb{R}^{n+1}} \frac{\exp\{\mathrm{i}(\boldsymbol{x} \bullet \boldsymbol{\xi} + y\eta)\}}{\boldsymbol{\xi}^2 + 2\boldsymbol{k} \bullet \boldsymbol{\xi} - \mathrm{i}\sigma\eta} \, \mathrm{d}\boldsymbol{\xi} \, \mathrm{d}\eta. \tag{6.2.15}$$

As $k_j \to \infty$, from equations (6.2.12) and (6.2.15), we see that

$$m(\boldsymbol{x}, y; \boldsymbol{k}_R, \boldsymbol{k}_I) = 1 + O\left(k_j^{-1}\right), \tag{6.2.16}$$

and

$$G(\boldsymbol{x}, y; \boldsymbol{k}_R, \boldsymbol{k}_I) = \frac{1}{(2\pi)^{n+1}} \frac{1}{2k_j} \iint_{\mathbb{R}^{n+1}} \exp\{\mathrm{i}(\boldsymbol{x} \cdot \boldsymbol{\xi} + y\eta)\} \frac{\mathrm{d}\boldsymbol{\xi}\, \mathrm{d}\eta}{\xi_j} + O\left(k_j^{-2}\right), \tag{6.2.17}$$

respectively and so equation (6.2.13) follows easily.

6.2.3 The Characterization Problem. Clearly, in general, the right-hand side of equation (6.2.13) will be a function of k_{R_i}, k_{I_i} for $i = 1, 2, \ldots, j-1, j+1, \ldots, n$. One possible way of characterizing admissible data would be to require that $T(\boldsymbol{k}_R, \boldsymbol{k}_I, \boldsymbol{\xi})$ to be such that the right-hand side of equation (6.2.13) are independent of these parameters, for all j. Such a requirement is analogous to the "miracle condition" referred to by Newton [1970, 1980, 1981, 1982a] for the time-independent problem. However, in this $\bar{\partial}$ formulation it is possible to go further and give conditions directly on $T(\boldsymbol{k}_R, \boldsymbol{k}_I, \boldsymbol{\xi})$. The importance of characterizing $T(\boldsymbol{k}_R, \boldsymbol{k}_I, \boldsymbol{\xi})$ directly not only has to do with understanding on which manifolds of $\boldsymbol{k}_R$, $\boldsymbol{k}_I$, one can hope to reconstruct the potential, but also may indicate how one could in principle measure data so as to produce local potentials in a stable manner.

For $n > 1$, the compatibility condition

$$\frac{\partial^2 m}{\partial \bar{k}_i \partial \bar{k}_j} = \frac{\partial^2 m}{\partial \bar{k}_j \partial \bar{k}_i}, \qquad \text{for} \qquad i \neq j, \tag{6.2.18}$$

leads to a nontrivial, nonlinear restriction on T:

$$\mathcal{L}_{ij}[T] = \mathrm{N}_{ij}[T], \tag{6.2.19a}$$

where

$$\begin{aligned} \mathcal{L}_{ij}[T] = (\xi_j - k_{R_j}) \left(\frac{\partial T}{\partial \bar{k}_i} + \frac{1}{2} \frac{\partial T}{\partial \xi_i} \right) \\ - (\xi_i - k_{R_i}) \left(\frac{\partial T}{\partial \bar{k}_j} + \frac{1}{2} \frac{\partial T}{\partial \xi_j} \right), \end{aligned} \tag{6.2.19b}$$

$$\begin{aligned} \mathrm{N}_{ij}[T](\boldsymbol{k}, \boldsymbol{\xi}) = \int_{\mathbb{R}^n} \left[(\eta_j - k_{R_j})(\xi_i - \eta_i) - (\eta_i - k_{R_i})(\xi_j - \eta_j) \right] \\ \times \delta(s(\boldsymbol{\eta}))\, T(\boldsymbol{k}_R, \boldsymbol{k}_I, \boldsymbol{\eta})\, T(\boldsymbol{\eta}, \boldsymbol{k}_I, \boldsymbol{\xi})\, \mathrm{d}\boldsymbol{\eta}. \end{aligned} \tag{6.2.19c}$$

To derive this, take $\partial / \partial \bar{k}_i$ of (6.2.7j), $\partial / \partial \bar{k}_j$ of (6.2.7i), then setting them equal yields

$$\begin{aligned} 0 &= \frac{\partial^2 m}{\partial \bar{k}_i \partial \bar{k}_j} - \frac{\partial^2 m}{\partial \bar{k}_j \partial \bar{k}_i}, \\ &= -\frac{C_n}{|\sigma_R|} \int_{\mathbb{R}^n} \left\{ \left[\mathrm{i}\,(\xi_j - k_{R_j}) \frac{\partial \beta}{\partial \bar{k}_i} - \mathrm{i}\,(\xi_i - k_{R_i}) \frac{\partial \beta}{\partial \bar{k}_j} \right] \mathrm{e}^{\mathrm{i}\beta} \right. \\ &\qquad \left. + \mathrm{e}^{\mathrm{i}\beta} \left[(\xi_j - k_{R_j}) \frac{\partial}{\partial \bar{k}_i} - (\xi_i - k_{R_i}) \frac{\partial}{\partial \bar{k}_j} \right] \right\} \\ &\qquad \times T(\boldsymbol{k}_R, \boldsymbol{k}_I, \boldsymbol{\xi}) \delta(s(\boldsymbol{\xi}))\, m(\boldsymbol{x}, y, \boldsymbol{\xi}, \boldsymbol{k}_I)\, \mathrm{d}\boldsymbol{\xi}. \end{aligned} \tag{6.2.20}$$

Since

$$\frac{\partial\beta}{\partial\overline{k}_j} = \tfrac{1}{2}\left\{-\left(x_j + 2yk_{I_j}/\sigma_R\right) + 2\mathrm{i}y\left(\xi_j - k_{R_j}\right)/\sigma_R\right\}$$
$$= -\frac{1}{2}\frac{\partial\beta}{\partial\xi_j} + \mathrm{i}y\left(\xi_j - k_{R_j}\right)/\sigma_R,$$

then

$$\mathrm{i}\left[\left(\xi_j - k_{R_j}\right)\frac{\partial\beta}{\partial\overline{k}_i} - \left(\xi_i - k_{R_i}\right)\frac{\partial\beta}{\partial\overline{k}_j}\right]$$
$$= -\frac{1}{2}\left[\left(\xi_j - k_{R_j}\right)\frac{\partial \mathrm{e}^{\mathrm{i}\beta}}{\partial\xi_i} - \left(\xi_i - k_{R_i}\right)\frac{\partial \mathrm{e}^{\mathrm{i}\beta}}{\partial\xi_j}\right],$$

and so integrating the first term in (6.2.20) by parts, noting the the contribution at ∞ is zero due to the fact that $\delta(s(\boldsymbol{\xi}))|_{\xi=\infty} = 0$ and

$$\frac{\partial}{\partial\xi_i}\left(\xi_j - k_{R_j}\right) - \frac{\partial}{\partial\xi_j}\left(\xi_i - k_{R_i}\right) = 0,$$

yields

$$0 = \int_{\mathbb{R}^n} \mathrm{e}^{\mathrm{i}\beta}\left[\left(\xi_j - k_{R_j}\right)\left(\frac{\partial}{\partial\overline{k}_i} + \frac{1}{2}\frac{\partial}{\xi_i}\right) - \left(\xi_i - k_{R_i}\right)\left(\frac{\partial}{\partial\overline{k}_j} + \frac{1}{2}\frac{\partial}{\xi_j}\right)\right]$$
$$\delta(s(\boldsymbol{\xi}))T(\boldsymbol{k}_R, \boldsymbol{k}_I, \boldsymbol{\xi})\, m(\boldsymbol{x}, y, \boldsymbol{\xi}, \boldsymbol{k}_I)\,\mathrm{d}\boldsymbol{\xi}$$
$$= \int_{\mathbb{R}^n} \mathrm{e}^{\mathrm{i}\beta}\mathcal{L}_{ij}[\delta(s(\boldsymbol{\xi}))T(\boldsymbol{k}_R, \boldsymbol{k}_I, \boldsymbol{\xi})\, m(\boldsymbol{x}, y, \boldsymbol{\xi}, \boldsymbol{k}_I)]\,\mathrm{d}\boldsymbol{\xi}$$
$$= \int_{\mathbb{R}^n} \mathrm{e}^{\mathrm{i}\beta}\delta(s(\boldsymbol{\xi}))\Big\{\mathcal{L}_{ij}[T(\boldsymbol{k}_R, \boldsymbol{k}_I, \boldsymbol{\xi})] m(\boldsymbol{x}, y, \boldsymbol{\xi}, \boldsymbol{k}_I)$$
$$+ T(\boldsymbol{k}_R, \boldsymbol{k}_I, \boldsymbol{\xi})\mathcal{L}_{ij}[m(\boldsymbol{x}, y, \boldsymbol{\xi}, \boldsymbol{k}_I)]\Big\}\mathrm{d}\boldsymbol{\xi},$$

where $\mathcal{L}_{ij}$ is given by (6.2.19b), since $\mathcal{L}_{ij}(\delta(s(\boldsymbol{\xi})) = 0$. However

$$\mathcal{L}_{ij}[m(\boldsymbol{x}, y, \boldsymbol{\xi}, \boldsymbol{k}_I)] =$$
$$\left[\left(\xi_j - k_{R_j}\right)\left(\frac{\partial}{\partial\overline{k}_i} + \frac{1}{2}\frac{\partial}{\xi_i}\right) - \left(\xi_i - k_{R_i}\right)\left(\frac{\partial}{\partial\overline{k}_j} + \frac{1}{2}\frac{\partial}{\xi_j}\right)\right] m(\boldsymbol{x}, y, \boldsymbol{\xi}, \boldsymbol{k}_I)$$
$$= \frac{1}{2}\left[\left(\xi_j - k_{R_j}\right)\left(\frac{\partial}{\partial\overline{\xi}_i} + \frac{\partial}{R_{I_i}}\right) - \left(\xi_i - k_{R_i}\right)\left(\frac{\partial}{\partial\overline{\xi}_j} + \frac{\partial}{R_{I_j}}\right)\right] m(\boldsymbol{x}, y, \boldsymbol{\xi}, \boldsymbol{k}_I)$$
$$\left[\left(\xi_j - k_{R_j}\right)\frac{\partial m}{\partial\overline{\zeta}_i}(\boldsymbol{x}, y, \boldsymbol{\xi}, \boldsymbol{k}_I) - \left(\xi_i - k_{R_i}\right)\frac{\partial m}{\partial\overline{\zeta}_j}(\boldsymbol{x}, y, \boldsymbol{\xi}, \boldsymbol{k}_I)\right],$$

where $\boldsymbol{\zeta} = \boldsymbol{\xi} + \mathrm{i}\boldsymbol{k}_I$ (note the location of $\boldsymbol{\xi}$ in m which is independent of $\boldsymbol{k}_R$), and so, using (6.2.7) we have

$$\mathcal{L}_{ij}[m(\boldsymbol{x}, y, \boldsymbol{\xi}, \boldsymbol{k}_I)]$$
$$= -\frac{C_n}{|\sigma_R|}\int_{\mathbb{R}^n} \exp\{\mathrm{i}\beta(\boldsymbol{x}, y; \boldsymbol{\xi}, \boldsymbol{k}_I, \boldsymbol{\eta})\}\left[(\xi_j - k_{R_j})(\eta_i - \xi_i) - (\xi_i - k_{R_i})(\eta_j - \xi_j)\right]$$
$$\times\, \delta(s(\boldsymbol{\xi}))T(\boldsymbol{\xi}, \boldsymbol{k}_I, \boldsymbol{\eta})\, m(\boldsymbol{x}, y, \boldsymbol{\eta}, \boldsymbol{k}_I)\,\mathrm{d}\boldsymbol{\eta}.$$

Thus we find that

$$\begin{aligned}&\int_{\mathbb{R}^n} \exp\{\mathrm{i}\beta(\boldsymbol{x},y;\boldsymbol{k}_R,\boldsymbol{k}_I,\boldsymbol{\xi})\}\delta(s(\boldsymbol{\xi}))\mathcal{L}_{ij}[T(\boldsymbol{k}_R,\boldsymbol{k}_I,\boldsymbol{\xi})]m(\boldsymbol{x},y,\boldsymbol{\xi},\boldsymbol{k}_I)\,\mathrm{d}\boldsymbol{\xi}\\
&= -\int_{\mathbb{R}^n} \exp\{\mathrm{i}\beta(\boldsymbol{x},y;\boldsymbol{k}_R,\boldsymbol{k}_I,\boldsymbol{\xi})\}\delta(s(\boldsymbol{\xi}))T(\boldsymbol{k}_R,\boldsymbol{k}_I,\boldsymbol{\xi})\mathcal{L}_{ij}[m(\boldsymbol{x},y,\boldsymbol{\xi},\boldsymbol{k}_I)]\,\mathrm{d}\boldsymbol{\xi}\\
&= \frac{C_n}{|\sigma_R|}\iint_{\mathbb{R}^{2n}} \exp\{\mathrm{i}\beta(\boldsymbol{x},y;\boldsymbol{k}_R,\boldsymbol{k}_I,\boldsymbol{\xi})\}\delta\left(\left(\boldsymbol{\xi}+\frac{\sigma_I}{\sigma_R}\boldsymbol{k}_I\right)^2+\left(\boldsymbol{k}_R+\frac{\sigma_I}{\sigma_R}\boldsymbol{k}_I\right)^2\right)\\
&\quad\times T(\boldsymbol{k}_R,\boldsymbol{k}_I,\boldsymbol{\xi})\exp\{\mathrm{i}\beta(\boldsymbol{x},y;\boldsymbol{k}_R,\boldsymbol{k}_I,\boldsymbol{\xi})\}\delta\left(\left(\boldsymbol{\eta}+\frac{\sigma_I}{\sigma_R}\boldsymbol{k}_I\right)^2+\left(\boldsymbol{\xi}+\frac{\sigma_I}{\sigma_R}\boldsymbol{k}_I\right)^2\right)\\
&\quad\times T(\boldsymbol{\xi},\boldsymbol{k}_I,\boldsymbol{\eta})\,m(\boldsymbol{x},y,\boldsymbol{\eta},\boldsymbol{k}_I)\,\left[(\xi_j-k_{R_j})(\eta_i-\xi_i)-(\xi_i-k_{R_i})(\eta_j-\xi_j)\right]\,\mathrm{d}\boldsymbol{\eta}\,\mathrm{d}\boldsymbol{\xi},\end{aligned}$$

and so by noting that

$$\beta(\boldsymbol{x},y;\boldsymbol{k}_R,\boldsymbol{k}_I,\boldsymbol{\xi})+\beta(\boldsymbol{x},y;\boldsymbol{\xi},\boldsymbol{k}_I,\boldsymbol{\eta})\}=\beta(\boldsymbol{x},y;\boldsymbol{k}_R,\boldsymbol{k}_I,\boldsymbol{\eta}),$$

and interchanging the variables $\boldsymbol{\xi}$ and $\boldsymbol{\eta}$ we finally obtain the restriction (6.2.19).

In fact, there is a change of variables which allows (6.2.19) to be simplified. Without loss of generality, we consider equations (6.2.19) with $i=1$ (the case $i\neq 1$, is obtained from the case $i=1$ by straightforward manipulation), and introduce new variables $(\boldsymbol{z},\boldsymbol{w},w_0)\in\mathbb{C}^{n-1}\times\mathbb{R}^n\times\mathbb{R}$ which parameterize the sphere $s(\boldsymbol{\xi})$, where $\boldsymbol{z}=(z_2,\ldots,z_n)$

$$k_{R_1}=\sum_{j=2}^{n} w_j z_{R_j}-\tfrac{1}{2}w_1-\frac{\sigma_I w_0 w_1}{2\boldsymbol{w}^2}, \tag{6.2.21a}$$

$$k_{R_j}=-w_1 z_{R_j}-\tfrac{1}{2}w_2-\frac{\sigma_I w_0 w_j}{2\boldsymbol{w}^2}, \tag{6.2.21b}$$

$$k_{I_1}=\sum_{j=2}^{n} w_j z_{I_j}+\frac{\sigma_R w_0 w_1}{2\boldsymbol{w}^2}, \tag{6.2.21c}$$

$$k_{I_j}=-w_1 z_{I_j}+\frac{\sigma_R w_0 w_j}{2\boldsymbol{w}^2}, \tag{6.2.21d}$$

$$\xi_1=\sum_{j=2}^{n} w_j z_{R_j}+\tfrac{1}{2}w_1-\frac{\sigma_I w_0 w_1}{2\boldsymbol{w}^2}, \tag{6.2.21e}$$

$$\xi_j=-w_1 z_{R_j}+\tfrac{1}{2}w_j-\frac{\sigma_I w_0 w_j}{2\boldsymbol{w}^2}, \tag{6.2.21f}$$

for $j\geq 2$ [i.e., $s(\boldsymbol{\xi})=0$ with (6.2.20)]. Thus if $w_1\neq 0$, there is a 1-1 map: $(\boldsymbol{k}_R,\boldsymbol{k}_I,\boldsymbol{\xi})\to(\boldsymbol{z},\boldsymbol{w},w_0)$ such that

$$\boldsymbol{w}=\boldsymbol{\xi}-\boldsymbol{k}_R,\qquad w_0=\frac{2}{\sigma_R}\boldsymbol{k}_I\bullet(\boldsymbol{\xi}-\boldsymbol{k}_R),\qquad \frac{\partial}{\partial\bar{z}_j}=\mathcal{L}_{1j},$$

which for $i=1$, $j=2,\ldots,n$ yields

$$\frac{\partial T}{\partial\bar{z}_j}=\mathrm{N}_{1j}[T](\boldsymbol{z},\boldsymbol{w},w_0),\qquad j=2,\ldots,n. \tag{6.2.22}$$

Again, using the generalized Cauchy formula we have

$$\begin{aligned} \mathrm{I}_j[T](\boldsymbol{z},\boldsymbol{w},w_0) \;&:= T(\boldsymbol{z},\boldsymbol{w},w_0) - \frac{1}{\pi}\int_{-\infty}^{\infty}\int_{-\infty}^{\infty} \frac{\mathrm{N}_{1j}[T](\boldsymbol{\zeta}^j,\boldsymbol{w},w_0)}{z_j-\zeta_j}\,\mathrm{d}\zeta_{R_j}\,\mathrm{d}\zeta_{I_j} \\ &= \hat{u}(\boldsymbol{w},w_0), \end{aligned} \tag{6.2.23}$$

with $\boldsymbol{\zeta}^j = (z_1,\ldots,z_{j-1},\zeta_j,z_{j+1}\ldots,z_n)$, where $\hat{u}(\boldsymbol{w},w_0) = \mathcal{F}(u(\boldsymbol{x},y))$ is the Fourier Transform of $u(\boldsymbol{x},y)$ with respect to $\boldsymbol{w}$, w_0. The term $\hat{u}(\boldsymbol{w},w_0)$ is the boundary value of $T(\boldsymbol{z},\boldsymbol{w},w_0)$ as $z_j\to\infty$. This can be seen from the definition of $T(\boldsymbol{z},\boldsymbol{w},w_0)$ [equation (6.2.6b)] and the fact that from equation (6.2.21), $z_j\to\infty$ implies $k_j\to\infty$ and hence $m\to 1$ from equation (6.2.16). Equation (6.2.23) leads both to admissibility criteria as well as the reconstruction of $u(\boldsymbol{x},y)$. Given $T(\boldsymbol{k}_R,\boldsymbol{k}_I,\boldsymbol{\xi})$, $\hat{u}(\boldsymbol{w},w_0)$ is computed by quadratures. We also reiterate the fact that equation (6.2.23) assumes that there are no homogeneous solutions to equation (6.2.3). This issue is nontrivial, for the case of the time-independent Schrödinger problem, such homogeneous solutions do arise (cf. Lavine and Nachman [1987]). Nachman and Ablowitz [1984a] conjecture that if $I_j[T]$ is independent of $\boldsymbol{z}$ and j and additionally has suitable decay properties for large $\boldsymbol{w}$, w_0, then T is *admissible*. The potential is reconstructed from

$$u(\boldsymbol{x},y) = \mathcal{F}^{-1}\left(\hat{u}(\boldsymbol{w},w_0)\right), \tag{6.2.24}$$

where $\mathcal{F}^{-1}$ denotes the inverse Fourier transform. Moreover, we see that the reconstruction follows purely by quadratures, given $T(\boldsymbol{k}_R,\boldsymbol{k}_I,\boldsymbol{\xi})$ on $s(\boldsymbol{\xi})=0$.

It turns out that the physically interesting cases of the time-dependent and time-independent Schrödinger equation in n dimensions arise as special cases of the above result. In the following two sections we discuss these cases as limits (reductions) of the above results and then briefly indicate how the formulae can be derived directly without recourse to any limit.

6.2.4 The Time-Dependent Schrödinger Equation. We consider the limit $\sigma\to\mathrm{i}$ in equation (6.2.1). In order to do this we first make the scaling $\hat{\boldsymbol{k}}_R = \boldsymbol{k}_R$, $\hat{\boldsymbol{k}}_I = (1/\sigma_R)\boldsymbol{k}_I$ and then set $\sigma_I = 1$ and let $\sigma_R \to 0-$ $(\sigma_R<0)$. Then $G(\boldsymbol{x},y;\boldsymbol{k}_R,\boldsymbol{k}_I)\to G_{\mathrm{L}}(\boldsymbol{x},y;\hat{\boldsymbol{k}}_R,\hat{\boldsymbol{k}}_I)$ (in the following we drop the symbol $\hat{}$), and

$$\begin{aligned} G_{\mathrm{L}}(\boldsymbol{x},y;\boldsymbol{k}_R,\boldsymbol{k}_I) = -\mathrm{i}C_n\mathrm{sgn}(y)\int_{\mathbb{R}^n} &\exp\{\mathrm{i}(\boldsymbol{x}\cdot\boldsymbol{\xi} - y(\boldsymbol{\xi}^2+2\boldsymbol{k}_R\cdot\boldsymbol{\xi})\} \\ &\times\Theta(y(\boldsymbol{\xi}^2+2(\boldsymbol{k}_R+\boldsymbol{k}_I)\cdot\boldsymbol{\xi}))\,\mathrm{d}\boldsymbol{\xi}. \end{aligned} \tag{6.2.25}$$

Equation (6.2.25) can be verified directly, i.e.

$$\mathcal{L}G_{\mathrm{L}}(\boldsymbol{x},y;\boldsymbol{k}_R,\boldsymbol{k}_I) = \delta(\boldsymbol{x})\delta(y), \tag{6.2.26}$$

where

$$\mathcal{L} \equiv \mathrm{i}\,\frac{\partial}{\partial y} + \sum_{j=1}^{n} \frac{\partial^2}{\partial x_j^2} + 2\mathrm{i}\boldsymbol{k}_R \bullet \nabla, \tag{6.2.27}$$

and hence $m \to m_{\mathrm{L}}$, where m_{L} satisfies

$$\mathcal{L} m_{\mathrm{L}} = -u m_{\mathrm{L}}, \tag{6.2.28a}$$

and

$$m_{\mathrm{L}} = 1 + \tilde{\mathrm{G}}_{\mathrm{L}}(u m_{\mathrm{L}}). \tag{6.2.28b}$$

Thus $G_{\mathrm{L}}(\boldsymbol{x}, y; \boldsymbol{k}_R, \boldsymbol{k}_I)$ provides a family of Green's functions parameterized by $\boldsymbol{k}_I$ which has the parameter entering via the "boundary conditions" (i.e., through the integral equation (6.2.28b) since $\mathcal{L}$ depends on $\boldsymbol{k}_R$ only). As $|k_{I_j}| \to \pm\infty$

$$G_{\mathrm{L}}(\boldsymbol{x}, y; \boldsymbol{k}_R, \boldsymbol{k}_I) \to -\mathrm{i}C_n \mathrm{sgn}(y) \int_{\mathbb{R}^n} \exp\{\mathrm{i}(\boldsymbol{x}\bullet\boldsymbol{\xi} - y(\boldsymbol{\xi}^2 + 2\boldsymbol{k}_R \bullet \boldsymbol{\xi})\}\Theta(\pm y \xi_j)\,\mathrm{d}\boldsymbol{\xi}, \tag{6.2.29}$$

hence

$$\lim_{k_{I_j} \to \pm\infty} G_{\mathrm{L}}(\boldsymbol{x}, y; \boldsymbol{k}_R, \boldsymbol{k}_I) = G_{\mathrm{L}}^{\pm j}\left(\boldsymbol{x}, y; \boldsymbol{k}_R, \boldsymbol{k}_I^j\right),$$

with $\boldsymbol{k}_I^j = (k_{I_1}, \ldots k_{I_{j-1}}, k_{I_{j+1}}, \ldots, k_{I_n}) \in \mathbb{R}^{n-1}$ and where $G_{\mathrm{L}}^{\pm j}\left(\boldsymbol{x}, y; \boldsymbol{k}_R, \boldsymbol{k}_I^j\right)$ are $(\mp)$ functions of k_{R_j}. Similarly

$$\lim_{k_{I_j} \to \pm\infty} m_{\mathrm{L}}(\boldsymbol{x}, y; \boldsymbol{k}_R, \boldsymbol{k}_I) = m_{\mathrm{L}}^{\pm j}\left(\boldsymbol{x}, y; \boldsymbol{k}_R, \boldsymbol{k}_I^j\right).$$

Then by direct calculation (or alternatively by taking limits)

$$\frac{\partial m_{\mathrm{L}}}{\partial k_{I_j}} = -2\mathrm{i}C_n \int_{\mathbb{R}^n} \exp\{\mathrm{i}\beta_{\mathrm{L}}(\boldsymbol{x}, y; \boldsymbol{k}_R, \boldsymbol{k}_I, \boldsymbol{\xi})\}\, T_{\mathrm{L}}(\boldsymbol{k}_R, \boldsymbol{k}_I, \boldsymbol{\xi}) \times (\xi_j - k_{R_j})\, \delta(s_{\mathrm{L}}(\boldsymbol{\xi}))\, m_{\mathrm{L}}(\boldsymbol{x}, y; \boldsymbol{\xi}, \boldsymbol{k}_I)\, \mathrm{d}\boldsymbol{\xi}, \tag{6.2.30a}$$

where

$$T_{\mathrm{L}}(\boldsymbol{k}_R, \boldsymbol{k}_I, \boldsymbol{\xi}) = \iint_{\mathbb{R}^{n+1}} \exp\{-\mathrm{i}\beta_{\mathrm{L}}(\boldsymbol{x}, y; \boldsymbol{k}_R, \boldsymbol{k}_I, \boldsymbol{\xi})\} \times u(\boldsymbol{x}, y)\, m(\boldsymbol{x}, y; \boldsymbol{k}_R, \boldsymbol{k}_I)\, \mathrm{d}\boldsymbol{x}\, \mathrm{d}y, \tag{6.2.30b}$$

$$\beta_{\mathrm{L}}(\boldsymbol{x}, y; \boldsymbol{k}_R, \boldsymbol{k}_I, \boldsymbol{\xi}) = (\boldsymbol{x} + 2y\boldsymbol{k}_I)\bullet(\boldsymbol{\xi} - \boldsymbol{k}_R), \tag{6.2.30c}$$

$$s_{\mathrm{L}}(\boldsymbol{\xi}) = (\boldsymbol{\xi} + \boldsymbol{k}_I)^2 - (\boldsymbol{k}_R + \boldsymbol{k}_I)^2. \tag{6.2.30d}$$

The reconstruction formula for m_{L} is then given by

$$m_{\mathrm{L}}(\boldsymbol{x}, y; \boldsymbol{k}_R, \boldsymbol{k}_I) = 1 - \frac{1}{2\pi\mathrm{i}} \int_{-\infty}^{\infty}\int_{-\infty}^{\infty} \left[\frac{\Theta\left(k_{I_j} - \kappa_{I_j}\right)}{k_{R_j} - \kappa_{R_j} - \mathrm{i}0} + \frac{\Theta\left(\kappa_{I_j} - k_{I_j}\right)}{k_{R_j} - \kappa_{R_j} + \mathrm{i}0}\right] \times \frac{\partial m}{\partial k_{I_j}}\left(\boldsymbol{x}, y; \boldsymbol{\kappa}_R^j, \boldsymbol{\kappa}_I^j\right) \mathrm{d}\kappa_{R_j}\, \mathrm{d}\kappa_{I_j}, \tag{6.2.31}$$

where $\boldsymbol{\kappa}_R^j = (k_{R_1}, \ldots, k_{R_{j-1}}, \kappa_{R_j}, k_{R_{j+1}}, \ldots, k_{R_n})$ and similarly for $\boldsymbol{\kappa}_I^j$, with (6.2.30) inserted into (6.2.31) (where $\Theta(\bullet)$ is the usual Heaviside function). Equation (6.2.31) can be derived directly by making use of the analytic properties of m_{L} at $k_{I_j} = \pm\infty$ (or follows by limits). To show this directly note that

$$\begin{aligned}\int_{-\infty}^{\infty} \frac{\partial m}{\partial k_{I_j}} \, \mathrm{d}k_{I_j} &= m(k_{I_j} = \infty) - m(k_{I_j} = -\infty) \\ &= m_j^-\left(\boldsymbol{x}, y; \boldsymbol{k}_R, \boldsymbol{k}_I^j\right) - m_j^+\left(\boldsymbol{x}, y; \boldsymbol{k}_R, \boldsymbol{k}_I^j\right). \qquad (6.2.32)\end{aligned}$$

Thus by projection

$$m_j^\pm\left(\boldsymbol{x}, y; \boldsymbol{k}_R, \boldsymbol{k}_I^j\right) = 1 - \mathcal{P}_j^\pm\left(\int_{-\infty}^{\infty} \frac{\partial m}{\partial k_{I_j}}(\boldsymbol{x}, y; \boldsymbol{k}_R, \boldsymbol{k}_I)\, \mathrm{d}k_{I_j}\right), \qquad (6.2.33a)$$

where

$$\mathcal{P}_j^\pm g(k_{R_j}) = \frac{1}{2\pi \mathrm{i}} \int_{-\infty}^{\infty} \frac{g(\kappa)}{\kappa - (k_{R_j} \pm \mathrm{i}0)}\, \mathrm{d}\kappa, \qquad (6.2.33b)$$

and the usual $\mathcal{P}_j^\pm$ projectors satisfy

$$\left(\mathcal{P}_j^+ - \mathcal{P}_j^-\right) g\left(k_{R_j}\right) = g\left(k_{R_j}\right).$$

Carrying out the integral in (6.2.32) and using the above boundary conditions at $k_{I_j} = -\infty$ yields

$$\begin{aligned}m(\boldsymbol{x}, y; \boldsymbol{k}_R, \boldsymbol{k}_I) &= 1 - \mathcal{P}_i^+ \int_{-\infty}^{\infty} \left(\frac{\partial m}{\partial k_{I_j}}\right) \mathrm{d}\kappa_{I_j} + \int_{-\infty}^{k_{I_j}} \left(\frac{\partial m}{\partial k_{I_j}}\right) \mathrm{d}\kappa_{I_j}, \qquad (6.2.34a) \\ &= 1 - \int_{-\infty}^{k_{I_j}} \mathcal{P}_j^- \left(\frac{\partial m}{\partial k_{I_j}}\right) \mathrm{d}\kappa_{I_j} - \int_{k_{I_j}}^{\infty} \mathcal{P}_j^+ \left(\frac{\partial m}{\partial k_{I_j}}\right) \mathrm{d}\kappa_{I_j} \qquad (6.2.34b)\end{aligned}$$

(with the obvious notation). Equation (6.2.34b) is equivalent to equation (6.2.31). The analogue of equation (6.2.19) is obtained from the compatibility condition (alternatively via limits)

$$\frac{\partial^2 m}{\partial k_{I_j} \partial k_{I_i}} = \frac{\partial^2 m}{\partial k_{I_i} \partial k_{I_j}}, \qquad i \neq j, \qquad (6.2.35a)$$

$$\mathcal{L}_{ij}^{\mathrm{L}}[T_{\mathrm{L}}] = \mathrm{N}_{ij}^{\mathrm{L}}[T_{\mathrm{L}}], \qquad (6.2.35b)$$

where

$$\mathcal{L}_{ij}^{\mathrm{L}}[T_{\mathrm{L}}] = \tfrac{1}{2}\mathrm{i}\left[(\xi_j - k_{R_j}) \frac{\partial T_{\mathrm{L}}}{\partial k_{I_j}} - (\xi_i - k_{R_i}) \frac{\partial T_{\mathrm{L}}}{\partial k_{I_i}}\right], \qquad (6.2.35c)$$

$$\begin{aligned}\mathrm{N}_{ij}^{\mathrm{L}}[T_{\mathrm{L}}] = -C_n \int_{\mathbb{R}^n} &\left[(\eta_j - k_{R_j})(\xi_i - \eta_i) - (\eta_i - k_{R_i})(\xi_j - \eta_j)\right] \\ &\times \delta(s_{\mathrm{L}}(\boldsymbol{\eta}))\, T_{\mathrm{L}}(\boldsymbol{k}_R, \boldsymbol{k}_I, \boldsymbol{\eta})\, T_{\mathrm{L}}(\boldsymbol{\eta}, \boldsymbol{k}_I, \boldsymbol{\xi})\, \mathrm{d}\boldsymbol{\eta}. \qquad (6.2.35d)\end{aligned}$$

Defining new variables $(\boldsymbol{z}, \boldsymbol{w}, w_0) \in \mathbb{C}^{n-1} \times \mathbb{R}^n \times \mathbb{R}$

$$
\begin{aligned}
k_{R_1} &= \sum_{j=2}^{n} w_j z_{R_j} - \tfrac{1}{2} w_1 - \frac{w_0 w_1}{2\boldsymbol{w}^2}, \\
k_{R_j} &= -w_1 z_{R_j} - \tfrac{1}{2} w_2 - \frac{w_0 w_j}{2\boldsymbol{w}^2}, \qquad 2 \le j \le n \\
k_{I_1} &= \sum_{j=2}^{n} w_j z_{I_j}, \qquad k_{I_j} = -w_1 z_{I_j}, \qquad 2 \le j \le n \\
\xi_1 &= \sum_{j=2}^{n} w_j z_{R_j} + \tfrac{1}{2} w_1 - \frac{\sigma_I w_0 w_1}{2\boldsymbol{w}^2}, \\
\xi_j &= -w_1 z_{R_j} + \tfrac{1}{2} w_j - \frac{\sigma_I w_0 w_j}{2\boldsymbol{w}^2}, \qquad 2 \le j \le n
\end{aligned}
$$

we have [taking $i = 1$ in (6.2.35)]

$$\frac{\partial T_{\mathrm{L}}}{\partial z_{I_j}} = -2\mathrm{i} N^{\mathrm{L}}_{I_j}[T_{\mathrm{L}}], \tag{6.2.36}$$

and the analogue of (6.2.23)

$$\hat{u}(\boldsymbol{w}, w_0) = T_{\mathrm{L}}(\boldsymbol{z}, \boldsymbol{w}, w_0) - \frac{1}{\pi} \int_{-\infty}^{\infty} \int_{-\infty}^{\infty} \left[\frac{\Theta\left(z_{I_j} - \zeta_{I_j}\right)}{z_{R_j} - \zeta_{R_j} - \mathrm{i}0} + \frac{\Theta\left(\zeta_{I_j} - z_{I_j}\right)}{z_{R_j} - \zeta_{R_j} + \mathrm{i}0} \right] \times \mathrm{N}_{1j}[T_{\mathrm{L}}](\boldsymbol{\zeta}, \boldsymbol{w}, w_0)\, \mathrm{d}\zeta_{R_j}\, \mathrm{d}\zeta_{I_j}. \tag{6.2.37}$$

Admissible data T_{L} are such that the left-hand side of equation (6.2.37) will be independent of $\boldsymbol{z}$ and j, and have suitable decay in $\boldsymbol{w}$, w_0. Equation (6.2.37) can be derived directly using the fact that $\lim_{z_{I_j} \to \pm\infty} (T_{\mathrm{L}} - \hat{u})$ is, as a function of z_{R_j}, the boundary value of an analytic function of z_{R_j} in the lower $(z_{R_j} \to \infty)$/upper $(z_{R_j} \to -\infty)$ half-plane and tending to zero as $z_{R_j} \to \infty$. The argument is identical to that of reconstructing m_{L} above [i.e., equations (6.2.32-34)].

6.2.5 The Time-Independent Schrödinger Equation. The following results can be obtained either by a reduction of those for m_{L} or alternatively can be obtained directly using similar techniques to those described above and so will be omitted. The reduction assumes the $u(\boldsymbol{x}, y) = u(\boldsymbol{x})$ and $m(\boldsymbol{x}, y; \boldsymbol{k}_R, \boldsymbol{k}_I) = m(\boldsymbol{x}; \boldsymbol{k}_R, \boldsymbol{k}_I)$.

The "stationary" eigenfunction $m_{\mathrm{s}}(\boldsymbol{x}; \boldsymbol{k}_R, \boldsymbol{k}_I)$ satisfies

$$m_{\mathrm{s}}(\boldsymbol{x}; \boldsymbol{k}_R, \boldsymbol{k}_I) = 1 + G_{\mathrm{s}}\left(u m_{\mathrm{s}}\right), \tag{6.2.38}$$

where the Green's function is given by

$$G_{\mathrm{s}}(\boldsymbol{x}; \boldsymbol{k}_R, \boldsymbol{k}_I) = -C_n \int_{\mathbb{R}^n} \left[\frac{\Theta\left(\boldsymbol{k}_I \bullet \boldsymbol{\xi}\right)}{\boldsymbol{\xi}^2 + 2\boldsymbol{\xi} \bullet \boldsymbol{k}_R - \mathrm{i}0} + \frac{\Theta\left(-\boldsymbol{k}_I \bullet \boldsymbol{\xi}\right)}{\boldsymbol{\xi}^2 + 2\boldsymbol{\xi} \bullet \boldsymbol{k}_R + \mathrm{i}0} \right] \times \exp\{\mathrm{i}\boldsymbol{\xi} \bullet \boldsymbol{x}\}\, \mathrm{d}\boldsymbol{\xi}. \tag{6.2.39}$$

We shall assume that equation (6.2.38) has no homogeneous solutions. The questions of homogeneous solutions is discussed in the literature (see for example, Lavine and Nachman [1987]). By direct calculations it can be verified that G_s satisfies

$$\mathcal{L}_s G_s(\boldsymbol{x};\boldsymbol{k}_R,\boldsymbol{k}_I) = \delta(\boldsymbol{x}), \tag{6.2.40a}$$

where

$$\mathcal{L}_s \equiv \sum_{j=1}^{n} \frac{\partial^2}{\partial x_j^2} + 2\mathrm{i}\boldsymbol{k}_R \bullet \nabla, \tag{6.2.40b}$$

and m_s satisfies

$$\mathcal{L}_s m_s = u m_s, \tag{6.2.41a}$$

or, if $v_s(\boldsymbol{x};\boldsymbol{k}_R,\boldsymbol{k}_I) = m_s \exp\{\mathrm{i}\boldsymbol{k}_R \bullet \boldsymbol{x}\}$, v_s satisfies

$$\sum_{j=1}^{n} \frac{\partial^2 v_s}{\partial x_j^2} + \left\{\boldsymbol{k}_R^2 - u(\boldsymbol{x})\right\} v_s = 0. \tag{6.2.41b}$$

G_s is obtained from G_L by

$$G_s(\boldsymbol{x};\boldsymbol{k}_R,\boldsymbol{k}_I) = \int_{-\infty}^{\infty} G_L(\boldsymbol{x},y;\boldsymbol{k}_R,\boldsymbol{k}_I)\,\mathrm{d}y, \tag{6.2.42}$$

where we have used the identity

$$\frac{\Theta(x+y)}{x-\mathrm{i}0} + \frac{\Theta(-x-y)}{x+\mathrm{i}0} = \frac{\Theta(y)}{x-\mathrm{i}0} + \frac{\Theta(-y)}{x+\mathrm{i}0}.$$

Indeed the Green's function $G_s(\boldsymbol{x};\boldsymbol{k}_R,\boldsymbol{k}_I)$ turns out to be the same as that of Faddeev [1965, 1966, 1976].

The analogue of equation (6.2.7) is

$$\frac{\partial m_s}{\partial k_{I_j}} = -2\pi\mathrm{i}C_n \int_{\mathbb{R}^n} \exp\{\mathrm{i}(\boldsymbol{x}+2y\boldsymbol{k}_R)\bullet(\boldsymbol{\xi}-\boldsymbol{k}_R)\}\, T_s(\boldsymbol{k}_R,\boldsymbol{k}_I,\boldsymbol{\xi})\,(\xi_j - k_{R_j})$$
$$\times\, \delta(\boldsymbol{\xi}^2 - \boldsymbol{k}_R^2)\,\delta(\boldsymbol{k}_I\bullet(\boldsymbol{\xi}-\boldsymbol{k}_R))\, m(\boldsymbol{x};\boldsymbol{\xi},\boldsymbol{k}_I)\,\mathrm{d}\boldsymbol{\xi}, \tag{6.2.43a}$$

where

$$T_s(\boldsymbol{k}_R,\boldsymbol{k}_I,\boldsymbol{\xi}) = \int_{\mathbb{R}^n} \exp\{-\mathrm{i}\boldsymbol{x}\bullet(\boldsymbol{\xi}-\boldsymbol{k}_R)\}\, u(\boldsymbol{x})\, m_s(\boldsymbol{x};\boldsymbol{k}_R,\boldsymbol{k}_I)\,\mathrm{d}\boldsymbol{x}. \tag{6.2.43b}$$

It should be remarked that the reduction of T_L to T_s obeys

$$T_L(\boldsymbol{k}_R,\boldsymbol{k}_I,\boldsymbol{\xi}) = \pi\delta\left(\boldsymbol{k}_I\bullet(\boldsymbol{\xi}-\boldsymbol{k}_R)\right) T_s(\boldsymbol{k}_R,\boldsymbol{k}_I,\boldsymbol{\xi}).$$

The reconstruction equation for $m_s(\boldsymbol{x};\boldsymbol{\xi},\boldsymbol{k}_I)$ follows directly (noting that as $k_{I_j}\to\pm\infty$, m_s is a $(\mp)$ function of k_{R_j}):

$$m_s(\boldsymbol{x},y;\boldsymbol{k}_R,\boldsymbol{k}_I)=1-\frac{1}{2\pi\mathrm{i}}\int_{-\infty}^{\infty}\int_{-\infty}^{\infty}\left[\frac{\Theta\left(k_{I_j}-\kappa_{I_j}\right)}{k_{R_j}-\kappa_{R_j}-\mathrm{i}0}+\frac{\Theta\left(\kappa_{I_j}-k_{I_j}\right)}{k_{R_j}-\kappa_{R_j}+\mathrm{i}0}\right]$$
$$\times\frac{\partial m_s}{\partial k_{I_j}}(\boldsymbol{x},y;\boldsymbol{\kappa}_R^j,\boldsymbol{\kappa}_I^j)\,\mathrm{d}\kappa_R\,\mathrm{d}\kappa_I,\qquad(6.2.44)$$

using (6.2.43).

Taking the restriction $w_0=\boldsymbol{k}_I\bullet(\boldsymbol{\xi}-\boldsymbol{k}_R)=0$ into account, then the compatibility condition

$$\frac{\partial^2 m_s}{\partial k_{I_i}\partial k_{I_j}}=\frac{\partial^2 m_s}{\partial k_{I_j}\partial k_{I_i}},\qquad i\neq j,$$

gives

$$\mathcal{L}^{s}_{ij}[T_s]=\pi\mathrm{N}^{s}_{ij}[T_s],\qquad\text{on}\qquad w_0=0,\qquad(6.2.45a)$$

$$\mathcal{L}^{s}_{ij}[T_s]=\tfrac{1}{2}\mathrm{i}\left(\left(\xi_j-k_{R_j}\right)\frac{\partial T_s}{\partial k_{I_i}}-\left(\xi_i-k_{R_i}\right)\frac{\partial T_s}{\partial k_{I_j}}\right),\qquad(6.2.45b)$$

$$\mathrm{N}^{s}_{ij}[T_s]=-C_n\int_{\mathbb{R}^n}\left[\left(\xi_j-k_{R_j}\right)\left(\xi_i-\eta_i\right)-\left(\xi_i-k_{R_i}\right)\left(\xi_j-\eta_j\right)\right]$$
$$\times\,\delta\left(\boldsymbol{\eta}^2-\boldsymbol{k}_R^2\right)\delta(\boldsymbol{k}_I\bullet(\boldsymbol{\eta}-\boldsymbol{k}_R))\,T_s(\boldsymbol{k}_R,\boldsymbol{k}_I,\boldsymbol{\eta})\,T_s(\boldsymbol{\eta},\boldsymbol{k}_I,\boldsymbol{\xi})\,\mathrm{d}\boldsymbol{\eta}.\quad(6.2.45c)$$

The change of variables

$$k_{R_1}=\sum_{j=2}^{n}w_j z_{R_j}-\tfrac{1}{2}w_1,\qquad k_{R_j}=w_1 z_{R_j}-\tfrac{1}{2}w_j,\qquad(6.2.46a)$$

$$k_{I_1}=\sum_{j=2}^{n}w_j z_{I_j},\qquad k_{I_j}=-w_1 z_{I_j},\qquad(6.2.46b)$$

$$\xi_1=\sum_{j=2}^{n}w_j z_{R_j}+\tfrac{1}{2}w_1,\qquad \xi_j=-w_1 z_{R_j},\qquad(6.2.46c)$$

for $2\le j\le n$, gives a transformation $(\boldsymbol{k}_R,\boldsymbol{k}_I,\boldsymbol{\xi})\to(\boldsymbol{z},\boldsymbol{w})$ from $3n-2$ variables to $3n-2$ variables (note that the restrictions $\boldsymbol{\xi}^2=k_R^2$ and $w_0=\boldsymbol{k}_I\bullet(\boldsymbol{\xi}-\boldsymbol{k}_R)=0$ are incorporated into this transformation). Hence equations (6.2.45-46) imply that [setting $i=1$ in (6.2.45)]

$$\frac{\partial T_s}{\partial z_{I_j}}=-2\pi\mathrm{i}\mathrm{N}_{ij}[T_s],$$

and then by integration

$$\hat{u}(\boldsymbol{w})=T_s(\boldsymbol{z},\boldsymbol{w})-\int_{-\infty}^{\infty}\int_{-\infty}^{\infty}\left[\left(\xi_j-k_{R_j}\right)\left(\xi_i-\eta_i\right)-\left(\xi_i-k_{R_i}\right)\left(\xi_j-\eta_j\right)\right]$$
$$\times\,\mathrm{N}_{1j}[T_s](\boldsymbol{\zeta},w)\,\mathrm{d}\zeta_{R_j}\,\mathrm{d}\zeta_{I_j},\qquad(6.2.47)$$

where $\hat{u}(\boldsymbol{w})$ is the Fourier transform of $u(\boldsymbol{x})$ with respect to $\boldsymbol{w}$. Equation (6.2.47) plays the role of characterizing suitable data $T_s(\boldsymbol{z}, \boldsymbol{w})$ as well as reconstructing the potential in analogy with equation (6.2.23). Namely, the left-hand side of (6.2.47) must be independent of $\boldsymbol{z}$ and j and have appropriate decay in $\boldsymbol{w}$. Again we note that $\lim_{z_{I_j} \to \pm\infty}(T_s - \hat{u}(\boldsymbol{w}))$ is, as a function of z_{R_j}, the boundary value of an analytic function in the lower (∞)/upper $(-\infty)$ half-plane and vanishing at $z_{R_j} = \infty$. This allows (6.2.47) to be obtained using similar ideas to those discussed previously.

6.2.6 The Relationship between the Inverse Data and the Scattering Data. In this section we show how the inverse data ($T(\boldsymbol{k}_R, \boldsymbol{k}_I, \boldsymbol{\xi})$, $T_L(\boldsymbol{k}_R, \boldsymbol{k}_I, \boldsymbol{\xi})$, $T_s(\boldsymbol{k}_R, \boldsymbol{k}_I, \boldsymbol{\xi})$ for the general, limit and reduction cases respectively), can be related to the scattering data. In the limit (L) and reduction (S) cases, scattering theory has a clear physical meaning. In the general case we shall define formal scattering and show how the time-dependent and time-independent physical scattering can be recovered as special cases. Naturally one can derive these special cases directly; however, since such an analysis is essentially identical to the general case, it will be omitted. We also remark that such formulae for the time-independent case was originally derived by Faddeev [1965, 1966, 1976].

Define a "left-Volterra" operator in terms of a Green's function

$$G_v(\boldsymbol{x}, y; \boldsymbol{k}) = C_n \sigma^{-1}\, \Theta(y) \int_{\mathbb{R}^n} \exp\left\{ i\boldsymbol{x} \bullet \boldsymbol{\xi} + (\boldsymbol{\xi}^2 + 2\boldsymbol{k} \bullet \boldsymbol{\xi}) y/\sigma \right\} d\boldsymbol{\xi}, \tag{6.2.48}$$

where $\boldsymbol{k} = \boldsymbol{k}_R + i\boldsymbol{k}_I$ and we require $\sigma_R < 0$ for convenience. Then for functions $u(\boldsymbol{x}, y)$ of compact support in both $\boldsymbol{x}$ and y (a much wider class of functions is allowed in the limit and reduction cases):

$$m_v(\boldsymbol{x}, y; \boldsymbol{k}) = 1 + G_v(u m_v)(\boldsymbol{x}, y; \boldsymbol{k}). \tag{6.2.49}$$

The scattering function is defined by the limit $y \to \infty$ of equation (6.2.49) (as $y \to -\infty$, $m_v \to 1$):

$$m_v(\boldsymbol{x}, y; \boldsymbol{k}) = 1 + \int_{\mathbb{R}^n} \exp\left\{ i\boldsymbol{x} \bullet \boldsymbol{\xi} + (\boldsymbol{\xi}^2 + 2\boldsymbol{k} \bullet \boldsymbol{\xi}) y/\sigma \right\} \tilde{S}(\boldsymbol{k}_R, \boldsymbol{k}_I, \boldsymbol{\xi})\, d\boldsymbol{\xi}, \tag{6.2.50a}$$

$$\tilde{S}(\boldsymbol{k}_R, \boldsymbol{k}_I, \boldsymbol{\xi}) = C_n \sigma^{-1} \iint_{\mathbb{R}^{n+1}} \exp\left\{ -i\boldsymbol{x} \bullet \boldsymbol{\xi} + (\boldsymbol{\xi}^2 + 2\boldsymbol{k} \bullet \boldsymbol{\xi}) y/\sigma \right\} \times u(\boldsymbol{x}, y)\, m_v(\boldsymbol{x}, y; \boldsymbol{k})\, d\boldsymbol{x}\, dy, \tag{6.2.50b}$$

or, by changing variables $\boldsymbol{\xi} \to \boldsymbol{\xi} - \boldsymbol{k}_R$,

$$\hat{S}(\boldsymbol{k}_R, \boldsymbol{k}_I, \boldsymbol{\xi}) = C_n \sigma^{-1} \iint_{\mathbb{R}^{n+1}} \exp\{-\hat{\beta}(\boldsymbol{x}, y; \boldsymbol{k}_R, \boldsymbol{k}_I, \boldsymbol{\xi})\} \times u(\boldsymbol{x}, y)\, m_v(\boldsymbol{x}, y; \boldsymbol{k})\, d\boldsymbol{x}\, dy, \tag{6.2.50c}$$

where

$$\hat{\beta}(\boldsymbol{x},y;\boldsymbol{k}_R,\boldsymbol{k}_I,\boldsymbol{\xi}) = \mathrm{i}\boldsymbol{x}\bullet(\boldsymbol{\xi}-\boldsymbol{k}_R) + (\boldsymbol{\xi}^2 - \boldsymbol{k}_R^2 + 2\mathrm{i}\boldsymbol{k}_I\bullet(\boldsymbol{\xi}-\boldsymbol{k}_R))y/\sigma. \qquad (6.2.50\text{d})$$

The eigenfunctions m and m_{v} are related as follows:

$$\begin{aligned}(m - m_{\mathrm{v}}) &= \tilde{\mathrm{G}}(um) - \tilde{\mathrm{G}}_{\mathrm{v}}(um_{\mathrm{v}}) \\ &= \tilde{G}_{\mathrm{v}}(u(m-m_{\mathrm{v}})) + (\tilde{\mathrm{G}} - \tilde{\mathrm{G}}_{\mathrm{v}})(um). \qquad (6.2.51)\end{aligned}$$

Using equations (6.2.4) and (6.2.48), we have

$$(G - G_{\mathrm{v}})(\boldsymbol{x},y;\boldsymbol{k}_R,\boldsymbol{k}_I) = -C_n\sigma^{-1}\int_{\mathbb{R}^n} \exp\{\hat{\beta}(\boldsymbol{x},y;\boldsymbol{k}_R,\boldsymbol{k}_I,\boldsymbol{\xi})\}\,\Theta(-s(\boldsymbol{\xi}))\,\mathrm{d}\boldsymbol{\xi}, \quad (6.2.52\text{a})$$

hence

$$(G - G_{\mathrm{v}})(um) = -\int_{\mathbb{R}^n} \exp\{\hat{\beta}(\boldsymbol{x},y;\boldsymbol{k}_R,\boldsymbol{k}_I,\boldsymbol{\xi})\}\,\Theta(-s(\boldsymbol{\xi}))\,\hat{T}(\boldsymbol{k}_R,\boldsymbol{k}_I,\boldsymbol{\xi})\,\mathrm{d}\boldsymbol{\xi}, \quad (6.2.52\text{b})$$

where

$$\begin{aligned}\hat{T}(\boldsymbol{k}_R,\boldsymbol{k}_I,\boldsymbol{\xi}) = -C_n\sigma^{-1}\iint_{\mathbb{R}^{n+1}} &\exp\{-\hat{\beta}(\boldsymbol{x},y;\boldsymbol{k}_R,\boldsymbol{k}_I,\boldsymbol{\xi})\} \\ &\times u(\boldsymbol{x},y)\,m(\boldsymbol{x},y;\boldsymbol{k}_R,\boldsymbol{k}_I)\,\mathrm{d}\boldsymbol{x}\,\mathrm{d}y. \qquad (6.2.52\text{c})\end{aligned}$$

Note that when

$$s(\boldsymbol{\xi}) = \left(\boldsymbol{\xi} + \frac{\sigma_I}{\sigma_R}\boldsymbol{k}_R\right)^2 - \left(\boldsymbol{k}_I + \frac{\sigma_I}{\sigma_R}\boldsymbol{k}_R\right)^2 = 0,$$

then

$$\begin{aligned}\hat{\beta}(\boldsymbol{x},y;\boldsymbol{k}_R,\boldsymbol{k}_I,\boldsymbol{\xi}) &= \beta(\boldsymbol{x},y;\boldsymbol{k}_R,\boldsymbol{k}_I,\boldsymbol{\xi}), \\ \hat{T}(\boldsymbol{k}_R,\boldsymbol{k}_I,\boldsymbol{\xi}) &= T(\boldsymbol{k}_R,\boldsymbol{k}_I,\boldsymbol{\xi}),\end{aligned}$$

[see equation (6.2.6)].

Then, using the symmetry condition for G_{v}

$$\exp\{-\hat{\beta}(\boldsymbol{x},y;\boldsymbol{k}_R,\boldsymbol{k}_I,\boldsymbol{\xi})\}G_{\mathrm{v}}(\boldsymbol{x},y;\boldsymbol{k}_R,\boldsymbol{k}_I) = G_{\mathrm{v}}(\boldsymbol{x},y;\boldsymbol{\xi},\boldsymbol{k}_I) \qquad (6.2.53)$$

(which is verified directly), we have from equations (6.2.49,51-53)

$$\begin{aligned}(m - m_{\mathrm{v}})(\boldsymbol{x},y;\boldsymbol{k}_R,\boldsymbol{k}_I) = \int_{\mathbb{R}^n} &\exp\{\hat{\beta}(\boldsymbol{x},y;\boldsymbol{k}_R,\boldsymbol{k}_I,\boldsymbol{\xi})\} \\ &\times \Theta(-s(\boldsymbol{\xi}))\,\hat{T}(\boldsymbol{k}_R,\boldsymbol{k}_I,\boldsymbol{\xi})\,m_{\mathrm{v}}(\boldsymbol{x},y;\boldsymbol{\xi},\boldsymbol{k}_I)\,\mathrm{d}\boldsymbol{\xi}. \quad (6.2.54)\end{aligned}$$

Now multiply (6.2.54) by $C_n\sigma^{-1}\exp\{-\hat{\beta}(\boldsymbol{x},y;\boldsymbol{k}_R,\boldsymbol{k}_I,\boldsymbol{\xi})\}$ and take $\iint \mathrm{d}\boldsymbol{x}\,\mathrm{d}y$, then using the definitions of $\hat{T},\hat{S}$ [equations (6.2.52c), (6.2.50c)] we see that

$$\hat{T}(\boldsymbol{k}_R,\boldsymbol{k}_I,\boldsymbol{\xi}) + \hat{S}(\boldsymbol{k}_R,\boldsymbol{k}_I,\boldsymbol{\xi}) + \int_{\mathbb{R}^n}\Theta(-s(\boldsymbol{\eta}))\,\hat{T}(\boldsymbol{k}_R,\boldsymbol{k}_I,\boldsymbol{\eta})\,\hat{S}(\boldsymbol{\eta},\boldsymbol{k}_I,\boldsymbol{\xi})\,\mathrm{d}\boldsymbol{\eta} = 0. \quad (6.2.55)$$

Using equation (6.2.55), we can obtain the "inverse data" $T(\boldsymbol{k}_R, \boldsymbol{k}_I, \boldsymbol{\xi})$ given the "scattering data" $S(\boldsymbol{k}_R, \boldsymbol{k}_I, \boldsymbol{\xi})$.

The limit and reduction cases now follow immediately. For $\sigma \to \mathrm{i}$, $\sigma_R < 0$

$$\hat{\beta} \to \hat{\beta}_{\mathrm{L}}(\boldsymbol{x}, y; \boldsymbol{k}_R, \boldsymbol{\xi}) = \mathrm{i}\left(\boldsymbol{x} \bullet (\boldsymbol{\xi} - \boldsymbol{k}_R) - (\boldsymbol{\xi}^2 - \boldsymbol{k}_R^2)y\right), \tag{6.2.56a}$$

$$\hat{S} \to \hat{S}_{\mathrm{L}}(\boldsymbol{k}_R, \boldsymbol{\xi}) = -\mathrm{i}C_n \iint_{\mathbb{R}^{n+1}} \exp\{-\hat{\beta}_{\mathrm{L}}(\boldsymbol{x}, y; \boldsymbol{k}_R, \boldsymbol{\xi})\} \times u(\boldsymbol{x}, y)\, m_{\mathrm{v}}^{\mathrm{L}}(\boldsymbol{x}, y; \boldsymbol{k}_R)\, \mathrm{d}\boldsymbol{x}\, \mathrm{d}y. \tag{6.2.56b}$$

Note that in equation (6.2.56a),

$$\hat{\beta}_{\mathrm{L}}(\boldsymbol{x}, y; \boldsymbol{k}_R, \boldsymbol{\xi}) = \beta_{\mathrm{L}}(\boldsymbol{x}, y; \boldsymbol{k}_R, \boldsymbol{k}_I, \boldsymbol{\xi}) = \mathrm{i}(\boldsymbol{x} + 2y\boldsymbol{k}_I) \bullet (\boldsymbol{\xi} - \boldsymbol{k}_R),$$

on the "shell" $s_{\mathrm{L}}(\boldsymbol{\xi}) = (\boldsymbol{\xi} + \boldsymbol{k}_I)^2 - (\boldsymbol{k}_R + \boldsymbol{k}_I)^2 = 0$. In equation (6.2.56b), $m_{\mathrm{v}}^{\mathrm{L}}$ is defined by

$$m_{\mathrm{v}}^{\mathrm{L}}(\boldsymbol{x}, y; \boldsymbol{k}_R) = 1 + \tilde{\mathrm{G}}_{\mathrm{v}}^{\mathrm{L}}(u m_{\mathrm{v}}^{\mathrm{L}})(\boldsymbol{x}, y; \boldsymbol{k}_R), \tag{6.2.56c}$$

where

$$\tilde{\mathrm{G}}_{\mathrm{v}}^{\mathrm{L}}(\boldsymbol{x}, y; \boldsymbol{k}_R) = -\mathrm{i}C_n \Theta(y) \int_{-\infty}^{\infty} \exp\{\mathrm{i}\boldsymbol{x} \bullet \boldsymbol{\xi} - \mathrm{i}(\boldsymbol{\xi}^2 + 2\boldsymbol{k}_R \bullet \boldsymbol{\xi})y\}\, \mathrm{d}\boldsymbol{\xi}. \tag{6.2.56d}$$

$m_{\mathrm{v}}^{\mathrm{L}}$ is not a function of $\boldsymbol{k}_I$, hence neither is $\hat{S} = \hat{S}(\boldsymbol{k}_R, \boldsymbol{\xi})$. Similarly:

$$\hat{T} \to \hat{T}_{\mathrm{L}}(\boldsymbol{k}_R, \boldsymbol{k}_I, \boldsymbol{\xi}) = \mathrm{i}C_n \iint_{\mathbb{R}^{n+1}} \exp\{-\hat{\beta}_{\mathrm{L}}(\boldsymbol{x}, y; \boldsymbol{k}_R, \boldsymbol{\xi})\}\, (u m_{\mathrm{v}}^{\mathrm{L}})(\boldsymbol{x}, y; \boldsymbol{k}_R)\, \mathrm{d}\boldsymbol{x}\, \mathrm{d}y. \tag{6.2.56e}$$

Thus we have the scattering relationship

$$\hat{T}_{\mathrm{L}}(\boldsymbol{k}_R, \boldsymbol{\xi}) + \hat{S}_{\mathrm{L}}(\boldsymbol{k}_R, \boldsymbol{\xi}) + \int_{\mathbb{R}^n} \Theta(-s_{\mathrm{L}}(\boldsymbol{\eta}))\, \hat{T}_{\mathrm{L}}(\boldsymbol{k}_R, \boldsymbol{\eta})\, \hat{S}_{\mathrm{L}}(\boldsymbol{\eta}, \boldsymbol{\xi})\, \mathrm{d}\boldsymbol{\eta} = 0. \tag{6.2.57}$$

As above, given the scattering data $\hat{S}_{\mathrm{L}}(\boldsymbol{k}_R, \boldsymbol{\xi})$ we can obtain (in principle) the inverse data $\hat{T}_{\mathrm{L}}(\boldsymbol{k}_R, \boldsymbol{k}_I, \boldsymbol{\xi})$ using equation (6.2.57) and this equals $T_{\mathrm{L}}(\boldsymbol{k}_R, \boldsymbol{k}_I, \boldsymbol{\xi})$ on $s_{\mathrm{L}}(\boldsymbol{\xi}) = 0$ [see equation (6.2.30)].

For the time-independent (reduction) problem, we make the observation

$$\begin{aligned} G_{\mathrm{v}}^{\mathrm{s}}(x; \boldsymbol{k}_R) &= \int_{-\infty}^{\infty} G_{\mathrm{v}}^{\mathrm{L}}(\boldsymbol{x}, y; \boldsymbol{k}_R)\, \mathrm{d}y \\ &= -C_n \int_{\mathbb{R}^n} \frac{\exp\{\mathrm{i}\boldsymbol{x} \bullet \boldsymbol{\xi}\}}{\boldsymbol{\xi}^2 + 2\boldsymbol{k}_R \bullet \boldsymbol{\xi} - \mathrm{i}0}\, \mathrm{d}\boldsymbol{\xi} \equiv G_+(\boldsymbol{x}; \boldsymbol{k}_R). \end{aligned} \tag{6.2.58a}$$

Namely $G_{\mathrm{v}}^{\mathrm{s}}(\boldsymbol{x}; \boldsymbol{k}_R)$ is identical to the standard outgoing Green's function (which is also analytic in the upper half $s = |\boldsymbol{k}_R|$ plane). Thus

$$m_{\mathrm{v}}^{\mathrm{s}}(\boldsymbol{x}; \boldsymbol{k}_R) = m_+(\boldsymbol{x}; \boldsymbol{k}_R). \tag{6.2.58b}$$

Using the fact that

$$\int_{-\infty}^{\infty} \exp\{-\mathrm{i}y(\boldsymbol{\xi}^2 - \boldsymbol{k}_R^2)\}\, \mathrm{d}y = 2\pi\delta(\boldsymbol{\xi}^2 - \boldsymbol{k}_R^2),$$

we have the following identifications when $u(\boldsymbol{x}, y) = u(\boldsymbol{x})$

$$\hat{S}_{\mathrm{L}}(\boldsymbol{k}_R, \boldsymbol{\xi}) = -C_{n-1}\delta(\boldsymbol{\xi}^2 - \boldsymbol{k}_R^2)\, A(\boldsymbol{k}_R, \boldsymbol{\xi}), \tag{6.2.58c}$$

$$\hat{T}_{\mathrm{L}}(\boldsymbol{k}_R, \boldsymbol{\xi}) = \mathrm{i}C_{n-1}\delta(\boldsymbol{\xi}^2 - \boldsymbol{k}_R^2)\, T_{\mathrm{s}}(\boldsymbol{k}_R, \boldsymbol{k}_I, \boldsymbol{\xi}), \tag{6.2.58d}$$

where

$$A(\boldsymbol{k}_R, \boldsymbol{\xi}) = \int_{\mathbb{R}^n} \exp\{-\mathrm{i}\boldsymbol{x} \bullet (\boldsymbol{\xi} - \boldsymbol{k}_R)\}\, (um_+)(\boldsymbol{x}; \boldsymbol{k}_R)\, \mathrm{d}\boldsymbol{x}, \tag{6.2.58e}$$

and $T_s(\boldsymbol{k}_R, \boldsymbol{k}_I, \boldsymbol{\xi})$ is defined by equation (6.2.43b). Then equation (6.2.57) reduces to

$$T_{\mathrm{s}}(\boldsymbol{k}_R, \boldsymbol{k}_I, \boldsymbol{\xi}) = A(\boldsymbol{k}_R, \boldsymbol{\xi}) + \mathrm{i}C_{n-1} \int_{\mathbb{R}^n} \Theta\,(\boldsymbol{k}_I \bullet (\boldsymbol{k}_R - \boldsymbol{\xi}))\, \delta(\boldsymbol{\eta}^2 - \boldsymbol{k}_R^2)\, T_{\mathrm{s}}(\boldsymbol{k}_R, \boldsymbol{k}_I, \boldsymbol{\eta})\, A(\boldsymbol{\eta}, \boldsymbol{\xi})\, \mathrm{d}\boldsymbol{\eta}, \tag{6.2.59}$$

on $\boldsymbol{\xi}^2 = \boldsymbol{k}_R^2$. Equation (6.2.59) was obtained by Faddeev [1965, 1966, 1976] in his study of the time-dependent Schrödinger problem and serves to relate the physical outgoing scattering amplitude to the "inverse data" $T_{\mathrm{s}}(\boldsymbol{k}_R, \boldsymbol{k}_I, \boldsymbol{\xi})$ (also called the nonphysical scattering amplitude). We reiterate the fact that the derivation above could be carried out directly on the time-dependent/independent Schrödinger operator without any recourse to "generalized scattering" as discussed above.

6.2.7 Further Remarks

(i) $T(\boldsymbol{k}_R, \boldsymbol{k}_I, \boldsymbol{\xi})$ [$T_{\mathrm{L}}(\boldsymbol{k}_R, \boldsymbol{k}_I, \boldsymbol{\xi})$, $T_{\mathrm{s}}(\boldsymbol{k}_R, \boldsymbol{k}_I, \boldsymbol{\xi})$] satisfies a quadratically nonlinear differential-integral equation when $n > 1$; i.e. equation (6.2.19) [(6.2.35), (6.2.45)]. The fundamental feature of this equation is that it leads to a characterization (admissibility criteria) for the inverse data. However, at the same time the nonlinear nature of the condition generally precludes the existence of a simple time evolution of the inverse data, for example $T(\bullet, t) \neq T(\bullet, 0)\,\mathrm{e}^{\mathrm{i}\omega t}$. Such simple flows are associated with the KdV, KP equations etc.. Hence this result provides further explanation as to the difficulty in associating local nonlinear evolution equations with the multidimensional scattering problem (6.2.1).

(ii) Equations (6.2.23) [together with the limit and reduction cases, equations (6.2.37) and (6.2.47)], provides characterization/admissibility criteria for the inverse data and a reconstruction formula for the potential in the same formula. Even for the classical problem of the time-independent Schrödinger operator (cf. Faddeev [1965, 1966, 1976]; Newton [1970, 1980, 1981, 1982a,b,c, 1983, 1989]), equation (6.2.47) above gives some novel information: it shows that Faddeev's characterization (with which it is essentially equivalent, cf. Nachman and Ablowitz [1984a]), naturally arises as an integral equation for T, a somewhat more convenient condition to verify than Faddeev's analyticity requirement; it also shows that once T is known then the potential can be found by quadratures. The scattering data are related to the inverse data via equations (6.2.55), (6.2.57) and (6.2.59). Given admissible inverse data, then solving (6.2.57) and (6.2.59) yield the scattering amplitudes which should be physically relevant and, in principle measurable. It is an open and important problem regarding how one could measure the scattering amplitude and at the same time ensure that the inverse data resulting from equations (6.2.57) and (6.2.59) will still be admissible even when small errors are present. Namely, how can one adjust errors in data in order to ensure admissibility.

6.3 Multidimensional Inverse Scattering for First-Order Systems.

In this section we briefly discuss inverse scattering associated with multidimensional generalization of the $N \times N$ one-dimensional scattering problem

$$\frac{\partial \boldsymbol{v}}{\partial y} + \sigma \sum_{p=1}^{n} \mathbf{J}_p \frac{\partial \boldsymbol{v}}{\partial x_p} = \mathbf{Q}(\boldsymbol{x}, y)\boldsymbol{v}, \tag{6.3.1}$$

with $\boldsymbol{x} = (x_1, x_2, \ldots, x_n) \in \mathbb{R}^n$ and $y \in \mathbb{R}$, where the $\mathbf{J}_p$ are constant, real, diagonal $N \times N$ matrices (we denote the diagonal entries of $\mathbf{J}_p$ by $J_p^1, J_p^2, \ldots, J_p^N$ and assume that $J_p^i \neq J_p^j \neq 0$, for $i \neq j$, $i,j = 1,2,\ldots,N$, $p = 1,2,\ldots,n$); $\mathbf{Q}(\boldsymbol{x}, y) = (Q^{ij}) \in \mathbb{C}^{N\times N}$ is an off-diagonal matrix containing the potentials and $\sigma = \sigma_R + \mathrm{i}\sigma_I$ is a complex parameter. (For further details see Nachman and Ablowitz [1984b]; Fokas [1986a,b]; Nachman, Fokas and Ablowitz [1986]. We caution the reader that the aforementioned papers use differing notations.)

6.3.1 The Direct and Inverse Scattering Problems. In order to study (6.3.1), we look for a fundamental matrix solution $\Psi(\boldsymbol{x}, y; \boldsymbol{k})$ (an $N \times N$ matrix valued function) in the form

$$\Psi(\boldsymbol{x}, y; \boldsymbol{k}) = \mathrm{m}(\boldsymbol{x}, y; \boldsymbol{k}) \exp\left\{\mathrm{i} \sum\nolimits_{p=1}^{n} k_p (x_p - \sigma y \mathbf{J}_p)\right\}, \tag{6.3.2}$$

with $\boldsymbol{k} = (k_1, k_2, \ldots, k_n) \in \mathbb{C}^n$, then the $N \times N$ matrix valued eigenfunction $\mathrm{m}(\boldsymbol{x}, y; \boldsymbol{k})$ satisfies the equation

$$\frac{\partial \mathrm{m}}{\partial y} + \sigma \sum_{p=1}^{n} \mathbf{J}_p \frac{\partial \mathrm{m}}{\partial x_p} + \mathrm{i}\sigma \sum_{p=1}^{n} k_p \left[\mathbf{J}_p, \mathrm{m}\right] = \mathbf{Q}(\boldsymbol{x}, y)\mathrm{m}. \tag{6.3.3}$$

We seek a solution of this equation, bounded for all values of $\boldsymbol{k} = \boldsymbol{k}_R + \mathrm{i}\boldsymbol{k}_I$ and tending to the unit matrix $\mathbf{I}$ as $|\boldsymbol{k}| \to \infty$, which is given by the solution $\mathbf{m} = (m^{ij})$ (the ijth component of $\mathbf{m}$) of the following linear integral equation

$$\mathbf{m} = \mathbf{I} + \tilde{\mathbf{G}}(\mathbf{Q}\mathbf{m}), \tag{6.3.4a}$$

that is

$$m^{ij}(\boldsymbol{x}, y; \boldsymbol{k}) = \delta_{ij} + \iint_{\mathbb{R}^{n+1}} G^{ij}(\boldsymbol{x} - \boldsymbol{x}', y - y'; \boldsymbol{k}) \times \left((\mathbf{Q}(\boldsymbol{x}', y')\,\mathbf{m}(\boldsymbol{x}', y'; \boldsymbol{k}))^{ij}\right) \mathrm{d}\boldsymbol{x}'\,\mathrm{d}y'. \tag{6.3.4b}$$

By taking a Fourier transform of the equation which $G^{ij}(\boldsymbol{x}, y; \boldsymbol{k})$ satisfies

$$\frac{\partial G^{ij}}{\partial y} + \sigma \sum_{p=1}^{n} J_p^i \frac{\partial G^{ij}}{\partial x_p} + \mathrm{i}\sigma \sum_{p=1}^{n} k_p (J_p^i - J_p^j) G^{ij} = \delta(\boldsymbol{x})\delta(y),$$

we find that the Green's function is given by

$$G^{ij}(\boldsymbol{x}, y; \boldsymbol{k}) = \frac{-\mathrm{i}}{(2\pi)^{n+1}} \iint_{\mathbb{R}^{n+1}} \frac{\exp\{\mathrm{i}(\boldsymbol{x} \bullet \boldsymbol{\xi} + y\eta)\}}{\eta + \sigma \sum_{p=1}^{n} \left[J_p^i \xi_p + k_p \left(J_p^i - J_p^j\right)\right]} \,\mathrm{d}\boldsymbol{\xi}\,\mathrm{d}\eta. \tag{6.3.4c}$$

For simplicity, we shall assume that $\mathbf{Q}$ is such that this integral equation has a bounded solution $\mathbf{m}$ for all $\boldsymbol{k} \in \mathbb{C}^n$.

$\mathbf{G}(\boldsymbol{x}, y; \boldsymbol{k})$ can be evaluated explicitly by contour integration to yield

$$G^{ij}(\boldsymbol{x}, y; \boldsymbol{k}) = \frac{\operatorname{sgn}(\sigma_I J_1^i)}{2\pi\mathrm{i}(x_1 - \sigma J_1^i y)} \exp\left\{\mathrm{i}\alpha^{ij}(\boldsymbol{x}, y; \boldsymbol{k})\right\} \prod_{p=2}^{n} \delta\left(x_p - \frac{J_p^i}{J_1^i} x_1\right), \tag{6.3.5a}$$

where

$$\alpha^{ij}(\boldsymbol{x}, y; \boldsymbol{k}) = \sum_{p=1}^{n} \frac{J_p^i - J_p^j}{\sigma_I} \left(|\sigma|^2 y k_{I_p} - \frac{x_p}{J_p^i} (\sigma k_p)_I\right). \tag{6.3.5b}$$

The next step is to express $\partial\mathbf{m}/\partial\bar{k}$ in terms $\mathbf{m}$. Taking $\partial/\partial\bar{k}_p$ of equation (6.3.4a) gives

$$\frac{\partial \mathbf{m}}{\partial \bar{k}_p} = \frac{\partial \tilde{\mathbf{G}}}{\partial \bar{k}_p}(\mathbf{Q}\mathbf{m}) + \tilde{\mathbf{G}}\left(\mathbf{Q}\frac{\partial \mathbf{m}}{\partial k_p}\right), \tag{6.3.6a}$$

where

$$\left(\frac{\partial \tilde{\mathbf{G}}}{\partial \bar{k}_p}(\mathbf{Q}\mathbf{m})\right)^{ij} = \frac{\overline{\sigma}(J_p^i - J_p^j)}{2\mathrm{i}|\sigma_I|(2\pi)^n} \int_{\mathbb{R}^n} \delta\left(\textstyle\sum_{\ell=1}^{n} J_\ell^i \lambda_\ell\right) \times \exp\left\{\mathrm{i}\beta^{ij}(\boldsymbol{x}, y; \boldsymbol{k}, \boldsymbol{\lambda})\right\} T^{ij}(\boldsymbol{k}, \boldsymbol{\lambda})\,\mathrm{d}\boldsymbol{\lambda}, \tag{6.3.6b}$$

with

$$\beta^{ij}(\boldsymbol{x},y;\boldsymbol{k},\boldsymbol{\lambda}) = \alpha^{ij}(\boldsymbol{x},y;\boldsymbol{k}) + \sum_{\ell=1}^{n}(x_\ell - \sigma_R J_\ell^i y)\lambda_\ell, \tag{6.3.6c}$$

and

$$T^{ij}(\boldsymbol{k},\boldsymbol{\lambda}) = \iint_{\mathbb{R}^{n+1}} \exp\left\{-\mathrm{i}\beta^{ij}(\boldsymbol{\xi},\eta;\boldsymbol{k},\boldsymbol{\lambda})\right\}(\mathbf{Q}(\boldsymbol{\xi},\eta)\mathbf{m}(\boldsymbol{\xi},\eta;\boldsymbol{k}))^{ij}\,\mathrm{d}\boldsymbol{\xi}\,\mathrm{d}\eta. \tag{6.3.6d}$$

The calculation of $\partial \mathbf{m}/\partial \bar{k}$ is then based on the following crucial symmetry property of the Green's function:

$$\exp\left\{-\mathrm{i}\beta^{ij}(\boldsymbol{x},y;\boldsymbol{k},\boldsymbol{\lambda})\right\} G^{rj}(\boldsymbol{x},y;\boldsymbol{k}) = G^{ri}\left(\boldsymbol{x},y;\hat{\boldsymbol{k}}^{ij}(\boldsymbol{k},\boldsymbol{\lambda})\right), \tag{6.3.7a}$$

whenever

$$\sum_{\ell=1}^{n} J_\ell^i \lambda_\ell = 0; \tag{6.3.7b}$$

where $\hat{\boldsymbol{k}}^{ij}(\boldsymbol{k},\boldsymbol{\lambda})$ is the point in $\mathbb{C}^n$ whose pth component is

$$\left(\hat{\boldsymbol{k}}^{ij}(\boldsymbol{k},\boldsymbol{\lambda})\right)_p = \frac{J_p^j - J_p^i}{\sigma_I J_p^i}(\sigma k_p)_I + k_p + \lambda_p. \tag{6.3.7c}$$

Once equation (6.3.7) has been established, we introduce an $N \times N$ matrix valued eigenfunction $\mathbf{M}_{ij}(\boldsymbol{x},y;\boldsymbol{k},\boldsymbol{\lambda})$ which satisfies an equation obtained from equation (6.3.4a) by replacing $\mathbf{I}$ by $\exp\left\{\mathrm{i}\beta^{ij}(\boldsymbol{x},y;\boldsymbol{k},\boldsymbol{\lambda})\right\}\mathbf{E}_{ij}$ where $\mathbf{E}_{ij}$ is the $N \times N$ matrix with entries

$$(\mathbf{E}_{ij})^{rs} = \delta_{ir}\,\delta_{js} = \begin{cases} 1, & \text{if } i = r \text{ and } j = s, \\ 0, & \text{otherwise;} \end{cases}$$

i.e., $\mathbf{M}_{ij}$ satisfies the integral equation

$$\mathbf{M}_{ij} = \exp\left\{\mathrm{i}\beta^{ij}\right\}\mathbf{E}_{ij} + \tilde{\mathbf{G}}(\mathbf{Q}\mathbf{M}_{ij}).$$

Hence

$$\begin{aligned} M_{ij}^{\ell j}(\boldsymbol{x},y;\boldsymbol{k},\boldsymbol{\lambda}) &= \exp\left\{\mathrm{i}\beta^{ij}(\boldsymbol{x},y;\boldsymbol{k},\boldsymbol{\lambda})\right\}\delta_{\ell j} \\ &\quad + \iint_{\mathbb{R}^{n+1}} G^{\ell j}(\boldsymbol{x}-\boldsymbol{x}',y-y';\boldsymbol{k})\left(\mathbf{Q}(\boldsymbol{x}',y')\,\mathbf{M}(\boldsymbol{x}',y';\boldsymbol{k})\right)_{ij}^{\ell j}\,\mathrm{d}\boldsymbol{x}'\,\mathrm{d}y', \end{aligned}$$

with $G^{\ell j}(\boldsymbol{x},y;\boldsymbol{k})$ as in equation (6.3.5). Multiplying this by $\exp\left\{-\mathrm{i}\beta^{ij}(\boldsymbol{x},y;\boldsymbol{k},\boldsymbol{\lambda})\right\}\delta_{ij}$, then, after some manipulation, we obtain the "symmetry condition"

$$M_{ij}^{\ell j}(\boldsymbol{x},y;\boldsymbol{k},\boldsymbol{\lambda}) = m^{\ell i}\left(\boldsymbol{x},y;\hat{k}^{ij}(\boldsymbol{k},\boldsymbol{\lambda})\right)\exp\left\{\mathrm{i}\beta^{ij}(\boldsymbol{x},y;\boldsymbol{k},\boldsymbol{\lambda})\right\}.$$

Using this (and assuming that (6.3.4) has no homogeneous solutions), it may be shown that (6.3.6) implies

$$\frac{\partial \mathbf{m}}{\partial \bar{k}_p}(\boldsymbol{x}, y; \boldsymbol{k}) = \sum_{i,j=1}^{n} \frac{\bar{\sigma}(J_p^i - J_p^j)}{2\mathrm{i}|\sigma_I|(2\pi)^n} \int_{\mathbb{R}^n} \delta\left(\sum\nolimits_{\ell=1}^{n} J_\ell^i \lambda_\ell\right) T^{ij}(\boldsymbol{k}, \boldsymbol{\lambda})$$
$$\times \exp\left\{\mathrm{i}\beta^{ij}(\boldsymbol{x}, y; \boldsymbol{k}, \boldsymbol{\lambda})\right\} \mathbf{m}\left(\boldsymbol{x}, y; \hat{\boldsymbol{k}}^{ij}(\boldsymbol{k}, \boldsymbol{\lambda})\right) \mathbf{E}_{ij}\, \mathrm{d}\boldsymbol{\lambda}. \quad (6.3.8p)$$

Now fix all k_ℓ, $\ell \neq p$, and apply the (one-dimensional) generalized Cauchy integral formula

$$f(z) = \frac{1}{2\pi \mathrm{i}} \int_{|\zeta|=R} \frac{f(\zeta)}{\zeta - z} \mathrm{d}\zeta + \frac{1}{2\pi \mathrm{i}} \iint_{|\zeta| \leq R} \frac{\partial f}{\partial \bar{\zeta}}(\zeta) \frac{\mathrm{d}\zeta \wedge \mathrm{d}\bar{\zeta}}{\zeta - z}, \quad (6.3.9)$$

to the variable k_p. Then we can convert (6.3.8p) into an integral equation. Noting that $\mathbf{m}(\boldsymbol{x}, y; \boldsymbol{k}) \sim \mathbf{I}$ as $|k_p| \to \infty$ and denoting $\boldsymbol{\kappa}^p = (k_1, \ldots, k_{p-1}, \kappa_p, k_{p+1}, \ldots, k_n)$, we have

$$\mathbf{m}(\boldsymbol{x}, y; \boldsymbol{k}) = \mathbf{I} - \frac{\mathrm{i}\bar{\sigma}}{|\sigma_I|(2\pi)^{n+1}} \sum_{i,j=1}^{n} (J_p^i - J_p^j) \iiint_{\mathbb{R}^{n+2}} \frac{\delta(\sum_{\ell=1}^{n} J_\ell^i \lambda_\ell)}{k_p - \kappa_p} T^{ij}(\boldsymbol{\kappa}^p, \boldsymbol{\lambda})$$
$$\times \exp\left\{\mathrm{i}\beta^{ij}(\boldsymbol{x}, y; \boldsymbol{\kappa}^p, \boldsymbol{\lambda})\right\} \mathbf{m}\left(\boldsymbol{x}, y; \hat{\boldsymbol{k}}^{ij}(\boldsymbol{\kappa}^p, \boldsymbol{\lambda})\right) \mathbf{E}_{ij}\, \mathrm{d}\boldsymbol{\lambda}\, \mathrm{d}\kappa_{R_p}\, \mathrm{d}\kappa_{I_p}. \quad (6.3.10p)$$

[More generally, one can use equation (6.3.9) with $\mathbf{f}(z) = \mathbf{m}(\boldsymbol{x}, y; \boldsymbol{k} + z\boldsymbol{v})$, $z \in \mathbb{C}$, with $\boldsymbol{k}$ fixed and with an arbitrary $\boldsymbol{v} \in \mathbb{C}^n$ which is not perpendicular to any of the vectors $\boldsymbol{J}^i - \boldsymbol{J}^j$, $i \neq j$, where $\boldsymbol{J}^i := (J_1^i, J_2^i, \ldots, J_n^i)$]. The matrix-valued function $\mathbf{T}(\boldsymbol{k}, \boldsymbol{\lambda})$ defined in equation (6.3.6d) is the *inverse data* and equation (6.3.10p) is the *scattering equation*, i.e., given $\mathbf{T}(\boldsymbol{k}, \boldsymbol{\lambda})$, one can reconstruct $\mathbf{m}$ from (6.3.10p). Once $\mathbf{m}$ has been found, the potential is reconstructed via the formula

$$\mathbf{Q}(\boldsymbol{x}, y) = \frac{\mathrm{i}\sigma}{\pi} \left[\mathbf{J}_p, \int_{-\infty}^{\infty} \int_{-\infty}^{\infty} \frac{\partial \mathbf{m}}{\partial \bar{k}_p}(\boldsymbol{x}, y; \boldsymbol{\kappa}^p)\, \mathrm{d}\kappa_{R_p}\, \mathrm{d}\kappa_{I_p}\right]. \quad (6.3.11)$$

(Equation (6.3.11) is easily obtained by considering the large k_p asymptotics of equations (6.3.4) and (6.3.10) and comparing the k_p^{-1} terms.)

6.3.2 The Characterization Problem. Equation (6.3.10p) indicates that there exist n inversion formulae for m. Furthermore equation (6.3.11) shows that, unless the inverse data $T^{ij}(\boldsymbol{k}, \boldsymbol{\lambda})$ are appropriately constrained, the reconstructed Q will depend on $\boldsymbol{k}$. Fokas [1986b] shows that Q being independent of $\boldsymbol{k}$ is equivalent to the equality of all the inversion formulae.

For equation (6.3.10p), the compatibility condition

$$\frac{\partial^2 \mathbf{m}}{\partial \bar{k}_r \partial \bar{k}_p} = \frac{\partial^2 \mathbf{m}}{\partial \bar{k}_p \partial \bar{k}_r},$$

yields (using the same technique as that used in §6.2) the following characterization equations for $\mathbf{T}$:

$$\mathcal{L}^{ij}_{pr}[\mathbf{T}](\boldsymbol{k},\boldsymbol{\lambda}) = \mathrm{N}^{ij}_{pr}[\mathbf{T}](\boldsymbol{k},\boldsymbol{\lambda}), \tag{6.3.12$^{ij}_{pr}$}$$

where

$$\begin{aligned}\mathcal{L}^{ij}_{pr}[\mathbf{T}](\boldsymbol{k},\boldsymbol{\lambda}) &:= (J^i_p - J^j_p)\frac{\partial T^{ij}}{\partial \bar{k}_r} - (J^i_r - J^j_r)\frac{\partial T^{ij}}{\partial \bar{k}_p} \\ &+ \frac{\mathrm{i}\bar{\sigma}}{2\sigma_I}(J^i_p - J^j_p)(J^i_r - J^j_r)\left(\frac{1}{J^i_r}\frac{\partial T^{ij}}{\partial \lambda_r} - \frac{1}{J^i_p}\frac{\partial T^{ij}}{\partial \lambda_p}\right),\end{aligned} \tag{6.3.13a}$$

and

$$\begin{aligned}\mathrm{N}^{ij}_{pr}[\mathbf{T}](\boldsymbol{k},\boldsymbol{\lambda}) &:= \frac{\mathrm{i}\bar{\sigma}}{2|\sigma_I|(2\pi)^n}\sum_{m=1}^{N}\Bigg\{\left[(J^m_p - J^j_p)(J^i_r - J^m_r) - (J^m_r - J^j_r)(J^i_p - J^m_p)\right] \\ &\times \int_{\mathbb{R}^n}\delta\left(\sum_{\ell=1}^{n} J^m_\ell \mu_\ell\right) T^{mj}(\boldsymbol{k},\boldsymbol{\mu})\, T^{im}\left(\hat{\boldsymbol{k}}^{mj}(\boldsymbol{k},\boldsymbol{\mu}), \hat{\boldsymbol{\lambda}}^{mi}\right)\mathrm{d}\boldsymbol{\mu}\Bigg\},\end{aligned} \tag{6.3.13b}$$

with $\boldsymbol{\mu} = (\mu_1, \mu_2, \ldots, \mu_n)$ and where $\hat{\boldsymbol{\lambda}}^{mi}$ is the vector such that

$$\left(\hat{\boldsymbol{\lambda}}^{mi}\right)_\ell = \lambda_\ell - (J^\kappa_\ell / J^i_\ell)\mu_\ell, \qquad \ell = 1, 2, \ldots, n.$$

For compatibility, equation (6.3.12) need only hold whenever $\sum_{\ell=1}^{n} J^i_\ell \lambda_\ell = 0$; however, one can verify that if T is defined by equation (6.3.6c) then (6.3.12) is satisfied everywhere.

As with the Schrödinger operator, it turns out to be useful to recast (6.3.12) into integral form. [It is sufficient to consider only (6.3.12$^{ij}_{p1}$).] We parametrize the hyperplane $\{(\boldsymbol{k},\boldsymbol{\lambda}) \in \mathbb{C}^n \times \mathbb{R}^n : \sum_{\ell=1}^{n} J^i_\ell \lambda_\ell = 0\}$ by new variables $(\boldsymbol{z}, w_0, \boldsymbol{w}) \in \mathbb{C}^{n-1} \times \mathbb{R} \times \mathbb{R}^n$, so that, in the new coordinates $\mathcal{L}^{ij}_{p1} \equiv \partial/\partial \bar{z}_p$, for $2 \le p \le n$, and $\beta^{ij}(\boldsymbol{x}, y; \boldsymbol{k}, \boldsymbol{\lambda}) = y w_0 + \boldsymbol{x} \bullet \boldsymbol{w}$. These requirements determine (up to a translation of $\boldsymbol{z}$) the following (invertible) map

$$k_p = (J^j_1 - J^i_1) z^{ij}_p, \qquad 2 \le p \le n, \tag{6.3.14a}$$

$$k_1 = \sum_{\ell=2}^{n}\left\{(J^i_\ell - J^j_\ell) z^{ij}_\ell + \frac{1}{J^i_1 - J^i_1}\left(\frac{\bar{\sigma}}{|\sigma|^2} w^{ij}_0 + \sum_{\ell=1}^{n} w^{ij}_\ell\right)\right\}, \tag{6.3.14b}$$

$$\lambda_p = \frac{(J^j_1 - J^i_1)(J^i_p - J^j_p)}{\sigma_I J^i_p}(\sigma z^{ij}_p)_I + w^{ij}_p, \qquad 2 \le p \le n, \tag{6.3.14c}$$

$$\lambda_1 = \sum_{\ell=2}^{n}\left\{\frac{(J^i_1 - J^j_1)(J^i_\ell - J^j_\ell)}{\sigma_I J^i_1}(\sigma z^{ij}_\ell)_I - \frac{J^i_\ell}{J^i_1} w^{ij}_\ell\right\}. \tag{6.3.14d}$$

For convenience, we shall usually suppress the superscripts i,j in w_0, w_p, z_p, $p = 2,\dots,n$; and let $\boldsymbol{w} = (w_1,\dots,w_n)$ and $\boldsymbol{z} = (z_2,\dots,z_n)$. In these new coordinates, equations $(6.3.12^{ij}_{p1})$ and (6.3.6d) become

$$\frac{\partial T^{ij}}{\partial \bar{z}_p}(w_0,\boldsymbol{w},\boldsymbol{z}) = \mathrm{N}^{ij}_{p1}[\mathbf{T}](w_0,\boldsymbol{w},\boldsymbol{z}), \tag{6.3.15a}$$

and

$$T^{ij}(w_0,\boldsymbol{w},\boldsymbol{z}) = \iint_{\mathbb{R}^{n+1}} \exp\{-\mathrm{i}(w_0\eta + \boldsymbol{w}\bullet\boldsymbol{\xi})\} \times (\mathbf{Q}(\boldsymbol{\xi},\eta)\,\mathbf{m}(\boldsymbol{\xi},\eta,w_0,\boldsymbol{w};\boldsymbol{z}))^{ij}\,\mathrm{d}\eta\,\mathrm{d}\boldsymbol{\xi}, \tag{6.3.15b}$$

respectively.

We now need to consider the limit of $T^{ij}(w_0,\boldsymbol{w},\boldsymbol{z})$ as $|z_p|\to\infty$ (with z_ℓ, $\ell\neq p$, w_0, $\boldsymbol{w}$ fixed); it turns out that this depends on whether we have

$$(J_1^r - J_1^j)(J_p^i - J_p^j) = (J_1^i - J_1^j)(J_p^r - J_p^j), \tag{6.3.16}$$

for some $r\neq i,j$.

For brevity we shall consider only two cases (which both arise in the study of the N-wave interaction equations in multidimensions): case I where (6.3.16) does *not* hold for any distinct i, j, r and any $p\neq 1$; and case II where relation (6.3.16) holds for all i, j, p, r (in other words, the vectors $\boldsymbol{J}^1, \boldsymbol{J}^2, \dots, \boldsymbol{J}^1$, all lie on the same line in $\mathbb{R}^n$). First, in the generic case I, we have

$$(J_1^r - J_1^j)(J_p^i - J_p^j) \neq (J_1^i - J_1^j)(J_p^r - J_p^j), \tag{6.3.17}$$

for *any* distinct i, j, r and any $p\neq 1$. Now define $m_i^{\ell j}$ and $\mu_i^{\ell j}$ as follows:

$$m_i^{\ell j} := m^{\ell j}(\boldsymbol{x},y;w_0^{ij},\boldsymbol{w}^{ij},\boldsymbol{z}^{ij}), \tag{6.3.18}$$

$$\mu_i^{\ell j} = \lim_{|z_p|\to\infty} m_i^{\ell j}. \tag{6.3.19}$$

It follows from equations (6.3.4) that $\mu_i^{\ell j}$ satisfy the following equations:

$$\mu_i^{ij}(\boldsymbol{x},y;w_0,w_1) = \frac{\mathrm{sgn}(\sigma_I J_1^i)}{\mathrm{i}(2\pi)^n}\iiint_{\mathbb{R}^{2n}} \frac{\exp\left[\mathrm{i}\{(y-\eta)w_0 + (\boldsymbol{x}-\boldsymbol{\xi})\bullet\boldsymbol{w}\}\right]}{(x_1 - x_1') - \sigma J_1^i(y-\eta)} \times Q^{ij}(\boldsymbol{\xi},\eta)\,\mu_i^{jj}(\boldsymbol{\xi},\eta;w_0,w_1)\,\mathrm{d}\boldsymbol{\xi}\,\mathrm{d}\eta\,\mathrm{d}\hat{\mathbf{w}}, \tag{6.3.20a}$$

$$\mu_i^{jj}(\boldsymbol{x},y;w_0,w_1) = 1 + \frac{\mathrm{sgn}(\sigma_I J_1^j)}{\mathrm{i}(2\pi)^n}\iiint_{\mathbb{R}^{2n}} \frac{Q^{ij}(\boldsymbol{\xi},\eta)\mu_i^{ij}(\boldsymbol{\xi},\eta;w_0,w_1)}{(x_1-\xi_1) - \sigma J_1^j(y-\eta)} \times \mathrm{d}\boldsymbol{\xi}\,\mathrm{d}\eta\,\mathrm{d}\hat{\mathbf{w}}, \tag{6.3.20b}$$

$$\mu_i^{\ell j} = 0 \qquad \text{for all} \qquad \ell,\ \ \ell\neq i,\ \ \ell\neq j, \tag{6.3.20c}$$

where $d\hat{\mathbf{w}} = dw_2\,dw_3\,\dots\,dw_n$. Then we have

$$\begin{aligned}\hat{A}^{ij}(w_0,\boldsymbol{w}) &:= \lim_{|z_p|\to\infty} T^{ij}(w_0,\boldsymbol{w},\boldsymbol{z})\\ &= \iint_{\mathbb{R}^{n+1}} \exp\{-\mathrm{i}(w_0\eta + \boldsymbol{w}\bullet\boldsymbol{\xi})\}Q^{ij}(\boldsymbol{\xi},\eta;w_0,\boldsymbol{w})\,\mathrm{d}\boldsymbol{\xi}\,\mathrm{d}\eta,\end{aligned} \tag{6.3.21}$$

and the characterization equation (6.3.13b) becomes

$$\begin{aligned}\mathrm{I}_p^{ij}[\mathbf{T}](w_0,\boldsymbol{w},\boldsymbol{z}) &:= \hat{A}^{ij}(w_0,\boldsymbol{w})\\ &= T^{ij}(w_0,\boldsymbol{w},\boldsymbol{z}) - \frac{1}{\pi}\int_{-\infty}^{\infty}\int_{-\infty}^{\infty}\frac{\mathrm{N}_{p1}^{ij}[\mathbf{T}](w_0,\boldsymbol{w},\boldsymbol{\zeta}^p)}{z_p-\zeta_p}\,\mathrm{d}\zeta_{R_p}\,\mathrm{d}\zeta_{I_p},\end{aligned} \tag{6.3.22}$$

where $\boldsymbol{\zeta}^p = (z_2\dots,z_{p-1},\zeta_p,z_{p+1},\dots,z_n)$. Fokas [1986b] shows that $\hat{A}^{ij}(w_0,\boldsymbol{w})$ arises in a reduced first order linear system of equations in two independent variables, from which reconstruction formulae can be ascertained. Equation (6.3.22) provides an explicit characterization of the inverse data $T^{ij}(w_0,\boldsymbol{w},\boldsymbol{z})$. Nachman and Ablowitz [1984b] conjecture that the main condition needed to characterize the scattering data is that the $\hat{I}_p^{ij}[\mathbf{T}](w_0,\boldsymbol{w},\boldsymbol{z})$ are independent of $\boldsymbol{z}$ and p and have suitable decay properties in $(w_0,\boldsymbol{w})$. Furthermore, given a $\mathbf{T}$ which passes this admissibility test we can (re)construct a local potential $\mathbf{Q}$ by carrying out the inverse scattering of a reduced first order system (cf. Fokas [1986b]). However, we note that $I_p^{ij}[\mathbf{T}]$ does *not* equal the Fourier transform of Q^{ij} and hence we are unable to obtain a closed form solution for Q^{ij} in terms of $I_p^{ij}[\mathbf{T}]$. This demonstrates a significant difference between inverse scattering associated with equation (6.3.1) and that for the n-dimensional Schrödinger equation (§6.2), for which $\hat{A}$ is the Fourier transform of Q [see equations (6.2.23)]. (Note that Fokas [1986a] corrects Nachman and Ablowitz [1984b], who state that $\hat{A}^{ij}$ is the Fourier transform of Q^{ij}.)

In case II we suppose that the $\mathbf{J}_p$'s satisfy the constraint

$$(J_1^r - J_1^j)(J_p^i - J_p^j) = (J_1^i - J_1^j)(J_p^r - J_p^j), \tag{6.3.23}$$

for all $p = 1,\dots,n$, $i,j,r = 1,\dots,N$. Then equation (6.3.12) implies that

$$\frac{\partial T^{ij}}{\partial \bar{z}_p} = 0, \tag{6.3.24}$$

for all p, $2 \le p \le n$, and hence (using Liouville's theorem), we have the characterization equation

$$T^{ij}(w_0,\boldsymbol{w},\boldsymbol{z}) = B^{ij}(w_0,\boldsymbol{w}). \tag{6.3.25}$$

Additionally, if the $\mathbf{J}_p$'s satisfy the constraint (6.2.23), then there exists a single parameter κ defined by

$$\sum_{p=1}^{n}\left(J_p^j - J_p^j\right)k_p = \left(J_1^i - J_1^j\right)\kappa, \tag{6.3.26}$$

for all $i, j = 1, \ldots, N$. (This is easily proved by multiplying (6.3.23) by k_p and summing over p.) This suggests that the eigenvalue problem associated with this case should be reducible to two dimensions. Fokas [1986a] indeed shows that this is the case, i.e., if the $\mathbf{J}_p$'s satisfy the constraint (6.3.23), then there exists a suitable change of variables such that the system (6.3.1) can be reduced to two dimensions. (Nachman and Ablowitz [1984b] showed that if the $\mathbf{J}_p$'s satisfy the constraint (6.3.23), then the system (6.3.1) is reducible to at most three dimensions.) Furthermore, he reduces the general problem of reconstructing an $N \times N$ potential $\mathbf{Q}$ in $n+1$ dimensions to one of reconstructing a 2×2 potential with entries q^{ij}, q^{ji} in two dimensions. The inverse data required for this reconstruction is precisely $\hat{A}^{ij}$, $\hat{A}^{ji}$. This reduction makes crucial use of the existence of redundant scattering parameters. In this sense it is the analog of the Born approximation for the Schrödinger equation (Newton [1982b]).

6.3.3 The Hyperbolic Limit. Formally substituting $\sigma = -1$ in equation (6.3.4c) shows that, away from the hyperplanes $\Sigma_{ij} = \{k \in \mathbb{C}^n : \sum_{p=1}^n (J_p^i - J_p^j) k_{I_p} = 0\}$, the eigenfunction $\mathbf{m}(\boldsymbol{x}, y; \boldsymbol{k})$ is well-defined and holomorphic. Thus it suggests that the inverse problem in the hyperbolic case may be viewed as a Riemann-Hilbert problem with data on the hyperplanes Σ_{ij}, $1 \le i < j \le N$. In order to consider the limit as $\sigma \to -1$ of the above, we consider the limit of $\mathbf{m}(\boldsymbol{x}, y; \boldsymbol{k}_R, \sigma_I \boldsymbol{k}_I)$, with $\boldsymbol{k}_I$ playing the role of a parameter. Therefore, if

$$\lim_{\sigma \to -1+\mathrm{i}0} \mathbf{m}(\boldsymbol{x}, y; \boldsymbol{k}_R, \sigma_I \boldsymbol{k}_I) = \mathbf{m}_{\mathrm{L}}(\boldsymbol{x}, y; \boldsymbol{k}_R, \boldsymbol{k}_I), \tag{6.3.27}$$

then $\mathbf{m}_{\mathrm{L}}(\boldsymbol{x}, y; \boldsymbol{k}_R, \boldsymbol{k}_I)$ is a solution of

$$\frac{\partial \mathbf{m}_{\mathrm{L}}}{\partial y} - \sum_{p=1}^{n} \mathbf{J}_p \frac{\partial \mathbf{m}_{\mathrm{L}}}{\partial x_p} - \mathrm{i} \sum_{p=1}^{n} k_{R_p} [\mathbf{J}_p, \mathbf{m}_{\mathrm{L}}] = \mathbf{Q}\mathbf{m}_{\mathrm{L}}, \tag{6.3.28}$$

and also solves the integral equation

$$\mathbf{m}_{\mathrm{L}} = \mathbf{I} + \tilde{\mathbf{G}}_{\mathrm{L}}(\mathbf{Q}\mathbf{m}_{\mathrm{L}}), \tag{6.3.29a}$$

i.e.,

$$m_{\mathrm{L}}^{ij}(\boldsymbol{x}, y; \boldsymbol{k}_R, \boldsymbol{k}_I) = \delta_{ij} + \iint_{\mathbb{R}^{n+1}} G_{\mathrm{L}}^{ij}(\boldsymbol{x} - \boldsymbol{x}', y - y'; \boldsymbol{k}_R, \boldsymbol{k}_I) \times \left(\mathbf{Q}(\boldsymbol{x}', y')\, \mathbf{m}_{\mathrm{L}}(\boldsymbol{x}', y'; \boldsymbol{k}_R, \boldsymbol{k}_I)\right)^{ij} \, \mathrm{d}\boldsymbol{x}' \, \mathrm{d}y', \tag{6.3.29b}$$

where

$$G_{\mathrm{L}}^{ij}(\boldsymbol{x}, y; \boldsymbol{k}_R, \boldsymbol{k}_I) = \lim_{\sigma \to -1+\mathrm{i}0} G^{ij}(\boldsymbol{x}, y; \boldsymbol{k}_R, \sigma_I \boldsymbol{k}_I)$$

$$= \frac{-\mathrm{i}}{(2\pi)^{n+1}} \iint_{\mathbb{R}^{n+1}} \left\{ \frac{\Theta\left(\sum_{p=1}^{n} \left[J_p^i\lambda_p + (k_{R_p} - k_{I_p})(J_p^i - J_p^j)\right]\right)}{\eta - \sum_{p=1}^{n}\left[J_p^i\lambda_p + k_{R_p}(J_p^i - J_p^j)\right] + \mathrm{i}0} + \frac{\Theta\left(-\sum_{p=1}^{n} \left[J_p^i\lambda_p + (k_{R_p} - k_{I_p})(J_p^i - J_p^j)\right]\right)}{\eta - \sum_{p=1}^{n}\left[J_p^i\lambda_p + k_{R_p}(J_p^i - J_p^j)\right] - \mathrm{i}0} \right\} \times \exp\{\mathrm{i}(y\eta + \boldsymbol{x} \bullet \boldsymbol{\lambda})\} \mathrm{d}\eta\, \mathrm{d}\boldsymbol{\lambda}, \tag{6.3.29c}$$

with $\Theta(\bullet)$ the Heaviside function.

The inverse data is given by

$$T_{\mathrm{L}}^{ij}(\boldsymbol{k}_R, \boldsymbol{k}_I, \boldsymbol{\lambda}) = \iint_{\mathbb{R}^{n+1}} \exp\left\{-\mathrm{i}\beta_{\mathrm{L}}^{ij}(\boldsymbol{x}, y; \boldsymbol{k}_R, \boldsymbol{k}_I, \boldsymbol{\lambda})\right\} \times \left(\mathbf{Q}(\boldsymbol{\xi}, \eta)\mathbf{m}_{\mathrm{L}}(\boldsymbol{x}, y; \boldsymbol{k}_R, \boldsymbol{k}_I)\right)^{ij} \mathrm{d}\boldsymbol{\xi}\, \mathrm{d}\eta, \tag{6.3.30a}$$

with

$$\beta_{\mathrm{L}}^{ij}(\boldsymbol{x}, y; \boldsymbol{k}_R, \boldsymbol{k}_I, \boldsymbol{\lambda}) = \sum_{p=1}^{n} \left\{ (J_p^i - J_p^j)\left[y k_{I_p} - \frac{x_p}{J_p^i}(k_{R_p} - k_{I_p})\right] + (x_p + J_p^i y)\lambda_p \right\}. \tag{6.3.30b}$$

By taking the appropriate limits in equation (6.3.8), we find the reconstruction equation for m_{L}:

$$\mathbf{m}_{\mathrm{L}}(\boldsymbol{x}, y; \boldsymbol{k}_R, \boldsymbol{k}_I) = \mathbf{I} + \frac{\mathrm{i}}{(2\pi)^{n+1}} \sum_{i,j=1}^{n} (J_p^i - J_p^j) \times \iiint_{\mathbb{R}^{n+2}} \left[\frac{\Theta\left(k_{I_p} - \kappa_{I_p}\right)}{k_{R_p} - \kappa_{R_p} + \mathrm{i}0} + \frac{\Theta\left(\kappa_{I_p} - k_{I_p}\right)}{k_{R_p} - \kappa_{R_p} - \mathrm{i}0} \right] \delta\left(\sum_{\ell=1}^{n} J_\ell^i \lambda_\ell\right) \times T_{\mathrm{L}}^{ij}(\boldsymbol{k}_R, \boldsymbol{k}_I, \boldsymbol{\lambda}) \exp\left\{\mathrm{i}\beta_{\mathrm{L}}^{ij}(\boldsymbol{x}, y; \boldsymbol{k}_R, \boldsymbol{k}_I, \boldsymbol{\lambda})\right\} \times \mathbf{m}_{\mathrm{L}}\left(\boldsymbol{x}, y; \hat{\boldsymbol{k}}_{\mathrm{L}}^{ij}(\boldsymbol{\kappa}_R, \boldsymbol{\kappa}_I, \boldsymbol{\lambda})\right) \mathbf{E}_{ij}\, \mathrm{d}\boldsymbol{\lambda}\, \mathrm{d}\kappa_{R_p}\, \mathrm{d}\kappa_{I_p}, \tag{6.3.31}$$

where the real and imaginary parts of the pth component of $\hat{\boldsymbol{k}}_{\mathrm{L}}^{ij}$ are given by

$$\left(\hat{\boldsymbol{k}}_{\mathrm{L}}^{ij}(\boldsymbol{k}_R, \boldsymbol{k}_I, \boldsymbol{\lambda})\right)_{R_p} = \frac{J_p^j}{J_p^i} k_{R_p} + \frac{J_p^i - J_p^j}{J_p^i} k_{I_p} + \lambda_p, \qquad \left(\hat{\boldsymbol{k}}_{\mathrm{L}}^{ij}\right)_{I_p} = k_{I_p},$$

respectively.

In order to write the characterization equations for T_{L}^{ij}, we introduce new variables [suggested by the limit of (6.3.14)] $(\boldsymbol{z}, w_0, \boldsymbol{w}) \in \mathbb{C}^{n-1} \times \mathbb{R} \times \mathbb{R}^n$ to parametrize the hyperplane $\sum_{p=1}^{n} J_p^i \lambda_p = 0 \in \mathbb{R}^{3n}$ as follows:

$$k_{R_p} = (J_1^j - J_1^i) z_{R_p}^{ij}, \qquad p \geq 2, \tag{6.3.32a}$$

$$k_{R_1} = \sum_{\ell=2}^{n} \left\{ \left(J_\ell^i - J_\ell^j\right) z_{R_\ell}^{ij} + \frac{1}{J_1^i - J_1^j} \left(w_0^{ij} - \sum_{\ell=1}^{n} J_\ell^i w_\ell^{ij} \right) \right\}; \tag{6.3.32b}$$

$$k_{I_p} = (J_1^j - J_1^i) z_{I_p}^{ij}, \qquad p \geq 2, \tag{6.3.32c}$$

$$k_{I_1} = \sum_{\ell=2}^{n} \left(J_\ell^i - J_\ell^j\right) z_{I_\ell}^{ij} + \frac{w_0^{ij}}{J_1^i - J_1^j}; \tag{6.3.32d}$$

$$\lambda_p = \frac{(J_1^j - J_1^i)(J_p^i - J_p^j)}{J_p^i} \left(z_{R_p}^{ij} - z_{I_p}^{ij} \right), \qquad p \geq 2, \tag{6.3.32e}$$

$$\lambda_1 = \sum_{\ell=2}^{n} \left\{ \frac{(J_1^i - J_1^j)(J_\ell^i - J_\ell^j)}{J_1^i} \left(z_{R_\ell}^{ij} - z_{I_\ell}^{ij} \right) - \frac{J_\ell^i}{J_\ell^j} w_\ell \right\}. \tag{6.3.32f}$$

Then provided that

$$(J_1^r - J_1^j)(J_p^i - J_p^j) \neq (J_1^i - J_1^j)(J_p^r - J_p^j) \tag{6.3.33}$$

for any distinct i, j, r and any $p \neq 1$, the limit of (6.3.21) yields

$$T_{\mathrm{L}}^{ij}(\boldsymbol{z}_R, \boldsymbol{z}_I, w_0, \boldsymbol{w}) - \frac{1}{\pi} \int_{-\infty}^{\infty} \int_{-\infty}^{\infty} \left[\frac{\Theta(z_{I_p} - \zeta_{I_p})}{z_{R_p} - \zeta_{R_p} + \mathrm{i}0} + \frac{\Theta(\zeta_{I_p} - z_{I_p})}{z_{R_p} - \zeta_{R_p} - \mathrm{i}0} \right]$$
$$\times \, \mathrm{N}_{p1}^{ij}[\mathbf{T}_{\mathrm{L}}](\boldsymbol{\zeta}, w_0, \boldsymbol{w}) \, \mathrm{d}\zeta_{R_p} \, \mathrm{d}\zeta_{I_p} =: \hat{T}_{\mathrm{L}}^{ij}(w_0, \boldsymbol{w}), \tag{6.3.34}$$

with $\mathrm{N}_{p1}[T_{\mathrm{L}}]$ given by a slight modification of equation (6.3.12b). If the J_p's satisfy the constraint

$$(J_1^r - J_1^j)(J_p^i - J_p^j) = (J_1^i - J_1^j)(J_p^r - J_p^j), \tag{6.3.35}$$

then

$$T_{\mathrm{L}}^{ij}(\boldsymbol{z}_R, \boldsymbol{z}_I, w_0, \boldsymbol{w}) = T_{\mathrm{L}}^{ij}(w_0, \boldsymbol{w}). \tag{6.3.36}$$

We remark that once the Green's function G_{L} has been found, all the above results can be obtained directly without recourse to the limiting arguments. [$\nabla_{\boldsymbol{k}_I} m_{\mathrm{L}}$ can be expressed in terms of m_{L} using the appropriate symmetry property of G_{L}, and the analytic behavior of m_{L} for $\boldsymbol{k}_I$ large — needed to establish equation (6.3.31) — follows from equation (6.3.29c). These analyticity properties, together with the compatibility conditions

$$\frac{\partial^2 \mathbf{m}_{\mathrm{L}}}{\partial k_{I_r} \partial k_{I_p}} = \frac{\partial^2 \mathbf{m}_{\mathrm{L}}}{\partial k_{I_p} \partial k_{I_r}}, \qquad 1 \leq p, r \leq n,$$

imply equations (6.3.34) and (6.3.36) (see Nachman and Ablowitz [1984b]; Fokas [1986a]).]

For the case $\sigma = -1$ we now derive a relations between $\mathbf{T}_{\mathrm{L}}$ and the scattering operator. To fix notation we sketch an elementary definition of the scattering operator. When $\mathbf{Q}(\boldsymbol{x}, y) \equiv 0$, given $\boldsymbol{f}(\boldsymbol{x}) : \mathbb{R}^n \to \mathbb{R}^N$, the solution of the Cauchy problem

$$\frac{\partial \boldsymbol{u}}{\partial y} + \sigma \sum_{p=1}^{n} \mathbf{J}_p \frac{\partial \boldsymbol{u}}{\partial x_p} = 0, \qquad \boldsymbol{u}(\boldsymbol{x}, 0) = \boldsymbol{f}(\boldsymbol{x}),$$

is

$$u_j(\boldsymbol{x}, y) = f^j(x_1 + yJ_1^j, x_2 + yJ_2^j, \ldots, x_n + yJ_n^j), \qquad 1 \le j \le N,$$

which we write as $\boldsymbol{u}(\boldsymbol{x}, y) = \boldsymbol{f}(\boldsymbol{x} + y\boldsymbol{J})$.

When $\mathbf{Q}(\boldsymbol{x}, y)$ is, say, smooth and of compact support, given and for reasonable $\boldsymbol{f}(\boldsymbol{x}) : \mathbb{R}^n \to \mathbb{R}^N$, there is a unique solution $\boldsymbol{u}(\boldsymbol{x}, y)$ with $\boldsymbol{u}(\boldsymbol{x}, y) = \boldsymbol{f}(\boldsymbol{x} + y\boldsymbol{J})$ for $y \ll 0$; furthermore there is a unique solutions $\boldsymbol{g}(\boldsymbol{x})$ such that $\boldsymbol{u}(\boldsymbol{x}, y) = \boldsymbol{g}(\boldsymbol{x} + y\boldsymbol{J})$ when $y \gg 0$. Write $\boldsymbol{g} = \mathbf{S}\boldsymbol{f}$, then using analogous ideas to those for the scalar case in §6.2.7, we find that on the Fourier transform side $\mathbf{S}$ can be written as

$$\left(\hat{\mathbf{S}}\boldsymbol{f}\right)(\boldsymbol{\xi}) = \boldsymbol{f}(\boldsymbol{\xi}) + \frac{1}{(2\pi)^n} \int_{\mathbb{R}^n} \mathbf{S}(\boldsymbol{\xi}, \boldsymbol{k}_R) \hat{\boldsymbol{f}}(\boldsymbol{k}_R)\, \mathrm{d}\boldsymbol{k}_R.$$

The question we want to address is how to recover the "inverse data" $\mathbf{T}_{\mathrm{L}}$ given the "scattering data" $\mathbf{S}(\boldsymbol{\xi}, \boldsymbol{k}_R)$. To relate $\mathbf{T}_{\mathrm{L}}$ and $\mathbf{S}(\boldsymbol{\xi}, \boldsymbol{k}_R)$ it transpires that we need to relate $\mathbf{m}_{\mathrm{L}}$ and the eigenfunction $\mathbf{m}_{\mathrm{v}}(\boldsymbol{x}, y; \boldsymbol{k}_R)$ corresponding to the "Volterra" Green's function

$$G_{\mathrm{v}}^{ij}(\boldsymbol{x}, y; \boldsymbol{k}_R) = \Theta(y) \exp\left\{-\mathrm{i} \sum\nolimits_{p=1}^{n} \left(x_p + yJ_p^j\right) k_{R_p}\right\} \prod_{\ell=1}^{n} \delta\left(x_\ell + yJ_\ell^i\right).$$

We start we the identity

$$\mathbf{m}_{\mathrm{L}} - \mathbf{m}_{\mathrm{v}} = (\mathbf{G}_{\mathrm{L}} - \mathbf{G}_{\mathrm{v}})(\mathbf{Q}\mathbf{m}_{\mathrm{L}}) + \mathbf{G}(\mathbf{Q}(\mathbf{m}_{\mathrm{L}} - \mathbf{m}_{\mathrm{v}}));$$

write $G_{\mathrm{L}}^{ij} - G_{\mathrm{V}}^{ij}$ as a superposition of $\exp\left(\mathrm{i}\beta_{\mathrm{L}}^{ij}\right)$ and use a suitable symmetry property of $\mathbf{G}$. The main result is the following *linear* equation for the "inverse data" $\mathbf{T}_{\mathrm{L}}$ given the "scattering data" $\mathbf{S}(\boldsymbol{k}_R, \boldsymbol{\lambda})$:

$$T_{\mathrm{L}}^{ij}(\boldsymbol{k}_R, \boldsymbol{k}_I, \boldsymbol{\lambda}) = S^{ij}\left(\hat{\boldsymbol{k}}_{\mathrm{L},R}^{ij}(\boldsymbol{k}_R, \boldsymbol{k}_I, \boldsymbol{\lambda}), \boldsymbol{k}_R\right) - \frac{1}{(2\pi)^n} \sum_{\ell=1}^{N} \int_{\mathbb{R}^n} \Theta\left(\sum\nolimits_{p=1}^{n} J_p^\ell \lambda_p\right)$$
$$\times S^{ij}\left(\hat{\boldsymbol{k}}_{\mathrm{L},R}^{ij}(\boldsymbol{k}_R, \boldsymbol{k}_I, \boldsymbol{\lambda}), \hat{\boldsymbol{k}}_{\mathrm{L},R}^{\ell j}(\boldsymbol{k}_R, \boldsymbol{k}_I, \boldsymbol{\mu})\right) T_{\mathrm{L}}^{\ell j}(\boldsymbol{k}_R, \boldsymbol{k}_I, \boldsymbol{\mu})\, \mathrm{d}\boldsymbol{\mu}, \tag{6.3.37}$$

where $\hat{\boldsymbol{k}}_{\mathrm{L},R}^{ij}(\boldsymbol{k}_R, \boldsymbol{k}_I, \boldsymbol{\lambda})$ denotes the real part of $\hat{\boldsymbol{k}}_{\mathrm{L}}^{ij}(\boldsymbol{k}_R, \boldsymbol{k}_I, \boldsymbol{\lambda})$.

We note that if the $\mathbf{J}_p$'s satisfy the constraint (6.3.35), then, as for the general σ case considered above, there exists a suitable change of variables such that the system

(6.3.28) can be reduced to two (spatial) dimensions and the reconstruction of the $N \times N$ potential $\mathbf{Q}$ in $N+1$ dimensions is reduced to the reconstruction of a 2×2 potential in two dimensions.

Finally, we remark that the physically interesting hyperbolic case $\sigma = -1$ can also be considered directly, without recourse to the limit procedure used above (see Fokas [1986b]). Fokas [1986b] obtains the following results: (a), the characterization problem of the inverse data is explicitly solved using the analyticity of the underlying eigenfunctions; (b), the problem of reconstructing $\mathbf{Q}$ is again reduced to reconstructing a 2×2 potential matrix in two dimensions; and (c), a simple relationship is found between the inverse data $\mathbf{T}$ and the scattering data $\mathbf{S}$. Actually if $N = 3$, then $\mathbf{T} = \mathbf{S}$.

6.3.4 The N-wave Interaction Equations. The multidimensional N-wave interaction equations are

$$\frac{\partial Q^{ij}}{\partial t} = \sigma^{-1} a_{ij} \frac{\partial Q^{ij}}{\partial y} + \sum_{p=1}^{n} (a_{ij} J_p^i - B_p^i) \frac{\partial Q^{ij}}{\partial x_p} + \sigma^{-1} \sum_{p=1}^{n} (a_{ip} - a_{pj}) Q^{ip} Q^{pj}, \quad (6.3.38)$$

with

$$a_{ij} = \frac{B_p^i - B_p^j}{J_p^j - J_p^i}, \qquad 1 \le p \le n, \quad (6.3.39)$$

for some real B_p^i, $1 \le p \le n$, $1 \le i \le n$. The equations (6.3.38) are the compatibility conditions of the Lax pair:

$$\frac{\partial \boldsymbol{v}}{\partial y} + \sigma \sum_{p=1}^{n} \mathbf{J}_p \frac{\partial \boldsymbol{v}}{\partial x_p} = \mathbf{Q}(\boldsymbol{x}, y) \boldsymbol{v}, \quad (6.3.40)$$

$$\frac{\partial \boldsymbol{v}}{\partial t} + \sum_{p=1}^{n} \mathbf{B}_p \frac{\partial \boldsymbol{v}}{\partial x_p} = \mathbf{A}(\boldsymbol{x}, y, t) \boldsymbol{v}; \quad (6.3.41)$$

where the matrices $\mathbf{B}_p$, $1 \le p \le n$, are constant, real, diagonal and

$$A^{ij}(\boldsymbol{x}, y, t) = a_{ij} \sigma^{-1} Q^{ij}(\boldsymbol{x}, y, t),$$

with a_{ij} given by equation (6.3.39). Equation (6.3.39) imposes restrictions on the choice of matrices $\mathbf{J}_p$, $1 \le p \le n$. If equation (6.3.39) holds then the matrix (a_{ij} is symmetric and

$$a_{ip} - a_{pj} = (a_{ij} - a_{pj}) \frac{J_\ell^i - J_\ell^i}{J_\ell^p - J_\ell^i}, \quad (6.3.42)$$

for every ℓ and for all i, j, p distinct. (Conversely, if (6.3.42) holds with (a_{ij}) symmetric, then $1 \le p \le n$, can be found so that (6.3.39) is satisfied.) Note that if $a_{ip} \ne a_{pj}$, then (6.3.42) implies that $\boldsymbol{J}^i$, $\boldsymbol{J}^j$, $\boldsymbol{J}^p$ are collinear. There are two cases:

(i) $\quad a_{ip} = a_{pj}$ *for all distinct* i, j, p.

In this case (6.3.42) imposes no restriction upon $\mathbf{J}_\ell$; in particular, they may be chosen so that

$$(J_1^r - J_1^j)(J_p^i - J_p^j) \neq (J_1^i - J_1^j)(J_p^r - J_p^j) \tag{6.3.43}$$

for any distinct i, j, r and $p \neq 1$. In this case the system (6.3.38) is linear.

(ii) *For some* i_0, j_0, p_0 *distinct*, $a_{i_0 p_0} \neq a_{p_0 j_0}$.

In this case we show that the vectors $\boldsymbol{J}^1, \dots, \boldsymbol{J}^N$ must *all* be collinear. From (6.3.42) we know that $\boldsymbol{J}^{i_0}$, $\boldsymbol{J}^{j_0}$, $\boldsymbol{J}^{p_0}$ are collinear. For any $r \neq i_0, j_0, p_0$, one of the following must be true: (a), $a_{i_0 r} \neq a_{r j_0}$; (b), $a_{r i_0} \neq a_{i_0 p_0}$; (c), $a_{r j_0} \neq a_{j_0 p_0}$ (since if not, then

$$a_{i_0 p_0} = a_{r j_0} = a_{r j_0} = a_{p_0 j_0},$$

which contradicts our assumption). In any of these three possibilities, $\boldsymbol{J}^r$ lies on the line passing through $\boldsymbol{J}^{i_0}$, $\boldsymbol{J}^{p_0}$, $\boldsymbol{J}^{j_0}$; this being true for any r, $1 \le r \le N$. Hence $\boldsymbol{J}^1, \dots, \boldsymbol{J}^N$ are collinear. (Conversely, given $\boldsymbol{J}^1, \boldsymbol{J}^2, \dots, \boldsymbol{J}^N$ collinear with $\boldsymbol{J}^i \neq \boldsymbol{J}^j$, then we can construct (a_{ij}) symmetric such that $a_{i_0 p_0} \neq a_{p_0 j_0}$ for some distinct i_0, j_0, p_0 and equation (6.3.42) holds.)

Hence the $\mathbf{J}_p$'s satisfy the constraint

$$(J_1^r - J_1^j)(J_p^i - J_p^j) = (J_1^i - J_1^j)(J_p^r - J_p^j), \tag{6.3.44}$$

for all $p = 1, \dots n$, $i, j, r = 1, \dots N$. Therefore by the above (§§6.3.3, 6.3.4), in this case the N-wave interaction equations in $n+1$ spatial dimensions can always be reduced to two spatial dimensions by an appropriate change of variables.

Thus, depending upon whether the $\mathbf{J}_p$'s satisfy the constraint (6.3.44), the N-wave interaction equations in $n+1$ spatial dimensions are either a linear system or are reducible to two spatial dimensions. Consequently, a genuine nonlinear evolution equation in three (or more) spatial dimensions, related to the $N \times N$ multidimensional first order scattering problem, remains to be found.

In equations (6.3.40), (6.3.41), let

$$\mathbf{v}(\boldsymbol{x}, y, t) = \mathbf{m}(\boldsymbol{x}, y, t; \boldsymbol{k}) \exp\left\{\mathrm{i} \sum\nolimits_{p=1}^{n} k_p (x_p - \sigma y \mathbf{J}_p - t\mathbf{B}_p)\right\}, \tag{6.3.45}$$

then m satisfies

$$\frac{\partial \mathbf{m}}{\partial y} + \sigma \sum_{p=1}^{n} \mathbf{J}_p \frac{\partial \mathbf{m}}{\partial x_p} + \mathrm{i}\sigma \sum_{p=1}^{n} k_p [\mathbf{J}_p, \mathbf{m}] = \mathbf{Q}\mathbf{m}, \tag{6.3.46}$$

and

$$\mathrm{M}_{\boldsymbol{k}} \mathbf{m} := \frac{\partial \mathbf{m}}{\partial t} + \sum_{p=1}^{n} \mathbf{B}_p \frac{\partial \mathbf{m}}{\partial x_p} + \mathrm{i} \sum_{p=1}^{n} k_p [\mathbf{B}_p, \mathbf{m}] - \mathbf{A}\mathbf{m} = 0. \tag{6.3.47}$$

Applying the operator M_k to equation (6.3.8), we find that (when $\sigma_I \neq 0$)

$$\frac{\partial T^{ij}}{\partial t}(t,\boldsymbol{k},\boldsymbol{\lambda}) = i\sum_{p=1}^{n}\left[B_p^j k_p - B_p^i \hat{k}_p^{ij}(\boldsymbol{k},\boldsymbol{\lambda})\right] T^{ij}(t,\boldsymbol{k},\boldsymbol{\lambda}). \tag{6.3.48}$$

In the case $\sigma = -1$, the equations (obtained as limits of (6.3.47) or by a parallel derivation) are

$$\frac{\partial T_{\mathrm{L}}^{ij}}{\partial t}(t,\boldsymbol{k}_R,\boldsymbol{k}_I,\boldsymbol{\lambda}) = \mathrm{i}\sum_{p=1}^{n}\left[B_p^j k_{R_p} - B_p^i \hat{k}_{R_p}^{ij}(\boldsymbol{k}_R,\boldsymbol{k}_I,\boldsymbol{\lambda})\right] T_{\mathrm{L}}^{ij}(t,\boldsymbol{k}_R,\boldsymbol{k}_I,\boldsymbol{\lambda}). \tag{6.3.49}$$

Therefore when $\sigma = -1$, we can apply the inverse scattering procedure together with equation (6.3.47) in order to construct the solution to the Cauchy problem for equations (6.3.38). Note that $\mathbf{T}_{\mathrm{L}}(t,\boldsymbol{k}_R,\boldsymbol{k}_I,\boldsymbol{\lambda})$ as given by (6.3.49) satisfies the characterization equations if $\mathbf{T}_{\mathrm{L}}(0,\boldsymbol{k}_R,\boldsymbol{k}_I,\boldsymbol{\lambda})$ does.

6.4 The Generalized Wave and Generalized Sine-Gordon Equations.

6.4.1 Introduction. The Sine-Gordon equation

$$u_{tt} - u_{xx} + \kappa \sin u = 0, \tag{6.4.1}$$

where κ is a constant, has attracted the interest both of mathematicians and physicists. The initial value problem associated with (6.4.1) was first solved by I.S.T. by Ablowitz, Kaup, Newell and Segur [1973a]. In mathematics it has arisen classically in the study of differential geometry and in physics it has arisen in the study of Josephson junctions, models of particle physics, stability of fluid motions etc. (see Scott, Chu and McLaughlin [1973]; Gibbon [1985] and the references therein for details). In this section we describe work by Ablowitz, Beals and Tenenblat [1986] on the I.S.T. scheme for solving certain nonlinear n-dimensional generalizations of the Sine-Gordon and wave equations ($\kappa = 0$) which naturally arise in theory of differential geometry.

As mentioned in §6.1, the Sine-Gordon equation arises in the study of pseudospherical surfaces M^2 (that is, surfaces of constant negative curvature) embedded in Euclidean space $\mathbb{R}^3$. There is an intimate relationship between such surfaces and solutions of the equation, which led Bäcklund in about 1875 to discover the transformation which now bears his name (Bäcklund [1876, 1880, 1882, 1905]). Let M^2 and $\bar{M}^2$ be surfaces in $\mathbb{R}^3$ and $\ell : M^2 \to \bar{M}^2$ be a diffeomorphism such that for any point p in M^2 and corresponding $\bar{p} = \ell(p)$ one has the following: (a), the line determined by p and $\bar{p}$ is tangent to M^2 and $\bar{M}^2$ at p and $\bar{p}$ respectively; (b), the distance $d(p,\bar{p}) = r > 0$ is a constant independent of p; and (c), the angle between the normal vectors $\boldsymbol{N}(p)$ and $\bar{\boldsymbol{N}}(\bar{p})$ to the surfaces is a constant θ independent of p.

Bäcklund proved that under these conditions, the surfaces M^2 and $\bar{M}^2$ have constant Gaussian curvature $\kappa = \bar{\kappa} = -r^{-2}\sin^2\theta$ which can be normalized to be -1. Moreover he showed that given any surface $M^2 \subset \mathbb{R}^3$ with curvature $\kappa = -1$, there exists a two-parameter family of surfaces $\bar{M}^2$ with curvature $\bar{\kappa} = -1$ related to M^2 by diffeomorphisms which satisfy (a)–(c).

The analytic interpretation of these results originated in what is now called a Bäcklund transformation, which provides new solutions of the Sine-Gordon equation from a given one. In order to give the Bäcklund transformation for the Sine-Gordon equation, consider the Sine-Gordon equation in characteristic coordinates

$$w_{\xi\tau} = \sin w, \tag{6.4.2}$$

where

$$\xi = \tfrac{1}{2}(x+t), \qquad \tau = \tfrac{1}{2}(t-x).$$

Then the Bäcklund transformation for (6.4.2) is given by

$$u_\xi + v_\xi = 2a\sin\left[\tfrac{1}{2}(u-v)\right], \tag{6.4.3a}$$

$$u_\tau - v_\tau = \frac{2}{a}\sin\left[\tfrac{1}{2}(u+v)\right], \tag{6.4.3b}$$

where a is an arbitrary constant. It is easily verified by cross-differentiation that (6.4.3) are consistent if and only if $u(\xi,\tau)$ and $v(\xi,\tau)$ satisfy (6.4.2). Later Bianchi [1918, 1922] obtained a permutability theorem for surfaces which provides superposition formulae for the Sine-Gordon equation.

Motivated in part by the work of Chern and Terng [1980], Tenenblat and Terng [1980] and Terng [1980] obtained the natural geometrical generalizations of these results by considering hyperbolic (constant sectional curvature equal to -1) n-dimensional submanifolds M^n of the Euclidean space $\mathbb{R}^{2n-1}$. The geometrical results for hyperbolic manifolds M^n contained in $\mathbb{R}^{2n-1}$ were extended by Tenenblat [1985] to manifolds M^n of constant sectional curvature $\kappa < 1$ (respectively $\kappa < -1$) contained in the unit spheres S^{2n-1} (respectively hyperbolic space H^{2n-1}). In particular, the zero-curvature submanifolds of the unit sphere correspond to solutions of a generalized wave (GW) equation which is a homogeneous version [$\epsilon = 0$ in (4.1.9)] of the generalized Sine-Gordon (GSG) equation associated with embeddings in Euclidean space. (For further details on Bäcklund transformations see, for example, the monographs by Anderson and Ibragimov [1979], Rogers and Shadwick [1982] and Rogers and Ames [1989] and the references therein.)

The higher dimensional version of Bäcklund's results takes the following form

$$\mathrm{d}\mathbf{X} + \mathbf{X}\mathbf{A}^T\mathbf{X} = \mathbf{A} - \mathbf{X}\mathbf{B}, \tag{6.4.4}$$

where

$$\mathrm{d}\mathbf{X} = \sum_{j=1}^{n} \frac{\partial \mathbf{X}}{\partial x_j}\,\mathrm{d}x_j, \tag{6.4.5a}$$

$$A_{ij} = b_i(k)\,a_{ij}\,\mathrm{d}x_j, \tag{6.4.5b}$$

$$B_{ij} = \frac{1}{A_{1i}}\frac{\partial a_{1j}}{\partial x_i}\,\mathrm{d}x_j - \frac{1}{A_{1j}}\frac{\partial a_{1i}}{\partial x_j}\,\mathrm{d}x_i, \tag{6.4.5c}$$

for $1 \le i,j \le n$ and $\mathbf{a} = (a_{ij}) \in \mathbb{R}^{n\times n}$. Equations (6.4.4,5) reduce to the Bäcklund transformation for the generalized Sine-Gordon (GSG) equation when

$$b_i(k) = \frac{k^2 + (2\delta_{i1} - 1)}{2k} \tag{6.4.6a}$$

and for the generalized wave (GW) equation when

$$b_i(k) = -\frac{1-k^2}{2k} \equiv \lambda(k). \tag{6.4.6b}$$

The compatibility condition required for the existence of solutions to these Bäcklund transformations results in a system of second order partial differential equations for an orthogonal $n \times n$ matrix $\mathbf{a} = (a_{ij})$ in (6.4.4) which is a function of n independent variables $\mathbf{a} = \mathbf{a}(\boldsymbol{x})$, where $\boldsymbol{x} = (x_1, x_2, \ldots, x_n)$. The equation has the form

$$\begin{aligned} \frac{\partial}{\partial x_i}\left(\frac{1}{a_{1i}}\frac{\partial a_{1j}}{\partial x_i}\right) &+ \frac{\partial}{\partial x_j}\left(\frac{1}{a_{1j}}\frac{\partial a_{1i}}{\partial x_j}\right) \\ &+ \sum_{k\neq i,j} \frac{1}{a_{ij}^2}\frac{\partial a_{1i}}{\partial x_k}\frac{\partial a_{1k}}{\partial x_j} = \epsilon a_{1i}a_{1j}, \qquad i \neq j, \\ \frac{\partial}{\partial x_k}\left(\frac{1}{a_{1j}}\frac{\partial a_{1i}}{\partial x_j}\right) &= \frac{1}{a_{1k}a_{1j}}\frac{\partial a_{1i}}{\partial x_k}\frac{\partial a_{1k}}{\partial x_j}, \qquad i,j,k \quad \text{distinct}, \\ \frac{\partial a_{jk}}{\partial x_i} &= \frac{a_{ji}}{a_{1j}}\frac{\partial a_{1k}}{\partial x_i}, \qquad i \neq k. \end{aligned} \tag{6.4.7}$$

where $\epsilon = 1$ for the GSG equation and $\epsilon = 0$ for the GW equation.

Note that when $n = 2$ and $\epsilon = 1$ (GSG), the orthogonal matrix $\mathbf{a} = (a_{ij})$:

$$\mathbf{a} = \begin{pmatrix} \cos\left(\frac{1}{2}u\right) & \sin\left(\frac{1}{2}u\right) \\ -\sin\left(\frac{1}{2}u\right) & \cos\left(\frac{1}{2}u\right) \end{pmatrix} \tag{6.4.8}$$

for the function $u = u(x,t)$ reduces the GSG equation to the classical Sine-Gordon equation (6.4.1). On the other hand, when $n = 2$ and $\epsilon = 0$, then (6.4.1) reduces the GW equation to the wave equation [(6.4.1) with $\kappa = 0$]. When $n \ge 3$, the GSG and GW equations are *not* obtained by replacing u_{xx} in (6.4.1) by $\nabla^2 u$, where $\nabla^2 \equiv \sum_{j=1}^{n} \partial^2/\partial x_j^2$ (the generalization of the wave equation is nonlinear).

The Bäcklund transformations (6.4.4) described above are in fact matrix Riccati equations. Linearizations of such equations can be performed in straightforward manner (see, for example, Winternitz [1983]). Introducing the transformation,

$$\mathbf{X}(\boldsymbol{x}) = \mathbf{U}(\boldsymbol{x})\mathbf{V}^{-1}(\boldsymbol{x}), \tag{6.4.9}$$

where $\mathbf{U}(\boldsymbol{x})$, $\mathbf{V}(\boldsymbol{x})$ are $n \times n$ matrix functions of $\boldsymbol{x} = (x_1, x_2, \ldots, x_n)$ the following linear system is obtained

$$\begin{pmatrix} \mathrm{d}\mathbf{U} \\ \mathrm{d}\mathbf{V} \end{pmatrix} = \begin{pmatrix} \mathbf{0} & \mathbf{A}(\boldsymbol{x}) \\ \mathbf{A}^T(\boldsymbol{x}) & \mathbf{B}(\boldsymbol{x}) \end{pmatrix} \begin{pmatrix} \mathbf{U} \\ \mathbf{V} \end{pmatrix}, \tag{6.4.10}$$

with the components of $\mathbf{A}(\boldsymbol{x})$, $\mathbf{B}(\boldsymbol{x})$ as given in (6.4.5). Compatibility ensures that the orthogonal matrix $\mathbf{a} = (a_{ij})$ satisfies the GSG equation with (6.4.6a) and GW equation with (6.4.6b). Alternatively, if we set $\boldsymbol{\psi} = (\mathbf{U}, \mathbf{V})$, the following system of $2n$ ordinary differential equations are obtained:

$$\frac{\partial \boldsymbol{\psi}}{\partial x_j} = \lambda(k)\tilde{\mathbf{A}}_j(\boldsymbol{x}; k)\boldsymbol{\psi} + \mathbf{C}_j(\boldsymbol{x})\boldsymbol{\psi}, \tag{6.4.11}$$

where $\tilde{\mathbf{A}}_j(\boldsymbol{x}; k)$, $\mathbf{C}_j(\boldsymbol{x})$ are $2n \times 2n$ matrices with the block structure

$$\tilde{\mathbf{A}}_j(\boldsymbol{x}; k) = \begin{pmatrix} \mathbf{0} & \tilde{\mathbf{a}}_j(\boldsymbol{x}; k) \\ \tilde{\mathbf{a}}_j^T(\boldsymbol{x}; k) & \mathbf{0} \end{pmatrix}, \qquad \mathbf{C}_j(\boldsymbol{x}) = \begin{pmatrix} \mathbf{0} & \mathbf{0} \\ \mathbf{0} & \mathbf{c}_j(\boldsymbol{x}; k) \end{pmatrix}, \tag{6.4.12}$$

where $\tilde{\mathbf{a}}_j(\boldsymbol{x}; k), \mathbf{c}_j(\boldsymbol{x}; k)$ are $n \times n$ matrices with the following structure

$$\tilde{\mathbf{a}}_j(\boldsymbol{x}; k) = \left(\frac{\delta}{\lambda} - 1\right) \mathbf{e}_1 \mathbf{a}_j(\boldsymbol{x}) + \mathbf{a}_j(\boldsymbol{x}), \tag{6.4.13a}$$

$$\mathbf{a}_j(\boldsymbol{x}) = \mathbf{a}(\boldsymbol{x})\mathbf{e}_j, \tag{6.4.13b}$$

$$(\mathbf{c}_j)_{k\ell} = (1 - \delta_{kj})\frac{1}{a_{1k}}\frac{\partial a_{1j}}{\partial x_k}\delta_{j\ell} - (1 - \delta_{\ell j})\frac{1}{a_{1\ell}}\frac{\partial a_{1j}}{\partial x_\ell}\delta_{kj}, \tag{6.4.14}$$

and $\mathbf{e}_j = (\mathbf{e}_j)_{ik}$ is the $n \times n$ unit matrix

$$(\mathbf{e}_j)_{rs} = \delta_{jr}\delta_{js} = \begin{cases} 1, & \text{if } r = s = j, \\ 0, & \text{otherwise.} \end{cases}$$

In (6.4.13), $\mathbf{a}(\boldsymbol{x}) : \mathbb{R}^n \to \mathrm{SO}(n)$ [the space of orthogonal $n \times n$ matrices with determinant equal to 1] is a diagonal matrix associated with the GW equation when $\delta = \lambda$ and with the GSG equation when

$$\delta = \tfrac{1}{2}(k + k^{-1}), \qquad \lambda = \tfrac{1}{2}(k - k^{-1}), \tag{6.4.15}$$

and $\mathbf{c}_j$ is the matrix (6.4.14): $\mathbb{R}^n \to \mathrm{M}_n(\mathbb{R})$ [the space of $n \times n$ matrices over $\mathbb{R}$], $\mathbf{c}_j + \mathbf{c}_j^T = 0$. Although $\mathbf{c}_j$ is determined by $\mathbf{a}$, it is convenient is treat $(\mathbf{a}, \mathbf{c}) :=$

$(\mathbf{a}, \mathbf{c}_1, \mathbf{c}_2, \ldots, \mathbf{c}_n)$ as the data. Then both (6.4.8) and (6.4.14) arise as the compatibility conditions for the scattering problem (6.4.11).

6.4.2 The Direct and Inverse Scattering Problems for the GW Equation. We shall consider the scattering problem associated with the GW equation by transforming it to a standard form and then associating to it a Riemann-Hilbert factorization problem as in Beals and Coifman [1984].

The scattering problem for the GW equation for $\psi = \psi(\boldsymbol{x}; k)$ is given by

$$\frac{\partial \psi}{\partial x_j} = k\mathbf{A}_j(\boldsymbol{x})\psi + \mathbf{C}_j(\boldsymbol{x})\psi, \tag{6.4.16}$$

where $\mathbf{A}_j(\boldsymbol{x})$, $\mathbf{C}_j(\boldsymbol{x})$ are $2n \times 2n$ matrices with the block structure

$$\mathbf{A}_j(\boldsymbol{x}) = \begin{pmatrix} \mathbf{0} & \mathbf{a}_j(\boldsymbol{x}) \\ \mathbf{a}_j^T(\boldsymbol{x}) & \mathbf{0} \end{pmatrix}, \qquad \mathbf{C}_j(\boldsymbol{x}) = \begin{pmatrix} \mathbf{0} & \mathbf{0} \\ \mathbf{0} & \mathbf{c}_j \end{pmatrix}, \tag{6.4.17}$$

with $\mathbf{a}_j$ and $\mathbf{c}_j$ as given by equations (6.4.13b) and (6.4.14), respectively.

In (6.4.16) we make the transformation

$$\tilde{\psi}(\boldsymbol{x}; k) = \mathbf{U}^{-1}(\boldsymbol{x})\psi(\boldsymbol{x}; k), \tag{6.4.18}$$

then $\tilde{\psi}(\boldsymbol{x}; k)$ satisfies

$$\frac{\partial \tilde{\psi}}{\partial x_j} = k\mathbf{J}_j\tilde{\psi} + \mathbf{Q}_j(\boldsymbol{x})\tilde{\psi}, \tag{6.4.19}$$

where

$$\mathbf{J}_j = \begin{pmatrix} \mathbf{e}_j & \mathbf{0} \\ \mathbf{0} & -\mathbf{e}_j \end{pmatrix}, \qquad \mathbf{Q}_j(\boldsymbol{x}) = \mathbf{U}_2^{-1} \begin{pmatrix} \mathbf{b}_j(\boldsymbol{x}) & \mathbf{0} \\ \mathbf{0} & \mathbf{c}_j(\boldsymbol{x}) \end{pmatrix} \mathbf{U}_2; \tag{6.4.20a}$$

and

$$\mathbf{U}_2 = \frac{1}{\sqrt{2}} \begin{pmatrix} \mathbf{I} & -\mathbf{I} \\ \mathbf{I} & \mathbf{I} \end{pmatrix}, \qquad \mathbf{b}_j = -\mathbf{a}^T \frac{\partial \mathbf{a}}{\partial x_j}. \tag{6.4.20b}$$

(6.4.19) is virtually the same problem as that considered in §4.6 above. Now, we look for a fundamental matrix solution of (6.4.19) in the form

$$\tilde{\Psi}(\boldsymbol{x}; k) = \mathbf{m}(\boldsymbol{x}; k) \exp\left(k\boldsymbol{x} \bullet \boldsymbol{J}\right), \qquad \boldsymbol{x} \bullet \boldsymbol{J} := \sum_{j=1}^{n} x_j \mathbf{J}_j, \tag{6.4.21}$$

then $\mathbf{m}(\boldsymbol{x}; k)$ satisfies

$$\frac{\partial \mathbf{m}}{\partial x_j} = k[\mathbf{J}_j, \mathbf{m}] + \mathbf{Q}_j\mathbf{m}. \tag{6.4.22}$$

These equations imply that $\det \mathbf{m}$ is constant, hence we look for $\mathbf{m}$ such that

$$\mathbf{m}(\boldsymbol{x}; k) \quad \text{is bounded,} \qquad \det \mathbf{m}(\boldsymbol{x}; k) \equiv 1. \tag{6.4.23}$$

We study the problem (6.4.22-23) by restricting to lines in $\mathbb{R}^n$. Let $\boldsymbol{w}$ be a unit vector in $\mathbb{R}^n$ and $\boldsymbol{y}$ a vector orthogonal to $\boldsymbol{w}$. Along the line

$$L(\boldsymbol{w},\boldsymbol{y}) = \{\boldsymbol{x} = \boldsymbol{y} + s\boldsymbol{w} : s \in \mathbb{R}\}, \tag{6.4.24}$$

we consider the restriction of $\mathbf{m}$:

$$\tilde{\mathbf{m}}(s;k) \equiv \mathbf{m}(s,k;\boldsymbol{w},\boldsymbol{y}) \equiv \mathbf{m}(\boldsymbol{y}+s\boldsymbol{w};k). \tag{6.4.25}$$

Then from equation (6.4.22)

$$\frac{\partial \tilde{\mathbf{m}}}{\partial s} = k[\mathbf{J}_w, \tilde{\mathbf{m}}] + \mathbf{Q}\tilde{\mathbf{m}}, \tag{6.4.26a}$$

where

$$\frac{\partial}{\partial s} = \sum_{j=1}^{n} \frac{\partial x_j}{\partial s}\frac{\partial}{\partial x_j} = \sum_{j=1}^{n} w_j \frac{\partial}{\partial x_j}, \tag{6.4.26b}$$

$$\mathbf{J}_w \equiv \boldsymbol{w} \bullet \boldsymbol{J} \equiv \sum_{j=1}^{n} w_j \mathbf{J}_j = \operatorname{diag}(w_1,\ldots,w_n,-w_1,\ldots,-w_n), \tag{6.4.26c}$$

$$\mathbf{Q}(s) \equiv \mathbf{Q}(s,\boldsymbol{w},\boldsymbol{y}) \equiv \sum_{j=1}^{n} w_j \mathbf{Q}_j(\boldsymbol{y}+s\boldsymbol{w}). \tag{6.4.26d}$$

Now if $\mathbf{Q}(s,\boldsymbol{w},\boldsymbol{y}) \to 0$ sufficiently rapidly as $|s| \to \infty$, in fact if $\mathbf{Q}(s)$ satisfies

$$\int_{-\infty}^{\infty} \|\mathbf{Q}(s,\boldsymbol{y},\boldsymbol{w})\|\, \mathrm{d}s \le c < 1, \tag{6.4.27}$$

for some constant c and all $\boldsymbol{y}$ orthogonal to $\boldsymbol{w}$, then Ablowitz, Beals and Tenenblat [1986] prove (using the previous results of Beals and Coifman [1984]), that for each $k \in \mathbb{C}\backslash i\mathbb{R}$ there is a unique $\mathbf{m}(\,\bullet\,;k)$ which solves (6.4.22,23) and satisfies the asymptotic condition

$$\lim_{s\to-\infty} \mathbf{m}(\boldsymbol{y}+s\boldsymbol{w};k) = \mathbf{I}, \tag{6.4.28}$$

for all $\boldsymbol{y}$ orthogonal to $\boldsymbol{w}$. Moreover $\mathbf{m}$ is bounded, $\mathbf{m}(\boldsymbol{x};\,\bullet)$ is holomorphic on $\mathbb{C}\backslash i\mathbb{R}$ and the limits

$$\mathbf{m}_\pm(\boldsymbol{x};k) = \lim_{\epsilon\to 0} \mathbf{m}(\boldsymbol{x};k\pm\epsilon), \tag{6.4.29}$$

exist and are smooth functions on $\mathbb{R}^n \times i\mathbb{R}$. Also

$$\lim_{|k|\to\infty} \mathbf{m}(\boldsymbol{x};k) = \mathbf{I}, \tag{6.4.30a}$$

uniformly with respect to $\boldsymbol{x}$. Furthermore, there exists a matrix valued function $\mathbf{V}(k) : i\mathbb{R} \to \mathrm{SL}(2n,\mathbb{C})$ [the space of $n \times n$ matrices with determinant equal to 1], which is the *scattering data* associated to $(\mathbf{a},\mathbf{c})$, such that

$$\mathbf{m}_+(\boldsymbol{x};k) = \mathbf{m}_-(\boldsymbol{x};k)\exp(k\boldsymbol{x}\bullet\boldsymbol{J})\,\mathbf{V}(k)\exp(-k\boldsymbol{x}\bullet\boldsymbol{J}), \tag{6.4.30b}$$

for all $\boldsymbol{x} \in \mathbb{R}^n$, for all $k \in \mathbb{R}$.

We now discuss the inverse problem for the GW equation. It essentially follows from the results of Beals and Coifman [1984] that given a matrix valued function $\mathbf{V} : i\mathbb{R} \to SL(2n, \mathbb{C})$, then for each $\boldsymbol{x} \in \mathbb{R}^n$, there exists a unique matrix valued function $\mathbf{m}(\boldsymbol{x}; \bullet)$ which is bounded and holomorphic on $\mathbb{C}\backslash i\mathbb{R}$, with continuous limits $\mathbf{m}_\pm$ on $\mathbb{R}$, satisfying (6.4.30). (Note that because the normalization (6.4.28) depends upon the choice of a direction $\boldsymbol{w}$, then the solution $\mathbf{m}$ and the associated scattering data $\mathbf{V}$ depend on $\boldsymbol{w}$.) The function $\mathbf{m}$ is smooth on $\mathbb{R}^n \times (\mathbb{C}\backslash i\mathbb{R})$ and satisfies

$$\frac{\partial \mathbf{m}}{\partial x_j} = k[\mathbf{J}_j, \mathbf{m}] + \mathbf{Q}_j(\boldsymbol{x})\mathbf{m}, \tag{6.4.31a}$$

where $\mathbf{Q}_j + \mathbf{Q}_j^T \equiv 0$, $\mathbf{Q}_j(\boldsymbol{x})$ real and

$$\mathbf{Q}_j(\boldsymbol{x}) = \mathbf{U}_2^{-1} \begin{pmatrix} \mathbf{b}_j(\boldsymbol{x}) & 0 \\ 0 & \mathbf{c}_j(\boldsymbol{x}) \end{pmatrix} \mathbf{U}_2. \tag{6.4.31b}$$

Also there exists a function $a(\boldsymbol{x}) : \mathbb{R}^n \to SO(n)$, which is unique up to left multiplication by a fixed element of $SO(n)$ such that

$$\mathbf{b}(s) = -\mathbf{a}^T \frac{\partial \mathbf{a}}{\partial s}, \tag{6.4.32}$$

or

$$\frac{\partial \mathbf{a}}{\partial s} = -\mathbf{a}\mathbf{b}(s), \tag{6.4.33}$$

where $\mathbf{a} \to \mathbf{I}$ as $s \to -\infty$. If $\mathbf{a}$ is any such solution then $(\mathbf{a}, \mathbf{c})$ are the *inverse data* for the function $\mathbf{V}$ (and satisfy the GW equation).

The inverse data can be related explicitly to the asymptotics of $\mathbf{m}$ as $|k| \to \infty$. $\mathbf{m}$ has an asymptotic expansion

$$\mathbf{m}(\boldsymbol{x}; k) \sim \sum_{p=0}^{\infty} \mathbf{m}_p(\boldsymbol{x}) k^{-p}, \qquad \text{as} \quad k \to \infty. \tag{6.4.34}$$

Hence substituting in (6.4.31) and equating powers of k we have

$$\frac{\partial \mathbf{m}_p}{\partial x_j} = [\mathbf{J}_j, \mathbf{m}_{p+1}] + \mathbf{Q}_j \mathbf{m}_p. \tag{6.4.35}$$

In particular, $\mathbf{m}_0 \equiv \mathbf{I}$ and so we obtain

$$\begin{aligned} \mathbf{Q}_j(\boldsymbol{x}) &= -[\mathbf{J}_j, \mathbf{m}_1(\boldsymbol{x})] \\ &= -\lim_{k \to \infty} k[\mathbf{J}_j, \mathbf{m}(\boldsymbol{x}; k)]. \end{aligned} \tag{6.4.36}$$

Therefore by comparing equations (6.4.31b) and (6.4.36) we see that the inverse data $(\mathbf{a}, \mathbf{c})$ can be recovered from the asymptotics of $\mathbf{m}$ as $k \to \infty$.

We are now able to solve an "initial value" problem for the GW equation. Suppose that $\boldsymbol{w}$ is a direction (unit vector) in $\mathbb{R}^n$ and $\boldsymbol{y}$ is orthogonal to $\boldsymbol{w}$, then we parameterize the line $L(w,y)$ by $s \to \boldsymbol{y} + s\boldsymbol{w}$. Without loss of generality we translate coordinates and take $\boldsymbol{y} = 0$. Specializing (6.4.30) to the line $L(\boldsymbol{w},0)$ [see (6.4.24)] we have

$$\mathbf{m}_+(s\boldsymbol{w};k) = \mathbf{m}_-(s\boldsymbol{w};k)\exp(ks\mathbf{J}_w)\,\mathbf{V}(k)\exp(-ks\mathbf{J}_w), \qquad (6.4.37a)$$

$$\mathbf{m} \equiv \mathbf{I}, \qquad \text{as} \quad k \to \infty, \qquad (6.4.37b)$$

with $\mathbf{J}_w$ as defined in equation (6.4.26c). Factorization of the function

$$\exp(k\boldsymbol{x}\bullet\boldsymbol{J})\,\dot{\mathbf{V}}(k)\exp(-k\boldsymbol{x}\bullet\boldsymbol{J}),$$

provides an extension of the function $\tilde{\mathbf{m}}$ to $\mathbb{R}$. Then there exists an associated solution $\mathbf{m}$ of the Riemann-Hilbert problem (6.4.30) such that

$$\tilde{\mathbf{m}}(s;k) = \mathbf{m}(s\boldsymbol{w},k). \qquad (6.4.38)$$

Suppose $\tilde{\mathbf{a}} : L(\boldsymbol{w},0) \to \mathrm{SO}(n)$ and $\tilde{\mathbf{c}} : L(\boldsymbol{w},0) \to \mathrm{M}_n(\mathbb{R})$ are smooth mappings, $\tilde{\mathbf{c}}^T + \tilde{\mathbf{c}} \equiv 0$ and if $\tilde{\mathbf{b}} = -\tilde{\mathbf{a}}^T\partial\tilde{\mathbf{a}}/\partial s$ with

$$\int_{-\infty}^{\infty} \|\tilde{\mathbf{b}}(s)\|\,\mathrm{d}s < 3 - \sqrt{2},$$

then there exists a unique small solution of the GW equation such that

$$\tilde{\mathbf{a}}(s) = \mathbf{a}(s\boldsymbol{w}), \qquad (6.4.39a)$$

$$\tilde{\mathbf{c}}(s) = \sum_{j=1}^{n} w_j\mathbf{c}_j(s\boldsymbol{w}). \qquad (6.4.39b)$$

The "time evolution" is given by

$$\mathbf{V}(k;\boldsymbol{y}) = \exp(k\boldsymbol{y}\bullet\boldsymbol{J})\,\mathbf{V}(k;0)\exp(-k\boldsymbol{y}\bullet\boldsymbol{J}). \qquad (6.4.40)$$

Hence we have the Riemann-Hilbert boundary value problem

$$\mathbf{m}_+(s\boldsymbol{w},\boldsymbol{y};k) = \mathbf{m}_-(s\boldsymbol{w},\boldsymbol{y};k)\exp\{k(s\boldsymbol{w}+\boldsymbol{y})\bullet\boldsymbol{J}\}$$
$$\times\,\mathbf{V}(k;0)\exp\{-k(s\boldsymbol{w}+\boldsymbol{y})\bullet\boldsymbol{J}\}, \qquad (6.4.41a)$$

$$\lim_{|k|\to\infty}\mathbf{m}(\boldsymbol{x};k) = \mathbf{I}, \qquad (6.4.41b)$$

or

$$\mathbf{m}_+(\boldsymbol{x};k) = \mathbf{m}_-(\boldsymbol{x};k)\exp(k\boldsymbol{x}\bullet\boldsymbol{J})\,\mathbf{V}(k;0)\exp(-k\boldsymbol{x}\bullet\boldsymbol{J}), \qquad (6.4.42a)$$

$$\lim_{|k|\to\infty}\mathbf{m}(s\boldsymbol{w},\boldsymbol{y};k) = \mathbf{I}, \qquad (6.4.42b)$$

The solution $\mathbf{m}$ of this Riemann-Hilbert boundary value problem satisfies

$$\frac{\partial \mathbf{m}}{\partial x_j} = k[\mathbf{J}_j, \mathbf{m}] + \mathbf{Q}_j \mathbf{m},$$

and the associated scattering data satisfy the GW equation by compatibility.

6.4.3 The Direct and Inverse Scattering Problems for the GSG Equation. Similar ideas to those used in §6.4.2 above for the GW equation are applicable to the scattering and inverse scattering problems for the GSG equation (that is, to consider the problem on oblique lines), so we shall only sketch a brief outline of the method (see Ablowitz, Beals and Tenenblat [1986] for further details).

The scattering problem for the GSG equation for $\boldsymbol{\psi} = \boldsymbol{\psi}(\boldsymbol{x}; k)$ is given by

$$\frac{\partial \boldsymbol{\psi}}{\partial x_j} = \tfrac{1}{2} k \mathbf{A}_j(\boldsymbol{x}) \boldsymbol{\psi} + \tfrac{1}{2} k^{-1} \mathbf{B}_j(\boldsymbol{x}) \boldsymbol{\psi} + \mathbf{C}_j(\boldsymbol{x}) \boldsymbol{\psi}, \tag{6.4.43}$$

where

$$\mathbf{A}_j(\boldsymbol{x}) = \begin{pmatrix} \mathbf{0} & \mathbf{a}_j(\boldsymbol{x}) \\ \mathbf{a}_j^T(\boldsymbol{x}) & 0 \end{pmatrix}, \qquad \mathbf{B}_j(\boldsymbol{x}) = \begin{pmatrix} \mathbf{0} & \mathbf{u}\mathbf{a}_j(\boldsymbol{x}) \\ \mathbf{a}_j^T(\boldsymbol{x})\mathbf{u} & 0 \end{pmatrix}, \tag{6.4.44a}$$

$$\mathbf{C}_j(\boldsymbol{x}) = \begin{pmatrix} \mathbf{0} & \mathbf{0} \\ \mathbf{0} & \mathbf{c}_j(\boldsymbol{x}) \end{pmatrix}, \tag{6.4.44b}$$

with $\mathbf{u} = \operatorname{diag}(1, -1, \ldots, -1) \in \mathrm{M}_n(\mathbb{R})$. Unlike the the scattering problem for the GW equation, the scattering problem (6.4.43) cannot easily be transformed into a single standard form. Nevertheless, it is still possible to associate a factorization problem of Riemann-Hilbert type.

We look for a fundamental matrix solution of (6.4.43) in the form

$$\boldsymbol{\Psi}(\boldsymbol{x}; k) = \mathbf{U}(\boldsymbol{x}) \mathbf{m}(\boldsymbol{x}; k) \exp\{\boldsymbol{x} \bullet \boldsymbol{J}(k)\}, \tag{6.4.45}$$

where

$$\mathbf{U}(\boldsymbol{x}) = \frac{1}{\sqrt{2}} \begin{pmatrix} \mathbf{a} & -\mathbf{a} \\ \mathbf{I} & \mathbf{I} \end{pmatrix}, \qquad \boldsymbol{x} \bullet \boldsymbol{J}(k) = \sum_{j=1}^{n} x_j \hat{\mathbf{J}}_j(k), \tag{6.4.46a}$$

$$\hat{\mathbf{J}}_j(k) = \tfrac{1}{2}(k \mathbf{J}_j + k^{-1} \mathbf{J}_j^{\#}), \qquad \mathbf{J}_1^{\#} = \mathbf{J}_1, \qquad \mathbf{J}_j^{\#} = -\mathbf{J}_j, \quad j \geq 2. \tag{6.4.46b}$$

Then $\mathbf{m}(\boldsymbol{x}; k)$ satisfies

$$\frac{\partial \mathbf{m}}{\partial x_j} = [\hat{\mathbf{J}}_j(k), \mathbf{m}] + \mathbf{Q}_j(\boldsymbol{x}) \mathbf{m}, \tag{6.4.47}$$

where

$$\mathbf{Q}_j(\boldsymbol{x}) = \mathbf{U}_2^{-1} \begin{pmatrix} \mathbf{b}_j(\boldsymbol{x}) & \mathbf{0} \\ \mathbf{0} & \mathbf{c}_j(\boldsymbol{x}) \end{pmatrix} \mathbf{U}_2 + \tfrac{1}{2} k^{-1} \left(\mathbf{U}^{-1} \mathbf{B}_j(\boldsymbol{x}) \mathbf{U} - \mathbf{J}_j^{\#} \right), \tag{6.4.48a}$$

and

$$\mathbf{b}_j(\boldsymbol{x}) = -\mathbf{a}^T \frac{\partial \mathbf{a}}{\partial x_j}, \qquad \mathbf{U}_2 = \frac{1}{\sqrt{2}} \begin{pmatrix} \mathbf{I} & -\mathbf{I} \\ \mathbf{I} & \mathbf{I} \end{pmatrix}. \tag{6.4.48b}$$

Now we consider the line

$$L(\boldsymbol{w}, \boldsymbol{y}) = \{\boldsymbol{x} = \boldsymbol{y} + s\boldsymbol{w} : s \in \mathbb{R}\}. \tag{6.4.49}$$

Then on this line, (6.4.47) becomes

$$\frac{\partial \mathbf{m}}{\partial s} = k[\mathbf{J}_w(k), \mathbf{m}] + \mathbf{Q}(s)\mathbf{m}, \tag{6.4.50a}$$

where

$$\frac{\partial}{\partial s} = \sum_{j=1}^{n} \frac{\partial x_j}{\partial s} \frac{\partial}{\partial x_j} = \sum_{j=1}^{n} w_j \frac{\partial}{\partial x_j}, \tag{6.4.50b}$$

$$\mathbf{J}_w(k) = \operatorname{diag}(w_1\delta(k), w_2\lambda(k), \ldots, w_n\lambda(k), - w_1\delta(k), -w_2\lambda(k), \ldots, -w_n\lambda(k)), \tag{6.4.50c}$$

$$\mathbf{Q}(s) \equiv \mathbf{Q}(s, \boldsymbol{w}, \boldsymbol{y}) \equiv \sum_{j=1}^{n} w_j \mathbf{Q}_j(\boldsymbol{y} + s\boldsymbol{w}), \tag{6.4.50d}$$

where $\delta(k)$ and $\lambda(k)$ are defined in equation (6.4.15).

Define Σ by

$$\Sigma = \mathrm{i}\mathbb{R} \cup \{k : |k| = 1\} \tag{6.4.51}$$

(i.e., the union of the imaginary axis and the unit circle). If $\mathbf{Q}(s)$ decays sufficiently rapidly as $|s| \to \infty$, then for each $k \in \mathbb{C}\backslash\Sigma$, there is a unique $\mathbf{m}(\bullet; k)$ satisfying (6.4.47) and such that for each $\boldsymbol{y}$ which is orthogonal to $\boldsymbol{w}$

$$\lim_{s \to -\infty} \mathbf{m}(\boldsymbol{y} + s\boldsymbol{w}; k) = \mathbf{I}. \tag{6.4.52}$$

Moreover, $\mathbf{m}$ is bounded, $\mathbf{m}(\boldsymbol{x}; \bullet)$ is holomorphic on $\mathbb{C}\backslash\Sigma$ and $\mathbf{m}(\boldsymbol{x}; \bullet)$ has continuous limits on Σ from each of the five components of $\mathbb{C}\backslash\Sigma$. Furthermore, there exists a matrix valued function $\mathbf{V} : \Sigma\backslash\{\pm \mathrm{i}\} \to \mathrm{SL}(2n, \mathbb{C})$ which is the *scattering data* associated to $(\mathbf{a}, \mathbf{c})$ and the direction $\boldsymbol{w}$, such that

$$\mathbf{m}_+(\boldsymbol{x}; k) = \mathbf{m}_-(\boldsymbol{x}; k) \exp\{\boldsymbol{x} \bullet \boldsymbol{J}(k)\} \mathbf{V}(k) \exp\{-\boldsymbol{x} \bullet \boldsymbol{J}(k)\}, \tag{6.4.53a}$$

$$\lim_{|k| \to \infty} \mathbf{m}(\boldsymbol{x}; k) = \mathbf{I}. \tag{6.4.53b}$$

We shall now discuss the inverse problem for the GSG equation. It essentially follows from the results of Beals and Coifman [1984] that given a matrix valued function $\mathbf{V} : \Sigma \to \mathrm{SL}(2n, \mathbb{C})$, then for each $\boldsymbol{x} \in \mathbb{R}^n$, there exists a unique matrix valued function $\mathbf{m}(\boldsymbol{x}; \bullet)$ which is bounded and holomorphic on $\mathbb{C}\backslash\Sigma$, with continuous limits

$\mathbf{m}$ on Σ, satisfying (6.4.53). Furthermore, there exists a function $\mathbf{a} : \mathbb{R}^n \to \mathrm{SO}(n)$, which is unique up to left multiplication by a fixed element of $\mathrm{SO}(n)$ such that

$$\mathbf{b}(s,y) = -\mathbf{a}^T \frac{\partial \mathbf{a}}{\partial s}, \tag{6.4.54}$$

or

$$\frac{\partial \mathbf{a}}{\partial s} = -\mathbf{a}\mathbf{b}(s,y), \tag{6.4.55}$$

where $\mathbf{a} \to \mathbf{I}$ as $|s| \to \infty$. If $\mathbf{a}$ is any such solution then $(\mathbf{a}, \mathbf{c})$ are the *inverse data* for the function $\mathbf{V}$ (and satisfy the GSG equation).

As for the GW equation, the inverse data $(\mathbf{a}, \mathbf{c})$ can be related explicitly to the asymptotics of $\mathbf{m}$ as $|k| \to \infty$, as in (6.4.36). $\mathbf{m}$ has an asymptotic expansion

$$\mathbf{m}(\boldsymbol{x}; k) = \sum_{p=0}^{\infty} \mathbf{m}_p(\boldsymbol{x}) k^{-p}. \tag{6.4.56}$$

In particular, $\mathbf{m}_0(\boldsymbol{x}) \to \mathbf{I}$ and then it follows from (6.4.47) that

$$\begin{aligned}\mathbf{Q}_j(\boldsymbol{x}) &= \lim_{|k|\to\infty} \mathbf{Q}_j(\boldsymbol{x}; k) = \mathbf{U}_2^{-1} \begin{pmatrix} \mathbf{b} & \mathbf{0} \\ \mathbf{0} & \mathbf{c}_j \end{pmatrix} \mathbf{U}_2 \\ &= \tfrac{1}{2}[\mathbf{J}_j, \mathbf{m}_1(\boldsymbol{x})]. \end{aligned} \tag{6.4.57}$$

Therefore the inverse data $(\mathbf{a}, \mathbf{c})$ can be recovered from the asymptotics of $\mathbf{m}$ as $|k| \to \infty$.

We are now able to solve an "initial value" problem for the GSG equation. As for the GW equation, suppose that $\boldsymbol{w}$ is a direction (unit vector) in $\mathbb{R}^n$ and $\boldsymbol{y}$ is orthogonal to $\boldsymbol{w}$, then we parameterize the line $L(\boldsymbol{w}, \boldsymbol{y})$ by $s \to \boldsymbol{y} + s\boldsymbol{w}$. Without loss of generality we translate coordinates and take $\boldsymbol{y} = 0$. Specializing (6.4.53) to the line $L(\boldsymbol{w}, 0)$ we have

$$\tilde{\mathbf{m}}]_+(s;k) = \tilde{\mathbf{m}}_-(s;k) \exp\{s\mathbf{J}_w(k)\} \mathbf{V}(k;0) \exp\{-s\mathbf{J}_w(k)\}, \tag{6.4.58a}$$

$$\mathbf{m} \to \mathbf{I}, \qquad \text{as} \quad |k| \to \infty. \tag{6.4.58b}$$

Suppose $\tilde{\mathbf{a}} : L(\boldsymbol{w}, 0) \to \mathrm{SO}(n)$ and $\tilde{\mathbf{c}} : L(\boldsymbol{w}, 0) \to \mathrm{M}_n(\mathbb{R})$ are smooth mappings, $\tilde{\mathbf{c}}^T + \tilde{\mathbf{c}} \equiv 0$ and if $\tilde{\mathbf{b}} = -\tilde{\mathbf{a}}^T \partial\tilde{\mathbf{a}}/\partial s$ with

$$\int_{-\infty}^{\infty} \|\tilde{\mathbf{b}}(s)\| \, \mathrm{d}s < k_0,$$

where k_0 is a sufficiently small constant, then there exists a unique small solution of the GW equation such that

$$\mathbf{a}(s) = \mathbf{a}(s\boldsymbol{w}), \tag{6.4.59a}$$

$$\tilde{\mathbf{c}}(s) = \sum_{j=1}^{n} w_j \mathbf{c}_j(s\boldsymbol{w}). \tag{6.4.59b}$$

The "time evolution" is given by

$$\mathbf{V}(k;y) = \exp\{\boldsymbol{y} \bullet \boldsymbol{J}(k)\}\mathbf{V}(k;0)\exp\{-\boldsymbol{y} \bullet \boldsymbol{J}(k)\}, \tag{6.4.60}$$

and hence we have the Riemann-Hilbert boundary value problem

$$\mathbf{m}_+(\boldsymbol{x};k) = \mathbf{m}_-(\boldsymbol{x};k)\exp\{\boldsymbol{x} \bullet \boldsymbol{J}(k)\}\mathbf{V}(k)\exp\{-\boldsymbol{x} \bullet \boldsymbol{J}(k)\}, \tag{6.4.61a}$$

$$\lim_{|k|\to\infty} \mathbf{m} = \mathbf{I}, \tag{6.4.61b}$$

the solution of which satisfies a system of ordinary differential equations and the associated scattering data satisfy the GSG equation.

6.4.4 Further Remarks. Recently Beals and Tenenblat [1988] have extended the above results in connection with inverse scattering for the GSG and GW equations. In particular, they prove global existence results for the associated inverse problems and derive relationships between geometric Bäcklund transformations and the scattering data.

Motivated by the results of Ablowitz, Beals and Tenenblat [1986] on the GW and GSG equations, Ablowitz, Costa and Tenenblat [1987] developed a method for obtaining solutions of multidimensional generalizations of other well-known nonlinear systems, in particular for the multidimensional extension of the anti-self-dual Yang-Mills equation for gauge fields in gl$(n, \mathbb{C})$,

$$S_j \frac{\partial}{\partial \bar{z}_{j+1}}\left(\boldsymbol{\Omega}^{-1}\frac{\partial \boldsymbol{\Omega}}{\partial z_i}\right) - S_i \frac{\partial}{\partial \bar{z}_{i+1}}\left(\boldsymbol{\Omega}^{-1}\frac{\partial \boldsymbol{\Omega}}{\partial z_j}\right) = 0, \tag{6.4.62a}$$

$$\sum_{\substack{l=1 \\ l\neq j}}^{n} S_{jl}\frac{\partial}{\partial \bar{z}_l}\left(\boldsymbol{\Omega}^{-1}\frac{\partial \boldsymbol{\Omega}}{\partial z_i}\right) - \sum_{\substack{k=1 \\ k\neq i}}^{n} S_{ik}\frac{\partial}{\partial \bar{z}_k}\left(\boldsymbol{\Omega}^{-1}\frac{\partial \boldsymbol{\Omega}}{\partial z_j}\right) = 0, \tag{6.4.62b}$$

$$r_j \frac{\partial}{\partial \bar{z}_{j-r_j(n/2)}}\left(\boldsymbol{\Omega}^{-1}\frac{\partial \boldsymbol{\Omega}}{\partial z_i}\right) - r_i \frac{\partial}{\partial \bar{z}_{i-r_i(n/2)}}\left(\boldsymbol{\Omega}^{-1}\frac{\partial \boldsymbol{\Omega}}{\partial z_j}\right) = 0, \tag{6.4.62c}$$

where n is an even integer

$$S_j = (-1)^j, \qquad S_{jk} = \begin{cases} 1, & \text{if } j > k, \\ -1, & \text{if } j < k \end{cases}, \qquad r_j = \begin{cases} -1, & \text{if } j \leq n/2, \\ 1, & \text{if } j > n/2 \end{cases}$$

and $\boldsymbol{\Omega}(\boldsymbol{z})$ is an $N \times N$ matrix valued function of $\boldsymbol{z} = (z_1, \ldots, z_n) \in \mathbb{C}^n$. Note that when $n = 2$, these equations reduces the the anti-self-dual Yang-Mills equation (cf. Pohlymeyer [1980])

$$\frac{\partial}{\partial \bar{z}_1}\left(\boldsymbol{\Omega}^{-1}\frac{\partial \boldsymbol{\Omega}}{\partial z_1}\right) + \frac{\partial}{\partial \bar{z}_2}\left(\boldsymbol{\Omega}^{-1}\frac{\partial \boldsymbol{\Omega}}{\partial z_2}\right) = 0, \tag{6.4.63}$$

where $\boldsymbol{\Omega}(\boldsymbol{z})$ is an $N \times N$ matrix valued function of $\boldsymbol{z} = (z_1, z_2) \in \mathbb{C}^2$.

6.5 The Self-dual Yang-Mills Equations.

6.5.1 Introduction. One of the most interesting and important multidimensional integrable equations is the self-dual Yang-Mills (SDYM) equations (Ward [1977]; Belavin and Zakharov [1978]). These four-dimensional equations arise in the study of field theory (see, for example, Chau [1984]) and relativity (Woodhouse and Mason [1988]; Mason and Newman [1989]). The SDYM equations are regarded as being "integrable" as a consequence of the so-called "twistor" construction, which relate solutions of the equations to certain holomorphic vector bundles (cf. Atiyah and Ward [1977]; Corrigan, Fairlie, Yates and Goddard [1978]; Ward [1981]), or equivalently being able to construct arbitrarily many solutions via a Riemann-Hilbert problem.

It is standard to define the Yang-Mills field as

$$F_{\mu\nu} = \frac{\partial \mathcal{A}_\nu}{\partial x_\mu} - \frac{\partial \mathcal{A}_\mu}{\partial x_\nu} - [\mathcal{A}_\mu, \mathcal{A}_\nu], \tag{6.5.1}$$

where x_μ, $\mu = 0,1,2,3$, are the co-ordinates in Euclidian space E^4 (Euclidean space is taken for convenience). $\mathcal{A}_\mu$ and $\mathcal{A}_\nu$ are the Yang-Mills potentials and take values in some Lie algebra g (associated with a Lie group $\mathcal{G}$) generally, but also can be algebras related to diffeomorphisms, where [,] is the associated Lie bracket. For example, if the Lie group $\mathcal{G}$ is $\mathrm{GL}(N,\mathbb{C})$ (the Lie group of $N \times N$ matrices over the complex field), then the associated Lie algebra is $\mathrm{gl}(N,\mathbb{C})$ with Lie bracket defined by the commutation $[\mathbf{A},\mathbf{B}] = \mathbf{AB} - \mathbf{BA}$; the Lie algebra $\mathrm{gl}(N,\mathbb{C})$, or one of its subalgebras, is frequently used.

Consider the standard null co-ordinates in Euclidean space is given by

$$x_\alpha = t + \mathrm{i}z, \qquad x_{\bar\alpha} = t - \mathrm{i}z, \qquad x_\beta = x + \mathrm{i}y, \qquad x_{\bar\beta} = x - \mathrm{i}y, \tag{6.5.2a}$$

the metric being $\mathrm{d}s^2 = \mathrm{d}x_\alpha\,\mathrm{d}x_{\bar\alpha} + \mathrm{d}x_\beta\,\mathrm{d}x_{\bar\beta}$. Furthermore we have

$$\mathcal{A}_\alpha = \mathcal{A}_t + \mathrm{i}\mathcal{A}_z, \qquad \mathcal{A}_{\bar\alpha} = \mathcal{A}_t - \mathrm{i}\mathcal{A}_z, \tag{6.5.2b}$$

$$\mathcal{A}_\beta = \mathcal{A}_x + \mathrm{i}\mathcal{A}_y, \qquad \mathcal{A}_{\bar\beta} = \mathcal{A}_x - \mathrm{i}\mathcal{A}_y, \tag{6.5.2c}$$

and

$$\frac{\partial}{\partial x_\alpha} \equiv \frac{\partial}{\partial t} - \mathrm{i}\frac{\partial}{\partial z}, \qquad \frac{\partial}{\partial x_{\bar\alpha}} \equiv \frac{\partial}{\partial t} + \mathrm{i}\frac{\partial}{\partial z},$$
$$\frac{\partial}{\partial x_\beta} \equiv \frac{\partial}{\partial x} - \mathrm{i}\frac{\partial}{\partial y}, \qquad \frac{\partial}{\partial x_{\bar\beta}} \equiv \frac{\partial}{\partial x} + \mathrm{i}\frac{\partial}{\partial y}.$$

For the coordinates (6.5.2), the SDYM field equations are given by

$$F_{\alpha\beta} = 0, \tag{6.5.3a}$$

$$F_{\bar\alpha\bar\beta} = 0, \tag{6.5.3b}$$

$$F_{\alpha\bar\alpha} + F_{\beta\bar\beta} = 0. \tag{6.5.3c}$$

This is a system of three partial differential equations for $\mathcal{A}_\alpha$, $\mathcal{A}_{\bar{\alpha}}$, $\mathcal{A}_\beta$ and $\mathcal{A}_{\bar{\beta}}$. We note that the SDYM equations are invariant under the gauge transformation

$$\mathcal{A}_\mu \rightarrow \mathcal{F}\mathcal{A}_\mu\mathcal{F}^{-1} - (\partial_\mu\mathcal{F})\mathcal{F}^{-1}, \tag{6.5.4}$$

with $\partial_\mu \equiv \partial/\partial x_\mu$, for any $\mathcal{F}(x_\mu)$. Frequently the gauge freedom is exploited (which we shall use later — for example, $\mathcal{A}_\alpha = 0$). The Lax pair for the SDYM equations (6.5.3) is given by

$$(\partial_\alpha + \zeta\partial_{\bar{\beta}})\Psi = (\mathcal{A}_\alpha + \zeta\mathcal{A}_{\bar{\beta}})\Psi, \tag{6.5.5a}$$

$$(\partial_\beta - \zeta\partial_{\bar{\alpha}})\Psi = (\mathcal{A}_\beta - \zeta\mathcal{A}_{\bar{\alpha}})\Psi, \tag{6.5.5b}$$

where ζ is the spectral parameter. The compatibility condition is expressed as a polynomial in ζ

$$(\partial_\alpha + \zeta\partial_{\bar{\beta}})(\mathcal{A}_\beta - \zeta\mathcal{A}_{\bar{\alpha}}) - (\partial_\beta - \zeta\partial_{\bar{\alpha}})(\mathcal{A}_\alpha + \zeta\mathcal{A}_{\bar{\beta}}) = [(\mathcal{A}_\alpha + \zeta\mathcal{A}_{\bar{\beta}}), (\mathcal{A}_\beta - \zeta\mathcal{A}_{\bar{\alpha}})].$$

Equating coefficients of powers of ζ yields (6.5.3).

An aspect of current 'soliton folklore' is the conjecture that every completely integrable equation arises as a reduction of some 'master equation'. Ward [1985] has conjectured that:

> "*many (and perhaps all?) of the ordinary and partial differential equations that are regarded as being integrable or solvable may be obtained from the self-dual gauge field equations (or its generalizations) by reduction*".

Are all, integrable equations, in some sense, special cases of the SDYM equations? Ward [1985, 1986] discusses this conjecture in some detail, though he also remarked that, at that time, it was not known whether either the Korteweg-de Vries (KdV) or Kadomtsev-Petviashvili (KP) equations arise as special cases of the SDYM equations. However recently, Mason and Sparling [1989] have shown that the KdV and nonlinear Schrödinger equations arise as reductions of the SDYM equations (see Example 6.5.5 below), thus considerably increasing the validity of the conjecture.

Moreover it is known that the Toda Molecule equation (a 2+1-dimensional version of the 1 + 1-dimensional Toda Lattice) is obtained by reductions (cf., equation (6.5.29) below). We show that by straightforward asymptotic analysis the Toda molecule reduces to the KP equation and from the KP equation it is also straightforward to obtain the Davey-Stewartson (DS) system by asymptotic reduction. Consequently from the SDYM equations one obtains the KP and DS equations via reduction and asymptotic limits. This strongly suggests that the KP and DS equations can be obtained from SDYM by exact reductions (i.e., no asymptotic limits).

In this section we give details of some reductions of the SDYM, to partial differential equations and to ordinary differential equations; we shall consider reductions of equations (6.5.3) or equivalently reductions of (6.5.6), choosing which is the more convenient for the example concerned. Here we shall not be concerned with the techniques for solving the equations, merely listing some of the equations which can be obtained as reductions of the SDYM equations. See Forgács and Manton [1980] for a general discussion of dimensional reduction using symmetries via gauge theory; see also Rosenhaus [1988] for a discussion of the symmetry of the solutions of the SDYM equations.

The SDYM equations are often written in the form due to Pohlmeyer [1976, 1980]

$$\partial_{\bar{\alpha}}(\mathbf{J}^{-1}\partial_\alpha \mathbf{J}) + \partial_{\bar{\beta}}(\mathbf{J}^{-1}\partial_\beta \mathbf{J}) = 0, \tag{6.5.6}$$

where $\mathbf{J} \in \mathrm{gl}(N,\mathbb{C})$. To show this, suppose that

$$\mathcal{A}_\alpha = \mathbf{H}^{-1}\partial_\alpha \mathbf{H}, \qquad \mathcal{A}_{\bar{\alpha}} = \mathbf{K}^{-1}\partial_{\bar{\alpha}}\mathbf{K}, \tag{6.5.7a}$$

$$\mathcal{A}_\beta = \mathbf{H}^{-1}\partial_\beta \mathbf{H}, \qquad \mathcal{A}_{\bar{\beta}} = \mathbf{K}^{-1}\partial_{\bar{\beta}}\mathbf{K}, \tag{6.5.7b}$$

with $\mathbf{H}$ and $\mathbf{K}$ in $\mathrm{gl}(N,\mathbb{C})$. It is readily seen that (6.5.3a,b) are identically satisfied [integrability conditions of (6.5.7)]. Substituting (6.5.7) in (6.5.3c) and setting $\mathbf{J} = \mathbf{H}\mathbf{K}^{-1}$ yields (6.5.6). However, as remarked by Ward [1985], whereas this is a neat form of the equations, it is not as general as the original form (6.5.3) since the SO(4)-invariance, which was present in (6.5.3), and the geometrical interpretation (in terms of connections and curvatures) have been lost [SO(4) is the Lie group of matrices $\mathbf{A}$ such that $\det \mathbf{A} = 1$ and $\mathbf{A}\mathbf{A}^T = 1$, where $\mathbf{A}^T$ is the transpose of $\mathbf{A}$]. Special solutions using inverse scattering formalism (direct methods) of (6.5.3) have been studied by Beals and Coifman [1985b, 1986] and Villarroel [1990b].

EXAMPLE 6.5.1 *The* SU(2) *SDYM Equations*

For the Lie algebra su(2) [the *special unitary* Lie algebra of skew-symmetric Hermitian 2×2 matrices $\mathbf{U}$ with zero trace ($\mathbf{U}^\dagger \mathbf{U} = \mathbf{I}$, where $\mathbf{U}^\dagger = \mathbf{U}^{*T}$)], if in (6.5.6) we take

$$\mathbf{J} = \frac{1}{u}\begin{pmatrix} 1 & w \\ v & u^2 + vw \end{pmatrix},$$

then we obtain

$$u(\nabla \bullet \bar{\nabla} u) = \nabla u \bullet \bar{\nabla} u - \nabla v \bullet \bar{\nabla} w, \tag{6.5.8a}$$

$$u(\nabla \bullet \bar{\nabla} v) = 2\nabla v \bullet \bar{\nabla} u, \tag{6.5.8b}$$

$$u(\nabla \bullet \bar{\nabla} w) = 2\nabla u \bullet \bar{\nabla} w, \tag{6.5.8c}$$

where $\nabla = (\partial_\alpha, \partial_{\bar{\alpha}})$ and $\bar{\nabla} = (\partial_\beta, \partial_{\bar{\beta}})$, which are Yang's equation "in R-gauge" (Yang [1977]; Pohlmeyer [1980]).

EXAMPLE 6.5.2 *Plebanski Equations*

A very interesting system arises by considering the SDYM equations with four translational symmetries and the gauge group of volume preserving diffeomorphisms of an auxiliary four-dimensional manifold M. Here we sketch briefly, how one can obtain the self-dual Einstein equations in a form which is generally known as Plebanski's first and second "Heavenly" equations (Plebanski [1975]), starting from the SDYM equations, following Chakravarty, Mason and Newman [1991].

Let $\mathcal{A}_\mu = \{\mathcal{A}_\alpha, \mathcal{A}_{\bar{\alpha}}, \mathcal{A}_\beta, \mathcal{A}_{\bar{\beta}}\}$ be the components of the Yang-Mills vector potential satisfying the following conditions: (a), the $\mathcal{A}_\mu$'s are totally independent of the coordinates (this is often called the *maximal symmetry condition*); (b), the $\mathcal{A}_\mu$'s are the generators of SDiff$(\mathcal{M})$ — the group of volume preserving diffeomorphisms of an auxiliary manifold $\mathcal{M}$ and can be regarded as four divergence free vector fields on $\mathcal{M}$.

Then from equations (6.5.3) without derivatives, the SDYM field equations are given by

$$[\mathcal{A}_{\bar{\alpha}}, \mathcal{A}_{\bar{\beta}}] = 0, \tag{6.5.9a}$$

$$[\mathcal{A}_\alpha, \mathcal{A}_{\bar{\alpha}}] + [\mathcal{A}_\beta, \mathcal{A}_{\bar{\beta}}] = 0, \tag{6.5.9b}$$

$$[\mathcal{A}_\alpha, \mathcal{A}_\beta] = 0, \tag{6.5.9c}$$

where the [,] represent Lie derivatives of vector fields on $\mathcal{M}$ defined as follows. If v and w are two vector fields on $\mathcal{M}$ defined in terms of local coordinates as

$$v = \sum_{\mu=1}^{4} v_\mu \frac{\partial}{\partial x_\mu}, \qquad w = \sum_{\mu=1}^{4} w_\mu \frac{\partial}{\partial x_\mu},$$

then their Lie derivative is given by

$$[v, w] = \sum_{\mu=1}^{4} [v, w]_\mu \frac{\partial}{\partial x_\mu},$$

where

$$[v, w]_\mu = \sum_{\lambda=1}^{4} \left(v_\lambda \frac{\partial w_\mu}{\partial x_\lambda} - w_\lambda \frac{\partial v_\mu}{\partial x_\lambda} \right).$$

In terms of local coordinates $x_\mu = \{p, q, r, s\}$ on $\mathcal{M}$, it is possible to choose the divergence free vector fields as (cf. Chakravarty, Mason and Newman [1991])

$$\mathcal{A}_{\bar{\alpha}} = \frac{\partial}{\partial p}, \qquad \mathcal{A}_{\bar{\beta}} = \frac{\partial}{\partial q}, \qquad \mathcal{A}_\alpha = f_r \frac{\partial}{\partial s} - f_s \frac{\partial}{\partial r}, \qquad \mathcal{A}_\beta = g_r \frac{\partial}{\partial s} - g_s \frac{\partial}{\partial r}, \tag{6.5.10}$$

for functions $f(x_\mu)$ and $g(x_\mu)$. Substituting (6.5.10) in (6.5.9), we find that (6.5.9a) is identically satisfied and after some appropriate rescaling of the functions f and g, we obtain from (6.5.9b) and (6.5.9c)

$$f_p + g_q = 0, \tag{6.5.11a}$$

and

$$f_r g_s - g_r f_s = 1, \tag{6.5.11b}$$

respectively. Equation (6.5.11a) implies the existence of a scalar function Ω such that

$$f = \Omega_q, \qquad g = -\Omega_p. \tag{6.5.12}$$

Then equation (6.5.12), using equation (6.5.11b), yields Plebanski's second Heavenly equation

$$\Omega_{pr}\Omega_{qs} - \Omega_{qr}\Omega_{ps} = 1. \tag{6.5.13}$$

One can also introduce a new se of coordinates defined by $x \equiv f$, $y \equiv g$ which replaces the pair (r, s) since equation (6.5.11b) is the nonvanishing Jacobian of such a transformation. The vector fields in equations (6.5.10) can then also be rewritten in terms of the new coordinates $X_\mu = \{p, q, x, y\}$ as

$$\mathcal{A}_{\bar{\alpha}} = \frac{\partial}{\partial p} + a_x\frac{\partial}{\partial y} - a_y\frac{\partial}{\partial x}, \tag{6.5.14a}$$

$$\mathcal{A}_{\bar{\beta}} = \frac{\partial}{\partial q} + b_x\frac{\partial}{\partial y} - b_y\frac{\partial}{\partial x}, \tag{6.5.14b}$$

$$\mathcal{A}_{\alpha} = \frac{\partial}{\partial y}, \tag{6.5.14c}$$

$$\mathcal{A}_{\beta} = -\frac{\partial}{\partial x}, \tag{6.5.14d}$$

where a and b are functions of X_μ. Substituting equations (6.5.14) in the field equations (6.5.9), we find that now equations (6.5.9c) is identically satisfied and proceeding similarly as above we obtain the integrability condition of a second scalar function q from equation (6.5.9b), given by

$$a = \Theta_x, \qquad b = \Theta_y. \tag{6.5.15}$$

Finally, using equation (6.5.15) in equation (6.5.9a) we obtain Plebanski's first Heavenly equation

$$\Theta_{yp} - \Theta_{xq} + \Theta_{xx}\Theta_{yy} - \Theta_{xy}\Theta_{yx} = 0. \tag{6.5.16}$$

The vector fields in equations (6.5.10) together with equation (6.5.13) [alternatively, equations (6.5.14) with equation (6.5.16)] define the self-dual Einstein metric on the auxiliary manifold $\mathcal{M}$, first found by Plebanski [1975]. Furthermore, it is interesting to observe that if $r = p$ and $s = q$ in equation (6.5.13), we obtain the Monge-Ampere equation

$$\Omega_{pp}\Omega_{qq} - \Omega_{pq}\Omega_{qp} = 1, \tag{6.5.17}$$

whilst if $\Theta = \Theta(x, y)$ in equation (6.5.16) then we obtain

$$\Theta_{xx}\Theta_{yy} - \Theta_{xy}\Theta_{yx} = 0. \tag{6.5.18}$$

6.5.2 Reductions to 2 + 1-dimensional Equations. In this section we discuss reductions of the SDYM equations to equations with three independent variables.

EXAMPLE 6.5.3 *A* $2+1$*-dimensional* SU(N) *Chiral Field Equation with Torsion*

Suppose that $\mathbf{J}$ lies in the Lie algebra su(N), then Ward [1988a,b] obtained a $2+1$-dimensional SU(2) chiral field equation with an additional 'torsion' (or 'Wess-Zumino-Witten') term (see equation (6.5.43) below for the $1+1$-dimensional chiral field equation).

$$-\left(\mathbf{J}^{-1}\mathbf{J}_t\right)_t+\left(\mathbf{J}^{-1}\mathbf{J}_x\right)_x+\left(\mathbf{J}^{-1}\mathbf{J}_y\right)_y+v_0\left\{\left(\mathbf{J}^{-1}\mathbf{J}_y\right)_x-\left(\mathbf{J}^{-1}\mathbf{J}_x\right)_y\right\}$$
$$+v_1\left\{\left(\mathbf{J}^{-1}\mathbf{J}_t\right)_y-\left(\mathbf{J}^{-1}\mathbf{J}_y\right)_t\right\}+v_2\left\{\left(\mathbf{J}^{-1}\mathbf{J}_x\right)_t-\left(\mathbf{J}^{-1}\mathbf{J}_t\right)_x\right\}=\mathbf{0}, \quad (6.5.19)$$

where $\boldsymbol{v}=(v_0,v_1,v_2)$ is a constant unit vector, which may be either timelike, i.e., $-v_0^2+v_1^2+v_2^2=-1$, or spacelike, i.e, $-v_0^2+v_1^2+v_2^2=+1$ (Ward [1988a,b]; Villarroel [1990a]). The Lax pair for equation (6.5.19), with $\boldsymbol{v}=(0,1,0)$, is given by

$$\mathrm{L}\Psi=[k\partial_x-\partial_y-\mathbf{A}]\Psi=\mathbf{0}, \quad (6.5.20a)$$
$$\mathrm{M}\Psi=[k\partial_t-\partial_x-\mathbf{B}]\Psi=\mathbf{0}, \quad (6.5.20b)$$

with $\mathbf{A}=\mathbf{J}^{-1}\mathbf{J}_y$, $\mathbf{B}=\mathbf{J}^{-1}\mathbf{J}_x$.

We make the following remarks:

1. If $\mathcal{G}$ is a group, then a 'natural' generalization of the $\mathcal{G}$ chiral equation (or $\mathcal{G}$ sigma equation) in $1+1$-dimensions (see equation (6.5.43) below), to $n+1$ dimensions is
$$\partial_0(\mathbf{J}^{-1}\partial_0\mathbf{J})-\sum_{m=1}^{n}\partial_m(\mathbf{J}^{-1}\partial_m\mathbf{J})=\mathbf{0}, \quad (6.5.21)$$
where $\partial_m=\partial/\partial x_m$, and $\mathbf{J}=\mathbf{J}(\boldsymbol{x})$ takes values in $\mathcal{G}$, and $\boldsymbol{x}=(x_0,x_1,x_2,\dots,x_n)$. However, this equation is thought to be *nonintegrable* for $n>1$; in particular, the $2+1$ dimensional chiral equation
$$(\mathbf{J}^{-1}\mathbf{J}_t)_t-(\mathbf{J}^{-1}\mathbf{J}_x)_x-(\mathbf{J}^{-1}\mathbf{J}_y)_y=\mathbf{0}, \quad (6.5.22)$$
appears to be nonintegrable since, for example, it does not satisfy the necessary requirements of the Painlevé test to be integrable (Ward [1988b]).
2. If in equation (6.5.19), $\boldsymbol{v}$ is timelike (i.e., $-v_0^2+v_1^2+v_2^2=-1$), then it is equivalent to the equation
$$\left(\mathbf{J}^{\dagger}\mathbf{J}_t\right)_t+\left(\mathbf{J}^{\dagger}\mathbf{J}_\xi\right)_{\bar\xi}=\mathbf{0}, \quad (6.5.23)$$
where $\xi=x+iy$, $\bar\xi=x-iy$ for which the I.S.T. scheme has been discussed by Manakov and Zakharov [1981] and Villarroel [1989]. The Lax pair for equation (6.5.23) is given by
$$\mathrm{L}\Psi=[k\partial_{\bar\xi}-k^{-1}(\partial_\xi+\mathbf{B})+\mathbf{A}]\Psi=\mathbf{0},$$
$$\mathrm{M}\Psi=[\partial_t+k\partial_{\bar\xi}+\mathbf{A}]\Psi=\mathbf{0},$$

with $\partial_\xi = \partial/\partial\xi$, $\partial_{\bar{\xi}} = \partial/\partial\bar{\xi}$, $\mathbf{A} = \mathbf{J}^\dagger \mathbf{J}_t$ and $\mathbf{B} = \mathbf{J}^\dagger \mathbf{J}_\xi$.

3. There are significant differences in the I.S.T. schemes for equations (6.5.19) and (6.5.23): (a), for equation (6.5.19) it is formulated as a Riemann-Hilbert problem on the real line (Villarroel [1990a]), whereas for equation (6.5.23) it is formulated as a Riemann-Hilbert problem on the unit circle (Villarroel [1989]); (b), for equation (6.5.19) the evolution of the scattering data is pure imaginary where for equation (6.5.23) the evolution is real exponential decay/growth.

EXAMPLE 6.5.4 *The Bogomolny Equations*

If in the SDYM equations (6.5.3) we take

$$\mathcal{A}_\alpha = \Phi - \mathrm{i}\mathcal{A}_3, \qquad \mathcal{A}_\beta = \mathcal{A}_1 - \mathrm{i}\mathcal{A}_2,$$
$$\mathcal{A}_{\bar{\alpha}} = \Phi + \mathrm{i}\mathcal{A}_3, \qquad \mathcal{A}_{\bar{\beta}} = \mathcal{A}_1 + \mathrm{i}\mathcal{A}_2,$$

and assume that $\mathcal{A}_1$, $\mathcal{A}_2$, $\mathcal{A}_3$ and Φ are independent of t, then we obtain the Bogomolny equations (in Euclidean coordinates)

$$\frac{\partial\Phi}{\partial x} + [\Phi, \mathcal{A}_1] + \frac{\partial\mathcal{A}_2}{\partial z} - \frac{\partial\mathcal{A}_3}{\partial y} + [\mathcal{A}_2, \mathcal{A}_3] = 0, \tag{6.5.24a}$$
$$\frac{\partial\Phi}{\partial y} + [\Phi, \mathcal{A}_2] + \frac{\partial\mathcal{A}_3}{\partial x} - \frac{\partial\mathcal{A}_1}{\partial z} + [\mathcal{A}_3, \mathcal{A}_1] = 0, \tag{6.5.24b}$$
$$\frac{\partial\Phi}{\partial z} + [\Phi, \mathcal{A}_3] + \frac{\partial\mathcal{A}_1}{\partial y} - \frac{\partial\mathcal{A}_2}{\partial x} + [\mathcal{A}_1, \mathcal{A}_2] = 0, \tag{6.5.24c}$$

which are relevant in the study of magnetic monopoles (Atiyah and Hitchen [1985a,b, 1988]; Hitchen [1987]).

EXAMPLE 6.5.5 *The Toda Molecule and Toda Lattice Equations*

Suppose that g is a simple Lie algebra of rank N [e.g., $\mathrm{sl}(N+1,\mathbb{C})$], with basis $\{H_i, E_i^+, E_i^-\}_{i=1}^N$, which satisfy the commutation relations

$$[H_i, H_j] = 0, \qquad [E_i^+, E_j^-] = \delta_{ij} H_j, \tag{6.5.25a}$$
$$[H_i, E_j^+] = K_N^{ji} E_j^+, \qquad [H_i, E_j^-] = -K_N^{ji} E_j^-, \tag{6.5.25b}$$

where $\mathbf{K}_N = [K_N^{ij}]$ is a *Cartan matrix* and δ_{ij} is the kronecker delta. A Cartan matrix $\mathbf{K}_N$ is an $N \times N$ matrix with the following properties:

(i) $\mathbf{K}_N$ is positive definite and of rank N;

(ii) $K^{ii} = 2$;

(iii) K^{ij} is either 0 or a negative integer;

(iv) $K^{ij} = 0$ if and only if $K^{ji} = 0$

(for further details on the Cartan matrix see, for example, Olive and Turok [1983]). Now require that $\mathcal{A}_\mu$ are functions only of x_α and $x_{\bar{\alpha}}$ (i.e., functions of t and z), and have the following forms

$$\mathcal{A}_\alpha = \sum_{i=1}^{N} f_i(x_\alpha, x_{\bar{\alpha}}) H_i, \qquad \mathcal{A}_{\bar{\alpha}} = \sum_{i=1}^{N} g_i(x_\alpha, x_{\bar{\alpha}}) H_i, \tag{6.5.26a}$$

$$\mathcal{A}_\beta = \sum_{i=1}^{N} u_i(x_\alpha, x_{\bar{\alpha}}) E_i^+, \qquad \mathcal{A}_{\bar{\beta}} = \sum_{i=1}^{N} u_i(x_\alpha, x_{\bar{\alpha}}) E_i^-, \tag{6.5.26b}$$

for some functions $f_i(x_\alpha, x_{\bar{\alpha}})$, $g_i(x_\alpha, x_{\bar{\alpha}})$, $u_i(x_\alpha, x_{\bar{\alpha}})$, $i = 1, 2, \ldots, N$. Substituting these into equation (6.5.3a) yields

$$F_{\alpha\beta} = \sum_{i,j=1}^{N} \left(\frac{\partial u_i}{\partial x_\alpha} - u_i f_j K_N^{ij} \right) E_i^+ = 0.$$

The $\{E_i^+\}_{i=1}^N$ are linearly independent so we have

$$\frac{\partial u_i}{\partial x_\alpha} = u_i \sum_{j=1}^{N} K_N^{ij} f_j. \tag{6.5.27a}$$

Similarly substituting (6.5.26) in equation (6.5.3b) yields

$$\frac{\partial u_i}{\partial x_{\bar{\alpha}}} = -u_i \sum_{j=1}^{N} K_N^{ij} g_j. \tag{6.5.27b}$$

Finally, equation (6.5.3c) yields

$$\sum_{i=1}^{N} \left\{ \frac{\partial g_i}{\partial x_\alpha} - \frac{\partial f_i}{\partial x_{\bar{\alpha}}} - u_i^2 \right\} H_i = 0. \tag{6.5.28}$$

Since the H_i are not linearly independent, we are unable to deduce that the expression in parentheses in (6.5.28) vanishes, though, by taking the Lie bracket with E_j^+, we deduce from equation (6.5.25) that

$$\sum_{i=1}^{N} K_N^{ji} \left\{ \frac{\partial g_i}{\partial x_\alpha} - \frac{\partial f_i}{\partial x_{\bar{\alpha}}} - u_i^2 \right\} E_j^+ = 0. \tag{6.5.29}$$

Hence, by differentiating equation (6.5.27) with respect to $x_{\bar{\alpha}}$, and equation (6.5.29) with respect to x_α, and then adding we obtain the Toda molecule equations

$$\Delta \phi_i = - \sum_{j=1}^{N} K_N^{ij} \exp(\phi_j), \qquad i = 1, 2, \ldots, N, \tag{6.5.30}$$

where $\Delta\phi \equiv \partial_\alpha\partial_{\bar{\alpha}}\phi = \phi_{zz} + \phi_{tt}$ and $u_i = \exp(\frac{1}{2}\phi_i)$. When N is infinite the Toda molecule equation (6.5.30) may be considered as a three-dimensional equation — having two continuous variables and one discrete variable. If in equation (6.5.30) we further assume that ϕ_i does not depend on z and consider the limit as $N \to \infty$, then we obtain the original Toda Lattice equations (Toda [1967, 1975, 1976]), which we discussed in §3.3.

The simplest case of equation (6.5.30) is when $N = 1$, with Lie algebra su(2), then the Cartan matrix is a scalar, $K_N^{11} = 2$ and we obtain the Liouville equation

$$\Delta\phi_1 = -2\exp(\phi_1). \tag{6.5.31}$$

For $N = 2$ and $N = 3$, examples of Cartan matrices are

$$\mathbf{K}_2 = \begin{pmatrix} 2 & -1 \\ -1 & 2 \end{pmatrix}, \qquad \mathbf{K}_3 \begin{pmatrix} 2 & -1 & 0 \\ -1 & 2 & -1 \\ 0 & -1 & 2 \end{pmatrix}, \qquad \mathbf{K}_4 = \begin{pmatrix} 2 & -1 & 0 & 0 \\ -1 & 2 & -1 & 0 \\ 0 & -1 & 2 & -1 \\ 0 & 0 & -1 & 2 \end{pmatrix},$$

and so from (6.5.30) we obtain the following systems of equations

$$\begin{aligned} \Delta\phi_1 &= -2\exp(\phi_1) + \exp(\phi_2), \\ \Delta\phi_2 &= \exp(\phi_1) - 2\exp(\phi_2), \end{aligned}$$

$$\begin{aligned} \Delta\phi_1 &= -2\exp(\phi_1) + \exp(\phi_2), \\ \Delta\phi_2 &= \exp(\phi_1) - 2\exp(\phi_2) + \exp(\phi_3), \\ \Delta\phi_3 &= \exp(\phi_2) - 2\exp(\phi_3), \end{aligned}$$

and

$$\begin{aligned} \Delta\phi_1 &= -2\exp(\phi_1) + \exp(\phi_2), \\ \Delta\phi_2 &= \exp(\phi_1) - 2\exp(\phi_2) + \exp(\phi_3), \\ \Delta\phi_3 &= \exp(\phi_2) - 2\exp(\phi_3) + \exp(\phi_4), \\ \Delta\phi_4 &= \exp(\phi_3) - 2\exp(\phi_4), \end{aligned}$$

respectively.

A similar result is obtained by taking a basis $\{H_i, E_i^+, E_i^-\}_{i=0}^N$, together with an *extended Cartan matrix* $\tilde{\mathbf{K}}_N$ and the same commutation relations (6.5.25) (with $\mathbf{K}_N$ replaced by $\tilde{\mathbf{K}}_N$). An extended Cartan matrix $\tilde{\mathbf{K}}_N$ is an $(N+1) \times (N+1)$ matrix with the following properties:

(i) $\tilde{\mathbf{K}}_N$ is of rank N;

(ii) $\tilde{\mathbf{K}}_N$ possesses a zero eigenvalue ($\det \tilde{\mathbf{K}}_N = 0$);

(iii) $\tilde{K}^{ii} = 2$;
(iv) $\tilde{K}^{ij}$ is either 0 or a negative integer;
(v) $\tilde{K}^{ij} = 0$ if and only if $\tilde{K}^{ji} = 0$

In this case one obtains the Toda Lattice equations

$$\Delta\phi_i = -\sum_{j=0}^{N} \tilde{K}_{N+1}^{ij} \exp(\phi_j), \qquad i = 0, 1, \ldots, N. \tag{6.5.32}$$

The simplest example is when $N = 1$, Lie algebra $\mathrm{sl}(2, \mathbb{C})$, with

$$\tilde{\mathbf{K}}_1 = \begin{pmatrix} 2 & -2 \\ -2 & 2 \end{pmatrix},$$

so that (6.5.32) becomes

$$\begin{aligned} \Delta\phi_1 &= -2\{\exp(\phi_1) - \exp(\phi_2)\}, \\ \Delta\phi_2 &= -2\{\exp(\phi_2) - \exp(\phi_1)\}, \end{aligned}$$

which on setting $\phi_2 = -\phi_1$ reduces to the sinh-Gordon equation

$$\Delta\phi_1 = -4\sinh(\phi_1). \tag{6.5.33}$$

Examples of extended Cartan matrices for $N = 2$ and $N = 3$ are

$$\tilde{\mathbf{K}}_2 = \begin{pmatrix} 2 & -1 & -1 \\ -1 & 2 & -1 \\ -1 & -1 & 2 \end{pmatrix}, \qquad \tilde{\mathbf{K}}_3 = \begin{pmatrix} 2 & -1 & 0 & -1 \\ -1 & 2 & -1 & 0 \\ 0 & -1 & 2 & -1 \\ -1 & 0 & 1 & 2 \end{pmatrix},$$

and so from (6.5.32) we obtain the following systems of equations

$$\Delta\phi_1 = -2\exp(\phi_1) + \exp(\phi_2) + \exp(\phi_3), \tag{6.5.34a}$$
$$\Delta\phi_2 = \exp(\phi_1) - 2\exp(\phi_2) + \exp(\phi_3), \tag{6.5.34b}$$
$$\Delta\phi_3 = \exp(\phi_1) + \exp(\phi_2) - 2\exp(\phi_3), \tag{6.5.34c}$$

and

$$\Delta\phi_1 = -2\exp(\phi_1) + \exp(\phi_2) + \exp(\phi_4), \tag{6.5.35a}$$
$$\Delta\phi_2 = \exp(\phi_1) - 2\exp(\phi_2) + \exp(\phi_3), \tag{6.5.35b}$$
$$\Delta\phi_3 = \exp(\phi_2) - 2\exp(\phi_3) + \exp(\phi_4), \tag{6.5.35c}$$
$$\Delta\phi_4 = \exp(\phi_1) + \exp(\phi_3) - 2\exp(\phi_4), \tag{6.5.35d}$$

respectively. Setting $\phi_1 = 3\psi - \theta$, $\phi_2 = 2\theta$, $\phi_3 = -3\psi - \theta$ in equations (6.5.34) yields

$$\Delta\theta = -e^{2\theta} + e^{-\theta}\cosh(3\psi),$$
$$\Delta\psi = -e^{-\theta}\sinh(3\psi),$$

which is known to be integrable (Fordy and Gibbons [1980b]) and have as a special case ($\theta = -\phi$, $\psi = 0$) the equation

$$\Delta\phi = -e^{\phi} + e^{-2\phi}. \tag{6.5.36}$$

Similarly, setting $\phi_1 = 2\theta + 2\phi$, $\phi_2 = -2\theta + 2\psi$, $\phi_3 = 2\theta - 2\phi$, $\phi_4 = -2\theta - 2\psi$ in equations (6.5.35) yields

$$\Delta\theta = -e^{2\theta}\cosh(2\phi) + e^{-2\theta}\cosh(2\psi),$$
$$\Delta\phi = -e^{2\theta}\sinh(2\phi),$$
$$\Delta\psi = -e^{-2\theta}\sinh(2\psi),$$

which were also shown to be integrable by Fordy and Gibbons [1980b].

In the general case, if we take

$$\tilde{\mathbf{K}}_N = \begin{pmatrix} 2 & -1 & 0 & \dots & 0 & -1 \\ -1 & 2 & -1 & & & 0 \\ 0 & \ddots & \ddots & \ddots & & \vdots \\ \vdots & & \ddots & \ddots & \ddots & 0 \\ 0 & & & -1 & 2 & -1 \\ -1 & 0 & \dots & 0 & -1 & 2 \end{pmatrix},$$

then we obtain the periodic Toda lattice

$$\Delta\phi_i = \exp(\phi_{i-1}) - 2\exp(\phi_i) + \exp(\phi_{i+1}), \qquad i = 0, 1, \dots, N,$$

with the identification $\phi_i = \phi_{i+N+1}$ (Flaschka [1974b]). For further details on the Toda Molecule and Toda Lattice equations see, for example, Olive and Turok [1983].

EXAMPLE 6.5.6 *Limits to the Kadomtsev-Petviashvili and Davey-Stewartson Equations*

It is possible to obtain the $2+1$-dimensional version of equation (6.5.30) directly as a reduction of the SDYM equations via and algebra of diffeomorphisms (Chakravarty and Ablowitz [1991]). Suppose that the $\mathcal{A}_\mu$'s are independent of x_β, $x_{\bar\beta}$ and take values in the Lie algebra A_∞, i.e., they can be expressed in terms of kernels of integral operators. Consider the continuous version of (6.5.26) given by

$$\mathcal{A}_\alpha(x_\alpha, x_{\bar\alpha}; \xi, \eta) = f(x_\alpha, x_{\bar\alpha}; \xi + h)\delta(\xi - \eta), \tag{6.5.37a}$$
$$\mathcal{A}_{\bar\alpha}(x_\alpha, x_{\bar\alpha}; \xi, \eta) = g(x_\alpha, x_{\bar\alpha}; \xi + h)\delta(\xi - \eta), \tag{6.5.37b}$$
$$\mathcal{A}_\beta(x_\alpha, x_{\bar\alpha}; \xi, \eta) = \exp\{\tfrac{1}{2}q(x_\alpha, x_{\bar\alpha}; \xi)\}\delta(\xi + h - \eta), \tag{6.5.37c}$$
$$\mathcal{A}_{\bar\beta}(x_\alpha, x_{\bar\alpha}; \xi, \eta) = \exp\{\tfrac{1}{2}q(x_\alpha, x_{\bar\alpha}; \xi)\}\delta(\xi - h - \eta), \tag{6.5.37d}$$

where ξ, η parametrize the Lie algebra and h is a parameter such that the continuum limit corresponds to the limit process $h \to 0$, $n \to \infty$ with $nh = \xi$ remaining fixed. The commutation relation between two elements of the Lie algebra A_∞ is defined to be

$$[F,G](\xi,\eta) = \int_{-\infty}^{\infty} \{F(\xi,\zeta)G(\zeta,\eta) - G(\xi,\zeta)F(\zeta,\eta)\}\,\mathrm{d}\zeta.$$

Substituting (6.5.37) into equations (6.5.3) yields

$$\frac{\partial q}{\partial x_\alpha}(\xi) = -2[f(\xi+h) - f(\xi)], \tag{6.5.38a}$$

$$\frac{\partial q}{\partial x_{\bar{\alpha}}}(\xi) = 2[g(\xi+h) - g(\xi)], \tag{6.5.38b}$$

$$\frac{\partial f}{\partial x_{\bar{\alpha}}}(\xi+h) - \frac{\partial g}{\partial x_{\bar{\alpha}}}(\xi+h) = \exp[q(\xi)] - \exp[q(\xi+h)], \tag{6.5.38c}$$

where, for convenience, the dependence on x_α and $x_{\bar{\alpha}}$ has been suppressed. Differentiating (6.5.38a) with respect to x_α, (6.5.38b) with respect to $x_{\bar{\alpha}}$, subtracting and using (6.5.38c) yields the continuous version of the Toda molecule equation

$$\frac{\partial^2 q}{\partial x_\alpha \partial x_{\bar{\alpha}}}(\xi) = \exp\{q(\xi+h)\} + \exp\{q(\xi-h)\} - 2\exp\{q(\xi)\}. \tag{6.5.39}$$

It is natural to ask what asymptotic equations might be contained in (6.5.39). Letting $h \to 0$ and expanding the right-hand side as a power series in h yields

$$q_{tt} + q_{zz} = h^2 q_{\xi\xi} + \tfrac{1}{12}h^4[q_{\xi\xi\xi\xi} + (q^2)_{\xi\xi}] + O(h^6), \tag{6.5.40}$$

where the derivatives are expressed in the usual cartesian coordinates. Next we make the change of variables

$$X = \xi + ht, \qquad T = \tfrac{1}{24}ht, \qquad Y = \tfrac{1}{2}h^2 z, \qquad Q = \tfrac{1}{2}h^2 q,$$

then in the limit $h \to 0$, equation (6.5.40) becomes the KP equation

$$(Q_T - 6QQ_X - Q_{XXX})_X + Q_{YY} = 0.$$

Thus one has obtained the 2 + 1-dimensional KP equation in the asymptotic limit $h \to 0$ as a reduction of SDYM equation with the A_∞ gauge algebra!

Furthermore, as shown in §5.5.8 above, the Davey-Stewartson (DS) equations

$$\begin{aligned} &iu_T + \tfrac{1}{2}(u_{XX} + \sigma^2 u_{YY}) + |u|^2 u = \phi u, \\ &\phi_{XX} - \sigma^2 \phi_{YY} = 2(|u|^2)_{XX}, \end{aligned}$$

may be obtained from the KP equation through a multiscales expansion (Zakharov and Kuznetzov [1986]; Ablowitz, Manakov and Schultz [1990]); hence the DS equations also arise as a reduction of the SDYM equations in an asymptotic limit.

It is important to recognize that these results show that the SDYM equations contain not only the classical 1 + 1-dimensional soliton equations (see §6.5.3 below), but also the classical 2 + 1-dimensional soliton equations, at least in an asymptotic limit.

6.5.3 Reductions to $1+1$-dimensional Equations. In this section we discuss reductions of the SDYM equations to partial differential equations in $1+1$-dimensions.

EXAMPLE 6.5.7 *N-wave Interaction Equations*

In this example, following Chakravarty and Ablowitz [1991], we suppose that the $\mathcal{A}_\mu$'s are independent of $x_{\bar\alpha}$ and $x_{\bar\beta}$ and take values in the Lie algebra su$(n,\mathbb{R})$. Equation (6.5.3b) then reduces to $[\mathcal{A}_{\bar\alpha},\mathcal{A}_{\bar\beta}]=0$ and so we take

$$\mathcal{A}_{\bar\alpha}=\mathrm{diag}(a_1,a_2,\dots,a_n),\qquad \mathcal{A}_{\bar\beta}=\mathrm{diag}(b_1,b_2,\dots,b_n),$$

where $a_i(x_\alpha,x_\beta)$, $b_i(x_\alpha,x_\beta)$, $i=1,2,\dots,n$, are functions to be determined. Equation (6.5.3c) yields

$$\frac{\partial\mathcal{A}_{\bar\alpha}}{\partial x_\alpha}-\frac{\partial\mathcal{A}_{\bar\beta}}{\partial x_\beta}=[\mathcal{A}_\alpha,\mathcal{A}_{\bar\alpha}]+[\mathcal{A}_\beta,\mathcal{A}_{\bar\beta}].$$

The left-hand-side of this equation is a diagonal matrix whereas the right-hand-side is off-diagonal. Hence taking

$$\mathcal{A}_\alpha=[A^{ij}(x_\alpha,x_\beta)],\qquad \mathcal{A}_\beta=[B^{ij}(x_\alpha,x_\beta)],$$

to take values in the Lie algebra so$(n,\mathbb{R})$, we have

$$\frac{\partial a_i}{\partial x_\alpha}-\frac{\partial b_i}{\partial x_\beta}=0, \tag{6.5.41a}$$

$$A^{ij}=-\lambda_{ij}B^{ij},\qquad \lambda_{ij}=-\frac{b_i-b_j}{a_i-a_j}. \tag{6.5.41b}$$

We may suppose the a_i, b_i, $i=1,2,\dots,n$, are constants, without loss of generality (if not, then we make an appropriate gauge transformation). Finally equation (6.5.3b), using equation (6.5.41), yields

$$\frac{\partial B^{ij}}{\partial x_\alpha}-\lambda_{ij}\frac{\partial B^{ij}}{\partial x_\beta}=\sum_{k=1}^{n}(\lambda_{ik}-\lambda_{kj})B^{ik}B^{kj}, \tag{6.5.42}$$

which are the N-wave interaction equations (cf. Ablowitz and Haberman [1975b]). Similarly, the Lax pair obtained from (6.5.5) is the usual Lax pair. In particular, when $n=3$, if one chooses $[A^{ij}]$ and $[B^{ij}]$ to be anti-Hermitian matrices, then one obtains the well-known three-wave interaction equations arising in fluid dynamics and plasma physics (Zakharov and Manakov [1973]).

EXAMPLE 6.5.8 *The Korteweg-de Vries and Nonlinear Schrödinger Equations*

In this important example (due to Mason and Sparling [1989]), we suppose that $\mathcal{A}_\mu$ is independent of x_β and z and are in sl$(2,\mathbb{C})$ [the *special linear* Lie algebra of 2×2

matrices with zero trace over $\mathbb{C}$]; furthermore, we impose the gauge condition $\mathcal{A}_\alpha = 0$. For convenience, set $\mathcal{A}_{\bar\alpha} = \mathcal{A}$, $\mathcal{A}_\beta = \mathcal{B}$, $\mathcal{A}_{\bar\beta} = \mathcal{C}$ [$\mathcal{A}$, $\mathcal{B}$, $\mathcal{C} \in \mathrm{sl}(2,\mathbb{C})$], $x_\alpha = x_{\bar\alpha} = \xi$ and $x_{\bar\beta} = \tau$, then the SDYM equations (6.5.3) become

$$\mathcal{B}_\xi = \mathbf{0}, \tag{6.5.43a}$$

$$\mathcal{C}_\xi - \mathcal{A}_\tau - [\mathcal{A},\mathcal{C}] = \mathbf{0}, \tag{6.5.43b}$$

$$\mathcal{A}_\xi - \mathcal{B}_\tau - [\mathcal{B},\mathcal{C}] = \mathbf{0}. \tag{6.5.43c}$$

Now if in equations (6.5.43) we choose

$$\mathcal{A} = \begin{pmatrix} w & 1 \\ w_\xi - w^2 & -w \end{pmatrix},$$

$$\mathcal{B} = \begin{pmatrix} 0 & 0 \\ 1 & 0 \end{pmatrix},$$

$$\mathcal{C} = \tfrac{1}{2}\begin{pmatrix} (w_\xi - w^2)_\xi & -2w_\xi \\ \tfrac{1}{4}(w_{\xi\xi\xi} - 4ww_{\xi\xi} - 2w_\xi^2 + 4w^2 w_\xi) & -(w_\xi - w^2)_\xi \end{pmatrix},$$

then $u = -w_\xi$ satisfies the KdV equation

$$4u_\tau = u_{\xi\xi\xi} + 12uu_\xi. \tag{6.5.44}$$

The associated Lax pair obtained from equations (6.5.5) is given by

$$\phi_\xi = (\mathcal{A} - \lambda\mathcal{B})\phi = \begin{pmatrix} w & 1 \\ w_\xi - w^2 - \lambda & -w \end{pmatrix}\phi, \tag{6.5.45a}$$

$$\phi_\tau = (\mathcal{C} - \lambda\mathcal{A} + \lambda^2\mathcal{B})\phi, \tag{6.5.45b}$$

where $\lambda = 1/\zeta$. We remark that this Lax pair is different from the usual one for the KdV equation [recall equations (1.9.10,20) or (2.2.1)], though from (6.5.45a) we can obtain the Schrödinger equation

$$\phi_{1,\xi\xi} + (2u + \lambda)\phi_1 = 0.$$

Alternatively, if in equations (6.5.43) we choose

$$\mathcal{A} = \begin{pmatrix} 0 & q \\ -r & 0 \end{pmatrix}, \qquad \mathcal{B} = \kappa\begin{pmatrix} 1 & 0 \\ 0 & -1 \end{pmatrix}, \qquad \kappa\mathcal{C} = \tfrac{1}{2}\begin{pmatrix} qr & q_\xi \\ r_\xi & -qr \end{pmatrix}, \tag{6.5.46}$$

then q and r satisfy

$$2\kappa q_\tau = q_{\xi\xi} + 2q^2 r,$$

$$2\kappa r_\tau = -r_{\xi\xi} - 2qr^2,$$

Hence setting $\kappa = \frac{1}{2}\mathrm{i}$ and $r = \pm q^*$, we obtain the nonlinear Schrödinger equation

$$\mathrm{i}q_\tau = q_{\xi\xi} \pm 2|q|^2 q. \tag{6.5.47}$$

We remark that the Lax pair obtained from equations (6.5.5) [i.e., equation (6.5.45a) with $\mathcal{A}$, $\mathcal{B}$ given by equation (6.5.46)], is the standard Lax pair for the nonlinear Schrödinger equation [recall equations (1.9.10,1.9.19)]. Recently Strachan [1991] has shown how the method developed by Forgács, Horvath and Palla [1983] to contruct the monopole solutions to the SDYM equations may be used to give the soliton soltuions of the nonlinear Schrödinger equation (6.5.47).

EXAMPLE 6.5.9 *The Sine-Gordon Equation*

In Example 6.5.4 it was shown that the Sinh-Gordon equation (6.5.33) arises as a special case of the Toda lattice equation (6.5.32) and hence it is a reduction of the SDYM equations. In this example we give details of how the Sine-Gordon equation

$$\Delta\phi = \sin\phi, \tag{6.5.48}$$

with $\Delta\phi \equiv \phi_{xx} + \phi_{tt}$, can be directly obtained from the SDYM equations (6.5.3). Suppose that $\mathcal{A}_\mu$ lies in the Lie algebra sl(2, $\mathbb{C}$) and depends on x and t only (i.e., $x_\alpha = x_{\bar{\alpha}} = t$ and $x_\beta = x_{\bar{\beta}} = x$). Choosing

$$\mathcal{A}_\alpha = \tfrac{1}{4}\begin{pmatrix} -\phi_x & -2\sin(\frac{1}{2}\phi) \\ 2\sin(\frac{1}{2}\phi) & \phi_x \end{pmatrix},$$

$$\mathcal{A}_{\bar{\alpha}} = \tfrac{1}{4}\begin{pmatrix} -\phi_x & 2\sin(\frac{1}{2}\phi) \\ -2\sin(\frac{1}{2}\phi) & \phi_x \end{pmatrix}$$

$$\mathcal{A}_\beta = \tfrac{1}{4}\begin{pmatrix} \phi_t & -2\cos(\frac{1}{2}\phi) \\ -2\cos(\frac{1}{2}\phi) & -\phi_t \end{pmatrix},$$

$$\mathcal{A}_{\bar{\beta}} = \tfrac{1}{4}\begin{pmatrix} \phi_t & 2\cos(\frac{1}{2}\phi) \\ 2\cos(\frac{1}{2}\phi) & -\phi_t \end{pmatrix},$$

then equations (6.5.3a,b) are satisfied provided that ϕ satisfies the Sine-Gordon equation (6.5.48), whilst equation (6.5.3c) is satisfied identically. We remark that the Sine-Gordon equation (6.5.48) arises as a reduction of the Bogomolny equations (6.5.24).

An alternative method of deriving the Sine-Gordon equation from the SDYM equations is given by Chakravarty and Ablowitz [1991]. Suppose that the $\mathcal{A}_\mu$'s depend only on x_α and $x_{\bar{\alpha}}$ and take values in the Lie algebra su(2) which is generated by the Pauli matrices $\boldsymbol{\sigma}_i$, $i = 1, 2, 3$, given by

$$\boldsymbol{\sigma}_1 = \begin{pmatrix} 0 & 1 \\ 1 & 0 \end{pmatrix}, \qquad \boldsymbol{\sigma}_2 = \begin{pmatrix} 0 & -\mathrm{i} \\ \mathrm{i} & 0 \end{pmatrix}, \qquad \boldsymbol{\sigma}_3 = \begin{pmatrix} 1 & 0 \\ 0 & -1 \end{pmatrix}$$

which satisfy $[\boldsymbol{\sigma}_i, \boldsymbol{\sigma}_j] = 2\mathrm{i}\sum_{k=1}^{3} \epsilon_{ijk}\boldsymbol{\sigma}_k$. By using an appropriate gauge freedom we can set $\mathcal{A}_{\bar{\alpha}} = 0$. The resulting Lax pair is given by

$$\partial_\alpha\Psi = (\mathcal{A}_\alpha + \zeta\mathcal{A}_{\bar{\beta}})\Psi, \qquad \zeta\partial_{\bar{\alpha}}\Psi = -\mathcal{A}_\beta\Psi. \tag{6.5.49}$$

The remaining gauge potentials are then chosen as follows:

$$\mathcal{A}_\alpha = c\boldsymbol{\sigma}_3, \qquad \mathcal{A}_\beta = a\boldsymbol{\sigma}_2 + b\boldsymbol{\sigma}_2, \qquad \mathcal{A}_{\bar{\beta}} = \boldsymbol{\sigma}_1,$$

where a, b and c are functions of x_α and $x_{\bar{\alpha}}$. Substituting these into the Lax pair (6.5.49) yields the following three first order equations for a, b and c

$$\partial_\alpha a = -2\mathrm{i}bc, \tag{6.5.50a}$$

$$\partial_\alpha b = 2\mathrm{i}ac, \tag{6.5.50b}$$

$$\partial_{\bar{\alpha}} c = 2\mathrm{i}b. \tag{6.5.50c}$$

Equations (6.5.50a,b) are easily solved to yield $a^2 + b^2 = \text{constant} = 1$; choosing $a = \cos\phi$, $b = \sin\phi$ one obtains $c = \frac{1}{2}\mathrm{i}\partial_\alpha\phi$, which when substituted equation (6.5.50c) yields the Sine-Gordon equation (6.5.48) with $\Delta\phi \equiv \frac{1}{4}\partial_\alpha\partial_{\bar\alpha}\phi = \frac{1}{4}(\phi_{zz} + \phi_{tt})$.

It is possible to obtained (6.5.48) by making yet another choice of the $\mathcal{A}_\mu$'s in (6.5.49): $\mathcal{A}_\alpha = q\boldsymbol{\sigma}_2$, $\mathcal{A}_\beta = F\boldsymbol{\sigma}_1 + G\boldsymbol{\sigma}_3$ and $\mathcal{A}_{\bar\beta} = \boldsymbol{\sigma}_3$, where q, F and G are functions of x_α and $x_{\bar\alpha}$. Then from the integrability conditions for (6.5.49) and after calculations similar to those above, one obtains $q = -\frac{1}{2}\partial_\alpha\phi$, $F^2 - G^2 = 1$, and finally equation (6.5.48) with $\Delta \equiv \frac{1}{4}\partial_\alpha\partial_{\bar\alpha} = \frac{1}{4}(\partial_z^2 + \partial_t^2)$. Note that with this particular choice of the $\mathcal{A}_\mu$'s, equations (6.5.49) are the standard Lax pair for the Sine-Gordon equation (Ablowitz, Kaup, Newell and Segur [1974]).

EXAMPLE 6.5.10 *The Ernst Equations*

In this example we assume that $\mathbf{J}$ only depends on z and $\rho = \sqrt{x^2 + y^2}$, then equation (6.5.6) becomes

$$\rho^{-1}\left(\rho\mathbf{J}^{-1}\mathbf{J}_\rho\right)_\rho + \left(\mathbf{J}^{-1}\mathbf{J}_z\right)_z = 0. \tag{6.5.51}$$

Now we suppose that the Lie algebra is su(2, $\mathbb{R}$) (note the additional constraint that each entry in the 2×2 matrix $\mathbf{J}$ is real valued has been imposed), then (6.5.51) is (a form of) the Ernst equation of general relativity. Choosing

$$\mathbf{J} = \frac{1}{F}\begin{pmatrix} 1 & G \\ G & F^2 + G^2 \end{pmatrix},$$

yields

$$F\nabla^2 F = (\nabla F)^2 - (\nabla G)^2, \tag{6.5.52a}$$

$$F\nabla^2 G = 2(\nabla F)\bullet(\nabla G), \tag{6.5.52b}$$

where ∇ is the axisymmetric gradient operator, which are equivalent to the stationary axisymmetric vacuum Einstein equations (Ernst [1968]; Hauser and Ernst [1980]; Cosgrove [1980]). ($F = F(\rho, z)$, $\nabla^2 F = F_{\rho\rho} + \rho^{-1}F_\rho + F_{zz}$, $|\nabla F|^2 = F_\rho^2 + F_z^2$. For further discussion on the reduction of the SDYM equations to the Ernst/stationary axisymmetric Einstein-Maxwell equation see, for example, Witten [1979]; Gürses and Xanthopoulos [1980]; Ward [1983]; Woodhouse, [1987, 1990]; Woodhouse and Mason [1988].)

EXAMPLE 6.5.11 *The Chiral Field Equation*

If in (6.5.6) we suppose that $\mathbf{J}$ is independent of z and y (i.e., $x_\alpha = x_{\bar\alpha} = t$ and $x_\beta = x_{\bar\beta} = x$), then

$$\left(\mathbf{J}^{-1}\mathbf{J}_t\right)_t + \left(\mathbf{J}^{-1}\mathbf{J}_x\right)_x = 0. \tag{6.5.53}$$

In turn this equation reduces to other equations frequently studied in physics; for example, if we require that $\mathbf{J}$ satisfies $\mathbf{J}^2 = \mathbf{I}$ so that it can be written as $\mathbf{J} = \mathbf{I} - 2\mathbf{P}$ with $\mathbf{P}^2 = \mathbf{P}$, then (6.5.53) becomes $[\mathbf{P}, \mathbf{P}_{xx} + \mathbf{P}_{tt}]$, which is a form of the CP^n equations (see, for example Din, Horvath and Zakrzewski [1984]).

6.5.4 Reductions to Ordinary Differential Equations. In this section we consider reductions to ordinary differential equations.

EXAMPLE 6.5.12 *The Euler-Arnold Top*

In this example, following Chakravarty, Ablowitz and Clarkson [1990, 1991], we assume that $\mathcal{A}_\mu$ depends only on one variable x_α [i.e., $\mathcal{A}_\mu = \mathcal{A}_\mu(x_\alpha)$], and that $\mathcal{A}_\mu$ take values in the Lie algebra $\mathrm{gl}(n, \mathbb{C})$. In particular, we take

$$\mathcal{A}_{\bar{\alpha}} = \mathrm{diag}(b_1^2, \ldots, b_n^2), \qquad \mathcal{A}_{\bar{\beta}} = \mathrm{diag}(b_1, \ldots, b_n),$$

with b_i, $i = 1, \ldots, n$, constants, to be in the Cartan subalgebra of $\mathrm{gl}(n, \mathbb{C})$; furthermore

$$\mathcal{A}_\alpha = [A^{ij}(x_\alpha)], \qquad \mathcal{A}_\beta = [B^{ij}(x_\alpha)],$$

are chosen to be in the subalgebra $\mathrm{so}(n)$ [the *special orthogonal* Lie algebra of skew-symmetric $n \times n$ matrices]. Since $\mathcal{A}_{\bar{\alpha}}$ and $\mathcal{A}_{\bar{\beta}}$ are constant, diagonal matrices, then (6.5.3b) is identically satisfied. Equation (6.5.3c) reduces to

$$[\mathcal{A}_\alpha, \mathcal{A}_{\bar{\alpha}}] + [\mathcal{A}_\beta, \mathcal{A}_{\bar{\beta}}] = 0,$$

which, by considering the ijth component, defines A^{ij} in terms of B^{ij}

$$A^{ij} = -\frac{B^{ij}}{b_i + b_j}. \tag{6.5.54}$$

Finally using this, equation (6.5.3a) yields

$$\frac{\mathrm{d}B^{ij}}{\mathrm{d}x_\alpha} = \sum_{k=1}^{n} \left\{ \frac{1}{b_j + b_k} - \frac{1}{b_k + b_i} \right\} B^{ik} B^{kj}. \tag{6.5.55}$$

Equations (6.5.55) are the Euler-Arnold equations for free motion of an n-dimensional rigid body about a fixed point (Arnold [1978]; Manakov [1976]). Specifically when $n = 3$, taking

$$B_{12} = L_3, \qquad B_{23} = L_1, \qquad B_{31} = L_2,$$
$$b_1 + b_2 = I_3, \qquad b_2 + b_3 = I_1, \qquad b_3 + b_1 = I_2,$$

one recovers from (6.5.55) the Euler equation for a free, spinning top about a fixed point:

$$\frac{\mathrm{d}L_1}{\mathrm{d}x_\alpha} = \gamma_1 L_2 L_3, \qquad \frac{\mathrm{d}L_2}{\mathrm{d}x_\alpha} = \gamma_2 L_3 L_1, \qquad \frac{\mathrm{d}L_3}{\mathrm{d}x_\alpha} = \gamma_3 L_1 L_2, \tag{6.5.56a}$$

with

$$\gamma_1 = \frac{I_2 - I_3}{I_2 I_3}, \qquad \gamma_2 = \frac{I_3 - I_1}{I_3 I_1}, \qquad \gamma_3 = \frac{I_1 - I_2}{I_1 I_2}, \tag{6.5.56b}$$

or alternatively

$$\frac{\mathrm{d}\boldsymbol{L}}{\mathrm{d}x_\alpha} + \boldsymbol{w} \wedge \boldsymbol{L} = 0, \qquad \boldsymbol{L} = \mathbf{I}\boldsymbol{w}, \qquad \mathbf{I} = \mathrm{diag}(I_1, I_2, I_3).$$

The L_i's are the usual angular momentum components in the body frame, and the I_i's are the principal moments of inertia and 'x_α' is the complex 'time' variable.

The Lax pair for equation (6.5.55) can be written as

$$\frac{\mathrm{d}\mathrm{L}_I}{\mathrm{d}x_\alpha} = [\mathrm{L}_I, \mathrm{M}_I], \tag{6.5.57a}$$

with

$$\mathrm{L}_I = \mathcal{A}_\beta - \zeta \mathcal{A}_{\bar{\alpha}}, \qquad \mathrm{M}_I = -(\mathcal{A}_\alpha + \zeta \mathcal{A}_{\bar{\beta}}). \tag{6.5.57b}$$

One can check from equations (6.5.57) that

$$\partial_\alpha \,\mathrm{Tr}(\mathrm{L}_I^n) = 0, \qquad n = 0, 1, 2, \dots\ ,$$

which when expanded in a power series in ζ yields the constants of motion. The non-trivial constants of motion are obtained from (a), the coefficient of ζ^0 in

$$\mathrm{Tr}(\mathrm{L}_I^2) = \mathrm{Tr}(\mathcal{A}_\beta^2) = 2\sum_{i=1}^{3} L_i^2, \tag{6.5.58a}$$

and (b), the coefficient of ζ in

$$\mathrm{Tr}(\mathrm{L}_I^3) = -3\,\mathrm{Tr}(\mathcal{A}_{\bar{\alpha}} \mathcal{A}_\beta^2). \tag{6.5.58b}$$

(6.5.58a) yields the square of the total angular momentum and the Hamiltonian is then obtained from (6.5.58b).

EXAMPLE 6.5.13 *The Kovalevskaya top and its Generalizations*

In this reduction, following Chakravarty, Ablowitz and Clarkson [1990, 1991], we take $x_\alpha = x_{\bar\alpha} = t$, and $\mathcal{A}_\mu(t)$ in the algebra so(p,q) [the Lie algebra of $n \times n$ matrices $\mathbf{A}$, with $n = p+q$, such that $\mathbf{A}^T \mathbf{g}_{pq} + \mathbf{g}_{pq}\mathbf{A} = \mathbf{0}$, where $\mathbf{g}_{pq}$ is a block matrix of the form

$$\mathbf{g}_{pq} = \begin{pmatrix} \mathbf{I}_p & \mathbf{O}_{pq} \\ \mathbf{O}_{qp} & -\mathbf{I}_q \end{pmatrix},$$

with $\mathbf{I}_p$ the $p \times p$ identity matrix, and $\mathbf{O}_{pq}$ the $p \times q$ null matrix]. $\mathcal{A}_z = (\mathcal{A}_\alpha - \mathcal{A}_{\bar\alpha})/(2\mathrm{i})$ and $\mathcal{A}_t + \mathrm{i}\mathcal{A}_z = \mathcal{A}_\alpha$ are chosen so that they lie in the maximal sub-algebra $\mathcal{K} =$ so$(p)\oplus$so(q) whereas $\mathcal{A}_\beta$ and $\mathcal{A}_{\bar\beta}$ are in the orthogonal complement to $\mathcal{K}$. The classical Kovalevskaya system is obtained by considering the algebra so$(3,1)$. The maximal subalgebra is so(3) and the $\mathcal{A}_\mu$'s are represented by the following matrices:

$$\mathcal{A}_\beta = \begin{pmatrix} 0 & 0 & 0 & g_1 \\ 0 & 0 & 0 & g_2 \\ 0 & 0 & 0 & g_3 \\ g_1 & g_2 & g_3 & 0 \end{pmatrix}, \qquad \mathcal{A}_{\bar\beta} = -2\begin{pmatrix} 0 & 0 & 0 & c_1 \\ 0 & 0 & 0 & c_2 \\ 0 & 0 & 0 & c_3 \\ c_1 & c_2 & c_3 & 0 \end{pmatrix},$$

$$\mathrm{i}\mathcal{A}_z = \begin{pmatrix} 0 & L_3 & -L_2 & 0 \\ -L_3 & 0 & L_1 & 0 \\ L_2 & -L_1 & 0 & 0 \\ 0 & 0 & 0 & 0 \end{pmatrix}, \qquad \mathcal{A}_t + \mathrm{i}\mathcal{A}_z = \begin{pmatrix} 0 & \omega_3 & -\omega_2 & 0 \\ -\omega_3 & 0 & \omega_1 & 0 \\ \omega_2 & -\omega_1 & 0 & 0 \\ 0 & 0 & 0 & 0 \end{pmatrix},$$

with $\boldsymbol{g} = (g_1, g_2, g_3)^T$ the field vector, $\boldsymbol{c} = (c_1, c_2, c_3)^T$ the center of mass vector, where L_k are the components of angular momentum and $L_i = I_i\omega_i$, with I_i the principal moments of inertia. Making the above choices in the Lax pair of the SDYM equation (6.5.5) yields, after some manipulation,

$$\frac{\mathrm{dL}_{II}}{\mathrm{d}t} = [\mathrm{L}_{II}, \mathrm{M}_{II}], \tag{6.5.59a}$$

with

$$\mathrm{L}_{II} = \mathcal{A}_\beta + 2\mathrm{i}\zeta\mathcal{A}_z + \zeta^2\mathcal{A}_{\bar\beta}, \qquad \mathrm{M}_{II} = -(\mathcal{A}_t + \mathrm{i}\mathcal{A}_z + \zeta\mathcal{A}_{\bar\beta}), \tag{6.5.59b}$$

from which, after expanding in powers of ζ, we obtain

$$\frac{\mathrm{d}\boldsymbol{L}}{\mathrm{d}t} = \boldsymbol{L}\wedge\boldsymbol{\omega} + \boldsymbol{g}\wedge\boldsymbol{c}, \qquad \frac{\mathrm{d}\boldsymbol{g}}{\mathrm{d}t} = \boldsymbol{g}\wedge\boldsymbol{\omega}, \tag{6.5.60}$$

where $\boldsymbol{L} = (L_1, L_2, L_3)^T$ and $\boldsymbol{\omega} = (\omega_1, \omega_2, \omega_3)^T$. Equation (6.5.60) describes, in the moving frame, the motion of a 'heavy top' rotating about a fixed point. However to derive (6.5.60) from (6.5.59) one has to impose certain conditions on the I_i's depending on the choice for the center of mass vector $\boldsymbol{c}$. When $\boldsymbol{c}$ is along one of the principal axes of the moment of inertia ellipsoid, these conditions essentially coincide with those of Kovalevskaya [1889]. For so(p,q) one can obtain from (6.5.59) generalizations of the Kovalevskaya case to a p-dimensional rigid body moving in q homogeneous fields (Adler and van Moerbeke [1988]; Bobenko, Reyman and Semenov-Tian-Shansky [1989]; Fomenko and Trofimov [1987]). As in the previous example, the integrals of motion for the so(p,q) systems are obtained by expanding $\mathrm{Tr}(\mathrm{L}_{II}^n)$, $n = 0, 1, 2, \ldots$, as a power series in ζ, in particular, the coefficient of ζ^2 in $\mathrm{Tr}(\mathrm{L}_{II}^2)$ yields the Hamiltonian; the remaining nontrivial integrals of motion are obtained from the traces of higher exponents of L_{II}.

EXAMPLE 6.5.14 *The Nahm and Chazy equations*

In this example, following Chakravarty, Ablowitz and Clarkson [1990, 1991], we take $\mathcal{A}_\mu = \mathcal{A}_\mu(t)$ as in the previous case. Recall that the SDYM equations (6.5.3) are invariant under the gauge transformation (6.5.3); we set $\mathcal{A}_t = 0$ by utilizing this gauge freedom. Then from (6.5.3) we find that the remaining $\mathcal{A}_\mu$'s satisfy

$$\frac{\mathrm{d}\mathcal{A}_1}{\mathrm{d}t} = [\mathcal{A}_2, \mathcal{A}_3], \qquad \frac{\mathrm{d}\mathcal{A}_2}{\mathrm{d}t} = [\mathcal{A}_3, \mathcal{A}_1], \qquad \frac{\mathrm{d}\mathcal{A}_3}{\mathrm{d}t} = [\mathcal{A}_1, \mathcal{A}_2], \tag{6.5.61}$$

where $\mathcal{A}_1 = \mathcal{A}_x$, $\mathcal{A}_2 = \mathcal{A}_y$, $\mathcal{A}_3 = \mathcal{A}_z$. The Nahm and Chazy equations are obtained from (6.5.61) by taking the $\mathcal{A}_i$'s in the Lie algebra su(2). It is convenient to work with the representation of su(2) in S^3 and express the $\mathcal{A}_i$'s in terms of the left invariant vector fields:

$$\begin{aligned}
\mathrm{X}_1 &= \cos\psi\frac{\partial}{\partial\theta} + \sin\psi\,\mathrm{cosec}\,\theta\frac{\partial}{\partial\phi} - \cot\theta\sin\psi\frac{\partial}{\partial\psi},\\
\mathrm{X}_2 &= -\sin\psi\frac{\partial}{\partial\theta} + \cos\psi\,\mathrm{cosec}\,\theta\frac{\partial}{\partial\phi} - \cot\theta\cos\psi\frac{\partial}{\partial\psi},\\
\mathrm{X}_3 &= \frac{\partial}{\partial\psi},
\end{aligned}$$

where θ, ϕ and ψ are the Euler angles and $[\mathrm{X}_i, \mathrm{X}_j] = \sum_{k=1}^{3} \epsilon_{ijk}\mathrm{X}_k$. Setting $\mathcal{A}_i = \omega_i(t)\mathrm{X}_i$, $i = 1, 2, 3$, in (6.5.61) one obtains the Nahm equations

$$\frac{\mathrm{d}\omega_1}{\mathrm{d}t} = \omega_2\omega_3, \qquad \frac{\mathrm{d}\omega_2}{\mathrm{d}t} = \omega_3\omega_1, \qquad \frac{\mathrm{d}\omega_3}{\mathrm{d}t} = \omega_1\omega_2 \tag{6.5.62}$$

(Nahm [1981, 1982]; Hitchen [1983, 1987]).

Next, consider the adjoint action in su(2) given by

$$\mathcal{A}_i \to \tilde{\mathcal{A}}_i = \mathcal{F}\mathcal{A}_i\mathcal{F}^{-1}, \qquad \mathcal{F} \in \mathrm{su}(2).$$

It is well known that this automorphism induces a homomorphism $\mathbf{\Phi}(\mathcal{F}) \in \mathrm{so}(3)$ and the $\mathcal{A}_i$ transforms like a triad in so(3) as

$$\mathcal{A}_i \to \tilde{\mathcal{A}}_i = \sum_{j=1}^{3} O^{ij}\mathcal{A}_j = \sum_{j=1}^{3} O^{ij}\omega_j(t)\mathrm{X}_j,$$

$\mathbf{O} \in \mathrm{so}(3)$ is expressed in terms of the Euler angles in the usual manner, i.e.

$$\mathbf{O} = \begin{pmatrix} \cos\psi\cos\phi - \cos\theta\sin\phi\sin\psi & -\sin\psi\cos\phi - \cos\theta\sin\phi\cos\psi & \sin\theta\sin\phi \\ \cos\psi\sin\phi + \cos\theta\cos\phi\sin\psi & -\sin\psi\sin\phi + \cos\theta\cos\phi\cos\psi & -\sin\theta\cos\phi \\ \sin\theta\sin\psi & \sin\theta\cos\psi & \cos\theta \end{pmatrix}$$

(see, for example, Goldstein [1965]). Substituting the $\tilde{\mathcal{A}}_i$'s in (6.5.61) yields

$$\sum_{l=1}^{3} O^{il}\frac{\mathrm{d}\mathcal{A}_l}{\mathrm{d}t} = \tfrac{1}{2}\sum_{j,k,r,s=1}^{3} \epsilon_{ijk}O^{jr}O^{ks}[\mathcal{A}_r,\mathcal{A}_s] + \sum_{j,k,r,s=1}^{3} \epsilon_{ijk}O^{jr}\mathcal{A}_r(O^{ks})\mathcal{A}_s.$$

The second term of the right-hand side is the extra contributions from the Lie bracket [due the action of the vector fields X_i's on $O^{ij}(\theta,\phi,\psi)$] and is given by

$$\mathrm{X}_i(O^{jk}) = \sum_{p=1}^{3}\epsilon_{ikp}O^{jp}.$$

Multiplying both sides of (6.5.61) by O^{in} and noting the $\mathbf{O}$ is an orthogonal matrix, we finally obtain

$$\frac{\mathrm{d}\omega_1}{\mathrm{d}t} = \omega_2\omega_3 - \omega_1(\omega_2+\omega_3), \tag{6.5.63a}$$

$$\frac{\mathrm{d}\omega_2}{\mathrm{d}t} = \omega_3\omega_1 - \omega_2(\omega_3+\omega_1), \tag{6.5.63b}$$

$$\frac{\mathrm{d}\omega_3}{\mathrm{d}t} = \omega_1\omega_2 - \omega_3(\omega_1+\omega_2), \tag{6.5.63c}$$

where we have used the following

$$\sum_{i,j,k=1}^{3}\epsilon_{ijk}O^{ip}O^{jq}O^{kr} = \epsilon_{pqr}, \qquad \sum_{i=1}^{3}\epsilon_{ijk}\epsilon_{imn} = \tfrac{1}{2}(\delta_{jm}\delta_{kn} - \delta_{jn}\delta_{km}).$$

Indeed, equation (6.5.63) arises as the self-dual reduction of Bianchi-IX comological models (Gibbons and Pope [1979]; Hitchen [1987]). In a diagonal gauge the complex Bianchi-IX metric can be written as

$$\mathrm{d}s^2 = (abc)^2\mathrm{d}t^2 + a^2\sigma_1^2 + b^2\sigma_2^2 + c^2\sigma_3^2,$$

where a, b and c are functions of complex 'time' t, are related to the ω_i's mentioned above by $\omega_1 = bc$, $\omega_2 = ca$, $\omega_3 = ab$, and the σ_i's are the left invariant 1-forms dual to the X_i's satisfying the Maurer-Cartan relations

$$\frac{\mathrm{d}\sigma_i}{\mathrm{d}t} = \tfrac{1}{2}\sum_{j,k=1}^{3}\epsilon_{ijk}\sigma_j\wedge\sigma_k.$$

The field equations, when expressed in terms of the orthonormal frame $e_0 = abc\mathrm{d}t$, $e_1 = a\sigma_1$, $e_2 = b\sigma_2$, $e_3 = c\sigma_3$, can be cast in the form of (6.5.61). This is a special case of what has been shown to be true for general half-flat metrics (Ashtekar, Jacobson and Smolin [1988]; Mason and Newman [1989]).

For Bianchi-IX, the orthonormal frame is acted on by the isometry group SU(2). The left action of SU(2) is trivial since it leaves the σ_i's invariant and the field equation (6.5.61) yield the Nahm equations (6.5.62). Under the adjoint action of SU(2), the σ_i's transform as $\sum_{j=1}^{3} O^{ij}\sigma_j$ and equations (6.5.61) yield (6.5.63).

An interesting feature of these models is that they yield gravitational instantons if one considers real solutions ('t Hooft [1976a,b]; Gibbons and Hawking [1979]). Then the associated real metric is Riemannian and positive-definite. Here we shall be concerned with the general solution of equations (6.5.63) in the complex t-plane.

By taking $y = -2(\omega_1 + \omega_2 + \omega_3)$, (6.5.63) yields a third order, nonlinear ordinary differential equation in $y(t)$

$$\frac{\mathrm{d}^3 y}{\mathrm{d}t^3} = 2y\frac{\mathrm{d}^2 y}{\mathrm{d}t^2} - 3\left(\frac{\mathrm{d}y}{\mathrm{d}t}\right)^2, \tag{6.5.64}$$

which is the classical Chazy equation [1910, 1911].

Equation (6.5.64) can be solved by introducing a new variable s such that

$$y(t(s)) = 6\frac{\mathrm{d}}{\mathrm{d}t}\ln z_1, \qquad t(s) = \frac{z_2}{z_1},$$

where z_1 and z_2 are two independent solution of the hypergeometric equation

$$s(1-s)\frac{\mathrm{d}^2 z}{\mathrm{d}s^2} + \left(\frac{1}{2} - \frac{7s}{6}\right)\frac{\mathrm{d}z}{\mathrm{d}s} - \frac{z}{144} = 0. \tag{6.5.65}$$

Given a solution of $y(t)$, one can recover the ω_i's in a straightforward way as follows. With the above definition of $y(t)$, using (6.5.63) we obtain

$$\begin{aligned}\frac{\mathrm{d}y}{\mathrm{d}t} &= 2(\omega_1\omega_2 + \omega_2\omega_3 + \omega_3\omega_1),\\ \frac{\mathrm{d}^2 y}{\mathrm{d}t^2} &= -12\omega_1\omega_2\omega_3.\end{aligned}$$

Thus y and its derivatives are related to the symmetric products of the ω_i's which are given be the solutions of the following cubic equation

$$\omega^3 + \frac{1}{2}y\omega^2 + \frac{1}{2}\frac{\mathrm{d}y}{\mathrm{d}t}\omega + \frac{1}{12}\frac{\mathrm{d}^2 y}{\mathrm{d}t^2} = 0. \tag{6.5.66}$$

As long as the ω_i's are distinct they can be obtained uniquely from (6.5.66) since the Jacobian of the transformation between $(y, \mathrm{d}y/\mathrm{d}t, \mathrm{d}^2y/\mathrm{d}t^2)$ and $(\omega_1, \omega_2, \omega_3)$ is given

FIGURE 6.5.1 *The triangle group* $(\frac{1}{3}, \frac{1}{2}, 0)$

by $(\omega_1 - \omega_2)(\omega_2 - \omega_3)(\omega_3 - \omega_1)$. This can be also seen from the equations that the difference of the ω_i's satisfy

$$\frac{\mathrm{d}}{\mathrm{d}t}(\omega_1 - \omega_2) = -2\omega_3(\omega_1 - \omega_2),$$
$$\frac{\mathrm{d}}{\mathrm{d}t}(\omega_1 - \omega_3) = -2\omega_2(\omega_1 - \omega_3).$$

Thus if $\omega_1 - \omega_2$ and $\omega_1 - \omega_3$ are nonzero at the initial time, then they remain nonzero at any later time. On the other hand if these differences vanish initially, they remain zero and the resulting system can be trivially integrated to yield

$$y(t) = \frac{-6}{t - t_0} + \frac{A}{(t - t_0)^2}, \tag{6.5.67}$$

where t_0 and A are arbitrary constants of integration determined by the initial conditions.

The mapping $t = t(s)$ takes the upper half s-plane into the interior of a spherical triangle with angles 0, $\pi/2$ and $\pi/3$ (see, for example, Nehari [1952]). The mapping $t(s)$ can be analytically continued by successive reflections of the upper half s-plane about the real line and by corresponding inversions of the fundamental triangle across its sides to complimentary triangles. However, these triangles do not cover the whole t-plane but (in the limit of successive inversions) 'tile' a region of the t-plane bounded by a circle C (see Figure 6.5.1). The inverse mapping $S(0, \pi/2, \pi/3; t)$ (which is the

Schwarzian triangle function) is single valued (for these choices of parameters), meromorphic and possesses a natural boundary as it can not be analytically continued beyond C (see, for example, Nehari [1952]). The vertex $t(\infty) = t_\infty$ of the fundamental triangle lies on C and is an isolated essential singular point of $S(t)$ as we show below. As any point on C is arbitrarily close to an image of t_∞ in the limit of successive inversions, C forms a dense set of essential singular points of $S(t)$ and is called the natural boundary of $S(t)$. Since $S(t)$ is invariant under the fractional linear or Möbius transformation: $S((at+b)/(ct+d)) = S(t)$, then this natural boundary C is movable. Any solution $y(t(s))$ of (6.5.64) also inherits the same natural boundary C with radius and center specified by the initial conditions. Specifically, let y, $\mathrm{d}y/\mathrm{d}t$ and $\mathrm{d}^2y/\mathrm{d}t^2$ be given as some initial time t_0. We define the mapping

$$t(s) = \zeta_2(s)/\zeta_1(s), \tag{6.5.68}$$

where

$$\zeta_1(s) = az_1(s) + bz_2(s), \qquad \zeta_2(s) = cz_1(s) + dz_2(s). \tag{6.5.69}$$

$z_1(s)$ and $z_2(s)$ are two special linearly independent solutions of the hypergeometric equation (6.5.65) given by

$$z_1(s) = F(\tfrac{1}{12}, \tfrac{1}{12}, \tfrac{1}{2}; s), \qquad z_2(s) = s^{1/2}F(\tfrac{7}{12}, \tfrac{7}{12}, \tfrac{3}{2}; s),$$

near $s = 0$, where $F(\alpha, \beta, \gamma; s)$ is the usual hypergeometric function (cf. Nehari [1952]). a, b, c and d are constants to be determined from the initial data t_0 and y, $\mathrm{d}y/\mathrm{d}t$ and $\mathrm{d}^2y/\mathrm{d}t^2$ at t_0. The fact that ζ_1 and ζ_2 are also linearly independent solutions of equation (6.5.65) is ensured by the normalization

$$ad - bc = 1. \tag{6.5.70}$$

The solutions to equation (6.5.64) is now given by

$$y(t(s)) = \frac{6}{\zeta_1}\frac{\mathrm{d}\zeta_1}{\mathrm{d}t} = \frac{6\zeta_1}{W}\frac{\mathrm{d}\zeta_1}{\mathrm{d}s},$$

where the Wronskian $W = s^{-1/2}(1-s)^{-2/3}$ and

$$\frac{\mathrm{d}\zeta_1}{\mathrm{d}s} = \frac{Wy}{6\zeta_1}. \tag{6.5.71}$$

Differentiating y with respect to t once again and using (6.5.65) one can solve for ζ_1 as

$$\zeta_1 = \frac{24^{1/4}}{(1-s_0)^{1/12}}\left(\frac{\mathrm{d}y}{\mathrm{d}t} - \tfrac{1}{6}y^2\right)^{1/4}. \tag{6.5.72}$$

By differentiating again and after some manipulations, one can solve for s_0 in terms of the initial data

$$s_0 = \frac{\frac{3}{8}K^2}{1+\frac{3}{8}K^2}, \tag{6.5.73a}$$

where

$$K = \left[\frac{d^2y}{dt^2} - y\frac{dy}{dt} + \tfrac{1}{9}y^3\right]\left(\frac{dy}{dt} - \tfrac{1}{6}y^2\right)^{-3/2}. \tag{6.5.73b}$$

Using (6.5.73) in equations (6.5.71,6.5.72), ζ_1 and $d\zeta_1/ds$ can be solved in terms of initial data and then substituting them in equation (6.5.69) yields the following expressions for a and b

$$a = \frac{1}{W}\left(\frac{dz_2}{ds}\zeta_1 - \frac{d\zeta_2}{ds}z_2\right), \qquad b = \frac{1}{W}\left(\frac{d\zeta_2}{ds}z_1 - \frac{dz_1}{ds}\zeta_1\right) \tag{6.5.74}$$

Finally using equations (6.5.70,6.5.72,6.5.74a,b) in equation (6.5.68) one obtains c as

$$c = at_0 - z_2/\zeta_1. \tag{6.5.74c}$$

Thus we can reconstruct the mapping $t(s)$ given by (6.5.68) from the initial data of the third order equation (6.5.64) together with the normalization condition (6.5.70). Equation (6.5.68) can be expressed as

$$t(s) = \frac{c+d\tau}{a+b\tau},$$

where $\tau(s) = z_2/z_1$, i.e., the mappings $t(s)$ and $\tau(s)$ are related by a Möbius transformation. For $\tau(s)$ one can check that the center of the natural boundary C is at $\tau(0)$ and the radius R_τ is given by the well-known gamma functions

$$R_\tau = |\tau(\infty) - \tau(0)| = \tfrac{1}{2}[\Gamma(\tfrac{5}{12})\Gamma(\tfrac{1}{12})]^2.$$

Under the Möbius transformation, the orthogonal circle C of the τ-plane is mapped into the orthogonal circle $\tilde{C}$ of the t-plane whose center and radius are given by

$$t(0) = c/a, \qquad R_t = \frac{R_\tau}{|a(a+ibR_\tau)|}. \tag{6.5.75}$$

The nature of the singularity and the asymptotic behavior of the inverse mapping $s(t)$ and the solution $y(t(s))$ near the natural boundary, or more specifically near t_∞ in the fundamental triangle, can now be analyzed. Substituting the expressions for the special solutions z_1 and z_2 for $|s| \to \infty$ and Im $s > 0$ in equation (6.5.69), we have (Caratheodory [1954])

$$\zeta_1(s) = A\phi_1(s) + B\phi_2(s), \qquad \zeta_2(s) = C\phi_1(s) + D\phi_2(s), \tag{6.5.76a}$$

with

$$\phi_1(s) = (-s)^{-1/12} F(\tfrac{7}{12}, \tfrac{1}{12}, 1; 1/s), \tag{6.5.76b}$$
$$\phi_2(s) = (-s)^{-1/12} \ln(-s) F(\tfrac{7}{12}, \tfrac{1}{12}, 1; 1/s) + g(1/s), \tag{6.5.76c}$$

where $F(a,b,c;s)$ is the hypergeometric function mentioned above and $G(1/s)$ is analytic in its argument with $G(0) = 0$. The coefficients A, B, C and D are constants which depend on a, b, c, d and gamma functions of various arguments and we note that $t_\infty = D/B$, $B \neq 0$. Using equations (6.5.76) in equation (6.5.68) and inverting the resulting relation, the asymptotic form of the inverse map $S(t)$ near t_∞ is

$$S(t) \sim -\exp\{h(t)\} \left[1 + \tfrac{41}{72} \exp\{-\tfrac{11}{12} h(t)\} P_1(\exp\{-h(t)\})\right], \tag{6.5.77}$$

with $h(t) = (C - At)/(Bt - D)$ and where P_1 is a series in $\exp\{-h(t)\}$ with $P_1(0) = 1$ and $\mathrm{Re}[h(t)] \to \infty$. Hence $S(t)$ has an essential singularity at $t_\infty = D/B$. In a similar fashion, the asymptotic behavior of $y(t(s))$ can be calculated to obtain

$$y(t(s)) \sim \frac{-6}{t - t_\infty} - \frac{1 - \frac{7}{12} \exp\{-h(t)\} P_2(\exp\{-h(t)\})}{2B^2 (t - t_\infty)^2}, \tag{6.5.78}$$

where P_2 is series in $\exp\{-h(t)\}$ with $P_2(0) = 1$.

$t(s)$ is one of a variety of such mappings related to Fuchsian differential equations which play an important role in the uniformization of the Riemann surfaces (Zograf and Takhtajan [1988a,b]). The simplest example of such a mapping is where $t(s)$ corresponds to Fuchsian equations with two regular singular points ($s = 0$ and $s = \infty$). In our case the corresponding Fuchsian equations has three regular singular points at $s = 0$, 1, and ∞. Thus the mapping associated with the Chazy equation provides a natural generalization of such mapping functions.

It is also possible to obtain a "Bäcklund transformation" for the Chazy equation (6.5.64). One solution of the Chazy equation is given by

$$y_2(t) = 6 \frac{\mathrm{d}}{\mathrm{d}t} (\ln z_1), \qquad t = z_2/z_1. \tag{6.5.79}$$

A second (general) solution is given by

$$\begin{aligned} y_1(T) &= 6 \frac{\mathrm{d}}{\mathrm{d}T} \ln(c z_2 + d z_1), \\ T &= \frac{a z_2 + b z_1}{c z_2 + d z_1} = \frac{at + b}{ct + d} = \gamma(t). \end{aligned} \tag{6.5.80}$$

Since

$$\frac{\mathrm{d}}{\mathrm{d}T} = (ct + d)^2 \frac{\mathrm{d}}{\mathrm{d}t},$$

then

$$y_1(T) = 6(ct+d)^2 \frac{\mathrm{d}}{\mathrm{d}t}\left[\ln z_1 + \ln(ct+d)\right]. \tag{6.5.81}$$

Hence we obtain the "Bäcklund transformation"

$$y_2(t) = (ct+d)^{-2} y_1(\gamma(t)) - \frac{6c}{ct+d}. \tag{6.5.82}$$

Now define

$$N_t(y) = \frac{\mathrm{d}^3 y}{\mathrm{d}t^3} - 2y\frac{\mathrm{d}^2 y}{\mathrm{d}t^2} + \left(\frac{\mathrm{d}y}{\mathrm{d}t}\right)^2, \tag{6.5.83}$$

then it is easily shown that

$$N_t(y_2) = (ct+d)^8 N_T(y_1), \tag{6.5.84}$$

where $y_1(T)$ and $y_2(t)$ are related by the "Bäcklund transformation" (6.5.82), and so if $N_T(y_1) = 0$, then necessarily $N_t(y_2) = 0$. Suppose we take as the starting solution $y_1(t) = 0$, then the "Bäcklund transformation" (6.5.82) yields

$$y_2(t) = -\frac{6c}{ct+d}, \tag{6.5.85}$$

i.e.,

$$y_2(t) = -\frac{6}{t-t_0}, \qquad t_0 = -d/c. \tag{6.5.86}$$

Similarly if we take $y_2(t) = -6/(t-t_0)$, then the "Bäcklund transformation" (6.5.82) yields

$$y_3(t) = -\frac{6}{t-\tilde{t}_0}, \qquad \tilde{t}_0 = -\frac{b-dt_0}{a-ct_0}. \tag{6.5.87}$$

We also remark that the solution of Chazy's equation can be written in terms of modular functions [i.e., functions belonging to the modular group $\mathrm{PSL}(2,\mathbb{Z})$], in an elegant formula (Takhtajan, Ablowitz and Chakravarty, private communication 1991). Namely

$$y(t) = \tfrac{1}{2}\frac{\mathrm{d}}{\mathrm{d}t}\ln \Delta(t), \tag{6.5.88}$$

where $\Delta(t)$ is the discriminant modular form (cf., Koblitz [1984]), satisfying

$$\Delta(t) = \sum_{n=1}^{\infty} \tau(n) q^n = q \prod_{n=1}^{\infty} (1-q^n)^{24}, \tag{6.5.89}$$

where $q = \mathrm{e}^{2\pi i t}$, $\tau(n)$ are the well known Ramanujan coefficients and

$$\Delta\left(\frac{at+b}{ct+d}\right) = (ct+d)^{12}\Delta(t),$$

with a, b, c and d integers such that $ad-bc=1$. Moreover, $\Delta(s(t))$ and the Schwarzian triangle function are related via the following. It is standard to call the inverse of $t(s) = z_2(s)/z_1(s)$, $\mathcal{J} = \mathcal{J}(t) = S(0, \pi/2, \pi/3; t)$ (i.e., the Schwarzian triangle function, $\mathcal{J}(t)$ is referred to as the Klein-Haupt function — after a suitable Möbius transformation to put $\mathcal{J}(t)$ in canonical form), $\mathcal{J}(t)$ satisfies

$$\mathcal{J}\left(\frac{at+b}{ct+d}\right) = \mathcal{J}(t), \qquad ad-bc=1,$$

and

$$\Delta(t) = \frac{(\mathcal{J}')^6}{\mathcal{J}^3(\mathcal{J}-1)^4}. \tag{6.5.90}$$

A similar type of solution can be written for the $w_j(t)$ satisfying equation (6.5.63).

Rankine [1956] shows (apart from a misprint) that $\Delta(t)$ satisfies

$$\begin{aligned}\Delta^3\frac{\mathrm{d}^4\Delta}{\mathrm{d}t^4} - 5\frac{\mathrm{d}^3\Delta}{\mathrm{d}t^3}\left(\frac{\mathrm{d}\Delta}{\mathrm{d}t}\right)^2 - \tfrac{3}{2}\Delta^2\left(\frac{\mathrm{d}^2\Delta}{\mathrm{d}t^2}\right)^2 \\ + 12\Delta\frac{\mathrm{d}^2\Delta}{\mathrm{d}t^2}\left(\frac{\mathrm{d}\Delta}{\mathrm{d}t}\right)^2 - \tfrac{13}{2}\left(\frac{\mathrm{d}\Delta}{\mathrm{d}t}\right)^4 = 0.\end{aligned} \tag{6.5.91}$$

Note that since $\Delta(t)$ has no zeros or poles, it is a natural analogue to the well known Theta function solutions of nonlinear wave equations; i.e., it is a "tau"-function. It is interesting to observe that it satisfies a quartic nonlinear equation, i.e., not a bilinear equation.

EXAMPLE 6.5.15 *The Painlevé Equations*

In the following Chapter we shall discuss the six Painlevé equations $\mathrm{P_I} - \mathrm{P_{VI}}$. It is well-known that some of these arise as similarity reductions of some of the equations which themselves arise as reductions of the SDYM equations. Hence, the Painlevé equations are further examples of integrable systems which are reductions of the SDYM equations. Below we list some of these reductions:

(1) The N-wave interaction equations (6.5.42) possess similarity reductions to $\mathrm{P_{III}}$, $\mathrm{P_{IV}}$, $\mathrm{P_V}$ and $\mathrm{P_{VI}}$ (see Fokas, Leo, Martina and Soliani [1986]; Leo, Martina and Soliani [1986]; Leo, Martina, Soliani and Tondo [1986]; Martina and Winternitz [1989]).

(2) The KdV equation (6.5.44) possesses similarity reductions to $\mathrm{P_I}$ and $\mathrm{P_{II}}$ (see Fokas and Ablowitz [1982]).

(3) The nonlinear Schrödinger equation (6.5.47) possesses similarity reductions to $\mathrm{P_{II}}$ and $\mathrm{P_{IV}}$ (see Tajiri [1984]).

(4) The Liouville equation (6.5.31), Sinh-Gordon equation (6.5.33), equation (6.5.36) and the Sine-Gordon equation (6.5.48) possess similarity reductions to $\mathrm{P_{III}}$ (see Clarkson, McLeod, Olver and Ramani [1986]).

(5) The Ernst equations (6.5.52) possess similarity reductions to P_{III} and P_V (see Chandrasekhar [1986]; Léauté and Marcilhacy [1979, 1982, 1983, 1984, 1985]; Marcilhacy [1979]; Persides and Xanthopoulus [1988]; van den Bergh and Wils [1983]; Wils [1989a,b]).

6.5.5 The SDYM Hierarchy. The SDYM equations are given by the compatibility of the Lax pair (6.5.5). Using the gauge freedom we suppose that: (a), $\mathcal{A}_{\bar{\alpha}}$ and $\mathcal{A}_{\beta}$ are constant diagonal matrices; and (b), $\mathcal{A}_{\alpha}$ are off-diagonal matrices. Then the resulting SDYM equations are

$$\partial_{\alpha}\mathcal{A}_{\beta} - \partial_{\beta}\mathcal{A}_{\alpha} = [\mathcal{A}_{\alpha}, \mathcal{A}_{\beta}], \tag{6.5.92a}$$

$$\partial_{\bar{\beta}}\mathcal{A}_{\beta} + \partial_{\bar{\alpha}}\mathcal{A}_{\alpha} = [\mathcal{A}_{\bar{\beta}}, \mathcal{A}_{\beta}], +[\mathcal{A}_{\bar{\alpha}}, \mathcal{A}_{\alpha}]. \tag{6.5.92b}$$

We can generate a hierarchy of the SDYM equations by keeping $D_1 \equiv \partial_{\alpha} + \zeta\partial_{\bar{\beta}}$ and $\mathcal{A}_1 \equiv \mathcal{A}_{\alpha} + \zeta\mathcal{A}_{\bar{\beta}}$ fixed and by *prolonging* $D_2 \equiv \partial_{\beta} - \zeta\partial_{\bar{\alpha}}$ and $\mathcal{A}_2 \equiv \mathcal{A}_{\beta} - \zeta\mathcal{A}_{\bar{\alpha}}$ as follows: let

$$D_2^{(k)} \equiv \partial_{\beta} - \zeta^k\partial_{\bar{\alpha}}, \qquad \mathcal{A}_2^{(k)} \equiv \sum_{r=0}^{k} \mathbf{F}_{k-r}\zeta^r, \tag{6.5.93}$$

for $k \geq 1$. [Note that for $k = 1$, the resulting equations coincide with the SDYM equations (6.5.92).] The hierarchy is obtained from the following compatibility condition

$$D_1\mathcal{A}_2^{(k)} - D_2^{(k)}\mathcal{A}_1 = [\mathcal{A}_1, \mathcal{A}_2^{(k)}]. \tag{6.5.94}$$

Equating coefficients of powers of ζ yields

$$\partial_{\bar{\beta}}\mathbf{F}_0 = [\mathcal{A}_{\bar{\beta}}, \mathbf{F}_0], \tag{6.5.95a}$$

$$\partial_{\alpha}\mathbf{F}_0 + \partial_{\bar{\beta}}\mathbf{F}_1 + \partial_{\bar{\alpha}}\mathcal{A}_{\alpha} = [\mathcal{A}_{\alpha}, \mathbf{F}_0] + [\mathcal{A}_{\bar{\beta}}, \mathbf{F}_1], \tag{6.5.95b}$$

$$\partial_{\alpha}\mathbf{F}_{k-r} + \partial_{\bar{\beta}}\mathbf{F}_{k-r+1} = [\mathcal{A}_{\alpha}, \mathbf{F}_{k-r}] + [\mathcal{A}_{\bar{\beta}}, \mathbf{F}_{k-r+1}], \quad 2 \leq r \leq k-1, \tag{6.5.95c}$$

$$\partial_{\alpha}\mathbf{F}_{k-2} + \partial_{\bar{\beta}}\mathbf{F}_{k-1} = [\mathcal{A}_{\alpha}, \mathbf{F}_{k-2}] + [\mathcal{A}_{\bar{\beta}}, \mathbf{F}_{k-1}], \tag{6.5.95d}$$

$$\partial_{\alpha}\mathbf{F}_{k-1} + \partial_{\bar{\beta}}\mathbf{F}_{k} = [\mathcal{A}_{\alpha}, \mathbf{F}_{k-1}] + [\mathcal{A}_{\bar{\beta}}, \mathbf{F}_{k}], \tag{6.5.95e}$$

$$\partial_{\alpha}\mathbf{F}_{k} - \partial_{\beta}\mathcal{A}_{\alpha} = [\mathcal{A}_{\alpha}, \mathbf{F}_{k}], \tag{6.5.95f}$$

We note that: (a), these equations can be solved successively for $\mathbf{F}_r$, commencing with equation (6.5.95a) which we can take to be $\mathbf{F}_0 = -\mathcal{A}_{\bar{\alpha}}$; (b), the coefficients of ζ^r for $1 \leq r \leq k+1$ do not involve any x_{β} (which may be thought of as a "time" derivative) and will be identified with a function $\mathbf{F}(x_{\mu}, \zeta)$ having an asymptotic expansion

$$\mathbf{F}(x_{\mu}, \zeta) \sim \sum_{n=0}^{\infty} \mathbf{F}_n(x_{\mu})\zeta^n, \qquad \text{as} \qquad \zeta \to \infty,$$

with $\mathbf{F}_0 = -\mathcal{A}_{\bar{\alpha}}$, and which satisfies the following equation

$$\mathrm{D}_1\mathbf{F} = [\mathcal{A}_1, \mathbf{F}] - \partial_{\bar{\alpha}}\mathcal{A}_{\alpha}. \tag{6.5.96}$$

From the solution of (6.5.96) at a fixed "time" (i.e., $x_\beta = x_{\beta_0}$), one can construct the hierarchy (6.5.94) by defining

$$\mathcal{A}_2^{(k)} \equiv (\zeta^k \mathbf{F})|_+, \tag{6.5.97}$$

where $G|_+$ denotes the Taylor series part of the expansion.

The solution of the homogeneous equation

$$\mathrm{D}_1\mathbf{\Phi} = [\mathcal{A}_1, \mathbf{\Phi}], \tag{6.5.98}$$

with $\mathbf{\Phi}_0 = -\mathcal{A}_{\bar{\alpha}}$, is given by $\mathbf{\Phi} = -\mathbf{G}\mathcal{A}_{\bar{\alpha}}\mathbf{G}^{-1}$ where $\mathbf{G}$ satisfies equation (6.5.5a) and has the asymptotic behavior $\mathbf{G} \to \mathbf{I}$, as $\zeta \to \infty$. The solution of the inhomogenous equation is obtained naturally by using variation of parameters and defining $\mathbf{F} = \mathbf{GH}(x_\mu, \zeta)\mathbf{G}^{-1}$, with $\mathbf{H}_0 = -\mathcal{A}_{\bar{\alpha}}$. Then from equation (6.5.96) we obtain

$$\mathrm{D}_1\mathbf{F} - [\mathcal{A}_1, \mathbf{F}] = \mathbf{G}(\mathrm{D}_1\mathbf{H})\mathbf{G}^{-1},$$

and so

$$\mathrm{D}_1\mathbf{H} = \mathbf{G}^{-1}(\partial_{\bar{\alpha}}\mathcal{A}_\alpha)\mathbf{G}. \tag{6.5.99}$$

To obtain the SDYM hierarchy, we proceed as follows. First, solve equation (6.5.5a) with given "initial data" $\mathcal{A}_\alpha(x_\alpha, x_{\bar{\alpha}}, x_{\beta_0}, x_{\bar{\beta}})$. Next we construct $\mathbf{F}$ [and hence $\mathcal{A}_2^{(k)}$ through equation (6.5.97)], by solving equation (6.5.96) or, essentially, by solving equation (6.5.99) given $\partial_{\bar{\alpha}}\mathcal{A}_\alpha$. Finally, using the $\mathcal{A}_2^{(k)}$ thus obtained in equation (6.5.94), all the equations given by the coefficients of ζ^r for $1 \le r \le k+1$ are ***identically satisfied*** and thus we obtain the evolution form of the SDYM hierarchy given by

$$\partial_{\beta_k}\mathcal{A}_\alpha = (\partial_\alpha - \mathrm{ad}\mathcal{A}_\alpha)\mathbf{F}_k \tag{6.5.100}$$

(which is the coefficient of ζ^0).

Is it easily checked that for $k = 1$, we obtain the original SDYM equations; equation (6.5.5b) is identically satisfied (note that $\mathbf{F}_1 = \mathcal{A}_\beta$ and $\mathbf{F}_0 = -\mathcal{A}_{\bar{\alpha}}$) and equation (6.5.92a) is equation (6.5.100) with $k = 1$. Furthermore, from equation (6.5.96) we see that the $\mathbf{F}_k$'s satisfy the following recursion relations for $k > 1$

$$(\partial_\alpha - \mathrm{ad}\mathcal{A}_\alpha)\mathbf{F}_k = -(\partial_{\bar{\beta}} - \mathrm{ad}\mathcal{A}_{\bar{\beta}})\mathbf{F}_{k+1}. \tag{6.5.101}$$

If we define $\mathrm{L}_1 \equiv \partial_\alpha - \mathrm{ad}\mathcal{A}_\alpha$ and $\mathrm{L}_2 \equiv -(\partial_{\bar{\beta}} - \mathrm{ad}\mathcal{A}_{\bar{\beta}})$, then, formally, (6.5.101) yields

$$\mathbf{F}_n = \mathrm{R}^n\mathbf{F}_0, \qquad n > 1, \tag{6.5.102}$$

where $\mathrm{R} \equiv \mathrm{L}_2^{-1}\mathrm{L}_1$, and so equation (6.5.100) can be written in the form

$$\partial_{\beta_k}\mathcal{A}_\alpha = \mathrm{L}_1(\mathrm{R}^k\mathbf{F}_0), \qquad k > 1. \tag{6.5.103}$$

The Lax pair for the hierarchy can be rewritten as

$$\mathrm{D}_1\mathbf{G} = \mathcal{A}_1\mathbf{G}, \tag{6.5.104a}$$

$$\begin{aligned}\partial_{\beta_k}\mathbf{G} &= (\mathcal{A}_2^{(k)} + \zeta^k\partial_{\bar{\alpha}})\mathbf{G} \\ &= [\mathbf{F}_k + \zeta\mathbf{F}_{k-1} + \zeta^2\mathbf{F}_{k-2} + \ldots + \zeta^k(\mathbf{F}_0 + \partial_{\bar{\alpha}})]\mathbf{G},\end{aligned} \tag{6.5.104b}$$

whose compatibility condition is equation (6.5.100). Moreover this hierarchy contains the well known $1+1$ hierarchy when we have reductions from four variables to two, x_α, x_β.

Chapter Seven

The Painlevé Equations

7.1 Historical Origins and Physical Applications.

7.1.1 Singularities of Ordinary Differential Equations. Singularities of ordinary differential equations are of two types: *fixed* and *movable*. We begin by reviewing some facts about *linear* ordinary differential equations (cf. Ince [1956, Chapter 15]). Consider the nth order equation in the complex domain

$$\frac{\mathrm{d}^n y}{\mathrm{d}x^n} + p_{n-1}(x)\frac{\mathrm{d}^{n-1} y}{\mathrm{d}x^{n-1}} + \ldots + p_1(x)\frac{\mathrm{d}y}{\mathrm{d}x} + p_0(x)y = 0. \qquad (7.1.1)$$

If the n coefficients $p_0(x), \ldots, p_{n-1}(x)$ are all analytic in the neighborhood of a point x_0, and x_0 is a *regular* point of the ordinary differential equation, then there exists a unique solution of (7.1.1) such that it and its first $n-1$ derivatives assume arbitrarily assigned values at $x = x_0$. This solution is expressible as a Taylor series in $(x - x_0)$, convergent at least within the circle centered at x_0 and passing through that singular point of the coefficients lying nearest to x_0. Therefore the singular points of solutions of the equation (which are either *regular* or *irregular singular points*), can be located only at singularities of the coefficients. These singularities are said to be *fixed*, since their locations do not vary with the particular solution chosen but depend only on the equation. For example, solutions of the hypergeometric equation

$$x(1-x)\frac{\mathrm{d}^2 y}{\mathrm{d}x^2} + \{c - (a+b+1)x\}\frac{\mathrm{d}y}{\mathrm{d}x} - aby = 0,$$

can only have singularities at those of the coefficients, namely 0, 1, ∞. It is a general property of linear ordinary differential equations that any singularities of their solutions are fixed.

Nonlinear ordinary differential equations lose this property. Consider the very simple ordinary differential equation

$$\frac{\mathrm{d}y}{\mathrm{d}x} + y^2 = 0.$$

Its general solution is

$$y(x) = (x - x_0)^{-1},$$

where x_0 is the constant of integration, and also denotes the location of the singularity. This singularity, a pole, is said to be *movable* since its location depends on the constant of integration x_0. Solutions of nonlinear ordinary differential equations can have both movable and fixed singularities.

A solution of an ordinary differential equation can have various kinds of singularities. Anything other than a pole, of whatever order, is called a *critical point*, and may be an essential singularity or a branch point. Branch points may have a finite number (greater than or equal to two) of branches, or infinitely many. For example, consider the equations

$$\frac{\mathrm{d}y}{\mathrm{d}x} + y^3 = 0,$$
$$y\frac{\mathrm{d}^2y}{\mathrm{d}x^2} - \frac{\mathrm{d}y}{\mathrm{d}x} + 1 = 0,$$
$$ay\frac{\mathrm{d}^2y}{\mathrm{d}x^2} + (1-a)\left(\frac{\mathrm{d}y}{\mathrm{d}x}\right)^2 = 0,$$
$$\left[y\frac{\mathrm{d}^2y}{\mathrm{d}x^2} - \left(\frac{\mathrm{d}y}{\mathrm{d}x}\right)^2\right]^2 + 4x\left(\frac{\mathrm{d}y}{\mathrm{d}x}\right)^3 = 0,$$
$$(1+y^2)\frac{\mathrm{d}^2y}{\mathrm{d}x^2} + (1-2y)\left(\frac{\mathrm{d}y}{\mathrm{d}x}\right)^2 = 0,$$

the general solutions of which are, respectively

$$y(x) = [2(x-x_0)]^{-1/2}, \tag{7.1.2a}$$
$$y(x) = (x-x_0)\ln(x-x_0) + \alpha(x-x_0), \tag{7.1.2b}$$
$$y(x) = \alpha(x-x_0)^a, \tag{7.1.2c}$$
$$y(x) = \alpha\exp\{(x-x_0)^{-1}\}, \tag{7.1.2d}$$
$$y(x) = \tan\{\ln[\alpha(x-x_0)]\}, \tag{7.1.2e}$$

where α and x_0 are the constants of integration. These singularities are, respectively, a movable algebraic branch point, a movable logarithmic branch point, a transcendental singular point unless a is rational, a movable isolated essential singularity, and a non-isolated movable essential singularity. In (7.1.2e), $y(x)$ has no limit (even allowing infinity) as $x \to x_0$ unless x is taken along special paths. In fact, an infinite number of distinct branches originate from the point x_0, which is a limit point of poles, a branch point and an essential singularity.

If the differential equation is put in normal form, i.e., solved for its highest derivative

$$\frac{\mathrm{d}^ny}{\mathrm{d}x^n} = F\left(\frac{\mathrm{d}^{n-1}y}{\mathrm{d}x^{n-1}}, \ldots, \frac{\mathrm{d}y}{\mathrm{d}x}, y, x\right), \tag{7.1.3}$$

then its singularities are those of F. Values of x at which F is singular independently of the other variables are fixed singularities, the others are movable.

7.1.2 First Order Ordinary Differential Equations. Painlevé [1888] proved that for first order equations of the form

$$G\left(\frac{\mathrm{d}y}{\mathrm{d}x}, y, x\right) = 0,$$

with G a polynomial in $\mathrm{d}y/\mathrm{d}x$ and y, and analytic in x, the movable singularities of the solutions are poles and/or algebraic branch points (for proof see Hille [1976] or Ince [1956]). In fact, somewhat earlier, Fuchs [1884] showed that the only equation of the form

$$\frac{\mathrm{d}y}{\mathrm{d}x} = F(y,x) = \frac{P(y,x)}{Q(y,x)}, \tag{7.1.4}$$

where P and Q are polynomials in y with coefficients analytic in x, the critical points of whose solutions are all fixed, is the generalized Riccati equation

$$\frac{\mathrm{d}y}{\mathrm{d}x} = p_2(x)y^2 + p_1(x)y + p_0(x), \tag{7.1.5}$$

with $p_2(x)$, $p_1(x)$, $p_0(x)$ analytic functions (for proof see Hille [1976, p. 89] or Ince [1956, p. 293]). If $p_2(x) \equiv 0$, then (7.1.5) is just a linear equation. Otherwise make the transformation

$$y(x) = -\frac{1}{p_2(x)\,w}\frac{\mathrm{d}w}{\mathrm{d}x},$$

whereupon (7.1.5) becomes the second order linear equation

$$p_2(x)\frac{\mathrm{d}^2w}{\mathrm{d}x^2} - \left[\frac{\mathrm{d}p_2}{\mathrm{d}x} + p_1(x)p_2(x)\right]\frac{\mathrm{d}w}{\mathrm{d}x} + p_0(x)p_2^2(x)w = 0.$$

Its general solution has the form

$$w(x) = \alpha_1 w_1(x) + \alpha_2 w_2(x),$$

where $w_1(x)$ and $w_2(x)$ are any two linearly independent solutions and α_1 and α_2 are arbitrary constants. Therefore the general solution of the Riccati equation (7.1.5) is

$$y(x) = -\frac{\alpha_1\dfrac{\mathrm{d}w_1}{\mathrm{d}x} + \alpha_2\dfrac{\mathrm{d}w_2}{\mathrm{d}x}}{p_2(x)(\alpha_1 w_1 + \alpha_2 w_2)}.$$

7.1.3 The Work of Sophie Kowalevski. Sophie Kowalevski's classic work was on the theory of the motion of a rigid body (a top) rotating about a fixed point (Kowalevski [1889a,b]). She was awarded the Bordin prize of the Paris Academy of Science in 1888 for her work on this problem. In fact, her work was considered so outstanding that the prize money was almost trebled. (Note that the spelling of Kowalevski's name is not consistent, her name is sometimes spelt Sofya Kovalevskaya — we shall use the spelling given in her most famous work [1889a,b].)

Hamilton's equations describing the motion of a top spinning about a fixed point are a system of six nonlinear first-order coupled ordinary differential equations (called Euler's equations):

$$A\frac{\mathrm{d}p}{\mathrm{d}t} = (B-C)qr + Mg(\gamma y_0 - \beta z_0), \qquad \frac{\mathrm{d}\alpha}{\mathrm{d}t} = \beta r - \gamma q, \tag{7.1.6a}$$

$$B\frac{\mathrm{d}q}{\mathrm{d}t} = (C-A)pr + Mg(\alpha z_0 - \gamma x_0), \qquad \frac{\mathrm{d}\beta}{\mathrm{d}t} = \gamma p - \alpha r, \tag{7.1.6b}$$

$$C\frac{\mathrm{d}r}{\mathrm{d}t} = (A-B)pq + Mg(\beta x_0 - \alpha y_0), \qquad \frac{\mathrm{d}\gamma}{\mathrm{d}t} = \alpha q - \beta p, \tag{7.1.6c}$$

where, with respect to a moving Cartesian coordinate system based on the principal axes of inertia with origin at its fixed point, (p,q,r) are the components of angular velocity, (α,β,γ) the direction cosines of the direction of gravity, (A,B,C) the moments of inertia, and (x_0,y_0,z_0) the position of the center of mass of the system; also M is the mass of the top and g the acceleration due to gravity.

In order to solve the equations (7.1.6), it suffices to find six functionally independent integrals ("constants") of the motion. However, since the system of equations (7.1.6) is autonomous (the equations do not depend explicitly on t), it is suffices to seek only five time-independent integrals, the time variation being obtained thereafter by a quadrature. Additionally equations (7.1.6) represent a divergence-free flow in the six-dimensional phase space of points $(p,q,r,\alpha,\beta,\gamma)$, essentially by Liouville's theorem on the preservation of phase space volume for Hamiltonian systems. Indeed, the divergence

$$\frac{\partial}{\partial p}\left(\frac{\mathrm{d}p}{\mathrm{d}t}\right) + \frac{\partial}{\partial q}\left(\frac{\mathrm{d}q}{\mathrm{d}t}\right) + \frac{\partial}{\partial r}\left(\frac{\mathrm{d}r}{\mathrm{d}t}\right) + \frac{\partial}{\partial \alpha}\left(\frac{\mathrm{d}\alpha}{\mathrm{d}t}\right) + \frac{\partial}{\partial \beta}\left(\frac{\mathrm{d}\beta}{\mathrm{d}t}\right) + \frac{\partial}{\partial \gamma}\left(\frac{\mathrm{d}\gamma}{\mathrm{d}t}\right) = 0,$$

is satisfied, since each term vanishes individually. With this property, just four integrals suffice, because they permit the elimination of four variables and reduction to a flow on a two-variable phase space that preserves a known measure (inherited from the original phase space volume) and can consequently be integrated by quadrature. For further details see the discussion of the theory of postmultipliers in Golubov [1953,§§1.11,1.12] (see also Cooke [1984, Appendix 6]).

So it turns out that one only requires four integrals of motion; three such integrals are immediate, the energy

$$Ap^2 + Bq^2 + Cr^2 - 2Mg(\alpha x_0 + \beta y_0 + \gamma z_0) = C_1, \tag{7.1.7a}$$

the angular momentum about the gravitational direction

$$A\alpha p + \beta q + C\gamma r = C_2, \tag{7.1.7b}$$

and the geometric integral

$$\alpha^2 + \beta^2 + \gamma^2 = 1 \tag{7.1.7c}$$

(see Golubov [1953]).

The quest to determine a fourth integral was a popular problem for eighteenth and nineteenth century mathematicians. However, it was only solved in three cases:

1. $A = B = C$ with integral $px_0 + qy_0 + rz_0 = K_1$;
2. $x_0 = y_0 = z_0$, (due to Euler), with integral $A^2p^2 + B^2q^2 + C^2r^2 = K_2$;
3. $A = B$ and $x_0 = y_0 = 0$ (due to Lagrange), with integral $Cr = K_3$,

with K_1, K_2, K_3 constants. For each of these cases the solution of equations (7.1.6) was obtained in terms of elliptic functions and so the solutions were meromorphic functions of t, and the question was whether the solutions of the equations (7.1.6) were meromorphic in any other case.

The approach of Kowalevski was to use the mathematically powerful, yet seemingly unphysical, methods of complex function theory. Apparently motivated by the work of Fuchs [1884] on first order equations, which we have discussed in §7.1.2, Kowalevski determined for which values of the parameters A, B, C, x_0, y_0, z_0, the solution of the system of equations (7.1.6) has no movable critical points. In order to do this, she sought solutions in the form

$$p = t^{-m_1}\sum_{j=0}^{\infty} p_j t^j, \qquad q = t^{-m_2}\sum_{j=0}^{\infty} q_j t^j, \qquad r = t^{-m_3}\sum_{j=0}^{\infty} r_j t^j, \tag{7.1.8a}$$

$$\alpha = t^{-n_1}\sum_{j=0}^{\infty} \alpha_j t^j, \qquad \beta = t^{-n_2}\sum_{j=0}^{\infty} \beta_j t^j, \qquad \gamma = t^{-n_3}\sum_{j=0}^{\infty} \gamma_j t^j, \tag{7.1.8b}$$

with m_1, m_2, m_3, n_1, n_2, n_3 positive integers. For (7.1.8) to represent a general solution of (7.1.6), these series should contain five arbitrary constants (here t can be replaced by $t - t_0$, where t_0 is a sixth arbitrary constant).

Kowalevski found that solutions of (7.1.6) of the form (7.1.8), containing five arbitrary constants, exist for only four special combinations of the parameters A, B, C, x_0, y_0, z_0. Namely, in addition to the three known cases mentioned above there is one new one,

$$A = B = 2C, \qquad z_0 = 0, \tag{7.1.9}$$

which is commonly referred to as the Kowalevski top. For this case, Kowalevski was able to determine the fourth integral, though it is considerably more difficult to find than in the three other cases and presents major mathematical difficulties. In the previously known integrable cases, the integrals were expressed in terms of elliptic integrals (with elliptic functions as inverses), but for the Kowalevski top, the integrals are expressed in terms of hyperelliptic integrals. Whereas elliptic functions are meromorphic functions, hyperelliptic functions in general are not. However, Jacobi [1832, 1835] observed that certain symmetric combinations of hyperelliptic integrals actually do have meromorphic inverses (a result subsequently proved by Göpel [1847] and Rosenhain [1850] — cf. Golubov [1953]; Cooke [1984]). This happens to be the case for the Kowalevski top and so Kowalevski was able to solve equations (7.1.6) in terms of meromorphic functions. For further details of the work of Kowalevski on the motion of a rigid body, the reader is recommended to consult Kowalevski [1889a,b]; Golubov [1953]; Cooke [1984]; Kruskal and Clarkson [1991] — an account of Kowalevski's life story may be found in Kovalevskaya [1978]. Recent mathematical studies of Kowalevski's top include Adler and van Moerbeke [1982, 1988, 1989]; Bobenko, Reyman and Semernov-Tian-Shansky [1989]; Bobenko and Kuznetzov [1988]; Haine and Horozov [1987]; Horozov and van Moerbeke [1989]; Komarov [1987]; Komarov and Kuznetsov [1990]; van Moerbeke [1985].

7.1.4 Second Order Ordinary Differential Equations. We now consider ordinary differential equations of the form

$$\frac{\mathrm{d}^2 y}{\mathrm{d}x^2} = F\left(\frac{\mathrm{d}y}{\mathrm{d}x}, y, x\right), \tag{7.1.10}$$

where F is a rational function of $\mathrm{d}y/\mathrm{d}x$ and y, and an analytic function of x. In 1887, Picard posed the problem of determining which equations of the form (7.1.10) have solutions whose movable singularities are only poles, i.e., the solutions have *no movable critical points* and so the locations of singularities of any of the solutions other than poles are independent of the particular solution chosen and thus dependent only on the equation. This property is now known as the "*Painlevé property*," and ordinary differential equations which possess it are said to be of "*Painlevé type*". The problem was solved by Painlevé and his colleagues (Picard [1887, 1889, 1890, 1893]; Painlevé [1888a,b,c 1893a,b,c,d,e,f,g, 1897, 1898a,b,c, 1899a,b,c,d, 1900, 1901, 1902a,b,c,d,e, 1906]; Fuchs [1905]; Gambier [1906a,b,c,d, 1907a,b, 1909] — a comprehensive review of their work is given in Ince [1956, Chapter 14]; see also Hille [1976, p. 439]). More recently the problem has been considered by Bureau [1980a,b, 1981a,b,c,d,e,f,g,h, 1982a,b,c, 1983a,b,c, 1984] (see also Kruskal [1991]; Kruskal and Clarkson [1991]). Whereas for first order equations of the form (7.1.4), the only possible movable singularities of solutions are poles and algebraic branch points, Painlevé showed that for second order equations of the form (7.1.10), the movable singularities may also be logarithmic branch points and essential singularities. The general method used to study this problem is known as the α-method and is due to Painlevé.

Painlevé's α-method then consists firstly of finding a set of necessary conditions for the absence of movable critical points and then verifying, by direct integration or otherwise, that these are also sufficient. The second part of the method is very tedious since it involves considering a large number of cases and then showing that the equations in each different case are either reducible to equations whose solutions are known to have no movable critical points or define new transcendents.

Painlevé *et al.* showed that out of all possible equations of the form (7.1.10), there are only fifty types which have the property of having no movable critical points. Further they showed that of these fifty equations, forty-four were integrable in terms of previously known functions (such as elliptic functions and linear equations) or were reducible to one of six new nonlinear ordinary differential equations. These fifty canonical types are generalizable by the Möbius transformation

$$Y(X) = \frac{a(x)y + b(x)}{c(x)y + d(x)}, \qquad X = \phi(x),$$

where a, b, c, d, and ϕ are analytic functions, and these new equations contain all second order ordinary differential equations of the form (7.1.10) whose general solutions have no movable critical points. The most interesting of the fifty canonical equations are those which define new transcendental functions. These six equations are

$$\frac{\mathrm{d}^2y}{\mathrm{d}x^2} = 6y^2 + x, \tag{$\mathrm{P_I}$}$$

$$\frac{\mathrm{d}^2y}{\mathrm{d}x^2} = 2y^3 + xy + \alpha, \tag{$\mathrm{P_{II}}$}$$

$$\frac{\mathrm{d}^2y}{\mathrm{d}x^2} = \frac{1}{y}\left(\frac{\mathrm{d}y}{\mathrm{d}x}\right)^2 - \frac{1}{x}\frac{\mathrm{d}y}{\mathrm{d}x} + \frac{1}{x}(\alpha y^2 + \beta) + \gamma y^3 + \frac{\delta}{y}, \tag{$\mathrm{P_{III}}$}$$

$$\frac{\mathrm{d}^2y}{\mathrm{d}x^2} = \frac{1}{2y}\left(\frac{\mathrm{d}y}{\mathrm{d}x}\right)^2 + \frac{3}{2}y^3 + 4xy^2 + 2(x^2 - \alpha)y + \frac{\beta}{y}, \tag{$\mathrm{P_{IV}}$}$$

$$\frac{\mathrm{d}^2y}{\mathrm{d}x^2} = \left\{\frac{1}{2y} + \frac{1}{y-1}\right\}\left(\frac{\mathrm{d}y}{\mathrm{d}x}\right)^2 - \frac{1}{x}\frac{\mathrm{d}y}{\mathrm{d}x} + \frac{(y-1)^2}{x^2}\left(\alpha + \frac{\beta}{y}\right) + \frac{\gamma y}{x} + \frac{\delta y(y+1)}{y-1}, \tag{$\mathrm{P_V}$}$$

$$\begin{aligned}\frac{\mathrm{d}^2y}{\mathrm{d}x^2} &= \frac{1}{2}\left\{\frac{1}{y} + \frac{1}{y-1} + \frac{1}{y-x}\right\}\left(\frac{\mathrm{d}y}{\mathrm{d}x}\right)^2 - \left\{\frac{1}{x} + \frac{1}{x-1} + \frac{1}{y-x}\right\}\frac{\mathrm{d}y}{\mathrm{d}x}\\ &\quad + \frac{y(y-1)(y-x)}{x^2(x-1)^2}\left\{\alpha + \frac{\beta x}{y^2} + \frac{\gamma(x-1)}{(y-1)^2} + \frac{\delta x(x-1)}{(y-x)^2}\right\},\end{aligned} \tag{$\mathrm{P_{VI}}$}$$

where α, β, γ, and δ are arbitrary constants, these are known as the *Painlevé equations* and they require the introduction of new transcendental functions for their solution. These functions are called *Painlevé transcendents* (the six equations are often referred to as the Painlevé transcendents, though we shall solely use this term for their solutions). The first three equations, $\mathrm{P_I}$, $\mathrm{P_{II}}$ and $\mathrm{P_{III}}$, were discovered by Painlevé, the fourth and fifth, $\mathrm{P_{IV}}$ and $\mathrm{P_V}$, were added by Gambier and the general form of the sixth, P_{VI}, was given by Fuchs [1905]. The solutions of P_I, P_{II} and P_{IV} are meromorphic functions of x. If the substitution $x = e^t$ is made in P_{III} and P_V (i.e., $x = 0$ is a fixed singular point), then the solutions become meromorphic functions of t. However, for P_{VI}, 0, 1 and ∞ are fixed critical points (in fact, they are all essential singularities), and hence the general solution of P_{VI} is *not* meromorphic throughout the finite complex plane; furthermore, in contrast to the case for P_{III} and P_V above, since there are three critical points, there exists no transformation which removes all of them from the finite complex plane). In reality, equation P_{VI} contains the other five, since these may be obtained from it by passage to the limit and coalescence (Painlevé [1906]; cf. Ince [1956, p. 345]). Since Painlevé proved that solutions of P_I define new transcendental functions (Painlevé [1900, 1902b,c,d,e]), then one expects that the solutions of $P_{II}, \ldots, P_{VI}$ also do, except for special values of the constants α, β, γ and δ. Recently, Joshi and Kruskal [1990] have given a simple proof that solutions of the Painlevé equations have no movable essential singularities, which previously was the difficult part of the proof that an equation possessed the Painlevé property (cf. Ince [1956, p. 347]), and in practice was rarely done. Furthermore, Kruskal [1991] has developed a method for detecting movable essential singularities which involves expansions in terms of large (or small) exponentials.

7.1.5 Third and Higher Order Ordinary Differential Equations. Many physical problems are modeled by systems of first order ordinary differential equations of the form

$$\frac{d\boldsymbol{x}}{dt} = \boldsymbol{F}(\boldsymbol{x}, t), \tag{7.1.11}$$

with $\boldsymbol{x} = (x_1, x_2, \ldots, x_n)$ and $\boldsymbol{F} = (F_1, F_2, \ldots, F_n)$ and where each F_k is analytic in t and rational with respect to its other arguments. Examples include the motion of stars in the galaxy, the three-body problem, nonlinear lattices, turbulence, competition of species in biology, wave interactions and many other examples (cf. Guckenheimer and Holmes [1983]; MacKay and Meiss [1987]). These systems describe the time evolution of a set of dynamical variables $x_k(t)$ and recently there has been considerable interest in determining when such systems are integrable, often using Painlevé analysis (see, for example, the recent reviews of Hietarinta [1987a]; Kruskal, Ramani and Grammaticos [1990]; Ramani, Grammaticos and Bountis [1989]; Steeb and Euler [1988]). Systems of higher order equations

$$\frac{d^n x}{dt^n} = F\left(\frac{d^{n-1}x}{dt^{n-1}}, \ldots, \frac{dx}{dt}, x, t\right), \tag{7.1.12}$$

can easily be rewritten in the form (7.1.11).

One system which has been extensively studied is that due to Lorenz [1963],

$$\frac{dx}{dt} = \sigma(y - x), \tag{7.1.13a}$$

$$\frac{\mathrm{d}y}{\mathrm{d}t} = -xz + \rho x - y, \tag{7.1.13b}$$

$$\frac{\mathrm{d}z}{\mathrm{d}t} = xy - \beta z, \tag{7.1.13c}$$

with σ, ρ, β parameters. Segur [1980] used Painlevé analysis to determine for which values of the parameters σ, ρ and β, the system (7.1.13) possesses the Painlevé property (see also Bountis, Ramani, Grammaticos, and Dorizzi [1984]; Levine and Tabor [1988]; Sen and Tabor [1990]; Tabor [1989]; Tabor and Weiss [1981]; Weiss [1982]).

Painlevé's α-method, which enabled a complete discussion of second order equations of the form (7.1.10), can also be applied to equations of third (and higher) order of the form (7.1.3). As for the second order case, the procedure divides into two parts, first the determination of necessary conditions for an equation to have no movable critical points, and then the proof that these conditions are also sufficient. The determination of many necessary conditions presents few difficulties: however, the proof that these conditions are also sufficient increases in difficulty with the order of the equations under investigation.

At present there are no comprehensive results for third and higher order equations. Partial classifications have been given by Chazy [1911]; Garnier [1912]; Bureau [1964a,b, 1972b, 1987]; Martynov [1985]. We remark that neither are there comprehensive results for the algebraic differential equation of degree n (for $n \geq 2$) given by

$$G\left(\frac{\mathrm{d}^2y}{\mathrm{d}x^2}, \frac{\mathrm{d}y}{\mathrm{d}x}, y, x\right) = 0,$$

where G is polynomial of degree n in $\mathrm{d}^2y/\mathrm{d}x^2$, rational in $\mathrm{d}y/\mathrm{d}x$ and y and analytic in x (see Bureau [1972a]; Bureau, Garret, and Goffar [1972]; Chalkley [1987] and the references therein for some results). We remark that there have been several investigations in recent years on the relationship between the Painlevé property and algebraic complete integrability (cf. Adler and van Moerbeke [1982, 1988, 1989]; Ercolani and Siggia [1986, 1989, 1991]; Haine [1984]). Furthermore, Grammaticos, Ramani and colleagues (and many others) have extensively studied the integrability of Hamiltonian and non-Hamiltonian systems using Painlevé analysis (Bountis, Ramani, Grammaticos and Dorizzi [1984]; Dorizzi, Grammaticos and Ramani [1983, 1984]; Dorizzi, Grammaticos, Ramani and Winternitz [1985]; Grammaticos, Dorizzi and Ramani [1983, 1984]; Grammaticos, Moulin-Ollangnier, Ramani, Strelcyn and Wojciechowski [1990]; Hietarinta, Grammaticos, Dorizzi and Ramani [1984]; Kruskal, Ramani and Grammaticos [1990]; Ramani, Dorizzi and Grammaticos [1982]; Ramani, Dorizzi, Grammaticos and Bountis [1984]; Ramani, Grammaticos and Bountis [1989]; Ranada, Ramani, Dorizzi and Grammaticos [1985]; Yoshida, Grammaticos, and Ramani [1987, 1988]; Yoshida, Grammaticos, Ramani and Hietarinta [1987]).

A major difficulty at third order is the occurrence of movable *natural boundaries*. A *natural boundary* is a line of singularities through which the solution cannot be analytically continued (cf. Hille [1976]; Nehari [1952]). For example, consider the ordinary differential equation

$$\frac{\mathrm{d}^3 y}{\mathrm{d}x^3} = 2y\frac{\mathrm{d}^2 y}{\mathrm{d}x^2} - 3\left(\frac{\mathrm{d}y}{\mathrm{d}x}\right)^2, \tag{7.1.14}$$

which was studied by Chazy [1909, 1910, 1911]. Every solution of the Chazy equation (7.1.14) is analytic either in a punctured plane, or in a domain bounded by a straight line or a circle; the location of which is dependent on the constants of integration and the solution cannot be analytically continued outside the domain. Recall that in §6.5, it was shown that the Chazy equation (7.1.14) arises as a reduction of the SDYM equations.

As discussed in §6.5 above, the Chazy equation (7.1.14) can be solved in terms of the special case of the hypergeometric equation

$$s(1-s)\frac{\mathrm{d}^2 z}{\mathrm{d}s^2} + \left(\frac{1}{2} - \frac{7s}{6}\right)\frac{\mathrm{d}z}{\mathrm{d}s} - \frac{z}{144} = 0; \tag{7.1.15}$$

given any two independent solutions of this equation, $z_1(s)$ and $z_2(s)$, define the function $x(s)$ by

$$x(s) = \frac{z_2(s)}{z_1(s)}, \qquad \text{i.e.} \qquad \frac{\mathrm{d}x}{\mathrm{d}s} = \frac{W(z_1, z_2)}{z_1^2}, \tag{7.1.16a}$$

where

$$W(z_1, z_2) := z_1\frac{\mathrm{d}z_2}{\mathrm{d}s} - z_2\frac{\mathrm{d}z_1}{\mathrm{d}s} = C(s-1)^{-2/3}s^{-1/2}, \tag{7.1.16b}$$

with C an arbitrary constant. Then $x(s)$ maps the upper half s-plane into the interior of a spherical triangle with angles 0, $\frac{1}{2}\pi$, $\frac{1}{3}\pi$ (cf. Nehari [1952, p. 206]). As is shown below, the inverse function $s(x)$ [the *Schwarzian triangle function* $S(x; 0, \frac{1}{2}, \frac{1}{3})$], has a straight line or a circle as a natural boundary. A solution of the Chazy equation (7.1.14) is then given by

$$y(s(x)) = \frac{6}{z_1}\frac{\mathrm{d}z_1}{\mathrm{d}x} = \frac{6}{z_1}\frac{\mathrm{d}z_1}{\mathrm{d}s}\frac{\mathrm{d}s}{\mathrm{d}x} = \frac{6z_1}{W(z_1, z_2)}\frac{\mathrm{d}z_1}{\mathrm{d}s},$$

as is easily verified by directly substituting this into (7.1.14) and using (7.1.16). Furthermore $y(s(x))$ has the same natural boundary as $s(x)$. The general solution is then obtained by making the transformation $z_1 \to az_1 + bz_2$ and $z_2 \to cz_1 + dz_2$, with a, b, c, d arbitrary constants such that $ad - bc \neq 0$.

Indeed a class of equations whose general solutions have movable natural boundaries can be constructed using these ideas. Given two linearly independent solutions, $z_1(s)$ and $z_2(s)$, of the general hypergeometric equation

$$s(1-s)\frac{\mathrm{d}^2 z}{\mathrm{d}s^2} + [c - (1 + a + b)s]\frac{\mathrm{d}z}{\mathrm{d}s} - abz = 0,$$

then, as above, the function $x(s)$ defined by

$$x(s) = \frac{z_2(s)}{z_1(s)}, \tag{7.1.17}$$

maps the upper half-plane $\mathrm{Im}(s) > 0$ into the interior of a *spherical triangle* with angles $\pi\alpha$, $\pi\beta$, $\pi\gamma$ where $\alpha = 1 - c$, $\beta = a - b$, $\gamma = c - a - b$. A *spherical triangle* is one whose sides are comprised of arcs of circles (or in the limiting case straight lines) which meet with angles $\pi\alpha$, $\pi\beta$ and $\pi\gamma$ at the vertices, where $\alpha + \beta + \gamma < 1$; any such triangle possesses an *orthogonal circle* C, which is a circle intersecting each of the three circles/straight lines making up the triangle orthogonally.

$x(s)$ may be analytically continued by successive inversions (reflections) of the upper half-plane $\mathrm{Im}(s) > 0$ with respect to the real axis. Since the s-plane is punctured at $0, 1, \infty$, then there exist two cuts, say from 0 to $-\infty$ and from 1 to ∞ on the real s-axis, and so there are three alternative inversions. Correspondingly $s(x)$ may be analytically continued by successive inversions (reflections) of the spherical triangle for which there are three possibilities (one in each of the three sides). An inversion of a spherical triangle, Δ_0, in any one of the sides leads to a new spherical triangle, Δ_1, whose sides are also orthogonal to C, the orthogonal circle associated with Δ_0 (since the inversion is angle preserving), that is the circles which make up the triangle also intersect C orthogonally, so that Δ_0 and Δ_1 have the same orthogonal circle (i.e., C is invariant). Furthermore Δ_1 is entirely contained within the interior of the orthogonal circle C and the mapping $s(x)$ is single-valued if and only if $\alpha = 1/m$, $\beta = 1/n$ and $\gamma = 1/p$, with m, n and p positive integers. If we continue in this manner, then all the spherical triangles obtained by successive inversions of the original spherical triangle Δ_0 have the same orthogonal circle C and are situated in the interior of C. This shows that no point outside C can be reached by successive inversions and so $s(x)$ can not be analytically continued beyond C. On the other hand, every point within C lies in some spherical triangle Δ_N obtained by N successive inversions of Δ_0, for some N and so the domain of definition of $s(x)$ coincides with the interior of the orthogonal circle C.

Therefore if

$$\alpha = \frac{1}{m}, \qquad \beta = \frac{1}{n} \qquad \gamma = \frac{1}{p}, \qquad \text{where} \qquad \frac{1}{m} + \frac{1}{n} + \frac{1}{p} < 1, \tag{7.1.18}$$

with m, n, p positive integers, then $s(x)$ is *meromorphic* in the interior of the circle C and every point on C is a limit point of poles and so C forms a *natural boundary* for the function $s(x)$ which cannot be analytically continued beyond it. Hence, provided that (7.1.18) holds, then $x(s)$ defined by (7.1.17) maps the Riemann surface of the s-plane punctured at $0, 1, \infty$, into a region of the x-plane bounded by a circle C (or in the limit a straight line).

$x(s)$ also satisfies the third order ordinary differential equation

$$\{x,s\} = \frac{1}{2}\left\{\frac{1-\alpha^2}{s^2} + \frac{1-\gamma^2}{(1-s)^2} + \frac{1-\alpha^2-\gamma^2+\beta^2}{s(1-s)}\right\},$$

where $\{x,s\}$ is the Schwarzian derivative defined by

$$\{x,s\} := \frac{x'''}{x'} - \frac{3}{2}\left(\frac{x''}{x'}\right)^2 = \frac{\mathrm{d}}{\mathrm{d}s}\left(\frac{x''}{x'}\right) - \frac{1}{2}\left(\frac{x''}{x'}\right)^2,$$

with $' := \mathrm{d}/\mathrm{d}s$. However, our interest is in the inverse function $s(x)$ [which is termed the Schwarzian triangle function $S(x;\alpha,\beta,\gamma)$]. $s(x)$ satisfies the ordinary differential equation

$$\{s,x\} + \frac{1}{2}\left\{\frac{1-\alpha^2}{s^2} + \frac{1-\gamma^2}{(1-s)^2} + \frac{1-\alpha^2-\gamma^2+\beta^2}{s(1-s)}\right\}\left(\frac{\mathrm{d}s}{\mathrm{d}x}\right)^2 = 0, \qquad (7.1.19)$$

since

$$\{x,s\} = -\{s,x\}\left(\frac{\mathrm{d}x}{\mathrm{d}s}\right)^2.$$

As shown above, $s(x) = S(x;\alpha,\beta,\gamma)$ is meromorphic in the interior of a circle C associated with the fundamental triangle and C forms a natural boundary for $s(x)$. Since the Schwarzian function $\{s,x\}$ is invariant under a Möbius transformation, that is

$$\left\{\frac{as+b}{cs+d}, x\right\} = \{s,x\},$$

with a, b, c, d arbitrary constants, then the location of the boundary C depends on the constants of integration and so is movable. Hence all ordinary differential equations of the form (7.1.19), satisfying (7.1.18), have movable natural boundaries and have general solutions which are meromorphic in the domain in which they exist. Additionally this shows that the solution $y(x) = y(s(x))$ of the Chazy equation (7.1.14) also is meromorphic in a domain bounded by a circle (or in the limiting case a straight line), the location of which depends on the constants of integration.

Recent studies on the occurrence of movable natural boundaries in systems of ordinary differential equations include Bessis and Chafee [1986]; Chang, Greene, Tabor and Weiss [1983]; Chang, Tabor and Weiss [1982]; Thual and Frisch [1984].

7.1.6 Physical Applications of the Painlevé Equations. Although the Painlevé equations were first discovered from strictly mathematical considerations, they have recently appeared in several physical applications, including the theories of: the spin-spin correlation function of the two-dimensional Ising Model (Barouch, McCoy and Wu [1973]; Wu, McCoy, Tracy and Barouch [1976]; McCoy, Tracy and Wu [1977]); quantum gravity (Brézin and Kazakov [1990]; Fokas, Its and Kitaev [1991]; Gross and

Migdal [1990a,b]; Its and Kitaev [1990]; Moore [1990]); a spherical electric probe in a continuum plasma (de Boer and Ludford [1975]); the asymptotic behavior of solutions of the mKdV equation (Ablowitz and Segur [1977a]; Segur and Ablowitz [1981]); the one-particle reduced density matrix of the one-dimensional Bose gas (Jimbo, Miwa, Mori and Sato [1981]; Creamer, Thacker and Wilkinson [1982, 1986]); natural convective flows with viscous dissipation (Turcotte, Spence and Bau [1982]; Holmes and Spence [1984]); time-dependent correlation functions of the one dimensional XY model with spin $\frac{1}{2}$ (McCoy, Perk and Schrock [1983a,b]); and Görtler vortices in boundary layers (Bassom and Hall [1989]; Hall [1982]; Hall and Lakin [1988]); the study of polyelectrolytes (McCaskill and Fackerill [1988]); general relativity (Chandrasekar [1986]; Léauté and Marcilhacy [1979, 1982, 1983, 1984, 1987]; MacCallum [1983]; Marcilhacy [1979]; Persides and Xanthopoulos [1988]; van den Bergh and Wils [1983]; Wils [1989a,b]); and nonlinear optics (Giannini and Joseph [1989]).

We remark that discrete analogs of the Painlevé equations have recently been derived, particularly in connection with two-dimensional quantum gravity (cf. Fokas, Its and Kitaev [1991]; Nijhoff and Papageorgiou [1991]; Quispel, Roberts and Thompson [1988, 1989]; Ramani, Grammaticos and Hietarinta [1991]).

7.2 The Painlevé Tests.

7.2.1 The Relationship between the Painlevé Equations and Inverse Scattering. Ablowitz and Segur [1977b] demonstrated a close connection between completely integrable partial differential equations and the Painlevé equations. Consider the following examples. The mKdV equation

$$u_t - 6u^2 u_x + u_{xxx} = 0, \tag{7.2.1}$$

is completely integrable (Wadati [1972]; Ablowitz, Kaup, Newell and Segur [1974]) and has a self-similar solution

$$u(x,t) = (3t)^{-1/3} w(z), \qquad z = x/(3t)^{1/3},$$

where $w(z)$ satisfies the second Painlevé equation

$$w'' = 2w^3 + zw + \alpha, \tag*{$\mathrm{P_{II}}$}$$

with $' := \mathrm{d}/\mathrm{d}z$ and α a constant of integration. Ablowitz and Segur also noted that since the Miura transformation (1.4.1) relates solutions of the mKdV equation (7.2.1) to solutions of the KdV equation

$$u_t + 6uu_x + u_{xxx} = 0,$$

then solutions of the self-similar reduction of the KdV equation

$$u(x,t) = (3t)^{-2/3} v(z), \qquad z = x/(3t)^{1/3},$$

where $v(z)$ satisfies

$$v''' + 6vv' - (2v + zv') = 0, \tag{7.2.2}$$

are also expressible in terms of solutions of the second Painlevé equation. Actually, there exists a one-to-one correspondence between solutions of equation (7.2.2) and those of the second Painlevé equation $\mathrm{P_{II}}$ given by

$$v = -w' - w^2, \qquad w = \frac{v' + \alpha}{2v - z}$$

(cf. Fokas and Ablowitz [1982]). The KdV equation also has the similarity reduction

$$u(x,t) = w(z) - \lambda t, \qquad z = x + 3\lambda t^2,$$

with λ an arbitrary constant, where $w(z)$ is solvable in terms of the first Painlevé equation

$$\frac{\mathrm{d}^2 W}{\mathrm{d}Z^2} = 6W^2 + Z \tag{$\mathrm{P_I}$}$$

(cf. Fokas and Ablowitz [1982]; Olver [1986]).

The Sine-Gordon equation

$$u_{xt} = \sin u,$$

is completely integrable (Ablowitz, Kaup, Newell and Segur [1974]) and has a self-similar solution

$$u(x,t) = v(z), \qquad z = xt;$$

after making the transformation $w = \mathrm{e}^{\mathrm{i}v}$, the equation becomes

$$z\left[ww'' - (w')^2\right] + ww' = w(w^2 - 1),$$

which is a special case of the third Painlevé equation $\mathrm{P_{III}}$.

The Boussinesq equation

$$u_{tt} = u_{xx} - 6(u^2)_{xx} + u_{xxxx}, \tag{7.2.3}$$

is completely integrable (Zakharov and Shabat [1974]). It has the travelling wave solution

$$u(x,t) = w(z), \qquad z = x - ct \tag{7.2.4}$$

with c a constant, where $w(z)$ satisfies

$$w'' = 6w^2 + (c^2 - 1)w + \alpha z + \beta, \qquad ('= \mathrm{d}/\mathrm{d}z),$$

with α, β constants of integration. Depending upon whether $\alpha = 0$, $w(z)$ is expressible in terms of either elliptic functions or solutions of the first Painlevé equation P_I.

We digress to remark that the classical method for determining similarity reductions uses Lie groups of infinitesimal transformations (cf. Bluman and Cole [1974]; Olver [1986]; Bluman and Kumei [1989] — also see equations (7.2.20) below), for which symbolic manipulation programs have been developed, in MACSYMA (Rosenau and Schwarzmeier [1979]; Champagne and Winternitz [1985]; Champagne, Hereman and Winternitz [1990]), REDUCE (Schwarz [1985, 1988]) and MUMATH (Head [1990]). Applying this method to the Boussinesq equation (7.2.3) gives the travelling wave solution $u(x,t) = w_1(z)$ with $z = x - ct$, and the scaling reduction $u(x,t) = t^{-1}w_2(z)$ with $z = x/t^{1/2}$ and λ a constant, where $w_1(z)$ and $w_2(z)$ are solvable in terms of the first and fourth Painlevé equations, respectively (cf. Nishitani and Tajiri [1982]; Rosenau and Schwarzmeier [1986]). However the Boussinesq equation also possesses the similarity reduction

$$u(x,t) = w_3(z) - 4\lambda^2 t^2, \qquad z = x + \lambda t^2,$$

where $w_3(z)$ is solvable in terms of the second Painlevé equation (Quispel, Nijhoff and Capel [1982]; Nishitani and Tajiri [1982]; Rosenau and Schwarzmeier [1986]; Olver and Rosenau [1987]), which is *not* obtained using the classical Lie group method.

Recently Clarkson and Kruskal [1989] developed a new method of determining similarity reductions which involves seeking solutions in the form

$$u(x,t) = U(x,t,w(z(x,t))).$$

Substituting this into the Boussinesq equation (7.2.3) and demanding that the result be an ordinary differential equation for $w(z)$ imposes conditions upon U, z and their derivatives in the form of an overdetermined system of equations (since there are more equations than functions to be determined), whose solution yields the desired similarity reductions. Clarkson and Kruskal [1989] show that the general similarity reduction of the Boussinesq equation (7.2.3) is given by

$$u(x,t) = \theta^2(t)w(z) - \frac{1}{12\theta^2(t)}\left(x\frac{\mathrm{d}\theta}{\mathrm{d}t} + \frac{\mathrm{d}\sigma}{\mathrm{d}t}\right)^2 + \frac{1}{12}, \qquad z = x\theta(t) + \sigma(t), \tag{7.2.5a}$$

where $\theta(t)$, $\sigma(t)$ are any solutions of

$$\frac{\mathrm{d}^2\theta}{\mathrm{d}t^2} = A\theta^5, \qquad \frac{\mathrm{d}^2\sigma}{\mathrm{d}t^2} = (A\sigma + B)\theta^4, \tag{7.2.5b}$$

with A, B constants, and $w(z)$ satisfies

$$w'''' - 12\left[ww'' + (w')^2\right] - (Az + B)w' - 2Aw = \tfrac{1}{6}(Az + B)^2, \tag{7.2.5c}$$

with $' := \mathrm{d}/\mathrm{d}z$. Equation (7.2.5c) is solvable in terms of (i), the first Painlevé equation $\mathrm{P_I}$, if $A = 0$, $B = 0$; (ii), the second Painlevé equation $\mathrm{P_{II}}$, if $A = 0$, $B \neq 0$; and (iii), the fourth Painlevé equation $\mathrm{P_{IV}}$, if $A \neq 0$ (see Clarkson and Kruskal [1989] for details). There are six types of similarity reductions

$$u_1(x,t) = w_1(z), \qquad z = x + \lambda_1 t, \tag{7.2.6a}$$

$$u_2(x,t) = t^2 w_2(z) + \frac{t^2 - x^2}{12t^2}, \qquad z = xt, \tag{7.2.6b}$$

$$u_3(x,t) = w_3(z) + \frac{1 - 4\lambda_3 t^2}{12}, \qquad z = x + \lambda_3 t^2, \tag{7.2.6c}$$

$$u_4(x,t) = t^2 w_4(z) + \frac{t^2 - (x + 6\lambda_4 t^5)^2}{12t^2}, \qquad z = xt + \lambda_4 t^6, \tag{7.2.6d}$$

$$u_5(x,t) = t^{-1} w_5(z) + \frac{4t^2 - (x - 3\lambda_5 t^2)^2}{48t^2}, \qquad z = xt^{-1/2} + \lambda_5 t^{3/2}, \tag{7.2.6e}$$

$$u_6(x,t) = \wp^{-1}(t;0,g_3)\left[w_6(z) - \frac{1}{48}\left(z\frac{\mathrm{d}\wp}{\mathrm{d}t}(t;0,g_3) + 2\lambda_6 \wp^{3/2}(t;0,g_3)\right)^2\right] + \frac{1}{12},$$
$$z = \wp^{-1/2}(t;0,g_3)\left[x + \lambda_6 \zeta(t;0,g_3)\right], \tag{7.2.6f}$$

where $\wp(t;g_2,g_3)$ and $\zeta(t;g_2,g_3)$, with g_2, g_3 constants, are the Weierstrass elliptic and zeta functions, respectively (cf. Whittaker and Watson [1927]), and $\lambda_1, \ldots, \lambda_6$ are arbitrary constants; $w_1(z)$ and $w_2(z)$ are solvable in terms of solutions of $\mathrm{P_I}$, $w_3(z)$ and $w_4(z)$ in terms of $\mathrm{P_{II}}$, and $w_5(z)$ and $w_6(z)$ in terms of $\mathrm{P_{IV}}$. (Note that only the similarity reductions u_1 and u_5 in the special case $\lambda_5 = 0$ can be obtained using the classical Lie group method.) Levi and Winternitz [1989] subsequently showed that the above similarity reductions of the Boussinesq equation (7.2.6) found by Clarkson and Kruskal [1989] can also be obtained using the "nonclassical method" due to Bluman and Cole [1969].

7.2.2 The Painlevé ODE Test. Inspired by the observation of Ablowitz and Segur [1977], Ablowitz, Ramani and Segur [1978, 1980a] and Hastings and McLeod [1980] formulated the *Painlevé Conjecture*, or *Painlevé ODE test*;

Every ordinary differential equation which arises as a similarity reduction of a completely integrable partial differential equation is of Painlevé type, perhaps after a transformation of variables.

This conjecture, if true, would provide a useful *necessary* condition to test whether a given partial differential equation might be completely integrable. Ablowitz, Ramani and Segur [1980b] and McLeod and Olver [1983] have proved weakened versions of the Painlevé ODE test. These proofs are based on the fact that if a partial differential equation is completely integrable, then its solutions are expressible in terms of the solutions of the Gel'fand-Levitan-Marchenko equation, which is a *linear* integral equation. We shall now briefly describe the results of McLeod and Olver [1983] which are somewhat more general than those of Ablowitz, Ramani and Segur [1980b].

Consider the system of partial differential equations

$$\Delta(t, x, u) = 0, \tag{7.2.7a}$$

where $x \in \mathbb{R}$, $t \in \mathbb{R}$ and $u(x,t) \in \mathbb{R}$, together with the initial condition

$$u(x, 0) = f(x), \tag{7.2.7b}$$

where $f(x)$ is in a suitable Banach space $\mathcal{B}$ (usually a space of functions which decay sufficiently rapidly as $|x| \to \infty$, or a space of periodic functions).

DEFINITION 7.2.1

The system of partial differential equations (7.2.7) *is said to be <u>completely integrable relative to</u>* $\mathbf{Q}(u)$ *<u>in the Banach space</u>* $\mathcal{B}$ *if there exists a Gel'fand-Levitan-Marchenko equation*

$$\mathbf{K}(x, y; t) + \mathbf{F}(x, y; t) + \int_x^\infty \mathbf{K}(x, z; t)\, \mathbf{H}(z, y; t)\, \mathrm{d}z = 0, \tag{7.2.8}$$

satisfying the following properties:

1. $\mathbf{F}$, $\mathbf{H}$, $\mathbf{K}$ *are* $N \times N$ *matrices;*
2. $\mathbf{F}$ *and* $\mathbf{H}$ *are uniquely determined from the initial data* (7.2.7*b*)*;*
3. *For initial data in* $\mathcal{B}$, *and for all real* x, y, *all complex* ϵ, *and* t *in some domain* $\Omega \subset \mathbb{C}$, *the functions* $\mathbf{F}(x - \epsilon t, y - \epsilon t; t)$ *and* $\mathbf{H}(x - \epsilon t, y - \epsilon t; t)$ *are analytic in* ϵ, t, *and there exists a Banach space* $\mathcal{B}^*$ *(not necessarily the same as* $\mathcal{B}$*) for which* $\mathbf{F}(x - \epsilon t, y - \epsilon t; t) \in \mathcal{B}^*$ *as a function of* y *and the operator*

   $$\mathbf{T}(x,t)\mathbf{G}(y) = \int_x^\infty \mathbf{G}(z)\, \mathbf{H}(z - \epsilon t, y - \epsilon t; t)\, \mathrm{d}z,$$

   is a compact operator in $\mathcal{B}^*$*;*
4. *The Gel'fand-Levitan-Marchenko equation has a unique solution (in* $\mathcal{B}^*$*) for all* x *and at least one* $t \in \Omega$*;*
5. *The solution* u *of the original system* (7.2.7) *can be recovered from the solution of the Gel'fand-Levitan-Marchenko equation through a relation of the form*

   $$\mathbf{Q}[u(x,t)] = P[\mathbf{K}(x, x; t)]$$

where $\mathbf{Q}$ *is some function (usually linear) of* u *and its spatial derivatives*

In order to study the Gel'fand-Levitan-Marchenko equation (7.2.8), consider the family of integral operators given by

$$\mathbf{T}(x,t)\mathbf{G}(y) := \int_x^\infty \mathbf{G}(z)\,\mathbf{H}(z,y;t)\,\mathrm{d}z. \tag{7.2.9}$$

Then the Gel'fand-Levitan-Marchenko equation becomes

$$\mathbf{K}(x,y;t) = -[\mathbf{I} + \mathbf{T}(x,t)]^{-1}\mathbf{F}(x,y;t)$$

(it is assumed in Definition 7.2.1 above that the inverse exists for at least one $t \in \Omega$). By studying this integral operator (7.2.9), McLeod and Olver [1983] used a theorem due to Steinberg [1969] in order to prove the following result.

THEOREM 7.2.2

If a system of partial differential equations (7.2.7) *is completely integrable relative to* $\mathbf{Q}$ *in the Banach space* $\mathcal{B}$ *and if the initial data are in* $\mathcal{B}$, *then the function* $\mathbf{Q}[u(x,t)]$ *is meromorphic in* (x,t) *for* $x \in \mathbb{C}$, $t \in \Omega$.

Now suppose that $\mathcal{G}$ is a one-parameter symmetry group of the system (7.2.7), which acts on some domain $D_0 \subset \mathbb{C} \times \Omega$, then the $\mathcal{G}$-invariant solutions defined over D_0 are determined by integrating a system of ordinary differential equations, $\Delta/\mathcal{G} = 0$, defined over the image M_0 of D_0. Then, using Theorem 7.2.2, McLeod and Olver [1983] proved the following theorem.

THEOREM 7.2.3

Suppose that $\Delta = 0$ *is a completely integrable relative to* $\mathbf{Q}$ *in the Banach space* $\mathcal{B}$ *and* $\mathcal{G}$ *is an analytic, regular, projectable, one-parameter symmetry group. If* $u = w(x,t)$ *is a* $\mathcal{G}$*-invariant solution of* $\Delta = 0$ *with initial data in* $\mathcal{B}$, *then the combination corresponding to* $\mathbf{Q}$ *of the solution of the reduced system of ordinary differential equations is meromorphic in* M_0.

Essentially this theorem shows that if a system of partial differential equations is completely integrable and $\mathcal{G}$ is a symmetry group of this system, then the reduced system of ordinary differential equations must necessarily be of Painlevé type. However, since the initial data for the $\mathcal{G}$-invariant solutions must lie in $\mathcal{B}$, then only these particular solutions necessarily give ordinary differential equations of Painlevé type.

Nevertheless, despite the absence of a proof of the Painlevé Conjecture, in a more general setting, there is considerable evidence which suggests that it is true (cf. Ablowitz, Ramani and Segur [1978, 1980a]; Ablowitz and Segur [1977b, 1981]; Lakshmanan and Kaliappan [1983]; McLeod and Olver [1983]).

Since the formulation of the Painlevé ODE test, there has been considerable interest in using the Painlevé property as a means of determining whether given equations, both partial and ordinary differential equations, are integrable. The first person to use the Painlevé property as a method of identifying possible integrable equations was Sophie Kowalevski [1889a,b], whose work on the rotation of a top about a fixed point we discussed in §7.1.3 (and consequently we should perhaps call the method the "Kowalevski test").

7.2.3 Applications of the Painlevé ODE Test. The Painlevé ODE test may be applied as follows: if a given partial differential equation is reducible to an ordinary differential equation which is *not* of Painlevé type (even after allowing for a possible change of variables), then the Painlevé ODE test predicts that the partial differential equation is not completely integrable. There is a systematic, algorithmic method for determining similarity reductions of a given system of partial differential equations by the use of one-parameter Lie groups of infinitesimal transformations (cf. Bluman and Cole [1974]; Olver [1986]; Bluman and Kumei [1989]). (Though it should be noted that this method does *not* give all possible similarity reductions for some partial differential equations, such as the Boussinesq equation (7.1.4) — cf. Clarkson and Kruskal [1989].) However, it is often straightforward to determine simple such reductions, for example travelling wave solutions and self-similar (scaling) solutions, by considering "obvious" symmetries of a given partial differential equation. Then one can determine whether the resulting ordinary differential equation is of Painlevé type by using the algorithm due to Ablowitz, Ramani and Segur [1980a]. We shall now consider three examples to demonstrate the application of the Painlevé ODE test.

EXAMPLE 7.2.4 *A Generalized Korteweg-de Vries equation*

Consider the generalized KdV equation

$$u_t + 6uu_x + u_{xxx} + \frac{\alpha}{t}\,u = 0, \tag{7.2.10}$$

where α is a constant. This equation has the scaling similarity solution

$$u(x,t) = (3t)^{-2/3}w(z), \qquad z = x/(3t)^{1/3},$$

where $w(z)$ satisfies

$$w''' + 6ww' = zw' + (2-3\alpha)w, \tag{7.2.11}$$

with $' \equiv \mathrm{d}/\mathrm{d}z$. In the vicinity of a pole at z_0, using the algorithm due to Ablowitz, Ramani and Segur [1980a], it can be shown that as $z \to z_0$

$$\begin{aligned} w(z) = -\frac{2}{(z-z_0)^2} - \tfrac{1}{6}z_0 + \tfrac{1}{2}\alpha(z-z_0) + c_1(z-z_0)^2 + \frac{(3\alpha-2)z_0}{36}(z-z_0)^3 \\ + \left\{c_2 + \frac{3\alpha(1-2\alpha)}{28}\ln(z-z_0)\right\}(z-z_0)^4 + o((z-z_0)^4), \end{aligned}$$

with c_1, c_2 arbitrary constants. Therefore $w(z)$ has a movable logarithmic branch point, and so (7.2.11) is not of Painlevé type, unless either $\alpha = 0$ or $\alpha = \frac{1}{2}$, in which case (7.2.10) is either the KdV or cylindrical KdV equation respectively, both of which are known to be completely integrable (Gardner, Greene, Kruskal and Miura [1967]; Calogero and Degasperis [1978], respectively). Hence the Painlevé ODE test predicts that the generalized KdV equation (7.2.10) is not completely integrable unless either $\alpha = 0$ or $\alpha = \frac{1}{2}$. (See also Example 7.2.7 below.)

EXAMPLE 7.2.5 *Nonlinear Klein-Gordon equations*

Consider the nonlinear Klein-Gordon equation in characteristic coordinates

$$u_{xt} = f(u), \tag{7.2.12}$$

where $f(u)$ is a linear combination of exponentials $\exp(b_j u)$. McLeod and Olver [1983] showed that the only partial differential equations of this form which are reducible to ordinary differential equations of Painlevé type by travelling wave solutions

$$u(x,t) = w(z), \qquad z = x - ct,$$

are those of the form

$$u_{xt} = a_2 e^{2\beta u} + a_1 e^{\beta u} + a_{-1} e^{-\beta u} + a_{-2} e^{-2\beta u}, \tag{7.2.13}$$

for some (possibly complex) constants β, a_2, a_1, a_{-1}, a_{-2}. (In fact, the singularities of u are not really poles, but rather "pure logarithms" in the sense that u_x, u_t and $\exp(\beta u)$ have only poles — this possibility is allowed by the final clause of the Painlevé ODE test given above.) However, a partial differential equation which is included in the form (7.2.13) is the double Sine-Gordon equation

$$u_{xt} = \sin u + \sin 2u, \tag{7.2.14}$$

which is not thought to be completely integrable since numerical evidence suggests that the interaction of solitary waves for (7.2.14) is inelastic and so do not behave like solitons (Ablowitz, Kruskal and Ladik [1979]).

Suppose now we consider the similarity reduction of (7.2.13) given by

$$u(x,t) = v(z), \qquad z = xt/\beta;$$

then after making the transformation $w = \exp(\beta v)$ (to put the equation into rational form), $w(z)$ satisfies

$$z\left[ww'' - (w')^2\right] + ww' = a_2 w^4 + a_1 w^3 + a_{-1} w + a_{-2},$$

with $' := \mathrm{d}/\mathrm{d}z$. This equation is of Painlevé type if and only if $a_2 a_1 = 0$ and $a_{-2}a_{-1} = 0$ (cf. Clarkson, McLeod, Olver and Ramani [1986]). Therefore the Painlevé ODE test predicts that the if a partial differential equation of the form (7.2.12) is completely integrable, then it necessarily is one of the three standard forms,

$$u_{xt} = \sin u, \tag{7.2.15a}$$

$$u_{xt} = \mathrm{e}^u - \mathrm{e}^{-2u}, \tag{7.2.15b}$$

$$u_{xt} = \mathrm{e}^u, \tag{7.2.15c}$$

possibly after scaling and/or translation [which rules out the double Sine-Gordon equation (7.2.14)]. Equation (7.2.15a) is the Sine-Gordon equation which is known to be completely integrable (Ablowitz, Kaup, Newell and Segur [1974]. Equation (7.2.15b) is also known to be completely integrable (Fordy and Gibbons [1980b]; Mikhailov [1979, 1981]). Equation (7.2.15c) is the Liouville equation is reducible using a Bäcklund transformation to the linear partial differential equation $v_{xt} = 0$ (cf. Clarkson, McLeod, Olver and Ramani [1986]). Therefore, we conclude from the Painlevé ODE test that the only partial differential equations of the form (7.2.12) which might be completely integrable are equivalent to one of the three equations (7.2.15).

Equation (7.2.15b) also highlights an important aspect in the application of the Painlevé ODE test. It reduces under the similarity reduction $u(x,t) = \ln w(z)$, with $z = xt$, to the equation

$$z\left[ww'' - (w')^2\right] + ww' = w^3 - 1. \tag{7.2.16}$$

This second order equation is of Painlevé type yet is *not* one of the fifty canonical equations on the list in Ince [1956], which were given by Painlevé and his colleagues! However, if we make the transformation

$$w(z) = z^{-1/4}\eta(\xi), \qquad \xi = \tfrac{4}{3}z^{3/4},$$

then (7.2.16) is transformed into a special case of the third Painlevé equation $\mathrm{P_{III}}$. This shows the problem of using the list in Ince superficially, that is without considering possible transformations. It also shows that it is often necessary to consider more than one similarity reduction.

Next consider the n-dimensional Klein-Gordon equation

$$\sum_{j=1}^{n} \frac{\partial^2 u}{\partial x_j^2} - \frac{\partial^2 u}{\partial t^2} = \begin{cases} 2\sinh u, \\ \mathrm{e}^u, \\ \mathrm{e}^u - \mathrm{e}^{-2u}, \end{cases} \tag{7.2.17}$$

i.e., the n-dimensional generalizations of equations (7.2.15), all of which we have seen are completely integrable if $n = 1$. We make the similarity reduction

$$u(x_1, x_2, \ldots, x_n, t) = v(z), \qquad z = \tfrac{1}{4}(x_1^2 + x_2^2 + \ldots + x_n^2 - t^2),$$

then after making the transformation $w = e^v$, $w(z)$ satisfies

$$z\left[ww'' - (w')^2\right] + \tfrac{1}{2}(n+1)ww' = \begin{cases} w^3 - w, \\ w^3, \\ w^3 - 1, \end{cases} \tag{7.2.18}$$

respectively. In all three cases, it can be shown that in the vicinity of a pole at z_0, as $z \to z_0$

$$w(z) = \frac{2z_0}{(z-z_0)^2} - \frac{n-1}{z-z_0} + c_1 + \frac{n(n-1)}{3z_0}\ln(z-z_0) + o(1),$$

where c_1 is an arbitrary constant. Hence $w(z)$ has a movable logarithmic branch point and so (7.2.18) is not of Painlevé type unless either $n = 0$ or $n = 1$. Therefore the Painlevé ODE test predicts that the n-dimensional Klein-Gordon equations (7.2.17) are *not* completely integrable if $n \geq 2$.

We remark that the crucial question is not whether a given partial differential equation is reducible to one of the Painlevé equations, but rather whether it is reducible to an ordinary differential equation which is of Painlevé type. If a given partial differential equation is reducible to an ordinary differential equation of Painlevé type, even to one of the Painlevé equations, then the Painlevé ODE test says nothing about whether the partial differential equation might be integrable. The Painlevé ODE test requires that every reduction should give an ordinary differential equation of Painlevé type, however this is not sufficient.

EXAMPLE 7.2.6 *The Modified Benjamin-Bona-Mahoney Equation*

Consider the Modified Benjamin-Bona-Mahoney (MBBM) equation

$$u_t + u_x + u^2u_x - u_{xxt} = 0, \tag{7.2.19}$$

In order to determine similarity reductions of (7.2.19), we shall use the classical Lie group method of infinitesimal transformations (cf. Bluman and Cole [1974]). Consider the one-parameter (ϵ) Lie group of infinitesimals in (x,t,u) defined by

$$X = x + \epsilon\xi(x,t,u) + O(\epsilon^2), \tag{7.2.20a}$$

$$T = t + \epsilon\tau(x,t,u) + O(\epsilon^2), \tag{7.2.20b}$$

$$U = u + \epsilon\eta(x,t,u) + O(\epsilon^2), \tag{7.2.20c}$$

The MBBM equation (7.2.19) is invariant under this transformation if

$$U_T + U_X + U^2U_X - U_{XXT} = 0. \tag{7.2.19*}$$

Substituting (7.2.20) into (7.2.19*) yields an overdetermined system of homogeneous linear equations for the infinitesimals $\xi(x,t,u)$, $\tau(x,t,u)$, and $\eta(x,t,u)$ which have solution

$$\xi(x,t,u) = \alpha, \qquad \tau(x,t,u) = \beta, \qquad \eta(x,t,u) = 0,$$

with α, β arbitrary constants (see Clarkson [1989] for further details). The similarity reductions are then obtained by solving the characteristic equation

$$\frac{\mathrm{d}x}{\xi(x,t,u)} = \frac{\mathrm{d}t}{\tau(x,t,u)} = \frac{\mathrm{d}u}{\eta(x,t,u)}.$$

Hence the only non-constant similarity reduction of the MBBM equation obtained by this method is the travelling wave solution

$$u(x,t) = w(z), \qquad z = x - ct,$$

where $w(z)$ satisfies

$$\tfrac{1}{2}(1-c)w^2 + \tfrac{1}{12}w^4 + \tfrac{1}{2}c\,(w')^2 = Aw + B,$$

with $' := \mathrm{d}/\mathrm{d}z$ and A, B arbitrary constants. This is solvable in terms of elliptic functions and so is of Painlevé type. Hence the MBBM equation (7.2.19) satisfies the necessary conditions of the Painlevé ODE test to be completely integrable, since every ordinary differential equation which arises as a similarity reduction of the equation obtained by the classical Lie group method is of Painlevé type. Furthermore, the travelling wave solution is the only similarity reduction obtained using the method due to Clarkson and Kruskal [1989] mentioned above with respect to similarity reductions for the Boussinesq equation (see Clarkson [1989] for details). However, numerical evidence (Makhankov [1978]) suggests that the interaction of solitary waves for (7.2.19) is inelastic and so do not behave like solitons.

This example, whereas it does not contradict the assertion of the Painlevé ODE test, strongly suggests that the Painlevé ODE test may *not* be generalized so as to provide a *necessary and sufficient* condition for a partial differential equation to be completely integrable.

Now consider the Symmetric Regularized Long Wave (SRLW) equation

$$u_{tt} + u_{xx} + uu_{xt} + u_x u_t + u_{xxtt} = 0, \tag{7.2.21}$$

which arises in several physical applications including ion sound waves in a plasma (cf. Seyler and Fenstermacher [1984]). Using the classical Lie group method for determining similarity solutions, which we used above for the MBBM equation, it is easily seen that the only non-constant similarity reduction arising from this method is the travelling wave solution

$$u(x,t) = v(z), \qquad z = x - ct,$$

where $v(z)$ satisfies

$$c^2 v'' = \tfrac{1}{2}cv^2 - (1+c^2)v + Az + B,$$

with $' := \mathrm{d}/\mathrm{d}z$ and A, B arbitrary constants (Rosenau and Schwarzmeier [1986]). The solution of this equation is expressible either in terms of elliptic functions or the first Painlevé equation, depending upon the choice of constants. However the SRLW equation also possesses the similarity reduction

$$u(x,t) = t^{-1}w(z) - \lambda t, \qquad z = x + \lambda^{-1}\ln t,$$

and λ ($\neq 0$) is a constant, where $w(z)$ satisfies the equation

$$w'''' + 3\lambda w''' + (2\lambda^2+1)w'' + 3\lambda w' + 2\lambda^2 w = 2\lambda^2 ww' + \lambda\left[ww'' + (w')^2\right] \qquad (7.2.22)$$

(Clarkson [1989]). In the neighborhood of a pole at z_0, it is easily shown that

$$w(z) = \frac{12}{\lambda(z-z_0)^2} - \frac{12}{5(z-z_0)} + \frac{25-\lambda^2}{25\lambda} - \frac{(125+\lambda^2)}{125}(z-z_0) \\ + \left\{w_4 - \tfrac{6}{25}\lambda\ln(z-z_0)\right\}(z-z_0)^2 + o((z-z_0)^2),$$

with w_4 an arbitrary constant (Clarkson [1989] — this similarity reduction was obtained by using the method due to Clarkson and Kruskal [1989]). At higher orders of $z - z_0$, higher and higher powers of $\ln(z-z_0)$ are required which indicates that the solution has a movable logarithmic branch point and so equation (7.2.22) is not of Painlevé type. Therefore the Painlevé ODE test predicts that the SRLW equation (7.2.21) is not solvable by inverse scattering. This conclusion is supported by the fact that (a), numerical evidence shows that the interaction of solitary waves is inelastic (Bogolubsky [1977]; Makhankov [1978]; Seyler and Fenstermacher [1984]); and (b), it possesses only three independent polynomial conservation laws (Seyler and Fenstermacher [1984]). (Note that we would not have obtained this prediction from the Painlevé ODE test had we only considered similarity solutions obtained using the classical Lie group method.)

7.2.4 The Painlevé PDE Test. The major drawback in applying the Painlevé ODE test to a given partial differential equation is that one has to first reduce the partial differential equation to an ordinary differential equation (there may be several possible reductions) and then determine whether each resulting ordinary differential equation is of Painlevé type (after allowing for possible transformations). Despite the existence of a systematic method of finding similarity reductions (by exploiting the symmetries in the partial differential equation, though as we remarked, this method does not necessarily determine *all* possible similarity solutions of a given partial differential equation), the Painlevé ODE test is less effective for partial differential equations which have a limited number of such symmetries, such as the MBBM equation (7.2.19) and if a partial differential equation has no symmetries, then the Painlevé ODE test is ineffective. In principal it would also be better if one could directly test the partial differential equation. Weiss, Tabor and Carnevale [1983] introduced the ***Painlevé property for partial differential equations***, or ***Painlevé PDE Test***, as a method of applying the Painlevé ODE test directly to a given partial differential equation without having to reduce it to an ordinary differential equation. A partial differential equation is said to ***possess the Painlevé property*** if solutions of the partial differential equation are "single-valued" in the neighborhood of non-characteristic, movable singularity manifolds. Weiss, Tabor and Carnevale [1983] proposed a method of applying the Painlevé PDE Test to a given partial differential equation which is analogous to the algorithm given by Ablowitz, Ramani and Segur [1980a] to determine whether a given ordinary differential equation might be of Painlevé type; by seeking a solution of a given partial differential equation in the form of a Laurent series

$$u(z_1, z_2, \ldots, z_n) = u(\boldsymbol{z}) = \phi^{-p}(\boldsymbol{z}) \sum_{j=0}^{\infty} u_j(\boldsymbol{z})\phi^j(\boldsymbol{z}), \tag{7.2.23}$$

where $u_j(\boldsymbol{z})$, $j = 0, 1, \ldots$, are analytic functions of $\boldsymbol{z} = (z_1, z_2, \ldots, z_n)$, with $u_0(\boldsymbol{z}) \not\equiv 0$, in the neighborhood of a non-characteristic, movable singularity manifold defined by $\phi(\boldsymbol{z}) = 0$, where $\phi(\boldsymbol{z})$ is an analytic function of $z_1, z_2, \ldots, z_n$. Substituting (7.2.23) into the equation and equating coefficients of like powers of ϕ determines p and defines recursion relations for u_n, for $n \geq 1$, of the form

$$(n - \beta_1)(n - \beta_2)\ldots(n - \beta_N)u_n = F_n(u_0, u_1, \ldots, u_{n-1}, \phi, \boldsymbol{z}),$$

where N is the order of the equation, for some functional F_n. This defines u_n unless $n = \beta_j$ for some j, $1 \leq j \leq N$. $n = \beta_1, \beta_2, \ldots, \beta_N$ are the *resonances* (commonly $n = -1$ is a resonance and it is usually associated to the singularity manifold defined by $\phi = 0$ being arbitrary). For each positive integer resonance there is a compatibility condition (i.e., $F_\beta = 0$) which must be identically satisfied for the partial differential equation to have a solution of the form (7.2.23) and then $u_\beta(\boldsymbol{z})$ is an arbitrary function. Essentially, in order for a given partial differential equation to pass the Painlevé PDE test, it is required that p is an integer and there are $N-1$ consistent recursion relations (i.e., all the compatibility conditions are satisfied), so that the series (7.2.23) contains the requisite number of arbitrary functions as required by the Cauchy-Kowalevski theorem ($\phi(\boldsymbol{z})$ is the Nth arbitrary function) and thus corresponds to the general solution of the equation. We remark that recently there have been studies into the role of negative resonances suggesting that they are important (cf. Fordy and Pickering [1991]; Kruskal [1991]); in particular, it is easily shown that *all* the resonances for the Chazy equation (7.1.14) are negative (-1, -2 and -3).

Also, as for the Painlevé ODE test, it may be necessary to make a change of variables in order to apply the Painlevé PDE Test, for example, the Sine-Gordon equation

$$u_{xt} = \sin u, \tag{7.2.24}$$

is known to be completely integrable (Ablowitz, Kaup, Newell and Segur [1974]) and in order to apply the Painlevé PDE Test we must first make the transformation $v = e^{iu}$ to put (7.2.24) into rational form.

The singularity manifold must be non-characteristic, since on a characteristic manifold *any* type of singularity may propagate. For example, consider the linear wave equation

$$u_{tt} = u_{xx},$$

which has general solution

$$u(x,t) = f(x+t) + g(x-t), \tag{7.2.25}$$

where f and g are arbitrary, twice differentiable, functions. Clearly f and g can possess any type of singularity, for example essential singularities, so the general solution (7.2.25) will also possess the same singularities on the characteristic manifolds. Characteristic manifolds for partial differential equations are the analog of fixed singularities in ordinary differential equations since they are determined by the partial differential equation and not by the particular solution.

Weiss, Tabor and Carnevale [1983] did not attempt to prove any relationship between the Painlevé property and complete integrability, however a "proof" can be inferred from the "proof" of the Painlevé ODE test given by McLeod and Olver [1983]. As mentioned above, McLeod and Olver essentially proved that, under certain restrictions, the solution $u(z_1, z_2, \ldots, z_n)$ of a completely integrable partial differential equation is meromorphic when $z_1, z_2, \ldots, z_n$ assume complex values.

Despite being by no means foolproof (as we discuss further below — see also Clarkson [1989]; Kruskal [1991]; Kruskal and Clarkson [1991]; Pogrebkov [1989]), the Painlevé PDE test also provides a useful criterion for the identification of completely integrable partial differential equations. In addition to providing a valuable first test for whether a given partial differential equation is completely integrable, other important information relating to completely integrable equations can be obtained by use of Painlevé analysis, including Bäcklund transformations, Lax pairs, Hirota's bilinear representation, special and rational solutions, etc. (cf. Chudnovsky, Chudnovsky and Tabor [1983]; Conte [1988, 1989]; Flaschka, Newell and Tabor [1991]; Gibbon, Padmore, Tabor and Wood [1985]; Gibbon, Newell, Tabor and Zeng [1988]; Musette and Conte [1991]; Newell, Tabor and Zeng [1987]; Ramani, Grammaticos and Bountis [1989]; Steeb and Euler [1988]; Weiss [1983, 1984a,b, 1985a,b, 1986a,b, 1987, 1989]). Many of these results are obtained by seeking solutions of the partial differential equation in the form of a *truncated Laurent series expansion*

$$u(z) = u_0(z)\phi^{-p}(z) + u_1(z)\phi^{-p+1}(z) + \cdots + u_p(z). \tag{7.2.26}$$

We remark that equating the coefficient of each order of ϕ to zero can be too restrictive and a more general approach is required (cf. Newell, Tabor and Zeng [1987]).

If a compatibility condition is not satisfied for arbitrary $\phi(z)$ (i.e., $F_\beta \not\equiv 0$ for some β), then it is necessary to introduce terms of the form $\phi(z)^{\beta-p}\ln\phi(z)$ into the series (7.2.23) at this order to make the recursion relations consistent, thereby rendering it a multivalued logarithmic psi series. If for special choices of $\phi(z)$, all the compatibility conditions are identically satisfied (i.e., $\phi(z)$ satisfies a set of "*consistency conditions*" $F_\beta = 0$), then the equation is said the have the "*conditional Painlevé property*" (Weiss [1984b]). In these cases, useful information, such as special solutions, for non-integrable equations can be obtained by using truncated Laurent series expansions (7.2.26) (cf. Cariello and Tabor [1989]; Conte [1988, 1989]; Conte and Musette [1989]; Kudryashov [1991]; Nozaki [1987]; Weiss [1984b]). Additionally, even if the equation possesses neither the Painlevé property nor the conditional Painlevé property for any choice of $\phi(z)$, then analysis of the associated logarithmic psi series can still yield valuable insights (cf. Fournier, Levine and Tabor [1988]; Levine and Tabor [1988]).

7.2.5 Applications of the Painlevé PDE Test. We shall now discuss examples of the application of the Painlevé PDE test.

EXAMPLE 7.2.7 *A Generalized Korteweg-de Vries Equation*

In this example we consider the generalized KdV equation

$$u_t + 6uu_x + u_{xxx} + a(t)u = 0, \tag{7.2.27}$$

where $a(t)$ is analytic function, which arises in the description of solitary waves in a channel with slowly varying depth (cf. Grimshaw [1979a]), and determine the choices of the function $a(t)$ such that the equation (7.2.27) passes the Painlevé PDE test (note that this is a generalization of Example 7.2.4 above). Suppose the non-characteristic singularity manifold is defined

$$\phi(x,t) = x + \psi(t) = 0, \tag{7.2.28}$$

where $\psi(t)$ is an arbitrary analytic function (as suggested by Kruskal, private communication — see Jimbo, Kruskal and Miwa [1982]). We also note the analogy with the Laurent series expansion for ordinary differential equations in which one seeks a solution in the form

$$w(z) = (z-z_0)^{-p}\sum_{j=0}^{\infty} w_j(z-z_0)^j,$$

in the neighborhood of the point z_0. Since it is assumed that $\phi(x,t) = 0$ does not define a characteristic manifold, then necessarily $\phi_x(x,t) \neq 0$ and by use of the implicit function theorem, we may assume that $\phi(x,t)$ has the special form (7.2.28), without loss of generality. (The Laurent series expansion is purely a local expansion and so the implicit function theorem, which is intrinsically a local theorem, is applicable.) Then we seek a solution of (7.2.27) in the form

$$u(x,t) = \phi^{-p}(x,t)\sum_{j=0}^{\infty} u_j(t)\phi^j(x,t), \tag{7.2.29}$$

where $u_j(t)$, $j = 0, 1, \ldots$, are analytic functions of t, with $u_0 \not\equiv 0$, in the neighborhood of the singularity manifold defined by (7.2.28). [If we had assumed that the u_j were analytic functions of x and t, then these can be expanded in Taylor series in powers of x and then by collecting terms and renaming coefficients if necessary, we obtain the form (7.2.29).] To determine the dominant behavior we substitute $u \sim u_0\phi^{-p}$, into (7.2.27) and find that necessarily p satisfies $p+3 = 2p+1$, i.e., $p = 2$. Substituting (7.2.28-29) with $p = 2$ into (7.2.27) and equating coefficients of powers of ϕ determines the following recursion relations for the u_j;

$$\phi^{-5}: \qquad u_0 + 2 = 0,$$

$$\phi^{j-5}: \qquad (j+1)(j-4)(j-6)u_j + 3(j-4)\sum_{k=1}^{j-1} u_k u_{j-k} + a(t)u_{j-3} + \frac{\mathrm{d}u_{j-3}}{\mathrm{d}t} + (j-4)u_{j-2}\frac{\mathrm{d}\psi}{\mathrm{d}t} = 0, \tag{7.2.30}$$

for $j \geq 1$ (define $u_j = 0$ for $j < 0$). (7.2.30) defines u_j for $j \geq 1$ unless $j = 4$ or $j = 6$, which are the *resonances*. The *resonances* are the values of j at which arbitrary functions arise in the expansion (7.2.29) and for every positive resonance there is a compatibility condition which must be identically satisfied in order that (7.2.27) has a solution of the form (7.2.28-29). From (7.2.30) we obtain

$$u_1(t) = 0, \tag{7.2.31a}$$

$$u_2(t) = -\tfrac{1}{6}\frac{\mathrm{d}\psi}{\mathrm{d}t}, \tag{7.2.31b}$$

$$u_3(t) = \tfrac{1}{6}a(t), \tag{7.2.31c}$$

$$u_5(t) = -\frac{1}{36}\left(a(t)\frac{\mathrm{d}\psi}{\mathrm{d}t} + \frac{\mathrm{d}^2\psi}{\mathrm{d}t^2}\right), \tag{7.2.31d}$$

and the compatibility conditions

$$\frac{du_1}{dt} + a(t)u_1 = 0, \tag{7.2.32a}$$

$$\frac{da}{dt} + 2a^2 = 0. \tag{7.2.32b}$$

The first compatibility condition (7.2.32a) is identically satisfied from (7.2.31a). Solving the second compatibility condition (7.2.32b) for $a(t)$ (which is an ordinary differential equation of Painlevé type) gives either

$$a(t) = 0, \qquad \text{or} \qquad a(t) = \frac{1}{2(t-t_0)}, \tag{7.2.33}$$

with t_0 a constant (which we set to zero, without loss of generality). Therefore (7.2.27) can possess the Painlevé property only if $a(t)$ has one of the forms (7.2.33), which correspond to the KdV equation

$$u_t + 6uu_x + u_{xxx} = 0,$$

and the cylindrical KdV equation

$$u_t + 6uu_x + u_{xxx} + \frac{u}{2t} = 0,$$

respectively, both of which are known to be completely integrable (Gardner, Greene, Kruskal, and Miura [1967]; Calogero and Degasperis [1978], respectively). If $a(t)$ does not satisfy the compatibility condition (7.2.32b), then terms of the form

$$[u_6(t) + v_6(t)\ln\phi(x,t)]\phi^4(x,t),$$

are required at that order ($j = 6$) and at higher and higher orders in $\phi(x,t)$, higher and higher powers of $\ln\phi(x,t)$ are required. Such an expansion is not compatible with the Painlevé property. Therefore the Painlevé PDE Test suggests that the only partial differential equations of the form (7.2.27) which are completely integrable are the KdV and cylindrical KdV equations.

EXAMPLE 7.2.8 *A Generalized Nonlinear Schrödinger Equation*

In this example we consider the generalized nonlinear Schrödinger equation

$$\mathrm{i}u_t + u_{xx} - 2|u|^2 u = a(x,t)u + b(x,t), \tag{7.2.34}$$

where $a(x,t)$ and $b(x,t)$ are analytic functions, which arises in several physical applications (cf. Grimshaw [1979b]; Kaup and Newell [1978b]), and determine the choices of the functions $a(x,t)$ and $b(x,t)$ such that the equation (7.2.34) passes the Painlevé PDE test, following Clarkson [1988]. First, due to the non-analyticity of the nonlinear term, we complexify all variables and write (7.2.34) as the system

$$\mathrm{i}u_t + u_{xx} - 2u^2 v = a(x,t)u + b(x,t), \tag{7.2.35a}$$

$$-\mathrm{i}v_t + v_{xx} - 2uv^2 = \bar{a}(x,t)v + \bar{b}(x,t), \tag{7.2.35b}$$

in which u and v are treated as *independent* complex functions of the (complex) variables x and t, and where $\bar{a}(x,t)$ and $\bar{b}(x,t)$ are the formal complex conjugates of $a(x,t)$ and $b(x,t)$, respectively.

Now we seek solutions of (7.2.35) in the form

$$u(x,t) = \phi^p(x,t)\sum_{j=0}^{\infty} u_j(t)\phi^j(x,t), \qquad v(x,t) = \phi^q(x,t)\sum_{j=0}^{\infty} v_j(t)\phi^j(x,t), \tag{7.2.36}$$

with $\phi(x,t) = x+\psi(t)$, where $\psi(t)$ is an arbitrary analytic function and $u_j(t)$, $v_j(t)$, $j = 0,1,2,\ldots$, are analytic functions, such that $u_0 \not\equiv 0$ and $v_0 \not\equiv 0$, in the neighborhood of a non-characteristic movable singularity manifold defined by $\phi(x,t) = 0$. It is also necessary to expand the function $a(x,t)$ about the singularity manifold as follows:

$$a(x,t) = \sum_{j=0}^{\infty} a_j(t)\phi^j(x,t), \qquad a_j(t) := \frac{1}{j!}\frac{\partial^j a}{\partial x^j}(x,t)\bigg|_{x=-\psi(t)},$$

and similarly for $b(x,t)$, $\bar{a}(x,t)$, $\bar{b}(x,t)$.

Leading order analysis shows that $p = q = -1$ and $u_0 v_0 = 1$. Equating the coefficients of powers of ϕ^{j-3} yields the general recursion relation

$$\mathbf{Q}(j)\begin{pmatrix} u_j \\ v_j \end{pmatrix} \equiv \begin{pmatrix} j^2 - 3j - 2 & -2u_0^2 \\ -2v_0^2 & j^2 - 3j - 2 \end{pmatrix}\begin{pmatrix} u_j \\ v_j \end{pmatrix} = \begin{pmatrix} F_j \\ G_j \end{pmatrix}, \tag{7.2.37a}$$

where

$$\begin{aligned} F_j =& 2\sum_{k=1}^{j-1}\sum_{l=0}^{k} u_l u_{k-l} v_{j-k} + 2v_0 \sum_{l=0}^{j-1} u_l u_{j-l} \\ & - \mathrm{i}(j-2)u_{j-1}\frac{\mathrm{d}\psi}{\mathrm{d}t} - \mathrm{i}\frac{\mathrm{d}u_{j-2}}{\mathrm{d}t} + \sum_{k=0}^{j-2} a_k u_{j-k-2} + b_{j-3}, \end{aligned} \tag{7.2.37b}$$

for $j \geq 1$ (define $u_j = 0 = v_j$ for $j < 0$) and G_j is obtained from F_j by interchanging u_j and v_j and letting $\mathrm{i} \to -\mathrm{i}$. These recursion relations uniquely define u_j, v_j unless

$$\det \mathbf{Q}(j) = (j+1)j(j-3)(j-4) = 0.$$

Therefore the resonances are $j = -1, 0, 3, 4$. The resonances $j = -1$ and $j = 0$ correspond to the fact that $\psi(t)$ is arbitrary and that there is only one equation defining u_0 and v_0 (so one, say u_0, is arbitrary), respectively.

For $j = 1$ and $j = 2$, we find that

$$u_1 = -\tfrac{1}{2}\mathrm{i}u_0\frac{\mathrm{d}\psi}{\mathrm{d}t}, \qquad v_1 = \tfrac{1}{2}\mathrm{i}v_0\frac{\mathrm{d}\psi}{\mathrm{d}t}, \tag{7.2.38a}$$

$$u_2 = -\tfrac{1}{12}u_0\left(\frac{\mathrm{d}\psi}{\mathrm{d}t}\right)^2 - \tfrac{1}{12}u_0 w_0 + \tfrac{1}{6}u_0(\bar{a}_0 - 2a_0), \tag{7.2.38b}$$

$$v_2 = -\tfrac{1}{12}v_0\left(\frac{\mathrm{d}\psi}{\mathrm{d}t}\right)^2 + \tfrac{1}{12}v_0 w_0 + \tfrac{1}{6}u_0(a_0 - 2\bar{a}_0), \tag{7.2.38c}$$

where

$$w_0 = u_0\frac{\mathrm{d}v_0}{\mathrm{d}t} - v_0\frac{\mathrm{d}u_0}{\mathrm{d}t}. \tag{7.2.38d}$$

Setting $j=3$ in (7.2.37), we see that

$$-2(v_0u_3+u_0v_3)=F_3v_0,$$
$$-2(v_0u_3+u_0v_3)=G_3u_0,$$

therefore these are compatible if and only if $F_3v_0=G_3u_0$, i.e., using (7.2.37,38), if and only if

$$a_1-\bar{a}_1+v_0b_0-u_0\bar{b}_0=0.$$

Since u_0 is arbitrary and $v_0=1/u_0$, then necessarily

$$a_1=\bar{a}_1, \qquad b_0=\bar{b}_0\equiv 0.$$

By definition $b_0(t)=b(-\psi(t),t)$, therefore since $\psi(t)$ is arbitrary, then necessarily

$$b(x,t)\equiv 0, \qquad \bar{b}(x,t)\equiv 0.$$

Similarly the compatibility condition for $j=4$ is

$$v_0F_4+u_0G_4=0=-\tfrac{1}{2}(a_0-\bar{a}_0)^2+\tfrac{1}{2}\mathrm{i}\frac{\mathrm{d}}{\mathrm{d}t}(a_0-\bar{a}_0)+a_2+\bar{a}_2. \tag{7.2.39}$$

Then by supposing that

$$a(x,t)=\alpha(x,t)+\mathrm{i}\beta(x,t), \qquad \bar{a}(x,t)=\alpha(x,t)-\mathrm{i}\beta(x,t),$$

it is easily seen that (7.2.39) holds only if

$$\alpha(x,t)=x^2\left(\tfrac{1}{2}\frac{\mathrm{d}\beta}{\mathrm{d}t}-\beta^2\right)+x\alpha_1(t)+\alpha_0(t), \qquad \beta(x,t)=\beta(t),$$

where $\alpha_1(t)$, $\alpha_0(t)$ and $\beta(t)$ are arbitrary functions. Hence the Painlevé PDE test suggests that necessary conditions for the generalized nonlinear Schrödinger equation (7.2.34) to be completely integrable are that

$$a(x,t)=x^2\left(\tfrac{1}{2}\frac{\mathrm{d}\beta}{\mathrm{d}t}-\beta^2\right)+\mathrm{i}\beta(t)+x\alpha_1(t)+\alpha_0(t), \qquad b(x,t)\equiv 0, \tag{7.2.40}$$

where $\alpha_1(t)$, $\alpha_0(t)$ and $\beta(t)$ are arbitrary functions (see Clarkson [1988] for further details). Unless $a(x,t)$ and $b(x,t)$ take these special forms [together with analogous ones for $\bar{a}(x,t)$ and $\bar{b}(x,t)$], then the series expansions (7.2.36) are inconsistent and logarithmic terms $\ln\phi$ are required at the order where the compatibility condition is not satisfied; at higher powers of ϕ, higher and higher powers of $\ln\phi$ are required, an indication of non-Painlevé behavior.

Additionally, it can be shown that (7.2.40) are necessary and sufficient conditions for the generalized nonlinear Schrödinger equation (7.2.34):

(a), to be mapped into the standard nonlinear Schrödinger equation

$$\mathrm{i}\eta_\tau + \eta_{\xi\xi} - 2|\eta|^2\eta = 0,$$

the transformation being

$$u(x,t) = \eta(\xi(x,t),\tau(t))\exp\left\{-\tfrac{1}{2}\mathrm{i}x^2\beta(t) + \mathrm{i}xf(t) + \mathrm{i}g(t) + 2\int^t \beta(s)\,\mathrm{d}s\right\},$$

with

$$\xi(x,t) = x\exp\left\{2\int^t \beta(s)\,\mathrm{d}s\right\} + h(t), \qquad \tau(t) = \int^t \exp\left\{4\int^s \beta(s')\,\mathrm{d}s'\right\} ds,$$

where

$$f(t) = -\exp\left\{2\int^t \beta(s)\,\mathrm{d}s\right\}\int^t \alpha_1(s)\exp\left(-2\int^s \beta(s')\,\mathrm{d}s'\right)\mathrm{d}s,$$
$$g(t) = -\int^t \{\alpha_0(s) + f^2(s)\}\,ds,$$
$$h(t) = -2\int^t f(s)\exp\left(2\int^s \beta(s')\,\mathrm{d}s'\right)\mathrm{d}s;$$

(b), to possess special 'soliton' solutions of the form

$$u(x,t) = A(x,t)\exp\{-\mathrm{i}\sigma(x,t)\}\,\mathrm{sech}\,\Theta(x,t),$$

where $A(x,t)$, $\sigma(x,t)$ and $\Theta(x,t)$ are (for real x and t), sufficiently differentiable functions given by

$$A(x,t) = \tau(t), \qquad \sigma(x,t) = \tfrac{1}{4}x^2\frac{\mathrm{d}}{\mathrm{d}t}(\ln\tau) + x\frac{\mathrm{d}\mu}{\mathrm{d}t} + \gamma(t), \qquad \Theta(x,t) = x\tau(t) + \mu(t),$$

where $\tau(t)$, $\mu(t)$ and $\gamma(t)$ are related to $\beta(t)$, $\alpha_1(t)$ and $\alpha_0(t)$ as follows:

$$\beta(t) = \tfrac{1}{2}\frac{\mathrm{d}}{\mathrm{d}t}(\ln\tau), \qquad \alpha_1(t) = \frac{\mathrm{d}^2\mu}{\mathrm{d}t^2} - \frac{\mathrm{d}\mu}{\mathrm{d}t}\frac{\mathrm{d}}{\mathrm{d}t}(\ln\tau), \qquad \alpha_0(t) = \tau^2(t) - \frac{\mathrm{d}\mu}{\mathrm{d}t} + \frac{\mathrm{d}\gamma}{\mathrm{d}t}$$

(see Clarkson [1988] for further details).

EXAMPLE 7.2.9 *A Generalized Derivative Nonlinear Schrödinger Equation*

In this example we consider the generalized derivative nonlinear Schrödinger equation

$$\mathrm{i}u_t = u_{xx} + \mathrm{i}\alpha u u^* u_x + \mathrm{i}\beta u^2 u_x^* + \gamma u^3 u^{*2}, \tag{7.2.41}$$

with α, β and γ real parameters and where u^* is the complex conjugate of u, which in quantum field theory (cf. Clarkson and Tuszyński [1990]), weakly nonlinear dispersive water waves (cf. Johnson [1977]), and nonlinear optics (cf. Kodama [1985]; Hasegawa [1990]). Here we use a combination of the Painlevé ODE and PDE tests to determine the choices of the parameters α, β and γ such that the equation (7.2.41) is completely integrable (modulo the validity of the tests); this problem has been considered by Clarkson and Cosgrove [1987] (using a slightly different approach).

First, we consider the scaling reduction

$$u(x,t) = (2t)^{-1/4} R(z) \exp\{\mathrm{i}\theta(z)\}, \qquad z = x/(2t)^{1/2}.$$

Then $R(z)$ and $\theta(z)$ satisfy the system of equations

$$-\tfrac{1}{2}R - zR' = R\theta'' + 2R'\theta' + (\alpha+\beta)R^2 R', \tag{7.2.42a}$$

$$zR\theta' = R'' - R(\theta')^2 + (\beta-\alpha)R^3\theta' + \gamma R^5, \tag{7.2.42b}$$

with $' := \mathrm{d}/\mathrm{d}z$. Equation (7.2.42a) can be integrated once after first multiplying by R, which expresses θ' in terms of R^2. Substituting for θ' in (7.2.42b) and defining $w := R^2$ gives the second order equation

$$ww'' = (w')^2 + \tfrac{3}{2}\delta w^4 - \tfrac{1}{2}z^2 w^2 + \tfrac{1}{2}(\beta-\alpha)zw^3 + A(\alpha-3\beta)w^2 + A^2, \tag{7.2.43a}$$

with

$$\delta = \tfrac{1}{12}(5\beta^2 - 3\alpha^2 + 2\alpha\beta - 16\gamma), \tag{7.2.43b}$$

and where A is an arbitrary constant. It is easily shown that this equation is of Painlevé type if and only if

$$\gamma = \tfrac{1}{4}\beta(2\beta-\alpha), \tag{7.2.44}$$

in which case a suitable rescaling of the variables yields the fourth Painlevé equation. Therefore, the Painlevé ODE test suggests that (7.2.44) is a *necessary* condition for equation (7.2.41) to be completely integrable.

Next, we apply the Painlevé PDE test to the equation

$$\mathrm{i}u_t = u_{xx} + \mathrm{i}\alpha u u^* u_x + \mathrm{i}\beta u^2 u_x^* + \tfrac{1}{4}\beta(2\beta-\alpha)u^3 u^{*2}, \tag{7.2.45}$$

by considering the system

$$\mathrm{i}u_t = u_{xx} + \mathrm{i}\alpha uvu_x + \mathrm{i}\beta u^2 v_x + \tfrac{1}{4}\beta(2\beta-\alpha)u^3v^2, \tag{7.2.46a}$$

$$-\mathrm{i}v_t = v_{xx} + \mathrm{i}\alpha vuv_x + \mathrm{i}\beta v^2 u_x + \tfrac{1}{4}\beta(2\beta-\alpha)v^3u^2, \tag{7.2.46b}$$

and seeking solutions in the form

$$u(x,t) = \phi^p(x,t)\sum_{j=0}^{\infty} u_j(t)\phi^j(x,t), \qquad v(x,t) = \phi^q(x,t)\sum_{j=0}^{\infty} v_j(t)\phi^j(x,t), \tag{7.2.47}$$

with $\phi(x,t) = x + \psi(t)$, where $\psi(t)$ is an arbitrary analytic function and $u_j(t)$, $v_j(t)$, $j = 0, 1, 2, \ldots$, are analytic functions, such that $u_0 \not\equiv 0$ and $v_0 \not\equiv 0$, in the neighborhood of a non-characteristic movable singularity manifold defined by $\phi = 0$. Leading order analysis shows that

$$p = \frac{\alpha}{\alpha - \beta}, \qquad q = \frac{-\beta}{\alpha - \beta}, \qquad u_0 v_0 = \frac{-2\mathrm{i}}{\alpha - \beta},$$

provided that $\alpha \neq \beta$, and the general recursion relations take the form

$$\mathbf{Q}(j)\begin{pmatrix} u_j \\ v_j \end{pmatrix} = \begin{pmatrix} F_j(u_0, v_0, u_1, v_1, \ldots, u_{j-1}, v_{j-1}, \phi) \\ G_j(u_0, v_0, u_1, v_1, \ldots, u_{j-1}, v_{j-1}, \phi) \end{pmatrix},$$

for some functions F_j and G_j, where $\det \mathbf{Q}(j) = (j+1)j(j-2)(j-3)$, for $j \geq 1$. Therefore the resonances are $j = -1, 0, 2, 3$. For the resonances $j = 0$, $j = 2$ and $j = 3$, it turns out that there is only one independent equation defining u_j and v_j and so one of them is arbitrary (see Clarkson and Cosgrove [1987] for further details). Therefore equations (7.2.46) have solutions of the form (7.2.47) with the requisite number of arbitrary functions and so pass the Painlevé PDE test. [Clarkson and Cosgrove [1987] applied the Painlevé PDE test directly on equation (7.2.41), after complexifying all variables, and found that equation (7.2.44) is a necessary condition for solutions of the form (7.2.47) to exist.)]

We remark that the only restriction upon α and β is $\alpha \neq \beta$, so p may take *any* real value, rational or irrational. Therefore, unless p is an integer, the variables u and v exhibit non-Painlevé behavior at leading order (movable fractional or irrational powers according as to whether p is rational or irrational). However this non-Painlevé behavior is trivially removable, for example, by transforming to the R and θ polar variables and so has no bearing on the question of integrability. Furthermore, we note that the leading order powers of u and v are *not* equal, in general, since we are not, for the purposes of Painlevé analysis, restricting attention to the original situation where u and v are mutually complex conjugate functions of real variables.

Therefore the Painlevé tests suggest that the generalized nonlinear Schrödinger equation (7.2.41) can be completely integrable only if it takes the special form (7.2.45). This equation is a generalization of the derivative nonlinear Schrödinger (DNLS) equation considered by Kaup and Newell [1978a]

$$\mathrm{i}u_t = u_{xx} + \mathrm{i}\beta(|u|^2 u)_x, \tag{7.2.48a}$$

which we call DNLSI; that considered by Chen, Lee and Liu [1979]

$$iu_t = u_{xx} + i\alpha uu^* u_x, \tag{7.2.48b}$$

which we call DNLSII; and that considered by Gerjikov and Ivanov [1982] (cf. Kundu [1984])

$$iu_t = u_{xx} + +i\beta u^2 u_x^* + \tfrac{1}{2}\beta^2 u^3 u^{*2}, \tag{7.2.48c}$$

which we call DNLSIII. In equation (7.2.45), if we make the transformation

$$u(x,t) = U(x,t)\exp\{ik\Theta(x,t)\}, \qquad u^*(x,t) = U^*(x,t)\exp\{-ik\Theta(x,t)\}, \tag{7.2.49a}$$

where

$$\Theta_x(x,t) := u(x,t)u^*(x,t), \tag{7.2.49b}$$

then

$$\begin{aligned} iU_t = U_{xx} &+ i(\alpha + 2k)UU^*U_x + i(\beta + 2k)U^2U_x^* \\ &+ \tfrac{1}{4}(\beta + 2k)(2\beta - \alpha + 2k)U^3U^{*2}. \end{aligned} \tag{7.2.50}$$

Hence choosing $k = -\frac{1}{2}\beta$, $k = \frac{1}{2}\alpha - \beta$ and $k = -\frac{1}{2}\alpha$ in (7.2.50) gives DNLSI, DNLSII and DNLSIII, respectively, provided that $\alpha \neq \beta$; in the case $\alpha = \beta$, choosing $k = -\frac{1}{2}\alpha$ transforms (7.2.45) into a linear partial differential equation. This shows that the DNLS equation (7.2.45) is a trivial generalization since it may be mapped into DNLSI, DNLSII and DNLSIII through a gauge transformation, provided that $\alpha \neq \beta$ (see also Kundu [1984, 1987]), and so these four equations are equivalent. [The equivalence of DNLSI and DNLSII was apparently first noticed by Wadati and Sogo [1983]; although, the transformation is implied in the work of Kaup and Newell [1978a] — see their equations (11) and (12).] Since DNLSI was solved by inverse scattering by Kaup and Newell [1978a] (see also Kawata and Inoue [1978]; Kawata, Sakai and Kobayashi [1980]), then equation (7.2.45) is also completely integrable since it is equivalent to DNLSI (7.2.48a).

Clarkson and Cosgrove [1987] also showed that the generalized mixed nonlinear Schrödinger equation

$$iu_t = u_{xx} + i\alpha uu^*u_x + i\beta u^2u_x^* + \gamma u^3u^{*2} + \kappa u^2u^*, \tag{7.2.51}$$

with α, β, γ, κ real parameters, can also pass the Painlevé PDE test only if (7.2.44) is satisfied. The point transformation

$$u(x,t) = \eta(\xi,\tau)\exp\{i(kx + k^2t)\}, \qquad u^*(x,t) = \eta^*(\xi,\tau)\exp\{-i(kx + k^2t)\},$$

with

$$\xi := x + 2kt, \qquad \tau := t, \qquad k = \frac{\kappa}{\alpha - \beta}, \qquad \alpha \neq \beta,$$

maps (7.2.51) into

$$\mathrm{i}\eta_\tau = \eta_{\xi\xi} + \mathrm{i}\alpha\eta\eta^*\eta_\xi + \mathrm{i}\beta\eta^2\eta_\xi^* + \gamma\eta^3\eta^{*2},$$

which is equation (7.2.41). [In particular, this shows that the mixed nonlinear Schrödinger equation (cf. Wadati, Konno and Ichikawa [1979a])

$$\mathrm{i}u_t = u_{xx} + \mathrm{i}\beta(u^2u^*)_x + \kappa u^2u^*,$$

is equivalent to DNLSI (7.2.48a).] Hence we conclude that the Painlevé tests suggest that the generalized mixed nonlinear Schrödinger equation is completely integrable only if (7.2.44) is satisfied.

EXAMPLE 7.2.10 *A generalized variable-coefficient Kadomtsev-Petviashvili equation*

In this example we consider the generalized variable-coefficient Kadomtsev-Petviashvili equation (GKP)

$$\begin{aligned}(u_t + uu_x + u_{xxx})_x + a(y,t)u_x + b(y,t)u_y \\ + c(y,t)u_{yy} + d(y,t)u_{xy} + e(y,t)u_{xx} = 0,\end{aligned} \tag{7.2.52}$$

where $a(y,t)$, $b(y,t)$, $c(y,t)$, $d(y,t)$ and $e(y,t)$ are analytic functions, which arises in the description of surface waves in straits or channels with varying depth and width (cf. David, Levi and Winternitz [1987a, 1989]). Here we determine the choices of the functions $a(y,t)$, $b(y,t)$, $c(y,t)$, $d(y,t)$ and $e(y,t)$ such that the equation (7.2.52) passes the Painlevé PDE test. We seek solutions of (7.2.52) in the form

$$u(x,y,t) = \phi^p(x,y,t)\sum_{j=0}^{\infty} u_j(y,t)\phi^j(x,y,t), \qquad \phi(x,y,t) = x + \psi(y,t), \tag{7.2.53}$$

where $\psi(y,t)$ is an arbitrary function and $u_j(y,t)$, $j = 0,1,2,\ldots$, are analytic functions with $u_0 \not\equiv 0$, in the neighborhood of a non-characteristic movable singularity manifold defined by $\phi(x,y,t) = 0$. Leading order analysis shows that $p = -2$ and $u_0 = -12$; equating coefficients of like powers of ϕ yields the general recursion relation

$$(j+1)(j-4)(j-5)(j-6)u_j + F_j = 0, \tag{7.2.54a}$$

where

$$\begin{aligned}F_j = \tfrac{1}{2}(j-4)(j-5)\sum_{k=1}^{j-1} u_k u_{j-k} + (j-4)(j-5)u_{j-2}(\psi_t + c\psi_y^2 + d\psi_y + e) \\ + (j-5)u_{j-3}(a + b\psi_y + c\psi_{yy}) + (j-5)[u_{j-3,t} + (2c\psi_y + d)u_{j-3,y}] \\ + bu_{j-4,y} + cu_{j-4,yy},\end{aligned} \tag{7.2.54b}$$

for $j \geq 1$ (define $u_j = 0$ for $j < 0$). Thus the resonances are $j = -1, 4, 5, 6$; the recursion relations (7.2.54) are consistent provided that $F_j \equiv 0$ for $j = 4, 5, 6$ (the compatibility conditions). From (7.2.54) we obtain

$$u_1 = 0, \qquad u_2 = -(\psi_t + c\psi_y^2 + d\psi_y + e), \qquad u_3 = a + b\phi_y + c\psi_{yy}, \tag{7.2.55}$$

and it is easily seen that $F_4 \equiv 0$ and $F_5 \equiv 0$. The compatibility condition $F_6 \equiv 0$, after using (7.2.55), is given by

$$\begin{aligned}(a_t + 2a^2 + da_y - be_y - ce_{yy}) + (b_t + 4ab - bd_y - cd_{yy} + 2ca_y + b_y d)\psi_y \\ + (c_t + 4ac - 2cd_y + dc_y)\psi_{yy} + (2b^2 - bc_y - cc_{yy} + 2cb_y)\psi_y^2 \\ + 2c(2b - c_y)\psi_y\psi_{yy} = 0.\end{aligned}$$

Since $\psi(y,t)$ is an arbitrary function, then the derivatives of ψ are independent and so a, b, c, d and e must satisfy the system of equations

$$a_t + 2a^2 + da_y - be_y - ce_{yy} = 0, \tag{7.2.56a}$$
$$b_t + 4ab - bd_y - cd_{yy} + 2ca_y + b_y d = 0, \tag{7.2.56b}$$
$$c_t + 4ac - 2cd_y + dc_y = 0, \tag{7.2.56c}$$
$$2b^2 - bc_y - cc_{yy} + 2cb_y = 0, \tag{7.2.56d}$$
$$2c(2b - c_y) = 0. \tag{7.2.56e}$$

(7.2.56e) implies that either (i), $c = 0$; or (ii), $b = \frac{1}{2}c_y$.

Case (i). If $c = 0$, then equation (7.2.6d) implies $b = 0$ and the problem is equivalent to Example 7.2.6 above (we may set $d = e = 0$, without loss of generality, and so y only plays the role of a parameter).

Case (ii). If $b = \frac{1}{2}c_y$, then equation (7.2.56d) is identically satisfied and equation (7.2.56b) is the (partial) y-derivative of equation (7.2.56c). Hence we require that a, c, d and e satisfy

$$a_t + 2a^2 + da_y - \tfrac{1}{2}c_y e_y - ce_{yy} = 0, \tag{7.2.57a}$$
$$c_t + 4ac - 2cd_y + dc_y = 0. \tag{7.2.57b}$$

Therefore $a(y,t)$ and $c(y,t)$ can be chosen arbitrarily and $d(y,t)$ and $e(y,t)$ are given by

$$\begin{aligned}d(y,t) &= c^{1/2}(y,t)[2f(y,t) + g(y,t) + d_0(t)],\\ e(y,t) &= f^2(y,t) + \int_{y_0}^{y}\left[f_y(\xi,t)g(\xi,t) + f_t(\xi,t)c^{-1/2}(\xi,t)\right]\mathrm{d}\xi\\ &\qquad + f(y,t)\mathrm{d}_0(t) + e_1(t)\int_{y_0}^{y} c^{-1/2}(\xi,t)\,\mathrm{d}\xi + e_0(t),\end{aligned}$$

where

$$f(y,t) = \int_{y_0}^{y} a(\xi,t)c^{-1/2}(\xi,t)\,\mathrm{d}\xi,$$
$$g(y,t) = \tfrac{1}{2}\int_{y_0}^{y} c^{-3/2}(\xi,t)c_t(\xi,t)\,\mathrm{d}\xi,$$

with y_0 an arbitrary constant and $d_0(t)$, $e_1(t)$ and $e_0(t)$ arbitrary functions. If otherwise, the series expansion (7.2.53) is inconsistent and logarithmic terms are required, an indication of non-Painlevé behaviour.

Furthermore, it can be shown that the GKP equation (7.2.52) can be transformed into the standard KP equation

$$(u_t + 6uu_x + u_{xxx})_x + 3\sigma^2 u_{yy} = 0,$$

only if equations (7.2.56) are satisfied. These results strongly suggest that the GKP equation (7.2.52) is not be completely integrable unless (7.2.57) is satisfied (for further details see Clarkson [1990a]).

Neither the MBBM equation (7.2.19) nor the SRLW equation (7.2.21) pass the Painlevé PDE test (Clarkson [1989]), which might suggest that the Painlevé PDE test may provide a better criterion for complete integrability than the Painlevé ODE test (which was inconclusive for the MBBM equation). However, it is not sufficient to merely seek solutions in the form of a Laurent series (7.2.23a), since such an expansion can miss some essential singularities.

EXAMPLE 7.2.11

Consider the partial differential equation

$$u_t = (1+u^2)u_{xx} + (1-2u)u_x^2. \tag{7.2.58}$$

We seek a solutions of this in the form

$$u(x,t) = \phi^p(x,t)\sum_{j=0}^{\infty} u_j(t)\phi^j(x,t), \qquad \phi(x,t) = x + \psi(t), \tag{7.2.59}$$

in the neighborhood of a non-characteristic singularity manifold defined by $\phi(x,t)=0$, where $\psi(t)$, $u_j(t)$, $j=0,1,\ldots$, are analytic functions, with $u_0 \not\equiv 0$. By leading order analysis we find that there are two choices for p, namely $p=0$ and $p=-1$. The choice $p=0$ leads to the regular solution and for $p=-1$, by equating coefficients of powers of ϕ, we find that $u_0(t)$ is arbitrary and $u_j(t)$, for $j \geq 1$, is given by

$$j(j+1)u_0^2 u_j = F_j(u_0, u_1, \ldots, u_{j-1}, \psi),$$

for some function F_j. Therefore equation (7.2.58) has solutions of the form (7.2.59) with the requisite number of arbitrary functions. However, time-independent solutions, $u(x,t) = y(x)$, satisfy

$$(1+y^2)\frac{\mathrm{d}^2y}{\mathrm{d}x^2} + (1-2y)\left(\frac{\mathrm{d}y}{\mathrm{d}x}\right)^2 = 0. \tag{7.2.60}$$

This has general solution

$$y(x) = \tan\{\ln(\alpha x + \beta)\},$$

with α and β arbitrary constants and so has a movable essential singularity at $x = -\beta/\alpha$, consequently (7.2.60) is not of Painlevé type. It is clearly possible to make a transformation to equation (7.2.60) in order to make it of Painlevé type, however no such transformation was required in order to ensure that the partial differential equation (7.2.58) has solutions of the form (7.2.59). Recently Weiss [1990] has shown that equation (7.2.58) also possesses solutions of the form

$$u(x,t) = \pm\mathrm{i} + u_0(t)\phi^{2\mathrm{i}} + \dots\,, \qquad \phi(x,t) = x + \psi(t),$$

where $u_0(t)$ is arbitrary in the neighborhood of the singularity manifold defined by $\phi = 0$ (the resonances are $j = -1$ and $j = 0$).

Additionally, equation (7.2.58) is not "exactly solvable" in the sense given by Fokas [1980] (i.e., a symmetry approach), since the most general partial differential equation of the form

$$u_t = g(u)u_{xx} + f(u, u_x),$$

which has "higher polynomial symmetries" is

$$u_t = g(u)u_{xx} + \left(\frac{gg''}{g'} - \frac{g'}{2}\right)u_x^2 + \alpha g(u)u_x,$$

where $' := \mathrm{d}/\mathrm{d}u$, $f(u)$ is any smooth function of u and α is a constant (Fokas and Yortsos [1982]). It is easily seen that (7.2.58) is not of this form (see also Clarkson [1985]).

Finally, there also exist partial differential equations which are reduced by a similarity reduction to an ordinary differential equation whose general solution possesses a movable natural boundary. For example, consider the equation

$$u_t = u_{xxx} + 3u_x^2 - 2uu_{xx}; \tag{7.2.61}$$

time-independent solutions $u(x,t) = y(x)$, satisfy the Chazy equation

$$\frac{\mathrm{d}^3y}{\mathrm{d}x^3} = 2y\frac{\mathrm{d}^2y}{\mathrm{d}x^2} - 3\left(\frac{\mathrm{d}y}{\mathrm{d}x}\right)^2, \tag{7.2.62}$$

the general solution of which possesses a movable natural boundary, as shown in §§6.5 and 7.1.5 above. It is an open question as to the influence of the movable natural boundary on solutions of equation (7.2.61). One possible method in attempting to understand this would be to study the equation

$$u_t = u_{xxx} + 3u_x^2 - \beta u u_{xx}, \tag{7.2.63}$$

with β a parameter (note that if $\beta = 0$, then equation (7.2.63) is the potential KdV equation). This equation possesses the travelling wave solution

$$u(x,t) = A\tanh\{\kappa(x-ct)+x_0\},$$

with

$$A = \frac{6\kappa}{3-2\beta}, \qquad \kappa^2 = \frac{c(2\beta+3)}{4(\beta+3)}.$$

A numerical study of the interaction of two solitary waves for various values of β might provide some insight as to the effect of the movable natural boundary.

7.2.6 Quasilinear Partial Differential Equations and the Painlevé Tests. Most of the well-known completely integrable partial differential equations are of the form

$$u_t = u_{nx} + f(u, u_x, u_{xx}, \ldots, u_{(n-1)x}), \qquad n \geq 2, \quad u_{mx} := \frac{\partial^m u}{\partial x^m},$$

and then the equation is said to be *semilinear*. However there also exist completely integrable partial differential equations of the form

$$u_t = g(u)u_{nx} + f(u, u_x, u_{xx}, \ldots, u_{(n-1)x}), \qquad n \geq 2, \tag{7.2.64}$$

where $\mathrm{d}g/\mathrm{d}u \not\equiv 0$, in which case the equation is said to be *quasilinear*. Additionally, we say that two partial differential equations are *equivalent* if one can be obtained from the other by a transformation involving the dependent variables $u = \phi(v)$ and/or the introduction of a potential variable ($u = v_x$ or $u_x = v$).

The application of the Painlevé tests to quasilinear partial differential equations appears not to be as straightforward as to semilinear partial differential equations since some completely integrable quasilinear partial differential equations appear to be only "*weak-Painlevé*" in the sense that they have expansions in the form

$$u(\boldsymbol{z}) = \phi^{p/q}(\boldsymbol{z}) \sum_{j=0}^{\infty} u_j(\boldsymbol{z})\phi^{j/q}(\boldsymbol{z}),$$

where $\phi(\boldsymbol{z})$ and $u_j(\boldsymbol{z})$, $j = 0, 1, \ldots$, are analytic functions of $\boldsymbol{z} = (z_1, z_2, \ldots, z_n)$, with $u_0 \not\equiv 0$ and p, q are integers determined by leading order analysis, in the neighborhood of a non-characteristic, movable singularity manifold defined by $\phi(\boldsymbol{z}) = 0$. For example, the Dym equation (Kruskal [1975])

$$u_t + 2(u^{-1/2})_{xxx} = 0, \tag{7.2.65}$$

which is known to be completely integrable (Wadati, Konno and Ichikawa [1979] — see also Calogero and Degasperis [1982]), is "weak-Painlevé" since it has solutions of the form

$$u(x,t) = \phi^{-4/3}(x,t)\sum_{j=0}^{\infty} u_j(t)\phi^{j/3}(x,t), \qquad \phi(x,t) = x + \psi(t),$$

with $u_0 \not\equiv 0$, in the neighborhood of a non-characteristic, movable singularity manifold defined by $\phi(x,t) = 0$ and so has a movable cube root (Weiss [1983]). We remark that the Dym equation (7.2.65) may be transformed into both the KdV equation (Weiss [1983]; Levi, Ragnisco and Sym [1984]) and the mKdV equation (Kawamoto [1985]) using an *extended hodograph transformation*, i.e., a transformation involving the interchange of independent and dependent variables of the form

$$\tau = t, \qquad \xi = \int^x f(u(y,t))\,\mathrm{d}y, \qquad \eta(\xi,\tau) = x, \tag{7.2.66}$$

which is a generalization of the standard hodograph transformation (cf. Clarkson, Fokas and Ablowitz [1989]).

However the higher nonlinear KdV equation

$$u_t + u^3 u_x + u_{xxx} = 0, \tag{7.2.67}$$

is also "weak Painlevé" (Weiss [1986a]), yet it is thought *not* to be completely integrable since (a), it has only three independent polynomial conservation laws (Miura [1976]); (b), the interaction of solitary wave solutions is inelastic (Fornberg and Whitham [1978]); and (c), it appears not to be solvable by inverse scattering (McLeod and Olver [1983]). Therefore the "weak-Painlevé" concept does not appear to distinguish between integrable and non-integrable partial differential equations.

Recall that the Painlevé tests require that an integrable partial differential equation possesses the Painlevé property *possibly only after a transformation of variables*, so that we may first have to make a change of variables before applying the tests. An open question remains as to what kind of transformations are allowable in the application of the Painlevé tests (i.e., which transformations does one have to check?). It seems that completely integrable quasilinear partial differential equations such as the Dym equation (7.2.65), which are "weak-Painlevé" can be transformed into a partial differential equation which have the "full-Painlevé" property through an appropriate hodograph transformation (for further examples, see Clarkson and Cosgrove [1987]; Clarkson, Fokas and Ablowitz [1989]).

Recently, Clarkson, Fokas and Ablowitz [1989] developed an algorithmic method for transforming a quasilinear partial differential equations into semilinear partial differential equations. For equations of the form (7.2.64), this method involves two steps; first one puts the equation into its potential canonical form

$$v_t = v_x^{-n} v_{nx} + F(v_x, v_{xx}, \ldots, v_{(n-1)x}), \tag{7.2.68}$$

by making the transformation $v_x = g^{-1/n}(u)$; and then one applies a *pure hodograph transformation*, i.e. a transformation of the form

$$\tau = t, \qquad \xi = u(x,t), \qquad \eta(\xi,\tau) = x \tag{7.2.69}$$

[note that this is a special case of the extended hodograph transformation (7.2.66)]. If the equation (7.2.68) is transformable to a semilinear equation, then after making the transformation (7.2.69) it will be of the form

$$\eta_\tau = \eta_{n\xi} + H(\eta_\xi, \eta_{\xi\xi}, \ldots, \eta_{(n-1)\xi}). \tag{7.2.70}$$

Since this is a semilinear, then one can usually apply the Painlevé tests in a straightforward manner.

Using this approach, Clarkson, Fokas and Ablowitz [1989] prove that the most general equation of the form (7.2.64) which can be transformed via an extended hodograph transformation to a semilinear equation of the form (7.2.70) is given by

$$\begin{aligned} u_t = g(u)u_{nx} &+ \left(\frac{g''}{g'} - \frac{n+1}{n}\frac{g'}{g}\right) B(u, u_x, \ldots, u_{(n-2)x})u_x \\ &+ B_u u_x + \sum_{r=1}^{n-2} B_{u_{rx}} u_{(r+1)x} + \left(\frac{gg''}{g'} - \frac{g'}{n}\right) u_x u_{(n-1)x}, \end{aligned}$$

where prime denotes derivative with respect to u, and $g(u)$ and $B(u, u_x, \ldots, u_{(n-2)x})$ are arbitrary functions.

EXAMPLE 7.2.12

Consider the equation

$$u_t + 2(u^{-1/2})_{xxx} + f'(u^{1/2})u_x = 0, \tag{7.2.71}$$

where f is a rational function and prime denotes differentiation with respect to the argument. The objective is to determine for which choices of f is (7.2.71) integrable [note that if $f' \equiv 0$, then (7.2.71) is the Dym equation (7.2.65)]. First we put (7.2.71) into canonical form by making the transformation $v_x = u^{1/2}$, hence we obtain

$$v_t = v_x^{-3}v_{xxx} - \tfrac{3}{2}v_x^{-4}v_{xx}^2 - f(v_x).$$

Applying a pure hodograph transformation (7.2.69) to this gives

$$\eta_\tau = \eta_{\xi\xi\xi} - \tfrac{3}{2}\eta_{\xi\xi}^2\eta_\xi^{-1} + \eta_\xi f(\eta_\xi^{-1}),$$

which has the non-potential form ($w = \eta_\xi$)

$$w_\tau = w_{\xi\xi\xi} - \tfrac{3}{2}\left(w^{-1}w_\xi^2\right)_\xi + g(w)w_\xi, \tag{7.2.72a}$$

where

$$g(w) := \frac{\mathrm{d}}{\mathrm{d}w}\left[wf(w^{-1})\right]. \tag{7.2.72b}$$

It can be shown that (7.2.72) can pass the Painlevé tests if and only if

$$g(w) = 3\alpha w^2 + \beta + \gamma w^{-2}, \tag{7.2.73}$$

hence,

$$f(w) = \alpha w^{-2} + \beta - \gamma w^2 + \delta w, \tag{7.2.74}$$

with α, β, γ and δ arbitrary constants. Equation (7.2.72) with $g(w)$ as given by (7.2.73) is equivalent to the Calogero-Degasperis-Fokas equation

$$v_\tau = v_{\xi\xi\xi} - \tfrac{1}{8}v_\xi^3 + (ae^v + be^{-v} + c)v_\xi,$$

with a, b, c constants, if either $\alpha \neq 0$ or $\gamma \neq 0$ [let $w = \exp(v/2)$], which is known to be completely integrable (Calogero and Degasperis [1981]), whilst if $\alpha = \gamma = 0$ and $q = w_\xi/w$, then q satisfies the mKdV equation. Therefore we conclude equation (7.2.72) with $g(w)$ as given by (7.2.73) is completely integrable and so the most general equation of the form (7.2.71) which is linearizable is given by

$$u_t + 2(u^{-1/2})_{xxx} - 2\gamma u^{1/2}u_x - 2\alpha u^{-3/2}u_x + \delta u_x = 0.$$

To summarize, Painlevé ODE and PDE tests have proven to provide useful criterion for determining whether a given partial differential equation might be completely integrable. However, at present, they can only be regarded as "rules of thumb" since there are several aspects of their application which are not satisfactorily resolved. One of the most important of these is the question of transformations. It is known that the Painlevé property is not invariant to a transformation, so how do we know if it is necessary to make a transformation in order to apply either of the Painlevé tests, and if so, what is the desired transformation? What transformations do we have to check? We believe that extended hodograph transformations and the notion of equivalence may be useful tools in this direction. Another important question concerns single-valuedness, especially in connection with equations such as

$$\left[y\frac{\mathrm{d}^2y}{\mathrm{d}x^2} - \left(\frac{\mathrm{d}y}{\mathrm{d}x}\right)^2\right]^2 + 4x\left(\frac{\mathrm{d}y}{\mathrm{d}x}\right)^3 = 0,$$

which has the single-valued general solution $y(x) = \alpha\exp\{(x-x_0)^{-1}\}$, and the Chazy equation

$$\frac{\mathrm{d}^3y}{\mathrm{d}x^3} = 2y\frac{\mathrm{d}^2y}{\mathrm{d}x^2} - 3\left(\frac{\mathrm{d}y}{\mathrm{d}x}\right)^2,$$

whose solution is also single-valued in its domain of definition. Strictly, neither of these two equations possesses the Painlevé property, yet both, in some sense, may be regarded as being "integrable". Despite the "success" of the Painlevé tests, the lack of rigor and absence of precise definitions are notable limitations in their effectiveness. For further discussion on the limitations and flexibility required in the application of the Painlevé tests, the reader is advised to consult, for example, Kruskal [1991]; Kruskal and Clarkson [1991]; Kruskal and Joshi [1991].

7.3 Inverse Problems for the Painlevé Equations.

In this section we discuss inverse problems associated with the Painlevé equations. The paradigm is the second Painlevé equation ($\mathrm{P_{II}}$)

$$\frac{\mathrm{d}^2\eta}{\mathrm{d}\xi^2} = 2\eta^3 + \xi\eta + \mu, \tag{7.3.1}$$

where μ is a constant, as the prototype example. There are two approaches in order to develop inverse problems for the Painlevé equations, either through a Gel'fand-Levitan-Marchenko integral equation, or through the Inverse Monodromic Transform method (I.M.T.) which can be viewed as the nonlinear analog of Laplace transform method. Since $\mathrm{P_{II}}$ arises through the self-similar reduction

$$u(x,t) = (3t)^{-1/3}\eta(\xi), \qquad \xi = x/(3t)^{1/3}, \tag{7.3.2}$$

of the Modified KdV (mKdV) equation

$$u_t - 6u^2 u_x + u_{xxx} = 0, \tag{7.3.3}$$

both approaches may be derived from the I.S.T. method for the mKdV equation we shall begin by briefly outlining this.

7.3.1 Inverse Scattering for the Modified KdV Equation. The mKdV equation (7.3.3) arises as the compatibility condition of the following linear systems

$$\boldsymbol{v}_x = \begin{pmatrix} -\mathrm{i}k & u \\ u & \mathrm{i}k \end{pmatrix} \boldsymbol{v}, \tag{7.3.4a}$$

$$\boldsymbol{v}_t = \begin{pmatrix} -4\mathrm{i}k^3 - 2\mathrm{i}u^2k & 4k^2u + 2\mathrm{i}ku_x - u_{xx} + 2u^3 \\ 4k^2u - 2\mathrm{i}ku_x - u_{xx} + 2u^3 & 4\mathrm{i}k^3 + 2\mathrm{i}u^2k \end{pmatrix} \boldsymbol{v}, \tag{7.3.4b}$$

where $\boldsymbol{v} = (v_1, v_2)^T$. We review the I.S.T. method for the mKdV equation via a slightly different approach than discussed in §3.1.

The eigenfunctions $\boldsymbol{M}(x;k)$, $\bar{\boldsymbol{M}}(x;k)$, $\boldsymbol{N}(x;k)$, $\bar{\boldsymbol{N}}(x;k)$ satisfy the following asymptotic relations as $|x| \to \infty$

$$\begin{aligned}
\boldsymbol{M}(x;k) &= \phi(x;k)\mathrm{e}^{\mathrm{i}kx} &&\sim \begin{pmatrix}1\\0\end{pmatrix} &&\text{as} && x \to -\infty,\\
\bar{\boldsymbol{M}}(x;k) &= -\bar{\phi}(x;k)\mathrm{e}^{-\mathrm{i}kx} &&\sim \begin{pmatrix}0\\1\end{pmatrix} &&\text{as} && x \to -\infty,\\
\boldsymbol{N}(x;k) &= \psi(x;k)\mathrm{e}^{-\mathrm{i}kx} &&\sim \begin{pmatrix}0\\1\end{pmatrix} &&\text{as} && x \to \infty,\\
\bar{\boldsymbol{N}}(x;k) &= \bar{\psi}(x;k)\mathrm{e}^{\mathrm{i}kx} &&\sim \begin{pmatrix}1\\0\end{pmatrix} &&\text{as} && x \to \infty,
\end{aligned}$$

where the linearly independent solutions of (7.3.4a) satisfy

$$\phi(x;k) = a(k)\bar{\psi}(x;k) + b(k)\psi(x;k), \tag{7.3.5a}$$

$$\bar{\phi}(x;k) = -\bar{a}(k)\psi(x;k) + \bar{b}(k)\bar{\psi}(x;k). \tag{7.3.5b}$$

From the associated integral equations, it is easily shown that $\boldsymbol{M}(x;k)$, $\boldsymbol{N}(x;k)$ and $a(k)$ are analytic for $\operatorname{Im} k > 0$ and $\bar{\boldsymbol{M}}(x;k)$, $\bar{\boldsymbol{N}}(x;k)$ and $\bar{a}(k)$ are analytic for $\operatorname{Im} k < 0$ (see equations (3.1.7,16) in §3.1). Therefore (7.3.5) is equivalent to the following matrix Riemann-Hilbert boundary value problem

$$\mathbf{M}_{+}(x;k) = \begin{pmatrix} 1+\rho(k)\bar{\rho}(k) & \rho(k)\mathrm{e}^{2\mathrm{i}kx} \\ \bar{\rho}(k)\mathrm{e}^{-2\mathrm{i}kx} & 1 \end{pmatrix} \mathbf{M}_{-}(x;k), \tag{7.3.6a}$$

$$\equiv \mathbf{G}(x;k)\mathbf{M}_{-}(x;k), \tag{7.3.6b}$$

where $\mathbf{M}_{\pm} \to \mathbf{I}$, the 2×2 identity matrix, as $|k| \to \infty$,

$$\rho(k) := \frac{a(k)}{b(k)}, \qquad \bar{\rho}(k) := \frac{\bar{a}(k)}{\bar{b}(k)},$$

and the 2×2 matrices $\mathbf{M}_{\pm}$ are made up of the above column vectors

$$\mathbf{M}_{+}(x;k) = \left(\frac{\boldsymbol{M}(x;k)}{a(k)}, \boldsymbol{N}(x;k)\right)^{T}, \qquad \mathbf{M}_{-}(x;k) = \left(\bar{\boldsymbol{N}}(x;k), \frac{\bar{\boldsymbol{M}}(x;k)}{\bar{a}(k)}\right)^{T},$$

where T denotes the transpose. Directly from this we can obtain a linear integral equation for the matrix $\mathbf{M}_{-}(x;k)$, where as $|k| \to \infty$

$$a, \bar{a} \to 1, \qquad \boldsymbol{M}, \bar{\boldsymbol{N}} \to \begin{pmatrix}1\\0\end{pmatrix}, \qquad \bar{\boldsymbol{M}}, \boldsymbol{N} \to \begin{pmatrix}0\\1\end{pmatrix}$$

(we shall assume that a and $\bar{a}$ do not vanish). Subtracting $\mathbf{M}_{-}(x;k)$ from both sides of equation (7.3.6) and taking a minus projection yields

$$\mathbf{M}_{-}(x;k) = \mathbf{I} + \frac{1}{2\pi\mathrm{i}} \int_{-\infty}^{\infty} \frac{(\mathbf{G}(x;\zeta) - \mathbf{I})\mathbf{M}_{-}(x;\zeta)}{\zeta - (k - \mathrm{i}0)}\, \mathrm{d}\zeta. \tag{7.3.7}$$

Formally a linear Fredholm integral equation can be obtained corresponding to any Riemann-Hilbert boundary value problem such as (7.3.6). Calling

$$\mathbf{M}_{\pm}(x;k) - \mathbf{I} = \mathbf{\Psi}_{\pm}(x;k),$$

then (7.3.6) takes the form

$$\mathbf{\Psi}_{+}(k) = \mathbf{G}(k)\mathbf{\Psi}_{-}(k) + \mathbf{F}(k), \tag{7.3.8}$$

where $\mathbf{F}(k) := \mathbf{G}(k) - \mathbf{I}$. We note that $\mathbf{G}(k) \to \mathbf{I}$ as $|k| \to \infty$. The appropriate plus and minus projections (of $\mathbf{\Psi}_{\pm}(k) \to \mathbf{0}$ as $|k| \to \infty$) satisfy

$$\mathbf{\Psi}_{-}(k) = -\frac{1}{\pi \mathrm{i}} ⨍_{-\infty}^{\infty} \frac{\mathbf{\Psi}_{-}(\zeta)}{\zeta - k}\, \mathrm{d}\zeta, \tag{7.3.9a}$$

$$\mathbf{\Psi}_{+}(k) = -\frac{1}{\pi \mathrm{i}} ⨍_{-\infty}^{\infty} \frac{\mathbf{\Psi}_{+}(\zeta)}{\zeta - k}\, \mathrm{d}\zeta, \tag{7.3.9b}$$

where k lies on the contour and $⨍_{-\infty}^{\infty}$ refers to the usual principal value integral. Using (7.3.8) in (7.3.9b) yields

$$\mathbf{\Psi}_{-}(k) + \mathbf{G}^{-1}(k)f(k) = \frac{1}{\pi \mathrm{i}} ⨍_{-\infty}^{\infty} \frac{\mathbf{G}^{-1}(k)\mathbf{G}(\zeta)\mathbf{\Psi}_{-}(\zeta) + \mathbf{G}^{-1}(k)\mathbf{F}(\zeta)}{\zeta - k}\, \mathrm{d}\zeta. \tag{7.3.9c}$$

Adding (7.3.9a,c) yields the Fredholm integral equation

$$\mathbf{\Psi}_{-}(k) - \frac{1}{2\pi \mathrm{i}} ⨍_{-\infty}^{\infty} \frac{\mathbf{K}(k,\zeta)\mathbf{\Psi}_{-}(\zeta)}{\zeta - k}\, \mathrm{d}\zeta = \mathbf{H}(k), \tag{7.3.10a}$$

where

$$\mathbf{K}(k,\zeta) = \frac{\mathbf{G}^{-1}(k)\mathbf{G}(\zeta) - \mathbf{I}}{\zeta - k}, \tag{7.3.10b}$$

$$\begin{aligned} \mathbf{H}(k) &= \mathbf{G}^{-1}(k)\left(-\tfrac{1}{2}\mathbf{F}(k) + \frac{1}{2\pi \mathrm{i}} ⨍_{-\infty}^{\infty} \frac{\mathbf{F}(\zeta)}{\zeta - k}\, \mathrm{d}\zeta\right) \\ &= \mathbf{G}^{-1}(k)\mathbf{F}_{-}(k). \end{aligned} \tag{7.3.10c}$$

An alternative to this approach is to convert the Riemann-Hilbert boundary value problem (7.3.6) and projection (7.3.7) into a Gel'fand-Levitan-Marchenko integral equation by looking for triangular representations of the eigenfunctions. In particular, it is easily shown that if $F(x;t)$ is the Fourier transform of $\rho(k;t)$, i.e.

$$F(x;t) = \frac{1}{2\pi} \int_{-\infty}^{\infty} \rho(k;0) \exp\{\mathrm{i}(kx + k^3 t)\}\, \mathrm{d}k, \tag{7.3.11}$$

then $F(x;t)$ also satisfies the partial differential equation

$$\frac{\partial F}{\partial t}+\frac{\partial^3 F}{\partial x^3}=0, \tag{7.3.12}$$

and $K(x,y;t)$ is the solution of the the linear integral equation

$$\begin{aligned} K(x,y;t) &= F\left(\frac{x+y}{2};t\right) \\ &+\frac{1}{4}\int_x^\infty\int_x^\infty K(x,z;t)\,F\left(\frac{z+s}{2};t\right)F\left(\frac{z+y}{2};t\right)\mathrm{d}z\,\mathrm{d}s, \end{aligned} \tag{7.3.13}$$

then the solution of the mKdV equation is given by

$$u(x,t)=K(x,x;t) \tag{7.3.14}$$

(see Ablowitz, Kaup, Newell and Segur [1974] for details).

7.3.2 The Gel'fand-Levitan-Marchenko Integral Equation Method. In this section we shall (formally) derive a linear integral equation whose solution satisfies $\mathrm{P_{II}}$ (7.3.1). Now we suppose that F and K in equations (7.3.11-7.3.13) have self-similar solutions of the form

$$\begin{aligned} K(x,y;t) &= (3t)^{-1/3}\mathcal{K}(\xi,\zeta), \\ F\left(\frac{x+y}{2};t\right) &= (3t)^{-1/3}\mathcal{F}\left(\frac{\xi+\zeta}{2}\right), \end{aligned}$$

with $\xi := x/(3t)^{1/3}$, $\zeta = y/(3t)^{1/3}$. Substituting these into (7.3.13) gives

$$\mathcal{K}(\xi,\zeta)=\mathcal{F}\left(\frac{\xi+\zeta}{2}\right)+\frac{1}{4}\int_\xi^\infty\int_\xi^\infty \mathcal{K}(\xi,\lambda)\mathcal{F}\left(\frac{\lambda+\mu}{2}\right)\mathcal{F}\left(\frac{\mu+\zeta}{2}\right)\mathrm{d}\lambda\,\mathrm{d}\mu. \tag{7.3.15}$$

Additionally, (7.3.12), after integrating once, becomes

$$\frac{\mathrm{d}^2\mathcal{F}}{\mathrm{d}\xi^2}-\xi\mathcal{F}(\xi)=C_1,$$

with C_1 a constant. Furthermore, if $\mathcal{F}(\xi)\to 0$ as $\xi\to\infty$, then necessarily $C_1=0$ and so $\mathcal{F}(\xi)$ is a multiple of the Airy function $\mathcal{F}(\xi)=k\,\mathrm{Ai}(\xi)$. Therefore the integral equation (7.3.15) becomes

$$\begin{aligned} \mathcal{K}(\xi,\zeta) =& k\,\mathrm{Ai}\left(\frac{\xi+\zeta}{2}\right) \\ &+\frac{k^2}{4}\int_\xi^\infty\int_\xi^\infty \mathcal{K}(\xi,\lambda)\,\mathrm{Ai}\left(\frac{\lambda+\mu}{2}\right)\mathrm{Ai}\left(\frac{\mu+\zeta}{2}\right)\mathrm{d}\lambda\,\mathrm{d}\mu. \end{aligned} \tag{7.3.16}$$

The Airy function, Ai(x), is defined to be the solution of

$$\mathrm{Ai}''(x) - x\,\mathrm{Ai}(x) = 0, \qquad ' \equiv \mathrm{d}/\mathrm{d}x,$$

satisfying

$$\mathrm{Ai}(x) \sim \tfrac{1}{2}\pi^{-1/2}x^{-1/4}\exp\left(-\tfrac{2}{3}x^{3/2}\right), \qquad \text{as} \quad x \to \infty,$$
$$\mathrm{Ai}(x) \sim \pi^{-1/2}|x|^{-1/4}\cos\left(\tfrac{2}{3}|x|^{3/2} - \tfrac{1}{4}\pi\right), \qquad \text{as} \quad x \to -\infty.$$

Meanwhile if we apply the same self-similar reduction on the mKdV equation (7.3.3)

$$u(x,t) = (3t)^{1/3}\eta(\xi), \qquad \xi = x/(3t)^{1/3},$$

then, after integrating once $\eta(\xi)$ satisfies

$$\frac{\mathrm{d}^2\eta}{\mathrm{d}\xi^2} - 2\eta^3 - \xi\eta = C_2, \tag{7.3.17}$$

with C_2 another constant, which is $\mathrm{P_{II}}$. Furthermore, equation (7.3.14) reduces to

$$\eta(\xi) = \mathcal{K}(\xi,\xi).$$

Since $C_1 = 0$, then necessarily $C_2 = 0$ also and so (7.3.17) is just the special of $\mathrm{P_{II}}$ (7.3.1). Therefore we have shown that the solution of

$$\frac{\mathrm{d}^2\eta}{\mathrm{d}\xi^2} - 2\eta^3 - \xi\eta = 0, \tag{7.3.18a}$$

satisfying the boundary condition

$$\eta(\xi) \to 0, \qquad \text{as} \qquad \xi \to \infty, \tag{7.3.18b}$$

can be obtained from the solution of the integral equation (7.3.16) on the diagonal $\xi = \zeta$. (Two different proofs of this result are given in Ablowitz and Segur [1977b] and Ablowitz, Ramani and Segur [1980b].)

Using the integral equation (7.3.16), it is possible to derive many properties of solutions of (7.3.18), including global existence and uniqueness of solutions (see Ablowitz and Segur [1977b]; Ablowitz, Ramani and Segur [1980a,b]; Segur and Ablowitz [1981]; Hastings and McLeod [1980]). In §7.4 below, we use the integral equation (7.3.16) to derive connection formulae for (7.3.18), following Clarkson and McLeod [1988].

The third and fourth Painlevé equation ($\mathrm{P_{III}}$ and $\mathrm{P_{IV}}$) have also been studied using through Gel'fand-Levitan-Marchenko equation integral equations (Ablowitz, Ramani and Segur [1980b]; Bassom, Clarkson, Hicks and McLeod [1991]; Clarkson and McLeod [1991]). In particular, Bassom, Clarkson, Hicks and McLeod [1991] obtained "bound state" solutions (exponentially decaying solutions as $\xi \to \pm\infty$) for the non-linear harmonic oscillator

$$\frac{\mathrm{d}^2\eta}{\mathrm{d}\xi^2} = 3\eta^5 + 2\xi\eta^3 + (\tfrac{1}{4}\xi^2 - \nu - \tfrac{1}{2})\eta, \tag{7.3.19}$$

satisfying

$$\eta(\infty) = 0.$$

Equation (7.3.19) is a special case of $\mathrm{P_{IV}}$ since if we make the transformation

$$\eta(\xi) = 2^{-3/4}\sqrt{w(z)}, \qquad z = \xi/\sqrt{2},$$

then $w(z)$ satisfies $\mathrm{P_{IV}}$ with $\alpha = 2\nu + 1$ and $\beta = 0$.

7.3.3 The Inverse Monodromy Transform Method: Introduction. The similarity reduction from the modified KdV equation to $\mathrm{P_{II}}$ also leads the way towards finding compatible *linear* systems which form the basis of the monodromy method. Scaling the independent and dependent variables as follows:

$$u(x,t) = (3t)^{-1/3}\eta(\xi), \qquad \xi = x/(3t)^{1/3},$$
$$\boldsymbol{v}(x,t;k) = \boldsymbol{w}(\xi;\zeta), \qquad \zeta = k(3t)^{1/3},$$

in the linear system for the mKdV equation [i.e., equations (7.3.4)], Flaschka and Newell [1980] obtained the following system

$$\boldsymbol{w}_\xi = \begin{pmatrix} -\mathrm{i}\zeta & \eta \\ \eta & \mathrm{i}\zeta \end{pmatrix}\boldsymbol{w}, \tag{7.3.20a}$$

$$\boldsymbol{w}_\zeta = \begin{pmatrix} -\mathrm{i}(4\zeta^2 + 2\eta^2 + \xi) & 4\zeta\eta + 2\mathrm{i}\dfrac{\mathrm{d}\eta}{\mathrm{d}\xi} + \dfrac{\mu}{\zeta} \\ 4\zeta\eta - 2\mathrm{i}\dfrac{\mathrm{d}\eta}{\mathrm{d}\xi} + \dfrac{\mu}{\zeta} & \mathrm{i}(4\zeta^2 + 2\eta^2 + \xi) \end{pmatrix}\boldsymbol{w}, \tag{7.3.20b}$$

where $\boldsymbol{w} = (w_1, w_2)^T$. These equations are compatible (i.e., $\boldsymbol{w}_{\xi\zeta} = \boldsymbol{w}_{\zeta\xi}$) provided that $\eta(\xi)$ satisfies $\mathrm{P_{II}}$

$$\frac{\mathrm{d}^2\eta}{\mathrm{d}\xi^2} - 2\eta^3 - \xi\eta = \mu, \tag{7.3.21}$$

The linear system (7.3.20b) provides the key to solving $\mathrm{P_{II}}$, for all initial data. Flaschka and Newell [1980] investigated (7.3.20b) and found that the associated monodromy data around the regular singular point $\zeta = 0$ and the irregular singular point $\zeta = \infty$ are constant if $\eta(\xi)$ satisfies (7.3.21). Therefore $\mathrm{P_{II}}$ is a monodromy preserving deformation of an irregular monodromy problem.

We now make some remarks on the classical work of Fuchs [1907], Garnier [1912] and Schlesinger [1912], who considered the Painlevé equations as the isomonodromic conditions for suitable linear systems with rational coefficients possessing regular and irregular singular points.

Consider the following system of linear ordinary differential equations

$$\frac{\mathrm{d}\boldsymbol{y}}{\mathrm{d}z} = \sum_{j=1}^{n} \frac{\mathbf{A}_j}{z - a_j}\,\boldsymbol{y}(z), \tag{7.3.22}$$

where a_j are complex constant scalars and $\mathbf{A}_j$ constant $m \times m$ matrices. The fundamental matrix solution $\mathbf{Y}(z)$ is (in general) a multivalued analytic function with branch points at a_j, $j = 1, 2, \ldots, n$. If z moves along a contour that encircles one of the singular points a_j, then the fundamental solution $\mathbf{Y}(z)$ changes to $\mathbf{Y}(a_j + (z - a_j)\mathrm{e}^{2\pi\mathrm{i}})$, and while also being a solution of (7.3.22), is, in general, not equal to $\mathbf{Y}(z)$, but rather may be expressed as a linear combination of the original solutions

$$\mathbf{Y}(a_j + (z - a_j)\mathrm{e}^{2\pi\mathrm{i}}) = \mathbf{Y}(z)\,\mathbf{M}_j, \tag{7.3.23}$$

where $\mathbf{M}_j$ is the associated *monodromy matrix*.

In connection with (7.3.22), Riemann [1892] formulated the following inverse problem, which is known as the "*Riemann monodromy problem*": *given branch points* $a_1, a_2, \ldots, a_n$ *and matrices* $\mathbf{M}_1, \mathbf{M}_2, \ldots, \mathbf{M}_n$, *find* $\boldsymbol{y}(z)$. This problem was reduced by Riemann to what is now considered to be a discontinuous matrix Riemann-Hilbert problem on the contour joining the branch points; the jumps across this contour are uniquely determined in terms of the monodromy matrices. The Riemann-Monodromy problem has been considered by several authors including Hilbert [1924], Birkhoff [1909, 1910], Plemelj [1964] and more recently by Sato, Miwa and Jimbo [1977, 1979a,b,c, 1980a,b,c]. The so-called isomonodromy deformation problem associated with (7.3.22) is: *how do the matrices* $\mathbf{A}_j$ *depend upon the location of the poles* a_k, $k = 1, 2, \ldots, n$, *such that the monodromy matrices* $\mathbf{M}_j$ *remain fixed?*

Fuchs [1907] considered a special case of the system (7.3.22) with $m = 2$, namely

$$\frac{\partial^2 y}{\partial z^2} = p(z;\tau) y, \tag{7.3.24}$$

where $p(z;\tau)$ is a rational function of z and τ, with four regular singular points, three of which can be located at the fixed points 0, 1, ∞ and the fourth at the variable point τ. He showed that the monodromy matrices are independent of τ if and only if z as function of τ satisfies the sixth Painlevé equation ($\mathrm{P_{VI}}$)

$$\begin{aligned}\frac{\mathrm{d}^2 z}{\mathrm{d}\tau^2} &= \frac{1}{2}\left\{\frac{1}{\tau} + \frac{1}{\tau - 1} + \frac{1}{z - \tau}\right\}\left(\frac{\mathrm{d}z}{\mathrm{d}\tau}\right)^2 - \left\{\frac{1}{\tau} + \frac{1}{\tau - 1} + \frac{1}{z - \tau}\right\}\frac{\mathrm{d}z}{\mathrm{d}\tau} \\ &\quad + \frac{z(z-1)(z-\tau)}{\tau^2(\tau-1)^2}\left\{\alpha + \frac{\beta\tau}{z^2} + \frac{\gamma(\tau-1)}{(z-1)^2} + \frac{\delta\tau(\tau-1)}{(z-\tau)^2}\right\},\end{aligned}$$

with α, β, γ and δ constants. In fact P_{VI} is the compatibility condition of (7.3.24) and the equation

$$\frac{\partial y}{\partial \tau} = B(z;\tau)y + A(z;\tau)\frac{\partial y}{\partial z}, \tag{7.3.25}$$

for appropriate $A(z;\tau)$ and $B(z;\tau)$ which are rational functions of z and τ. The basic idea is to complement equation (7.3.24) with an equation of the form (7.3.25) and then require that $y_{zz\tau} = y_{\tau zz}$. This is the analogue of a *Lax pair* for P_{VI}.

Subsequently, Garnier [1912] generalized the results of Fuchs by considering equations with N regular singular points and also to the case with irregular singular points (though seemingly only on a purely formal level). He showed that the first five Painlevé equations P_I, P_{II}, P_{III}, P_{IV} and P_V, arise as isomonodromy conditions (or compatibility conditions of the Lax pairs) for equations of the form (7.3.24, 25) with appropriate $p(z;\tau)$, $A(z;\tau)$ and $B(z;\tau)$, though he did not explore this connection in order to solve the Painlevé equations.

Schlesinger [1912] showed that the general isomonodromy condition for the system (7.3.22) is that the matrices $\mathbf{A}_j$ satisfy the nonlinear differential equations

$$\frac{\partial \mathbf{A}_j}{\partial a_k} = \frac{[\mathbf{A}_j, \mathbf{A}_k]}{a_j - a_k}, \quad j \neq k, \qquad \sum_{j=1}^{n} \frac{\partial \mathbf{A}_j}{\partial a_j} = 0, \tag{7.3.26}$$

which are now known as the *Schlesinger equations*. In the same paper Schlesinger claimed to prove that these equations possess the Painlevé property. In the 1920's various special cases of the Schlesinger equations were integrated in terms of the Painlevé equations (see Okamoto [1986a]).

Since the development of the I.S.T. method for solving the initial value problem partial differential equations, there has been renewed interest in expressing the Painlevé equations as isomonodromic conditions for suitable linear systems possessing both regular and irregular singular points (cf. Flaschka and Newell [1980]; Jimbo, Miwa and Ueno [1981]; Jimbo and Miwa [1981a,b]). The emphasis of this work is on equations with irregular singular points when the monodromy matrices effectively are replaced by the *Stokes multipliers*.

Flaschka and Newell [1980] demonstrated that P_{II} was the deformation equation of equations (7.3.20b). Solving the initial value problem for P_{II} amounts to solving an "inverse problem" of $\boldsymbol{v}(\xi;z)$, i.e., using analytical properties of $\boldsymbol{w}(\xi;z)$ (in z with $z \in \mathbb{C}$) one can show that given appropriate monodromy data, this can in principle be used to reconstruct $\boldsymbol{w}(\xi;z)$. Obtaining this monodromy data is classical and analogous to the direct problem in I.S.T..

Fokas and Ablowitz [1983a] showed that the inverse problem for P_{II} could be formulated in terms of a Riemann-Hilbert boundary value problem on a complicated contour (six semi-rays intersecting at the origin). They mapped this Riemann-Hilbert problem through a series of transformations to three different standard Riemann-Hilbert problems (which need to be solved sequentially). Each of these Riemann-Hilbert problems can be solved through a system of two Fredholm integral equations for a region of parameter space $0 \leq \alpha < 0$; the entire parameter space is covered by using the "Bäcklund" transformations of P_{II} (cf. Airault [1980]; Fokas and Ablowitz [1982]). In a special case, using these Riemann-Hilbert problems, Lebeau and Lochak [1987] were able to obtain a connection formula for P_{II}. Recently Fokas and Zhou [1991] have rigorously studied the Riemann-Hilbert problems associated with P_{II} (and also P_{IV}). They show that, in general, the Cauchy problems for P_{II} (and P_{IV}) possess meromorphic global solutions.

Riemann-Hilbert problems associated to P_{IV} and P_V have been studied by Fokas, Mugan and Ablowitz [1987] (see also Mugan [1987]) and for P_{VI} by Cosgrove (private communication, 1986). Fokas, Mugan and Ablowitz provide a scheme similar to that of Fokas and Ablowitz [1983a] and map the solution of P_{IV} and P_V to a sequence of certain standard Riemann-Hilbert problems in a certain region of parameter sequence (e.g., the extra parameters in the equation, such as in P_{II}). They cover the entire region of parameter space by using the concept of Schlesinger transformations (first introduced for P_{VI} by Cosgrove, private communication). Fokas, Mugan and Ablowitz [1987] also point out the striking similarity of the inverse monodromy method to I.S.T. and refer to it as I.M.T.. In some sense the method is the analogue of the Laplace method for linear ordinary differential equations just as I.S.T. is the analogue of Fourier analysis for linear problems.

We further remark that McCoy and Tang [1986a,b,c] have derived connection formulae for P_{II}, P_{III} and P_V as an application of the monodromy preserving deformation technique due to Jimbo, Miwa and Ueno [1981] and Jimbo and Miwa [1981b,c]. Additionally, Jimbo [1982] and Cosgrove (private communication, 1986) have obtained connection formulae for P_{VI}.

The monodromy preserving deformation method of solution for the Painlevé equations (concentrating primarily on P_{II} and P_{III}) is reviewed in detail in a monograph by Its and Novokshenov [1986]. As an application of the technique, Its and Novokshenov derive and prove connection formulae for P_{II} and P_{III}. Furthermore, they list the deformation equations and various connection formulae for the six Painlevé equations. In fact, Soviet investigators have used the monodromy preserving deformation method to obtain properties of P_I (Kapaev [1989a]), P_{II} (Its and Kapaev [1988]; Kapaev [1989b]; Kapaev and Novokshenov [1986]; Suleimanov [1987]), P_{III} (Kitaev [1989]; Mugan and Fokas [1990]; Novokshenov [1985a,b]) and P_{IV} (Kitaev [1985]).

7.3.4 The Inverse Monodromy Transform Method: Direct Problem. The essence of the direct problem is to establish the analytic structure of $\mathbf{W}(\xi;\zeta)$, a fundamental matrix solution of the system (7.3.20b), with respect to the parameter ζ, in the entire complex ζ-plane. This is similar to the I.S.T. procedure for partial differential equations. However we note that, in contrast to the analogous problem in the I.S.T. for partial differential equations, this task is straightforward. Equation (7.3.20b) is a *linear* system of ordinary differential equations, therefore its analytic structure is completely determined by its singular points; for $\mu \neq 0$ it has a regular singular point at $\zeta = 0$ (if $\mu = 0$ then the origin is a regular point) and a third-order irregular singular point at $\zeta = \infty$. The formal asymptotic expansion of two linearly independent solutions of (7.3.20b) as $\zeta \to \infty$ is

$$\tilde{\boldsymbol{w}}^{(1)}(\xi;\zeta) \sim \mathrm{e}^{-\theta}\left\{\begin{pmatrix}1\\0\end{pmatrix} + \frac{\mathrm{i}}{2\zeta}\begin{pmatrix}\zeta\eta^2+\eta^4-(\eta')^2+2\mu\eta\\ \eta\end{pmatrix} + O(\zeta^{-2})\right\}, \tag{7.3.27a}$$

$$\tilde{\boldsymbol{w}}^{(2)}(\xi;\zeta) \sim \mathrm{e}^{\theta}\left\{\begin{pmatrix}0\\1\end{pmatrix} - \frac{\mathrm{i}}{2\zeta}\begin{pmatrix}\eta\\ \zeta\eta^2+\eta^4-(\eta')^2+2\mu\eta\end{pmatrix} + O(\zeta^{-2})\right\}, \tag{7.3.27b}$$

where $\theta(\xi;\zeta) := \mathrm{i}(\frac{4}{3}\zeta^3 + \xi\zeta)$. Since $\zeta = \infty$ is an irregular singular point then the actual asymptotic behavior of $\mathbf{W}(\xi;\zeta)$ is different in different sectors of the complex ζ-plane. These sectors are determined by $\mathrm{Re}\left[\mathrm{i}\left(\frac{4}{3}\zeta^3 + \xi\zeta\right)\right] = 0$; hence for large ζ, the boundaries of these sectors, Σ_j, are asymptotic to the rays $\arg\zeta = j\pi/3$, for $j = 0, 1, \ldots, 6$. Define the sector S_j as follows:

$$S_j := \{\zeta | (j-1)\pi/3 < \arg\zeta < j\pi/3\}.$$

The formal asymptotic solution $\tilde{\boldsymbol{w}}^{(1)}(\xi;\zeta)$ ($\tilde{\boldsymbol{w}}^{(2)}(\xi;\zeta)$) is exponentially growing (decaying) in the sectors S_1, S_3, S_5 and exponentially decaying (growing) in the sectors S_2, S_4, S_6. Suppose $\mathbf{W}_j(\xi;\zeta) = (\boldsymbol{w}_j^{(1)}(\xi;\zeta), \boldsymbol{w}_j^{(2)}(\xi;\zeta))$ is the fundamental matrix solution of (7.3.20b) in S_j, for $j = 0, 1, \ldots, 6$, such that $\boldsymbol{w}_j^{(i)}(\xi;\zeta)$ is asymptotic to $\tilde{\boldsymbol{w}}_j^{(i)}(\xi;\zeta)$ in S_j, for $i = 1, 2$.

The solutions are related as follows,

$$\mathbf{W}_2(\xi;\zeta) = \mathbf{W}_1(\xi;\zeta)\mathbf{G}_1, \qquad \tfrac{1}{3}\pi \le \arg\zeta < \tfrac{2}{3}\pi, \tag{7.3.28a}$$

$$\mathbf{W}_3(\xi;\zeta) = \mathbf{W}_2(\xi;\zeta)\mathbf{G}_2, \qquad \tfrac{2}{3}\pi \le \arg\zeta < \pi, \tag{7.3.28b}$$

$$\mathbf{W}_4(\xi;\zeta) = \mathbf{W}_3(\xi;\zeta)\mathbf{G}_3, \qquad \pi \le \arg\zeta < \tfrac{4}{3}\pi, \tag{7.3.28c}$$

$$\mathbf{W}_5(\xi;\zeta) = \mathbf{W}_4(\xi;\zeta)\mathbf{G}_4, \qquad \tfrac{4}{3}\pi \le \arg\zeta < \tfrac{5}{3}\pi, \tag{7.3.28d}$$

$$\mathbf{W}_6(\xi;\zeta) = \mathbf{W}_5(\xi;\zeta)\mathbf{G}_5, \qquad \tfrac{5}{3}\pi \le \arg\zeta < 2\pi, \tag{7.3.28e}$$

$$\mathbf{W}_7(\xi;\zeta) = \mathbf{W}_6(\xi;\zeta)\mathbf{G}_6, \qquad 2\pi \le \arg\zeta < \tfrac{7}{3}\pi, \tag{7.3.28f}$$

where $\mathbf{G}_j$, $j = 1, 2, \ldots, 6$, are constant matrices, called *Stokes multipliers*, given by

$$\mathbf{G}_1 = \begin{pmatrix}1&0\\a&1\end{pmatrix}, \qquad \mathbf{G}_2 = \begin{pmatrix}1&b\\0&1\end{pmatrix}, \qquad \mathbf{G}_3 = \begin{pmatrix}1&0\\c&1\end{pmatrix}, \tag{7.3.29a}$$

$$\mathbf{G}_4 = \begin{pmatrix}1&d\\0&1\end{pmatrix}, \qquad \mathbf{G}_5 = \begin{pmatrix}1&0\\e&1\end{pmatrix}, \qquad \mathbf{G}_6 = \begin{pmatrix}1&f\\0&1\end{pmatrix} \tag{7.3.29b}$$

(cf. Wasow [1987]). The parameters $\{a,b,c,d,e,f\}$ are the *monodromy data*, which are independent of ξ (cf. Flaschka and Newell [1980]). We remark that each $\mathbf{W}_j(\xi;\zeta)$, $j=1,2,\ldots,7$, satisfies the x-equation (7.3.20a). The fundamental matrix $\mathbf{W}_7(\xi;\zeta)$ is defined on a sector including the positive real ζ-axis; since the $\mathbf{W}_1(\xi;\zeta)$ are single-valued at $\zeta=\infty$, then necessarily

$$\mathbf{W}_7(\xi;\zeta)\equiv\mathbf{W}_1(\xi;\zeta e^{2\pi i}). \tag{7.3.30}$$

As $\zeta\to 0$, two linearly independent solutions of (7.3.20b), for $\alpha\neq n+\frac{1}{2}$, with n an integer, are given by

$$\mathbf{\Phi}(\xi;\zeta)=\left(\frac{\zeta^{\alpha}\exp\{-q(\xi)\}}{\sqrt{2}}\left\{\begin{pmatrix}1\\-1\end{pmatrix}-\frac{i\zeta(2\eta'+2\eta^2+\xi)}{1+2\alpha}\begin{pmatrix}1\\1\end{pmatrix}+O(\zeta^2)\right\},\right.$$
$$\left.\frac{\exp\{q(\xi)\}}{\sqrt{2}\,\zeta^{\alpha}}\left\{\begin{pmatrix}1\\1\end{pmatrix}+O(\zeta)\right\}\right). \tag{7.3.31}$$

where $q'(\xi)=\eta(\xi)$. This leads to the following monodromy matrix $\mathbf{J}_\mu$ associated with $\zeta=0$

$$\mathbf{\Phi}(\xi;\zeta e^{2\pi i})=\mathbf{\Phi}(\xi;\zeta)\mathbf{J}_\mu,\qquad \mathbf{J}_\mu=\begin{pmatrix}e^{2\pi i\mu}&0\\0&e^{-2\pi i\mu}\end{pmatrix},\qquad \mu\neq n+\tfrac{1}{2} \tag{7.3.32}$$

(note that $\mathbf{J}_\mu=\mathbf{I}$, if $\mu=n$). The matrices $\mathbf{W}_1(\xi;\zeta)$ and $\mathbf{\Phi}(\xi;\zeta)$ are related by

$$\mathbf{W}_1(\xi;\zeta)=\mathbf{\Phi}(\xi;\zeta)\mathbf{E},\qquad \mathbf{E}:=\begin{pmatrix}\alpha&\beta\\\gamma&\delta\end{pmatrix},\qquad \det\mathbf{E}=1. \tag{7.3.33}$$

($\mathbf{E}$ is referred to as the *connection* matrix.)

Using certain symmetries in the differential system (7.3.20b), we can obtain relationships between the monodromy data and the matrix $\mathbf{E}$ which are useful in the following. If $\mathbf{W}(\xi;\zeta)$ is a solution of equation (7.3.20b) then so is $\boldsymbol{\sigma}_1\mathbf{W}(\xi;-\zeta)$, where

$$\boldsymbol{\sigma}_1=\begin{pmatrix}0&1\\1&0\end{pmatrix}.$$

Hence it is easily shown that the $\mathbf{G}_j$, $j=1,2,\ldots,6$, satisfy the symmetry conditions

$$\mathbf{G}_{j+3}=\boldsymbol{\sigma}_1\mathbf{G}_j\boldsymbol{\sigma}_1,\qquad j=1,2,3, \tag{7.3.34}$$

and so we see that

$$a=d,\qquad b=e,\qquad c=f. \tag{7.3.35}$$

For equations (7.3.28,30,32) to be consistent, then necessarily

$$\mathbf{G}_1\mathbf{G}_2\mathbf{G}_3\mathbf{G}_4\mathbf{G}_5\mathbf{G}_6=\mathbf{E}^{-1}\mathbf{J}_\mu\mathbf{E}, \tag{7.3.36}$$

where $\mathbf{J}_\mu$ and $\mathbf{E}$ are as defined in (7.3.32,33) (note that $\mathbf{E}^{-1}\mathbf{J}_\mu\mathbf{E} = \mathbf{I}$, if $\mu = n$). Therefore from equations (7.3.29,35,36) we have the relationships (for $\mu \neq n + \frac{1}{2}$)

$$1 + ab = 2\alpha\gamma\cos(\mu\pi), \tag{7.3.37a}$$

$$1 + bc = -2\beta\delta\cos(\mu\pi), \tag{7.3.37b}$$

$$b = -\alpha\delta \mathrm{e}^{-\mu\pi\mathrm{i}} - \beta\gamma\mathrm{e}^{\mu\pi\mathrm{i}}, \tag{7.3.37c}$$

$$a + b + c + abc = 2\mathrm{i}\sin(\mu\pi). \tag{7.3.37d}$$

We note that if $\mu = n$, then these simplify to

$$1 + ab = 2(-1)^n\alpha\gamma, \tag{7.3.38a}$$

$$1 + bc = 2(-1)^{n+1}\beta\delta, \tag{7.3.38b}$$

$$b = (-1)^{n+1}(\alpha\delta + \beta\gamma), \tag{7.3.38c}$$

$$a + b + c + abc = 0. \tag{7.3.38d}$$

(For further details see Flaschka and Newell [1980].)

If in (7.3.20b) ξ, $\eta(\xi)$ and $\eta'(\xi)$ are real, if $\mathbf{W}(\xi;\zeta)$ is a solution of equation (7.3.20b) then so is $\boldsymbol{\sigma}_1\mathbf{W}^*(\xi;\zeta^*)$, which shows that

$$a = -c^*. \tag{7.3.39}$$

Hence, in the case when $\mu = 0$ (which we shall consider in §7.4.3), from (7.3.38d)

$$b = \frac{a^* - a}{1 - aa^*}, \tag{7.3.39}$$

provided that $aa^* \neq 1$, where the asterisk denotes the complex conjugate. Therefore we see that when we are considering real solutions of

$$\frac{\mathrm{d}^2\eta}{\mathrm{d}\xi^2} = 2\eta^3 + \xi\eta,$$

then the monodromy data is fully characterized by the complex parameter a.

7.3.5 The Inverse Monodromy Transform Method: Inverse Problem. The Riemann-Hilbert problems associated with I.M.T. possess several novelties. These are described in terms of the I.M.T. associated with $\mathrm{P_{II}}$. The inverse problem associated with $\mathrm{P_{II}}$ involves the reconstruction of a sectionally meromorphic function $\mathbf{W}(\xi;\zeta)$ with the following properties:

1. $\mathbf{W}(\xi;\zeta)$ has prescribed jumps on the contour defined by $\mathrm{Re}\left[\mathrm{i}\left(\frac{4}{3}\zeta^3 + \xi\zeta\right)\right] = 0$;
2. $\mathbf{W}(\xi;\zeta) \to \mathbf{I}$ as $|\zeta| \to \infty$ off the contour, but it oscillates on it;
3. $\mathbf{W}(\xi;\zeta) \sim \mathbf{W}_0(\xi;\zeta)\,\mathrm{diag}(\zeta^\mu, \zeta^{-\mu})$ as $\zeta \to 0$, where $\mathbf{W}_0(\xi;\zeta)$ is analytic at $\zeta = 0$.

The inverse problem associated with the scattering problem (7.3.20b) is: *given the monodromy data, reconstruct* $\mathbf{W}(\xi;\zeta)$. We shall sketch this derivation of this, the interested reader can find further details in Fokas and Ablowitz [1983a]. We remark that this inverse problem appears to be more complicated than those previously discussed in these notes with respect to nonlinear evolution equations solvable by I.S.T.. Furthermore, in order to obtain some properties of $\mathrm{P_{II}}$, for example connection formulae (see §7.4.3 below), the form of the inverse integral equations are not essential.

First define

$$\mathbf{\Psi}_j(\xi;\zeta) := \mathbf{W}_j(\xi;\zeta)\mathbf{\Theta}^{-1}(\xi;\zeta), \qquad \text{for} \quad j = 1,2,\ldots,6,$$

where $\mathbf{\Theta}(\xi;\zeta) = \mathrm{diag}(\mathrm{e}^{-\theta},\mathrm{e}^{\theta})$, with $\theta(\xi;\zeta) := \mathrm{i}(\frac{4}{3}\zeta^3 + \xi\zeta)$. Then equations (7.3.28) yield

$$\mathbf{\Psi}_{j+1}(\xi;\zeta) = \mathbf{\Psi}_j(\xi;\zeta)\mathbf{\Gamma}_j(\xi;\zeta), \qquad 1 \le j \le 6, \tag{7.3.40a}$$
$$\mathbf{\Psi}_7(\xi;\zeta) = \mathbf{\Psi}_1(\xi;\zeta\mathrm{e}^{2\pi\mathrm{i}}), \tag{7.3.40b}$$

where $\mathbf{\Gamma}_j(\xi;\zeta) = \mathbf{\Theta}\mathbf{G}_j\mathbf{\Theta}^{-1}$, i.e.,

$$\mathbf{\Gamma}_1(\xi;\zeta) = \begin{pmatrix} 1 & 0 \\ a\mathrm{e}^{2\theta} & 1 \end{pmatrix}, \qquad \mathbf{\Gamma}_2(\xi;\zeta) = \begin{pmatrix} 1 & b\mathrm{e}^{-2\theta} \\ 0 & 1 \end{pmatrix}, \tag{7.3.41a}$$
$$\mathbf{\Gamma}_3(\xi;\zeta) = \begin{pmatrix} 1 & 0 \\ c\mathrm{e}^{2\theta} & 1 \end{pmatrix}, \qquad \mathbf{\Gamma}_4(\xi;\zeta) = \begin{pmatrix} 1 & a\mathrm{e}^{-2\theta} \\ 0 & 1 \end{pmatrix}, \tag{7.3.41b}$$
$$\mathbf{\Gamma}_5(\xi;\zeta) = \begin{pmatrix} 1 & 0 \\ b\mathrm{e}^{2\theta} & 1 \end{pmatrix}, \qquad \mathbf{\Gamma}_6(\xi;\zeta) = \begin{pmatrix} 1 & c\mathrm{e}^{-2\theta} \\ 0 & 1 \end{pmatrix}. \tag{7.3.41c}$$

Also $\mathbf{\Psi}(\xi;\zeta)$ is analytic in ζ in the sector S_j and

$$\mathbf{\Psi}(\xi;\zeta) \to \mathbf{I}, \qquad \text{as} \quad \zeta \to \infty, \tag{7.3.42a}$$
$$\mathbf{\Psi}(\xi;\zeta) \to \mathbf{\Phi}(\xi;\zeta)\mathbf{E}\mathbf{\Theta}^{-1}, \qquad \text{as} \quad \zeta \to 0, \tag{7.3.42b}$$
$$\mathbf{\Psi}(\xi;\zeta\mathrm{e}^{2\pi\mathrm{i}}) = \mathbf{\Psi}(\xi;\zeta)\mathbf{\Theta}\mathbf{E}^{-1}\mathbf{J}_\mu\mathbf{E}^{-1}\mathbf{\Theta}^{-1}, \tag{7.3.42c}$$

where $\mathbf{\Phi}(\xi;\zeta)$ and $\mathbf{E}$ are as defined in equations (7.3.31,33) above.

Equations (7.3.40-42) define a discontinuous Riemann-Hilbert problem for the sectionally analytic, matrix-valued function $\mathbf{\Psi}(\xi;\zeta)$, which has different representation $\mathbf{\Psi}_1(\xi;\zeta), \mathbf{\Psi}_2(\xi;\zeta), \ldots, \mathbf{\Psi}_6(\xi;\zeta)$ in the different sectors $S_1, S_2, \ldots, S_6$. The boundary values of these different representations are related through equations (7.3.40). The Riemann-Hilbert problem is discontinuous (at $\zeta = 0$) since $\mathbf{\Gamma}_{j+3}(0) \neq \mathbf{\Gamma}_j(0)$, for $j = 1,2,3$.

Define C_j, $j = 1,3,5$, to be the contour from 0 to ∞ along the line $\arg\zeta = (j-1)\pi/3$ and C_j, $j = 2,4,6$, the contour from ∞ to 0 along the line $\arg\zeta = (j-1)\pi/3$.

Fokas and Ablowitz [1983a] show that the Riemann-Hilbert problem (7.3.40-42) can be mapped through a series of transformations solved sequentially to the following three different standard Riemann-Hilbert problems (each defined on two rays, which provide the continuation of each other)

$$\mathbf{P}_1(\xi;\zeta) = \mathbf{P}_2(\xi;\zeta)\mathbf{H}_1(\xi;\zeta), \qquad \zeta \text{ on } C_1 + C_4, \tag{7.3.43a}$$

$$\mathbf{P}_1(\xi;\zeta) \to \mathbf{I}, \qquad \text{as } \zeta \to \infty, \quad \text{in } S_1 + S_2 + S_3, \tag{7.3.43b}$$

$$\mathbf{Q}_1(\xi;\zeta) = \mathbf{Q}_2(\xi;\zeta)\mathbf{H}_2(\xi;\zeta), \qquad \zeta \text{ on } C_2 + C_5, \tag{7.3.44a}$$

$$\mathbf{Q}_1(\xi;\zeta) \to \mathbf{I}, \qquad \text{as } \zeta \to \infty, \quad \text{in } S_5 + S_6 + S_1, \tag{7.3.44b}$$

$$\mathbf{R}_1(\xi;\zeta) = \mathbf{R}_2(\xi;\zeta)\mathbf{H}_3(\xi;\zeta), \qquad \zeta \text{ on } C_3 + C_6, \tag{7.3.45a}$$

$$\mathbf{R}_1(\xi;\zeta) \to \mathbf{I}, \qquad \text{as } \zeta \to \infty, \quad \text{in } S_6 + S_1 + S_2, \tag{7.3.45b}$$

with

$$\mathbf{H}_1(\xi;\zeta) = \begin{cases} \mathbf{\Gamma}_6(\xi;\zeta), & \zeta \text{ on } C_1, \\ \mathbf{A}\mathbf{G}_1\mathbf{G}_2(\mathbf{G}_1\mathbf{G}_2\mathbf{\Gamma}_3)^{-1}\mathbf{A}^{-1}, & \zeta \text{ on } C_4, \end{cases}$$

where $\mathbf{A}$ is defined by

$$\mathbf{A}\mathbf{G}_1\mathbf{G}_2(\mathbf{G}_1\mathbf{G}_2\mathbf{G}_3)^{-1}\mathbf{A}^{-1} = \mathbf{G}_6,$$

$\mathbf{H}_2$ is expressed in terms of $\{\mathbf{G}_j,\ \mathbf{\Gamma}_j\}_{j=1}^{6}$, and $\mathbf{P}_1$, $\mathbf{P}_2$, and $\mathbf{H}_3$ in terms of $\{\mathbf{G}_j,\ \mathbf{\Gamma}_j\}_{j=1}^{6}$, $\mathbf{P}_1$, $\mathbf{P}_2$, and $\mathbf{Q}_1$, $\mathbf{Q}_2$. By solving these three Riemann-Hilbert problems in succession, we obtain $\mathbf{P}_1(\xi;\zeta)$, $\mathbf{P}_2(\xi;\zeta)$, then $\mathbf{Q}_1(\xi;\zeta)$, $\mathbf{Q}_2(\xi;\zeta)$, and finally $\mathbf{R}_1(\xi;\zeta)$, $\mathbf{R}_2(\xi;\zeta)$. From these solutions one can then construct $\mathbf{\Phi}(\xi;\zeta)$.

Each of the above Riemann-Hilbert problems can be solved through a system of Fredholm integral equations. For example, the Riemann-Hilbert problem (7.3.43) is equivalent to

$$\mathbf{P}_2(\xi;\zeta) = \mathbf{I} + \frac{1}{2\pi\mathrm{i}}\int_{C_4+C_1} \frac{\mathbf{P}_2(\xi;\tau)[\mathbf{H}_1(\xi;\tau)-\mathbf{I}]}{\tau-\zeta+\mathrm{i}0}\,\mathrm{d}\tau, \tag{7.3.46a}$$

$$\mathbf{P}_1(\xi;\zeta) = \mathbf{I} + \frac{1}{2\pi\mathrm{i}}\int_{C_4+C_1} \frac{\mathbf{P}_2(\xi;\tau)[\mathbf{H}_1(\xi;\tau)-\mathbf{I}]}{\tau-\zeta-\mathrm{i}0}\,\mathrm{d}\tau. \tag{7.3.46b}$$

We remark that the Riemann-Hilbert problem (7.3.40-42) can be written in the integral form

$$\mathbf{\Phi}(\xi;\zeta) = \mathbf{I} - \frac{1}{2\pi\mathrm{i}}\int_{C} \frac{\mathbf{\Phi}^{+}(\xi;\tau)[\mathbf{\Gamma}(\xi;\tau)-\mathbf{I}]}{\tau-\zeta}\,\mathrm{d}\tau, \tag{7.3.47}$$

where $\mathbf{\Gamma}(\xi;\zeta) = \mathbf{\Gamma}_j(\xi;\zeta)$, for ζ on C_{j+1} ($C_7 = C_1$), and $C = C_1 + C_2 + \cdots + C_6$. By taking the limit of (7.3.47) as ζ approaches C from S_1, S_3, S_5, one may obtain a system of linear integral equations. The above construction involving the three

Riemann-Hilbert problems may be interpreted as a method of mapping (7.3.47) to regular Fredholm integral equations.

Finally to reconstruct $\eta(\xi)$ we take the limit of (7.3.47) as $\zeta \to \infty$, and then using (7.3.27) we have

$$\eta(\xi) = \frac{1}{\pi} \int_C \left(\mathbf{\Phi}^+(\xi;\tau)[\mathbf{\Gamma}(\tau) - \mathbf{I}]\right)^{21} \, \mathrm{d}\tau. \tag{7.3.48}$$

7.4 Connection Formulae for the Second Painlevé Equation.

7.4.1 Introduction. In this section we discuss connection formulae for a special case of the second Painlevé transcendent

$$\frac{\mathrm{d}^2 y}{\mathrm{d}x^2} = xy + 2y^3, \tag{7.4.1a}$$

satisfying

$$y(x) \to 0, \qquad \text{as} \qquad x \to \infty. \tag{7.4.1b}$$

Hastings and McLeod [1980] have proved the following theorem in connection with (7.4.1) (see also Figure 7.4.1).

THEOREM 7.4.1

Any solution of (7.4.1a) *satisfying* (7.4.1b) *is asymptotic to* $k\,\mathrm{Ai}(x)$*, for some* k*, and conversely, for any* k*, there is a unique solution of* (7.4.1a) *asymptotic to* $k\,\mathrm{Ai}(x)$*. Let this solution be* $y_k(x)$*, then if* $|k| < 1$*, as* $x \to -\infty$*,*

$$y_k(x) \sim d|x|^{-1/4} \sin\left\{\tfrac{2}{3}|x|^{3/2} - \tfrac{3}{4}d^2 \ln|x| - \theta_0\right\}, \tag{7.4.2}$$

where the constants d *and* θ_0 *are dependent on* k*. If* $|k| = 1$*, as* $x \to -\infty$*,*

$$y_k(x) \sim \mathrm{sgn}(k)(-\tfrac{1}{2}x)^{1/2},$$

and if $|k| > 1$*,* $y_k(x)$ *has a pole at a finite* x_0*, dependent on* k*,*

$$y_k(x) \sim \mathrm{sgn}(k)(x - x_0)^{-1}, \qquad \text{as} \qquad x \downarrow x_0.$$

Since (7.4.1a) is left unchanged by the transformation $y \to -y$, we shall assume without loss of generality that $k > 0$. It has become a matter of some interest to establish the exact dependence of the constants d, θ_0 and x_0 on k and in this section we discuss the following results:

$$d^2(k) = -\pi^{-1} \ln(1 - k^2). \tag{7.4.3}$$

$$\theta_0(k) = \tfrac{3}{2}d^2 \ln 2 + \arg\left\{\Gamma(1 - \tfrac{1}{2}\mathrm{i}d^2)\right\} - \tfrac{1}{4}\pi, \tag{7.4.4}$$

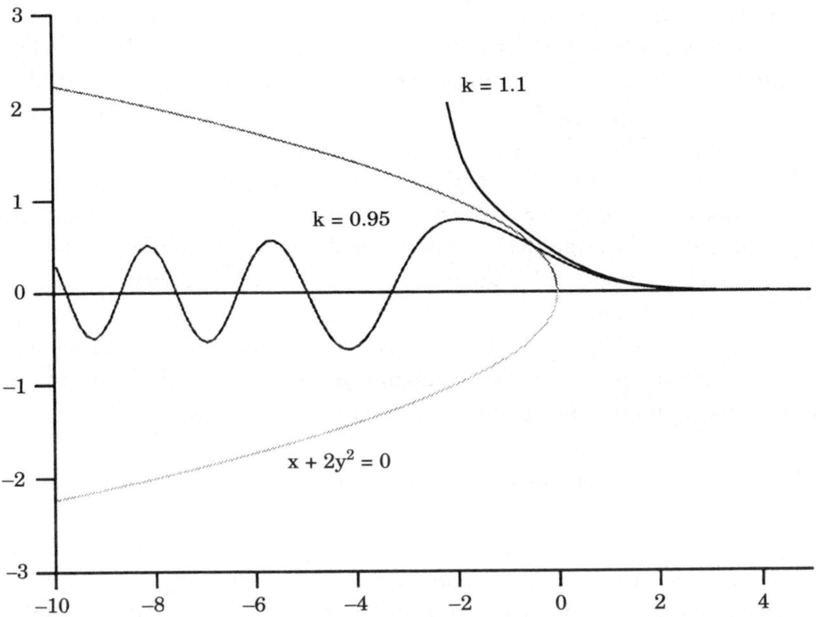

FIGURE 7.4.1 *Solutions of equation* (7.4.1) *with* $k = 1.1$ *and* $k = 0.95$

where $\Gamma(z)$ is the Gamma function. These results were first conjectured, found heuristically and subsequently verified numerically, by Ablowitz and Segur [1977a] and Segur and Ablowitz [1981], respectively. Subsequently Clarkson and McLeod [1988] have given a rigorous proof of (7.4.3).

(7.4.3) and (7.4.4) are examples of nonlinear *connection problems*, since we are relating the asymptotic behavior of the solution of (7.4.1) as $x \to \infty$ to the asymptotic behavior as $x \to -\infty$. Linear connection problems for ordinary differential equations have been extensively studied for over a hundred years, however nonlinear connection problems are rare. The usual method for linear equations is to consider x as a complex variable and pass from $x \to \infty$ to $x \to -\infty$ along a large semicircle in the complex x-plane. Provided that the coefficients in the equation have a reasonably simple asymptotic behavior as $|x| \to \infty$, then it is usually possible to construct an asymptotic expansion for the solution at all points on the semicircle, and so relate the asymptotic behavior as $x \to \infty$ to the asymptotic behavior as $x \to -\infty$.

However, this method fails in general for nonlinear equations since the solutions may be very complicated as $|x| \to \infty$. However, for nonlinear equations such as the Painlevé equations which have the Painlevé property (and in particular those equations such as the second Painlevé equation (7.4.1) whose solutions are meromorphic in the finite complex plane), this method is feasible. Indeed Boutroux [1913, 1914] (see also Ince [1956]; Hille [1976]; Bender and Orszag [1978]), has studied the asymptotics of the first Painlevé equation in considerable detail and remarks that his ideas can be extended to the second Painlevé equation also. Essentially, the solutions behave asymptotically like elliptic functions, at least locally, and although not considered by Boutroux, the solution of the connection problem is a matter of matching different elliptic functions in different sectors on the large semi-circle in the complex plane. Whereas this method is theoretically feasible for the second Painlevé equation (7.4.1), it certainly involves considerable technical difficulties and the connection problems are solvable by a different method. Recently, Joshi [1986] (Joshi, Kruskal, private communication — see also Joshi and Kruskal [1988, 1991]), in her thesis discusses how one can extend the ideas of Boutroux and use singular perturbative techniques, to obtain connection formulae for the first and second Painlevé equations.

7.4.2 The Gel'fand-Levitan-Marchenko Integral Equation Approach. In this section we discuss the proof of (7.4.3), following Clarkson and McLeod [1988]. This proof exploits the fact that, because of the relationship between the second Painlevé equation (7.4.1) and the mKdV equation, the inverse scattering method of solution for the mKdV equation leads to an expression for $y_k(x)$ in terms of the solution of a linear integral equation, as we discussed above. Ablowitz and Segur [1977b] (see also Ablowitz, Ramani and Segur [1980b]), show that if the integral equation

$$K(x,y) = k\,\mathrm{Ai}\left(\frac{x+y}{2}\right) + \frac{k^2}{4}\int_x^\infty\int_x^\infty K(x,s)\,\mathrm{Ai}\left(\frac{s+t}{2}\right)\mathrm{Ai}\left(\frac{t+y}{2}\right)\mathrm{d}s\,\mathrm{d}t, \tag{7.4.5}$$

is considered for x sufficiently large (depending on k), say $x > x_1$,

$$y_k(x) = K(x,x). \tag{7.4.6}$$

Ablowitz and Segur [1977b] further show that $K(x,y)$ satisfies the differential equation

$$\left(\frac{\partial}{\partial x} + \frac{\partial}{\partial y}\right)^2 K(x,y) = \left(\frac{x+y}{2}\right) K(x,y) + 2\left[K(x,x)\right]^2 K(x,y), \tag{7.4.7}$$

for $y \geq x$. Hastings and McLeod [1980] subsequently proved that (7.4.5-7) hold for all x and y if $0 < k < 1$.

If in (7.4.7) we write $u = -\frac{1}{2}(y+x)$, $v = \frac{1}{2}(y-x)$, $F(u,v) = K(x,y)$, then $F(u,v)$ satisfies

$$\frac{\partial^2 F}{\partial u^2}(u,v) = -uF(u,v) + 2[y_k(-u-v)]^2 F(u,v), \tag{7.4.8}$$

which is a *linear ordinary differential equation* for $F(u,v)$, where v is regarded as a parameter.

To prove (7.4.3), the technique is to investigate the asymptotics of $K(x,y)$ as $x \to -\infty$, with $(y-x)$ bounded away from zero (that is the parameter v in (7.4.8) is bounded away from zero). There are two ways of doing this, either through the integral equation (7.4.5), in which case the asymptotic behavior is expressed in terms of k, or through the asymptotic behavior of the differential equation (7.4.6), which in turn depends upon the asymptotic behavior as $u \to \infty$ of $y_k(-u-v)$ and this involves d. A comparison of the two resultant expressions for the asymptotic behavior of $K(x,y)$ yields (7.4.3).

First consider the differential equation

$$\frac{\mathrm{d}^2 w}{\mathrm{d}u^2} + \{u - 2[y_k(-u-v)]^2\} w(u) = 0. \tag{7.4.9}$$

If we make the transformation

$$\eta(u) = q^{1/4}(u)\, w(u), \qquad \xi(u) = \int_0^u q^{1/2}(t)\,\mathrm{d}t,$$

where $q(u) = \{u - 2[y_k(-u-v)]^2\}$, then (7.4.9) becomes

$$\frac{\mathrm{d}^2\eta}{\mathrm{d}\xi^2} + \left\{1 - \left(\frac{q''}{4q^2} - \frac{5(q')^2}{16q^3}\right)\right\}\eta = 0,$$

where $' \equiv \mathrm{d}/\mathrm{d}u$. Using (7.4.2), one then shows that as $\xi \to -\infty$

$$\frac{\mathrm{d}^2\eta}{\mathrm{d}\xi^2} + \left\{1 - \frac{2d^2}{3\xi}\cos\left[2\xi + 2v\left(\tfrac{3}{2}\xi\right)^{1/3} + \tfrac{1}{3}d^2\ln|\xi| - c_1\right] + O(|\xi|^{-4/3})\right\}\eta = 0, \tag{7.4.10}$$

where c_1 is a constant and

$$\xi(u) = -\tfrac{2}{3}u^{3/2} + \tfrac{1}{2}d^2\ln u + c_2 + O(u^{-1}),$$

as $u \to \infty$, where c_2 is another constant. Then, provided that the parameter v in (7.4.10) is bounded away from zero, one can show that $\eta(\xi)$ behaves asymptotically like a linear combination of $\cos\xi$ and $\sin\xi$. Hence we have:

THEOREM 7.4.2

If $y - x \geq \delta > 0$, for some positive constant δ, then as $x \to -\infty$,

$$\begin{aligned}|x+y|^{1/4}K(x,y) \sim\ & A\left(\frac{y-x}{2}\right)\sin\left\{\frac{2}{3}\left|\frac{x+y}{2}\right|^{3/2} - \frac{d^2}{2}\ln\left|\frac{x+y}{2}\right|\right\} \\ &+ B\left(\frac{y-x}{2}\right)\cos\left\{\frac{2}{3}\left|\frac{x+y}{2}\right|^{3/2} - \frac{d^2}{2}\ln\left|\frac{x+y}{2}\right|\right\},\end{aligned} \tag{7.4.11}$$

for some functions A and B.

Now we shall discuss the integral equation (7.4.5). Define the operator L_x, for any $f \in L^2(x,\infty)$, by

$$(\mathrm{L}_x f)(y) := \tfrac{1}{2}\int_x^\infty \mathrm{Ai}\left(\frac{y+s}{2}\right) f(s)\,\mathrm{d}s, \tag{7.4.12}$$

then the integral equation (1.8) can be written in the form

$$K(x,y) = k\,\mathrm{Ai}\left(\frac{y+s}{2}\right) + k^2(\mathrm{L}_x^2 K)(x,y). \tag{7.4.13}$$

The operator L_x has the following properties (for further details, see Ablowitz and Segur [1977b]; Ablowitz, Ramani and Segur [1980b]; Hastings and McLeod [1980]),

(i) L_x is a compact, self-adjoint operator on $L^2(x,\infty)$,

(ii) L_x^2 is a compact, positive, self-adjoint operator on $L^2(x,\infty)$,

(iii) $|||\mathrm{L}_x||| \leq 1$ for all finite x and $|||\mathrm{L}_x|||$ increases monotonically as $x \to -\infty$, where $|||\bullet|||$ denotes the operator norm (largest eigenvalue) of L_x in $L^2(x,\infty)$.

(iv) At least in a formal sense, as $x \to -\infty$, $\mathrm{L}_x^2 \to \mathrm{I}$, where I is the identity operator in $L^2(-\infty,\infty)$, that is for $f \in L^2(-\infty,\infty)$

$$f(y) = \tfrac{1}{4}\int_{-\infty}^\infty \mathrm{Ai}\left(\frac{y+s}{2}\right)\left\{\int_{-\infty}^\infty \mathrm{Ai}\left(\frac{s+t}{2}\right) f(t)\,\mathrm{d}t\right\}\mathrm{d}s. \tag{7.4.14}$$

Since $|||\mathrm{L}_x||| \leq 1$, then for $0 < k < 1$, it is possible to express $K(x,y)$ as the sum of the Neumann series

$$K(x,y) = k\sum_{n=0}^\infty \left(\tfrac{1}{2}k\right)^{2n} I_{2n}(x,y), \tag{7.4.15a}$$

with

$$I_0(x,y) := \mathrm{Ai}\left(\frac{x+y}{2}\right), \tag{7.4.15b}$$

$$I_m(t_0,t_{m+1}) := \int_x^\infty \overset{m}{\cdots} \int_x^\infty \prod_{j=0}^\infty \mathrm{Ai}\left(\frac{t_j+t_{j+1}}{2}\right)\mathrm{d}t_1 \ldots \mathrm{d}t_m. \tag{7.4.15c}$$

The integrals in (7.4.15) are over $[x,\infty)$, but our interest is in what happens as $x \to -\infty$. By exploiting (7.4.14) (i.e., by using the fact that $\mathrm{L}_x^2 \to \mathrm{I}$ as $x \to -\infty$), it can be shown for $0 < k < 1$ that

$$K(x,y) = \frac{k(2-k^2)}{2(1-k^2)}\sum_{n=0}^\infty \left(\frac{-k^2}{1-k^2}\right)^n A_n(x,y), \tag{7.4.16a}$$

with

$$A_0(x,y) := \mathrm{Ai}\left(\frac{x+y}{2}\right), \tag{7.4.16b}$$

$$A_n(x,y) := (-1)^n \int_1^\infty \cdot\overset{n}{\cdots}\cdot \int_1^\infty \mathrm{Ai}(x\lambda_n) \prod_{r=1}^{n} W(x\lambda_r, x\lambda_{r-1})\,\mathrm{d}\lambda_n \ldots \mathrm{d}\lambda_1, \tag{7.4.16c}$$

where

$$W(s,t) := \frac{\mathrm{Ai}(s)\,\mathrm{Ai}'(t) - \mathrm{Ai}'(s)\,\mathrm{Ai}(t)}{s-t}, \qquad \lambda_0 := \frac{x+y}{2x}. \tag{7.4.16d}$$

For $x < 0$, the $A_n(x,y)$ are multiple integrals over $(-\infty, x]$. From (7.4.16) it follows that $K(x,y)$ satisfies the integral equation

$$\begin{aligned} K(x,y) = {} & \frac{k(2-k^2)}{2(1-k^2)}\,\mathrm{Ai}\left(\frac{x+y}{2}\right) \\ & - \left(\frac{k^2}{1-k^2}\right)\int_{-\infty}^{x} K(x,t)\,W\left(\frac{x+t}{2}, \frac{x+y}{2}\right)\mathrm{d}t, \end{aligned} \tag{7.4.17}$$

for $0 < k < 1$.

Define the operator W_x, for any $f \in L^2(-\infty, x)$, by

$$(\mathrm{W}_x f)(y) := \int_{\infty}^{x} W\left(\frac{x+t}{2}, \frac{x+y}{2}\right) f(t)\,\mathrm{d}t,$$

then it can be shown that W_x is a self-adjoint, symmetric and positive operator in $L^2(-\infty, x)$.

Using the known asymptotic behavior of $\mathrm{Ai}(x)$ and $\mathrm{Ai}'(x)$ as $x \to -\infty$, i.e.,

$$\begin{aligned} \mathrm{Ai}(x) &\sim \pi^{-1/2}|x|^{-1/4}\cos\left\{\tfrac{2}{3}|x|^{3/2} - \tfrac{1}{4}\pi\right\}, \\ \mathrm{Ai}'(x) &\sim \pi^{-1/2}|x|^{1/4}\sin\left\{\tfrac{2}{3}|x|^{3/2} - \tfrac{1}{4}\pi\right\}, \end{aligned}$$

it can be shown that if $\mu > 1 + \varepsilon/|x|$, for some positive constant ε, then as $x \to -\infty$

$$\begin{aligned} \int_1^\infty \mathrm{Ai}(x\lambda)\,W(x\lambda, x\mu)\,\mathrm{d}\lambda &\sim -\tfrac{1}{2}\,\mathrm{Ai}(x\mu) \\ &+ \tfrac{1}{2}\pi^{-3/2}|x|^{-1/4}\mu^{-1/4}\sin\left\{\tfrac{2}{3}|x|^{3/2}\mu^{3/2} - \tfrac{1}{4}\pi\right\}\ln\left(\frac{\mu^{1/2}+1}{\mu^{1/2}-1}\right) \end{aligned} \tag{7.4.18}$$

(the logarithmic term arises from the integration of $\cos^2(x\lambda) = \frac{1}{2}(1+\cos(2x\lambda))$.

Now define $B_{n,m}(x,y)$ by

$$\begin{aligned} B_{n,m}(x,y) := {} & \frac{(-1)^{n+m}}{m!\,|x|^{1/4}\sqrt{\pi}} \int_1^\infty \cdot\overset{n}{\cdots}\cdot \int_1^\infty \prod_{r=1}^{n} W(x\lambda_r, x\lambda_{r-1}) \\ & \times \mathrm{Re}\left[\lambda_n^{-1/4}\exp(\mathrm{i}\theta_n) \prod_{p=0}^{m-1}\left\{\frac{1}{2\pi\mathrm{i}}\ln\left(\frac{\lambda_n^{1/2}+1}{\lambda_n^{1/2}-1}\right) + p\right\}\right]\mathrm{d}\lambda_n \ldots \mathrm{d}\lambda_1, \end{aligned} \tag{7.4.19a}$$

where

$$\theta_n := \tfrac{2}{3}|x|^{3/2}\lambda_n^{3/2} - \tfrac{1}{4}\pi, \tag{7.4.19b}$$

then we obtain the recurrence relation

$$A_n(x,y) = \tfrac{1}{2}A_{n-1}(x,y) + B_{n-1,1}(x,y) + R_n(x,y), \tag{7.4.20}$$

with $R_n(x,y)$ a remainder term. Similarly, by considering the first integral in $B_{n,m}$, we obtain a second recurrence relation

$$B_{n,m}(x,y) = B_{n-1,m}(x,y) + B_{n-1,m+1}(x,y) + R_{n,m}(x,y), \tag{7.4.21}$$

with $R_{n,m}(x,y)$ another remainder term.

A_n and $B_{n,m}$ are n-fold integrals, however by use of the recurrence relations (7.4.20) and (7.4.21) and doing a summation, we reduce (7.4.16) to a single integral. After doing this final integration we find that if $y - x$ is bounded away from zero, then as $x \to -\infty$

$$\begin{aligned} K(x,y) = \mathrm{Re}\Bigg[&\frac{k}{\sqrt{\pi(1-k^2)}}\left|\frac{x+y}{2}\right|^{-1/4} \\ &\times \exp\left\{\mathrm{i}\left(\frac{2}{3}\left|\frac{x+y}{2}\right|^{3/2} - \frac{\pi}{4}\right)\right\}(1-k^2)^{-(\ln X)/(2\pi\mathrm{i})}\Bigg], \end{aligned} \tag{7.4.22a}$$

where

$$X = \frac{1+\sqrt{\frac{1}{2}|x+y|}}{1-\sqrt{\frac{1}{2}|x+y|}}. \tag{7.4.22b}$$

Since

$$(1-k^2)^{-(\ln X)/(2\pi\mathrm{i})} = X^{-[\ln(1-k^2)]/(2\pi\mathrm{i})} \sim (4|x|/v)^{-[\ln(1-k^2)]/(2\pi\mathrm{i})},$$

as $x \to -\infty$, where $v = \frac{1}{2}(y-x)$, then we have the (7.4.11) with $d^2 = -\pi^{-1}\ln(1-k^2)$.

In order to rigorously prove that (7.4.22) actually is the leading term in the asymptotic behavior of $K(x,y)$ as $x \to -\infty$, we define

$$\mathrm{Ci}(t) := \mathrm{Ai}(\omega t), \qquad \omega = \exp\left(\tfrac{2}{3}\pi\mathrm{i}\right), \tag{7.4.23}$$

and, for $0 > y > x$,

$$\tilde{K}_1(x,y) = \mathrm{Re}\left(2k\exp\left(-\tfrac{1}{3}\pi\mathrm{i}\right)\mathrm{Ci}(z)\left(\frac{|z/x|^{1/2}-1}{|z/x|^{1/2}+1}\right)^{\frac{1}{2}\mathrm{i}d^2}\right), \tag{7.4.24a}$$

$$\tilde{K}_2(x,y) = \mathrm{Re}\left(2k\mathrm{i}\exp\left(-\tfrac{1}{3}\pi\mathrm{i}\right)z^{-1/2}\,\mathrm{Ci}(z)\left(\frac{|z/x|^{1/2}-1}{|z/x|^{1/2}+1}\right)^{\frac{1}{2}\mathrm{i}d^2}\right), \tag{7.4.24b}$$

with $z = \frac{1}{2}(x+y)$, and $d^2 \equiv \pi^{-1}\ln(1-k^2)$, then it can be shown that

$$\begin{aligned}\tilde{K}_1(x,y) &+ \frac{k^2}{1-k^2}\int_{-\infty}^{x} \tilde{K}_1(x,t)\, W\left(\frac{x+t}{2},\frac{x+y}{2}\right) dt \\ &= \frac{k(2-k^2)}{2(1-k^2)}\,\mathrm{Ai}\left(\frac{x+y}{2}\right) + \frac{B}{x+y}\,\mathrm{Ai}'\left(\frac{x+y}{2}\right) + \varepsilon_1(x,y), \qquad (7.4.25a)\\ \tilde{K}_2(x,y) &+ \frac{k^2}{1-k^2}\int_{-\infty}^{x} \tilde{K}_2(x,t)\, W\left(\frac{x+t}{2},\frac{x+y}{2}\right) dt \\ &= \frac{C}{x+y}\,\mathrm{Ai}'\left(\frac{x+y}{2}\right) + \varepsilon_2(x,y), \qquad (7.4.25b)\end{aligned}$$

with B, C real constants, and ε_1, ε_2 satisfy the order relations

$$\varepsilon_j(x,y) = \begin{cases} o(|x+y|^{-3/4}) + O(|x(y-x)|^{-1}|x+y|^{-1/4}\ln|x|), & \text{if } |y-x| \geq |x|^{-1/2}, \\ O(|x+y|^{-1/4}\ln|x-y|), & \text{if } |y-x| \leq |x|^{-1/2}. \end{cases}$$

(For $x < y < 0$, we make similar definitions for $\tilde{K}_1(x,y)$ and $\tilde{K}_2(x,y)$.) Therefore if we define

$$\tilde{K}(x,y) = \tilde{K}_1(x,y) - (B/C)\,\tilde{K}_2(x,y),$$

then

$$\left(\mathrm{I} + \frac{k^2}{1-k^2}\,\mathrm{W}_x\right)\tilde{K}(x,y) = \frac{k(2-k^2)}{2(1-k^2)}\,\mathrm{Ai}\left(\frac{x+y}{2}\right) + \varepsilon(x,y), \qquad (7.4.26)$$

where

$$\varepsilon(x,y) = \varepsilon_1(x,y) - (B/C)\,\varepsilon_2(x,y).$$

Since W_x is a self-adjoint, symmetric and positive operator on $L^2(-\infty,x)$, then it follows that (7.4.24a) represents the leading term in the asymptotic behavior of $K(x,y)$ as $x \to -\infty$, provided $y-x$ is bounded away from zero.

We shall now make some comments on the "proof" of (7.4.4), the following being just a heuristic derivation of the result. In the proof of (7.4.3), we obtained an expression for the asymptotic behavior of $K(x,y)$ as $x \to -\infty$ with $y-x$ bounded away from zero. As $y \to x$, there is a boundary layer effect which makes the evaluation of $K(x,y)$, on the diagonal $y = x$, directly from the integral equation (7.4.5) very difficult. To derive (7.4.4), we use the known asymptotic behavior for $K(x,y)$ on the edge of the boundary layer region (i.e., (7.4.11), to "guess" the form of the asymptotic behavior of $K(x,y)$ in the boundary layer region. By requiring that this is a solution of the differential equation (7.4.6) which satisfies (7.4.2) on the diagonal $y = x$, then matching with the known behavior on the edge of the boundary layer region yields (7.4.4).

In the above, it was shown that if $y - x$ is bounded away from zero, then the solutions $K(x,y)$ of the integral equation (7.4.5) has the following asymptotic behavior as $x \to -\infty$ [which is equivalent to (7.4.24a)]

$$K(x,y) \sim k\pi^{-1/2} \left|\frac{x+y}{2}\right|^{-1/4} \times$$
$$\mathrm{Re}\left[\exp\left\{\mathrm{i}\left(\frac{2}{3}\left|\frac{x+y}{2}\right|^{3/2} - \frac{\pi}{4}\right)\right\}\left(\frac{\sqrt{|x+y|}-\sqrt{2|x|}}{\sqrt{|x+y|}+\sqrt{2|x|}}\right)^{\frac{1}{2}\mathrm{i}d^2}\right]. \quad (7.4.27)$$

A close examination of the proof of Theorem 7.4.3 shows that (7.4.27) is valid unless $(y-x) = O(|x|^{-1/2})$, as $x \to -\infty$. (Note that (7.4.27) tends to no limit as $y \to x$.)

In (7.4.27) let $s = \frac{1}{2}(y-x)\sqrt{|x|}$, then as $x \to -\infty$,

$$K(x,y) \sim k\pi^{-1/2}|x|^{-1/4}\mathrm{Re}\left[\exp\left\{\mathrm{i}\left(\tfrac{2}{3}|x|^{3/2} - \tfrac{1}{4}\pi\right)\right\}\mathrm{e}^{\mathrm{i}s}\left(\tfrac{3}{4}s|x|^{-3/2}\right)^{\frac{1}{2}\mathrm{i}d^2}\right], \quad (7.4.28)$$

provided that $s = o(|x|^{3/2})$, as $x \to -\infty$.

Ablowitz and Segur [1977b] showed that $K(x,y)$ satisfies the differential equation (7.4.6). In (7.4.6), let $u = -\frac{1}{2}(x+y)$, $v = \frac{1}{2}(y-x)$, $K(x,y) = F(u,v)$, then

$$\frac{\partial^2 F}{\partial u^2}(u,v) = -uF(u,v) + 2[F(u+v,0)]^2 F(u,v). \quad (7.4.29)$$

Also from (7.4.28) we see that as $u \to \infty$ (i.e., as $x \to -\infty$)

$$F(u,v) \sim k\pi^{-1/2}(u+v)^{-1/4}\cos\{\tfrac{2}{3}(u+v)^{3/2} - \tfrac{3}{4}d^2\ln(u+v)$$
$$+ s + \tfrac{1}{2}d^2\ln s - d^2\ln 2 - \tfrac{1}{4}\pi\}, \quad (7.4.30)$$

with $s = -v\sqrt{(u+v)}$. If, v is a parameter which is bounded away from zero (i.e., effectively assuming the $v = O(1)$ as $x \to -\infty$), then (7.4.30) is valid for $|s|$ sufficiently large, dependent on u (and hence on x). Note that equations (7.4.28) [and therefore also (7.4.30)] is only valid provided that as $x \to -\infty$, $s = O(|x|^\alpha)$ for an α such that $0 < \varepsilon \le \alpha \le \frac{3}{2} - \varepsilon$, for a constant ε (we are assuming that $\alpha = \frac{1}{2}$). We now wish to determine the asymptotic behavior of $F(u,v)$ which is valid for all s.

Suppose that the leading order term in the asymptotic expansion of $F(u,v)$, in the boundary layer region, has the form

$$F_0(u,v) = (u+v)^{-1/4}\left\{f(s)\cos\theta(u,v) + g(s)\sin\theta(u,v)\right\}, \quad (7.4.31a)$$
$$\theta(u,v) = \tfrac{2}{3}(u+v)^{3/2} + \tfrac{3}{4}d^2\ln(u+v) - \theta_0, \quad (7.4.31b)$$

as $u \to -\infty$, where θ_0 is the constant we require to determine and for real s, $f(s)$ and $g(s)$ are real-valued functions. Substituting (7.4.31) into (7.4.29), then to leading order, it is seen that f and g satisfy

$$\frac{\mathrm{d}f}{\mathrm{d}s} = g, \qquad s\frac{\mathrm{d}g}{\mathrm{d}s} + (s+d^2)f = 0,$$

i.e.,

$$s\frac{\mathrm{d}^2 f}{\mathrm{d}s^2} + (s + d^2) f = 0. \tag{7.4.32}$$

We require the real solution (for real s) of this satisfying the initial conditions

$$f(0) = 0, \qquad \frac{\mathrm{d}f}{\mathrm{d}s}(0) = d,$$

in order that (7.4.2) is satisfied. (7.4.32) is a special case of Whittaker's equation

$$\frac{\mathrm{d}^2 W}{\mathrm{d}z^2} = \left(\frac{1}{4} - \frac{k}{z} + \frac{m^2 - \frac{1}{4}}{z^2}\right) W,$$

with $m = \frac{1}{2}$, $k = -\frac{1}{2}\mathrm{i}d^2$, $z = 2\mathrm{i}s$ (see Chapter 16 of Whittaker and Watson [1927]).

Therefore

$$f(s) = -\tfrac{1}{2}\mathrm{i}dM_{-\frac{1}{2}\mathrm{i}d^2,\frac{1}{2}}(2\mathrm{i}s), \tag{7.4.33}$$

with $M_{k,m}(z)$ Whittaker's function (cf. Whittaker and Watson [1927]). Hence using the known asymptotic behavior of $M_{\mathrm{i}\kappa,\mu}(\mathrm{i}\xi)$ for real κ, μ, ξ as $\xi \to \infty$

$$M_{\mathrm{i}\kappa,\mu}(\mathrm{i}\xi) \sim \frac{2\exp\left\{\frac{1}{4}\pi\left[2\kappa + \mathrm{i}(1+2\mu)\right]\right\}}{\left|\Gamma\left(\mu + \frac{1}{2} + \mathrm{i}\kappa\right)\right|} \cos\left\{-\kappa \ln\xi + \tfrac{1}{2}\xi + \delta - \tfrac{1}{4}\pi(1+2\mu)\right\},$$

where $\delta = \arg\left\{\Gamma\left(\mu + \frac{1}{2} + \mathrm{i}\kappa\right)\right\}$ (cf. Buchholz [1969, p.92]), we see that as $s \to \infty$,

$$f(s) \sim k\pi^{-1/2} \cos\left\{s + \tfrac{1}{2}d^2 \ln(2s) + \beta - \tfrac{1}{2}\pi\right\}, \tag{7.4.34a}$$

$$g(s) \sim -k\pi^{-1/2} \sin\left\{s + \tfrac{1}{2}d^2 \ln(2s) + \beta - \tfrac{1}{2}\pi\right\}, \tag{7.4.34b}$$

with $\beta = \arg\left\{\Gamma(1 - \frac{1}{2}\mathrm{i}d^2)\right\}$. Substituting (7.4.34) into (7.4.31) shows that as $s \to -\infty$

$$\begin{aligned} F_0(u,v) \sim k\pi^{-1/2}(u+v)^{-1/4} \cos\{&\tfrac{2}{3}(u+v)^{3/2} - \tfrac{3}{4}d^2 \ln(u+v) \\ &- \theta_0 + s + \tfrac{1}{2}d^2\ln(2s) + \beta - \tfrac{1}{2}\pi\}. \end{aligned} \tag{7.4.35}$$

A comparison of (7.4.35) and (7.4.30) formally yields (7.4.4), that is

$$\theta_0 = \tfrac{3}{2}d^2 \ln 2 + \arg\left\{\Gamma(1 - \tfrac{1}{2}\mathrm{i}d^2)\right\} - \tfrac{1}{4}\pi.$$

A rigorous proof that (7.4.31) with $f(s)$ as given by (7.4.33) is the leading term in the asymptotic expansion of $K(x,y)$ as $x \to -\infty$ in the boundary layer region, should be obtained by considering the integral equation (7.4.17) [as was the case for the proof of (7.4.3)].

7.4.3 The Inverse Monodromy Transform Approach. Flaschka and Newell [1980] proved that the monodromy data for the second Painlevé equation

$$\frac{d^2y}{dx^2} = 2y^3 + xy, \tag{7.4.40}$$

did not depend upon x. Therefore it is possible (theoretically at least) to compute the connection formulae for this equation using the asymptotic solution of the direct problem of I.M.T. for the associated system

$$\boldsymbol{w}_\zeta = \begin{pmatrix} -i(4\zeta^2 + 2y^2 + x) & 4\zeta y + 2iy' \\ 4\zeta y - 2iy' & i(4\zeta^2 + 2y^2 + x) \end{pmatrix} \boldsymbol{w}, \tag{7.4.41}$$

with $' \equiv d/dx$. That is, to determine the monodromy data $\{a, b, c\}$, given specific asymptotic behaviors of $y(x)$ as $x \to \pm\infty$ (which are obtained from a qualitative analysis of equation (7.4.40) — independent of the I.M.T.). The asymptotic solution of the direct problem for the system (7.4.41) leads to the expression of the monodromy data in two different ways: (i), by means the leading term of the asymptotics of the solution $y(x)$ as $x \to \infty$

$$y(x) \sim k\,\mathrm{Ai}(x) \sim \tfrac{1}{2} k \pi^{-1/2} x^{-1/4} \exp\left(-\tfrac{2}{3} x^{3/2}\right), \tag{7.4.42}$$

for some constant k; (ii), the leading term of the asymptotics of the same solution as $x \to -\infty$

$$y(x) \sim d|x|^{-1/4} \sin\left\{\tfrac{2}{3}|x|^{3/2} - \tfrac{3}{4} d^2 \ln|x| - \theta_0\right\}, \tag{7.4.43}$$

for some constants d and θ_0 (recall Theorem 7.4.1 above). Since the monodromy data does not depend on x, then we are able to determine the connection formulae (7.4.3,4). As mentioned in §7.3.4 above, the monodromy data for real solutions of equation (7.4.40) is fully characterized by the complex parameter a and the Stokes multiplier matrix

$$\mathbf{G}_1 = \begin{pmatrix} 1 & 0 \\ a & 1 \end{pmatrix}. \tag{7.4.44}$$

First, following Flaschka and Newell [1980], we consider the asymptotic behavior as $x \to \infty$, i.e., (7.4.42). The objective is to determine the relationship between the monodromy data a and the parameter k. The exponential decay of (7.4.42) as $x \to \infty$ ensures that scattering theory can be applied to the equation

$$\boldsymbol{w}_x = \begin{pmatrix} -i\zeta & y \\ y & i\zeta \end{pmatrix} \boldsymbol{w}. \tag{7.4.45}$$

In particular, it follows that there exists an eigenfunction $\boldsymbol{w}(x;\zeta)$ of (7.4.45) with asymptotic behavior as $x \to \infty$

$$\boldsymbol{w}(x;\zeta) \sim \exp(i\zeta x)\begin{pmatrix} 0 \\ 1 \end{pmatrix},$$

which exists for x sufficiently large and has a triangular representation

$$\boldsymbol{w}(x;\zeta)\exp(-\mathrm{i}\zeta x) = \begin{pmatrix}0\\1\end{pmatrix} + \int_x^\infty K(x,s)\exp[\mathrm{i}\zeta(s-x)]\,\mathrm{d}s. \tag{7.4.46}$$

By repeated integration of (7.4.46) by parts, one can derive the asymptotic expansion

$$\boldsymbol{w}(x;\zeta)\exp(-\mathrm{i}\zeta x) \sim \begin{pmatrix}0\\1\end{pmatrix} + \zeta^{-1}\boldsymbol{v}_1, \tag{7.4.47}$$

which is valid in the upper half ζ-plane. $\boldsymbol{w}^{(2)}(x;\zeta) = \boldsymbol{w}(x;\zeta)\exp\left(-\frac{4}{3}\mathrm{i}\zeta^2\right)$ is precisely the solution of equation (7.4.41) on $\arg\zeta = 0$ and, since the expansion (7.4.47) is valid in S_1, S_2 and S_3, then the Stokes parameter b vanishes, and so the Stokes parameter a is real [recall equation (7.3.39)]. In this case the following integral equation can be derived

$$\begin{pmatrix}0 & 1\\1 & 0\end{pmatrix}\boldsymbol{w}^{(2)}(x;\zeta)\exp\{\theta(x;\zeta)\} = \begin{pmatrix}1\\0\end{pmatrix} + \frac{a}{2\pi\mathrm{i}}\int_{-\infty}^{\infty}\frac{\boldsymbol{w}^{(2)}(\xi;\zeta)\exp\{\theta(\xi;\zeta)\}}{\zeta+\xi}\,\mathrm{d}\xi, \tag{7.4.48}$$

where $\theta(x;\zeta) = \mathrm{i}(\frac{4}{3}\zeta^3 + x\zeta)$. Substituting (7.4.46) into (7.4.48) and taking a Fourier transform yields the Gel'fand-Levitan-Marchenko equation equation (7.4.5) with k replaced by a. Hence the parameter k in the asymptotic form (7.4.42) equals the Stokes parameter a.

Next, following Suleimanov [1987], we consider the asymptotics as $x \to -\infty$. The scheme of the proof is to replace y and y' in (7.4.41) by the leading terms in their asymptotic expansions as $x \to -\infty$, i.e. (7.4.43), and thus obtain a singularly perturbed system of differential equation in the small parameter $(-x)^{-1}$. Letting this parameter tend to zero expresses d and θ_0 in terms of the Stokes parameter a.

For convenience, we set $\xi = \zeta(-x)^{-1/2}$ and $\varepsilon = (-x)^{-3/4}$ in equations (7.4.41), then substituting (7.4.43) into this system yields

$$\varepsilon^2\frac{\mathrm{d}\boldsymbol{w}}{\mathrm{d}\xi} = \mathbf{T}(\xi,\varepsilon)\boldsymbol{w}, \tag{7.4.49a}$$

where

$$\mathbf{T} = -\mathrm{i}[4\xi^2 - 1 + 2\varepsilon^2 d^2\sin^2\theta]\,\boldsymbol{\sigma}_3 + \varepsilon[4\xi d\sin\theta\,\boldsymbol{\sigma}_1 + 2d\cos\theta\,\boldsymbol{\sigma}_2], \tag{7.4.49b}$$

with

$$\theta = \tfrac{2}{3}|x|^{3/2} - \tfrac{3}{4}d^2\ln|x| - \theta_0, \tag{7.4.49c}$$

$$\boldsymbol{\sigma}_1 = \begin{pmatrix}0 & 1\\1 & 0\end{pmatrix}, \qquad \boldsymbol{\sigma}_2 = \begin{pmatrix}0 & -\mathrm{i}\\\mathrm{i} & 0\end{pmatrix}, \qquad \boldsymbol{\sigma}_3 = \begin{pmatrix}1 & 0\\0 & -1\end{pmatrix}. \tag{7.4.49d}$$

Next we consider the asymptotic expansion of (7.4.49) as $\varepsilon \to 0$ in a region where ξ is bounded away from $\frac{1}{2}$. Using the standard WKB method, we seek a solution of (7.4.49) in the form

$$\hat{\mathbf{W}} = \exp(-S\boldsymbol{\sigma}_3)\mathbf{V}_0 + \varepsilon\exp(S\boldsymbol{\sigma}_3)\mathbf{V}_1, \tag{7.4.50}$$

where $S = \mathrm{i}(\frac{4}{3}\zeta^3 + \zeta x) = \mathrm{i}(\frac{4}{3}\xi^3 - \xi)/\varepsilon^2$. Substituting (7.4.50) into (7.4.49) and equating coefficients of powers of ε yields, at order ε

$$\mathbf{V}_1 = \frac{(2\xi d\sin\theta\,\boldsymbol{\sigma}_2 - d\cos\theta\,\boldsymbol{\sigma}_1)}{4\xi^2 - 1}\mathbf{V}_0, \tag{7.4.51}$$

and at order ε^2

$$\frac{\mathrm{d}\mathbf{V}_0}{\mathrm{d}\xi} = \frac{2\mathrm{i}d^2}{4\xi^2 - 1}\mathbf{V}_0. \tag{7.4.52}$$

Hence the WKB approximation gives

$$\hat{\mathbf{W}} = \left[\exp(-S\boldsymbol{\sigma}_3) + \frac{\varepsilon(2\xi d\sin\theta\,\boldsymbol{\sigma}_2 - d\cos\theta\,\boldsymbol{\sigma}_1)}{4\xi^2 - 1}\exp(S\boldsymbol{\sigma}_3)\right] \times \exp\left\{\tfrac{1}{2}\mathrm{i}d^2\,\boldsymbol{\sigma}_3 \ln\left(\frac{\xi - \frac{1}{2}}{\xi + \frac{1}{2}}\right)\right\}, \tag{7.4.53}$$

which is defined on the complex ξ-plane with a cut along the segment $[-\frac{1}{2}, \frac{1}{2}]$ (so that it tends to the unit matrix at infinity). Consequently, Suleimanov [1987] proves the following theorem

THEOREM 7.4.3

If $\varepsilon/|4\xi^2 - 1| = o(1)$, then the matrix defined by (7.4.53) is a fundamental solution of the equation

$$\varepsilon^2\frac{\mathrm{d}\hat{\mathbf{W}}}{\mathrm{d}\xi} = [\mathbf{T}(\xi,\varepsilon) + \varepsilon^3\mathbf{R}(\xi,\varepsilon)]\hat{\mathbf{W}}, \tag{7.4.54}$$

where the elements R^{ij} of the matrix $\mathbf{R}$ satisfy

$$R^{ij} = \begin{cases} O(\varepsilon|4\xi^2 - 1|^{-1}[1 + |4\xi^2 - 1|^{-2})], & \text{for } i = j, \\ O((1 + |\xi|)|4\xi^2 - 1|^{-1}[1 + |4\xi^2 - 1|^{-1})], & \text{for } i \neq j. \end{cases} \tag{7.4.55}$$

This result is proved by comparing equations (7.4.49) and (7.4.54) and, essentially, writing an integral equation for the difference. In particular, in the course of the proof it is shown that

$$|\hat{\mathbf{W}}(2,\varepsilon) - \mathbf{W}_1(2,\varepsilon)| \le M\varepsilon, \tag{7.4.56a}$$

$$|\hat{\mathbf{W}}^{(1)}(0,\varepsilon) - \mathbf{W}_2^{(1)}(0,\varepsilon)| \le M\varepsilon, \tag{7.4.56b}$$

where $\mathbf{W}_j$ is the fundamental matrix defined in §7.3.4 above, the index $^{(i)}$ indicates the ith column of the matrix and M is a constant.

The next step is to derive approximate expressions for $\mathbf{W}_1(0,\varepsilon)$ and we can then use the inequality (7.4.56b) to calculate the first column of $\mathbf{G}_1$ by letting $\varepsilon \to 0$ in the expression $[\mathbf{W}_1(0,\varepsilon)]^{-1}\mathbf{W}_2^{(1)}(0,\varepsilon)$.

An approximation for the fundamental matrix $\mathbf{W}_1(0,\varepsilon)$ can be obtained from the system (7.4.49). In fact, the value of this matrix at $\xi = 2$ is known from (7.4.56); therefore, it is sufficient to construct a matrix satisfying the initial condition for $\xi = 2$ and the system (7.4.49), asymptotically with respect to ε. The complication is that the matrix constructed above leads to a remainder with a non-removable singularity at $\xi = \frac{1}{2}$. Henceforth we shall assume that $\xi \in [-2,2]$ and that all matrices are bounded in norm equal to the maximum of the moduli of their elements.

THEOREM 7.4.4

If $\varepsilon^\kappa + \frac{1}{2} \le \xi \le 2$ *and* $0 < \kappa < 1$, *then*

$$\mathbf{W}_1(\xi,\varepsilon) = \hat{\mathbf{W}}(\xi,\varepsilon)[\mathbf{I} + O(\varepsilon^{1-\kappa})], \tag{7.4.57}$$

and if $0 \le \xi \le \frac{1}{2} - \varepsilon^\kappa$ *and* $0 < \kappa < 1$, *then*

$$\mathbf{W}_1(\xi,\varepsilon) = \hat{\mathbf{W}}(\xi,\varepsilon)[\mathbf{I} + O(\varepsilon^{1-\kappa})]\exp\left(\tfrac{1}{2}\pi d^2\,\boldsymbol{\sigma}_3\right)\mathbf{K}(\varepsilon), \tag{7.4.58}$$

where $\mathbf{K}(\varepsilon)$ *is a non-singular matrix.*

This result is proved by writing the system (7.4.49) as a system of integral equations.

Consequently, the question of the value of $\mathbf{W}_1(0,\varepsilon)$ is thus reduced to the estimation of $\mathbf{K}(\varepsilon)$ in the limit as $\varepsilon \to 0$. Prior to doing this, we shall examine the behavior of a solution of (7.4.49) in the neighborhood of $\xi = \frac{1}{2}$. Introducing the variable $\lambda = (\xi - \frac{1}{2})/\varepsilon$, then equations (7.4.49) become

$$\frac{\mathrm{d}\boldsymbol{v}}{\mathrm{d}\lambda} = \{-\mathrm{i}[4\lambda+\varepsilon(4\lambda^2+d^2\sin\theta)]\,\boldsymbol{\sigma}_3+4\varepsilon\lambda d\sin\theta\,\boldsymbol{\sigma}_1+2d(\sin\theta\,\boldsymbol{\sigma}_1+\cos\theta\,\boldsymbol{\sigma}_2)\}\boldsymbol{v}. \tag{7.4.59}$$

Dropping the $O(\varepsilon)$ term gives

$$\frac{\mathrm{d}\boldsymbol{v}}{\mathrm{d}\lambda} = \begin{pmatrix} -4\mathrm{i}\lambda & -2\mathrm{i}d\exp(\mathrm{i}\theta) \\ 2\mathrm{i}d\exp(\mathrm{i}\theta) & 4\mathrm{i}\lambda \end{pmatrix}\boldsymbol{v}, \tag{7.4.60}$$

which has the fundamental solution

$$\tilde{\mathbf{W}} = \begin{pmatrix} D_{\frac{1}{2}\mathrm{i}d^2}(\sqrt{8\mathrm{i}}\,\lambda) & -\dfrac{d\mathrm{e}^{\mathrm{i}\theta}}{\sqrt{2\mathrm{i}}}D_{-\frac{1}{2}\mathrm{i}d^2-1}(\sqrt{-8\mathrm{i}}\,\lambda) \\ -\dfrac{d\mathrm{e}^{\mathrm{i}\theta}}{\sqrt{-2\mathrm{i}}}D_{\frac{1}{2}\mathrm{i}d^2-1}(\sqrt{8\mathrm{i}}\,\lambda) & D_{-\frac{1}{2}\mathrm{i}d^2}(\sqrt{-8\mathrm{i}}\,\lambda) \end{pmatrix}, \tag{7.4.61}$$

where $D_\nu(z)$ is the parabolic cylinder function (cf. Whittaker and Watson [1927]). Using the known asymptotics of the parabolic cylinder function, i.e.

$$D_\nu(z) \sim z^\nu \exp\left(-\tfrac{1}{4}z^2\right), \qquad |\arg z| < \tfrac{3}{4}\pi,$$

yields, as $\lambda \to +\infty$

$$\begin{aligned}\tilde{\mathbf{W}}(\lambda) = \exp\left\{\left[\tfrac{1}{2}\mathrm{i}d^2 \ln|\lambda| - 2\mathrm{i}\lambda^2 + \tfrac{3}{4}\mathrm{i}d^2 \ln 2\right] \boldsymbol{\sigma}_3 - \tfrac{1}{8}\pi d^2\right\} \\ \times [\mathbf{I} + O(|\lambda|^{-1})],\end{aligned} \tag{7.4.62}$$

and as $\lambda \to -\infty$

$$\begin{aligned}\tilde{\mathbf{W}}(\lambda) &= \exp\left\{\left[\tfrac{1}{2}\mathrm{i}d^2 \ln|\lambda| - 2\mathrm{i}\lambda^2 + \tfrac{3}{4}\mathrm{i}d^2 \ln 2\right] \boldsymbol{\sigma}_3 + \tfrac{3}{8}\pi d^2\right\} \\ &\times \begin{pmatrix} 1 & \dfrac{2\sqrt{\mathrm{i}\pi}\exp\left(-\frac{1}{4}\pi d^2\right)}{d\Gamma\left(\frac{1}{2}\mathrm{i}d^2\right)\exp(-\mathrm{i}\theta)} \\ \dfrac{2\sqrt{-\mathrm{i}\pi}\exp\left(-\frac{1}{4}\pi d^2\right)}{d\Gamma\left(-\frac{1}{2}\mathrm{i}d^2\right)\exp(\mathrm{i}\theta)} & 1 \end{pmatrix} \\ &\times [\mathbf{I} + O(|\lambda|^{-1})].\end{aligned} \tag{7.4.63}$$

By comparing equations (7.4.59) and (7.4.60) and using equations (7.4.62,63), it can be shown that the fundamental solution of equation (7.4.59) also satisfies an integral equation and using this the following result can be proved.

THEOREM 7.4.5

In the domain where $\varepsilon|\lambda|^3 = o(1)$ we have

$$\mathbf{W}_1(\xi, \varepsilon) = \tilde{\mathbf{W}}(\lambda)[\mathbf{I} + O(\varepsilon(1 + |\lambda|^3))]\mathbf{C}(\varepsilon), \tag{7.4.64}$$

where $\mathbf{C}(\varepsilon)$ is a non-singular matrix and $\tilde{\mathbf{W}}(\lambda)$ is given by (7.4.63).

The conditions

$$\varepsilon|\lambda|^3 = o(1), \qquad \varepsilon^\kappa + \tfrac{1}{2} \le \xi \le 2,$$

for $0 < \kappa < 1$, are both satisfied when

$$\varepsilon^\kappa \le \xi - \tfrac{1}{2} \le 2\varepsilon^\kappa, \qquad \tfrac{2}{3} < \kappa < 1.$$

For these values of ξ, equations (7.4.57) and (7.4.64) also hold. Moreover $|\xi - \frac{1}{2}| = o(1)$ and $\lambda^{-1} = o(1)$ and equations (7.4.53) and (7.4.57) imply that

$$\begin{aligned}\mathbf{W}_1(\xi, \varepsilon) = \exp\left\{\left[\tfrac{1}{2}\mathrm{i}d^2 \ln|\lambda| - 2\mathrm{i}\lambda^2 + \tfrac{3}{4}\mathrm{i}d^2 \ln 2\right] \boldsymbol{\sigma}_3\right\} \\ \times [\mathbf{I} + O(\varepsilon^{1-\kappa}) + O(\varepsilon^{3\kappa-2})],\end{aligned} \tag{7.4.65}$$

and equations (7.4.62) and (7.4.64) imply that

$$\begin{aligned} \mathbf{W}_1(\xi,\varepsilon) = \exp\left\{\left[\tfrac{1}{2}\mathrm{i}d^2\ln|\lambda| - 2\mathrm{i}\lambda^2 + \tfrac{3}{4}\mathrm{i}d^2\ln 2\right]\boldsymbol{\sigma}_3 - \tfrac{1}{8}\pi d^2\right\} \\ \times [\mathbf{I} + O(\varepsilon^{1-\kappa}) + O(\varepsilon^{3\kappa-2})]\mathbf{C}(\varepsilon). \end{aligned} \tag{7.4.66}$$

Since $\xi - \frac{1}{2} = \varepsilon\lambda$, we deduce from equations (7.4.65,66) that for $\nu > 0$

$$\mathbf{C}(\varepsilon) = \exp\left\{\left[\tfrac{1}{2}\mathrm{i}d^2\ln|\varepsilon| + \tfrac{1}{3}\mathrm{i}\varepsilon^{-2} - \tfrac{3}{4}\mathrm{i}d^2\ln 2\right]\boldsymbol{\sigma}_3 + \tfrac{1}{8}\pi d^2\right\}[\mathbf{I} + O(\varepsilon^{\nu})]. \tag{7.4.67}$$

Therefore by matching the asymptotic expansions for $\mathbf{W}_1$ for ξ bounded away from $\frac{1}{2}$ and ξ close to $\frac{1}{2}$, we can determine the asymptotic expansion for $\mathbf{C}(\varepsilon)$ as $\varepsilon \to 0$. Similarly, it can be shown that the elements of the matrix $\mathbf{K}(\varepsilon)$ have the following asymptotic representations (with ν a strictly positive constant)

$$K^{11} = K^{22}[1 + O(\varepsilon^{\nu})] = \exp\left(\tfrac{1}{2}\pi d^2\right)[1 + O(\varepsilon^{\nu})], \tag{7.4.68a}$$

$$\begin{aligned} K^{21} = K^{12*}[1 + O(\varepsilon^{\nu})] = \exp\left(\tfrac{1}{4}\pi d^2 - \mathrm{i}\theta + \tfrac{2}{3}\mathrm{i}\varepsilon^{-2} - \tfrac{3}{2}\mathrm{i}d^2\ln 2\right) \\ \times \frac{\exp\left(\mathrm{i}d^2\ln\varepsilon\right)2\sqrt{-\mathrm{i}\pi}}{d\Gamma\left(-\tfrac{1}{2}\mathrm{i}d^2\right)}[1 + O(\varepsilon^{\nu})], \end{aligned} \tag{7.4.68b}$$

where * denotes the complex conjugate.

From equations (7.4.56b) and (7.4.58)

$$\begin{pmatrix} 1 \\ a \end{pmatrix} = \lim_{\varepsilon\to 0}\left\{\frac{\exp\left(-\tfrac{1}{2}\pi d^2\right)}{\det\mathbf{K}(\varepsilon)}\begin{pmatrix} K^{11}(\varepsilon) \\ K^{21}(\varepsilon)\end{pmatrix}\right\}, \tag{7.4.69}$$

and so from equations (7.4.68) we have

$$\begin{aligned} -a &= \lim_{\varepsilon\to 0}\frac{d\exp\left(-\tfrac{3}{4}\pi d^2\right)}{d^2\exp\left(\tfrac{1}{2}\pi d^2\right) - 4\pi|\Gamma\left(-\tfrac{1}{2}\mathrm{i}d^2\right)|^{-2}} \\ &\quad\times\frac{2\sqrt{-\mathrm{i}\pi}}{\Gamma\left(-\tfrac{1}{2}\mathrm{i}d^2\right)}\exp\left(\mathrm{i}d^2\ln\varepsilon - \mathrm{i}\theta + \tfrac{2}{3}\mathrm{i}\varepsilon^{-2} - \tfrac{3}{2}\mathrm{i}d^2\ln 2\right), \\ &= \lim_{\varepsilon\to 0}\frac{2\sqrt{-\mathrm{i}\pi}}{d\Gamma\left(-\tfrac{1}{2}\mathrm{i}d^2\right)}\exp\left(\mathrm{i}d^2\ln\varepsilon - \tfrac{1}{4}\pi d^2 - \mathrm{i}\theta + \tfrac{2}{3}\mathrm{i}\varepsilon^{-2} - \tfrac{3}{2}\mathrm{i}d^2\ln 2\right), \end{aligned} \tag{7.4.70}$$

where we have used the result

$$|\Gamma\left(-\tfrac{1}{2}\mathrm{i}d^2\right)|^{-2} = \frac{d^2\sinh(\tfrac{1}{2}\pi d^2)}{2\pi}.$$

Taking absolute values in (7.4.70) yields (7.4.3) and comparing the arguments yields (7.4.4).

7.5 Properties of the Painlevé equations.

Boutroux, in two long memoirs [1913, 1914], studied the asymptotics of solutions of the first Painlevé equation in considerable detail. For the first Painlevé equation

$$\frac{d^2y}{dx^2} = 6y^2 + x, \tag{7.5.1}$$

if we make the transformation

$$y(x) = x^{1/2}Y(X), \qquad \text{with} \qquad X = \tfrac{4}{5}x^{5/4},$$

then $Y(X)$ satisfies

$$\frac{d^2Y}{dX^2} = 6Y^2 + 1 - \frac{1}{X}\frac{dY}{dX} + \frac{4Y}{25X^2}.$$

This equation may be compared with the equation

$$\frac{d^2W}{dX^2} = 6W^2 + 1,$$

which has general solution

$$W(X) = \wp(X - k; -2, g_3),$$

where $\wp(X; g_2, g_3)$ is the Weierstrass elliptic function, with k, g_2, g_3 constants (cf. Whittaker and Watson [1927]). This comparison suggests that for X sufficiently large

$$Y(X) \sim \wp(X - k; -2, g_3)$$

(for suitable k and g_3), that is for the first Painlevé transcendent

$$y(x) \sim x^{1/2}\wp\left(\tfrac{4}{5}x^{5/4}; -2, g_3\right), \qquad \text{as} \qquad |x| \to \infty.$$

Boutroux [1913, 1914] studied this relationship extensively, and determined the regions where it is valid. Similarly, for the second Painlevé equation

$$\frac{d^2y}{dx^2} = 2y^3 + xy + \alpha, \tag{7.5.2}$$

if we make the transformation

$$y = x^{1/2}Y(X), \qquad \text{with} \qquad X = \tfrac{2}{3}x^{3/2},$$

then we obtain

$$\frac{d^2Y}{dX^2} = 2Y^3 + Y - \frac{1}{X}\frac{dY}{dX} + \frac{Y}{9X^2} + \frac{2\alpha}{3X}. \tag{7.5.3}$$

Therefore the second Painlevé transcendent is asymptotically related to the Jacobian elliptic function sn(x) (which satisfies (7.5.3) omitting the last three terms). This property is analogous to the situation which arises for classical transcendental functions. For example, the Bessel function, $J_n(x)$, which is a solution of Bessel's equation

$$x^2\frac{\mathrm{d}^2 J_n}{\mathrm{d}x^2} + x\frac{\mathrm{d}J_n}{\mathrm{d}x} + (x^2 - n^2)J_n = 0,$$

and has the asymptotic expansion

$$J_n(x) \sim 2^{1/2}\pi^{-1/2}x^{-1/2}\cos\left(x - \tfrac{1}{2}n\pi - \tfrac{1}{4}\right).$$

In the literature there are a number of interesting results regarding properties of the Painlevé equations (many results of the Soviet school are reviewed by Erugin [1976a,b], where 127 references are cited). Although the Painlevé equations are irreducible in general, for special values of the parameters α, β, γ and δ, the equations $\mathrm{P_{II}}, \dots, \mathrm{P_{VI}}$ have certain special solutions and special relationships between solutions.

1. They have rational solutions and admit one-parameter families of solutions expressible in terms of the classical transcendental functions:
   $\mathrm{P_{II}}$ — Airy functions (Gambier [1909]);
   $\mathrm{P_{III}}$ — Bessel functions (Lukashevich [1965, 1967b]);
   $\mathrm{P_{IV}}$ — Weber-Hermite functions (Lukashevich [1965, 1967a]);
   $\mathrm{P_{V}}$ — Whittaker functions (Lukashevich [1968]);
   $\mathrm{P_{VI}}$ — hypergeometric functions (Lukashevich and Yablonskii [1967] and Fokas and Yortsos [1981]).
2. They admit transformations which map solutions of a given Painlevé equation to solutions of the same Painlevé equation (sometimes referred to as *Bäcklund transformations*), but with different values of the parameters (Lukashevich [1971]; Gromak [1975, 1976]; Fokas and Yortsos [1981]; Kitaev [1991]).
3. They can be written as a Hamiltonian system (Okamoto [1981]). The transformations and special solutions can be obtained from the I.M.T. method as well (see, for example, Fokas, Mugan and Ablowitz [1987]).

(See also Airault [1980]; Clarkson [1990b]; Fokas and Ablowitz [1982]; Gromak [1977]; Murata [1985].)

For example, consider $\mathrm{P_{II}}$.

(i) If $\alpha = n$, with n an integer, then $\mathrm{P_{II}}$ possesses rational solutions; examples of these are given in Table 7.5.1.

(ii) If $\alpha = n + \frac{1}{2}$, with n an integer, then $\mathrm{P_{II}}$ admits a one-parameter family of solutions characterized by

$$\frac{\mathrm{d}y}{\mathrm{d}x} + y^2 + \tfrac{1}{2}x = 0.$$

Setting $y = \phi'(x)/\phi(x)$ yields the Airy equation

$$\frac{d^2\phi}{dx^2} + \tfrac{1}{2}x\phi = 0; \qquad (7.5.4)$$

examples of these are given in Table 7.5.2.

(iii) If $y(x;\alpha)$ is a solutions of P_{II},

$$\tilde{y}(x;\tilde{\alpha}) = -y(x;\alpha) - \frac{1+2\alpha}{2y^2+2y'+x}, \qquad ' = d/dx, \qquad (7.5.5)$$

is also a solution of P_{II}, where $\tilde{\alpha} = \alpha + 1, \alpha \neq -\frac{1}{2}$.

(iii) P_{II} can be written as the Hamiltonian system

$$\frac{dy}{dx} = \frac{\partial H}{\partial p}, \qquad \frac{dp}{dx} = -\frac{\partial H}{\partial y},$$

with the Hamiltonian $H(x;p,y)$ given by

$$H(x;p,y) = \tfrac{1}{2}p^2 - (y^2 + \tfrac{1}{2}x)p - (\alpha + \tfrac{1}{2})y.$$

Table 7.5.1 — Rational Solutions of Painlevé II

α	$y(x;\alpha)$
0	0
1	$-\dfrac{1}{x}$
2	$\dfrac{1}{x} - \dfrac{3x^2}{x^3+4}$
3	$-\dfrac{6x^5+60x^2}{x^6+20x^3-80} + \dfrac{3x^2}{x^3+4}$
4	$-\dfrac{9x^8+360x^5}{x^9+60x^6+11200} + \dfrac{6x^5+60x^2}{x^6+20x^3-80} - \dfrac{1}{x}$
5	$-\dfrac{15x^{14}+1680x^{11}+25200x^8+470400x^5-9408000x^2}{x^{15}+140x^{12}+2800x^9+78400x^6-3136000x^3-627200} + \dfrac{9x^8+360x^5}{x^9+60x^6+11200} - \dfrac{1}{x}$

Table 7.5.2 — Airy solutions of Painlevé II

α	$y(x;\alpha)$
$\frac{1}{2}$	$-\phi'/\phi$
$\frac{3}{2}$	$\dfrac{2(\phi'/\phi)^3 + x(\phi'/\phi) - 1}{2(\phi'/\phi)^2 + x}$
$\frac{5}{2}$	$\dfrac{[4x(\phi'/\phi)^4 + 6(\phi'/\phi)^3 + 4x^2(\phi'/\phi)^2 + 3x(\phi'/\phi) + (x^3 - 1)]}{[2(\phi'/\phi)^2 + x][4(\phi'/\phi)^3 + 2x(\phi'/\phi) - 1]}$

where $\phi(x)$ is any solution of the Airy equation (7.5.4)

Chapter Eight

Further Remarks and Open Problems.

In this book we have collated a considerable amount information regarding nonlinear equations solvable by an appropriate I.S.T. scheme and related methods on the infinite line and in infinite space. We have quite deliberately considered neither the periodic problem nor *ad hoc* techniques. Additionally we have included a very large pool of references.

In this final chapter, we briefly discuss some research problems, which, to our knowledge are currently open. There is a long list of open problems in 'soliton theory', both in the pure and applied aspects of the subject. Indeed, the fact that soliton theory has attracted the interest from a rather diverse group of scientists (e.g., both algebraic geometers and experimental physicists), demonstrates a deep and fundamental relationship with both modern and core Mathematics (e.g., algebraic and differential geometry, analysis, group theory and chaos) together with many physically significant applications (e.g., in plasma physics, fluids, optics and relativity). This is a reflection of the enormous diversity of the subject! The open problems we identify in this section are some of the more important ones, in our opinion, and ones which are currently attracting considerable interest.

The origins of soliton theory as we know it today were in the late 1960's with the pioneering work of Zabusky and Kruskal [1965] and Gardner, Greene, Kruskal and Miura [1967].

During the 1970's, inverse scattering techniques were developed for the solution of nonlinear evolution equations in $1+1$-dimensions and some rigorous results were derived (e.g., for the Korteweg-de Vries (KdV) equation by Deift and Trubowitz [1979]). Additionally it was discovered that there exist nonlinear evolution equations in $2+1$-dimensions, integro-differential equations, differential-difference and partial-difference equations which are be solvable by inverse scattering. Furthermore, there was shown to be a deep and fundamental connection between nonlinear evolution equations solvable by inverse scattering and the Painlevé equations (Ablowitz and Segur [1977]).

During the 1980's, inverse scattering techniques were developed for the solution of nonlinear evolution equations in $2+1$-dimensions (including $2+1$-dimensional boundary value problems), integro-differential equations and the Painlevé equations. Some further rigorous results were obtained, in particular for equations in $2+1$-dimensions (cf. Beals and Coifman [1985a,b, 1987], Wickerhauser [1987]). Other developments included studies on the relationship between classical and quantum inverse scattering (see §8.5 below).

There is every reason to expect that during the 1990's there will a continuation of interest in soliton theory, inverse scattering techniques, related topics and applications. The field continues to be very active, and recent years has witnesses a considerable number of novel, interesting and important (both physically and mathematically) results. While some might suggest that the number of significant results being produced is slowing down; in fact, quite the contrary, there are indications that there is a broadening of activity and one can expect important developments during the next ten years. There are a considerable number of open questions and mathematical theories to be proved; below we identify a few of those we believe are significant.

8.1 Multidimensional Equations.

At present the most important open problem, in our opinion, is to find a genuinely nonlinear, $3+1$-dimensional evolution equation (i.e., an equation in three spatial and one temporal dimensions), which can be solved exactly (for example, via an appropriate I.S.T. scheme). As discussed above in Chapter 6, an important feature associated with inverse problems in three (and higher) spatial dimensions is the fact that whilst the potential $q(y, x_1, x_2)$ depends on three variables, the inverse data $T(k_R, k_I, \xi)$, $k_R \in \mathbb{R}^2$, $k_I \in \mathbb{R}^2$, $\xi \in \mathbb{R}^2$ is a function of six parameters with one restriction (i.e., it depends on five variables). This means that the inverse data must satisfy an appropriate constraint which takes the form of a quadratically nonlinear integro-differential equation and so provides a "characterization" of the inverse data. The main question is whether one can avoid these constraints or incorporate them into an I.S.T. solution in some way. At present, the self-dual Yang-Mills (SDYM) equations, which we discussed in §6.5 above, are the only example of an integrable equation in more than $2+1$-dimensions. However they are most naturally though of as a "$2+2$"-dimensional system. We recognize that the SDYM equations are very significant, especially in view of the conjecture (and partial verification) that they are a "master equation" (recall §6.5).

The existence of "redundancy" in the scattering data has important consequences. For the inverse scattering associated with the n-dimensional Schrödinger equation

$$\sigma \frac{\partial v}{\partial y} + \sum_{j=1}^{n} \frac{\partial^2 v}{\partial x_j^2} + u(x_1, x_2, \ldots, x_n, y) v = 0, \tag{8.1.1}$$

the existence of redundant scattering parameters is exploited to find a formula to reconstruct the potential $u(x, y)$ in *closed* form in terms of scattering amplitude function (Ablowitz and Nachman [1986]). Similarly for the inverse scattering associated with the $N \times N$ scattering problem

$$\frac{\partial v}{\partial y} + \sigma \sum_{\ell=1}^{n} J^{\ell} \frac{\partial v}{\partial x_{\ell}} = \mathbf{Q}(x_1, x_2, \ldots, x_n, y) v, \tag{8.1.2}$$

the existence of redundancy is exploited in such a way so as to reduce the problem of reconstructing the $N \times N$ matrix potential $\mathbf{Q}$ in $n+1$-dimensions to that of reconstructing a 2×2 matrix potential in two spatial dimensions (Fokas [1986a]). Such facts strongly suggest that solvable nonlinear evolution equations in higher dimensions will not be associated with (8.1.1,2).

For those scattering equations in $n+1$-dimensions which have been studied to date [i.e., equations (8.1.1) and (8.1.2)], the inverse data $T(k_R, k_I, \xi)$ has always been forced to satisfy constraints. $T(k_R, k_I, \xi)$ depends on $3n-1$ parameters and there exist $2n-2$ constraints in order that it is compatible with the potential which is to be reconstructed (i.e., the difference is $n+1$ parameters). Generically these constraints are quadratically nonlinear; however, based upon experience, the evolution of the scattering data has always been simple (i.e., of the form $\exp(i\omega t)$). In order to accommodate such both situations there would need to be still further restrictions upon the inverse data. Consequently with further restrictions one can expect that the potential (and hence the evolution equation) will possess associated symmetries so that there exists an appropriate change of variables which would reduce the evolution equation to one of lower-dimensions (cf. Nachman and Ablowitz [1984b]; Fokas [1986]).

A major open question is to find any scattering operators and associated (hopefully local), nonlinear evolution equations for which either a constraint on the scattering data does not exist or if it exists it still can be compatible with an associated $3+1$-dimensional nonlinear evolution equation.

8.2 Boundary Value Problems.

We have been primarily concerned with the method of solution of the initial value problem on the infinite line (in one spatial dimension, i.e., for $x \in (-\infty, \infty)$) given initial conditions which decay sufficiently rapidly as $|x| \to \infty$. An open problem is to generalize the I.S.T. method in order to solve the initial-boundary value problem for either (a) a finite interval (for example, $x \in [0,1]$), or (b), a semi-infinite interval (for example, $x \in [0, \infty)$).

Mathematically speaking, (a) involves constructing the solution of, for example, the KdV equation

$$u_t + 6uu_x + u_{xxx} = 0, \tag{8.2.1}$$

where $(x,t) \in [0,1] \times [0, \infty)$, with the given initial condition

$$u(x,0) = f(x), \tag{8.2.2}$$

and the boundary conditions

$$u(0,t) = g(t), \tag{8.2.3a}$$

$$u(1,t) = h(t), \tag{8.2.3b}$$

$$u_x(1,t) = j(t), \tag{8.2.3c}$$

for appropriate $f(x)$, $g(t)$, $h(t)$, $j(t)$. Other initial-boundary value problems may be considered as well.

The initial-boundary value problem for the KdV equation on the semi-infinite line (b) involves solving (8.2.1), where $(x,t) \in [0,\infty) \times [0,\infty)$, with given initial conditions (8.2.2) and boundary data

$$\alpha u(0,t) + \beta u_x(0,t) = g(t). \tag{8.2.4}$$

The solution of such an initial-boundary value problem on the semi-infinite line is of mathematical and physical interest. For example, the problem of wave generation by a moving model may be formulated as an initial-boundary value problem on the semi-infinite line. Chu, Xiang and Barnasky [1983] have considered the initial boundary value problem for the KdV equation (8.2.1) on the semi-infinite line for zero initial data [i.e., $f(x) \equiv 0$ in (8.2.2)] and the boundary condition

$$u(0,t) = g(t), \tag{8.2.5}$$

where $g(t)$ is a trapezoidal pulse; i.e., $g(t)$ has the form

$$g(t) = \begin{cases} at, & 0 < t < b, \\ ab, & b < t < c+b, \\ a(2b+c) - at, & c+b < t < c+2b, \\ 0, & t > c+2b. \end{cases}$$

They solved this problem numerically and found that a sequence of solitons were produced: the longer the pulse, the more solitons were produced.

Bona and Winther [1983] have proved existence and uniqueness theorems for the initial-boundary value problem for the KdV equation (8.2.1) on the semi-infinite line and shown that solution depends continuously (in strong norms) on both the initial and boundary data. However, the I.S.T. method to "solve" (i.e. linearize) this problem is currently unknown. Keller [1985] has discussed several aspects of this problem and suggested a possible method of solution using an analysis based upon the work of Lax and Levermore [1983a,b,c] and Venakides [1985a,b] who studied the KdV equation

$$u_t + 6uu_x + \epsilon u_{xxx} = 0,$$

in the limit as $\epsilon \to 0$.

Physically, the solution of initial-boundary value problems (either on a finite or semi-infinite interval) may be viewed as a partial differential equation to which external forces (for example, sources and/or sinks) are applied to the boundaries. Clearly solutions to such problems would greatly extend the scope and application of the I.S.T. method.

Similar questions such as those discussed above may be asked about other nonlinear evolution equations solvable by I.S.T. on the infinite line. Early work was done by Ablowitz and Segur [1975] on the nonlinear Schrödinger equation and other equations associated with the AKNS-ZS system. In their work certain symmetry conditions and continuity (oddness/evenness) on the potentials was assumed in order to show that the problem was solvable. In recent years a continuation of analytical and numerical studies of various semi-infinite problems have appeared for example for the Toda Lattice (Kaup [1984, 1987]; Kaup and Neuberger [1984]); the Sine-Gordon equation (Kaup [1985]); the nonlinear Schrödinger equation (Kaup [1985]; Kaup and Hansen [1985]; Bikbaev and Tarasov [1991]); and the forced nonlinear Schrödinger equation (Carroll [1990]). The forced Boussinesq equation has been studied numerically by Camacho and Lund [1985].

Recent studies by Fokas [1988, 1989a] discuss the problem of extending the I.S.T. scheme for the solution of the nonlinear Schrödinger equation

$$\mathrm{i}q_t = q_{xx} \pm 2|q|^2 q, \tag{8.2.6}$$

on the infinite line (which we described in §3.2 above), to solve the initial-boundary value problem for the nonlinear Schrödinger equation on $(x,t) \in [0,\infty) \times [0,\infty)$, subject the initial and boundary conditions

$$q(x,0) = f(x), \qquad \text{for} \quad x \geq 0, \tag{8.2.7a}$$

$$q_x(0,t) + \alpha q(0,t) = g(t), \qquad \text{for} \quad t \geq 0. \tag{8.2.7b}$$

where $h(x)$ decays sufficiently rapidly as $|x| \to \infty$, $\alpha$is a real constant and the given functions $g(t)$ and $h(x)$ have appropriate smoothness and satisfy the necessary compatibility conditions to ensure the existence of the solution at $x = 0$. The main difficulty encountered is that whereas the solution $q(x,t)$ of the nonlinear Schrödinger equation can be expressed as the solution of a *linear* integral equation uniquely defined in terms of the standard scattering data, $b(k,t)$, this satisfies a *nonlinear* integro-differential equation. For example, if $q(0,t)$ is given then $B(k,t) = b(k,t)/(4k)$ satisfies

$$B_t + \mathrm{i}kB = -q(0,t) + \frac{1}{2\pi}\int_{-\infty}^{\infty} \frac{B(k') - B(k)}{k' - k} \times \frac{\partial}{\partial t}\left\{ \mathrm{H}\left[\ln\left(1 \pm 16k^2|B(k')|^2\right)\right]\right\} \mathrm{d}k', \tag{8.2.8}$$

where H denotes the Hilbert transform, i.e.,

$$(\mathrm{H}f)(x) = \frac{1}{\pi} \int_{-\infty}^{\infty} \frac{f(y)}{y-x}\, \mathrm{d}y.$$

The existence and uniqueness of solutions of the nonlinear equation (8.2.8) still need to be established; furthermore, for practical as well as theoretical reasons, it is desirable to ascertain whether it can be linearized.

Fokas and Ablowitz [1989] discuss the I.S.T. schemes for the forced nonlinear Schrödinger (NLS) equation

$$\mathrm{i}q_t + q_{xx} \pm 2|q|^2 q = q(0,t)\delta(x), \tag{8.2.9a}$$

and the forced KdV equations

$$u_t + 6uu_x + u_{xxx} = u(0,t)\delta(x), \tag{8.2.9b}$$
$$u_t + 6uu_x + u_{xxx} = u(0,t)\delta'(x), \tag{8.2.9c}$$

with $\delta(x)$ the Dirac delta function. Analogous to the results for the NLS equation on the semi-infinite line mentioned above, they showed that the scattering data $b(k)$ satisfies the nonlinear integro-differential equation

$$\begin{aligned} B_t - 4\mathrm{i}\rho k^n B = -U(t) + &\frac{1}{2\pi} \int_{-\infty}^{\infty} \left\{ \frac{B(k') - B(k)}{k'-k} \right\} \\ &\times \frac{\partial}{\partial t} \left\{ \mathrm{H}\left[\ln\left(1 + f(k)|B(k')|^2\right)\right](k') \right\} \mathrm{d}k', \end{aligned} \tag{8.2.10}$$

where

(a),	$U(t) = q(0,t)$,	$B(k) = b(k)$,	$\rho = 1$,	$n = 2$,	$f(k) = 1$;
(b),	$U(t) = u(0,t)$,	$B(k) = kb(k)$,	$\rho = 2$,	$n = 3$,	$f(k) = 1/k^2$;
(c),	$U(t) = u(0,t)$,	$B(k) = b(k)$,	$\rho = 2$,	$n = 3$,	$f(k) = 1$,

respectively. We remark however that alternative "inverse" methods might prove to be problem. We also note that Ablowitz and de Lillo [1991] have shown that the Burgers equations with distributional forcing

$$u_t - 2uu_x - u_{xx} = f(t)\delta(x), \tag{8.2.12}$$

can be linearized via the semi-line methods developed for the Burgers equation by Calogero and de Lillo [1989a,b, 1991].

The inverse scattering problem for the Schrödinger equation

$$v_{xx} + u(x)v = \lambda v, \tag{8.2.11}$$

has been extensively studied in the literature for both a finite interval (for example, Isaacson and Trubowitz [1983]; Isaacson, McKean and Trubowitz [1984]; Dahlberg and Trubowitz [1984] — see also McLaughlin [1986]; Poschel and Trubowitz [1987] and the references therein) and on the semi-infinite line (when x is interpreted as a radial coordinate (see, for example, Gel'fand and Levitan [1955]; Faddeev [1963]; Chadan and Sabatier [1977] and the references therein). For further information and bibliography on the inverse scattering problem associated with (8.2.6) see Agronovich and Marchenko [1963] and Newton [1972]. We only remark that these studies must be connected with the time evolution of the data in an appropriate manner in order to be able to solve, say, the KdV equation.

In §1.11, we made some remarks about the periodic problem for the KdV equation; i.e., the solution of (8.2.1) with periodic boundary conditions of period T

$$u(x,t) = u(x+T,t),$$

given suitable initial data

$$u(x,0) = f(x).$$

Some of the other equations for which periodic problems have been considered are: the nonlinear Schrödinger equation (Its and Kotlyarov [1976]; Ma and Ablowitz [1981]); the Sine-Gordon equation (McKean [1981a]; Forest and McLaughlin [1982,1983]; Ercolani, Forest and McLaughlin [1984]; Ercolani and Forest [1985]); the Boussinesq equation (McKean [1981b,c]); and the Kadomtsev-Petviashvili equation (Krichever [1977a,b, 1980]; Krichever and Novikov [1980a,b] — see also Dubrovin [1981, 1982]). An intriguing open problem is to determine the precise relationship between the I.S.T. scheme for the periodic problem and that on the infinite interval. Various connections between the infinite-line and periodic I.S.T. schemes for the KdV equation are discussed by Osborne and Bergamasco [1986].

8.3 Ordinary Differential Equations.

As we discussed in Chapter 7, it is possible to use the ideas of inverse scattering to solve ordinary differential equations via certain associated Riemann-Hilbert problems; especially the classical Painlevé equations (see Ince [1956] for a discussion of the underlying mathematical properties of the Painlevé equations). There are a number of open questions such as rigorous analysis of these methods and development of a method to obtain the complete connection formulae for solvable nonlinear ordinary differential equations (some results in this direction have been obtained for the second and fourth Painlevé equations by Fokas and Zhou [1991]).

In §§6.5 and 7.1.4 above we discusses the third order ordinary differential equation due to Chazy [1909, 1910, 1911]

$$\frac{\mathrm{d}^3 y}{\mathrm{d}x^3} = 2y\frac{\mathrm{d}^2 y}{\mathrm{d}x^2} - \left(\frac{\mathrm{d}y}{\mathrm{d}x}\right)^2, \tag{8.3.1}$$

whose solution possesses a movable natural boundary. There is serious interest in this and similar equations, especially due to the fact that equation (8.3.1) arises as a reduction of the SDYM equation (Chakravarty, Ablowitz and Clarkson [1990]). Only recently has this system been connected with a physical problem despite the fact that the equation was first written down over eighty years ago (and an associated system over a hundred years ago by Halphen [1881a,b]). There are several open questions associated with equation (8.3.1), in particular: (i), in what sense is the equation is completely integrable; (ii), how can we identify *a priori* equations whose solutions have movable natural boundaries; (iii), what is the structure of higher order ordinary differential equations with movable natural boundaries; and (iv), if a given partial differential equation possesses a similarity reduction to an ordinary differential equation whose solution has a movable natural boundary, then what influence does the natural boundary have upon the solutions original partial differential equation?

In §4.3 we saw that N-soliton solutions of the Benjamin-Ono equation are closely related to solutions of the Calogero-Moser N-body problem, which itself is an integrable system (Calogero [1971]; Moser [1975]). More generally, there appears to be a close relationship between "pole expansion" solutions of complete integrable nonlinear evolution equations and finite dimensional dynamical systems of ordinary differential equations. Precisely how this relationship is manifest is not clear.

For example, there is a close relationship between the rational solutions of the KdV equation

$$u_t + 6uu_x + u_{xxx} = 0, \tag{8.3.4}$$

and solutions of the many body problem

$$\frac{\mathrm{d}q_j}{\mathrm{d}t} = p_j, \qquad \frac{\mathrm{d}p_j}{\mathrm{d}t} = 2\sum_{\substack{k=1\\k\neq j}}^{N} \frac{1}{(q_j - q_k)^3}, \qquad j = 1, 2, \ldots, N, \tag{8.3.5}$$

with a constraint. Airault, McKean and Moser [1977] (motivated by earlier work of Kruskal [1974] and Thickstun [1976] — see also Chudnovsky and Chudnovsky [1977]) showed that solutions of the KdV equation (8.3.4) of the form

$$u(x,t) = 2\sum_{j=1}^{N} \frac{1}{[x - q_j(t)]^2},$$

exist provided that the q_j satisfy the dynamical system (8.3.5) with initial conditions (constraints)

$$p_j(t_0) = 0, \qquad \sum_{\substack{k=1\\k\neq j}}^{N} \frac{1}{[q_j(t_0) - q_k(t_0)]^3} = 0,$$

and $N = \frac{1}{2}m(m+1)$, for $m = 1, 2, \ldots$. (The relationship between integrable dynamical systems and partial differential equations is further pursued in Calogero [1978] — see also Olshanetsky and Perelomov [1981, 1982] and the references therein.) That there is a solution to (8.3.5) with such constraints for all time is surprising itself.

We remark upon very recent work by Ruijsenaars and Schneider [1986] (see also Bruschi and Calogero [1987] and Ruijsenaars [1987a,b] and the references therein) on the connection between solitons of the Sine-Gordon equation and a relativistic invariant system of nonlinear ordinary differential equations obtained by a pole expansion associated with the Sine-Gordon equation.

Open problems include elucidations of the relationship between nonlinear evolution equations, their underlying dynamical systems and their respective linearizations; further understanding the scope and meaning of dynamical systems related to solvable evolution equations as well as the construction of other interesting integrable dynamical systems.

8.4 Functional Analysis and 2+1-dimensions.

Throughout this book, we have frequently neglected rigorous discussions about the direct and inverse scattering problems; in particular, we have not attempted to characterize the most general class of functions for which the scattering transform (i.e., I.S.T.) is bijective. Indeed, one of the principle reasons for this is that it is currently an active area of research and would take us outside the intended scope of this book.

In the I.S.T. method, both the direct and inverse problems involve linear integral equations. Therefore, it is required to establish necessary and sufficient conditions for the existence and uniqueness of the solution of these equations. That is, what Banach space should the initial conditions lie in so that we might solve the associated nonlinear evolution equation by an appropriate I.S.T. scheme.

A rigorous analytical discussion of the direct and inverse scattering problems for a given scattering operator should include the following:

(i) a formulation of the notion of scattering data V which is meaningful for (essentially) all potentials u in an appropriate Banach space (for example, $u \in L^1(\mathbb{R})$);

(ii) a proof that the mapping $u \to V$ is injective;

(iii) a characterization of the scattering data in terms of the algebraic, topological and analytical constraints such data must satisfy;

(iv) a proof that for (essentially) each set of data satisfying the constraints, there is a (unique) corresponding potential;

(v) a discussion of the relationship between the analyticity properties of u (for example, smoothness or decay at ∞) with the corresponding properties of V.

As we noted above, we have usually glossed over such questions in this paper. Therefore, strictly speaking (in particular in multidimensions), when we have stated that a nonlinear evolution equation has been "solved by I.S.T." often we should have said "formally solved by I.S.T.."

In the literature there are several analytical studies of the direct and inverse scattering of various systems of linear ordinary differential equations, which are associated with the solution of 1 + 1-dimensional nonlinear evolution equations. In particular, there have been a number of studies of the time independent Schrödinger equation

$$v_{xx} + u(x)v = \lambda v.$$

A controversy arose over the question of which Banach space the potential u should lie in, either P_1 or P_2, where

$$P_r = \{u : (1 + |x|^r)u \in L^1(\mathbb{R})\}.$$

Early on, Faddeev [1963] asserted that $u \in P_1$; however, as shown by Chadan and Sabatier [1977] and Deift and Trubowitz [1979], Faddeev's assumptions would only provide $u \in P_2$. Recently though, Marchenko (private communication, Faddeev) has shown that, with certain further conditions on the scattering data, indeed the condition $u \in P_1$ is sufficient (see also Melin [1985]; Marchenko [1986]).

The scattering and inverse scattering associated with the third order Sturm-Liouville equation

$$v_{xxx} + q(x)v_x + r(x)v = \lambda v, \tag{8.4.1}$$

has been studied in detail by Deift, Tomei and Trubowitz [1982] (see also Caudrey [1980]). In order to discuss (8.4.1) rigorously, Deift, Tomei and Trubowitz [1982] assumed that $q(x)$ and $r(x)$ were in the Schwarz space $\mathcal{S}(\mathbb{R})$ ($q \in \mathcal{S}(\mathbb{R})$ if there exist constants c_{pk} such that $|x^p q^{(k)}(x)| < c_{pk}$, for $p, k = 0, 1, \ldots$, and for all $x \in \mathbb{R}$, where $q^{(k)}(x) = \mathrm{d}^k q/\mathrm{d}x^k$). However, they note that their proofs could be extended to $q(x)$ and $r(x)$ with only a finite number of derivatives and a finite order of decay.) Beals [1985] (see also Beals, Deift and Tomei [1988]) has generalized this in a rigorous analytical study of scattering and inverse scattering associated with the nth order Sturm-Liouville equation

$$\frac{\mathrm{d}^n v}{\mathrm{d}x^n} + \sum_{j=0}^{n-2} q_j(x) \frac{\mathrm{d}^j v}{\mathrm{d}x^j} = \lambda v.$$

Beals [1985] reduced this problem to one studied earlier by Beals and Coifman [1984], who discussed the rigorous scattering and inverse scattering associated with the $N \times N$ AKNS-ZS scattering problem

$$\boldsymbol{v}_x = \mathrm{i}k\mathbf{J}\boldsymbol{v} + \mathbf{Q}\boldsymbol{v},$$

where $\mathbf{J} = \mathrm{diag}(J^1, J^2, \ldots, J^N)$, $J^i \neq J^j$ for $i \neq j$ and $Q^{ii} = 0$, under the condition that $Q^{jk} \in L^1(\mathbb{R})$ (see also Beals and Coifman [1985a,b, 1987, 1989]; Caudrey [1982]; Shabat [1975,1979]; Mikhailov [1981]).

Whereas in one spatial dimension there have been numerous rigorous analytical studies, in higher spatial dimensions there are very few results. Beals and Coifman [1985b] and Wickerhauser [1987] have given a rigorous analytic treatment for scattering operators associated with DSII and KPII, respectively (as briefly discussed in Chapter 5 above). That is, for the operators

$$\boldsymbol{v}_x = \mathbf{Q}\boldsymbol{v} + \mathbf{J}\boldsymbol{v}_y, \tag{8.4.2a}$$

where

$$\mathbf{Q} = \begin{pmatrix} 0 & q \\ q^* & 0 \end{pmatrix}, \qquad \mathbf{J} = \mathrm{diag}(-\mathrm{i}, \mathrm{i}), \tag{8.4.2b}$$

and

$$v_{xx} - v_y - u(x,y)v = 0, \tag{8.4.3}$$

respectively. Beals and Coifman [1985b] assumed that $q(x,y) \in L^{2-\epsilon}(\mathbb{R}^2) \cap L^{2+\epsilon}(\mathbb{R}^2)$, for some $\epsilon > 0$ and that the norms $\|q\|_{2-\epsilon}$ and $\|q\|_{2+\epsilon}$ were sufficiently small in order to develop a scattering theory for (8.4.2). Wickerhauser assumed $u(x,y) \in L^1(\mathbb{R}^2) \cap L^2(\mathbb{R}^2)$ and $\|u\|_1$ and $\|u\|_2$ sufficiently small. $\|u\|_p$ is the $L^p(\mathbb{R}^2)$ norm of u, i.e.

$$\|u\|_p = \left(\int_{-\infty}^{\infty} \int_{-\infty}^{\infty} |u(x,y)|^p \,\mathrm{d}x\,\mathrm{d}y \right)^{1/p}.$$

Sung and Fokas [1990] discuss some rigorous aspects of the I.S.T. for the N-wave equation (which is a 2 + 1-dimensional hyperbolic equation); Zhou [1990] discusses rigorous aspects of the I.S.T. for KPI, he employs the constraint $\int_{-\infty}^{\infty} u(x,y)\,\mathrm{d}x = 0$, (i.e., the non-generic case, see also §5.3). We also note that Coifman and Wickerhauser [1990] discuss rigorous aspects of the I.S.T. for the Benjamin-Ono equation (they analyze a different scattering problem than the one discussed in §4.3).

There are various technical assumptions made in the development of direct and inverse scattering theory in 2 + 1 (and higher) dimensions which have not yet been proved for "large" norm. For example, it was assumed that the integral equation for the scattering eigenfunction $m(x,y,t)$ [equation (5.3.7) in the I.S.T.associated to KPII and equation (6.2.3) in the multidimensional case] has no homogeneous solutions. If the integral equation has homogeneous solutions, this would suggest the existence of lump type solutions for KPII (which have never been shown to exist by construction). The existence (or more accurately the non-existence) of homogeneous solutions (and hence also bound states) remains an important open problem.

In recent work, Lavine and Nachman [1987] and Novikov and Khenkin [1987a,b], investigate the inverse scattering associated with the n-dimensional time independent Schrödinger equations (and also the acoustic equation: Novikov and Khenkin [1987b])

$$\sum_{j=1}^{n} \frac{\partial^2 v}{\partial x_j^2} - u(x_1, x_2, \ldots, x_n)v = \lambda v.$$

As shown in the work of Nachman and Ablowitz [1984a], Ablowitz and Nachman [1986], the scattering theory first developed by Faddeev [1965, 1966, 1976] could be recovered by $\bar{\partial}$ techniques. New ideas and methods of reconstruction were given. A serious issue regarding the existence of certain homogeneous solutions of the underlying integral equations remains largely open. Lavine and Nachman [1987] address this question and find certain conditions when there will/will not be homogeneous solutions and what one must do to incorporate them in an expanded theory. Khenkin and Novikov [1987a,b] rigorously discuss aspects involving the characterization of the data for the inverse problem.

In §3.4 we discussed the (discrete) I.S.T. scheme for differential-difference (and partial-difference) equations in 1 + 1-dimensions. An open problem is the development of general discrete I.S.T. techniques for differential- difference equations in $2+1$-dimension. Clearly, there are many possible discretizations in $2+1$-dimensions. Furthermore since the analysis for the continuous and discrete scattering problems in 1 + 1-dimensions were quite similar, we expect an analogous situation in 2 + 1-dimensions. (Some work in this direction has been done by Caudrey [1986] and Ragnisco, Santini, Chitlaru-Briggs and Ablowitz [1987].)

Another open problem in $2+1$-dimensions is the development of an I.S.T. scheme to solve $2+1$-dimensional integro-differential equations, such as a $2+1$-dimensional generalization of the Benjamin-Ono equation

$$u_t + 2uu_x + \mathrm{H}(u_{xx}) + 3\sigma^2\partial_x^{-1}(u_{yy}) = 0, \tag{8.4.4}$$

where $\mathrm{H}(f)$ represents the Hilbert transform [for a derivation of (8.4.4) see, for example, Ablowitz and Segur [1980]]. Furthermore, there exist other $2+1$-dimensional generalizations, for example

$$\begin{aligned}[u_t + 2(u\mathrm{H}u)_x + u_{xx}]_x &= -3\sigma^2 u_{yy},\\ [u_t + 2uu_x + \mathrm{H}(u_{xx})]_x &= [2uu_x + \mathrm{H}(u_{xx})]_y\end{aligned}$$

(cf. Ablowitz, Fokas and Kruskal [1987] — see also Matveev and Sall [1981]; Bobenko, Matveev and Sall [1982]).

8.5 Quantum Inverse Scattering and Statistical Mechanics.

An issue which has attracted considerable interest is to understand and make precise the underlying connection between integrable quantum field theories and their classical analogues (i.e., is their a direct relationship between quantum inverse scattering and classical I.S.T.?). Despite many important advances (which are reviewed in articles by Faddeev [1980]; Kulish and Sklyanin [1982]; Thacker [1982]), the analogy between the classical and quantum problems remains incomplete; questions that can be answered classically remain open for the analogous quantum problem and vice versa does every integrable system have a quantum analog which itself is solvable?

As an example to demonstrate this analogy, we shall consider the nonlinear Schrödinger equation, which is frequently referred to as the delta-function Bose gas model in quantum theories (see Segur [1986] for references and further details). The quantum problem is a one-dimensional non-relativistic field theory defined by the Hamiltonian

$$H_q = \int_{-\infty}^{\infty} (q_x^* q_x + cq^* q^* qq)\, \mathrm{d}x \tag{8.5.1}$$

and the commutation relations

$$[q(x), q^*(y)] = \delta(x-y), \tag{8.5.2a}$$
$$[q(x), q(y)] = 0 = [q^*(x), q^*(y)], \tag{8.5.2b}$$

where c is a real constant and q and q^* are annihilation and creation operators, respectively. A Hamiltonian naturally implies a dynamical (i.e., time-dependent) model. However, in quantum problems, the questions primarily considered are time-independent. Two questions of fundamental interest for quantum problems are (a), how to diagonalize the Hamiltonian hence obtain the eigenvalues and the relevant eigenfunctions and (b) how to compute the correlation functions.

The classical problem may also be represented by a Hamiltonian

$$H_c = \tfrac{1}{2}\mathrm{i} \int_{-\infty}^{\infty} (|u_x|^2 + c|u|^4)\, \mathrm{d}x, \tag{8.5.3}$$

where c is a real number and there is a Poisson Bracket (cf. §3.5.6), where u and u^* are treated as conjugate variables. Hence the evolution equation

$$\mathrm{i}u_t = u_{xx} - 2c|u|^2 u, \tag{8.5.4}$$

is a Hamiltonian system. The primary interest in the classical problem is the solution of the initial value problem for (8.5.4) on for example $-\infty < x < \infty$ (or periodic on $0 < x < L$), given initial data $u(x,0)$ such that, say

$$|c|^{1/2} \int_{-\infty}^{\infty} u(x,0)\, \mathrm{d}x < \infty.$$

The first step in the quantum problem is to diagonalize the Hamiltonian and find its eigenvalues and eigenfunctions. In a sense, the analogous procedure for the classical problem is to exploit the underlying linear equations (i.e., the Lax pair) whose compatibility implies (8.5.4). As we have discussed at length in these notes, one of these linear equations maps the solution of (8.5.4) into the scattering data, which evolve in a simple manner according to the other equation. The evolution is simple since the (canonical) transformation to the scattering data action-angle variables may be viewed as diagonalizing the Hamiltonian.

For the classical problem, the inverse transformation (i.e., the reconstruction of $u(x,t)$) is effected through a generalization of a Riemann-Hilbert boundary value problem for any finite c. In the quantum problem one is not really interested in $u(x,t)$ since it is an operator. Rather it is of interest to compute the correlation function. At present, correlation functions have only been computed for the nonlinear Schrödinger equation/delta-function Bose gas model (8.5.4) in the limit $c \to \infty$, which is known as the impenetrable boson model. A major open problem in quantum problems is the computation of correlation functions for finite values of c.

It is also an open question how to compute correlation functions for other quantum systems connected to integrable classical models.

A further indication of the relationship between the classical and quantum inverse scattering theories is the fact that the Painlevé equations arise in both contexts: as similarity reductions for nonlinear evolution equations solvable by classical I.S.T. and as the closed form solution of the correlation functions in the quantum case in a certain limit (i.e., at the critical value of the relevant parameter). For reductions of the nonlinear Schrödinger equation to the Painlevé equations see Boiti and Pempinelli [1980]; Lakshmanan and Kaliappan [1983]; Tajiri [1983]. For the calculation of the correlation functions in terms of the Painlevé equations for the impenetrable boson model see Jimbo, Miwa, Mori and Sato [1981]; Creamer, Thacker and Wilkinson [1982, 1986]; Thacker [1983]; McCoy and Tang [1986a,b,c].

Significantly, exactly solvable lattice models in statistical mechanics are closely related to the quantum inverse scattering method, prototype example being the six and eight vertex models of Baxter; a special case being the two-dimensional nearest neighbor Ising model which was solved by Onsager [1944]. Considerable progress in the solution of exactly solvable models in statistical mechanics is due to Baxter who pioneered the method of *commuting transfer matrices* and *corner transfer matrices* (cf. Baxter [1982, 1985, 1986] and the references therein — see also the reviews by Kulish and Sklyanin [1982]; Thacker [1981]; Tracy [1987]; Zamolodchikov [1980]).

One of the most interesting aspects of the quantum inverse scattering method is its relationship to the transfer matrix techniques developed in the solution of exactly solvable lattice models. The quantum inverse scattering method involves operators which are defined on continuous space and time whereas exactly solvable lattice models involve operators where are defined on discrete lattice sites. For example, the Baxter eight-vertex model (Baxter [1972]) is a solvable two-dimensional lattice model whose continuum limit is the relativitically invariant massive Thirring/Sine-Gordon field theory (Luther [1976] — see, for example, Thacker [1986] for a discussion of the relationship).

Most exactly solvable lattice models in statistical mechanics are two-dimensional whose solution is intimately associated to the *Yang-Baxter* (or *star-triangle*) equations. The only known example of a three-dimensional lattice model is given by Zamolodchikov [1981], which is associated with the three-dimensional Yang-Baxter equations. With regard to quantum field problems in higher dimensions, the only model so far studied is the Davey-Stewartson problem (see Schultz, Ablowitz, and BarYaacov [1987]).

Open problems associated with exactly solvable lattice models include: (i) the relationship between their solution and the classical inverse scattering method (for difference equations); and (ii), the solution of lattice models (and the associated Yang-Baxter equations) in higher dimensions. We note that there have been remarkable developments on the connection between lattice models and topology (polynomials) and a new class of groups termed quantum groups. These subjects are outside the scope of this book, but the interested reader can find many recent references to this material (cf. Faddeev [1990]; Faddeev and Takhtajan [1986]; Reshetkhin, Takhtajan and Faddeev [1989]; Moore and Reshetkhin [1989]; Kauffman [1990]).

8.6 Complete Integrability.

The study of the I.S.T. and the solution of associated nonlinear evolution equations raises the question of what it means for a given nonlinear evolution equation to be completely integrable; in the above we have discussed classical integrability. Indeed, such questions regarding integrability can be asked of both ordinary and partial differential equations. A fundamental question which has intrigued mathematicians for a long time is "given a system of equations, how can one determine *a priori* whether or not the system is integrable?" (by *a priori*, it is meant without first finding the method). Despite much current interest, this question remains to be completely answered, in fact some would argue that it is unanswerable! Determining the integrability of a given system of equations has, to date, often relied upon inspired guesswork, luck and/or extensive (and expensive) numerical investigations (which can never be conclusive). One major obstacle to answering this fundamental question is that at present there is not one definition of what it means for a system of equations to be integrable. If one asks a group of mathematicians (or physicists or any other scientists for that matter), what it means for a system of equations to be integrable, then it is probable that there will be several different definitions!

The Painlevé tests, as formulated by Ablowitz, Ramani and Segur [1978,1980a] and Weiss, Tabor and Carnevale [1983], which we discussed in §7.2, have proven to be useful criteria for the identification of completely integrable nonlinear evolution equations and have attracted considerable recent interest. Although this approach is by no means foolproof, it does appear to be capable of yielding other information such as Bäcklund transformations, Lax pairs, etc.. However it seems that the overriding issue is whether solutions of a given differential equation are *single valued*. As pointed out recently by Kruskal [1991], it appears that in the original investigations of Painlevé *et al.*, they were primarily concerned with the singlevaluedness of solutions and local essential singular points (and even natural boundaries) and fixed singularities are not regarded as being as "bad" (see for example the discussion in one of the definitive works of Painlevé himself [1902]).

Recently Kruskal has proposed a modification of the Painlevé tests (the "poly-Painlevé" test), in order to rectify some deficiencies in the original Painlevé tests (Kruskal and Clarkson [1991]). Kruskal's ideas are novel and we shall not attempt to describe them here; especially since these ideas have not yet been fully developed even for ordinary differential equations (although it seems highly probable that they can be extended to partial differential equations). Importantly the view point deals with the heart of the nature of integrability and may prove important in our question to understand the nature and inherent ramifications of integrability.

Appendix A

Riemann-Hilbert Boundary Value Problems

Riemann-Hilbert problems have played (explicitly or implicitly) an important role in soliton theory. In this section we review some basic elements of Riemann-Hilbert theory, as well as mentioning connections with the most important Riemann-Hilbert problems encountered in soliton theory.

A.1 Classical Riemann-Hilbert Problems for Hölder Functions.

We first introduce some definitions and results needed in the sequel.

1. Let C be a simple, smooth, closed contour dividing the complex z-plane into two regions D_+ and D_- (the positive direction of C will be taken as that for this D_+ is on the left; more than one contour could be considered but for simplicity here with consider only one). With suitable modifications we can also allow C to be a line separating the complex plane into two regions (e.g., $\text{Im} z = 0$, separates the complex plane into upper/lower half planes). The case of a "complicated" contour has, for example, many rays passing through the origin separating the complex plane into sections.

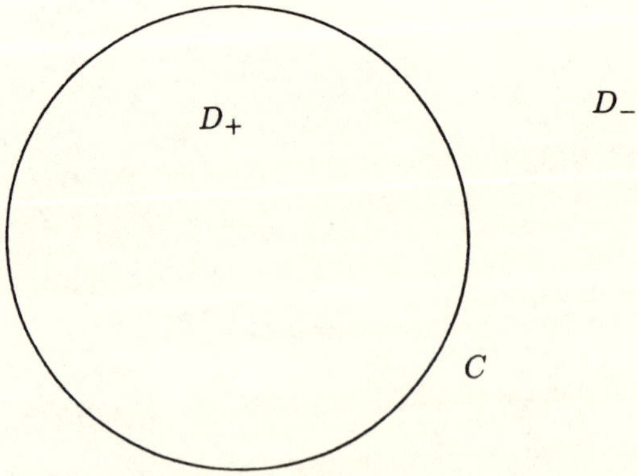

FIGURE A.1 *The contour C*

A scalar function $\phi(z)$ defined in the entire plane, except for points of C, will be called ***sectionally holomorphic*** if: (a), the function of $\phi(z)$ is holomorphic in each of the regions D_+ and D_- except, perhaps, at $z = \infty$; and (b), the function $\phi(z)$ is sectionally continuous with respect to C, i.e., as z approaches any point t on C along any path which lies wholly in either D_+ or D_-, the function $\phi(z)$ approaches a definite limiting value $\phi_+(t)$ or $\phi_-(t)$ respectively. (It then follows according to a theorem of Painlevé (cf. Hille [1976]) that $\phi(z)$ is continuous in the closed region $D_+ \cup C$, if it is assigned the value $\phi_+(t)$ on C and similarly for the region $D_- \cup C$.)

2. A function $\phi(t)$ defined on C will be called ***Hölder***, it for any two arbitrary points on C

$$|\phi(t_2) - \phi(t_1)| < A_1 |t_2 - t_1|^\lambda, \quad A > 0, \quad 0 < \lambda \leq 1 \tag{A.1}$$

(if $\lambda = 1$ then the Hölder condition reduces the Lipschitz condition).

3. The *index of a function $\phi(t)$ with respect to C* is the increment of its argument, in transversing the curve in the positive direction, divided by 2π, i.e.,

$$\text{ind}\, \phi(t) := \frac{1}{2\pi}[\arg \phi(t)]_C = \frac{1}{2\pi \text{i}} \int_C \text{d}(\ln \phi(t)). \tag{A.2}$$

4. The sectionally holomorphic function $\phi(z)$ has *finite degree at infinity* if for some finite integer k, $\lim_{z\to\infty} \left(\phi(z)|z|^{-k-1}\right) = 0$ and

$$\phi(z) \sim C_k z^k + O(z^{k-1}) \quad \text{as} \quad z \to \infty. \tag{A.3}$$

5. The following result, due to Plemelj, is of fundamental importance in the theory of Riemann-Hilbert problems: assume that $f(t)$ is Hölder on C, then the function

$$\phi(z) := \frac{1}{2\pi \text{i}} \int_C \frac{f(\tau)\, \text{d}\tau}{\tau - x}, \tag{A.4a}$$

is a sectionally holomorphic function vanishing at infinity. Furthermore the limits $\phi_+(t), \phi_-(t)$ as t approaches C are given by

$$\phi_\pm(t) = \pm\tfrac{1}{2} f(t) + \frac{1}{2\pi \text{i}} \int_C \frac{f(\tau)\, \text{d}\tau}{\tau - t}, \tag{A.4b}$$

where principal value integrals are assumed whenever needed. One consequence of the Plemelj formulae (A.4b) is that any function $f(t)$, Hölder on C, can be written as

$$\phi_+(t) - \phi_-(t) = f(t), \tag{A.5}$$

where $\phi_\pm(t)$ are the boundary values of the sectionally holomorphic function $\phi(z)$, uniquely defined in terms of $f(t)$ (assuming $\phi(z) \sim O(z^{-1})$ as $z \to \infty$).

6. All the above notions are easily extended to matrix and vector functions $\phi(z)$. So for example, the degree of $\phi(z)$ at infinity is the highest degree of any of its components. The Plemelj's formulae are also valid for matrix and vector functions.

Having introduced the above notions we can now formulate the simplest Riemann-Hilbert problems:

A.1.1 Scalar Homogeneous Riemann-Hilbert Problem. The scalar homogeneous Riemann-Hilbert problem is defined as follows: Given a closed (for convenience) contour C, and a function $G(t)$ which is Hölder on C and $G(t) \neq 0$ on C, find a sectionally holomorphic function $\phi(z)$, with finite degree at ∞, such that

$$\phi_+(t) = G(t)\phi_-(t), \qquad \text{on} \quad C \tag{A.6}$$

where $\phi_\pm(t)$ are the boundary values of $\phi(z)$ on C.

The solution of (A.6) can be given in closed form. This solution is characterized by the index of $G(t)$ and its behavior at infinity. Let

$$\kappa := \text{ind } G(t). \tag{A.7}$$

Then the sectionally holomorphic function $X(z)$ defined by

$$X(z) := \begin{cases} \exp\{\Gamma(z)\} & z \text{ in } D_+ \\ z^{-\kappa}\exp\{\Gamma(z)\} & z \text{ in } D_- \end{cases}; \qquad \Gamma(z) := \frac{1}{2\pi \mathrm{i}}\int_C \frac{\ln[\tau^{-\kappa}G(\tau)]}{\tau - z}\,\mathrm{d}\tau, \tag{A.8}$$

is called the *fundamental solution* of (A.6). Any solution $\phi(z)$ of (A.6) with finite degree at ∞ is given by

$$\Phi(z) = X(z)p(z), \tag{A.9}$$

where $p(z)$ is an arbitrary polynomial of z.

Remarks

(i) The above results can be easily obtained by multiplying $G(t)$ by $t^{-\kappa}$ and $\phi_-(t)$ by t^{κ}, taking the logarithm of (A.6) and then using Plemelj's formulae. The introduction of the term $\tau^{-\kappa}$ in (A.8) makes the function $\ln[\tau^{-\kappa}G(\tau)]$ Hölder on C. (For the infinite contour $\text{Im}k = 0$ we can multiply by factors $(t \pm \mathrm{i})^{\kappa}$, e.g., see Gakhov [1966].)

(ii) Riemann-Hilbert problems are intimately related to singular integral equations. For this and other reasons one is interested in *finding all solutions of* (A.6) *which vanish at infinity*. Equations (A.8), (A.9) imply that: (a), if $\kappa > 0$ then there exist κ linearly independent solutions of (A.6) vanishing at infinity; and (b), if $\kappa \leq 0$ then there exist no solutions of (A.6) vanishing at infinity.

(iii) Inhomogeneous Riemann-Hilbert problems

$$\phi_+(t) = g(t)\phi_-(t) + F(t), \tag{A.10}$$

where $F(t)$ is Hölder on C are solved in terms of homogeneous Riemann-Hilbert problems. In particular replacing $G(t)$ by $X_+(t)/X_-(t)$ in (A.10) and using Plemelj's formula it follows that

$$\phi(z) = X(z)[\psi(z) + p(z)]; \qquad \psi(z) := \frac{1}{2\pi \mathrm{i}} \int_C \frac{F(\tau)\,\mathrm{d}\tau}{X_+(\tau)(\tau - z)}. \tag{A.11}$$

Hence, if one is looking for solutions vanishing at infinity it follows that: (a), if $\kappa > 0$ then there exist κ linearly independent solutions given by (A.11) with $p(z) = P_{\kappa-1}(z)$; (b), if $\kappa = 0$, then there exist a unique solution $X(z)\psi(z)$; and (c), if $\kappa < 0$ then there exists a unique solution $X(z)\psi(z)$ provided the following orthogonality conditions are satisfied:

$$\int_C \frac{F(\tau)\tau^{n-1}}{X_+(\tau)}\,\mathrm{d}\tau = 0, \quad n = 1, 2, \ldots, -\kappa.$$

A.1.2 Vector homogeneous Riemann-Hilbert problems. The vector homogeneous Riemann-Hilbert problem is defined as follows:

Given a contour C, and an $n \times n$ matrix $\mathbf{G}(t)$ which is Hölder and non-singular on C (i.e., G_{ij} $i, j = 1, \ldots, n$ are Hölder and $\det \mathbf{G}(t) \neq 0$ *on C), find a sectionally holomorphic vector function $\phi(z)$, with finite degree at ∞, such that*

$$\phi_+(t) = \mathbf{G}(t)\phi_-(t). \tag{A.12}$$

The solution of (A.12) can not, in general, be found in closed form. It is characterized through a system of linear Fredholm integral equations. However, in contrast to the scalar case, where the existence or not of solutions is *a priori* determined in terms of the index, the existence and uniqueness of solutions of (A.12) must be investigated in terms of certain Fredholm equations. This difficulty arises fact that the solutions of (A.12) depend on the individual indices $\kappa_1, \ldots, \kappa_n$ which *cannot be a priori calculated.* Only their sum

$$\kappa = \kappa_1 + \cdots + \kappa_n = \text{ind } \det \mathbf{G}(t), \tag{A.13}$$

is *a priori* known.

Solving (A.12) means finding a *fundamental solution matrix* $\mathbf{X}(z)$ composed of solution vectors $\chi_1(z), \ldots, \chi_n(z)$. The individual index κ_ℓ is related to the behaviour of $\chi_\ell(z)$, $\ell = 1, \ldots, n$, as $z \to \infty$: Suppose we are looking for solutions of (A.12) whose degree at infinity does not exceed r. Among these solutions there exist some with the lowest possible degree $-\kappa_1$. Let χ_1 denote a solution with degree $-\kappa_1$. From the remaining solutions consider all those which cannot be obtained from χ_1 by $\phi(z) = p(z)\chi_1(z)$, where $p(z)$ is some polynomial. Among these solutions pick one with the lowest possible degree. Call χ_2 and $-\kappa_2$ the solution and degree respectively, etc.. It can be shown that the solution matrix $\mathbf{X}(z)$ constructed this way has the following two properties: (i), $\det \mathbf{X}(z) \neq 0$ for all finite z; and (ii), $\det(z^{\kappa_1}\chi_1, \dots, z^{\kappa_n}\chi_n) \neq 0$ at $z = \infty$. Furthermore any solution of (A.12) is given by

$$\phi(z) = \mathbf{X}(z)\boldsymbol{p}(z), \tag{A.14}$$

where $\boldsymbol{p}(z)$ is a polynomial vector.

The above non-constructive approach for determining $\mathbf{X}(z)$ can be turned into a constructive one provided that κ_ℓ, $\ell = 1, \dots, n$ are known. For example, suppose that $\kappa_1 = \kappa_2 = \cdots = \kappa_n = 0$ (a necessary but not sufficient condition for this is $\text{ind} \det G(t) = 0$). Then using (A.12) and

$$\tfrac{1}{2}\chi_{\ell -}(t) + \frac{1}{2\pi i}\int_C \frac{\chi_{\ell -}(\tau)}{\tau - t}\,d\tau - \boldsymbol{e}_\ell = 0 \tag{A.15}$$

(which is a necessary and sufficient condition for $\chi_{\ell -}(z)$ to be holomorphic in D and tend to the ℓth unit vector $\boldsymbol{e}_\ell$ as $z \to \infty$) it follows that $\chi_{\ell -}$ is found from the vector Fredholm integral equation

$$\chi_{\ell -}(t) - \frac{1}{2\pi i}\int_C \frac{\mathbf{G}^{-1}(t)\mathbf{G}(\tau) - \mathbf{I}}{\tau - t}\,\boldsymbol{x}_{\ell -}(\tau)\,d\tau = \boldsymbol{e}_\ell, \quad \ell = 1, \dots, n. \tag{A.16}$$

Conversely if the linear Fredholm integral equations (A.16) have unique solutions $\chi_{\ell -}$, $\ell = 1, \dots, n$ then all the individual indices κ_ℓ, $\ell = 1, \dots, n$ are zero. To find κ_ℓ in general one has to investigate integral equations of the type (A.16). We remark that in §7.3.1 above we derived the Fredholm integral equation (7.3.10) from the matrix Riemann-Hilbert boundary value problem (7.3.6) associated with the MKdV equation. This is an example of (A.16) (for general forcing) and shows how one may a Fredholm integral equation from a Riemann-Hilbert boundary value problem in a special case.

Remarks

(i) To establish the above results one introduces several Riemann-Hilbert problems related to (A.12) the so-called adjoint, accompanying or associated problems and then appropriately uses these problems and the known results of Fredholm theory (in particular one makes extensive use of Fredholm's alternative theorem).

(ii) The results expressed by remarks (ii) and (iii) of the scalar case are easily generalised to the vector case (for example $\mathbf{G}(t) = \mathbf{X}_+(t)(\mathbf{X}_-(t))^{-1}$).

(iii) The results for the vector problem can be extended to matrix problems, this is only a question of rotation.

A.1.3 A note of the history of Riemann-Hilbert problems. The Riemann-Hilbert boundary value problem (sometimes referred to in the literature as the "*Hilbert*" or "*Riemann*" problem) was first introduced by Riemann [] in connection with the so-called "*Riemann monodromy*" problem. This problem is defined roughly as follows:

Given points $\{a_1, \ldots, a_n\}$ in the complex plane and $m \times m$ matrices $\mathbf{M}_1, \ldots, \mathbf{M}_n$ characterizing the monodromy group of $a_1, \ldots, a_n$ find all linear ordinary differential equations

$$\frac{\mathrm{d}\boldsymbol{y}}{\mathrm{d}x} = \sum_{i=1}^{n} \frac{\mathbf{A}_i}{x - a_i} \boldsymbol{y}, \tag{A.17}$$

where $\mathbf{A}_i$ are $m \times m$ matrices, which possess the monodromy group generated by $\mathbf{M}_i$, for $i = 1, \ldots, n$.

The connection between the Riemann-Hilbert and Riemann-Monodromy problems as well as their implications to soliton theory is worth noting (recent references re soliton theory include: Chudnovsky [1980a,b]; Flaschka and Newell [1980]; Fokas and Ablowitz [1983a]). The Riemann-Monodromy problem was transformed to what we call a Riemann-Hilbert problem by Hilbert [1924] (see also Plemelj [1964]; Birkhoff [1909, 1910]). This results in a vector Riemann-Hilbert problem with "discontinuous" coefficients (see below). Plemelj used the results of Riemann-Hilbert problems to study the Riemann-Monodromy problem. We also mention that Lappo-Danilevskij [1957] studied the Riemann-Monodromy problem directly via hyperlogarithm functions (cf. Erugin [1966]). More recently Sato, Miwa and Jimbo [1977, 1979a,b,c, 1980a,b]). Jimbo, Miwa and Ueno [1981], and Jimbo and Miwa [1981a,b] using the equivalence between the Riemann-Monodromy problem and the so-called Schlesinger equations, studied the Riemann-Monodromy problem using classical operators of field theory.

Riemann made no attempt to solve the problem he formulated. The first solution of the homogeneous scalar Riemann-Hilbert problem (A.6) was given by Hilbert [1904] in terms of a Fredholm integral equation (Picard [1927] proceeded also along similar lines). Plemelj [1908a,b] gave the first closed form solution of (A.6) in the case that $\kappa = 0$. Carlman [1932] solved a related singular integral equation (as it was pointed out earlier singular integral equations are related to Riemann-Hilbert problems). Gakhov [1937] gave the full solution of the scaler Riemann-Hilbert problem (A.6). The vector Riemann-Hilbert problem (A.12) was considered by Plemelj (cf. [1964]), Gakhov (cf. [1966]), Muskhelishvili (cf. [1953]) and extensively by Vekua (cf. [1967]). One should also note that the work of Wiener and Hopf [1931] is related to Riemann-Hilbert problems.

Discontinuous Riemann-Hilbert problems, i.e., problems where $G(t)$ is Hölder on L except on a finite number of points where it is discontinuous, have also been considered in the literature. A complete solution of the scalar discontinuous Riemann-Hilbert problem was given independently by Gakhov [1941] and Muskhelishvili [1941]. These authors used two rather different methods; Gakhov's method was generalized by Vekua to the vector discontinuous Riemann-Hilbert problem (a complete bibliography on the above work is contained in the books by Gakhov [1966], Muskhelishvili [1953] and Vekua [1967]).

A.2 Some results of Krein and Gohberg.

After the classical work on Riemann-Hilbert problems for Hölder functions reviewed in §A.1, many investigators have considered Riemann-Hilbert problems for functions in various other spaces. An up to date review of these results can be found in Zabreyko, Koshelev, Krasnosel'skii, Mikhlin, Rakovshchik and Stet'senko [1975]. Here we only mention some results taken from the papers of Krein [1958] and Gohberg and Krein [1958]. These results:

(i) serve to illustrate further the notion of index;

(ii) contain the only, to our knowledge, general result providing an *a priori* determination of all individual indices.

We first introduce some notation. Let $L^1(\mathbb{R})$ denote the normed linear (vector) space of measurable, absolutely integrable functions with norm

$$||f||_1 := \int_{-\infty}^{\infty} |f(t)| \, \mathrm{d}t.$$

Let $F(\lambda)$ be the Fourier transform of $f(t)$ where $f(t) \in L^1(\mathbb{R})$ and R_0 the linear space associated with $F(\lambda)$. Let R_a denote the linear space of $F(\lambda; a)$, where $F(\lambda; a)$ denotes the Fourier transform of $a\delta(t) + f(t)$, with a some complex number, i.e., R_a is the set of all functions of the form

$$F(\lambda; a) = a + \int_{-\infty}^{\infty} f(t)\mathrm{e}^{\mathrm{i}\lambda t} \, \mathrm{d}t, \qquad \text{with} \quad f(t) \in L^1(\mathbb{R}). \tag{A.18}$$

R_a^+ is the subset of R_a consisting of all functions of the form

$$F_+(\lambda; a) = a + \int_0^{\infty} f(t)\mathrm{e}^{\mathrm{i}\lambda t} \, \mathrm{d}t.$$

It is clear that functions in R_a^+ are analytically continuable in the upper half λ-complex plane π_+. Similar considerations apply to R_a^- and π_-.

Gohberg and Krein [1958] were interested in the solution of

$$\boldsymbol{\chi}(t) - \int_0^{\infty} \mathbf{K}(t-s)\boldsymbol{\chi}(s) \, \mathrm{d}s = \boldsymbol{f}(t), \tag{A.19}$$

where $\boldsymbol{\chi}(t)$, $\boldsymbol{f}(t)$ are vectors, $\mathbf{K}(t)$ is a $n \times n$ matrix with elements in $L^1(\mathbb{R})$. Assume for concreteness that $\boldsymbol{f}(t) \in L^1_{(n\times 1)}(0,\infty)$ (i.e., $f_j(t) \in L^1(0,\infty)$, $j = 1, \ldots, n$, where

$\boldsymbol{f} = (f_1, f_2, \dots, f_n))$ and seek solutions $\boldsymbol{\chi}(t) \in L^1_{(n\times 1)}(0,\infty)$. The solution of (A.19) is intimately related to the vector Riemann-Hilbert problem (see below)

$$\phi_+(\lambda) = \mathbf{G}(\lambda)\phi_-(\lambda), \qquad \mathbf{G}(\lambda) \in R^{(n\times n)}, \qquad \det \mathbf{G}(\lambda) \neq 0, \qquad \lambda \in \mathbb{R}. \tag{A.20}$$

We note that (A.20) is not a special case of (A.12) since $\mathbf{G}(\lambda)$ is continuous but, in general, not Hölder. Gohberg and Krein have developed a substantial theory for analyzing (A.19), (A.20). Here we only touch upon this theory; in particular we concentrate only on some of the results concerning (A.20). For completeness we first review the scalar case (corresponding to (A.19) being a scalar equation).

A.2.1 Scalar Factorization Problem. Assume that the scalar function $G(\lambda) \in R$ is defined on the closed line $L : \{-\infty, \infty\}$ and $G \neq 0$ on L. The basic factorization problem associated with G, which as we shall see is related to a Riemann-Hilbert problem, can be formulated as follows: Find $G_\pm \in R_\pm$ such that

$$G = G_+ G_-, \quad G_\pm(\infty) = 1. \tag{A.21}$$

We call a factorization proper if either $G_+ \neq 0$ in π_+ or $G_- \neq 0$ in π_-. We call a factorization canonical if both $G_\pm \neq 0$ in $\pi_\pm$.

Krein [1958], using fundamental theorems of Wiener and Levy, proved that:

1. The function $G(\lambda)$ (specified above) admits a canonical factorization if and only if $G(\lambda) \neq 0$ and ind $G = 0$. If it admits a canonical factorization then it is the only proper one.
2. Suppose that $G \neq 0$ and ind $G = \kappa \neq 0$.
   (i), If $\kappa > 0$ then there exists a choice of points $\alpha_1, \dots, \alpha_m \in \pi_+$ and integers $p_1, \dots, p_m$ satisfying $p_1 + \cdots + p_m = \kappa$, such that there exists a function $G_+ \in R_+$ with zeros at $\alpha_1, \dots, \alpha_m$ and no other zeros in π_+. Furthermore, there exists a function $G_- \in R_-$ such that $G = G_+G_-$ and G_- has no zeros in π_-.
   (ii), If $\kappa < 0$ analogous statements are valid (of course now $p_1 + \cdots + p_m = -\kappa$).

Remarks

(a) Equation (A.21) can be written as

$$G_+ = G(G_-)^{-1}, \quad G_\pm(\infty) = 1. \tag{A.22}$$

Then assuming, for the sake of comparison, that G is also Hölder, equation (A.22) becomes a Riemann-Hilbert problem with the "jump" given along the real axis and the boundary condition of boundedness at infinity. Suppose that $\kappa > 0$, then G_- has no zeros in π_- and hence $(G_-)^{-1} \in R_-$. Thus $\phi_+(\lambda) = G_+(\lambda)$ and $\phi_-(\lambda) = (G_-(\lambda))^{-1}$. However, if $\kappa < 0$ then $(G_-)^{-1}$ is not holomorphic in π_- and the Riemann-Hilbert (A.22) has no solution bounded at infinity. This is consistent with the scalar Riemann-Hilbert theory.

(b) The functions $G_\pm(\lambda)$ admit the following representations

$$G_\pm(\lambda) = 1 + \int_0^\infty e^{i\lambda t}\gamma_\pm(t)\,dt, \qquad \lambda \in \pi_\pm,$$

where $\gamma_\pm \in L^1(\mathbb{R})$. The above formulae are useful in establishing various properties of G_+, however they have the disadvantage that they are not constructive. Krein [1975] also gives constructive formulae for $G_\pm$, which coincide with the ones obtained via Riemann-Hilbert theory.

A.2.2 A vector Riemann-Hilbert problem with all its indices positive. A fundamental result of Gohberg and Krein [1958] is a factorization theorem for matrices $\mathbf{G}(\lambda) \in R^{(n\times n)}$. In preparation for this result we first consider a vector Riemann-Hilbert problem with all its indices positive.

Consider the adjoint of the homogeneous equation corresponding to (A.19)

$$\boldsymbol{w}(t) - \int_0^\infty \mathbf{k}^T(t-s)\boldsymbol{w}(s)\,ds = 0. \tag{A.23}$$

Equations (A.19) and (A.23) are intimately related to the Riemann-Hilbert problem

$$\mathbf{F}_-(\lambda) = [\mathbf{I} - \mathbf{K}(\lambda)]\mathbf{F}_+(\lambda), \qquad \mathbf{F}_\pm(\lambda) \in R_\pm^{(n\times 1)} \tag{A.24}$$

where $\mathbf{K}(\lambda)$ is the Fourier transform of $\mathbf{k}(t)$ (recall that $\mathbf{k}(t) \in L^1_{(n\times n)}(\mathbb{R})$). If (A.23) has only the trivial solution then all the individual indices of the Riemann-Hilbert problem (A.24) are non-negative. Assuming that (A.23) has only the trivial solution, then one may prove the following:

1. the Riemann-Hilbert problem (A.24) has n solutions in $R_+^{(n\times 1)}$ whose values at ∞ are linearly independent;
2. the multiplicity of any zero λ_0, $\mathrm{Im}(\lambda_0) \geq 0$ of $\mathbf{F}_+(\lambda) \in R_+^{(n\times 1)}$ is $\leq \alpha$, where α is the number of linearly independent solutions of the homogeneous version of (A.19).

In the above case one may also introduce the notion of a *standard solution matrix* which is analogous to the fundamental solution matrix introduced in §A.1. This matrix $\mathbf{F}(\lambda)$ is defined as follows: (i), $\mathbf{F}_+(\lambda)$ solves (A.24); (ii), $\det \mathbf{F}_+(\infty) \neq 0$; (iii), assume that the jth column vector of $\mathbf{F}_+(\lambda)$ has a zero at $\lambda = i$ of multiplicity κ_j. Choose the columns of $\mathbf{F}_+(\lambda)$ in such an order that $\kappa_1 \geq \kappa_2 \geq \ldots \geq \kappa_n$. Define $\boldsymbol{\kappa} = (\kappa_1, \kappa_2, \ldots, \kappa_n)$ to be the index of (A.24). Because of the result 2 above, $\boldsymbol{\kappa}$ is bounded and hence from all solution matrices in $R_+^{(n\times n)}$ there must be one with the greatest index. This is the standard solution matrix.

The following results are proven by Gohberg and Krein [1958] about the above standard solution matrix.

3. Let $\mathbf{F}_+(\lambda)$ be the standard solution matrix of (A.24). Then $\det \mathbf{F}_+(\lambda) \neq 0$ for every $\lambda \in \pi_+$, with the exception of $\lambda = \mathrm{i}$ where the multiplicity of zero of $\det \mathbf{F}_+(\lambda)$ is equal to $\sum_{j=1}^{n} \kappa_j$. Furthermore

$$\sum_{j=1}^{n} \kappa_j = -\,\mathrm{ind}\ \det[\mathbf{I} - \mathbf{K}(\lambda)], \tag{A.25}$$

   and hence $\det \mathbf{F}_-(\lambda) \neq \mathbf{0}$ for every $\lambda \in \pi_-$.
4. Let $\mathbf{F}_+(\lambda) = (\boldsymbol{f}_1, \dots, \boldsymbol{f}_j, \dots, \boldsymbol{f}_n)$ be the standard solution matrix of (A.24). Then every solution of (A.24) has the form

$$\mathbf{F}(\lambda) = \sum_{j=1}^{n} \left[a_{j0} + a_{j1}(\lambda - \mathrm{i})^{-1} + \ldots + a_{j\kappa_j}(\lambda - \mathrm{i})^{-\kappa_j}\right] \boldsymbol{f}_j(\lambda).$$

A.2.3 A matrix factorization theorem.

1. Every non-singular matrix $\mathbf{G}(\lambda) \in R_{(n\times n)}$ possesses the following factorization

$$\mathbf{G}(\lambda) = \mathbf{G}_+(\lambda)\boldsymbol{\Delta}(\lambda)\mathbf{G}_-(\lambda), \tag{A.26}$$

   where $\mathbf{G}_\pm \in R^{\pm}_{(n\times n)}$, $\det \mathbf{G}_\pm(\lambda) \neq 0$ for $\lambda \in \pi_\pm$ and

$$\boldsymbol{\Delta}(\lambda) = \mathrm{diag}\left(\left(\frac{\lambda - \mathrm{i}}{\lambda + \mathrm{i}}\right)^{\kappa_1}, \dots, \left(\frac{\lambda - \mathrm{i}}{\lambda + \mathrm{i}}\right)^{\kappa_n}\right),$$

   for some integers $\kappa_1 \geq \kappa_2 \geq \cdots \geq \kappa_n$.
2. If a matrix possesses another factorization

$$\mathbf{G} = \tilde{\mathbf{G}}_+(\lambda)\tilde{\boldsymbol{\Delta}}(\lambda)\tilde{G}_-(\lambda),$$

   then

$$\tilde{\boldsymbol{\Delta}}(\lambda) = \boldsymbol{\Delta}(\lambda), \qquad \tilde{\mathbf{G}}_+(\lambda) = \mathbf{G}_+\mathbf{A}(\lambda), \qquad \tilde{\mathbf{G}}_-(\lambda) = \mathbf{A}^{-1}\mathbf{G}_-(\lambda),$$

   for some non-singular matrix $\mathbf{A}$.
3.

$$\sum_{1}^{n} \kappa_j = \mathrm{ind}\ \det \mathbf{G}(\lambda). \tag{A.27}$$

A.2.4 A theorem about indices. Let $\mathbf{G}^*$ denote the Hermitian conjugate of $\mathbf{G}$, i.e. $\mathbf{G}^* = \bar{\mathbf{G}}^T$, where $\mathbf{G}^T$ denotes the transpose of $\mathbf{G}$ and $\bar{\mathbf{G}}$ the complex conjugate. Define the real and imaginary parts of a matrix through

$$\mathbf{G}_R = \tfrac{1}{2}(\mathbf{G} + \mathbf{G}^*), \qquad \mathbf{G}_I = \tfrac{1}{2}(\mathbf{G} - \mathbf{G}^*). \tag{A.28}$$

A matrix $\mathbf{G}$ is definite if $\boldsymbol{x}^*\mathbf{G}\boldsymbol{x}$ is real and has only one sign for all $\boldsymbol{x} \neq 0$. The following result is valid (Gohberg and Krein [1958]).

If the real or imaginary part of $\mathbf{G}(\lambda) \in R_{(n\times n)}$ *is definite and* $\det \mathbf{G}(\lambda) \neq 0$ *then all its individual indices are zero.*

The above result is a corollary of the matrix factorization theorem presented in §A.2.3.

As an example consider the following matrix to be factorized:

$$\mathbf{G}(\lambda) = \begin{pmatrix} 1 - r(\lambda)\bar{r}(\lambda) & r(\lambda)\mathrm{e}^{2\mathrm{i}\lambda x} \\ -\bar{r}(\lambda)\mathrm{e}^{-2\mathrm{i}\lambda x} & 1 \end{pmatrix}, \tag{A. * 29}$$

where $\bar{r}(\lambda)$ is the complex conjugate of $r(\lambda)$. This arises in the scattering theory associated with the KdV and MKdV equation. We have

$$\mathbf{G}^*(\lambda) = \begin{pmatrix} 1 - r(\lambda)\bar{r}(\lambda) & -r(\lambda)\mathrm{e}^{2\mathrm{i}\lambda x} \\ \bar{r}(\lambda)\mathrm{e}^{-2\mathrm{i}\lambda x} & 1 \end{pmatrix}.$$

Consequently

$$\mathbf{G}_R(\lambda) = \begin{pmatrix} 1 - r(\lambda)\bar{r}(\lambda) & 0 \\ 0 & 1 \end{pmatrix},$$

which is definite whenever $|r(\lambda)\bar{r}(\lambda)| < 1$.

A.2.5 Stability of the indices. Finally we mention that Gohberg and Krein [1958] prove an important theorem about stability of indices. Namely corresponding to $G(\lambda)$ is the overall index $\kappa = \sum_{j=1}^{n} \kappa_j$ (A.25). If there are values q, r such that

$$\kappa_1 = \kappa_2 = \ldots \kappa_r = q + 1, \qquad \kappa_{r+1} = \kappa_{r+2} = \ldots = \kappa_n = q, \tag{A.29}$$

satisfying

$$\kappa = qn + r \qquad (0 \leq r < n). \tag{A.30}$$

Then the indices κ_j will not be altered by an arbitrary sufficiently small (in a suitable space) perturbation. It follows immediately that if $\kappa = 0$ then all the partial indices $\kappa_i = 0$, $i = 1, \ldots, n$, if and only ij $\kappa_i = 0$ satisfy (A.29-30) and they are stable (if $r = 0$ (A.30) implies, $q = 0$, hence $\kappa_i = 0$, $i = 1, \ldots, n$, if $q = 1, \ldots, n-1$ there is no solution to (A.30) with $\kappa = 0$; if $r = n$, $q = -1$, hence from (A.29) $\kappa_i = 0$, $i = 1, \ldots, n$).

These results mean that if the overall index κ is zero, then generically speaking, all the partial indices vanish as well; namely in any neighborhood of a "non-generic" kernel $\mathbf{G}(\lambda) = \mathbf{I} - \mathbf{K}(\lambda)$ of (A.24) are kernels for which $\kappa_i = 0$, $i = 1, \ldots, n$. Hence the constructions based upon all indices zero go through in a generic sense. In our studies of Riemann-Hilbert problems associated with nonlinear wave equations, we are primarily interested in the generic situation in any event.

A.3 Riemann-Hilbert problems appearing in inverse scattering transforms.

The following types of Riemann-Hilbert problems have been used in connection with the solution of nonlinear equations arising in soliton theory. We note that these problem are a generalization of a Riemann-Hilbert problem in that we allow our extended functions to be meromorphic (specifically to possess a finite number of poles) in the relevant region.

1. **Scalar Riemann-Hilbert problems with a "shift"**, e.g., the Riemann-Hilbert problem associated with the time independent Schrödinger equation is of the form (x, t are parameters, cf. Chapter 2)

$$m_+(k) - m_-(k) = V(k,x,t)m_-(-k), \qquad \text{on} \qquad \mathrm{Im}k = 0. \tag{A.31}$$

   It is clear that $m_-(-k)$ is a plus function, hence by considering (A.31) together with the equation obtained from (A.31) by letting $k \to -k$, we obtain a standard vector Riemann-Hilbert problem (with a $\mathbf{G}(k)$ such as (A.*29)). The Riemann-Hilbert problem associated with the Intermediate Long Wave equation is also of shift type.

2. **Vector Riemann-Hilbert problems**, e.g., the Riemann-Hilbert problem associated with the $n \times n$ scattering problem

$$\frac{\mathrm{d}\boldsymbol{v}}{\mathrm{d}x} = \lambda \mathbf{J}\boldsymbol{v} + \mathbf{Q}(x)\boldsymbol{v}, \tag{A.32}$$

   where all the entries of $\mathbf{J}$ are either real or purely imaginary, $\mathbf{Q}$ off-diagonal, is of the matrix form (cf. Chapter 3)

$$\mathbf{m}_+(k) - \mathbf{m}_-(k) = \mathbf{V}(k,x,t)\mathbf{m}_-(-k), \qquad \text{on} \qquad \mathrm{Im}k = 0, \tag{A.33}$$

   where $\mathbf{V}(k,x,t)$ is proportional to the scattering data.

3. **Vector Riemann-Hilbert problems on complicated contours**, e.g. the general case of the eigenvalue problem (A.32), where the entries of $\mathbf{J}$ are complex. In this case the Riemann-Hilbert problem is defined along rays (cf. Beals and Coifman [1984]), i.e., the matrix Riemann-Hilbert problem (A.33) on Σ, where Σ is a collection of semi-rays.

4. **Discontinuous vector Riemann-Hilbert problems on complicated contours**. Riemann-Hilbert problems associated with the initial value problem of the Painlevé equations , e.g. the second Painlevé equation (Fokas and Ablowitz [1983a], cf. Chapter 7). In this case the matrix $\mathbf{G}(k)$ is discontinuous at the origin and the jump conditions are defined on a set of rays.

5. **Scalar differential Riemann-Hilbert problems**. An example is the Riemann-Hilbert problem which plays the role of the direct scattering problem

(the time-independent part of the Lax pair) for the Benjamin-Ono equation (or more generally the Intermediate Long Wave equation, cf. Chapter 4)

$$\mathrm{i}v_x^+ + \lambda(v^+ - v^-) = -uv^+.$$

6. **Nonlocal scalar and vector Riemann-Hilbert problems.** The inverse problem of the Benjamin-Ono (Chapter 4) and KPI (Chapter 5) equations give rise to scalar nonlocal Riemann-Hilbert problems. Similarly the inverse problem associated with hyperbolic systems in the plane (Chapter 5) also yields nonlocal vector Riemann-Hilbert problem of the form

$$\boldsymbol{m}_+(k) - \boldsymbol{m}_-(k) = \int_{-\infty}^{\infty} \mathbf{F}(\ell, k; x, y, t)\boldsymbol{m}_-(\ell)\,\mathrm{d}\ell.$$

For further details of Riemann-Hilbert boundary value problems see, for example, Rodin [1987, 1988].

Appendix B

The DBAR Problem

In this section we give an elementary introduction to the DBAR ($\bar{\partial}$) problem, which may be viewed as a generalization of the Riemann-Hilbert problem discussed in Appendix A. There we dealt with complex functions on the plane, which were sectionally meromorphic, namely everywhere analytic with the exception of a discrete set of points (poles) and of a peace-wise continuous curve dividing the plane into sections. Here we consider the more general case of complex functions, whose domain of non-analyticity may be two-dimensional, and, therefore, they may be even nowhere analytic in the plane. In fact, the main mathematical tool here is the usual differential and integral calculus on the (x, y) plane. Some of the relevant formulae are given below for the benefit of the reader, and to establish the notation. In the present context it is convenient to replace the cartesian coordinates x and y with the complex variables z and $\bar{z}$. Though z and $\bar{z}$ are complex numbers, they should be regarded merely as a new coordinate system in the plane, the coordinate transformation being

$$z = x + \mathrm{i}y, \qquad \bar{z} = x - \mathrm{i}y, \tag{B.1a}$$

$$x = \tfrac{1}{2}(z + \bar{z}), \qquad y = -\tfrac{1}{2}\mathrm{i}(z - \bar{z}). \tag{B.1b}$$

By means of this transformation, all well-known formulae of calculus can be formally rewritten in the $(z, \bar{z})$ coordinate system. Thus, a complex function $g(x, y)$, defined in a domain R of the plane, can be also expressed as

$$g(x, y) = f(z, \bar{z}), \qquad z \in R \subset \mathbb{C}, \tag{B.2}$$

in terms of the two variables z and $\bar{z}$. Of course, the coordinate transformation (B.1) induces a relation among the partial differential operators, which reads

$$\partial \equiv \frac{\partial}{\partial z} = \frac{1}{2}\left(\frac{\partial}{\partial x} - \mathrm{i}\frac{\partial}{\partial y}\right), \qquad \bar{\partial} \equiv \frac{\partial}{\partial \bar{z}} = \frac{1}{2}\left(\frac{\partial}{\partial x} + \mathrm{i}\frac{\partial}{\partial y}\right), \tag{B.3}$$

$$\partial_x \equiv \frac{\partial}{\partial x} = \partial + \bar{\partial}, \qquad \partial_y \equiv \frac{\partial}{\partial y} = \mathrm{i}(\partial - \bar{\partial}), \tag{B.4}$$

In particular, the differential operator $\bar{\partial}$ is the so-called "DBAR" ($\bar{\partial}$) operator, its name originating from the fact that $\bar{\partial}$ is formally the complex conjugate of ∂, with the notational convention that a "bar" indicates the complex conjugation. This operator plays a special role in connection with the theory of analytic functions. In fact, if $f(z,\bar{z})$ is analytic in a domain R, then

$$\bar{\partial} f(z,\bar{z}) = 0, \qquad z \in R, \tag{B.5}$$

namely f does not depend on the coordinate $\bar{z}$. Indeed, if $f(z,\bar{z}) = u(x,y) + \mathrm{i}v(x,y)$, the differential equation (B.5) coincides with the Cauchy-Riemann conditions of analyticity expressed by the elliptic system of two partial differential equations

$$u_x = v_y, \qquad u_y = -v_x. \tag{B.6}$$

We will generally consider functions $f(z,\bar{z})$ which are not analytic, i.e. $\bar{\partial} f \neq 0$; for convenience, in the following we indicate with $f(z)$ *any* complex function in the plane, and, by abuse of notation, we will write equations such as $\bar{\partial} f(z) = \phi(z)$, with $\phi(z) \neq 0$.

In the present context, the Gauss-Green integral formulae

$$\iint_R \left(\frac{\partial P}{\partial x} + \frac{\partial Q}{\partial y} \right) \mathrm{d}x\, \mathrm{d}y = \oint_{\partial R} P\, \mathrm{d}y - Q\, \mathrm{d}x,$$

or in special cases

$$\oint_{\partial R} g(x,y)\, \mathrm{d}x = -\iint_R g_y(x,y)\, \mathrm{d}x \wedge \mathrm{d}y, \tag{B.7a}$$

$$\oint_{\partial R} g(x,y)\, \mathrm{d}y = \iint_R g_x(x,y)\, \mathrm{d}x \wedge \mathrm{d}y, \tag{B.7b}$$

are of basic importance; here R is a simply connected domain of the plane (for the sake of simplicity), its boundary ∂R being a piece-wise differentiable and counter-clockwise oriented curve, and $\mathrm{d}x \wedge \mathrm{d}y$ is the Lebesgue measure on the plane (we use the convenient wedge product notation, with its property of being skew-symmetric so that $\mathrm{d}x \wedge \mathrm{d}y = \frac{1}{2}\mathrm{i}\, \mathrm{d}z \wedge \mathrm{d}\bar{z}$, see (B.1b)). In the complex coordinates z and $\bar{z}$, the Gauss-Green formulae take the form

$$\oint_{\partial R} f(z)\, \mathrm{d}z = -\iint_R \bar{\partial} f(z)\, \mathrm{d}z \wedge \mathrm{d}\bar{z}, \tag{B.8a}$$

$$\oint_{\partial R} f(z)\, \mathrm{d}\bar{z} = \iint_R \partial f(z)\, \mathrm{d}z \wedge \mathrm{d}\bar{z}, \tag{B.8b}$$

where we have set $f(z) = g(x,y)$: and we note that the equation (B.8a) generalizes the Cauchy theorem to the class of non-analytic functions

$$\oint_{\partial R} f(z)\, \mathrm{d}z = 0,$$

for analytic functions, to which it reduces if $f(z)$ is analytic in R. Indeed, the formula (B.8a) clearly shows that the contribution to the closed curve integral in the left hand side originates from the "departure from analyticity" measured by the $\bar{\partial}$ derivative in the right hand side. The familiar residue theorem, which applies when $f(z)$ has simple pole singularities in R, can be recovered this way by computing the $\bar{\partial}$ derivative of such singular functions. However, in order to deal with this simple case, for convenience we introduce first the Dirac distribution in the plane. This is defined by the usual formula

$$\iint_R \delta(x-x_0)\,\delta(y-y_0)\,g(x,y)\,\mathrm{d}x\wedge\mathrm{d}y = g(x_0,y_0), \qquad (x_0,y_0)\in R, \tag{B.9}$$

which, in the coordinates $(z,\bar{z})$, reads

$$\iint_R \delta(z-z_0)\,f(z)\,\mathrm{d}z\wedge d\bar{z} = -2\mathrm{i}f(z_0), \tag{B.10}$$

where $g(x,y)=f(z)$, and we use the notation

$$\delta(z-z_0)=\delta(x-x_0)\,\delta(y-y_0). \tag{B.11}$$

Let now $F(z)$ be analytic in R, $\bar{\partial}F(z)=0$, and let us replace $f(z)$ with $F(z)/(z-z_0)$ in (B.8a), with $z_0\in R$; then the Gauss-Green formula reads

$$\oint_{\partial R}\frac{F(z)}{z-z_0}\,\mathrm{d}z = -\iint_R F(z)\bar{\partial}\left(\frac{1}{z-z_0}\right)\,\mathrm{d}z\wedge\mathrm{d}\bar{z}. \tag{B.12}$$

We note that the left hand side is the Cauchy integral representation of an analytic function, and it equals $2\pi\mathrm{i}F(z_0)$; therefore, comparing the resulting equality with the definition (B.10), we arrive at the useful formula

$$\bar{\partial}\left(\frac{1}{\pi(z-z_0)}\right)=\delta(z-z_0), \tag{B.13}$$

which shows that the simple pole expression $1/(\pi z)$ is the Green function, vanishing at infinity, of the operator $\bar{\partial}$. By using this result, we can now generalize the Cauchy integral representation formula to non-analytic functions; to this aim, replace again $f(z)$ with $F(z)/(z-z_0)$ ($z_0\in R$) in (B.8a), but without assuming that $F(z)$ be analytic in R. It is then easily found that the Gauss-Green relation (B.8a) yields the generalized Cauchy formula

$$F(z_0)=\frac{1}{2\pi\mathrm{i}}\oint_{\partial R}\frac{F(z)}{z-z_0}\,\mathrm{d}z+\frac{1}{2\pi\mathrm{i}}\iint_R\frac{\bar{\partial}F(z)}{z-z_0}\,\mathrm{d}z\wedge\mathrm{d}\bar{z}, \tag{B.14}$$

which is the basic result in investigating the $\bar{\partial}$ problem. A special case of particular relevance, for our application to inverse scattering, is obtained when the domain $R=$

R_∞ is the entire complex plane, $\partial R = C_\infty$ an arbitrarily large circle in the plane, and the function $F(z)$ is bounded everywhere, including the point at infinity, say

$$F(z) \to 1, \qquad |z| \to \infty. \tag{B.15}$$

Then the formula (B.14) yields

$$F(z_0) = 1 + \frac{1}{2\pi \mathrm{i}} \iint_{\mathbf{C}} \frac{\bar{\partial} F(z)}{z - z_0}\, \mathrm{d}z \wedge \mathrm{d}\bar{z}. \tag{B.16}$$

For future reference, we write below two more formulae. The first is the obvious identity (using equation (B.13))

$$\bar{\partial}\left[a(z) + \frac{1}{2\pi \mathrm{i}} \iint_R \frac{g(\zeta)}{\zeta - z}\, \mathrm{d}\zeta \wedge \mathrm{d}\bar{\zeta}\right] = g(z), \qquad z \in R, \tag{B.17}$$

where $a(z)$ is assumed to be analytic in R. The second formula deals with the $\bar{\partial}$ derivative of a sectionally analytic function, and it reads (again using equation (B.13))

$$\bar{\partial}\left[\frac{1}{2\pi \mathrm{i}} \int_\Gamma \frac{g(\zeta)}{\zeta - z}\, \mathrm{d}\zeta\right] = \tfrac{1}{2}\mathrm{i} \int_\Gamma g(\zeta)\, \delta(\zeta - z)\, \mathrm{d}\zeta, \tag{B.18}$$

where Γ is an oriented continuous curve and $g(\zeta)$ is Hölder on Γ (see appendix A). A special case of interest occurs when Γ coincides with the real axis, so that, if

$$f(z) = \frac{1}{2\pi \mathrm{i}} \int_{-\infty}^{\infty} \frac{g(\zeta)}{\zeta - z}\, \mathrm{d}\zeta, \tag{B.19a}$$

then (B.18) takes the form

$$\bar{\partial} f(z) = \tfrac{1}{2}\mathrm{i}[f_+(x) - f_-(x)]\, \delta(y), \tag{B.19b}$$

where we have used the notation (B.1), and $f_\pm(x) = f(x \pm \mathrm{i}0)$ are the boundary values. Recall from equation (A.4) that the Plemelj formulae yield

$$f_\pm(x) = \pm\tfrac{1}{2} g(x) + \frac{1}{2\pi \mathrm{i}} \int_{-\infty}^{\infty} \frac{g(\zeta)}{\zeta - x}\, \mathrm{d}\zeta,$$

i.e.,

$$f_+(x) - f_-(x) = g(x),$$
$$f_+(x) + f_-(x) = \frac{1}{\pi \mathrm{i}} \int_{-\infty}^{\infty} \frac{g(\zeta)}{\zeta - x}\, \mathrm{d}\zeta.$$

This equation gives the connection between the $\bar{\partial}$ derivative of a sectionally analytic functions and its jump discontinuity.

We are now in the position to consider the $\bar{\partial}$ equation

$$\bar{\partial} f(z) = g(z), \qquad z \in R, \tag{B.20}$$

where $g(z)$ is given in the open domain R, and $f(z)$ is the unknown function. This is an elliptic system of two coupled partial differential equations,

$$u_x - v_y = 2q, \qquad u_y + v_x = 2p, \tag{B.21}$$

with $f(z) = u(x, y) + iv(x, y)$ and $g(z) = q(x, y) + ip(x, y)$, in cartesian coordinates (see (B.3b)); the $\bar{\partial}$ formalism is clearly more convenient to represent such a system. In fact, from our previous discussion (in particular, equation (B.17)), the general solution of the differential equation (B.20) is given by

$$f(z) = a(z) + \frac{1}{2\pi \mathrm{i}} \iint_R \frac{g(\zeta)}{\zeta - z} \, \mathrm{d}\zeta \wedge \mathrm{d}\bar{\zeta}, \tag{B.22}$$

where $a(z)$ is any function that is analytic in R.

The associated $\bar{\partial}$ problem is the problem of finding the particular solution of the $\bar{\partial}$ equation (B.20), which satisfies the boundary condition

$$f(z) = h(z), \qquad z \in \partial R, \tag{B.23}$$

where $h(z)$ is a given function on the boundary of the domain R. The question here is whether an analytic function $a(z)$ can be found, such that the solution (B.22) satisfies the boundary condition (B.23). If $h(z)$ is an arbitrary function, then the $\bar{\partial}$ problem in general has no solution. In fact, suppose the solution $f(z)$ exist, then the generalized Cauchy formula (B.14) provides the explicit expression

$$f(z) = \frac{1}{2\pi \mathrm{i}} \oint_{\partial R} \frac{h(\zeta)}{\zeta - z} \, \mathrm{d}\zeta + \frac{1}{2\pi \mathrm{i}} \iint_R \frac{g(\zeta)}{\zeta - z} \, \mathrm{d}\zeta \wedge \mathrm{d}\bar{\zeta}, \qquad z \in R, \tag{B.24}$$

for the solution; by letting now z approach the boundary curve ∂R, and using the Plemelj formulae (A.3), we obtain for the given functions $h(z)$ and $g(z)$ of the $\bar{\partial}$ problem the following relationship

$$h(z) = \frac{1}{\pi \mathrm{i}} P \oint_{\partial R} \frac{h(\zeta)}{\zeta - z} \, \mathrm{d}\zeta + \frac{1}{\pi \mathrm{i}} \iint_R \frac{g(\zeta)}{\zeta - z} \, \mathrm{d}\zeta \wedge \mathrm{d}\bar{\zeta}, \qquad z \in \partial R, \tag{B.25}$$

which, for given $g(z)$, is an integral equation for the boundary value $h(z)$. Therefore, $h(z)$ cannot be arbitrary, but it has to satisfy the necessary condition (B.25). The relevant case in scattering theory, i.e. in the context of section 3, corresponds to $h(z) = 1$, $R = \mathbb{C}$ and $g(z)$ vanishing sufficiently fast as $|z| \to \infty$; therefore, the unknown function $f(z)$ has to satisfy the $\bar{\partial}$ equation (B.20) with the condition that

$$f(z) \to 1, \qquad |z| \to \infty. \tag{B.26}$$

It is easily seen that the condition (B.25) is satisfied (take R as a circle of radius ρ with $h(z) = 1$ on the boundary, and let $\rho \to \infty$). Consequently the solution of this $\bar{\partial}$ problem satisfies

$$f(z) = 1 + \frac{1}{2\pi \mathrm{i}} \iint_{\mathbf{C}} \frac{g(\zeta)}{\zeta - z} \, \mathrm{d}\zeta \wedge \mathrm{d}\bar{\zeta}. \tag{B.27}$$

We note that this is an explicit formula if $g(\zeta)$ is a known function, i.e., the $\bar{\partial}$ problem we have solved is non-homogeneous. However, in the application of scattering theory, the $\bar{\partial}$ problem is itself homogeneous, namely

$$\bar{\partial} f(z) = \eta(z) \, f(\gamma z), \tag{B.28}$$

(γz is an automorphism, e.g. $\gamma z = -z$) with (B.26) and for a given function $\eta(z)$, then the formula (B.27) is no longer explicit (such as in the KP equations, or a matrix system as in the Davey-Stewartson problem), but it transforms the differential equation (B.28) into the equivalent integral equation

$$f(z) = 1 + \frac{1}{2\pi \mathrm{i}} \iint_{\mathbf{C}} \frac{\eta(\zeta) f(\gamma\zeta)}{\zeta - z} \, \mathrm{d}\zeta \wedge \mathrm{d}\bar{\zeta}, \tag{B.29}$$

which turns out to be a more convenient means to investigate the solution $f(z)$. We finally notice that, if we allow the given function $g(z)$ in (B.20), or $\eta(z)$ in (B.28), to be a distribution, then a Riemann-Hilbert problem (see Appendix A) can be reformulated as a $\bar{\partial}$ problem. As an instance, consider the Riemann-Hilbert problem of finding a function $f(z)$, which is analytic in the upper and lower halves of the plane divided by the real axis, satisfying the asymptotic condition (B.26), and whose boundary values $f_{\pm}(x)$ on the real axis satisfy the equation (see (A.8))

$$f_{+}(x) = G(x) f_{-}(x) + h(x). \tag{B.30}$$

Then this problem is equivalent to the $\bar{\partial}$ problem of finding the solution of the $\bar{\partial}$ equation (see (B.19))

$$\bar{\partial} f(z) = \tfrac{1}{2}\mathrm{i}\{[G(x) - 1] \; f_{-}(x) + h(x)\}\delta(y), \tag{B.31}$$

with the condition (B.26).

References

Ablowitz M.J. [1977], *Nonlinear evolution equations — continuous and discrete*, SIAM Review, **19**, 663–684;

——— [1990], *Painlevé equations and the inverse scattering and inverse monodromy transforms*, in "Solitons in Physics, Mathematics and Nonlinear Optics," eds. P.J. Olver and D.H. Sattinger, IMA Series **25**, 27–43, Springer-Verlag, Berlin-Heidelberg-New York.

Ablowitz M.J., BarYaacov D. and Fokas A.S. [1983], *On the inverse scattering transform for the Kadomtsev-Petviashvili equation*, Stud. Appl. Math., **69**, 135–142.

Ablowitz M.J., Beals R. and Tenenblat K. [1986], *On the solutions of the generalized wave and generalized Sine-Gordon equations*, Stud. Appl. Math., **74**, 177–203; **77**, 101.

Ablowitz M.J., Costa D.G. and Tenenblat K. [1987], *Solutions of multidimensional extensions of the Anti-Self-Dual Yang-Mills equation*, Stud. Appl. Math., **77**, 37–46.

Ablowitz M.J. and de Lillo S. [1991], *Forced and semiline solutions of Burgers equation*, Phys. Lett., **156A**, 483–487.

Ablowitz M.J. and Fokas A.S. [1982], *A direct linearization associated with the Benjamin-Ono equation*, in "Mathematical Methods in Hydrodynamics and Integrability in Dynamical Systems," proceedings, La Jolla 1981, eds. M. Tabor and Y.M. Treve, AIP Conf. Proc., **88**, 229–236;

——— [1983], *Comments on the inverse scattering transform and related nonlinear evolution equations*, in "Nonlinear Phenomena," proceedings, Oaxtepec, Mexico 1982, ed. K.B. Wolf, Lect. Notes Phys., **189**, 3–24, Springer-Verlag, Berlin-Heidelberg-New York.

Ablowitz M.J., Fokas A.S. and Anderson R.L. [1983], *The direct linearizing transform and the Benjamin-Ono equation*, Phys. Lett., **93A**, 375–378.

Ablowitz M.J., Fokas A.S. and Kruskal M.D. [1987], *Note on solutions to class of nonlinear singular integro-differential equations*, Phys. Lett., **120A**, 215–218.

Ablowitz M.J., Fokas A.S., Satsuma J. and Segur H. [1982], *On the periodic intermediate long wave equation*, J. Phys. A: Math. Gen., **15**, 781–786.

Ablowitz M.J. and Haberman R. [1975a], *Nonlinear evolution equations — two and three dimensions*, Phys. Rev. Lett., **35**, 1185–1188;

——— [1975b], *Resonantly coupled nonlinear evolution equations*, J. Math. Phys., **16**, 2301–2305.

Ablowitz M.J. and Herbst B.M. [1990a], *On homoclinic structure and numerically induced chaos for the nonlinear Schrödinger equation*, SIAM J. Appl. Math., **50**, 339–351;

——— [1990b], *On homoclinic boundaries in the nonlinear Schrödinger equation*, in "Hamiltonian Systems, Transformation Groups and Spectral Transform," eds. J. Harnad and J. Marsden, 121–131, Les Publications de Centre Recherches Mathématiques, Montreal.

Ablowitz M.J., Herbst B.M. and Keiser J. [1990], *Nonlinear evolution equations, solitons, chaos and cellular automata*, in "Nonlinear Physics," Research Reports in Physics, eds. C. Gu, Y. Li and G. Tu, 166–189, Springer-Verlag, Berlin-Heidelberg-New York.

Ablowitz M.J., Kaup D.J. and Newell A.C. [1974], *Coherent pulse propagation, a dispersive, irreversible phenomenon*, J. Math. Phys., **15**, 1852–1858.

Ablowitz M.J., Kaup D.J., Newell A.C. and Segur H. [1973a], *Method for solving the Sine-Gordon equation*, Phys. Rev. Lett., **30**, 1262–1264;

——— [1973b], *Nonlinear evolution equations of physical significance*, Phys. Rev. Lett., **31**, 125–127;

——— [1974], *The inverse scattering transform — Fourier analysis for nonlinear problems*, Stud. Appl. Math., **53**, 249–315.

Ablowitz M.J. and Keiser J. [1990], *On particles and interaction properties of the Parity Rule Filter automata*, PAM preprint **68**, University of Colorado, Boulder.

Ablowitz M.J., Keiser J. and Takhtajan L.A. [1991a], *A class of stable multi-state time-reversible cellular automata with rich particle content*, PAM preprint **74**, University of Colorado, Boulder;

——— [1991b], *Stable multi-state time-reversible cellular automata with rich particle content*, PAM preprint **75**, University of Colorado, Boulder.

Ablowitz M.J. and Kodama Y. [1980], *Note on asymptotic solutions of the Korteweg-de Vries equation with solitons*, Stud. Appl. Math., **66**, 159–170.

Ablowitz M.J., Kruskal M.D. and Ladik J.F. [1979], *Solitary wave collisions*, SIAM J. Appl. Math., **36**, 428–437.

Ablowitz M.J., Kruskal M.D. and Segur H. [1979], *A note on Miura's transformation*, J. Math. Phys., **20**, 999–1003.

Ablowitz M.J. and Ladik J.F. [1975], *Nonlinear differential-difference equations*, J. Math. Phys., **16**, 598–603;

——— [1976a], *Nonlinear differential-difference equations and Fourier analysis*, J. Math. Phys., **17**, 1011–1018;

——— [1976b], *A nonlinear difference scheme and inverse scattering*, Stud. Appl. Math., **55**, 213–229;

——— [1977], *On the solution of a class of nonlinear partial difference equations*, Stud. Appl. Math., **57**, 1–12.

Ablowitz M.J., Manakov S.V. and Schultz C.L. [1990], *On the boundary conditions of the Davey-Stewartson equation*, Phys. Lett., **148A**, 50–52.

Ablowitz M.J. and Nachman A.I. [1986], *Multidimensional nonlinear evolution equations and inverse scattering*, Physica, **18D**, 223–241.

Ablowitz M.J., Ramani A. and Segur H. [1978], *Nonlinear evolution equations and ordinary differential equations of Painlevé type*, Lett. Nuovo Cim., **23**, 333–338;

——— [1980a,b], *A connection between nonlinear evolution equations and ordinary differential equations of P-type. I and II*, J. Math. Phys., **21**, 715–721; 1006–1015.

Ablowitz M.J. and Schultz C.L. [1989], *Strong coupling limit of certain multidimensional nonlinear wave equations*, Stud. Appl. Math., **80**, 229–238.

Ablowitz M.J. and Segur H. [1975], *The inverse scattering transform: semi-infinite interval*, J. Math. Phys., **16**, 1054–1056;

——— [1977a], *Asymptotic solutions of the Korteweg-de Vries equation*, Stud. Appl. Math., **57**, 13–44;

——— [1977b], *Exact linearization of a Painlevé transcendent*, Phys. Rev. Lett., **38**, 1103–1106;

——— [1979], *On the evolution of packets of water waves*, J. Fluid Mech., **92**, 691–715;

——— [1981], "*Solitons and the Inverse Scattering Transform*," SIAM, Philadelphia.

Ablowitz M.J. and Villarroel J. [1990a], *On the complete integrability of certain nonlinear evolution equations in one and two spatial dimensions*, in "*Chaos and Order*," eds. N. Joshi and R.L. Dewar, World Scientific, Singapore, to appear;

——— [1990b], *On the Kadomtsev-Petviashvili equation and associated constraints*, Stud. Appl. Math., to appear.

Abramowitz M. and Stegun I.A. [1972], "*Handbook of Mathematical Functions*," 10th edition, Dover, New York.

Adler M. and Moser J. [1978], *On a class of polynomials connected with the Korteweg-de Vries equation*, Commun. Math. Phys., **61**, 1–30.

Adler M. and van Moerbeke P. [1982], *The algebraic integrability of geodesic flow on* SO(4), Invent. Math., **67**, 297–331;

——— [1988], *The Kowalewski and Hénon-Heiles motions as Manakov geodesic flows on* SO(4) — *a two-dimensional family of Lax pairs*, Commun. Math. Phys., **113**, 659–700;

——— [1989], *The complex geometry of the Kowalewski-Painlevé analysis*, Invent. Math., **97**, 3–51.

Airault H. [1979], *Rational solutions of Painlevé equations*, Stud. Appl. Math., **61**, 31–53.

Airault H., McKean H.P. and Moser J. [1977], *Rational and elliptic solutions of the KdV equation and related many-body problems*, Commun. Pure Appl. Math., **30**, 95–198.

Agranovich Z. and Marchenko V. [1963], "*The Inverse Problem of Scattering Theory*," Gordon and Breach, New York.

Airy G.B. [1845], "*Tides and waves*," Encyclopedia Metropolotana, vol. **5**, 241–396, London.

Akhmediev N.N., Korneev V.I. and Mitskevich N.V. [1985], *Generation of periodic train of picosecond pulses in an optical fiber: exact solutions*, Sov. Phys. JETP, **62**, 894–899;

——— [1988], *N-modulation signals in a single-mode optical waveguide under nonlinear conditions*, Sov. Phys. JETP, **67**, 89–95.

Aktosun T. and van der Mee C. [1990], *Solution of the inverse scattering problem for the 3-D Schrödinger equation by Weiner-Hopf factorization of the scattering operator*, J. Math. Phys., **31**, 2172–2180;

——— [1991], *Solution of the inverse scattering problem for the three-dimensional Schrödinger equation using a Fredholm integral equation*, SIAM J. Math. Anal., **22**, 717–731.

Alpers W. and Salusti E. [1983], *Scylle and Charybdis observed from space*, J. Geophy. Res., **88**, 1800–1808.

Alonso L.M. [1985a], *Soliton motion in the case of a nonzero reflection coefficient*, Phys. Rev. Lett., **54**, 400–501;

——— [1985b], *Effect of the radiation component on soliton motion*, Phys. Rev., **32D**, 1459–1466;

——— [1985c], *Solitons as truncated asymptotic solutions of completely integrable models*, Phys. Lett., **112A**, 361–365.

Anderson R.L. and Ibragimov N.H. [1979], "*Lie-Bäcklund Transformations in Applications*," SIAM, Philadelphia.

Anderson R.L. and Taflin E. [1985], *Benjamin-Ono equation — recursivity of linearization maps — Lax pairs*, Lett. Math. Phys., **9**, 299–311.

Anker D. and Freeman N.C. [1978], *On the soliton solutions of the Davey-Stewartson equation for long waves*, Proc. Roy. Soc. London A, **360**, 529–540.

Antonowicz M. and Fordy A.P. [1987], *Coupled KdV equations with multi-Hamiltonian structures*, Physica, **28D**, 345–357;

——— [1989], *Super-extensions of energy dependent Schrödinger operators*, Commun. Math. Phys., **124**, 487–500.

Apel J.R. and Gonzalez F.I. [1983], *Nonlinear features of internal waves off Baja California as observed from the SEASAT imaging radar*, J. Geophys. Res., **88**, (C7), 4459–4466.

Apel J.R. and Holbrook J.R. [1980], *The Sulu sea internal soliton experiment, 1. Background and overview*, Eos Trans. AGU, **61**, 1009.

Arkadiev V.A., Pogrebkov A.K. and Polivanov M.C. [1989a], *Closed string-like solutions of the Davey-Stewartson equation*, Inverse Problems, **5**, L1–L6;

——— [1989b], *Inverse scattering transform method and soliton solutions for Davey-Stewartson II equation*, Physica, **36D**, 189–197.

Arnold V.I. [1978], "*Mathematical Methods of Classical Mechanics*," Graduate Texts Math., **60**, Springer-Verlag, Berlin-Heidelberg-New York.

Ashtekar A., Jacobsen E. and Smolin L. [1988], *A new characterization of half-flat solutions to Einstein's equation*, Commun. Math. Phys., **115**, 631–648.

Atiyah M.F., Drinfield V.G., Hitchen N.J. and Manin Yu.I. [1978], *Construction of instantons*, Phys. Lett., **65**, 185–187.

Atiyah M.F. and Hitchen N.J. [1985a], *Low energy scattering of non-Ableian monopoles*, Phys. Lett., **107A**, 21–25;

——— [1985b], *Low energy scattering of non-Ableian magnetic monopoles*, Phil. Trans. R. Soc. Lond. A, **315**, 459–469.

Atiyah M.F., Hitchen N.J. and Singer I.M. [1978], *Self-duality in four-dimensional Riemannian geometry*, Proc. Roy. Soc. London A, **362**, 425–461.

Atiyah M.F. and Ward R.S. [1977], *Instantons and Geometry*, Commun. Math. Phys., **55**, 117–124.

Bäcklund A.V. [1876], *Über Flächentransformationen*, Math. Ann., **9**, 207–320;

——— [1880], *Zür Theorie der partiellen Differentialgleichungen erster Ordnung*, Math. Ann., **17**, 285–328;

——— [1882], *Zür Theorie der Flächentransformationen*, Math. Ann., **19**, 387–422;

——— [1905], "Concerning Surfaces with Constant Negative Curvature," (trans. by E.M. Coddington) New Ear, Lancaster, PA.

Barouch E., Fokas A.S. and Papageorgiou V. [1988], *The bi-Hamiltonian formulation of the Landau-Lifshiftz equation*, J. Math. Phys., **29**, 2618–2633.

Barouch E., McCoy B.M. and Wu T.T. [1973], *Zero-field susceptibility of the two-dimensional Ising Model Near* T_c, Phys. Rev. Lett., **31**, 1490–1411.

Bassom A.P., Clarkson P.A., Hicks A.C. and McLeod J.B. [1991], *Integral equations and exact solutions for the fourth Painlevé equation*, preprint, University of Exeter, U.K..

Bassom A.P. and Hall P. [1989], *On the generation of mean flows by the interaction of Görtler vortices and Tollmien-Schlichting waves in curved channel flows*, Stud. Appl. Math., **81**, 185–219.

Baxter R.J. [1972], *Partition function of the eight vertex lattice model*, Ann. Phys., **70**, 193–228;

——— [1982], "Exactly Solved Models in Statistical Mechanics," Academic, New York;

——— [1985], *Exactly solved models in statistical mechanics*, in "Integrable Systems in Statistical Mechanics," eds. G.M. D'Ariano, P. Montorsi and M. Rasetti, Ser. Adv. Stat. Mech., **1**, World Scientific, Singapore.

——— [1986], *The Yang-Baxter equations and the Zamolodchikov model*, Physica, **18D**, 321–347.

Bazhanov V.V. [1987], *Integrable quantum systems and classical Lie algebras*, Commun. Math. Phys., **113**, 471–503.

Beals R. [1985], *The inverse problem for ordinary differential equations on the line*, Amer. J. Math., **107**, 281–366.

Beals R. and Coifman R.R. [1981, 1982], *Scattering, transformations spectrales et equations d'evolution nonlineare*, I and II, presented at Seminaire Goulaouic-Meyer-Schwartz 1980–1981, exp. 22 and 1981–1982, exp. 21, Ecole Polytechnique, Palaiseau;

——— [1984], *Scattering and inverse scattering for first order systems*, Commun. Pure Appl. Math., **37**, 39–90;

——— [1985a], *Inverse scattering and evolution equations*, Commun. Pure Appl. Math., **38**, 29–42;

——— [1985b], *Multidimensional inverse scattering and nonlinear partial differential equations*, Proc. Symp. Pure Math., **43**, Amer. Math. Soc., Providence, 45–70;

——— [1986], *The D-bar approach to inverse scattering and nonlinear evolutions*, Physica, **18D**, 242–249;

——— [1987], *Scattering and inverse scattering for first-order systems. II*, Inverse Problems, **3**, 577–593;

——— [1989], *Linear Spectral problems, nonlinear equations, and the $\bar{\partial}$-method*, Inverse Problems, **5**, 87–130.

Beals R., Deift P. and Tomei C. [1988], "*Direct and Inverse Scattering on the Line*," Math. Surv. Mono., **28**, American Mathematical Society, Providence, R.I..

Beals R. and Konopelchenko B. [1991], *On nontrivial interactions and complete integrability of soliton equations*, preprint.

Beals R., Rabelo M. and Tenenblat K. [1988], *Bäcklund transformations and inverse scattering and for some pseudospherical surface equations*, Stud. Appl. Math., **81**, 125–151.

Beals R. and Sattinger D.H. [1991], *On the complete integrability of completely integrable systems*, Commun. Math. Phys., **138**, 409–436.

Beals R. and Tenenblat K. [1988], *Inverse scattering and the Bäcklund transformation for the Generalized Wave and Generalized Sine-Gordon equations*, Stud. Appl. Math., **78**, 227–256.

Belavin A.A. and Zakharov V.E. [1978], *Yang-Mills equations as inverse scattering problem*, Phys. Lett., **73B**, 53–57.

Bender C.M. and Orszag S.A. [1978], "*Advanced Mathematical Methods for Scientists and Engineers*," McGraw-Hill, New York.

Benjamin T.B. [1967], *Internal waves of permanent form in fluids of great depth*, J. Fluid Mech., **29**, 559–592.

Benjamin T.B., Bona J.L. and Mahoney J.J. [1972], *Model equations for long waves in nonlinear dispersive systems*, Phil. Trans. Roy. Soc., **A272**, 47–78.

Benjamin T.B. and Feir J.F. [1967], *The disintegration of wave trains on deep water*, J. Fluid Mech., **27**, 417–430.

Benney D.J. [1966], *Long nonlinear waves in fluid flows*, J. Math. Phys. (Stud. Appl. Math.), **45**, 52–63.

Benney D.J. and Roskes G.J. [1969], *Waves instabilities*, Stud. Appl. Math., **48**, 377–385.

Bessis D. and Chafee N. [1986], *On the existence and non-existence of natural boundaries for non-integrable dynamical systems*, preprint.

Bianchi L. [1918], "*Lezioni sulla Teoria dei Gruppi Continui Finiti di Transformazioni*," Enrico Spoerri, Pisa;

——— [1922], "*Lezioni di Geometria Differenziale*," Vol. I. Enrico Spoerri, Pisa.

Bikbaev R.F. and Tarasov V.O. [1991], *Initial boundary value problem for the nonlinear Schrödinger equation*, J. Phys. A: Math. Gen., **24**, 2507–2516.

Birkhoff G.D. [1909], *Singular points of ordinary linear differential equations*, Trans. Amer. Math. Soc., **10**, 436–470;

——— [1910], *A simplified treatment of the regular singular point*, Trans. Amer. Math. Soc., **11**, 199–202.

Bishop A.R., Flesch R., Forest M.G., McLaughlin D.W. and Overman E.A. II [1990], *Correlations between chaos in a perturbed Sine-Gordon equations and in a truncated modal system*, SIAM J. Math. Anal., **21**, 1511–1536.

Bishop A.R., Forest M.G., McLaughlin D.W. and Overman E.A. II [1986], *A quasi-periodic route to chaos in a near-integrable PDE*, Physica, **23D**, 293–328;

——— [1988], *Quasi-periodic route to chaos in a near-integrable PDE*, Phys. Lett., **127**, 335–340.

Bishop A.R. and Lomdahl P.S. [1986], *Nonlinear dynamics in driven, damped Sine-Gordon systems*, Physica, **18D**, 54–66.

Blasak M. [1989], *Benjamin-Ono interacting solitons as field representations of Galilean point particles*, J. Phys. A: Math. Gen., **22**, 451–457.

Bluman G.W. and Cole J.D. [1969], *The general similarity solution of the heat equation*, J. Math. Mech., **18**, 1025–1042;

——— [1974], "*Similarity Methods for Differential Equations*," Appl. Math. Sci., **13**, Springer-Verlag, Berlin-Heidelberg-New York.

Bluman G.W. and Kumei S. [1989], "*Symmetries and Differential Equations*," Appl. Math. Sci., **81**, Springer-Verlag, Berlin-Heidelberg-New York.

Bobenko A.I. [1991], *Uniformization of Riemann surfaces and effectiveness of theta-functional formulae*, preprint.

Bobenko A.I. and Bordag L.A. [1989], *Periodic multiphase solutions of the Kadomtsev-Petviashvili equation*, J. Phys. A: Math. Gen., **22**, 1259–1274.

Bobenko A.I. and Kuznetsov V.B. [1988], *Lax representation and new formulae for the Goryachev-Chaplygin top*, J. Phys. A: Math. Gen., **21**, 999–2006.

Bobenko A.I., Matveev V.B. and Sall M.A. [1982], *Nonlocal Korteweg-de Vries and Kadomtsev-Petviashvili equations*, Sov. Phys. Dokl., **27**, 610–611.

Bobenko A.I., Reyman A.G. and Semenov-Tian-Shansky M.A. [1989], *The Kowalewski top 99 years later: a Lax pair, generalizations and explicit solutions*, Commun. Math. Phys., **122**, 321–354.

Bock T.L. and Kruskal M.D. [1979], *A two-parameter Miura transformation for the Benjamin-Ono equation*, Phys. Lett., **74A**, 173–176.

Bogdanov L.V. and Manakov S.V. [1988], *The non-local $\overline{\partial}$ problem and $(2+1)$-dimensional soliton equations*, J. Phys. A: Math. Gen., **21**, L537–L544.

Bogolubsky I.L. [1977], *Some examples of inelastic soliton interactions*, Comp. Phys. Commun., **13**, 149–155.

Bogolyubov N.N. and Prikarpat-skii A.K. [1982], *An inverse periodic problem for discrete approximation of the nonlinear Schrödinger equation*, Sov. Phys. Dokl., **27**, 113–116.

Bogoyavlensky O.I. [1988], *Five constructions of integrable dynamical systems connected with the Korteweg-de Vries equation*, Acta Appl. Math., **13**, 227–266.

Boiti M., Leon J.J-P., Manna M. and Pempinelli F. [1986], *On the spectral transform of a Korteweg-de Vries equation in two spatial dimensions*, Inverse Problems, **2**, 271–279;

——— [1987], *On a spectral transform of a KDV-like equation related to the Schrödinger operator in the plane*, Inverse Problems, **3**, 25–36.

Boiti M., Leon J.J-P., Martina L. and Pempinelli F. [1987], *On the recursion operator for the KP hierarchy in tow and three spatial dimensions*, Phys. Lett., **123A**, 340–344;

——— [1988a], *Integrable non-linear evolution in* $2+1$ *dimensions with non-analytic dispersion relations*, J. Phys. A: Math. Gen., **21**, 3611–3627.

——— [1988b], *Scattering of localized solitons in the plane*, Phys. Lett., **132A**, 432–439;

——— [1989], *Solitons in two dimensions*, in "Integrable Systems and Applications," proceedings, Oléron, France, 1988, eds. M. Balabane, P. Lochak and C. Sulem, 31–48, Lect. Notes Phys., **342**, Springer-Verlag, Berlin-Heidelberg-New York.

Boiti M., Leon J.J-P. and Pempinelli F. [1987a], *Integrable two-dimensional generalisation of the sine- and sinh-Gordon equations*, Inverse Problems, **3**, 37–49;

——— [1987b], *Spectral transform for a two spatial dimension extension of the dispersive long wave equation*, Inverse Problems, **3**, 371–387;

——— [1988], *Canonical and noncanonical recursion operators in multidimensions*, Stud. Appl. Math., **78**, 1–19;

——— [1989a], *Spectral transform and orthogonality relations for the Kadomtsev-Petviashvili I equation*, Phys. Lett., **141A**, 96–100;

——— [1989b], *A new spectral transform for the Davey-Stewartson I equation*, Phys. Lett., **141A**, 101–107;

——— [1990a], *On the spectral theory for the Davey-Stewartson equation*, in "Inverse Problems in Action," proceedings, Montpellier 1989, ed. P.C. Sabatier, 544–551, Springer-Verlag, Berlin-Heidelberg-New York;

——— [1990b], *Multidimensional solitons and their spectral transforms*, J. Math. Phys., **31**, 2612–2618;

——— [1990c], *Bifurcations of solitons in multidimensions*, Inverse Problems, **6**, 715–723;

——— [1991], *Waves in the Davey-Stewartson equation*, Inverse Problems, **7**, 175–185.

Boiti M. and Pempinelli F. [1980], *Nonlinear Schrödinger equation, Bäcklund transformation and Painlevé transcendents*, Nuovo Cim., **59B**, 40–58.

Bona J.L. and Smith R. [1975], *The initial value problem for the Korteweg-de Vries equation*, Philos. Trans. Roy. Soc. London Ser. A, **278**, 555–604.

Bona J. and Winther R. [1983], *The Korteweg-de Vries equation posed in quarter space*, SIAM J. Math. Anal., **14**, 1056–1106.

Bountis T.C., Ramani A., Grammaticos B. and Dorizzi B. [1984], *On the complete and partial integrability of non-Hamiltonian systems*, Physica, **128A**, 268–288.

Boussinesq M.J. [1871], *Théorie de l'intumescence appelée onde solitaire ou de translation se propageant dans un canal rectangulaire*, C.R. Acad. Sc. Paris, **72**, 755–759;

——— [1872], *Théorie des ondes et des remous qui se propagent le long d'un canal rectangulaire horizontal, en communiquant au liquide contenu dans ce canal des vitesses sensiblemant parielles de la surface au fond*, J. Math. Pures Appl. Ser. (2), **17**, 55–108.

Boutroux P. [1913, 1914], *Recherches sur les transcendents de M. Painlevé et l'étude asymptotique des équations différentielles du second ordre*, Ann. Ecole Norm. Super., **30**, 255–375; **31**, 99–159.

Brézin E. and Kazakov V.A. [1990], *Exactly solvable field theories of closed strings*, Phys. Lett., **236B**, 144–150.

Bruschi M. and Calogero F. [1987], *The Lax representation for an integrable class of relativistic dynamical systems*, Commun. Math. Phys., **109**, 481–492.

Buchholz H. [1969], "*The Confluent Hypergeometric Function*," Springer-Verlag, Berlin-Heidelberg-New York.

Bureau F. [1964a,b], *Differential equations with fixed critical points*, Ann. Mat. Pura Appl. (IV), **64**, 229–364; **66**, 1–116;

——— [1972a], *Equations différentielles du second ordre en $\ddot{y}$ et du second degré en y dont l'intégrale générale est à points critiques fixès*, Ann. Mat. Pura Appl. (IV), **91**, 163–281;

——— [1972b], *Integration of some nonlinear systems of ordinary differential equations*, Ann. Mat. Pura Appl. (IV), **94**, 344–360;

——— [1980a], *Systèmes differentiels à points critiques fixès. I*, Acad. Roy. Belg. Bull. Cl. Sc. (5), **66**, 672–683;

——— [1980b, 1981a,b,c,d], *Systèmes differentiels à points critiques fixès. II–VI. Points singuliers des intégrales*, Acad. Roy. Belg. Bull. Cl. Sc. (5), **66**, 762–775; **67**, 63–79; **67**, 241–247; **67**, 299–318; **67**, 447–454;

——— [1981e,f,g,h, 1982a,b,c], *Systèmes differentiels à points critiques fixès. VII–XIII. Les systèmes differentiels polynomiaux stables*, Acad. Roy. Belg. Bull. Cl. Sc. (5), **67**, 512–528; **67**, 637–665; **67**, 755–781; **67**, 942–957; **68**, 56–80; **68**, 220–248; **68**, 398–423;

——— [1983a], *Les équations différentielles du second ordre à points critiques fixès. I. Les intégrales de l'équation A2 de Painlevé*, Acad. Roy. Belg. Bull. Cl. Sc. (5), **69**, 80–104;

——— [1983b], *Les équations différentielles du second ordre à points critiques fixès. II. Les intégrales de l'équation A4 de Painlevé*, Acad. Roy. Belg. Bull. Cl. Sc. (5), **69**, 397–433;

——— [1983c], *Les équations différentielles du second ordre à points critiques fixès. III. Les intégrales de l'équation A3 de Painlevé*, Acad. Roy. Belg. Bull. Cl. Sc. (5), **69**, 614–640;

——— [1984], *Les équations différentielles du second ordre à points critiques fixès. IV. Les intégrales de l'équation A5 de Painlevé*, Acad. Roy. Belg. Bull. Cl. Sc. (5), **70**, 290–310;

——— [1987], *Sur des systèmes différentiels non linéaires du troisième ordre et les équations différentielles non linéaires associées*, Acad. Roy. Belg. Bull. Cl. Sc. (5), **73**, 335–353.

Bureau F., Garcet A. and Goffar J. [1972], *Transformées algébriques des équations du second ordre dont l'integrale générale est à points critiques fixès*, Ann. Mat. Pura Appl. (IV), **92**, 177–191.

Buslaev V.S., Faddeev L.D. and Takhtajan L.A. [1986], *Scattering theory for the Korteweg-de Vries (KdV) equation and its Hamiltonian interpretation*, Physica, **18D**, 255–266.

Calogero F. [1971], *Solution of the one-dimensional N-body problems with quadratic and/or inversely quadratic pair potentials*, J. Math. Phys., **12**, 419–436;

——— [1978], *Motion of poles and zeros of special solutions of nonlinear and linear partial differential equations, and related "solvable" many-body problems*, Nuovo Cim., **43B**, 177–421.

Calogero F. and Degasperis A. [1978], *Solution by the spectral transform method of a nonlinear evolution equation including as a special case the cylindrical KdV equation*, Lett. Nuovo Cim., **23**, 150–154;

——— [1981], *Reduction technique for matrix nonlinear evolution equations solvable by the spectral transform*, J. Math. Phys., **22**, 23–31;

——— [1982], "*Spectral Transform and Solitons I*," North-Holland, Amsterdam.

Calogero F. and de Lillo S. [1989b], *Burgers equation on the semi-infinite and finite intervals*, Nonlinearity, **2**, 37–43;

——— [1989b], *Burgers equation on the semiline*, Inverse Problems, **5**, L37–L40;

——— [1991], *Burgers equation on the semiline with general boundary conditions at the origin*, J. Math. Phys., **32**, 99–105.

Camacho C. and Lund F. [1985], *Generation of Boussinesq solitons via boundary conditions*, Phys. Lett., **108A**, 1–3.

Campbell D.K. and Peynard M. [1986], *Solitary wave collisions revisited*, Physica, **18D**, 47–53.

Caponi E.A., Saffman P.G. and Yuen H.C. [1982], *Instability and confined chaos in a nonlinear dispersive system*, Phys. Fluids, **25**, 2159–2166.

Caratheodory C. [1954], "*Theory of Functions of a Complex Variable*," Vol. 2, Chelsea, New York.

Cariello F. and Tabor M. [1989], *Painlevé expansions for nonintegrable evolution equations*, Physica, **39D**, 77–94.

Carleman T. [1932], *Sur la résolution de certaines équations intégrales*, Archiv för matematik astronomi och fysik, **16**, no. 26.

Carroll R. [1990], *On the forced nonlinear Schrödinger equation*, Japan J. Appl. Math., **7**, 321–344.

Carroll R. and Oharu S. [1990], *Some remarks on forced KdV*, Appl. Anal., **39**, 83–93.

Case K.M. [1978], *The N-soliton solution of the Benjamin-Ono equation*, Proc. Nat. Acad. Sci., **75**, 3562–3563;

——— [1979], *Benjamin-Ono-related equations and their solutions*, Proc. Nat. Acad. Sci., **76**, 1–3.

Case K.M. and Kac M. [1973], *A discrete version of the inverse scattering problem*, J. Math. Phys., **14**, 594–603.

Case K.M. and Monge A. [1989], *Explicitly time-dependent constant/symmetries of the higher-order KP equations*, J. Math. Phys., **30**, 1250–1253.

Caudrey P.J. [1980], *The inverse problem for the third order equation* $u_{xxx} + q(x)u_x + r(x)u = i\rho^3 u$, Phys. Lett., **79A**, 264–268;

——— [1982], *The inverse problem for a general $N \times N$ spectral equation*, Physica, **6D**, 51–66;

——— [1986], *Discrete and periodic spectral transforms related to the Kadomtsev-Petviashvili equation*, Inverse Problems, **2**, 281–292;

——— [1990], *Two-dimensional spectral transform*, in "Soliton Theory: A Survey of Results," ed. A.P. Fordy, 55–74, M.U.P., Manchester.

Caudrey P.J., Dodd R.K. and Gibbon J.D. [1976], *A new hierarchy of Korteweg-de Vries equations*, Proc. Roy. Soc. London A, **351**, 407–422.

Chadan K. and Sabatier P.C. [1977], "Inverse Problems in Quantum Scattering Theory," Springer-Verlag, Berlin-Heidelberg-New York.

Chakravarty S. and Ablowitz M.J. [1991], *On reductions of self-dual Yang-Mills equations*, in "Painlevé Transcendents, their Asymptotic and Physical Applications," proceedings, St. Adèle, Québec, Canada, 1990, eds., D. Levi and P. Winternitz, Plenum, New York, to appear.

Chakravarty S., Ablowitz M.J. and Clarkson P.A. [1990], *Reductions of self-dual Yang-Mills fields and classical systems*, Phys. Rev. Lett., **65**, 1085–1087;

——— [1991], *One dimensional reductions of self-dual Yang-Mills fields and classical systems*, in "Recent Advances in General Relativity: Essays in Honor of Ted Newman," proceedings, Pittsburgh, U.S.A., 1990, eds., A.I. Janis and J.R. Porter, Birkhäuser Boston, Basel, to appear.

Chakravarty S., Mason L. and Newman E.T. [1991], *Canonical structures on anti-self-dual four-manifolds and the diffeomorphism group*, J. Math. Phys., **32**, 1458–1464.

Chalkley R. [1987], *New contributions to the related work of Paul Appell, Lazarus Fuchs, George Hamel, and Paul Painlevé on nonlinear differential equations whose solutions are free of movable branch points*, J. Diff. Eqns., **68**, 72–117.

Champagne B., Hereman W. and Winternitz P. [1990], *The computer calculation of large systems of differential equations*, preprint CRM-**1689**, Centre de Recherches Mathématiques, Montréal, Canada.

Champagne B. and Winternitz P. [1985], *A MACSYMA program for calculating the symmetry group of a system of differential equations*, preprint CRM-**1278**, Centre de Recherches Mathématiques, Montréal, Canada;

——— [1988], *On the infinite-dimensional symmetry group of the Davey-Stewartson equations*, J. Math. Phys., **29**, 1–8.

Chandrasekhar S. [1986], *Cylindrical waves in general relativity*, Proc. Roy. Soc. London A, **408**, 209–232.

Chang Y.F., Greene J.M., Tabor M. and Weiss J. [1983], *The analytic structure of dynamical systems and self-similar natural boundaries*, Physica, **8D**, 183–207.

Chang Y.F., Tabor M. and Weiss J. [1982], *Analytic structure of the Hénon-Heiles Hamiltonian in integrable and nonintegrable regimes*, J. Math. Phys., **23**, 531–538.

Channel P.J. and Scovel C. [1990], *Symplectic integration of Hamiltonian systems*, Nonlinearity, **3**, 231–259;

——— [1991], *Integrators for Lie-Poisson dynamical systems*, Physica, **50D**, 80–88.

Chau L.L. [1984], *Integrability properties of supersymmetric Yang-Mills fields and relations with other nonlinear systems*, in "Group Theoretical Methods in Physics," proceedings of the XIIIth International Colloquium, ed. W.W. Zachary, 3–15, World Scientific, Singapore.

Chazy J. [1909], *Sur les équations différentielles dont l'intégrale générale est uniforme et admet des singularities essentielles mobiles*, C.R. Acad. Sc. Paris, **149**, 563–565;

——— [1910], *Sur les équations différentielles dont l'intégrale générale possède une coupure essentielle mobile*, C.R. Acad. Sc. Paris, **150**, 456–458;

——— [1911], *Sur les équations différentielles du troisième et d'ordre supérieur dont l'intégrale générale à ses points critiques fixés*, Acta Math., **34**, 317–385.

Chen H.H. [1975], *A Bäcklund transformation in two dimensions*, J. Math. Phys., **16**, 2382–2384.

Chen H.H. and Lee Y.C. [1979], *Internal wave solitons of fluids of finite depth*, Phys. Rev. Lett., **43**, 264–266.

Chen H.H., Lee Y.C. and Liu J.-E. [1979], *Integrability of nonlinear Hamiltonian systems by inverse scattering method*, Physica Scripta, **20**, 490–492;

——— [1983], *On a new hierarchy of symmetries for the Kadomtsev-Petviashvili equation*, Physica, **9D**, 439–445.

Chen H.H., Lee Y.C. and Pereira N.R. [1979], *Algebraic internal wave solitons and the integrable Calogero-Moser-Sutherland N-body problem*, Phys. Fluids, **22**, 187–188.

Chern S.S. and Terng C.L. [1980], *Analogue of Bäcklund's theorem in affine geometry*, Rocky Mountain Journal, **10**, 105–124.

Chudnovsky D.V. [1980a], *Riemann monodromy problem, isomonodromy deformation equations and completely integrable systems*, in "Bifurcation Phenomena in Mathematical Physics and Related Topics," proceedings, Cargese, 1979, eds. C. Bardos and D. Bessis, 385–447, Reidel, Dordrecht.

——— [1980b], *The generalized Riemann-Hilbert and the spectral interpretation*, in "Nonlinear Evolution Equations and Dynamical Equations," proceedings, Lecce 1979, eds. M. Boiti, F. Pempinelli and G. Soliani, Lect. Notes Phys., **120**, 103–149, Springer-Verlag, Berlin-Heidelberg-New York.

Chudnovsky D.V. and Chudnovsky G.V. [1977], *Pole expansions of nonlinear partial differential equations*, Nuovo Cim., **40B**, 339–353.

Chudnovsky D.V., Chudnovsky G.V. and Tabor M. [1983], *Painlevé property and multicomponent isospectral deformation equations*, Phys. Lett., **97A**, 268–274.

Chu C.K., Xiang L.W. and Baransky Y. [1983], *Solitary waves induced by boundary motion*, Commun. Pure Appl. Math., **36**, 495–504.

Clarkson P.A. [1985], *The Painlevé property and a partial differential equation with an essential singularity*, Phys. Lett., **109A**, 205–208;

——— [1986], *Remarks on the two-dimensional Sine-Gordon equation and the Painlevé tests*, Lett. Math. Phys., **10**, 297–299.

——— [1988], *Painlevé analysis for the damped, driven nonlinear Schrödinger equation*, Proc. Roy. Soc. Edin., **109A**, 109–126;

——— [1989], *New similarity solutions and Painlevé analysis for the Symmetric Regularized Long Wave and the Modified Benjamin-Bona-Mahoney equations*, J. Physics A: Math. Gen., **22**, 3821–3848;

——— [1990a], *Painlevé analysis and the complete integrability of a generalized variable-coefficient Kadomtsev-Petviashvili equation*, IMA J. Appl. Maths, **44**, 27–53;

——— [1990b], *New exact solutions of the Boussinesq equation*, Europ. J. Appl. Math., **1**, 279–300.

Clarkson P.A. and Cosgrove C.M. [1987], *The Painlevé property and a generalised derivative nonlinear Schrödinger equation*, J. Phys. A: Math. Gen., **20**, 2003–2024.

Clarkson P.A., Fokas A.S. and Ablowitz M.J. [1989], *Hodograph transformations on linearizable partial differential equations*, SIAM J. Appl. Math., **49**, 1188–1209.

Clarkson P.A. and Kruskal M.D. [1989], *New similarity solutions of the Boussinesq equation*, J. Math. Phys., **30**, 2201–2213.

Clarkson P.A. and McLeod J.B. [1983], *A connection formula for the second Painlevé transcendent*, in "*Ordinary and Partial Differential Equations*," proceedings, Dundee, Scotland, 1982, eds. W.N. Everitt and B.D. Sleeman, Lect. Notes Math., **964**, 135–142, Springer-Verlag, Berlin-Heidelberg-New York;

——— [1988], *A connection formula for the second Painlevé transcendent*, Arch. Rat. Mech. Anal., **103**, 97–138;

——— [1991], *Integral equations and connection formulae for the Painlevé equations*, in "*Painlevé Transcendents, their Asymptotic and Physical Applications*," proceedings, St. Adèle, Québec, Canada, 1990, eds., D. Levi and P. Winternitz, Plenum, New York, to appear.

Clarkson P.A., McLeod J.B., Ramani A. and Olver P.J. [1986], *Integrability of Klein-Gordon equations*, SIAM J. Math. Anal., **17**, 798–802.

Clarkson P.A. and Tuszyński J.A. [1990], *Exact solutions of the multidimensional derivative nonlinear Schrödinger equation for many-body systems near criticality*, J. Phys. A: Math. Gen., **23**, 4269–4288.

Clarkson P.A. and Winternitz P. [1991], *Nonclassical symmetry reductions for the Kadomtsev-Petviashvili equation*, Physica, **49D**, 257–272.

Coifman R.R. and Wickerhauser M.V. [1990], *The scattering transform for the Benjamin-Ono equation*, Inverse Problems, **6**, 825–861.

Cole J.D. [1951], *On a quasi-linear parabolic equation occuring in aerodynamics*, Q. Appl. Math., **9**, 225–236.

Constantin P., Lax P.D. and Majda A. [1985], *A simple one-dimensional model for the three-dimensional vorticity equation*, Commun. Pure Appl. Math., **38**, 715–724.

Conte R. [1988], *Universal invariance properties of Painlevé analysis and Bäcklund transformation in nonlinear partial differential equations*, Phys. Lett., **134A**, 100–104;

——— [1989], *Invariant Painlevé analysis of partial differential equations*, Phys. Lett., **140A**, 383–390.

Conte R. and Musette M. [1989], *Painlevé analysis and Bäcklund transformation in the Kuramoto-Sivashinsky equation*, J. Phys. A: Math. Gen., **22**, 169–177.

Cooke R. [1984], "*The Mathematics of Sonya Kovalevskaya*," Springer-Verlag, Berlin-Heidelberg-New York.

Cornille H. [1979], *Solutions of the nonlinear 3-wave equations in three spatial dimensions*, J. Math. Phys., **20**, 1653–1666.

Corrigan E.F., Fairlie D.B., Yates R.G. and Goddard P. [1978], *Construction of self-dual solutions to* SU(2) *gauge theory*, Commun. Math. Phys., **58**, 223–240

Cosgrove C.M. [1980], *Relationships between the group-theoretic and soliton-theoretic techniques for generating stationary axisymmetric gravitational solutions*, J. Math. Phys., **21**, 2417-2447.

Creamer D.B., Thacker H.B. and Wilkinson D. [1982], *Some exact results for the two-point function of an integrable quantum field theory*, Phys. Rev., **D23**, 3081–3084;

——— [1986], *A study of correlation functions for the delta-function bose gas*, Physica, **20D**, 155–186.

Curry J.H., Herring J.R., Loncaric J. and Orszag S.A. [1984], *Order and disorder in two- and three-dimensional Bénard convection*, J. Fluid Mech., **147**, 1–38.

Dahlberg B.E.J. and Trubowitz E. [1984], *The inverse Sturm-Liouville problem III*, Commun. Pure Appl. Math., **37**, 255–267.

Das A. [1989], "*Integrable Models*," World Scientific Lect. Notes Phys., **30**, World Scientific, Singapore.

Date E., Jimbo M., Kashiwara M. and Miwa T. [1981a], *Transformation groups for soliton equations. I. The τ-function of the Kadomtsev-Petviashvili equation*, Proc. Japan Acad. Ser. A, **57**, 342–347;

——— [1981b], *Transformation groups for soliton equations. II. Vertex operators and τ-functions*, Proc. Japan Acad. Ser. A, **57**, 387–392;

——— [1981c], *Transformation groups for soliton equations. III. Operator approach to the Kadomtsev-Petviashvili equation*, J. Phys. Soc. Japan, **50**, 3806–3812;

——— [1981d], *Transformation groups for soliton equations. IV. A new hierarchy of equations of KP-type*, Physica, **4D**, 343–365;

——— [1981e], *Transformation groups for soliton equations. VI. KP hierarchies of orthogonal and symplectic type*, J. Phys. Soc. Japan, **50**, 3813–3818;

——— [1982], *Transformation groups for soliton equations. V. Quasiperiodic solutions of the Kadomtsev-Petviashvili equation*, Publ. RIMS, Kyoto Univ., **18**, 1111–1119.

Davey A. and Stewartson K. [1974], *On three-dimensional packets of surface waves*, Proc. Roy. Soc. London A, **338**, 101–110.

David D., Kamran N., Levi D. and Winternitz P. [1986], *Symmetry reduction for the Kadomtsev-Petviashvili equation using a loop algebra*, J. Math. Phys., **27**, 1225–1237.

David D., Levi D. and Winternitz P. [1987a], *Integrable nonlinear equations for water waves in straits of varying width and depth*, Stud. Appl. Math., **76**, 133–168;

——— [1987b], *Bäcklund transformations and the infinite-dimensional symmetry group of the Kadomtsev-Petviashvili equation*, Phys. Lett., **118A**, 390–394;

——— [1989], *Solitons in shallow seas of variable depth and in marine straits*, Stud. Appl. Math., **80**, 1–23.

Davies B. [1981], *Difficulties with the inverse scattering transform method in quantum field theory*, J. Phys. A: Math. Gen., **14**, 145–153.

Davies R.E. and Acrivos A. [1967], *Solitary internal waves in deep water*, J. Fluid Mech., **29**, 593–607.

Davydov A.S. [1981], *The role of solitons in the energy and electron transfer in one-dimensional molecular systems*, Physica, **3D**, 1–22;

——— [Davydov A.S.], "*1985*," Solitons in Molecular SystemsReidel, Dordrecht.

de Boer P.C.T. and Ludford L.S.S. [1975], *Spherical electric probe in a continuum gas*, Plas. Phys., **17**, 29–43

Davydov A.S. [1982], "*Biology and Quantum Mechanics*," Pergamon, New York.

Degasperis A., Lebedev D., Olshanetsky M., Pakuliak S., Perelomov A. and Santini P.M. [1991a], *Recent developments for integrable integro-differential equations*, preprint, Roma University, Italy;

——— [1991b], *Nonlocal integrable partners to generalized MKdV and two-dimensional Toda Lattice equation in the formalism of a dressing method with quantized spectral parameter*, preprint, Roma University, Italy.

Degasperis A. and Sabatier P.C. [1987], *Extension of the one-dimensional scattering theory, and ambiguities*, Inverse Problems, **3**, 73–109.

Degasperis A. and Santini P.M. [1983], *Linear operator and conservation laws for a class of nonlinear integro-differential evolution equations*, Phys. Lett., **98A**, 240–244.

Degasperis A., Santini P.M. and Ablowitz M.J. [1985], *Nonlinear evolution equations associated with a Riemann-Hilbert scattering problem*, J. Math. Phys., **26**, 2469–2472.

Deift P., Tomei C. and Trubowitz E. [1982], *Inverse scattering and the Boussinesq equation*, Commun. Pure Appl. Math., **35**, 567–628.

Deift P. and Trubowitz E. [1979], *Inverse scattering on the line*, Commun. Pure Appl. Math., **32**, 121–251.

Din A.M., Horvath Z. and Zakrzewski W.J. [1984], *The Riemann-Hilbert problem and finite action* CP^{N-1} *model solutions*, Nucl. Phys., **B233**, 269–288.

Dodd R.K., Eilbeck J.C., Gibbon J.D. and Morris H.C. [1982], "*Solitons and Nonlinear Waves*," Academic, London.

Dodd R.K. and Fordy A.P. [1983], *The prolongation structures of quasi-polynomial flows*, Proc. Roy. Soc. London A, **385**, 389–429;

——— [1984], *Prolongation structures of complex quasi-polynomial evolution equations*, J. Phys. A: Math. Gen., **17**, 3249–3266.

Dodd R.K. and Gibbon J.D. [1977], *The prolongation structure of a higher order Korteweg-de Vries equation*, Proc. Roy. Soc. London A, **358**, 287–296.

Doering C.R., Gibbon J.D., Holm D.D. and Nicolaenko B. [1987], *Exact Lyapunov dimension of the universal attractor for the complex Landau-Ginzburg equation*, Phys. Rev. Lett., **59**, 2911–2914;

——— [1988], *Low-dimensional behaviour in the complex Ginzburg-Landau equation*, Nonlinearity, **1**, 279–309.

Dorizzi B., Grammaticos B. and Ramani A. [1983], *A new class of integrable systems*, J. Math. Phys., **24**, 2282–2288;

——— [1984], *Explicit integrability for Hamiltonian systems and the Painlevé conjecture*, J. Math. Phys., **25**, 481–485.

Dorizzi B., Grammaticos B., Ramani A. and Winternitz P. [1985], *Integrable Hamiltonian systems with velocity-dependent potentials*, J. Math. Phys., **26**, 3070–3079;
——— [1986], *Are all the equations of the Kadomtsev-Petviashvili hierarchy integrable*, J. Math. Phys., **27**, 2848–2852.

Drazin P.G. and Johnson R.S. [1989], "*Solitons: an Introduction*," Cambridge Texts in Applied Mathematics, C.U.P., Cambridge.

Drinfel'd V.G. [1983], *On constant, quasiclassical solutions of the Yang-Baxter quantum equation*, Sov. Math. Dokl., **28**, 667–671;
——— [1985], *Hopf algebras and the quantum Yang-Baxter equation*, Sov. Math. Dokl., **32**, 254–258.

Drinfel'd V.G. and Sokolov V.V. [1981], *Equations of KdV type and simple Lie algebras*, Sov. Math. Dokl., **23**, 457–462.

Dryuma V.S. [1974], *Analytic solution of the two-dimensional Korteweg-de Vries equation*, Sov. Phys. JETP Lett., **19**, 381–388;
——— [1983], *On the integration of the cylindrical Kadomtsev-Petviashvili equation by the method of the inverse problem of scattering theory*, Sov. Math. Dokl., **27**, 6–8.

Dubrovin B.A. [1981], *Theta functions and nonlinear equations*, Russ. Math. Surveys, **36**, 11–92;
——— [1982], *Multidimensional theta functions and their application to the integration of nonlinear equations*, Sov. Sci. Rev. C: Math. Phys. Rev., **3**, 83–150.

Dubrovin B.A., Matveev V.B. and Novikov S.P. [1976], *Nonlinear equations of Korteweg-de Vries type, finite zoned linear operators, and Ableian varieties*, Russ. Math. Surveys, **31**, 59–146.

Dubrovin B.A. and Novikov S.P. [1975], *Periodic and conditionally periodic analogues of the many-soliton solutions of the Korteweg-de Vries equation*, Sov. Phys. JETP, **40**, 1058–1063.

Easton R.W. [1984], *Computing the dependence on a parameter of a family of unstable manifolds: generalized Mel'nikov formulas*, Nonl. Anal., Theo., Meth. Appl., **8**, 1–4.

Eckhaus W. and van Harten A. [1981], "*The Inverse Scattering Transformation and the Theory of Solitons*," North-Holland, Amsterdam.

Eilbeck J.C., Gibbon J.D., Caudrey P.J. and Bullough R.K. [1973], *Solitons in nonlinear optics. I. A more accurate description of the 2π-pulse in self-induced transparency*, J. Phys. A: Math. Gen., **6**, 1337–1347.

Eilbeck J.C., Lombdahl P.S. and Scott A.C. [1985], *The discrete self-trapping equation*, Physica, **16D**, 318–338.

Ercolani N.M. and Forest M.G. [1985], *The geometry of real Sine-Gordon wavetrains*, Commun. Math. Phys., **99**, 1–49.

Ercolani N.M., Forest M.G. and McLaughlin D.W. [1984], *Modulations of two-phase Sine-Gordon wavetrains*, Stud. Appl. Math., **71**, 91–101;
——— [1986a], *Oscillations and instabilities in near integrable PDE's*, in "*Nonlinear Systems of Partial Differential Equations*," proceedings, SIAM-AMS Summer Seminar, Santa Fe, 1984, eds. B. Nicolaenko, D.D. Holm and J.M. Hyman, AMS Lect. Appl. Math., **23**, 3–46;

——— [1986b], *Modulation instabilities of periodic Sine-Gordon waves*, in "Nonlinear Systems of Partial Differential Equations," proceedings, SIAM-AMS Summer Seminar, Santa Fe, 1984, eds. B. Nicolaenko, D.D. Holm and J.M. Hyman, AMS Lect. Appl. Math., **23**, 149–166;

——— [1990a], *Geometry of modulational instability. I. Local Analysis*, Mem. Amer. Math. Soc., to appear;

——— [1990b], *Geometry of modulational instability. II. Global Analysis*, Mem. Amer. Math. Soc., to appear;

——— [1990c], *Geometry of modulational instability. III. Homoclinic orbits for th periodic Sine-Gordon equation*, Physica, **43D**, 349–384.

Ercolani N.M., Forest M.G., McLaughlin D.W. and Montgomery R. [1987], *Hamiltonian structure of the modulation equations for Sine-Gordon wavetrains*, Duke Math. J., **55**, 949–983.

Ercolani N. and Siggia E.D. [1986], *Painlevé property and integrability*, Phys. Lett., **119A**, 112–116;

——— [1989], *Painlevé property and geometry*, Physica, **34D**, 3303–346;

——— [1991], *Painlevé property and integrability*, in "What is Integrability," ed. V.E. Zakharov, 63–72, Springer-Verlag, Berlin-Heidelberg-New York.

Ernst F.J. [1968], *New formulation of the axially symmetric gravitational field problem*, Phys. Rev., **167**, 1175–1178.

Erugin N.P. [1966], "Linear Systems of Ordinary Differential Equations," Academic, New York.

——— [1976a,b], *Theory of nonstationary singularities of second-order equations*, I and II, Diff. Eqns, **12**, 267–289; **12**, 405–420.

Faddeev L.D. [1963], *The inverse problem in the quantum theory of scattering*, J. Math. Phys., **4**, 72–104;

——— [1965], *Increasing solutions of the Schrödinger equation*, Soviet Math. Dokl., **10**, 1033–1035;

——— [1966], *Factorization of the S-matrix for the multidimensional Schrödinger operator*, Soviet Phys. Dokl., **11**, 209–211.

——— [1976], *Inverse problem of quantum scattering theory, II*, J. Soviet Math., **5**, 334–396;

——— [1980], *A Hamiltonian interpretation of the inverse scattering method*, in "Solitons," eds. R.K. Bullough and P.J. Caudrey, Topics in Current Physics, **17**, 339–354, Springer-Verlag, Berlin-Heidelberg-New York;

——— [1980a], *Quantum completely integrable models in field theory*, Sov. Sci. Rev. C: Math. Phys. Rev., **1**, 107–155;

——— [1984], *Integrable models in* $(1+1)$*-dimensional quantum field theory*, in "Recent Advances in Field Theory and Statistical Mechanics," proceedings, Les Houches 1982, eds., J.-B. Zuber and R. Stora, 563–608, Elsevier, Amsterdam;

——— [1990], *From integrable models to quantum groups*, in "Fields and Particles," eds. H. Mitter and W. Schweiger, 89–115, Springer-Verlag, Berlin-Heidelberg-New York.

Faddeev L.D. and Takhtajan L.A. [1986], *Liouville model an the lattice*, in "*Quantum Gravity and Strings*," eds. H. de Vega and N. Sanchez, Lect. Notes Phys., **246**, 166–179, Springer-Verlag, Berlin-Heidelberg-New York;

——— [1987], "*Hamiltonian Methods in the Theory of Solitons*," Springer-Verlag, Berlin-Heidelberg-New York.

Fermi A., Pasta J. and Ulum S. [1955], *Studies of nonlinear problems. I*, Los Alamos Report LA1940. [Reproduced in "*Nonlinear Wave Motion*," proceedings, Potsdam, New York, 1972, ed. A.C. Newell, Lect. Appl. Math., **15**, 143–156, A.M.S., Providence, RI, (1974).]

Flaschka H. [1974a], *The Toda Lattice. I. Existence of integrals*, Phys. Rev., **B9**, 1924–1925;

——— [1974b], *On the Toda Lattice II. Inverse scattering solution*, Prog. Theo. Phys., **51**, 703–706.

Flaschka H. and Newell A.C. [1975], *Integrable systems of nonlinear evolution equations*, in "*Dynamical Systems Theory and Applications*," proceedings, Seattle 1974, ed. J. Moser, Lect. Notes Phys., **38**, 355–466, Springer-Verlag, Berlin-Heidelberg-New York;

——— [1980], *Monodromy- and spectrum preserving deformations. I*, Commun. Math. Phys., **76**, 65–116.

Flaschka H., Newell A.C. and Tabor M. [1991], *Integrability*, in "*What is Integrability*," ed. V.E. Zakharov, 73–114, Springer-Verlag, Berlin-Heidelberg-New York.

Flesch R., Forest M.G. and Sinha A. [1991], *Numerical inverse spectral transform for the periodic Sine-Gordon equation: theta function solutions and their linearized stability*, Physica, **48D**, 169–231.

Fokas A.S. [1980], *A symmetry approach to exactly solvable evolution equations*, J. Math. Phys., **21**, 1318–1325;

——— [1983], *Inverse scattering of first-order systems in the plane related to nonlinear multidimensional equations*, Phys. Rev. Lett., **51**, 3–6;

——— [1986a], *An inverse problem for multidimensional first-order systems*, J. Math. Phys., **27**, 1737–1746;

——— [1986b], *Inverse scattering and integrability in multidimensions*, Phys. Rev. Lett., **57**, 159–162;

——— [1987], *Symmetries and integrability*, Stud. Appl. Math., **77**, 253–299;

——— [1988], *The inverse scattering transform on the half-line: the nonlinear analogue of the sine transform inverse problems*, Montpellier, France, 1986, ed. P. Sabatier;

——— [1989a], *An initial-boundary value problem for the nonlinear Schrödinger equation*, Physica, **35D**, 167–185;

——— [1989b], *Linearized maps for the Davey-Stewartson I equation*, in "*Integrable Systems and Applications*," proceedings, Oléron, France, 1988, eds. M. Balabane, P. Lochak and C. Sulem, Lect. Notes Phys., **342**, 95–104, Springer-Verlag, Berlin-Heidelberg-New York.

Fokas A.S. and Ablowitz M.J. [1981], *Linearization of the Korteweg-de Vries and Painlevé II equations*, Phys. Rev. Lett., **47**, 1096–1100;

——— [1982], *On a unified approach to transformations and elementary solutions of Painlevé equations*, J. Math. Phys., **23**, 2033–2042;

——— [1983a], *On the initial value problem of the second Painlevé transcendent*, Commun. Math. Phys., **91**, 381–403;

——— [1983b], *The inverse scattering transform for the Benjamin-Ono equation — a pivot to multidimensional equations*, Stud. Appl. Math., **68**, 1–10;

——— [1983c], *On the inverse scattering and direct linearizing transforms for the Kadomtsev-Petviashvili equation*, Phys. Lett., **94A**, 67–70;

——— [1983d], *On the inverse scattering of the time-dependent Schrödinger equation and the associated Kadomtsev-Petviashvili equation (I) equation*, Stud. Appl. Math., **69**, 211–228;

——— [1983e], *Method of solution for a class of multidimensional nonlinear evolution equations*, Phys. Rev. Lett., **51**, 7–10;

——— [1983f], *The inverse scattering transform for multidimensional* (2+1) *problems*, in "*Nonlinear Phenomena*," proceedings, Oaxtepec, Mexico 1982, ed. K.B. Wolf, Lect. Notes Phys., **189**, 137–183, Springer-Verlag, Berlin-Heidelberg-New York;

——— [1984], *On the inverse scattering transform of multidimensional nonlinear evolution equations related to first-order systems in the plane*, J. Math. Phys., **25**, 2494–2505;

——— [1989], *Forced nonlinear evolution equations and the inverse scattering transform*, Stud. Appl. Math., **80**, 253–272.

Fokas A.S. and Anderson R.L. [1982], *On the use of isospectral eigenvalue problems for obtaining hereditary symmetries for Hamiltonian systems*, J. Math. Phys., **23**, 1066–1073.

Fokas A.S. and Fuchssteiner B. [1981], *The hierarchy of the Benjamin-Ono equation*, Phys. Lett., **86A**, 341.

Fokas A.S., Its A.R. and Kitaev A.V. [1991], *Discrete Painlevé equations and their appearance in quantum gravity*, INS preprint **164**, Clarkson University, Potsdam, New York.

Fokas A.S., Leo R.A., Martina L. and Soliani G. [1986], *The scaling reduction of the three-wave resonant system and the Painlevé VI equation*, Phys. Lett., **115A**, 329–332.

Fokas A.S. and Manakov S.V. [1990], *The dressing method, symmetries, and invariant solutions*, Phys. Lett., **150A**, 369–374.

Fokas A.S., Mugan U. and Ablowitz M.J. [1987], *A method of solution for Painlevé equations: Painlevé IV, V*, Physica, **30D**, 247–283.

Fokas A.S., Papadopoulou E.P. and Saridakis Y.G. [1990], *Soliton cellular automata*, Physica, **41D**, 297–321.

Fokas A.S., Papadopoulou E.P., Saridakis Y.G. and Ablowitz M.J. [1989], *Interaction of simple particles in soliton cellular automata*, Stud. Appl. Math., **81**, 153–180.

Fokas A.S. and Santini P.M. [1986], *The recursion operator of the Kadomtsev-Petviashvili equation and the squared eigenfunctions of the Schrödinger operator*, Stud. Appl. Math., **75**, 179–185;

—— [1988a], *Recursion operators and bi-Hamiltonian structures in multidimensions. II*, Commun. Math. Phys., **116**, 449–474;

—— [1988b], *Bi-Hamiltonian formulation of the Kadomtsev-Petviashvili and Benjamin-Ono equations*, J. Math. Phys., **29**, 604–617;

—— [1989], *Coherent structures in multidimensions*, Phys. Rev. Lett., **63**, 1329–1333;

—— [1990], *Dromions and a boundary value problem for the Davey-Stewartson I equation*, Physica, **44D**, 99–130.

Fokas A.S. and Yortsos Y.C. [1981], *The transformation properties of the sixth Painlevé equation and one-parameter families of solutions*, Lett. Nuovo Cim., **30**, 539–544;

—— [1982], *On the exactly solvable equation* $S_t = [(\beta S+\gamma)^{-2}S_x]_x + \alpha(\beta S+\gamma)^{-2}S_x$ *occuring in two-phase flow in porous media*, SIAM J. Appl. Math., **42**, 318–332.

Fokas A.S. and Zakharov V.E. [1990], *The dressing method and nonlocal Riemann-Hilbert problems*, INS preprint **137**, Clarkson University, Potsdam, New York.

Fokas A.S. and Zhou X. [1991], *On the solvability of Painlevé II and IV*, INS preprint **148**, Clarkson University, Potsdam, New York.

Fomenko A.T. and Trofimov V.V. [1987], "*Integrable Systems on Lie Algebras and Symmetric Spaces*," Gordon and Breach, New York.

Fordy A.P. [1984], *Derivative nonlinear Schrödinger equations and Hermitian symmetric spaces*, J. Phys. A: Math. Gen., **17**, 1235–1245;

—— [1990a], "*Soliton Theory: a Survey of Results*," (editor), M.U.P., Manchester;

—— [1990b], *Prolongation structures of nonlinear evolution equations*, in "*Soliton Theory: A Survey of Results*," ed. A.P. Fordy, 403–404, M.U.P., Manchester.

Fordy A.P. and Gibbons J. [1980a], *Some remarkable nonlinear transformations*, Phys. Lett., **75A**, 325;

—— [1980b], *Integrable nonlinear Klein-Gordon equations and Toda Lattices*, Commun. Math. Phys., **77**, 21–30.

Fordy A.P. and Kulish P.P. [1983], *Nonlinear Schrödinger equations and simple Lie algebras*, Commun. Math. Phys., **89**, 427–433.

Fordy A.P. and Pickering A. [1991], *Analysing negative resonances in the Painlevé test*, preprint, University of Leeds, U.K..

Forest E. and Ruth R.D. [1990], *Fourth-order symplectic integration*, Physica, **43D**, 105–117.

Forest M.G. [1988], *Correlations between chaos in the perturbed Sine-Gordon equation and finite modal equations*, in "*Nonlinear Evolutions*," ed. J.J.P. Leon, 537–557, World-Scientific, Singapore.

Forest M.G. and McLaughlin D.W. [1982], *Spectral theory for the periodic Sine-Gordon equation: a concrete viewpoint*, J. Math. Phys., **23**, 1248–1277;

—— [1983], *Modulations of Sinh-Gordon and Sine-Gordon wavetrains*, Stud. Appl. Math., **68**, 11–59.

Forgács P., Horvath Z. and Palla L. [1983], *Solution generating techniques for self-dual monopoles*, Nucl. Phys. B, **229**, 77–104.

Forgács P. and Manton N.S. [1980], *Space-time symmetries in gauge theories*, Commun. Math. Phys., **72**, 15–35.

Fornberg B. and Whitham G.B. [1978], *A numerical and theoretical study of certain nonlinear wave phenomena*, Phil. Trans. Roy. Soc. Lond. A, **289**, 373–404.

Forsyth A.R. [1906], "*Theory of Differential Equations. Part IV — Partial Differential Equations*," Vol. VI. C.U.P..

Fournier J.D., Levine G. and Tabor M. [1988], *Singularity clustering in the Duffing oscillator*, J. Phys. A: Math. Gen., **21**, 33–54.

Freeman N.C. [1984], *Soliton solutions of nonlinear evolution equations*, IMA J. Appl. Math., **32**, 125–145.

Freidlander S. and Vishik N.M. [1990], *Lax pair formulation for the Euler equation*, Phys. Lett., **148A**, 313–319.

Frisch U., Hasslacher B. and Pomeau Y. [1986], *Lattice gas automata for the Navier-Stokes equation*, Phys. Rev. Lett., **56**, 1505–1508.

Fu L.L. and Holt B. [1984], *Internal waves in the gulf of California: Observations spacebourne radar*, J. Geophy. Res., **89**, 2053–2060.

Fuchs L. [1884], *Über differentialgleichungen deren intégrale feste verzweigungspunkte besitzen*, Sitz. Akad. Wiss., Berlin, **32**, 669–720.

Fuchs R. [1905], *Sur quelques équations différentielles linéaires du second ordre*, C.R. Acad. Sc. Paris, **141**, 555–558;

——— [1907], *Über lineare homogene differentialgleichungen zweiter ordnung mit drei im endlich gelegene wesentlich singulären stellen*, Math. Ann., **63**, 301–321.

Fuchssteiner B. and Oevel G. [1989], *Geometry and action-angle variables of multi-soliton systems*, Rev. Math. Phys., **1**, 415–479.

Gabeskiria M.A. and Saveliev M.V. [1982], *On reduction of a two-dimensional generalized Toda lattice to an ordinary differential equations system*, Lett. Math. Phys., **6**, 109–112.

Gakhov F.D. [1937], *On the Riemann boundary problem*, Matem. sborn., **2**, 673–683;

——— [1941], *Boundary value problems of the theory of analytic functions and singular integral equations*, Doctor's thesis, Tbilisi;

——— [1966], "*Boundary Value Problems*," Pergamon, Oxford.

Gambier B. [1906a,b,c], *Sur les équations différentielles du second ordre dont l'intégrale générale est uniform*, C.R. Acad. Sc. Paris, **142**, 266–269; **142**, 1403–1406; **142**, 1497–1500;

——— [1906d, 1907a,b], *Sur les équations différentielles du second ordre et du premier degre dont l'intégrale générale est à points critiques fixés*, C.R. Acad. Sc. Paris, **143**, 741–743; **144**, 827–830; **144**, 962–964;

——— [1909], *Sur les équations différentielles du second ordre et du premeir degre dont l'intégrale générale est à points critiques fixés*, Acta Math., **33**, 1–55.

Ganesan S. and Lakshmanan M. [1987], *Singularity-structure analysis and Hirota's bilinearisation of the Davey-Stewartson equation*, J. Phys. A: Math. Gen., **20**, L1143–L1147.

Gaponov-Grekhov A.V. and Rabinovich M. [1990], *Disorder, dynamical chaos and structures*, Physics Today, **43**(7), 30–38.

Gardner C.S. [1971], *The Korteweg-de Vries equation and generalizations. IV. The Korteweg-de Vries equation as a Hamiltonian system*, J. Math. Phys., **12**, 1548–1551.

Gardner C.S., Greene J.M., Kruskal M.D. and Miura R.M. [1967], *Method for solving the Korteweg-de Vries equation*, Phys. Rev. Lett., **19**, 1095–1097;

——— [1974], *Korteweg-de Vries and generalizations. VI. Methods for exact solution*, Commun. Pure Appl. Math., **27**, 97–133.

Gardner C.S. and Morikawa G.K. [1960], *Similarity in the asymptotic behaviour of collision free hydrodynamic waves and water waves*, Courant Inst. Math. Sci. Res. Rep. NYO–9082, New York University, New York.

Garnier R. [1912], *Sur les équations différentielles du troisième ordre dont l'integral générale est uniforme et sur une classe d'équations nouvelles d'ordre superier dont l'integale générale à ses points critiques fixés*, Ann. Sci. Ecole Norm. Sup., **29**, 1–126;

——— [1926], *Solution du problème de Riemann pour les systèmes différentiels dont íntégrale générale est à point critiques fixès*, Ann. Sci. Ecole Norm. Sup., **43**, 177–307;

——— [1968], *Contribution à l'étude des solutions de l'équation (V) de Painlevé*, J. Math. Pures et Appl., **46**, 353–413.

Gel'fand I.M. and Dikii L.A. [1977], *Resolvants and Hamiltonian systems*, Func. Anal. Appl., **11**, 93–104.

Gel'fand I.M. and Levitan B.M. [1955], *On the determination of a differential equation from its spectral function*, Amer. Math. Soc. Transl., Ser. 2, **1**, 253–304.

Ghidaglia J.-M. and Saut J.-C. [1989], *Sur le problème de Cauchy pour les équations de Davey-Stewartson*, C.R. Acad. Sc. Paris, **308**, 115–120;

——— [1990], *On the initial value problem for the Davey-Stewartson systems*, Nonlinearity, **3**, 475–506.

Giannini J.A. and Joseph R.I. [1989], *The role of the second Painlevé transcendent in nonlinear optics*, Phys. Lett., **141A**, 417–419.

Gibbon J.D. [1985], *A survey of the origins and physical importance of soliton equations*, Phil. Trans. Roy. Soc. London A, **315**, 335–365.

Gibbon J.D., Caudrey P.J., Bullough R.K. and Eilbeck J.C. [1973], *An N-soliton solution of a nonlinear optics equation derived by a general inverse scattering method*, Lett. Nuovo Cim., **8**, 775–779.

Gibbon J.D., Newell A.C., Tabor M. and Zeng Y.B. [1988], *Lax pairs, Bäcklund transformations and special solutions for ordinary differential equations*, Nonlinearity, **1**, 481–490.

Gibbon J.D., Radmore P., Tabor M. and Wood D. [1985], *The Painlevé property and Hirota's method*, Stud. Appl. Math., **72**, 39–63.

Gibbon J.D. and Tabor M. [1985], *On one-and two-dimensional Toda Lattices and the Painlevé property*, J. Math. Phys., **26**, 1956–1960.

Gibbons G.W. and Pope C.N. [1979], *The positive action conjecture and asymptotically Euclidean metrics in quantum gravity*, Commun. Math. Phys., **66**, 267–290.

Gibbons J. and Kupershmidt B. [1980], *A linear scattering problem for the finite depth equation*, Phys. Lett., **79A**, 31–32.

Gohberg I.T. and Kreǐn M. [1958], *Systems of integral equations on the semiaxis with kernels depending on the difference of arguments*, Uspekhi Mat. Nauk., **13**, (2), 3–72.

Goldman M.V. [1986], *Langmuir wave solutions and spatial collapse in plasma physics*, Physica, **18D**, 67–76.

Goldstein H. [1965], "*Classical Mechanics*," Wiley, New York.

Golubov V.V. [1953], "*Lectures on the integration of the equations of motion of a rigid body about a fixed point*," State Publishing House, Moscow.

Grammaticos B., Dorizzi B. and Ramani A. [1983], *Hamiltonians with high-order integrals and the "weak-Painlevé" concept*, J. Math. Phys., **24**, 2289–2295;

——— [1984], *Integrability of Hamiltonians with third- and fourth-degree polynomial potentials*, J. Math. Phys., **25**, 3470–3473.

Grammaticos B., Moulin-Ollangnier J., Ramani A., Strelcyn J.-M. and Wojciechowski S. [1990], *Integrals of quadratic ordinary differential equations in* $\mathbb{R}^3$*: the Lotka-Volterra system*, Physica, **163A**, 683–722.

Grimshaw R. [1979a], *Slowly varying solitary waves. I. Korteweg-de Vries equation*, Proc. Roy. Soc. London A, **368**, 359–375;

——— [1979b], *Slowly varying solitary waves. II. Nonlinear Schrödinger equation*, Proc. Roy. Soc. London A, **368**, 377–388.

Grinevich P.G. and Novikov R.G. [1986], *Analogues of multisoliton potentials for the two-dimensional Schrödinger operator, and a nonlocal Riemann problem*, Sov. Math. Dokl., **33**, 9–12.

Gromak V.I. [1975], *Theory of Painlevé's equation*, Diff. Eqns., **11**, 285–287;

——— [1976], *Solutions of Painlevé's fifth equation*, Diff. Eqns., **12**, 519–521;

——— [1977], *One-parameter systems of solutions of Painlevé's equations*, Diff. Eqns., **14**, 1510–1513;

——— [1987], *Theory of the fourth Painlevé equation*, Diff. Eqns., **23**, 506–513.

Gross D.J. and Migdal A.A. [1990a], *Nonperturbative two-dimensional quantum gravity*, Phys. Rev. Lett., **64**, 127–130;

——— [1990b], *Nonperturbative solutions of the Ising model on a random surface*, Phys. Rev. Lett., **64**, 717–720.

Guckenheimer J. and Holmes P. [1983], "*Nonlinear Oscillations, Dynamical Systems and Bifurcations of Vector Fields*," Appl. Math. Sci., **42**, Springer-Verlag, Berlin-Heidelberg-New York.

Gürses M. and Xanthopoulus B.C. [1980], *Axially symmetric, static self-dual* SU(3) *gauge fields and stationary Einstein-Maxwell metrics*, Phys. Rev. D, **17**, 3243–3246.

Haine L. [1984], *The algebraic complete integrability of Geodesic flow on* SO(N), Commun. Math. Phys., **94**, 271–287.

Haine L. and Horozov E. [1987], *A Lax pair for Kowalevski's top*, Physica, **29D**, 173–180.

Hall P. [1982], *On the nonlinear evolution of Görtler vortices in non-parallel boundary layers*, IMA J. Appl. Math., **29**, 173–196.

Hall P. and Lakin W.D. [1988], *The fully nonlinear development of Görtler vortices in growing boundary layers*, Proc. Roy. Soc. London A, **415**, 421–444.

Halphen G. [1881a], *Sur un système d'équations différentielles*, C.R. Acad. Sc. Paris, **92**, 1101–1103 [Œuvres, Paris, **2** (1918) 475–477];

—— [1881b], *Sur certains systèmes d'équations différentielles*, C.R. Acad. Sc. Paris, **92**, 1404–1407 [Œuvres, Paris, **2** (1918) 478–481].

Hammack J.L. and Segur H. [1974], *The Korteweg-de Vries equation and water waves part 2. Comparison with experiments*, J. Fluid Mech., **65**, 289–314;

—— [1978], *The Korteweg-de Vries equation and water waves part 3. Oscillatory waves*, J. Fluid Mech., **84**, 337–358.

Hasegawa A. [1990], "*Optical Solitons in Fibers*," 2nd. Ed., Springer-Verlag, Berlin-Heidelberg-New York.

Hasegawa A. and Kodama Y. [1981], *Signal transmission by optical solitons in a monomode fiber*, Proc. IEEE, **69**, 1145–1150.

Hasimoto H. and Ono H. [1972], *Nonlinear modulation of gravity waves*, J. Phys. Soc. Japan, **33**, 805–811.

Hastings S.P. and McLeod J.B. [1980], *A boundary value problem associated with the second Painlevé transcendent and the Korteweg-de Vries equation*, Arch. Rat. Mech. Anal., **73**, 31–51.

Hauser I. and Ernst F.J. [1980], *A homogeneous Hilbert problem for the Kinnersley-Chitre transformations*, J. Math. Phys., **21**, 1126–1140.

Head A. [1990], *LIE: a muMATH program for the calculation of the Lie algebra of differential equations*, preprint, CSIRO Division of Material Sciences, Clayton, Australia.

Herbst B.M. and Ablowitz M.J. [1989a], *Numerically induced chaos in the nonlinear Schrödinger equation*, Phys. Rev. Lett., **62**, 2065–2068;

—— [1989b], *On numerical chaos in the nonlinear Schrödinger equation*, in "*Integrable Systems and Applications*," proceedings, Oléron, France, 1988, eds. M. Balabane, P. Lochak and C. Sulem, Lect. Notes Phys., **342**, 192–206, Springer-Verlag, Berlin-Heidelberg-New York;

—— [1990], *Mel'nikov analysis and numerically induced chaos*, in "*Chaos and Order*," eds. N. Joshi and R.L. Dewar, World Scientific, Singapore, to appear.

Hietarinta J. [1987a], *Direct methods for the search of the second invariant*, Phys. Rep., **147**, 87–154;

—— [1987b], *A search for bilinear equations passing Hirota's three-soliton condition. I. KdV-type bilinear equations*, J. Math. Phys., **28**, 1732–1742;

—— [1987c], *A search for bilinear equations passing Hirota's three-soliton condition. II. MKdV-type bilinear equations*, J. Math. Phys., **28**, 2094–2101;

—— [1987d], *A search for bilinear equations passing Hirota's three-soliton condition. III. Sine-Gordon-type bilinear equations*, J. Math. Phys., **28**, 2586–2592;

—— [1988], *A search for bilinear equations passing Hirota's three-soliton condition. IV. Complex bilinear equations*, J. Math. Phys., **29**, 628–635.

Hietarinta J., Grammaticos B., Dorizzi B. and Ramani A. [1984], *Coupling-constant metamorphosis and duality between integrable Hamiltonian systems*, Phys. Rev. Lett., **53**, 1707–1710.

Hilbert D. [1904], *Über eine anwendung der integralgleichungen auf ein problem der funktionentheorie*, in "*Verhandl. des. III Internat. Mathematiker Kongresses, Heidelberg*," 233-240, Teubner, Leipzig.

——— [1924], "*Grundzüge einer allgemeinen Theorie der linearen Integralgleichungen*," 2nd edition, Dritter Abschnitt, Leipzig-Berlin.

Hille E. [1976], "*Ordinary Differential Equations in the Complex Domain*," Wiley, New York.

Hirota R. [1971], *Exact solution of the Korteweg-de Vries equation for multiple collisions of solitons*, Phys. Rev. Lett., **27**, 1192–1194;

——— [1977a], *Nonlinear partial difference equations I. A difference analogue of the Korteweg-de Vries equation*, J. Phys. Soc. Japan, **43**, 1429–1433;

——— [1977b], *Nonlinear partial difference equations II. Discrete time Toda equation*, J. Phys. Soc. Japan, **43**, 2074–2078;

——— [1977c], *Nonlinear partial difference equations III. Discrete sine-Gordon equations*, J. Phys. Soc. Japan, **43**, 2079–2089;

——— [1980], *Direct methods in soliton theory*, in "*Solitons*," eds. R.K. Bullough and P.J. Caudrey, Topics in Current Physics, **17**, 157–176, Springer-Verlag, Berlin-Heidelberg-New York.

Hirota R. and Ramani A. [1980], *The Miura transformations of Kaup's equations and of Mikhailov's equation*, Phys. Lett., **76A**, 95–96.

Hirota R. and Satsuma J. [1981], *Soliton solutions of a coupled Korteweg-de Vries equation*, Phys. Lett., **85**, 407–408.

Hitchen N.J. [1983], *On the construction of monopoles*, Commun. Math. Phys., **89**, 145–190;

——— [1987], "*Monopoles, Minimal Surfaces and Algebraic Curves*," Séminaire de Mathématiques Supérieures, **105**, Montreal.

Hlavatý L. [1986], *The Painlevé analysis of damped KdV equation*, J. Phys. Soc. Japan, **55**, 1405–1406;

——— [1988], *Painlevé analysis of nonautonomous evolution equations*, Phys. Lett., **128A**, 335–338.

Hochstadt H. [1971], "*The Functions of Mathematical Physics*," Wiley, New York.

Hodnett P.F. and Moloney T.P. [1989], *On the structure during interaction of the two-soliton solution of the Korteweg-de Vries equation*, SIAM J. Appl. Math., **49**, 1174–1187.

Holbrook J.R., Apel J.R. and Tsai J.J. [1980], *The Sulu sea internal soliton experiment, 2. Observations of large-amplitude nonlinear waves*, Eos Trans. AGU, **61**, 1009.

Holm D.D., Kovacic G. and Sundaram B. [1990], *Chaotic laser-matter interaction*, preprint **LA-UR 90-2629**, Los Alamos, New Mexico.

Holm D.D., Marsden J.E., Ratiu T. and Weinstein A. [1985], *Nonlinear stability of fluid and plasma equilibria*, Phys. Rep., **123**, 1–116.

Holmes P.J. and Spence D. [1984], *On a Painlevé-type boundary-value problem*, Q. Jl. Mech. Appl. Math., **37**, 525–538

Hopf E. [1950], *The partial differential equation* $u_t + uu_x = \mu u_{xx}$, Commun. Pure Appl. Math., **3**, 201–230.

Hörmander L. [1973], "*An Introduction to Complex Analysis in Several Variables*," 2nd edition, North-Holland, Amsterdam.

Horozov E. and van Moerbeke P. [1989], *The full geometry of Kowalewski's top and* $(1,2)$*-Abelian surfaces*, Commun. Pure Appl. Math., **42**, 357–407.

Hughes B.A. and Gower J.F.R. [1983], *SAR imagery and surface truths comparisons of internal waves in Georgia Strait, British Columbia, Canada*, J. Geophy. Res., **88**, 1809–1824.

Hyman J.M., McLaughlin D.W. and Scott A.C. [1981], *On Davydov's alpha-helix soliton*, Physica, **3D**, 23–44;

Ince E.L. [1956], "*Ordinary Differential Equations*," Dover, New York.

Infeld E. and Rowlands G. [1990], "*Nonlinear Waves, Solitons and Chaos*," C.U.P., Cambridge.

Isaacson E.L., McKean H.P. and Trubowitz E. [1984], *The inverse Sturm-Liouville problem. II*, Commun. Pure Appl. Math., **37**, 1–12.

Isaacson E.L. and Trubowitz E. [1983], *The inverse Sturm-Liouville problem. I*, Commun. Pure Appl. Math., **36**, 767–784.

Ishimori Y. [1981], *On the modified Korteweg-de Vries equation and the loop soliton*, J. Phys. Soc. Japan, **50**, 2471–2472;

——— [1982], *A relationship between the Ablowitz-Kaup-Newell-Segur and Wadati-Konno-Ichikawa schemes of the inverse scattering method*, J. Phys. Soc. Japan, **51**, 3036–3041;

——— [1984], *Multi-vortex solutions of two-dimensional nonlinear wave equation*, Prog. Theo. Phys., **72**, 33–37.

Ito M. [1982], *Symmetries and conservation laws of a coupled nonlinear wave equation*, Phys. Lett., **91A**, 335–338.

Its A.R. [1986], *"Isomonodromy" solutions of equations of zero curvature*, Math. USSR Izvestiya, **26**, 497–529.

Its A.R. and Kapaev A.A. [1988], *The method of isomonodromy deformations and connection formulas for the second Painlevé transcendent*, Math. USSR Izvestiya, **31**, 193–207.

Its A.R. and Kitaev A.V. [1990], *Mathematical aspects of the non-perturbative 2D quantum gravity*, Mod. Phys. Lett., **5A**, 2079–2083.

Its A.R. and Kotlyarov V.P. [1976], *Explicit formulas for solutions of a nonlinear Schrödinger equation*, Dokl. Akad. Nauk Ukain. SSR Ser. A, **1976** (11), 965–968.

Its A.R. and Matveev V.B. [1975], *The periodic Korteweg-de Vries equation*, Func. Anal. Appl., **9**, 65–66.

Its A.R. and Novokshenov V.Yu. [1986], "*The Isomonodromic Deformation Method in the Theory of Painlevé Equations*," Lect. Notes Math., **1191**, Springer-Verlag, Berlin-Heidelberg-New York.

Jackiw R. [1977], *Quantum meaning of classical field theory*, Rev. Mod. Phys., **49**, 681–706.

Jaulent M. and Manna M. [1986a], *The spatial transform method: $\overline{\partial}$ derivation of the AKNS hierarchy*, Phys. Lett., **117A**, 62–66;

——— [1986b], *Connection between the KDV and AKNS spatial transforms*, Inverse Problems, **3**, L35–L41;

——— [1987a], *The spatial transform method for $\overline{\partial}$ equations of nth linear order order*, Inverse Problems, **3**, L15–L18;

——— [1987b], *The "spectral Wronskian" tool and the $\overline{\partial}$ investigation of the KdV hierarchy*, J. Math. Phys., **28**, 2338–2342.

Jaulent M., Manna M. and Martinez Alonso L. [1988], *$\overline{\partial}$ equations in the theory of integrable systems*, Inverse Problems, **4**, 123–150;

——— [1989], *A solvable hierarchy of $(N+1)$-dimensional non-linear evolution equations with constraints*, J. Phys. A: Math. Gen., **22**, L13–L18.

Jeffrey A. and Kakutani T. [1972], *Weak nonlinear dispersive waves, a discussion centered around the Korteweg-de Vries equation*, SIAM Rev., **14**, 582–643.

Jiang Z. [1989], *Construction of scattering data for a class of multidimensional scattering operators*, Inverse Problems, **5**, 349–374.

Jiang Z. and Bullough R.K. [1987], *Combined $\overline{\partial}$ and Riemann-Hilbert inverse methods for integrable nonlinear evolution equations in $(2+1)$ dimensions*, J. Phys. A: Math. Gen., **20**, L429–L435.

Jiang Z., Bullough R.K. and Manakov S.V. [1985], *Complete integrability of the Kadomtsev-Petviashvili equation in $2+1$ dimensions*, Physica, **18D**, 305–307.

Jimbo M. [1979], *An inverse scattering problem and the fifth Painlevé transcendent*, Prog. Theo. Phys., **61**, 359–360;

——— [1982], *Monodromy problem and the boundary condition for some Painlevé equations*, Publ. RIMS, Kyoto Univ., Japan, **18**, 1137–1161.

Jimbo M. Kruskal M.D. and Miwa T. [1982], *Painlevé test for the Self-dual Yang-Mills equation*, Phys. Lett., **92A**, 59–60.

Jimbo M., Miwa T., Mori Y. and Sato M. [1981], *Density matrix of an impenetrable bose gas and the fifth Painlevé transcendent*, Physica, **1D**, 80–139.

Jimbo M. and Miwa T. [1981a,b], *Monodromy preserving deformations of linear ordinary differential equations with rational coefficients. II and III*, Physica, **2D**, 407–448; **4D**, 26–46;

——— [1983], *Solitons and infinite dimensional Lie algebras*, Publ. RIMS, Kyoto Univ., **19**, 943–1001.

Jimbo M., Miwa T. and Ueno K. [1981], *Monodromy preserving deformation of linear ordinary differential equations with rational coefficients. I*, Physica, **2D**, 306–352.

Johnson R.S. [1977], *On the modulation of water waves in the neighbourhood of $kh \approx 1.363$*, Proc. Roy. Soc. London A, **357**, 131–141;

——— [1980], *Water waves and Korteweg-de Vries equations*, J. Fluid Mech., **97**, 701–719.

Joseph R.I. [1977], *Solitary waves in a finite depth fluid*, J. Phys. A: Math. Gen., **10**, L225–L227.

Joseph R.I. and Egri R. [1978], *Multi-soliton solutions in a finite depth fluid*, J. Phys. A: Math. Gen., **11**, L97–L102.

Joshi N. [1986], *The connection problem for the first and second Painlevé transcendents*, Ph.D. thesis, Princeton University.

——— [1987], *Painlevé property of general variable-coefficient versions of the Korteweg-de Vries and nonlinear Schrödinger equations*, Phys. Lett., **125A**, 456–460;

——— [1990], *An explicit one-soliton solution of the* $(2+1)$*-dimensional generalized Sine-Gordon equation*, Lett. Math. Phys., **20**, 261–270.

Joshi N. and Kruskal M.D. [1988], *An asymptotic approach to the connection problem for the first and second Painlevé equations*, Phys. Lett., **130A**, 129–137;

——— [1990], *A simple proof that solutions of the six Painlevé equations have no movable singularities except poles*, preprint, CMA-R06-90, Centre for Mathematical Analysis, Australian National University, Canberra;

——— [1991], *The Painlevé connection problem: an asymptotic approach*, Stud. Appl. Math., to appear.

Kac M. and van Moerbeke P. [1975], *On an explicitly soluble system of nonlinear differential equations related to certain Toda lattices*, Advances Math., **16**, 160–169.

Kadomtsev B.B. and Petviashvili V.I. [1970], *On the stability of solitary waves in weakly dispersive media*, Sov. Phys. Dokl., **15**, 539–541.

Kang F. and Meng-Zhao Q. [1987], *The symplectic methods for the computation of Hamiltonian equations*, in "Numerical Methods for Partial Differential Equations," eds. Zhu You-Ian and Guo Ben-yu, Lect. Notes Math., **1297**, 1–37, Springer-Verlag, Berlin-Heidelberg-New York.

Kapaev A.A. [1989a], *Asymptotics of solutions of the Painlevé equation of the first kind*, Diff. Eqns., **24**, 1107–1115;

——— [1989b], *Asymptotics of solutions for the second Painlevé equation*, Theo. Math. Phys., **77**, 1227–1234.

Kapaev A.A. and Novokshenov V.Yu. [1986], *Two-parameter family of real solutions of the second Painlevé equation*, Sov. Phys. Dokl., **31**, 719–721.

Kauffman L.H. [1990], *From knots to quantum groups (and back)*, in "Hamiltonian Systems, Transformation Groups and Spectral Transform," eds. J. Harnad and J. Marsden, 161–176, Les Publications de Centre Recherches Mathématiques, Montreal.

Kaup D.J. [1975], *A higher-order water-wave equation and the method for solving it*, Prog. Theo. Phys., **54**, 396–408;

——— [1976], *The three-wave interaction — a nondispersive phenomenon*, Stud. Appl. Math., **55**, 9–44;

——— [1980a], *The inverse scattering solution for the full three-dimensional three-wave resonant interaction*, Physica, **D1**, 45–67;

——— [1980b], *On the inverse scattering problem for the cubic eigenvalue problem of the class* $\phi_{xxx}+6Q\phi_x+6R\phi=\lambda\phi$, Stud. Appl. Math., **62**, 189–216;

——— [1980c], *The Estabrook-Wahlquist method with examples of application*, Physica, **1D**, 391–411;

——— [1981a], *The solution of the general initial value problem for the full three-dimensional three wave resonant interaction*, Physica, **3D**, 374–395;

——— [1981b], *The lump solutions and the Bäcklund transformation for the three-dimensional three-wave resonant interaction*, J. Math. Phys., **22**, 1176–1181;

——— [1984], *The forced Toda Lattice: an example of an almost integrable system*, J. Math. Phys., **25**, 277–281;

——— [1985], *Forced integrable systems — an overview*, in "Nonlinear Systems of Partial Differential Equations in Applied Mathematics," Part 1, proceedings, SIAM-AMS Summer Seminar, Santa Fe, 1984, eds. B. Nicolaenko, D.D. Holm and J.M. Hyman, AMS Lect. Appl. Math., **23**, 195–215.

Kaup D.J. and Hansen P.J. [1986], *The forced nonlinear Schrödinger equation*, Physica, **18D**, 77–84.

Kaup D.J. and Neuberger D.H. [1984], *The soliton birth rate in the forced Toda Lattice*, J. Math. Phys., **25**, 282–284.

Kaup D.J. and Newell A.C. [1978a], *An exact solution for a derivative nonlinear Schrödinger equation*, J. Math. Phys., **19**, 798–801;

——— [1978b], *Solitons as particles, oscillators, and in slowly varying media: a singular perturbation theory*, Proc. Roy. Soc. London A, **361**, 413–446.

Kawamoto S. [1985], *An exact transformation from the Harry Dym equation to the Modified KdV equation*, J. Phys. Soc. Japan, **54**, 2055–2056.

Kawata T. and Inoue H. [1978], *Exact solutions of the derivative nonlinear Schrödinger equation under the nonvanishing conditions*, J. Phys. Soc. Japan, **44**, 1968–1976.

Kawata T., Sahhai J. and Kobayashi N. [1980], *Inverse method for the mixed nonlinear Schrödinger equations and soliton solutions*, J. Phys. Soc. Japan, **48**, 1371–1379.

Kay I. and Moses H.E. [1956], *Reflectionless transmission through dielectrics and scattering potentials*, J. Appl. Phys., **27**, 1503–1508.

Keiser J. [1989], *On the computation of periodic particles for cellular automata*, Masters Thesis, Clarkson University, Potsdam, New York.

Keller J.B. [1985], *Soliton generation and nonlinear wave propagation*, Phil. Trans. R. Soc. Lond. A, **315**, 367–377.

Kenkre V.M. and Campbell D.K. [1986], *Self-trapping on a dimer: time-dependent solutions of a discrete nonlinear Schrödinger equation*, Phys. Rev. B, **34**, 4959–4961.

Kenkre V.M. and Tsironis G.P. [1987], *Nonlinear effects in quasilinear neutron scattering: exact line-shape calculation for a dimer*, Phys. Rev. B, **35**, 1473–1484.

Kitaev A.V. [1985], *Self-similar solutions of the modified nonlinear Schrödinger equation*, Theo. Math. Phys., **64**, 878–894;

——— [1989], *The method of isomonodromy deformations and the asymptotics of solutions of the "complete" third Painlevé equation*, Math. USSR Sbornik, **62**, 421–444;

——— [1990], *On similarity reductions of the three-wave resonant system to the Painlevé equations*, J. Phys. A: Math. Gen., **23**, 3543–3553;

——— [1991], *Quadratic transformations for the sixth Painlevé equation*, Lett. Math. Phys., **21**, 105–111.

Kivshar Y.S. and Malomed B.A. [1989], *Dynamics of solitons in nearly integrable systems*, Rev. Mod. Phys., **61**, 763–915.

Khan T.M. and Can M. [1989], *On the initial value problems for the second Painlevé equation*, Bull. Tech. Univ. Istanbul, **42**, 43–52.

Koblitz N. [1984], "Introduction to Elliptic Curves and Modular Forms," Grad. Texts Math., **97**, Springer-Verlag, Berlin-Heidelberg-New York.

Kodama Y. [1985], *Optical solitons in a monomode fiber*, J. Stat. Phys., **39**, 597–614;

——— [1988a], *A method for solving the dispersionless KP equation and its exact solutions*, Phys. Lett., **129A**, 223–226;

——— [1988b], *A solution method for the dispersionless KP equation and its exact solutions*, Prog. Theo. Phys. Supp., **94**, 184–194.

Kodama Y., Ablowitz M.J. and Satsuma J. [1982], *Direct and inverse scattering problems of the nonlinear intermediate long wave equation*, J. Math. Phys., **23**, 564–576.

Kodama Y. and Gibbons J. [1989], *A method for solving the dispersionless KP equation and its exact solutions. II*, Phys. Lett., **135A**, 167–170.

Kodama Y., Satsuma J. and Ablowitz M.J. [1981], *Nonlinear intermediate long-wave equation: Analysis and method of solution*, Phys. Rev. Lett., **46**, 687–690.

Komarov I.V. [1987], *A generalization of the Kovalevskaya top*, Phys. Lett., **123A**, 14–15.

Komarov I.V. and Kuznetsov V.B. [1990], *Kowalewski's top on the Lie algebras* o(4), e(3) *and* o(3, 1), J. Phys. A: Math. Gen., **23**, 841–846.

Konno K. and Jeffrey A. [1984], *The loop soliton*, in "Advances in Nonlinear Waves," ed. L. Debnath, Res. Notes Math., **95**, 162–183, Pitman, London.

Konopelchenko B. [1981], *The two dimensional matrix spectral problem: General structure of the integrable equations and their Bäcklund transformations*, Phys. Lett., **89A**, 346–350;

——— [1982a], *On the gauge-invariant description of the evolution equations integrable by Gel'fand-Dikij spectral problems*, Phys. Lett., **92A**, 323–327;

——— [1982b], *On the general structure of nonlinear evolution equations integrable by the two-dimensional matrix spectral problem*, Commun. Math. Phys., **87**, 105–125.

——— [1982c], *On the general structure of nonlinear evolution equations and their Bäcklund transformations connected with the matrix nonstationary Schrödinger spectral problem*, J. Phys. A: Math. Gen., **15**, 3425–3437;

——— [1987], "Nonlinear Integrable Equations," Lect. Notes Phys., **270**, Springer-Verlag, Berlin-Heidelberg-New York.

Konopelchenko B. and Dubrovsky V. [1984], *Some new integrable nonlinear evolution equations in* $2+1$ *dimensions*, Phys. Lett., **102A**, 15–17;

——— [1985], *Bäcklund-Calogero group and general form of integrable equations for the two-dimensional Gel'fand-Dikij-Zakharov-Shabat problem. Bilocal approach*, Physica, **16D**, 79–98;

——— [1990], *Localized spectral transform for the Kadomtsev-Petviashvili equation*, preprint;

——— [1991], *Inverse spectral transform for the modified Kadomtsev-Petviashvili equation*, Stud. Appl. Math., to appear.

Konopelchenko B.G. and Matkarimov B.T. [1989], *On the inverse scattering transform for the Ishimori equation*, Phys. Lett., **135A**, 183–189;

——— [1990a], *Inverse spectral transform for the nonlinear evolution equations generating the Davey-Stewartson and Ishimori equations*, Stud. Appl. Math., **82**, 319–359;

——— [1990b], *Inverse spectral transform for the Ishimori equation: I. Initial value problem*, J. Math. Phys., **31**, 2737–2746.

Korteweg D.J. and de Vries G. [1895], *On the change of form of long waves advancing in a rectangular canal, and on a new type of long stationary waves*, Philos. Mag. Ser. 5, **39**, 422–443.

Kovalevskaya S. [1978], "*A Russian childhood*," ed. B. Stillman, Springer-Verlag, Berlin-Heidelberg-New York.

Kowalevski S. [1889a], "*Sur le problème de la rotation d'un corps solide autour d'un point fixè*," Acta Math.12177–232;

——— [1889b], *Sur une propriété d'un système d'équations différentielles qui definit la rotation d'un corps solide autour d'un point fixè*, Acta Math., **14**, 81–93.

Krein M. [1958], *Integral equations on the half-line with a kernel depending on the difference of the arguments*, Uspekhi Mat. Nauk., **13,** (5), 3–120.

Krichever I.M. [1977a], *Methods of algebraic geometry in the theory of nonlinear equations*, Russ. Math. Surveys, **32,** (6), 185–213;

——— [1977b], *Integration of nonlinear equations by methods of algebraic geometry*, Russ. Math. Surveys, **11**, 11–2.

——— [1980], *Elliptic solutions of the Kadomtsev-Petviashvili equation and integrable systems of particles on a straight line*, Func. Anal. Appl., **14**, 282–290.

Krichever I.M. and Novokov S.P. [1980a], *Holomorphic bundles over algebraic curves and nonlinear equations*, Russ. Math. Surveys, **35,** (6), 53–79;

——— [1980b], *The inverse problem method and holomorphic bundles on Riemann surfaces*, Sov. Sci. Rev. C: Math. Phys. Rev., **1**, 5–26.

Krumhansl J.A. [1991], *Unity in the science of physics*, Physics Today, **44** (3), 33–38.

Kruskal M. [1974], *The Korteweg-de Vries equation and related evolution equations*, in "*Nonlinear Wave Motion*," proceedings, Potsdam, New York, 1972, ed. A.C. Newell, Lect. Appl. Math., **15**, 61–83, A.M.S., Providence, RI;

——— [1975], *Nonlinear wave equations*, in "*Dynamical Systems Theory and Applications*," proceedings, Seattle 1974, ed. J. Moser, Lect. Notes Phys., **38**, 310–354, Springer-Verlag, Berlin-Heidelberg-New York;

——— [1991], *Flexibility in applying the Painlevé test*, in "*Painlevé Transcendents, their Asymptotic and Physical Applications*," proceedings, St. Adèle, Québec, Canada, 1990, eds., D. Levi and P. Winternitz, Plenum, New York, to appear.

Kruskal M.D. and Clarkson P.A. [1991], *The Painlevé-Kowalevski and Poly-Painlevé tests for integrability*, Stud. Appl. Math., to appear.

Kruskal M.D. and Joshi N. [1991], *Soliton theory, Painlevé property and integrability*, preprint AM90/37, University of New South Wales, Australia.

Kruskal M.D., Ramani A. and Grammaticos B. [1990], *Singularity analysis and its relation to complete, partial and non-integrability*, in "*Partially Integrable Evolution*

Equations in Physics," proceedings, Les Houches, France 1989, ed. R. Conte and N. Boccara, NATO ASI Series C, **310**, 321–372, Kluwer, Dordrecht.

Kruskal M.D. and Zabusky N.J. [1963], *Progress on the Fermi-Pasta-Ulum nonlinear string problem*, Princeton Plasma Physics Laboratory Annual Rep. MATT-Q-21, Princeton, N.J., 301–308.

Kubota T., Ko D.R.S. and Dobbs L.D. [1978], *Weakly nonlinear long internal gravity waves in stratified fluids of finite depth*, AIAA J. Hydronautics, **12**, 157–165.

Kudryashov N.A. [1991], *On types of nonlinear nonintegrable equations with exact solutions*, Phys. Lett., **155A**, 269–275.

Kulish P.P. [1981], *Quantum difference nonlinear Schrödinger equation*, Lett. Math. Phys., **5**, 191–197.

Kulish P.P. and Lipovsky V.D. [1988], *Hamiltonian structure of the Davey-Stewartson equation and Poisson brackets of scattering data*, Phys. Lett., **127A**, 413–417;

——— [1989a], *Hamiltonian interpretation of the method of the inverse problem for the Davey-Stewartson equation*, J. Sov. Math., **46**, 2083–2093.

——— [1989b], *Scattering data for the nonstationary Dirac equation*, J. Sov. Math., **47**, 2488–2492.

Kulish P.P. and Sklyanin E.K. [1982], *Quantum spectral transform method. Recent developments*, in "*Integrable Quantum Field Theories*," proceedings, Tuarminne, Finland 1981, eds., J. Hietarinta and C. Montonen, Lect. Notes Phys., **151**, 61–119, Springer-Verlag, Berlin-Heidelberg-New York.

Kundu A. [1984], *Landau-Lifschitz and higher-order nonlinear systems gauge generated from nonlinear Schrödinger type equations*, J. Math. Phys., **25**, 3433–3438;

——— [1987], *Exact solutions to higher-order nonlinear equations through gauge transformations*, Physica, **25D**, 399–406.

Kuperschmidt B.A. [1984], *A super KdV equation: an integrable system*, Phys. Lett., **102A**, 213–215;

——— [1985], *Mathematics of dispersive water waves*, Commun. Math. Phys., **99**, 51–73.

Kuperschmidt B.A. and Wilson G. [1981], *Modifying Lax equations and the second Hamiltonian structure*, Invent. Math., **62**, 403–436.

Kuznetsov E.A. and Mikhailov A.V. [1977], *On the complete integrability of the two-dimensional classical Thirring model*, Theoret. Math. Phys., **30**, 193–200.

Lacombe H. and Richez C. [1982], *The regime if the straits of Gibralta*, in "*Hydrodynamics of Semi-Enclosed Seas*," ed. J.C.J. Nilhoud, 13–73, Elsevier, Amsterdam.

Lakshmanan M. [1977], *Continuum spin system as an exactly solvable dynamical system*, Phys. Lett., **61A**, 53–54.

Lakshmanan M. and Kaliappan P. [1983], *Lie transformations, nonlinear evolution equations, and Painlevé forms*, J. Math. Phys., **24**, 795–805.

Lamb G.L. [1973], *Phase variation in coherent-optical-pulse propagation*, Phys. Rev. Lett., **31**, 196–199;

——— [1980], "*Elements of Soliton Theory*," Wiley, New York.

Lamb H. [1932], "*Hydrodynamics*," Dover, New York.

Landman M.J., Le Mesurier B.J., Papanicolaou G.C., Sulem C. and Sulem P.L. [1989], *Singular solutions of the cubic Schrödinger equation*, in "*Integrable Systems and Applications*," proceedings, Oléron, France, 1988, eds. M. Balabane, P. Lochak and C. Sulem, Lect. Notes Phys., **342**, 207–217, Springer-Verlag, Berlin-Heidelberg-New York.

Landman M.J., Papanicolaou G.C., Sulem C. and Sulem P.L. [1988], *Rate of blowup for solutions of the nonlinear Schrödinger equation at critical dimension*, Phys. Rev. A, **38**, 3937–3843.

Lappo-Danilevskij I.A. [1957], "*An application of the functions on matrices to the theory of systems of linear ordinary differential equations*," Acad. Sci. USSR, Moscow.

Larson J.W. and Tracy E.R. [1988], *Spectral averaging of small-amplitude Sine-Gordon wave trains*, Phys. Rev. A, **38**, 4419–4430.

Lavine R. and Nachman A. [1987], *On the inverse scattering transform for the n-dimensional Schrödinger operator*, in "*Topics in Soliton Theory and Exactly Solvable Nonlinear Equations*," proceedings, Oberwolfach, W. Germany, 1986, eds. M.J. Ablowitz, B. Fuchssteiner and M.D. Kruskal, 33–44, World Scientific, Singapore.

Lax P.D. [1968], *Integrals of nonlinear equations of evolution and solitary waves*, Commun. Pure Appl. Math., **21**, 467–490;

——— [1975], *Periodic solutions of the Korteweg-de Vries equation*, Commun. Pure Appl. Math., **28**, 141–188;

——— [1978], *A Hamiltonian approach to the KdV and other equations*, in "*Nonlinear Evolution Equations*," ed. M. Crandall, 207–224, Academic, New York.

Lax P.D. and Levermore C.D. [1983a,b,c], *The small dispersion limit of the Korteweg-de Vries equation*, I, II, and III, Commun. Pure Appl. Math., **37**, 253–290; 571–593; 809–830.

Léauté B. and Marcilhacy G. [1979], *Sur certaines particulières transcendantes des équations d'Einstein*, Ann. Inst. H. Poincaré, **31**, 363–375;

——— [1982], *A new transcendent solution of Einstein's equations*, Phys. Lett., **87A**, 159–161;

——— [1983], *Transcendent self-dual solutions for* SU(2) *gauge fields on Euclidean four-dimensional space*, Phys. Lett., **93A**, 394–398;

——— [1984], *A new transcendent solution of Einstein-Maxwell equations*, Lett. Nuovo Cim., **40**, 102–106;

——— [1987], *A class of self-dual solutions for* SU(2) *gauge fields on Euclidean space*, J. Math. Phys., **28**, 774–776.

Lebowitz J.L., Orlandi E. and Presutti E. [1988], *Convergence of stochastic cellular automaton to Burgers' equation: fluctuation and stability*, Physica, **33D**, 165–188.

Lebeau G. and Lochak P. [1987], *On the second Painlevé equation*, J. Diff. Eqns., **68**, 344–372.

Lebedev R.R. and Radul A.O. [1983], *Generalized internal long wave equations: construction, Hamiltonian structure and conservation laws*, Commun. Math. Phys., **91**, 543–555.

Leclert G.P., Karney C.F.F., Bers A. and Kaup D.J. [1979], *Two-dimensional self modulation of lower hybrid waves in inhomogeneous plasmas*, Phys. Fluids, **22**, 1545–1553.

Leo R.A., Mancarella G., Soliani G. and Solombrino L. [1988], *On the Painlevé property of nonlinear field equations in* $2+1$ *dimensions: the Davey-Stewartson system*, J. Math. Phys., **29**, 2666–2671.

Le Mesurier B.J., Papanicolaou G.C., Sulem C. and Sulem P.L. [1988a], *Focusing and multi-focusing solutions of the nonlinear Schrödinger equation*, Physica, **31D**, 78–102;

Le Mesurier B.J., Papanicolaou G.C., Sulem C. and Sulem P.L. [1988b], *Local structure of the self-focusing singularity of the nonlinear Schrödinger equation*, Physica, **32D**, 210–226.

Leo R.A., Martina L. and Soliani G. [1986], *Group analysis of the three-wave resonant system in* $(2+1)$ *dimensions*, J. Math. Phys., **27**, 2623–2628.

Leo R.A., Martina L., Soliani G. and Tondo G. [1986], *On certain symmetry reduction systems of the three-wave interaction in* $(2+1)$ *dimensions*, Prog. Theo. Phys., **76**, 739–751.

Leon J.J.P. [1987], *General evolution of the spectral transform from the* $\overline{\partial}$*-approach*, Phys. Lett., **123A**, 65–70.

LeVeque R.J. [1987], *On the interaction of nearly equal solitons in the KdV equation*, SIAM J. Appl. Math., **47**, 254–262.

Levi D., Pilloni L. and Santini P.M. [1981], *Bäcklund transformations for* $(2+1)$*-dimensional integrable systems*, Phys. Lett., **81A**, 419–423.

Levi D., Ragnisco O. and Sym A. [1984], *The Bäcklund transformation for nonlinear evolution equations which exhibit exotic solitons*, Phys. Lett., **100A**, 7–10.

Levi D. and Winternitz P. [1989], *Nonclassical symmetry reduction: example of the Boussinesq equation*, J. Phys. A: Math. Gen., **22**, 2915–2924.

Levine G. and Tabor M. [1988], *Integrating the nonintegrable: analytic structure of the Lorenz system revisited*, Physica, **33D**, 189–210.

Levinson N. [1949], *On the uniqueness of the potential in a Schrödinger equation for a given asymptotic phase*, Danske Vid. Selsk. Math. Fys. Medd., **25**, 1–29.

Lipovsky V.D. [1986], *Hamiltonian structure of the Kadomtsev-Petviashvili-II equation in the class of decreasing Cauchy data*, Func. Anal. Appl., **20**, 282–291.

Lipovsky V.D., Matveev V.B. and Smirnov A.O. [1989], *On the connection of the Kadomtsev-Petviashvili equation and Johnson equation*, J. Sov. Math., **46**, 1609–1612.

Long R.R. [1964], *Solitary waves in the Westerlies*, J. Atmos. Sci., **21**, 156–179.

Lorenz E.N. [1963], *Deterministic non-periodic flow*, J. Atmos. Sci., **20**, 130–141.

Lukashevich N.A. [1965], *Elementary solutions of certain Painlevé equations*, Diff. Eqns., **1**, 561–564;

——— [1967a], *Theory of the fourth Painlevé equation*, Diff. Eqns., **3**, 395–399;

——— [1967b], *On the theory of the third Painlevé equation*, Diff. Eqns, **3**, 994–999;

——— [1968], *Solutions of the fifth equation of Painlevé equation*, Diff. Eqns., **4**, 732–735;

——— [1971], *The second Painlevé equation*, Diff. Eqns., **7**, 853–854.

Lukashevich N.A. and Yablonskii A.I. [1967], *On a set of solutions of the sixth Painlevé equation*, Diff. Eqns., **3**, 264–266.

Lund F. and Regge T. [1976], *Unified approach to strings and vortices with soliton solutions*, Phys. Rev., **D14**, 1524–1535.

Luther A. [1976], *Eigenvalue spectrum of interacting massive fermions in one dimension*, Phys. Rev., **B14**, 2153–2159.

Ma Y.-C. [1978], *The complete solution of the Long-Wave-Short-Wave Resonance Equations*, Stud. Appl. Math., **59**, 201–221.

Ma Y.-C. and Ablowitz M.J. [1981], *The periodic cubic Schrödinger equation*, Stud. Appl. Math., **65**, 113–158.

MacCallum M.A.H. [1983], *Static and stationary 'cylindrically symmetric' Einstein-Maxwell fields, and the solutions of van den Bergh and Wils*, J. Phys. A: Math. Gen., **16**, 3853–3866.

MacKay R.S. and Meiss J.D. [1987], "*Hamiltonian Dynamical Systems*," Adam Hilger, Bristol.

Magri F. [1978], *A simple model of the integrable Hamiltonian equation*, J. Math. Phys., **19**, 1156–1162;

——— [1980], *A geometrical approach to the nonlinear solvable equations*, in "*Nonlinear Evolution Equations and Dynamical Systems*," eds. M. Boiti, F. Pempinelli and G. Soliani, Lect. Notes Phys., **120**, 233–263, Springer-Verlag, Berlin-Heidelberg-New York.

Makhankov V.G. [1978], *Dynamics of classical solitons*, Phys. Rep., **35**, 1–128.

Manakov S.V. [1975], *Complete integrability and stochastization of discrete dynamical systems*, Sov. Phys. JETP, **40**, 269–274;

——— [1976], *Remarks on the integrals of the Euler equations of the n-dimensional heavy top*, Func. Anal. Appl., **10**, 93–94.

——— [1981], *The inverse scattering transform for the time-dependent Schrödinger equation and Kadomtsev-Petviashvili equation*, Physica, **3D**, 420–427.

Manakov S.V. and Zakharov V.E. [1981], *Three-dimensional model of relativistic-invariant theory, integrable by the inverse scattering transform*, Lett. Math. Phys., **5**, 247–253.

Manakov S.V., Zakharov V.E., Bordag L.A., Its A.R. and Matveev V.B. [1977], *Two-dimesional solitons of the Kadomtsev-Petviashvili equation and their interaction*, Phys. Lett., **63A**, 205–206.

Manzella G., Bohm E. and Salustry E. [1983], *Evidence of internal waves in the Archipelago of La Maddalena*, Nuovo Cim., **6C**, 381–400.

Marchenko V.A. [1986], "*Sturm-Liouville Operators and Applications*," Birkhauser-Verlag, Basel-Boston-Stuttgart.

Marcilhacy G. [1979], *New exact cylindrical solutions of the Einstein-Maxwell equations of general relativity*, Phys. Lett., **73A**, 157–158.

Marcus P.S. [1981], *Effects of truncation in modal representations of thermal convection*, J. Fluid Mech., **103**, 241–255.

Martina L. and Winternitz P. [1989], *Analysis and applications of the symmetry group of the multidimensional three-wave interaction problem*, Ann. Phys., **196**, 231–277.

Martynov I.P. [1985], *A particular class of 3rd-order equations without mobile critical points*, Diff. Eqns., **21**, 623–630.

Mason L.J. and Newman E.T. [1989], *A connection between the Einstein and Yang-Mills equations*, Commun. Math. Phys., **121**, 659–688.

Mason L.J. and Sparling G.A.J. [1989], *Nonlinear Schrödinger and Korteweg-de Vries are reductions of self-dual Yang-Mills*, Phys. Lett., **137A**, 29–33.

Matsuno Y. [1979a], *Exact multi-soliton solution of the Benjamin-Ono equation*, J. Phys. A: Math. Gen., **12**, 619–621;

——— [1979b], *Exact multi-soliton solution for nonlinear waves in a stratified fluid of finite depth*, Phys. Lett., **74a**, 233–235;

——— [1980], *Interaction of the Benjamin-Ono solitons*, J. Phys. A: Math. Gen., **13**, 1519–1536;

——— [1984], "*Bilinear Transformation Method*," Academic, London;

——— [1986], *N-soliton solutions for the Sine-Hilbert equation*, Phys. Lett., **119A**, 229–233;

——— [1987a], *Periodic problem for the Sine-Hilbert equation*, Phys. Lett., **120A**, 187–190;

——— [1987b], *Kinks of the the Sine-Hilbert equation and dynamical motions*, J. Phys. A: Math. Gen., **20**, 3587–3606.

Matveev V.B. and Sall M.A. [1981], *Nonlocal analogues of the Korteweg-de Vries and Kadomtsev-Petviashvili equations*, Sov. Math. Dokl., **24**, 536–540.

Matveev V.B., Salle M.A. and Rybin A.V. [1988], *Darboux transformations and coherent interaction of the light pulse with two-level media*, Inverse Problems, **4**, 173–183.

Maxworthy T. and Redekopp L. [1976], *Theory of the Great Red Spot and other observed features of the Jovian atmosphere*, Icarus, **29**, 261–271.

McCaskill J.S. and Fackerell E.D. [1988], *Painlevé solution of the Poisson-Boltzmann equation for a cylindrical polyelectrolyte in excess salt solution*, J. Chem. Soc., Faraday Trans. 2, **84**, 161–179.

McCoy B.M., Perk J.H.H. and Shrock R.E. [1983a], *Time-dependent correlation functions of the transverse Ising chain at the critical magnetic field*, Nucl. Phys., **B220 [FS8]**, 35–47;

——— [1983b], *Correlation functions of the transverse Ising chain at the critical field for large temporal and spatial separations*, Nucl. Phys., **B220 [FS8]**, 269–282.

McCoy B.M. and Tang S. [1986a], *Connection formulae for Painlevé functions*, Physica, **18D**, 190–196;

——— [1986b], *Connection formulae for Painlevé V functions*, Physica, **19D**, 42–72;

——— [1986c], *Connection formulae for Painlevé V functions. II. The δ function Bose gas problem*, Physica, **20D**, 187–216.

McCoy B.M. Tracy C.A. and Wu T.T. [1977], *Painlevé functions of the third kind*, J. Math. Phys., **18**, 1058–1092.

McKean H.P. [1981a], *The Sine-Gordon and Sinh-Gordon equations on the circle*, Commun. Pure Appl. Math., **34**, 197–257;

——— [1981b,c], *Boussinesq's equation on the circle*, Physica, **3D**, 294–305; Commun. Pure Appl. Math., **34**, 599–691.

McKean H.P. and Trubowitz E. [1976], *Hill's operator and hyperelliptic function theory in the presence of infinitely many branch points*, Commun. Pure Appl. Math., **29**, 143–226.

McKean H.P. and van Moerbeke P. [1975], *The spectrum of Hill's equation*, Invent. Math., **30**, 217–274.

McLaughlin J.R. [1986], *Analytical methods for recovering coefficients in differential equations from spectral data*, SIAM Review, **28**, 53–72.

McLeod J.B. and Olver P.J. [1983], *The connection between partial differential equations soluble by inverse scattering and ordinary differential equations of Painlevé type*, SIAM J. Math. Anal., **14**, 488–506.

Melin A. [1985], *Operator methods for inverse scattering on the real line*, Commun. Part. Diff. Eqns., **10**, 677–766.

Mel'nikov V.K. [1983], *On equations for wave interactions*, Lett. Math. Phys., **7**, 129–136;

——— [1986], *Wave emission and absorption in a nonlinear integrable system*, Phys. Lett., **118A**, 22–24;

——— [1987a], *Reflection of waves in nonlinear integrable systems*, J. Math. Phys., **28**, 2603–2609;

——— [1987b], *A direct method for deriving a multi-soliton solution for the problem of interaction of waves an the x, y plane*, Commun. Math. Phys., **112**, 639–652.

Menyuk C.R. [1989], *Hamiltonian deformations and optical fibers*, in "Integrable Systems and Applications," proceedings, Oléron, France, 1988, eds. M. Balabane, P. Lochak and C. Sulem, Lect. Notes Phys., **342**, 218–231, Springer-Verlag, Berlin-Heidelberg-New York.

Meyers S.D., Sommeria J. and Swinney H.L. [1989], *Laboratory study of the dynamics of Jovian-type vortices*, Physica, **37D**, 515–530.

Mikhailov A.V. [1976], *Integrability of the two-dimensional Thirring model*, JETP Lett., **23**, 320–323;

——— [1979], *Integrability of a two-dimensional generalization of the Toda chain*, Sov. Phys. JETP Lett., **30**, 414–418;

——— [1981], *The reduction problem and the inverse scattering method*, Physica, **3D**, 73–117;

——— [1982], *The Landau-Lifschitz equation and the Riemann boundary value problem on a torus*, Phys. Lett., **92A**, 51–55.

Mikhailov A.V., Shabat A.B. and Yamilov R.I. [1987], *The symmetry approach to the classification of non-linear equations. Complete lists of integrable systems*, Russ. Math. Surveys, **42**, 1–63;

——— [1988], *Extension of the module of invertible transformations. Classification of integrable systems*, Commun. Math. Phys., **115**, 1–19.

Miles J.W. [1978], *On the second Painlevé transcendent*, Proc. Roy. Soc. Lond., **A361**, 277–291;

——— [1980a], *The second Painlevé transcendent: a nonlinear Airy function*, in "Mechanics Today. Vol.5," 297–313, Pergamon, Oxford;

——— [1980b], *Solitary waves*, Ann. Rev. Fluid Mech., **12**, 11–43;

——— [1981], *The Korteweg-de Vries equation: an historical essay*, J. Fluid Mech., **106**, 131–147.

Miura R.M. [1968], *Korteweg-de Vries equations and generalizations I. A remarkable explicit nonlinear transformation*, J. Math. Phys., **9**, 1202–1204;

——— [1976], *The Korteweg-de Vries equation: a survey of results*, SIAM Rev., **18**, 412–459.

Miura R.M., Gardner C.S. and Kruskal M.D. [1968], *Korteweg-de Vries equations and generalizations. II. Existence of conservation laws and constants of motion*, J. Math. Phys., **9**, 1204–1209.

Mollenauer L.F. [1985], *Solitons in optical fibres and the soliton laser*, Phil. Trans. Roy. Soc. London A, **315A**, 437–450.

Mollenauer L.F. and Stolen R.H. [1982], *Solitons in optical fibres*, Laser Focus, **18**, 193–198.

Mollenauer L.F., Stolen R.H. and Gordon J.P. [1983], *Experimental observation of picosecond pulse narrowing and solitons in optical fibres*, Phys. Rev. Lett., **45**, 1045–1048.

Moloney T.P. and Hodnett P.F. [1986], *Soliton interactions (for the Korteweg-de Vries equation): a new prospective*, J. Phys. A: Math. Gen., **19**, L1129–L1135.

Moore G. [1990], *Geometry of the string equation*, Commun. Math. Phys., **133**, 261–304.

Moore G. and Reshetikhin N. [1989], *A comment on quantum group symmetry in conformal field theory*, Nucl. Phys., **B328**, 557–574.

Morris H.C. and Dodd R.K. [1980], *Geometric structures and inverse scattering problems for Painlevé transcendents*, Phys. Lett., **75A**, 249–253.

Moser J. [1975], *Three integrable Hamiltonian systems connected with isospectral deformations*, Adv. Math., **16**, 197–220.

Mugan U. [1987], *On the initial value problem of some Painlevé equations*, Ph.D. thesis, Clarkson University, Potsdam, New York.

Mugan U. and Fokas A.S. [1990], *On the initial value problem of the third Painlevé equation*, in "*Nonlinear Evolution Equations and Dynamical Systems*," Research Reports in Physics (Nonlinear Dynamics), eds. S. Carillo and O. Ragnisco, 82–86, Springer-Verlag, Berlin-Heidelberg-New York.

Murata Y. [1985], *Rational solutions of the second and the fourth Painlevé equations*, Funkcial. Ekvac., **28**, 1–32.

Murray J.D. [1984], "*Asymptotic Analysis*," 2nd. Ed., Springer-Verlag, Berlin-Heidelberg-New York.

Mushkeilishvili N.I. [1941], *An application of the Cauchy type integral to a class of singular integral equations*, Trudy Tbil. matem. inst., **10**, 1–43;

——— [1953], "*Singular Integral Equations*," Noordhoff, Groningen.

Musette M. and Conte R. [1991], *Algorithmic method for deriving Lax pairs from the invariant Painlevé analysis of nonlinear partial differential equations*, J. Math. Phys., **32**, 1450–1457.

Nachman A.I. and Ablowitz M.J. [1984a], *A multidimensional inverse scattering method*, Stud. Appl. Math., **71**, 243–250;

——— [1984b], *Multidimensional inverse scattering for first order systems*, Stud. Appl. Math., **71**, 251–262.

Nachman A.I., Fokas A.S. and Ablowitz M.J. [1986], *Note on the inverse problem for a class of first order multidimensional systems*, in "*Nonlinear Systems of Partial Differential Equations in Applied Mathematics*," Part 1, proceedings, SIAM-AMS Summer Seminar, Santa Fe, 1984, eds. B. Nicolaenko, D.D. Holm and J.M. Hyman, American Mathematical Society Lect. Appl. Math., **23**, 217–222.

Nahm W. [1981], *All self-dual multimonopoles for arbitrary gauge groups*, CERN preprint **3172**;

——— [1982], *The algebraic geometry of multimonopoles*, in "*Group Theoretical Methods in Physics*," eds. M. Serdaroglu and E. Inonu, Lect. Notes Phys., **180**, 456–466, Springer-Verlag, Berlin-Heidelberg-New York.

Nakamura A. [1979], *A direct method of calculating periodic wave solutions to nonlinear evolution equations I. Exact two-periodic wave solution*, J. Phys. Soc. Japan, **47**, 1701–1705;

——— [1981a], *Simple similarity-type multiple-decay-mode solution of the 2-dimensional Korteweg-de Vries equation*, Phys. Rev. Lett., **46**, 751–753;

——— [1981b], *Decay mode solution of the 2-dimensional KdV equation and the generalized Bäcklund transformation*, J. Math. Phys., **22**, 2456–2462;

——— [1981c], *One dimensionally aligned decay mode solutions of the 2-dimensional nonlinear Schrödinger equation*, J. Phys. Soc. Japan, **50**, 2469–2470;

——— [1981d], *A hierarchy of explode-mode solutions of the 2-dimensional KdV equation and their superpositions*, J. Phys. Soc. Japan, **51**, 19–20;

——— [1982], *Explode-decay mode lump solutions of a 2-dimensional nonlinear Schrödinger equation*, Phys. Lett., **88A**, 55–56;

——— [1983], *Exact explode-decay soliton solutions of a coupled nonlinear Schrödinger equation*, J. Phys. Soc. Japan, **52**, 37113–3721.

Nehari Z. [1952], "*Conformal Mapping*," McGraw-Hill, New York. [Reprinted by Dover, New York, (1975).]

Newell A.C. [1978], *The general structure of integrable evolution equations*, Proc. Roy. Soc. London A, **365**, 283–311;

Newell A.C., Tabor M. and Zeng Y.B. [1987], *A unified approach to Painlevé expansions*, Physica, **29D**, 1–68.

Newton R. [1970], *A new result on the inverse scattering problem in three dimensions*, Phys. Rev. Lett., **43**, 541–542;

——— [1980], *Inverse Scattering II. Three dimensions*, J. Math. Phys., **21**, 1698–1715; **22**, 631 (1981); **23**, 693, (1982);

——— [1981], *Inverse Scattering III. Three dimensions, continued*, J. Math. Phys., **22**, 2191–2200; **23**, 693, (1982);

——— [1982a], *Inverse Scattering IV. Three dimensions: Generalised Marchenko construction with bound states*, J. Math. Phys., **23**, 594–604;

——— [1982b], "*Scattering Theory of Waves and Particles*," 2nd ed., Springer-Verlag, Berlin-Heidelberg-New York.

——— [1982c], *On a generalized Hilbert problem*, J. Math. Phys., **23**, 2257–2265.

—— [1983], *The Marchenko and Gel'fand-Levitan methods in the inverse scattering problem in one and three dimensions*, in "*Conference on Inverse scattering: Theory and Application*," ed. J.B. Bednor, R. Redner, E. Robinson and A. Weglein, 1–74, SIAM, Philadelphia;

—— [1989], "*Inverse Schrödinger Scattering in Three Dimensions*," Springer-Verlag, Berlin-Heidelberg-New York.

Nicholson D.R. and Goldman M.V. [1978], *Virial theory of direct Langmuir collisions*, Phys. Rev. Lett., **41**, 406–410.

Nijhoff F.W. and Papageorgiou V.G. [1991], *Similarity reductions of integrable lattices and discrete analogues of Painlevé II equation*, Phys. Lett., **153A**, 337–344.

Nishitani T. and Tajiri M. [1982], *On similarity solutions of the Boussinesq equation*, Phys. Lett., **89A**, 379–380.

Nizhnik L.P. [1980], *Integration of multidimensional nonlinear equations by the method of the inverse problem*, Sov. Phys. Dokl., **25**, 706–708.

Novikov R.G. and Khenkin G.M. [1987a], *$\bar{\partial}$ equation in the multidimensional inverse scattering problem*, Russian Math. Surveys, **42**, 109–180;

—— [1987b], *Solution of a multidimensional inverse scattering problem on the basis of generalized dispersion relations*, Sov. Math. Dokl., **35**, 153–157.

Novikov S.P. [1974], *The periodic problem for the Korteweg-de Vries equation*, Func. Anal. Appl., **8**, 236–246.

Novikov S.P., Manakov S.V., Pitaevskii L.P. and Zakharov V.E. [1984], "*Theory of Solitons. The Inverse Scattering Method*," Plenum, New York.

Novokshenov V.Yu. [1990], *The modulated elliptic function as the solution of Painlevé's second equation in the complex plane*, Sov. Math. Dokl., **41**, 252–255.

Nozaki K. [1987], *Hirota's method and the singular manifold expansion*, J. Phys. Soc. Japan, **56**, 3052–3054.

Oevel W. and Fuchssteiner B. [1982], *Explicit formulas for symmetries and conservation laws of the Kadomtsev-Petviashvili equation*, Phys. Lett., **88A**, 323–327.

Oishi S. [1979], *Relationship between Hirota's method and the inverse spectral method — the Korteweg-de Vries equations's case*, J. Phys. Soc. Japan, **47**, 1037–1038.

Okamoto K. [1981], *On the τ-function of the Painlevé equations*, Physica, **2D**, 525–535;

—— [1986a], *Isomonodromic deformation and Painlevé equations, and the Garnier system*, J. Fac. Sci. Univ. Tokyo, **33**, 575–618;

—— [1986b], *Studies on the Painlevé equations. III. Second and fourth Painlevé equations,* P_{II} *and* P_{IV}, Math. Ann., **275**, 221–255.

Olive D. and Turok N. [1983], *The symmetries of Dynkin diagrams and the reduction of Toda field equations*, Nucl. Phys., **B215 [FS7]**, 470–494.

Olshanetsky M.A. and Perelomov A.M. [1981], *Classical integrable finite-dimensional systems related to Lie algebras*, Phys. Rep., **71**, 313–400;

—— [1982], *Integrable systems and Lie algebras*, Sov. Sci. Rev. C: Math. Phys. Rev., **3**, 151–220.

Olver F.W.J. [1974], "*Asymptotics and Special Functions*," Academic, New York.

Olver P.J. [1986], "*Applications of Lie Groups to Differential Equations*," Graduate Texts Math., **107**, Springer-Verlag, Berlin-Heidelberg-New York.

Olver P.J. and Rosenau P. [1986], *The construction of special solutions to partial differential equations*, Phys. Lett., **114A**, 107–112;

——— [1987], *Group-invariant solutions of differential equations*, SIAM J. Appl. Math., **47**, 263–278.

Omote M. [1988], *Infinite-dimensional symmetry algebras and an infinite number of conserved quantities of the* $(2+1)$*-dimensional Davey-Stewartson equation*, J. Math. Phys., **29**, 2599–2603.

Ono H. [1975], *Algebraic solitary waves in stratified fluids*, J. Phys. Soc. Japan, **39**, 1082–1091.

Onsager L. [1944], *Crystal statistics, I. A two-dimensional model with an order-disorder transition*, Phys. Rep., **65**, 117–149.

Orfandis S.J. [1978], *Sine-Gordon equations and nonlinear σ model on a lattice*, Phys. Rev., **D18**, 3828–3832;

——— [1980], *σ models of nonlinear evolution equations*, Phys. Rev., **D21**, 1513–1522.

Osborne A.R. [1990], *The generation and propagation of internal solitons in the Andaman Sea*, in "*Soliton Theory: A Survey of Results*," ed. A.P. Fordy, 152–173, M.U.P., Manchester.

Osborne A.R. and Bergamasco L. [1986], *The solitons of Zabusky and Kruskal revisited: perspective in terms of the periodic spectral transform*, Physica, **18D**, 26–46.

Osborne A.R. and Burch T.L. [1980], *Interval solitons in the Andaman Sea*, Science, **258**, 451–460.

Painlevé P. [1888a,b,c], *Sur les équations différentielles du premier ordre*, C.R. Acad. Sc. Paris, **107**, 221–224; 320–323; 724–726;

——— [1893a,b], *Sur les équations différentielles d'ordre supérieur dont l'intégrale n'admet qu'un nombre fini determinations*, C.R. Acad. Sc. Paris, **116**, 88–91; **116**, 173–176;

——— [1893c], *Sur les singularities essentielles des équations différentielles d'ordre supérieur*, C.R. Acad. Sc. Paris, **116**, 362–365;

——— [1893d], *Sur les transcendents definies par les équations différentielles du second ordre*, C.R. Acad. Sc. Paris, **116**, 566–569;

——— [1893e], *Sur les équations du second degre dont l'intégrale générale est uniforme*, C.R. Acad. Sc. Paris, **117**, 211–214;

——— [1893f,g], *Sur les équations du second ordre à points critiques fixés et sur la correspondence univoque entre deux surfaces*, C.R. Acad. Sc. Paris, **117**, 611–614; 686–688;

——— [1897], "*Leçons sur la théorie analytique des équations différentielles, profesées à Stockholm, 1895*," Hermann, Paris

——— [1898a], *Sur la determination explicite des équations différentielles du second ordre à points critiques fixés*, C.R. Acad. Sc. Paris, **126**, 1329–1332;

——— [1898b,c, 1899a,b,c,d], *Sur les équations différentielles du second ordre à points critiques fixés*, C.R. Acad. Sc. Paris, **126**, 1185–1188; **126**, 1699–1700; **127**, 541–544; **127**, 945–948; **129**, 750–753; **129**, 949–952;

——— [1900], *Memoire sur les équations différentielles dont l'intégrale générale est uniforme*, Bull. Soc. Math. France, **28**, 201–261;

——— [1901], *Sur les singularités essentielles des équations différentielles*, C.R. Acad. Sc. Paris, **133**, 910–913;

——— [1902a], *Sur les équations différentielles du second ordre et d'ordre supérieur dont l'intégrale générale est uniforme*, Acta Math., **25**, 1–85;

——— [1902b,c,d], *Sur l'irreducibilite de l'équation* $y'' = 6y^2 + x$, C.R. Acad. Sc. Paris, **135**, 411–415; **135**, 641–647; **135**, 1020–1025;

——— [1902e], *Sur les transcendantes uniformes definies par l'équation* $y'' = 6y^2 + x$, C.R. Acad. Sc. Paris, **135**, 757–761;

——— [1906], *Sur les équations différentielles du second ordre à points critiques fixès*, C.R. Acad. Sc. Paris, **143**, 1111–1117.

Papageorgiou V. [1988], *Analytical approaches to integrable and near-integrable dynamical systems*, Ph.D. Thesis, Clarkson University, Potsdam, New York.

Papageorgiou V.G., Nijhoff F.W. and Capel H.W. [1990], *Integrable mappings and nonlinear integrable lattice equations*, Phys. Lett., **147A**, 106–114.

Papatheodorou T.S., Ablowitz M.J. and Saridakis Y.G. [1988], *A rule for fast computation and analysis of soliton automata*, Stud. Appl. Math., **79**, 173–184.

Papatheodorou T.S. and Fokas A.S. [1989], *Evolution theory, periodic particles and solitons in cellular automata*, Stud. Appl. Math., **80**, 165–182.

Park J.H.H., Steiglitz K. and Thurston W. [1986], *Soliton-like behaviour in automata*, Physica, **19D**, 423–432.

Persides S. and Xanthopoulus B.C. [1988], *Some new stationary axisymmetric asymptotically flat space-times obtained from Painlevé transcendents*, J. Math. Phys., **29**, 674–680.

Picard E. [1887], *Sur un classe d'équations différentielles*, C.R. Acad. Sc. Paris, **104**, 41–43;

——— [1889], *Memoire sur la theorie des fonctions algebrique de deux variables*, J. de Math., **5**, 135–318;

——— [1890], *Sur un classe d'èquations différentielles dont l'intégrale générale est uniforme*, C.R. Acad. Sc. Paris, **110**, 877–880;

——— [1893], *Remarques sur les équations différentielles*, Acta Math., **17**, 297–300;

——— [1927], "Leçons sur quelques types simples d'équations aux dérivées partielles," Gauthier-Villars, Paris.

Pilloni L. and Levi D. [1982], *The inverse scattering transform for solving the discrete Sine-Gordon equation*, Phys. Lett., **92A**, 5–8.

Plemelj J. [1908a], *Ein ergänzungssatz zur Cauchyschen integralderstellung analytischer funktionen*, Monatschefte für Math. und Phys., **19**, 205–210;

——— [1908b], *Riemannsche funktionenscharen mit gegebener monodromiegruppe*, Monatschefte für Math. und Phys., **19**, 211–245;

——— [1964], "*Problems in the sense of Riemann and Klein*," Wiley, New York.

Pogorzelski W. [1966], "*Integral Equations and their Applications*," Vol.I., Pergamon, New York.

Pogrebkov A.K. [1989], *On the formulation of the Painlevé test as a criterion of complete integrability of partial differential equations*, Inverse Problems, **5**, L7–L10.

Pohlmeyer K. [1976], *Integrable Hamiltonian systems and interactions through quadratic constraints*, Commun. Math. Phys., **46**, 207–221;

——— [1980], *On the Lagrangian theory of anti-self-dual fields in four-dimensional Euclidean space*, Commun. Math. Phys., **72**, 37–47.

Porter D. and Stirling D.S.G. [1990], "*Integral Equations: a Practical Treatment, from Spectral Theory to Applications*," Cambridge Texts in Applied Mathematics, C.U.P., Cambridge.

Pöschel J. and Trubowitz E. [1987], "*Inverse Spectral Theory*," Academic, New York.

Quispel G.R.W., Nijhoff F.W. and Capel H.W. [1982], *Linearization of the Boussinesq equation and the modified Boussinesq equation*, Phys. Lett., **91A**, 143–145.

Quispel G.R.W., Roberts J.A.G. and Thompson C.J. [1988], *Integrable mappings and soliton equations*, Phys. Lett., **126A**, 419–421;

——— [1989], *Integrable mappings and soliton equations. II*, Physica, **34D**, 183–192.

Radul A.O. [1985], *The equations of a two-layer fluid is a completely integrable Hamiltonian system*, Soviet Math. Dokl., **32**, 96–99.

Ragnisco O., Santini P.M., Chitlaru-Briggs S. and Ablowitz M.J. [1987], *An example of $\overline{\partial}$ problem arising in a finite difference context: direct and inverse problems for the discrete analog of the equation $\phi_{xx} + u\phi = \phi_y$*, J. Math. Phys., **28**, 777–780.

Rajaraman R. [1982], "*Solitons and Instantons*," North-Holland, Amsterdam.

Ramani A., Dorizzi B. and Grammaticos B. [1982], *Painlevé conjecture revisited*, Phys. Rev. Lett., **49**, 1539–1541.

Ramani A., Dorizzi B., Grammaticos B. and Bountis T.C. [1984], *Integrability and the Painlevé property for low-dimensional systems*, J. Math. Phys., **25**, 878–883.

Ramani A., Grammaticos B. and Bountis T. [1989], *The Painlevé property and singularity analysis of integrable and non-integrable systems*, Phys. Reports, **180**, 159–245.

Ramani A., Grammaticos B. and Hietarinta J. [1991], *Discrete versions of Painlevé equations*, preprint.

Ranada A.F., Ramani A., Dorizzi B. and Grammaticos B. [1985], *The weak-Painlevé property as a criterion for the integrability of dynamical systems*, J. Math. Phys., **26**, 708–710.

Rasmussen J.J. and Rypdal K. [1986], *Blow-up in nonlinear Schroedinger equations — I a general review*, Physica Scripta, **33**, 481–497.

Rayleigh, Lord [1876], *On waves*, Philos. Mag. Ser.(5), **1**, 257–279.

Redekopp L. [1977], *On the theory of solitary Rossby waves*, J. Fluid Mech., **82**, 725–745.

Redekopp L. and Weidman P.D. [1977], *Solitary Rossby waves in zonal shear flows and their interactions*, J. Atmos. Sci., **35**, 790–804.

Reshetikhin N.Yu. and Semenov-Tian-Shansky M.A. [1990], *Central extensions of quantum current groups*, Lett. Math. Phys., **19**, 133–142.

Reshetikhin N.Yu., Takhtajan L.A. and Faddeev L.D. [1989], *Quantization of Lie groups and Lie algebras*, Algebra and Analysis, **1**, 178–206.

Rodin Yu.L. [1984], *The Riemann boundary problem in Riemann surfaces and the inverse scattering problem for the Landau-Lifschitz equation*, Physica, **11D**, 90–108;

——— [1987], *The Riemann boundary value problem on closed Riemann surfaces and integrable systems*, Physica, **24D**, 1–53;

——— [1988], "*The Riemann Boundary Problem on Riemann Surfaces*," Reidel, Dordrecht.

Rogers C. and Ames W.F. [1989], "*Nonlinear Boundary Value Problems in Science and Engineering*," Academic, New York.

Rogers C. and Shadwick W.F. [1982], "*Bäcklund Transformations and their Applications*," Academic, New York.

Rogers C. and Wong P. [1984], *On reciprocal Bäcklund transformations of inverse scattering schemes*, Physica Scripta, **30**, 10–14.

Rosales R.S. [1978a], *Exact solutions of some nonlinear evolution equations*, Stud. Appl. Math., **59**, 117–151;

——— [1978b], *The similarity solutions for the Korteweg-de Vries equation and the related Painlevé transcendent*, Proc. Roy. Soc. London A, **361**, 265–275.

Rosenau P.J. and Schwarzmeier J.L. [1986], *On similarity solutions of Boussinesq type equations*, Phys. Lett., **115A**, 75–77.

Rosenhaus V. [1988], *On the symmetry of solutions of the Yang-Mills equation*, J. Phys. A: Math. Gen., **21**, 1125–1135.

Rossby G.G. [1939], *Relation between variations in the intensity of the zonal circulation of the atmosphere*, J. Marine Res., **2**, 38–55.

Ruijsenaars S.N.M. [1987a], *Relativistic Calogero-Moser systems and solitons*, in "*Topics in Soliton Theory and Exactly Solvable Nonlinear Equations*," proceedings, Oberwolfach, W. Germany, 1986, eds. M.J. Ablowitz, B. Fuchssteiner and M.D. Kruskal, 182–190, World Scientific, Singapore;

——— [1987b], *Complete integrability of relativisitic Calogero-Moser systems and elliptic identities*, Commun. Math. Phys., **110**, 191–213.

Ruijsenaars S.N.M. and Schneider H. [1986], *A new class of integrable systems and its relation to solitons*, Ann. Phys., **170**, 370–405.

Russell J. Scott [1838], *Report of the committee on waves*, Report of the 7th Meeting of the British Association for the Advancement of Science, Liverpool, 417–496;

——— [1844], *Report on waves*, Report of the 14th Meeting of the British Association for the Advancement of Science, John Murray, London, 311–390.

Rypdal K. and Rasmussen J.J. [1986], *Blow-up in nonlinear Schroedinger equations — II similarity structure of the blow-up singularity*, Physica Scripta, **33**, 498–504.

Sabatier P.C. [1990], *Quest of multidimensional nonlinear equations with exponentially confined solutions*, Inverse Problems, **6**, L29–L32.

Santini P.M. [1981], *On the evolution of two-dimensional packets of water waves over an uneven bottom*, Lett. Nuovo Cim., **30**, 236–240;

——— [1986], *Nonlinear integral evolution equations associated with a Riemann-Hilbert scattering problem. Aspects of the initial value problem for the Sine-Hilbert class*, in "*Systèmes Dynamiques Non Linéaires Intégrabilité et Comportement Qualitatif,*" ed. P. Winternitz, SMS **102**, 240–259, University of Montreal, Montreal;

——— [1989a], *dimensional deformations of integrable systems: I. An approach to integrability in multidimensions*, Inverse Problems, **5**, 67–86;

——— [1989b], *Bi-Hamiltonian formulations of the Intermediate Long Wave Equation*, Inverse Problems, **5**, 203–225;

——— [1990], *Energy exchange of interacting coherent structures in multidimensions*, Physica, **41D**, 26–54.

Santini P.M., Ablowitz M.J. and Fokas A.S. [1984a], *On the limit from the intermediate long wave equation to the Benjamin-Ono equation*, J. Math. Phys., **25**, 892–899;

——— [1984b], *The direct linearization of a class of nonlinear evolution equations*, J. Math. Phys., **25**, 2614–2619;

——— [1987], *On the initial value problem for a class of nonlinear integral evolution equations including the Sine-Hilbert equation*, J. Math. Phys., **28**, 2310–2316.

Santini P.M. and Fokas A.S. [1988], *Recursion operators and bi-Hamiltonian structures in multidimensions. I*, Commun. Math. Phys., **115**, 375–419.

——— [1990], *The initial-boundary value problem for the Davey-Stewartson I equation; how to generate and drive localized coherent structures in multidimensions*, in "*Partially Integrable Evolution Equations in Physics*," proceedings, Les Houches, France 1989, ed. R. Conte and N. Boccara, NATO ASI Series C, **310**, 223–259, Kluwer, Dordrecht.

Santoleri R. [1983], *Preliminary observations of large amplitude tidal internal waves near the strait of Messina*, in "*Nonlinear Phenomena*," proceedings, Oaxtepec, Mexico 1982, ed. K.B. Wolf, Lect. Notes Phys., **189**, 428–437, Springer-Verlag, Berlin-Heidelberg-New York.

Sanz-Serna J.M. [1988], *Runge-Kutta schemes for Hamiltonian systems*, BIT, **28**, 877–883.

Sato M., Miwa T. and Jimbo M. [1977, 1979a,b,c, 1980a], *Holonomic quantum fields*, I, II, III, IV, V, Publ. RIMS, Kyoto Univ., **14**, 223–267; **15**, 201–278; **15**, 577–629; **15**, 871–972; **16**, 531–584;

——— [1980b], *Holonomic quantum fields — the unanticipated link between deformation theory of differential equations and quantum fields*, in "*Mathematical Problems in Theoretical Physics*," proceedings, Lausanne, Switzerland, 1979, ed. K. Osterwalder, Lect. Notes Phys., **116**, 119–142, Springer-Verlag, Berlin-Heidelberg-New York;

——— [1980c], *Aspects of Holonomic quantum fields — isomonodromic deformation and Ising model*, in "*Complex Analysis, Microlocal Calculus and Relativistic Quantum Theory*," proceedings, Les Houches, ed. D. Iagolnitzer, Lect. Notes Phys., **126**, 429–491, Springer-Verlag, Berlin-Heidelberg-New York.

Satsuma J. [1976], *N-soliton solution of the two-dimensional Korteweg-de Vries equation*, J. Phys. Soc. Japan, **40**, 286–290;

——— [1981], *Exact solutions of a nonlinear diffusion equation*, J. Phys. Soc. Japan, **50**, 1423–1424.

Satsuma J. and Ablowitz M.J. [1979], *Two-dimensional lumps in nonlinear dispersive systems*, J. Math. Phys., **20**, 1496–1503;

——— [1980], *Solutions of an internal wave equation describing a stratified fluid with finite depth*, in "*Nonlinear Partial Differential Equations in Engineering and Applied Science*," eds. R.L. Sternberg, A.J. Kalinowski and J.S. Papadakis, 397, Marcel Dekker, New York.

Satsuma J., Ablowitz M.J. and Kodama Y. [1979], *On an internal wave equation describing a stratified fluid with finite depth*, Phys. Lett., **73A**, 283–286.

Satsuma J. and Ishimori Y. [1979], *Periodic wave and rational soliton solutions of the Benjamin-Ono equation*, J. Phys. Soc. Japan, **46**, 681–687.

Satsuma J. and Kaup D.J. [1978], *A Bäcklund transformation for a higher order Korteweg-de Vries equation*, J. Phys. Soc. Japan, **43**, 692–697.

Satsuma J., Taha T.R. and Ablowitz M.J. [1984], *On a Bäcklund transformation and scattering problem for the modified intermediate long wave equation*, J. Math. Phys., **25**, 900–904.

Sattinger D.H. [1989], *Flat connections and integrable systems*, Int. J. Mod. Phys., **3B**, 211–233;

——— [1990], *Flat connections and scattering theory on the line*, SIAM J. Math. Anal., **21**, 729–756.

Sawada S. and Kotera T. [1974], *A method for finding N-soliton solutions of the KdV and KdV-like equation*, Prog. Theo. Phys., **51**, 1355–1367.

Schlesinger L. [1912], *Über eine klasse von differentialsystemen beliebiger ordnung mit festen kritischen punkten*, J. für Math., **141**, 96–145.

Schultz C.L. and Ablowitz M.J. [1989a], *Action-angle variables and trace formula for D-bar limit case of Davey-Stewartson I*, Phys. Lett., **135A**, 433–437;

——— [1989b], *Exact solutions of nonlinear evolution equations in the strong coupling limit*, INS preprint **115**, Clarkson University, Potsdam, New York.

Schultz C.L., Ablowitz M.J. and BarYaacov D. [1987], *Davey-Stewartson I System — a quantum* $2+1$*-dimensional integrable system*, Phys. Rev. Lett., **59**, 2825–2828.

Schuur P.C. [1986], "*Asymptotic Analysis of Soliton Problems*," Lect. Notes Math., **1232**, Springer-Verlag, Berlin-Heidelberg-New York.

Schwarz F. [1982], *Symmetries of the two-dimensional Korteweg-de Vries equation*, J. Phys. Soc. Japan, **51**, 2387–2388;

——— [1985], *Automatically determining symmetries of partial differential equations*, Computing, **34**, 91–106;

——— [1988], *Symmetries of differential equations: from Sophus Lie to computer algebra*, SIAM Rev., **30**, 450–481.

Scott A.C., Chu F.Y.F. and McLaughlin D.W. [1973], *The soliton — a new concept in applied science*, Proc. IEEE, **61**, 1443–1483.

Scovel C. [1990], *Symplectic numerical integrations of Hamiltonian systems*, in "*Geometry of Hamiltonian Systems*," to appear.

Segur H. [1973], *The Korteweg-de Vries equation and water waves. Solutions of the equation. I*, J. Fluid Mech., **59**, 721–736;

——— [1974], *An alternative method to solve the Korteweg-de Vries equation?*, in "*Nonlinear Wave Motion*," proceedings, Potsdam, New York, 1972, ed. A.C. Newell, Lect. Appl. Math., **15**, 165–171, A.M.S., Providence, RI, (1974).

——— [1982], *Comments on IS for the Kadomtsev-Petviashvili equation*, in "*Mathematical Methods in Hydrodynamics and Integrability in Dynamical Systems*," proceedings, La Jolla 1981, eds. M. Tabor and Y.M. Treve, AIP Conf. Proc., **88**, 211–228;

——— [1983], *Some physical applications of solitons*, in "*Nonlinear Phenomena*," proceedings, Oaxtepec, Mexico 1982, ed. K.B. Wolf, Lect. Notes Phys., **189**, 210–262, Springer-Verlag, Berlin-Heidelberg-New York;

——— [1986], *Some open problems*, Physica, **18D**, 1–12.

Segur H. and Ablowitz M.J. [1980], *Long internal waves in fluids of great depth*, Stud. Appl. Math., **62**, 249–262;

——— [1981], *Asymptotic solutions of nonlinear evolution equations and a Painlevé transcendent*, Physica, **3D**, 165–184.

Segur H. and Finkel A. [1985], *An analytical model of periodic waves in shallow water*, Stud. Appl. Math., **73**, 183–220.

Segur H. and Hammack J.L. [1982], *Soliton models of long internal waves*, J. Fluid Mech., **118**, 285–304.

Sen T. and Tabor M. [1990], *Lie symmetries of the Lorenz model*, Physica, **44D**, 313–339.

Seyler C.E. and Fenstermacher D.L. [1984], *A symmetric regularized long-wave equation*, Phys. Fluids, **27**, 4–7.

Shabat A.B. [1975], *Inverse-scattering problem for a system of differential equations*, Func. Anal. Appl., **9**, 244–247;

——— [1979], *An inverse scattering problem*, Diff. Eqns., **15**, 1299–1307.

Shen M.-M. and Nicholson R.R. [1987], *Stochasticity in numerical solutions of the nonlinear Schrödinger equation*, Phys. Fluids, **30**, 3150–3154.

Shimizu T. and Wadati M. [1980], *A new integrable nonlinear evolution equation*, J. Phys. Soc. Japan, **63**, 808–820.

Silverman J.H. [1986], "*The Arithmetic of Elliptic Curves*," Grad. Texts Math., **106**, Springer-Verlag, Berlin-Heidelberg-New York.

Sklyanin E.K. [1979], *On complete integrability of the Landau-Lifschitz equation*, preprint LOMI, E-3-79, Leningrad.

Soffer A. and Weinstein M.I. [1989], *Multichannel nonlinear scattering theory for nonintegrable equations*, in "*Integrable Systems and Applications*," proceedings, Oléron, France, 1988, eds. M. Balabane, P. Lochak and C. Sulem, Lect. Notes Phys., **342**, 312–327, Springer-Verlag, Berlin-Heidelberg-New York.

——— [1990], *Multichannel nonlinear scattering for nonintegrable equations*, Commun. Math. Phys., **133**, 119–146.

Sokolov V.V. and Shabat A.B. [1984], *Classification of integrable evolution equations*, Sov. Sci. Rev. C: Math. Phys. Rev., **4**, 221–280.

Steeb W.-H. and Euler N. [1988], "*Nonlinear Evolution Equations and Painlevé Test*," World Scientific, Singapore.

Steiglitz K., Kamal I. and Watson A. [1988], *Embedding computation in one-dimensional automata by phase coding solitons*, IEEE Trans. Comput., **37**, 138–145.

Steinberg S. [1969], *Meromorphic families of compact operators*, Arch. Rat. Mech. Anal., **31**, 372–379.

Stokes G. [1847], *On the theory of oscillatory waves*, Trans. Camb. Phil. Soc., **8**, 441–455.

Strachan I.A.B. [1991], *Self-dual Yang-Mills fields and the nonlinear Schrödinger equation*, Phys. Lett., **154A**, 123–126.

Stuart J.T. and DiPrima R.C. [1978], *The Eckhaus and Benjamin-Feir resonance mechanisms*, Proc. Roy. Soc. London A, **362**, 27–41.

Suleimanov B.I. [1987], *The relation between asymptotic properties of the second Painlevé equation in different directions towards infinity*, Diff. Eqns., **23**, 569–576.

Sung L.Y. and Fokas A.S. [1991a], *Inverse problems in multidimensions*, SIAM J. Math. Anal., to appear.

——— [1991b], *Inverse problem of $N \times N$ hyperbolic systems on the plane and N-wave interactions*, Commun. Pure Appl. Math., to appear.

Svinoupov S.I. and Sokolov V.V. [1983], *Evolution equations with nontrivial conservation laws*, Func. Anal. Appl., **16**, 317–319.

Svinoupov S.I., Sokolov V.V. and Yamiov R.I. [1983], *On Bäcklund transformations for integrable evolution equations*, Soviet Math. Dokl., **28**, 165–168.

Tabor M. [1984], *Modern dynamics and classical analysis*, Nature, **310**, 277–282;

——— [1989], "*Chaos and Integrability in Nonlinear Dynamics*," Wiley, New York.

Tabor M. and Gibbon J.D. [1986], *Aspects of the Painlevé property for partial differential equations*, Physica, **18D**, 180–189.

Tabor M. and Weiss J. [1981], *Analytic structure of the Lorenz system*, Phys. Rev., **24A**, 2157–2167.

Tagami Y. [1989], *Soliton-like solutions to a $(2+1)$-dimensional generalization of the KdV equation*, Phys. Lett., **141A**, 116–120.

Taha T.R. and Ablowitz M.J. [1984a], *Analytical and numerical aspects of certain nonlinear equations. I. Analytical*, J. Comp. Phys., **55**, 192–202;

——— [1984b], *Analytical and numerical aspects of certain nonlinear evolution equations. II. Numerical, nonlinear Schrödinger equation*, J. Comp. Phys., **55**, 203–230;

——— [1984c], *Analytical and numerical aspects of certain nonlinear evolution equations. III. Numerical, Korteweg-de Vries equation*, J. Comp. Phys., **55**, 231–253;

——— [1984c], *Analytical and numerical aspects of certain nonlinear evolution equations. IV. Numerical, modified Korteweg-de Vries equation*, J. Comp. Phys., **77**, 540–548.

Tajiri M. [1983], *Similarity reductions of the one and two dimensional nonlinear Schrödinger equations*, J. Phys. Soc. Japan, **52**, 1908–1917.

Tajiri M., Nishitani T. and Kawamoto S. [1982], *Similarity reductions of the Kadomtsev-Petviashvili equation*, J. Phys. Soc. Japan, **51**, 2350–2356.

Takhtajan L.A. [1977], *Integration of the continuous Heisenberg spin chain through the inverse scattering method*, Phys. Lett., **64A**, 235–237.

Taki M., Spatschek K.H., Fernandez J.C., Grauer R. and Reinisch G. [1989], *Breather dynamics in the nonlinear Schrödinger regime of perturbed Sine-Gordon systems*, Physica, **40D**, 65–82.

Tanaka S. [1972], *On the N-tuple wave solutions of the Korteweg-de Vries equation*, Publ. RIMS, Kyoto Univ., **8**, 419–428.

Tenenblat K. [1985], *Bäcklund's theorem for submanifolds of space forms*, Bol. Soc. Brasil. Math., **16**, 67–92.

Tenenblat K. and Terng C.L. [1980], *Bäcklund's theorem for N-dimensional submanifolds of* S^{2n-1}, Ann. Math., **111**, 477–490.

Terng C.L. [1980], *A higher dimensional generalisation of the Sine-Gordon equation and its soliton theory*, Ann. Math., **111**, 491–510.

Terrones G., McLaughlin D.W., Overman E.A. II and Pearlstein A.J. [1990], *Stability and bifurcation of spacialy coherent solutions of the damped-driven NLS equation*, SIAM J. Appl. Math., **50**, 791–818.

Thacker H.B. [1981], *Exact integrability in quantum field theories and statistical systems*, Rev. Mod. Phys., **53**, 253–285;

——— [1982], *The quantum inverse method and Green's functions for completely integrable field theories*, in "*Integrable Quantum Field Theories*," proceedings, Tuarminne, Finland 1981, eds., J. Hietarinta and C. Montonen, Lect. Notes Phys., **151**, 1–60, Springer-Verlag, Berlin-Heidelberg-New York.

Thickstun W. [1976], *A system of particles equivalent to solitons*, J. Math. Anal. Appl., **55**, 335–346.

't Hooft G. [1976a], *Symmetry breaking through Bell-Jackiw anomalies*, Phys. Rev. Lett., **37**, 8–11;

——— [1976b], *Computation of the quantum effects due to a four-dimensional pseudoparticle*, Phys. Rev. D, **14**, 3432–3450.

Thual O. and Frisch U. [1986], *Natural boundary in the Kuramoto model*, in "*Combustion and Nonlinear Phenomena*," proceedings, Les Houches, France, 1985, eds. P. Clavin, B. Larroutvrou and P. Pelce, 327–326, Les Ulis: Editions de Physique.

Toda M. [1967], *Vibration of a chain with nonlinear interaction*, J. Phys. Soc. Japan, **22**, 431–436;

——— [1975], *Studies of a nonlinear lattice*, Phys. Reports, **18**, 1-112;

——— [1976], *Exact treatment of nonlinear lattice waves*, Supp. Prog. Theo. Phys., **59**, 1–161.

Tracy C.A. [1987], *The emerging role of number theory in exactly solvable models in lattice statistical mechanics*, Physica, **25D**, 1–19;

——— [1989], *Monodromy preserving deformation theory of the Klein-Gordon equation in the hyperbolic plane*, Physica, **34D**, 347–365.

Tracy E.R. [1984], *Topics in nonlinear wave theory with applications*, Ph.D. Thesis, University of Maryland, College Park.

Tracy E.R. and Chen H.H. [1988], *Nonlinear self-modulational: an exactly solvable model*, Phys. Rev. A, **37**, 815–839;

Tracy E.R., Chen H.H. and Lee Y.C. [1984], *Study of quasiperiodic solutions of the nonlinear Schrödinger equation and the nonlinear modulational instability*, Phys. Rev. Lett., **53**, 218–211;

——— [1986], *A study of nonlinear modulations*, Physica, **18D**, 481–482.

Turcotte D.L., Spence D.A. and Bau H.H. [1982], *Multiple solutions for natural convectived flows in an internally heated, vertical channel with viscous dissipation and pressure work*, Int. J. Heat Mass Trans., **25**, 699–706.

van den Bergh N. and Wils P. [1983], *A class of stationary Einstein-Maxwell solutions with cylindrical symmetry*, J. Phys. A: Math. Gen., **16**, 3843–3851.

van Moerbeke P. [1985], *Algebraic geometrical methods in Hamiltonian mechanics*, Phil. Trans. Roy. Soc. London A, **315**, 379–380.

Vekua I.A. [1962], "*Generalized Analytic Functions*," Pergamon, Oxford.

Venakides S. [1985a], *The zero dispersion limit of the KdV equation with nontrivial reflection coefficient*, Commun. Pure Appl. Math., **38**, 125–155;

——— [1985b], *The generation of modulated wavetrains in the solution of the KdV equation*, Commun. Pure Appl. Math., **38**, 883–909.

Veselov A.P. and Novikov S.P. [1984a], *Finite-zone, two-dimensional, potential Schrödinger operators, explicit formulas and evolution equations*, Sov. Math. Dokl., **30**, 588–591;

——— [1984b], *Finite-zone, two-dimensional Schrödinger operators. Potential operators*, Sov. Math. Dokl., **30**, 705–708.

Villarroel J. [1989], *Scattering data for the self-duality equations*, Inverse Problems, **5**, 1157–1162;

——— [1990a], *The inverse problem for Ward's system*, Stud. Appl. Math., **83**, 211–222;

——— [1990b], *On the initial value problem for the self-dual Yang-Mills equations*, PAM preprint **12**, University of Colorado, Boulder;

——— [1991], *The DBAR problem and the Thirring model*, Stud. Appl. Math., **84**, 207–220.

Villarroel J. and Ablowitz M.J. [1991], *On the Hamiltonian formalism for the Davey-Stewartson system*, Inverse Problems, **7**, 451–460.

Wadati M. [1972], *The modified Korteweg-de Vries equation*, J. Phys. Soc. Japan, **32**, 1681.

Wadati M., Konno K. and Ichikawa Y.-H. [1979a], *A generalization of the inverse scattering method*, J. Phys. Soc. Japan, **46**, 1965–1966;

——— [1979b], *New integrable nonlinear evolution equations*, J. Phys. Soc. Japan, **47**, 1698–1700.

Wadati M. and Sogo K. [1983], *Gauge transformations in soliton theory*, J. Phys. Soc. Japan, **52**, 394–398.

Wadati M. and Toda M. [1972], *The exact N-soliton solution of the Korteweg-de Vries equation*, J. Phys. Soc. Japan, **22**, 431–436.

Wahlquist H.D. and Estabrook F.B. [1973], *Bäcklund transformations for solutions of the Korteweg-de Vries equation*, Phys. Rev. Lett., **23**, 1386–1389;

——— [1975, 1976], *Prolongation structures and nonlinear evolution equations*, J. Math. Phys., **16**, 1–7; **17**, 1403–1411.

Ward R.S. [1977], *On Self-dual gauge fields*, Phys. Lett., **61A**, 81–82.

——— [1981], *Ansätze for self-dual Yang-Mills fields*, Commun. Math. Phys., **80**, 563–574;

——— [1983], *Stationary axisymmetric space-times: a new approach*, Gen. Rel. Grav., **15**, 105–109;

——— [1984], *The Painlevé property for the self-dual gauge-field equations*, Phys. Lett., **102A**, 279–282;

——— [1985], *Integrable and solvable systems, and relations among them*, Phil. Trans. R. Soc. Lond. A, **315**, 451–457;

——— [1986], *Multi-dimensional integrable systems*, in "Field Theory, Quantum Gravity and Strings. II," proceedings, Meudon and Paris VI, France, 1985/86, eds. H.J. de Vega and N. Sanchez, Lect. Notes Phys., **280**, 106–110, Springer-Verlag, Berlin-Heidelberg-New York

——— [1988a], *Soliton solutions in an integrable chiral in* $2+1$ *dimensions*, J. Math. Phys., **29**, 386–389;

——— [1988b], *Integrability of the chiral equations with torsion term*, Nonlinearity, **1**, 671–679;

——— [1989], *Twistors in* $2+1$*-dimensions*, J. Math. Phys., **30**, 2246–2251.

Wasow W. [1987], "Asymptotic Expansions for Ordinary Differential Equations," Dover, New York.

Weder R. [1990], *Multidimensional inverse scattering: the reconstruction problem*, Inverse Problems, **6**, 267–298;

——— [1991], *Characterization of the scattering data in multidimensional inverse scattering theory*, Inverse Problems, **7**, 461–489.

Weideman J.A.C. [1986], *Computation of instability and recurrence phenomena in the nonlinear Schrödinger equation*, Ph.D. thesis, University of the Orange Free State, Bloemfontein.

Weideman J.A.C. and Herbst B.M. [1987], *Recurrence in semidiscrete approximations of the nonlinear Schrödinger equation*, SIAM J. Sci. Statist. Comput., **8**, 988–1004.

Weiner N. and Hopf E. [1931], *Über eine klasses singulärer integralgleichungen*, Sitz. Berliner Akad. Wiss., 696–706.

Weiss J. [1982], *Analytic structure of the Henon-Heiles system*, in "Mathematical Methods in Hydrodynamics and Integrability in Dynamical Systems," eds. M. Tabor and Y. Treve, A.I.P. Conf. Proc., **88**, 243–277;

——— [1983], *The Painlevé property for partial differential equations. II: Bäcklund transformation, Lax pairs, and the Schwarzian derivative*, J. Math. Phys., **24**, 1405–1413;

——— [1984a], *On classes of integrable systems and the Painlevé property*, J. Math. Phys., **25**, 13–24;

——— [1984b], *The Sine-Gordon equations: Complete and partial integrability*, J. Math. Phys., **25**, 2226–2235;

——— [1985a], *The Painlevé property and Bäcklund transformations for the sequence of Boussinesq equations*, J. Math. Phys., **26**, 258–269;

——— [1985b], *Modified equations, rational solutions, and the Painlevé property for the Kadomtsev-Petviashvili and Hirota-Satsuma equations*, J. Math. Phys., **26**, 2174–2180;

——— [1986a], *Bäcklund transformations and the Painlevé property*, J. Math. Phys., **27**, 1293–1305;

——— [1986b], *Periodic fixed points of Bäcklund transformations and the Korteweg-de Vries equation*, J. Math. Phys., **27**, 2647–2656;

——— [1987], *Periodic fixed points of Bäcklund transformations*, J. Math. Phys., **28**, 2025–2039.

——— [1990], *Bäcklund transformations and the Painlevé property*, in "*Partially Integrable Evolution Equations in Physics*," eds. R. Conte and N. Boccara, NATO ASI Series C: Mathematical and Physical Sciences, **310**, 375–411, Kluwer, Dordrecht.

Weiss J., Tabor M. and Carnevale G. [1983], *The Painlevé property for partial differential equations*, J. Math. Phys., **24**, 522–526.

Whitham G.B. [1965], *Nonlinear dispersive waves*, Proc. Roy. Soc. London A, **283**, 238–261;

——— [1974], "*Linear and Nonlinear Waves*," Wiley, New York.

Whittaker E.T. and Watson G.N. [1927], "*A Course of Modern Analysis*," 4th edition, C. U. P., Cambridge.

Wickerhauser M.V. [1987], *Inverse scattering for the heat operator and evolutions in* $2+1$ *variables*, Commun. Math. Phys., **108**, 67–89;

——— [1988], *Hamilton's form for the Kadomtsev-Petviashvili equation*, J. Math. Phys., **29**, 2300-2302.

Wiggins S. [1988], "*Global Bifurcations and Chaos*," Springer-Verlag, Berlin-Heidelberg-New York.

Wils P. [1989a], *A new Painlevé solution of the Ernst equations*, Phys. Lett., **135A**, 425–426;

——— [1989b], *Painlevé solutions in general relativity*, Class. Quantum Grav., **6**, 1231–1241.

Wilson G. [1982], *The affine Lie algebra* $C_2^{(1)}$ *and an equation of Hirota and Satsuma*, Phys. Lett., **89A**, 332–334.

Winternitz P. [1983], *Lie groups and solutions of nonlinear differential equations*, in "*Nonlinear Phenomena*," proceedings, Oaxtepec, Mexico 1982, ed. K.B. Wolf, Lect. Notes Phys., **189**, 263–331, Springer-Verlag, Berlin-Heidelberg-New York.

Witten L. [1979], *Static axially symmetric solutions of self-dual* SU(2) *gauge fields in Euclidean four-dimensional space*, Phys. Rev. D, **19**, 718–720.

Wolfram S. [1986], "*Theory and Applications of Cellular Automata*," World Scientific, Singapore.

Wood D. [1984], *The self-focusing singularity in the nonlinear Schrödinger equation*, Stud. Appl. Math., **71**, 103–115.

Woodhouse N.M.J. [1987], *Twistor description of the symmetries of Einstein's equations for stationary axisymmetric space times*, Class. Quan. Grav., **4**, 799–814;

——— [1990], *Ward's splitting construction for stationary axisymmetric solutions of the Einstein-Maxwell equations*, Class. Quan. Grav., **7**, 257–260.

Woodhouse N.M.J. and Mason L.J. [1988], *The Geroch group and non-Hausdorff twistor spaces*, Nonlinearity, **1**, 73–114.

Wu T.T., McCoy B.M., Tracy C.A. and Barouch E. [1976], *Spin-spin correlation functions for the two-dimensional Ising model-exact theory in the scaling region*, Phys. Rev., **13B**, 316–374

Yajima N. and Oikawa M. [1976], *Formation and interaction of sonic-Langmuir solitons — inverse scattering method*, Prog. Theo. Phys., **56**, 1719–1739.

Yan Y. and Caudrey P.J. [1990], *A discrete spectral transform in* $2+1$ *dimensions*, Inverse Problems, **6**, 153–163.

Yang C.N. [1977], *Condition of self-duality for* SU(2) *gauge fields on Euclidean four-dimensional space.*, Phys. Rev. Lett., **38**, 1377–1379.

Yoshida H., Grammaticos B. and Ramani A. [1987], *Painlevé resonances versus Kowalevski exponents: some exact results on singularity structure and integrability of dynamical systems*, Acta Appl. Math., **8**, 75–103;

——— [1988], *Non-integrability of the fourth-order truncated Toda Hamiltonian*, Physica, **30D**, 151–163.

Yoshida H., Grammaticos B., Ramani A. and Hietarinta J. [1987], *On the non-integrability of some generalized Toda lattices*, Physica, **144A**, 310–321.

Yuen H.C. and Ferguson W.E. [1978], *Relationship between Benjamin-Feir instability and recurrence in the nonlinear Schrödinger equation*, Phys. Fluids, **21**, 1275–1278.

Yuen H.C. and Lake B.M. [1975], *Instabilities of waves on deep water*, Phys. Fluids, **18**, 956–960.

Zabreyko P.P., Koshelev A.I., Krasnosel'skii M.A., Mikhlin S., Rakovskchik L.S. and Stet'senko V.Ya. [1975], "*Integral Equations — A Reference Text*," Noordhoff, Leyden.

Zabusky N.J. and Kruskal M.D. [1965], *Interaction of solitons in a collisionless plasma and the recurrence of initial states*, Phys. Rev. Lett., **15**, 240–243.

Zakharov V.E. [1971], *Kinetic equation for solitons*, Soviet Physics JETP, **33**, 538–541;

——— [1972], *Collapse of Langmuir waves*, Sov. Phys. JETP, **35**, 908–914;

——— [1974], *On stocastization of one-dimensional chains of nonlinear oscillations*, Sov. Phys. JETP, **38**, 108–110;

——— [1990], *On the dressing method*, in "*Inverse Methods in Action*," proceedings, Montpellier 1989, ed. P.C. Sabatier, 602–623, Springer-Verlag, Berlin-Heidelberg-New York.

Zakharov V.E. and Faddeev L.D. [1971], *Korteweg-de Vries equation, a completely integrable Hamiltonian system*, Func. Anal. Appl., **5**, 280–287.

Zakharov V.E. and Kuznetsov E.A. [1986], *Multiple-scale expansions in the theory of systems integrable by the inverse scattering transform*, Physica, **18D**, 455–463.

Zakharov V.E. and Manakov S.V. [1973], *Resonant interaction of wave packets in nonlinear media*, Sov. Phys. JETP Lett., **18**, 243–245;

——— [1974], *On the complete integrability of the nonlinear Schrödinger equation*, Theo. Math. Phys., **19**, 551–559;

——— [1976], *The theory of resonance interaction of wave packets in nonlinear media*, Sov. Phys. JETP, **42**, 842–850;

——— [1985], *The construction of multidimensional nonlinear integrable systems and their solutions*, Func. Anal. Appl., **19**, 89–101.

Zakharov V.E. and Mikhailov A.V. [1978], *Relativistically invariant two-dimensional models of field theory which are integrable by means of the inverse scattering problem method*, Sov. Phys. JETP, **47**, 1017–1027.

Zakharov V.E. and Schulman E.I. [1982], *To the integrability of the system of two coupled nonlinear Schrödinger equations*, Physica, **4D**, 270–274.

Zakharov V.E. and Shabat A.B. [1972], *Exact theory of two-dimensional self-focusing and one-dimensional of waves in nonlinear media*, Sov. Phys. JETP, **34**, 62–69;

——— [1974], *A scheme for integrating the nonlinear equations of mathematical physics by the method of the inverse scattering problem*, Func. Anal. Appl., **8**, 226–235;

——— [1979], *Integration of nonlinear equations of mathematical physics by the method of the inverse scattering. II*, Func. Anal. Appl., **13**, 166–174.

Zamolodchikov A.B. [1980], *Factorized S matrices and lattice statistical systems*, Sov. Sci. Rev.: Phys. Rev., **2**, ed. I.M. Khalantnikov, 1–40, Harwood, Switzerland;

——— [1981], *Tetrahedron equations and the relativistic S-matrix of straight strings in* $2+1$ *dimensions*, Commun. Math. Phys., **79**, 489–505.

Zhou X. [1989], *Direct and inverse scattering transforms with arbitrary spectral singularities*, Commun. Pure Appl. Math., **42**, 895–938;

——— [1990], *Inverse scattering transform for the time dependent Schrödinger equation with applications to the KPI equation*, Commun. Math. Phys., **128**, 551–564.

Zograf P.G. and Takhtadzhjan L.A. [1988a], *On Liouville's equation, accessory parameters, and the geometry of Teichmüller space for Riemann surfaces of genus*-0, Math. USSR Sbornik, **60**, 143–161;

——— [1988b], *On uniformization of Riemann surfaces and the Weil-Peterson metric on Teichmüller and Schottky-spaces*, Math. USSR Sbornik, **60**, 297–313.

INDEX

Page numbers in italics indicate definitions